BIOLOGICAL AND MEDICAL PHYSICS,
BIOMEDICAL ENGINEERING

BIOLOGICAL AND MEDICAL PHYSICS, BIOMEDICAL ENGINEERING

The fields of biological and medical physics and biomedical engineering are broad, multidisciplinary and dynamic. They lie at the crossroads of frontier research in physics, biology, chemistry, and medicine. The Biological and Medical Physics, Biomedical Engineering Series is intended to be comprehensive, covering a broad range of topics important to the study of the physical, chemical and biological sciences. Its goal is to provide scientists and engineers with textbooks, monographs, and reference works to address the growing need for information.

Books in the series emphasize established and emergent areas of science including molecular, membrane, and mathematical biophysics; photosynthetic energy harvesting and conversion; information processing; physical principles of genetics; sensory communications; automata networks, neural networks, and cellular automata. Equally important will be coverage of applied aspects of biological and medical physics and biomedical engineering such as molecular electronic components and devices, biosensors, medicine, imaging, physical principles of renewable energy production, advanced prostheses, and environmental control and engineering.

J. Fitter T. Gutberlet J. Katsaras

Neutron Scattering in Biology

Techniques and Applications

With 240 Figures

Springer

Dr. Jörg Fitter
Forschungszentrum Jülich GmbH
Abt. IBI-2
52425 Jülich, Germany
e-mail: j.fitter@fz-juelich.de

Dr. John Katsaras
National Research Council
Chalk River
KoJ 1Jo Ontario, Canada
e-mail: john.katsaras@nrc.gc.ca

Dr. Thomas Gutberlet
Laboratory of Neutron Scattering
Paul Scherrer Institut
5232 Villigen, Switzerland
e-mail: thomas.gutberlet@psi.ch

ISSN 1618-7210

ISBN 978-3-642-06707-5 e-ISBN 978-3-540-29111-4

Springer is a part of Springer Science+Business Media

springer.com

Softcover reprint of the hardcover 1st edition 2006

Cover concept by eStudio Calamar Steinen

Preface

"Certainly no subject or field is making more progress on so many fronts at the present moment, than biology, and if we were to name the most powerful assumption of all, which leads one on and on in an attempt to understand life, it is that all things are made of atoms, and that everything that living things do can be understood in terms of the jigglings and wigglings of atoms."

Richard P. Feynmann, from "Six easy pieces" (1963)

In 1932, James Chadwick discovered the neutron, but initially the only sources of neutrons were from the radioactive decay of unstable nuclei. It was not until 1942 when Enrico Fermi constructed the first nuclear reactor in the squash courts beneath the University of Chicago's Stagg Field, that a controlled and sustained nuclear chain reaction was achieved. After World War II, nuclear reactors became available for civilian research, and in 1945 Ernest Wollan set up a double-crystal diffractometer at ORNL's Graphite Reactor. This marks the beginning of neutron scattering.

Neutrons produced by present reactor- and accelerator-based sources, typically have wavelengths in the order of Ångstroms, and hence are well-suited for probing the structures and motions of molecules. For biological materials rich in hydrogen, the large difference in scattering cross-sections between hydrogen and deuterium provides the possibility of contrast variation, a powerful method achieved by selective deuteration for emphasizing, or not, the scattering from a particular portion of a molecule or molecular assembly. Using a variety of scattering methods, the structures and dynamics of biological systems can be determined.

The present compilation aims to provide the reader with some of the important applications of neutron scattering in structural biology, biophysics, and systems relevant to biology.

The location of hydrogen atoms in biomolecules such as, proteins, is – despite the high brilliance and power of third generation synchrotron sources – not readily available by X-ray crystallography or related physical techniques. In the case of hydrogens attached to electronegative atoms (e.g., O and N), even high resolution X-ray structures (resolution <1 Å) cannot unequivocally locate these H atoms. On the other hand, these atoms can effectively be located using high resolution crystallographic neutron diffraction methods. Radiation damage leading to changes in metal oxidation state and subsequent loss of hydrogens can also pose a problem with X-rays, but not so with neutrons. When good quality, large ($>1\,mm^3$) single crystals cannot be obtained, low resolution neutron diffraction offers an alternative technique in determining the hydrated structure of macromolecules and their various hydrogen-bonding patterns.

Small-angle neutron scattering (SANS) is probably the technique most often applied to biological materials as it can probe the size, shape and conformation of macromolecules and macromolecular complexes in aqueous solution on a length scale from ten to several thousand Ångstroms. The ability to scatter from materials in solution allows for biologically relevant conditions to be mimicked, and also permits for the study of samples that are either difficult or impossible to crystallize. In recent years, SANS has greatly benefited from the production of "cold neutrons" that have wavelengths 10–20 times larger than "thermal neutrons", allowing SANS to examine complex materials, such as living cells.

Over the past decade, neutron reflectometry has increasingly become an important technique for the characterization of biological and biomimetic thin films attached to a solid support, in contact with water. Advancements in sample environments, instrumentation, and data analysis now make it possible to obtain high resolution information about the composition of these materials along the axis perpendicular to the plane of the membrane or substrate. Most recently, a newly developed phase-sensitive neutron reflectometry technique also allows direct inversion of the reflectometry data to obtain unique compositional depth profiles of the films in question.

Studies exploring the relationship between the function and the dynamics of biological systems are still in their nascent stages. Incoherent neutron scattering (INS) techniques such as, elastic (EINS), quasielastic (QINS), and inelastic (IINS) neutron scattering, along with molecular dynamics (MD) simulations offer the real possibility of investigating the dynamics associated with a molecule's biological function(s). Using the large incoherent scattering cross-section intrinsic to naturally abundant hydrogen atoms, various INS type measurements can be carried out. These results, in conjunction with MD simulations, offer a glimpse of for example, a protein's internal structure on the picosecond time scale. Moreover, the current developments of intense pulsed neutron sources promise, in the near future, to accelerate our understanding of the relationship between a molecule's dynamics and its function.

The study of materials under difficult environmental conditions (such as high magnetic fields, high pressures, shear, and 100% relative humidity) is by no means straight forward and requires specialized equipment. In many cases, these experiments are better accommodated by the fact that neutrons interact weakly, thus nondestructively, with many commonly used materials (e.g., aluminum and its alloys) that are readily available and suitable for the construction of sample environments. The conditions created by these specialized environments provide us with a more detailed physical understanding of biologically relevant materials.

The present volume begins with a general introduction into the generation and properties of neutrons and is followed by a series of papers describing the various elastic and inelastic neutron scattering techniques used to study biological and biologically relevant systems. The reader is introduced to the basic principles of neutron crystallography, low resolution neutron diffraction, neutron small-angle scattering, neutron reflectometry, inelastic and quasielastic neutron scattering, and neutron spin echo spectroscopy. Papers describing sample environments and preparatory techniques, in addition to molecular dynamics simulations used to evaluate the neutron data, are also included. Finally, there are a series of papers describing recent neutron research that has elucidated the structure and dynamics of soluble proteins, membrane embedded proteins, and of complex biological aggregates.

The editors wish to express their great appreciation to all of the contributors whose diligence, efforts, and timeliness made this compilation possible.

Jülich
Villigen
Chalk River
Spring 2005

Jörg Fitter
Thomas Gutberlet
John Katsaras

Contents

List of Contributors

N.F. Berk
National Institute of Standards and Technology
Gaithersburg, MD 20899, USA
norman.berk@nist.gov

M. Blakeley
EMBL Grenoble
6 rue Jules Horowitz
BP 181, 38042 Grenoble, France
blakeley@embl-grenoble.fr

M. Chakrapani
National Research Council
Steacie Institute for Molecular Sciences
Chalk River Laboratories
Chalk River, ON, K0J 1J0
Canada

W. Doster
Technische Universität München
Physikdepartment E 13
85748 Garching, Germany
wdoster@ph.tum.de

J. Fitter
Research Center Jülich
IBI-2: Structural Biology
52425 Jülich, Germany
j.fitter@fz-juelich.de

V.T. Forsyth
Partnership for Structural Biology
Institut Laue-Langevin
6 rue Jules Horowitz
BP 156, 38042 Grenoble Cedex 9
France
and
Institute of Science and Technology in Medicine
Keele University Medical School
Staffordshire ST4 7QB, UK
forsyth@ill.fr

P. Fratzl
Max Planck Institute of Colloids and Interfaces
Department of Biomaterials
14424, Potsdam, Germany
fratzl@mpikg-golm.mpg.de

A.J.K. Gilboa
Department of Structural Biology
The Weizmann Institute
71600 Rehovot, Israel

R.B. Gregory
Department of Chemistry
Kent State University
Kent, OH 44242-0001, USA
rgregory@kent.edu

S.K. Gregurick
Department of Chemistry
and Biochemistry
University of Maryland
Baltimore County
1000 Hilltop Circle
Baltimore, MD 20850, USA
gregurick@umbc.edu

T. Gutberlet
Laboratory of Neutron Scattering
Paul Scherrer Institut
5232 Villigen, Switzerland
thomas.gutberlet@psi.ch

J. Habash
Department of Chemistry
University of Manchester
Manchester M13 9PL, UK

T.A. Harroun
National Research Council
Steacie Institute for Molecular
Sciences
Chalk River Laboratories
Chalk River, ON, K0J 1J0
Canada
harrount@aecl.ca

J.R. Helliwell
Department of Chemistry
University of Manchester
Manchester M13 9PL, UK
J.R.Helliwell@dl.ac.uk

J. Katsaras
National Research Council
Steacie Institute for Molecular
Sciences
Chalk River Laboratories
Chalk River, ON, K0J 1J0
Canada
john.katsaras@nrc.gc.ca

J.K. Krueger
Chemistry Department University
of North Cardina at Charlotte
9201, University City Blvd.
Charlotte, NC 28223-0001, USA
jkkruege@email.uncc.edu

S. Krueger
NIST Center for Neutron Research
National Institute of Standards
and Technology
NIST, 100 Bureau Drive
Gaithersburg, MD 20899-8562, USA
susan.krueger@nist.gov

D. Kuzmanovic
Geo-Centers, Inc.
Gunpowder Branch
P.O. Box 68
Aberdeen Proving Ground
MD 21010, USA

R.E. Lechner
Hahn-Meitner-Institut Berlin
Glienicker Strasse 100
14109 Berlin, Germany
lechner@hmi.de

U. Lehnert
Yale University
Department of Molecular Biophysics
& Biochemistry
266 Whitney Avenue
New Haven, CT 06520, USA
lehnert@csb.yale.edu

S. Longeville
Laboratoire Léon Brillouin
CEA Saclay
91191 Gif-sur-Yvette, France
longevil@llb.saclay.cea.fr

M. Lösche
Carnegie Mellon University
Department of Physics
Pittsburgh, PA 15213, USA
and CNBT Consortium, NIST
Center for Neutron Research
Gaithersburg, MD 20899
USA
quench@cmu.edu

J.R. Lu
Biological Physics Group
Department of Physics
UMIST Oxford Road, M13 9PL, UK
j.lu@manchester.ac.uk

C.F. Majkrzak
National Institute of Standards
and Technology
Gaithersburg, MD 20899, USA
charles.majkrzak@nist.gov

R.P. May
Institut Laue-Langevin
6 rue Jules Horowitz
BP 156, 38042 Grenoble, France
Roland.May@ill.fr

H.D. Middendorf
Clarendon Laboratory
University of Oxford
Oxford OX13PU, UK
hdm01@isise.rl.ac.uk

D.A. Myles
Center for Structural Molecular
Biology
Oak Ridge National Laboratory
Oak Ridge, TN 37831, USA
mylesda@ornl.gov

M.-P. Nieh
National Research Council
Steacie Institute for Molecular
Sciences
Chalk River Laboratories
Chalk River, ON, K0J 1J0 Canada
Mu-Ping.Nieh@nrc.gc.ca

N. Niimura
Ibaraki University & Japan Atomic
Energy Research
Institute (JAERI)
4-12-1 Naka-narusawa, Hitachi
Ibaraki 316-8511, Japan
niimura@mx.ibaraki.ac.jp

O. Paris
Institute of Metal Physics
University of Leoben,
and Erich Schmid Institute
of Materials Science
Austrian Academy of Sciences
8700 Leoben, Austria
Current address: Max Planck
Institute of Colloids and Interfaces
Dept. of Biomaterials
14424 Potsdam, Germany
oskar.paris@mpikg-golm.mpg.de

I.M. Parrot
Institut Laue-Langevin
6 rue Jules Horowitz
BP 156, 38042 Grenoble Cedex 9,
France
and
Institute of Science and
Technology in Medicine
Keele University Medical School
Staffordshire ST4 7QB, UK
parrot@ill.fr

U.A. Perez-Salas
NIST Center for Neutron Research
National Institute of Standards
and Technology
NIST, 100 Bureau Drive
Gaithersburg, MD 20899-8562, USA
ursula.perez-salas@nist.gov

J. Raftery
Department of Chemistry
University of Manchester
Manchester, M13 9PL, UK
jrtest@spec.ch.man.ac.uk

V.A. Raghunathan
Raman Research Institute
Bangalore, 560 080, India
raghu@rri.res.in

M.C. Rheinstädter
Institut Laue-Langevin
6 rue Jules Horowitz
BP 156, 38042 Grenoble, France
rheinstaedter@ill.fr

M.W. Roessle
EMBL-Outstation Hamburg
Notkestr. 85
22603 Hamburg, Germany
manfred.roessle@embl-hamburg.de

T. Salditt
Institut für Röntgenphysik
Friedrich-Hund-Platz 1
37077 Göttingen, Germany
tsaldit@gwdg.de

A.P. Sokolov
Department of Polymer Science
The University of Akron
Akron, OH 44325, USA
alexei@polymer.uakron.edu

M. Tarek
Equipe de dynamique des
assemblages membranaires
Unite mixte de recherché
Cnrs/Uhp 7565
Universite Henri Poincare
BP 239
54506 Vanduvre-les–Nancy Cedex
France
mtarek@edam.uhp-nancy.fr

P. Timmins
Institut Laue-Langevin
6 rue Jules Horowitz
BP 156, 38042 Grenoble, France
timmins@ill.fr

D.J. Tobias
Department of Chemistry
and Institute for Surface
and Interface Science
University of California
Irvine, CA 92697-2025, USA
dtobias@uci.edu

M.J. Watson
National Research Council
Steacie Institute for Molecular
Sciences
Chalk River Laboratories
Chalk River, ON, K0J 1J0
Canada
Mike.Watson@nrc.gc.ca

M. Weik
Institut de Biologie Structurale
41 rue Jules Horowitz
38027 Grenoble Cedex 1, France
martin.weik@ibs.fr

G.D. Wignall
Oak Ridge National Laboratory
Oak Ridge, TN 37830-6393, USA
wignallgd@ornl.gov

C.C. Wilson
Department of Chemistry
University of Glasgow
Glasgow, G12 8QQ, UK

ISIS Facility CCLRC Rutherford
Appleton Laboratory
Chilton, Didcot
Oxon OX11 0QX, UK
chick@chem.gla.ac.uk

1

Neutron Scattering for Biology

T.A. Harroun, G.D. Wignall, J. Katsaras

1.1 Introduction

The structure and dynamics of a specimen can be determined by measuring the changes in energy and momentum of neutrons scattered by the sample. For biological materials, the structures of interest may be complex molecular structures, membranes, crystal lattices of macromolecules (e.g., proteins), micellar dispersions, or various kinds of aggregates. These soft materials may exhibit various modes of motion, such as low-energy vibrations, undulations or diffusion.

Neutrons are non-charged particles that penetrate deeply into matter. Neutrons are isotope-sensitive, and as they possess a magnetic moment, scatter from magnetic structures. Neutron scattering can often reveal aspects of structure and dynamics that are difficult to observe by other probes, including X-ray diffraction, nuclear magnetic resonance, optical microscopy, and various spectroscopies. It is particularly powerful for the study of biologically relevant materials which often contain hydrogen atoms and must be held in precise conditions of pH, temperature, pressure, and/or hydration in order to reveal the behaviors of interest.

Neutron scattering is practiced at facilities possessing reactor-based and accelerator-based neutron sources, and to which researchers travel to undertake their scattering experiments with the help of local scientific and technical expertise. Compared to traditional "hard" materials, in biologically relevant materials the characteristic length-scales are larger and the energy levels are lower. As such, additional neutron scattering measurements are possible if the reactor or accelerator-based source includes a cold moderator that emits a large proportion of long wavelength, lower velocity neutrons, which are better suited to the typical structures and dynamics found in bio-materials.

This chapter will follow neutrons from their production in a fission or spallation event, into the specimen where they scatter and are subsequently detected in a way that discriminates changes in momentum and energy. The advantages of using neutron scattering for problems in biology will be outlined.

However, details of specific instruments and data analysis for the associated scattering methods will be left to subsequent chapters.

1.2 Production of Neutrons

The neutron is a neutral, subatomic, elementary particle that had been postulated by Rutherford, and discovered in 1932 by James Chadwick [1, 2]. It is found in all atomic nuclei except hydrogen (^{1}H), has a mass similar to the proton, a nuclear spin of 1/2, and a magnetic moment [3]. Neutron beams with intensities suitable for scattering experiments are presently being produced either by nuclear reactors (Fig. 1.1), where the fission of uranium nuclei results in neutrons of energies between 0.5 and 3 MeV [4], or by spallation sources (Fig. 1.2), where accelerated subatomic particles (e.g., protons) strike a heavy metal target (e.g., tungsten or lead), expelling neutrons from the target nuclei [5].

In Canada, for example, the 125 MW National Research Universal (NRU) reactor, located at Chalk River Laboratories, has a peak thermal flux of

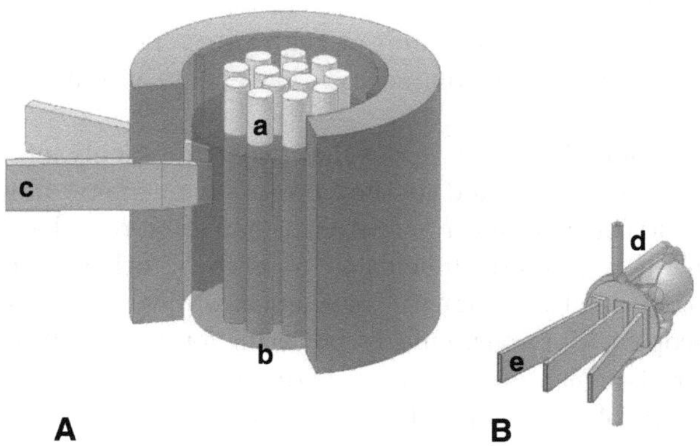

Fig. 1.1. Schematic of a nuclear reactor that produces thermal neutrons. Fuel rods (**a**) contain ^{235}U atoms which when they encounter moderated neutrons undergo fission producing ~2.5 high-energy neutrons/^{235}U atom. The probability of a fast (high energy) neutron interacting with a ^{235}U atom is small. To sustain the chain reaction, neutrons must be slowed down or thermalized by passing through a moderator. In practice, moderators such as H_2O, D_2O, graphite, or beryllium are used, filling the space in the reactor core around the fuel rods. For reasons of cost, H_2O is the most commonly used moderator (**b**) Thermal neutrons with a peak flux centered at ~1.2 Å can either be extracted directly from the reactor via a beam tube (**c**) or can be furthered slowed down by interaction with another, colder moderator, for example, a vessel of liquid hydrogen (**d**) These cold neutrons, with their Maxwellian distribution shifted toward lower energies, can be transported over many meters to the various spectrometers by ^{58}Ni-coated optically flat glass surfaces (**e**) through a process known as total external reflection

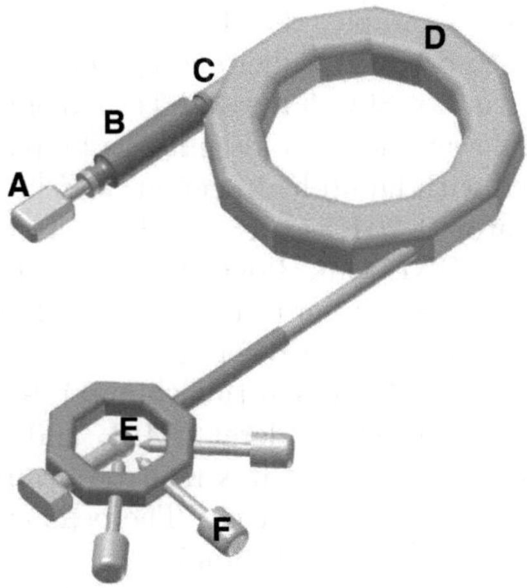

Fig. 1.2. Schematic of the Spallation Neutron Source (SNS) presently under construction at Oak Ridge National Laboratory. (**a**) H^- ions produced by an ion source are accelerated to 2.5 MeV (**b**) the H^- ion beam is then delivered to a Linac further accelerating the 2.5 MeV H^- ion beam to 1 GeV (**c**) prior to delivery from the Linac to the accumulator ring, H^- ions are stripped of all of their electrons by a stripper foil resulting in H^+ ions (**d**) these H^+ ions are bunched and intensified by the accumulator ring for delivery to the (**e**) liquid mercury target where a nuclear reaction takes place creating spallation neutrons for use at various spectrometers (**f**) the duration of the SNS proton pulse is 10^{-6} s and the repetition rate is 60 Hz. Not unlike reactor-based neutrons, spallation neutrons are moderated by either water or a liquid hydrogen source, giving rise to thermal or cold neutrons, respectively. The SNS chose mercury as the target for the proton pulses for the following reasons: (**i**) Unlike solid materials, liquid mercury does not experience radiation damage. (**ii**) Mercury is a high atomic number material resulting in many spallation neutrons (~20–30 neutrons/mercury atom). (**iii**) Compared to a solid target, a liquid target at room temperature better dissipates heat and withstands shock effects

3×10^{14} neutrons cm^{-2} s^{-1}. Fast MeV neutrons are produced from fission of ^{235}U atoms which are in turn thermalized, through successive collisions with deuterium atoms in a heavy water moderator at room temperature, to an average energy of ~0.025 eV. Neutron beams exiting the reactor have a Maxwellian distribution of energy, [4] and are usually monochromated using a crystal monochromator, and then used to study a variety of condensed matter.

For a thermal neutron reactor, such as the Institut Laue-Langevin (ILL, Grenoble, France) the Maxwell spectrum peak is centered at ~1 Å due to a 300 K D_2O moderator [6]. However, the peak of the spectrum can be shifted to higher energies (or shorter wavelengths) by allowing the thermal neutrons to equilibrate with a “hot source”, or shifted to lower energies with the use of a "cold source". For example, the ILL uses a self-heating graphite block hot-source at 2400 K to produce higher energy neutrons, [7] while the reactor at the National Institute of Standards and Technology (NIST, Gaithersburg, Maryland) produces lower energy cold neutrons by passing thermal neutrons through a vessel filled with liquid hydrogen at 40 K [8]. Similarly, a supercritical hydrogen moderator at 20 K is currently being installed at the Oak Ridge National Laboratory (ORNL, Oak Ridge, Tennessee) High Flux Isotope Reactor (HFIR) that will feed a suite of instruments, including a 35 m small-angle neutron scattering facility optimized for the study of biological systems (see contribution by Krueger and Wignall this volume) [9].

Presently, the heavy-water moderated ILL and light-water moderated ORNL reactors produce the highest flux neutron beams, operating at a thermal power of 58 and 85 MW, respectively. The peak core flux of both sources is $>10^{15}$ neutrons $cm^{-2}\ s^{-1}$. Since the ability to remove heat from the reactor core dictates the maximum power density, and thus the maximum neutron flux, it is unlikely that a reactor far exceeding the thermal flux characteristics of the ILL and ORNL high flux reactors will ever be constructed.

The notion of accelerator driven neutron sources dates back to the 1950s. In an accelerator-based pulsed neutron source, high energy subatomic particles, such as protons, are produced in a linear accelerator (Linac) [10–12]. These accelerated protons then impinge on a heavy metal target releasing neutrons from the nuclei of the target material. Since the Linac operation uses travelling electromagnetic waves, the arrival of the protons at the target are in pulsed bunches, and therefore the neutron beams produced are also pulsed. As with neutrons produced in a reactor, spallation neutrons have very high initial energies and must be slowed down from MeV to meV energies. However, their characteristic spectra differ considerably as the neutron spectrum from a spallation source contains both a high energy slowing component of incomplete thermalized neutrons, and a Maxwell distribution characteristic of the moderator temperature. Compared to reactor sources, the biggest advantage of spallation sources is that they create much less heat per neutron produced, translating into increased neutron fluxes. Nevertheless, since neutrons are produced in pulses, the time-averaged flux of even the most powerful pulsed source, that of ISIS (Oxford, UK), is less than that of a high flux reactor source (e.g., ILL). However, judicious use of time-of-flight techniques, which can utilize the many neutron wavelengths present in each pulse, can exploit the high brightness and can, for certain experiments, more than compensate for the time-averaged flux disadvantage.

The Spallation Neutron Source (SNS), presently being constructed at ORNL, will have a time-averaged flux comparable to a high-flux reactor but each pulse will contain neutron intensities between 50 and 100 times greater than the ILL or ORNL reactor-based sources. Moreover, the intense short-pulse neutron beams produced by accelerator-based neutron sources make it possible to perform time-of-flight experiments, and the study of kinetics and dynamics of various systems.

1.3 Elements of Neutron Scattering Theory

1.3.1 Properties of Neutrons

X-rays interact with charged subparticles of an atom, primarily with electrons [13]. On the other hand, neutrons, as mentioned previously, are non-charged subatomic particles having a mass (m) of 1.0087 atomic mass units (1.675×10^{-27} kg), spin of 1/2, and a magnetic moment (μ_n) of -1.9132 nuclear magnetons [6]. These properties of the neutron give rise to two principal modes of interaction which are different from those of X-rays.

As neutrons are zero charge particles, their interaction with matter, both nuclear and magnetic, is short ranged. As a result of this small interaction probability, neutrons can penetrate deep into condensed matter. Moreover, the interaction between the neutron and atomic nuclei involve complex nuclear interactions between the nuclear spins and magnetic moments. For this reason, there is no general trend throughout the periodic table of an atom's ability to scatter neutrons. This is quite unlike the X-ray atomic scattering factor which increases with atomic number [13, 14]. In addition, different isotopes of the same element may have very different abilities to scatter neutrons. This concept of a difference in scattering power, or *contrast*, between various components in a sample as a result of the different scattering properties of the various elements (particularly ^{1}H and ^{2}H) is the core principle of neutron scattering, and from which biology greatly benefits [14–16].

The second mode of interaction is the magnetic dipole interaction between the magnetic moments associated with unpaired electron spins in magnetic samples and the nuclear magnetic moment of the neutron. This type of neutron–atom interaction is of limited use to biology, and as such, for the purposes of this chapter only nuclear scattering will be considered. It should be noted that the interaction between the magnetic field of the X-ray and the orbital magnetic moments of the electron is not zero. However, compared to charge scattering, X-ray magnetic scattering is weak [13].

1.3.2 Energy and Momentum Transfer

In a scattering experiment the neutron undergoes a change in momentum after interacting with the sample. This means the neutron has a change in direction and/or velocity. The neutron's momentum is given by $\boldsymbol{p} = \hbar \boldsymbol{k}$, where $\hbar = h/2\pi$ is Planck's constant and $\boldsymbol{k}$ is the neutron wave vector, $|\boldsymbol{k}| = 2\pi/\lambda$. The wavelength, λ, of a neutron is given by

$$\frac{h^2}{2m\lambda^2} = 2k_{\mathrm{B}}T, \tag{1.1}$$

where k_{B} is Boltzmann's constant and T is the neutron moderator temperature.

The momentum change can be described by a momentum transfer vector or the *scattering vector*, $\boldsymbol{Q}$, and is defined as the vector difference between the incoming and scattered wave vectors,

$$\boldsymbol{Q} = \boldsymbol{k}_0 - \boldsymbol{k}_1, \tag{1.2}$$

where $\boldsymbol{k}_0$ and $\boldsymbol{k}_1$ are the incident and scattered wave vectors, respectively (Fig. 1.3). The change in the neutron's momentum is given by $\hbar\boldsymbol{Q}$.

Besides a change in direction, the magnitude of $\boldsymbol{k}$ can also change as energy between the incident neutron and the sample are exchanged. The law of energy conservation can be expressed as

$$E = E_0 - E_1 = \hbar^2\frac{\boldsymbol{k}_0^2}{2m} - \hbar^2\frac{\boldsymbol{k}_1^2}{2m} = \hbar\omega, \tag{1.3}$$

where E is the energy gained or lost by the neutron. Any process whereby the neutron is scattered from $\boldsymbol{k}_0$ to $\boldsymbol{k}_1$ is therefore associated with $\boldsymbol{Q}$ and E.

1.3.3 Diffraction

Scattering is totally *elastic* when $E = 0$. In this case, from Eq. 1.3 we must have $|\boldsymbol{k}_1| = |\boldsymbol{k}_0|$ and as such, from Eq. 1.2 we get $|\boldsymbol{Q}| = 2\boldsymbol{k}_0 \sin\theta$. For crystalline materials Bragg peaks appear at values $\boldsymbol{Q}$ equal to the reciprocal lattice spacing:

$$|\boldsymbol{Q}| = \frac{2\pi}{d}, \tag{1.4}$$

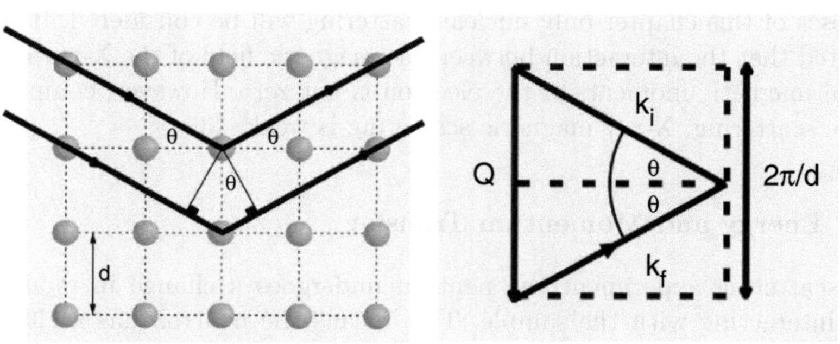

Fig. 1.3. Neutrons strike an array of atoms (green) from the left, and are scattered to the right. Horizontal planes of atoms are separated by distance d. Both the incident and diffracted neutron beams make an angle θ with respect to the planes of atoms (*left*). The change of the neutron's momentum, $\boldsymbol{Q}$, is given in Eq. 1.2 and is schematically represented schematically. In reciprocal space, when $\boldsymbol{Q}$ points along the reciprocal lattice of spacing $2\pi/d$, the Bragg condition for diffraction is met, and constructive interference leads to a diffraction peak or so-called Bragg maximum (*right*)

where d is the characteristic spacing of a set of crystal planes. Since $\boldsymbol{k}_0 = 2\pi/\lambda$, carrying out the appropriate substitutions leads to the now familiar Bragg formula:

$$\lambda = 2d \sin \theta. \tag{1.5}$$

Simply stated, this is the condition of constructive interference of waves with incident angle θ on a set of equidistant planes separated by a distance d.

The measurement of truly elastic scattering requires that both the incident and scattered neutrons have the same wavelength, i.e., $|\boldsymbol{k}_1| = |\boldsymbol{k}_0|$. However, in practice this type of elastic scattering experiment, using an analyzer crystal to choose the appropriate energy scattered neutron, is seldom performed and the inelastic contribution ($E \neq 0$) is usually not removed.

1.3.4 Scattering Length and Cross-Section

Neutron, X-ray, and light scattering all involve interference phenomena between the wavelets scattered by different elements in the system. In the simple case of neutron scattering from a single, fixed nucleus, incident neutrons can be represented as a plane wave, $\psi_0 = \exp \mathrm{i}\boldsymbol{k}_0 z$. The resulting scattered wave is a spherical wave, and is given by

$$\psi_1 = \frac{b}{r} \mathrm{e}^{\mathrm{i}\boldsymbol{k}_1 \cdot \boldsymbol{r}}, \tag{1.6}$$

where $\boldsymbol{r}$ is the location of the detector from the nucleus. The quantity b has the dimensions of length, and is the measure of the scattering ability of the atomic nucleus. It may be regarded as a real and known constant for a given nucleus or isotope.

A typical experiment involves counting the number of neutrons scattered in a particular direction, and in this simple case, without regard of any changes in energy. If the distance from the detector to the nucleus is assumed to be large, so that the small solid angle $\mathrm{d}\Omega$ subtended by the detector is well defined, we can then define the *differential cross-section* as

$$\frac{\mathrm{d}\sigma}{\mathrm{d}\Omega} = \frac{(\text{neutrons s}^{-1} \text{ scattered into } \mathrm{d}\Omega)}{\Phi \mathrm{d}\Omega}, \tag{1.7}$$

where Φ is the incident neutron flux (number of neutrons $\mathrm{cm}^{-2}\,\mathrm{s}^{-1}$). The *total scattering cross-section* is defined as the total number of neutrons scattered per second, normalized to the flux;

$$\sigma_{\mathrm{s}} = \int \left(\frac{\mathrm{d}\sigma}{\mathrm{d}\Omega}\right) \mathrm{d}\Omega, \tag{1.8}$$

where the integral is over all directions. For the single, fixed nucleus that we are considering, we can readily relate the total cross-section to b. If v is the velocity of the incident neutrons, then the number of neutrons passing through an area $\mathrm{d}S$ s^{-1} is

$$v\mathrm{d}S\,|\psi|^2 = v\mathrm{d}S\frac{b^2}{r^2} = vb^2\mathrm{d}\Omega. \tag{1.9}$$

From the definition of a neutron cross-section,

$$\frac{\mathrm{d}\sigma}{\mathrm{d}\Omega} = \frac{vb^2\mathrm{d}\Omega}{\Phi\mathrm{d}\Omega}, \tag{1.10}$$

where $\sigma_\mathrm{s} = 4\pi b^2$ [17].

From the above it is obvious that σ_s has the dimensions of area. Moreover, the magnitude of b is typically of the order 10^{-12} cm, giving rise to the usual unit for cross-section, commonly known as the barn (1 barn = $10^{-24}\,\mathrm{cm}^2$). [1] To a first approximation, the cross-section may be regarded as the effective area which the target nucleus presents to the incident beam of neutrons for the elastic scattering process and is usually referred-to as the *bound atom* cross-section, as the nucleus is considered fixed at the origin [18]. Where the atom is free to recoil, such as in the gaseous state, the *free atom* cross-section is applicable. The bound atom cross-section is generally relevant to biological studies which are virtually always conducted on samples of macroscopic dimensions in the solid or liquid state.

Neutrons are scattered isotropically from individual nuclei, whereas for X-ray scattering, the scattering originates in the electron cloud, which is very large compared to the X-ray wavelength. In the case of X-rays, the atomic form factors are Q-dependent. However, the variation in practice is small (<1% for $Q < 0.1\,\text{Å}^{-1}$), and usually neglected in the small angle region. The Thompson scattering amplitude of a classical electron is $r_\mathrm{T} = 0.282 \times 10^{-12}$ cm, so the X-ray scattering length of an atom, f, is proportional to the atomic number ($\mathrm{f} = r_\mathrm{T}Z$) and increases with the number of electrons/atom. For neutrons, values of b vary from isotope to isotope (Sect. 1.3.5). If the nucleus has a nonzero spin, it can interact with the neutron spin, and the total cross-section (σ_s) contains both, coherent and incoherent components.

1.3.5 Coherent and Incoherent Cross-Sections

Atomic nuclei are characterized by an incoherent and a coherent neutron scattering length b. The coherent scattering length is analogous to the atomic form factor in X-rays, f, while there is no X-ray analogue for the incoherent scattering length. For the purposes of this review, we will only consider the case where the nuclear moments of the material being probed with neutrons are completely disordered, giving rise to incoherent scattering.

---

[1] The origin of the barn unit is thought to lie in the colloquialism "as big as a barn", and was recommended in 1950 by the Joint Commission on Standards, Units and Constants of Radioactivity, because of its common usage in the USA.

When a neutron of spin 1/2 encounters a single isotope with nuclear spin I, the spin of the neutron–nucleus system can assume two values, $I \pm 1/2$. The scattering lengths of the two systems are denoted by b^+ and b^-, and the number of spin states associated with each are $2(I + 1/2) + 1 = 2I + 2$ and $2(I - 1/2) + 1 = 2I$, respectively. The total number of states is $4I + 2$. If the neutrons are unpolarized and the nuclear spins are randomly oriented, each spin state has the same probability. Thus the the frequency of the b^+ occurring is weighted by $(I+1)/(2I+1)$, and for b^-, $I/(2I+1)$. The coherent cross-section for each isotope is given as $\sigma_c = 4\pi\bar{b}^2$, where $\bar{b}$ represents the thermally averaged scattering length with + and − spin state populations. Similarly, the total scattering cross-section is given by $\sigma_s = 4\pi\overline{b^2}$. The average coherent scattering length is then given by

$$\bar{b} = \frac{1}{2I+1}\left[(I+1)\,b^+ + Ib^-\right], \tag{1.11}$$

$$\overline{b^2} = \frac{1}{2I+1}\left[(I+1)\,(b^+)^2 + I(b^-)^2\right], \tag{1.12}$$

The difference between σ_s and σ_c is the incoherent scattering cross-section, σ_i.

If the isotope has no spin (e.g., ^{12}C), then $b^2 = \bar{b}^2 = \overline{b^2}$ and there is no incoherent scattering. Only the coherent scattering cross-section contains information on interference effects arising from spatial correlations of the nuclei in the system, in other words, the structure of the sample. The incoherent cross-section contains no structural information or interference effects, and forms an isotropic (flat) background which must be subtracted off from the raw data (e.g., see J. Krueger et al. this volume). The incoherent component of the scattering does, however, contain information on the motion of single atoms which may be investigated via by studying the changes in energy of the scattered beam (e.g., see contributions by Lechner et al., Doster, Sokolov et al. or Fitter in this volume).

While most of the atoms encountered in neutron scattering of biologically relevant materials are mainly coherent scatterers, such as carbon and oxygen, there is one important exception. In the case of hydrogen (^{1}H) the spin-up and spin-down scattering lengths have opposite sign ($b^+ = 1.080 \times 10^{-12}$ cm; $b^- = -4.737 \times 10^{-12}$ cm). Since $I = 1/2$ we then have σ_c, σ_i, and σ_s equal to 1.76×10^{-24}, 79.7×10^{-24}, and 81.5×10^{-24} cm^2, respectively.

Unlike neutrons, for photons there is no strict analog of incoherent scattering. X-ray Compton scattering is similar in that it contains no information on interference effects, i.e., the structure of the sample, and contributes a background to the coherent signal. However, to a good approximation this background goes to zero in the limit $Q \to 0$ and in X-ray studies, is usually neglected. Table 1.1 gives the cross-sections and scattering lengths for atoms commonly encountered in synthetic, natural and biomaterials.

The cross-sections given previously for hydrogen refer to bound protons and neglect inelastic effects arising from the interchange of energy with the

Table 1.1. Bound atom scattering lengths and cross-sections for typical elements in synthetic and natural biomaterials

atom	nucleus	b_c (10^{-12} cm)	σ_c (10^{-24} cm^2)	σ_i (10^{-24} cm^2)	σ^a_{abs} (10^{-24} cm^2)	$f_{X\text{-ray}}$ (10^{-12} cm)
hydrogen	^{1}H	−0.374	1.76	79.7	0.33	0.28
deuterium	^{2}H	0.667	5.59	2.01	0	0.28
carbon	^{12}C	0.665	5.56	0	0	1.69
nitrogen	^{14}N	0.930	11.1	0	1.88	1.97
oxygen	^{16}O	0.580	4.23	0	0	2.25
fluorine	^{19}F	0.556	4.03	0	0	2.53
silicon	^{28}Si	0.415	2.16	0	0.18	3.94
phosphorous	P^b	0.513	3.31	0	0.17	4.22
chlorine	Cl^b	0.958	11.53	5.9	33.6	4.74

[a] Values of the absorption cross-section (σ_{abs}) are a function of wavelength and are given at $\lambda = 1.8$ Å. As $\sigma_{abs} \sim \lambda$, values at other wavelengths may be estimated by scaling by $\lambda/1.8$; $f_{X\text{-ray}}$ is given for $\theta = 0$

[b] Values are an average over the natural abundance of the various isotopes

neutron. For coherent scattering, which is a collective effect arising from the interference of scattered waves over a large correlation volume, this approximation is reasonable, especially at low Q where recoil effects are small. However, for incoherent scattering, which depends on the uncorrelated motion of individual atoms, inelastic effects become increasingly important for long wavelength neutrons. In most biological systems, the atoms are not rigidly bound, so due to effects of torsion, rotation, and vibration, the scattering generally contains an inelastic component. This has two consequences: Firstly, the scattering, which in the center-of-mass system is elastic, may induce a change of energy of the neutron in the laboratory frame. This gives rise to inelastic scattering which contains information about the motion of atoms in the sample (e.g., see Lechner et al.). Secondly, the effective total scattering cross-section in the laboratory system is wavelength-dependent, an effect that is particularly important for ^{1}H-containing samples, where the transmission is a function of both the incident neutron energy and temperature. This effect is important for H_2O, a common solvent for biomaterials, and for which the total scattering cross-section at 20°C is given by $\log \sigma = 4.45 + 0.46 \log \lambda$, where σ is expressed in barns [19]. For further discussion of such inelastic effects, see contribution by S. Krueger et al.

1.4 Neutron Diffraction and Contrast

Compared to synchrotron X-rays, the single biggest disadvantage of neutrons is that neutron fluxes from reactor, or even accelerator-based sources, are small. Effectively, this translates into neutron experiments taking much longer to achieve the same signal-to-noise values as ones performed with X-rays. Moreover, the availability of neutron sources is scant compared to the combined availability of the different types of X-ray sources, such as sealed tube and rotating anode X-ray generators, and synchrotron facilities. Nevertheless, as we have seen in a previous section, the many advantageous properties of neutrons, especially those of contrast variation and sensitivity to low Z atoms equally well as heavy ones, make neutrons a highly desirable probe.

1.4.1 Contrast and Structure

Contrast variation has been exploited in several ways. Here, we will only present a broad outline of how it is used to determine the structure and dynamics of biological macromolecules, and leave it to subsequent chapters to provide explicit detail and examples.

The scattering associated with coherent cross-section will have a spatial distribution, which is a function of the distribution of atoms in the sample. The amplitude of the scattered neutron wave is often called the structure factor, and is given by

$$S(\boldsymbol{Q}) = \sum_i b_i e^{\mathrm{i}\boldsymbol{Q}\cdot\boldsymbol{r}_i}, \tag{1.13}$$

where the sum is over all atoms in the sample. The measured intensity of neutrons is then proportional to the structure factor squared

$$I(\boldsymbol{Q}) \propto |S(\boldsymbol{Q})|^2 . \tag{1.14}$$

In a diffraction experiment, resolution is defined as $2\pi/Q_{\mathrm{max}}$, where Q_{max} is the maximum value of measured amplitude, $|\boldsymbol{Q}|$. When working at resolutions where individual atoms are not resolved (e.g., $\gtrsim$10 Å) [18, 20], it is valid to use the concept of a neutron refractive index or scattering length density, $\rho(\boldsymbol{r})$. Because each nucleus has a different scattering amplitude (ref. Table 1.2), the scattering length density (SLD) is defined as the sum of the coherent scattering lengths over all atoms within a given volume δV, divided by δV [18, 20] or

$$\rho(\boldsymbol{r})\delta V = \sum_i b_i. \tag{1.15}$$

SLD is the Fourier transform of the structure factor

$$\rho(\boldsymbol{r}) = \int S(\boldsymbol{Q})\mathrm{e}^{-\mathrm{i}\boldsymbol{Q}\cdot\boldsymbol{r}}\mathrm{d}\boldsymbol{Q}. \tag{1.16}$$

In the case of a single crystal, the integral in Eq. 1.16 is over all atoms in the unit cell, and techniques used in X-ray crystallography are entirely applicable. The goal in this case is to determine the scattering length density $\rho(\boldsymbol{r})$ over the unit cell, rather than the electron density. Whereas both methods yield the locations of the atoms, $\boldsymbol{r}_i$, in the case of neutrons hydrogen atoms with their negative b value (ref. Table 1.2) stand out in much more detail, whereas hydrogen is for all purposes invisible to X-rays.

In general, when solving a crystal structure from diffraction data one has to deal with the well-known *phase problem.* This problem arises from the fact that the structure factor is a complex function, however, the complex part, or the phase, is lost in the measured intensity. A technique devised to resolve the phase problem is *isomorphous replacement*, and involves the addition of an element which effectively changes the neutron or electron density contrast of the crystal. In X-ray crystallography, this usually means the incorporation of heavy atoms such as, Hg into the structure.

In the ideal case, isomorphic replacement does not alter the macromolecule's conformation or the unit cell parameters. This is not always the case when heavy atoms are used to change the sample contrast. On the other hand, the exchange of deuterium for hydrogen, whether in the solvent or explicitly on selected chemical groups, is as nearly perfect isomorphous replacement as possible. The scattering length density of the specific deuterium label, $\rho_{\mathrm{l}}(\boldsymbol{r})$, can be isolated by taking the measured structure factors from the protonated sample and subtracting them from the deuterated sample as follows

$$\rho_{\mathrm{l}}(\boldsymbol{r}) = \int [S_D(\boldsymbol{Q}) - S_H(\boldsymbol{Q})]\, \mathrm{e}^{-\mathrm{i}\boldsymbol{Q}\cdot\boldsymbol{r}} \mathrm{d}\boldsymbol{Q}. \tag{1.17}$$

This is analogous to a difference Fourier map in X-ray crystallography, but the possibility of altering the molecule's conformation has been greatly reduced [21].

Where neutron diffraction excels, is the study of samples which cannot be crystallized and display a high degree of disorder, and dispersions of particles in solution. In this case, the benefits of contrast variation are easily seen.

There would be no observable diffraction if particles of uniform scattering length density $\bar{\rho}$ were placed in a solvent where the SLD *matches*, $\rho_{\mathrm{s}} = \bar{\rho}$, and the contrast is zero. Instead, the effective scattering density of a particle whose SLD varies with $\boldsymbol{r}$ is $\rho(\boldsymbol{r}) - \rho_{\mathrm{s}}$. For a particle in solution, the measurable contribution of the particle against a backdrop of solvent is given by

$$S_{\mathrm{p}}(\boldsymbol{r}) = \int [\rho(\boldsymbol{r}) - \rho_{\mathrm{s}}]\, \mathrm{e}^{\mathrm{i}\boldsymbol{Q}\cdot\boldsymbol{r}} \mathrm{d}\boldsymbol{r}, \tag{1.18}$$

where the integral is over the the particle volume. The key to finding the particle's structure in solution is to separate $\rho(\boldsymbol{r})$ into the mean particle density at the match point, $\overline{\rho}_{\mathrm{m}}$, and fluctuations about the mean, $\rho_{\mathrm{f}}(\boldsymbol{r})$,

$$\rho(\boldsymbol{r}) = \overline{\rho}_{\mathrm{m}} + \rho_{\mathrm{f}}(\boldsymbol{r}), \tag{1.19}$$

where $\rho_{\mathrm{f}}(\boldsymbol{r})$ is normalized by

$$\int \rho_{\mathrm{f}} \mathrm{d}\boldsymbol{r} = 0. \tag{1.20}$$

The contrast in this situation is defined as $\rho_c = \overline{\rho}_{\mathrm{m}} - \rho_s$, which is adjusted by varying the amount of D_2O in the solvent (ref. Sect. 1.4.3). Therefore, contrast variation helps separate particle shape and internal structure contributions to the scattered amplitude. Because scattering from solution averages over all orientations of the particles, modelling of $\rho(\boldsymbol{r})$ and fitting of $S_{\mathrm{p}}(\boldsymbol{r})$ are usually performed to fully analyze the data. Although we have neglected the exchange of the molecule's labile H atoms with solvent D atoms, such exchange does take place and will be discussed in the following section.

1.4.2 Contrast and Dynamics

Using neutron spectroscopy to study the dynamics of biological molecules is a comparatively new and developing field. The analysis of inelastic neutron scattering data is complicated and beyond the scope of this introductory chapter. However, the lessons of contrast in structural determination are still applicable. It should be pointed-out that scattering length density is time-dependent, $\rho(\boldsymbol{r}, t)$, as the atoms are moving, giving rise to inelastic and incoherent scattering, as discussed. In the previous section, we were only concerned with the time-averaged values, $\langle \rho(\boldsymbol{r}) \rangle$, as we wanted to illustrate the importance of contrast in determining structural information. The dynamic structure factor $S(\boldsymbol{Q}, \omega)$ is in general more complicated, and is given by

$$G(\boldsymbol{r}, t) = \frac{1}{N} \int \langle \rho(\boldsymbol{r}', 0) \rho(\boldsymbol{r}' - \boldsymbol{r}, t) \rangle \, \mathrm{d}\boldsymbol{r}' \tag{1.21}$$

$$S(\boldsymbol{Q}, \omega) = \frac{1}{2\pi\hbar} \int \int G(\boldsymbol{r}, t) \mathrm{e}^{\mathrm{i}(\boldsymbol{Q} \cdot \boldsymbol{r} - \omega t)} \mathrm{d}\boldsymbol{r} \mathrm{d}t, \tag{1.22}$$

where $G(\boldsymbol{r}, t)$ is called the time dependent pair correlation function. Clearly, by matching the scattering length density of the solvent to parts of the molecule, one can isolate the relative motions of particular groups or molecules.

1.4.3 Contrast and Biology

It may be seen from Table 1.2 that there is a large difference in the coherent scattering length of deuterium (^{2}H) and hydrogen (^{1}H), and that the value for the latter, is negative. This arises from a change of phase of the scattered wave with respect to the incident wave, and as explained above, results in a marked difference in scattering power (contrast) between hydrogenous materials containing ^{2}H or ^{1}H. This has important consequences for the scattering lengths of commonly found biological groups.

Table 1.2 shows the relevant values of scattering cross-section for common biological molecules such as water, and the components of proteins, nucleic

Table 1.2. Bound atom scattering lengths for typical biological chemical groups

A. Amino acids and proteins

amino acid	exchangeable hydrogen	b_s H_2O (10^{-12} cm)	b_s D_2O (10^{-12} cm)	b_s deuterated (10^{-12} cm)	volume[a] (Å^3)
glycine	1	1.73	2.77	4.85	71.9
alanine	1	1.65	2.69	6.85	100.5
valine	1	1.48	2.52	10.85	150.8
leucine	1	1.40	2.44	12.85	179.0
isoleucine	1	1.40	2.44	12.85	175.7
phenylalanine	1	4.14	5.18	13.51	201.8
tyrosine	2	4.72	6.80	14.09	205.2
tryptophan	2	6.04	8.12	16.45	239.0
aspartic acid	1	3.85	4.89	8.01	124.2
glutamic acid	1	3.76	4.80	10.01	149.3
serine	2	2.23	4.31	7.43	100.6
threonine	2	2.14	4.23	9.43	127.7
asparagine	3	3.46	6.58	9.70	129.5
glutamine	3	3.37	6.50	11.70	155.9
lysine	4	1.59	5.75	15.12	181.0
arginine	6	3.47	9.72	17.00	211.6
histadine	1.5	4.96	6.52	11.73	163.2
methionine	1	1.76	2.81	11.13	175.4
cystine	2	1.93	4.01	7.14	122.0
proline	0	2.23	2.23	9.52	137.5

B. Nucleotides and nucleic acids

base		exchangeable hydrogen	b_s H_2O (10^{-12} cm)	b_s D_2O (10^{-12} cm)	b_s deuterated (10^{-12} cm)	volume[a] (Å^3)
adenine	RNA	3	11.24	14.36	22.69	314.0
	DNA	2	10.66	12.74	22.11	
guanine	RNA	4	11.82	15.98	23.27	326.3
	DNA	3	11.24	14.36	22.69	
cytosine	RNA	2	9.27	12.39	20.72	285.6
	DNA	3	8.69	10.77	20.14	
uracil	RNA	2	9.29	11.37	19.70	282.3
thymine	DNA	1	8.62	9.66	21.12	308.7

C. Water

	b_s (10^{-12} cm)	ρ (10^{-12} cm Å^{-3})
H_2O	−0.168	−0.00562
D_2O	1.915	0.06404

Table 1.2. *contd.*

D. Phosphatidylcholine lipids[b]

	b_s (10^{-12} cm)	ρ (10^{-12} cm Å^{-3})	b_s deut. (10^{-12} cm)	ρ deut. (10^{-12} cm Å^{-3})
CH_3	−0.458	−0.0085	2.67	0.0495
CH_3	−0.083	−0.0031	2.0	0.0744
headgroup	2.24	0.011	15.67	0.071

[a] Values are from Durchschlag and Zipper [22]. Number of exchangeable hydrogen are assumed for pH 7.
[b] Values are from Jacrot [19]

acids, and lipids. In nearly all neutron studies some deuteration is used, either for the water in solvation, or of the chemical group itself. When solvating water is replaced by heavy water, some of the hydrogens in the sample will be replaced by deuterium through exchange with the solvent, changing its scattering length density. In general, hydrogen bound to nitrogen or oxygen will be the most likely candidates for exchange. In Table 1.2 this has been taken into account.

Table 1.2 makes two important points. First is that common biological macromolecules have very different scattering lengths. For example, DNA and RNA have considerably larger scattering lengths than proteins, which in turn, are much larger than lipids. This is due to the fact that DNA/RNA have more nitrogen (high positive SLD) and fewer hydrogen (negative SLD) atoms than either, protein or lipid molecules. Lipids have the greatest number of hydrogens per molecule, thanks to their hydrocarbon chains and few exchangeable hydrogens. Thus in any complex, the effects of different molecular species can be highlighted with appropriate contrast matching.

As a simple example, consider the case of a two component particle, containing protein and DNA. In this case, $\rho(\boldsymbol{r}) = \rho_{\text{pro}}(\boldsymbol{r}) + \rho_{\text{dna}}(\boldsymbol{r})$. When $\rho_s = \overline{\rho_{\text{pro}}(\boldsymbol{r})}$, the scattering is dominated by the nucleic acid structure, and vice versa.

The second, and probably most important point that can be drawn from Table 1.2 and Fig. 1.4 is that D_2O has a larger scattering length density, and H_2O a lower scattering length density than any of the biological molecules listed. This means that an appropriate mixture of the two solvents can contrast match almost any biological molecule. This is represented graphically in Fig. 1.4, which shows the average scattering length density for model RNA, protein, and lipid membrane systems, as a function of the concentration of D_2O solvent. The points where the line for water crosses the lines for other molecules is called the solvent match point, where the contrast is zero (Fig. 1.4). For DNA and RNA this occurs ∼70% D_2O, while for protein, it

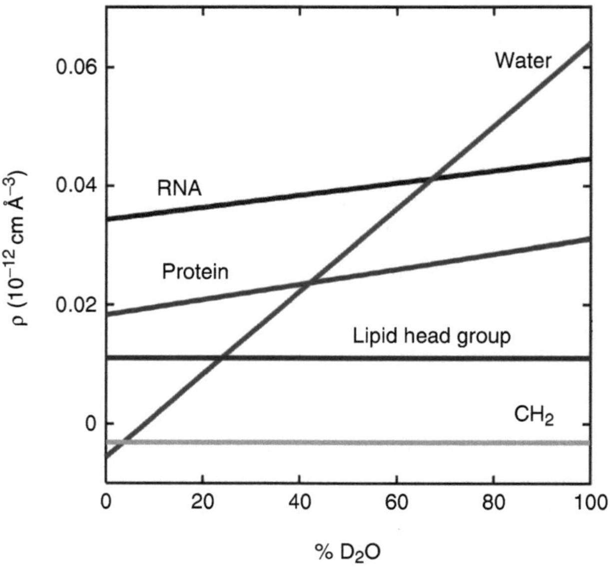

Fig. 1.4. The average scattering length density of typical biological macromolecules, as a function of D_2O concentration in the solvent. The figure is calculated from the data in Table 1.2. The number of exchanged hydrogen is assumed to be complete in 100% D_2O. The figure will depend of the solvent accessbile area and specific volume of the molecule, and each case is unique. Note that for water with 8% D_2O, $\rho = 0$. For protein, the line is calculated from the natural abundance of mammalian amino-acid weigthed average, and is $\rho = 0.0128 \cdot X + 0.0183$. RNA and DNA (not shown) are less sensitive to H/D exchange; $\rho = 0.0103 \cdot X + 0.0343$ for RNA and $\rho = 0.007 \cdot X + 0.0317$ for DNA

occurs closer to 40%. A more detailed description of the principles underlying contrast variation methods is given in the contribution by J. Krueger et al. (Chapter 8).

1.5 Conclusions

Neutrons are commonly thought of as a tool for hard materials, and for good reason. For the year 2002, published reports involving experiments classified as biological, made up only ~8% of all reports at the Hahn-Meitner Institut (Berlin, Germany) [23], and ~4% at NRC Chalk River [24]. In the 2003 JAERI annual report (Tokai, Japan) ~9% of reports dealt with biology, [25] while only about 6% of the beam time allocated at ILL in 2002 went to proposals in biology [7]. These numbers increase however, if one considers experiments involving so-called bio-materials, which are often classified under soft condensed matter, rather than biology. In this case, around one in eight instrument days at the ILL is devoted to science involving some form of biologically related material [7]. More importantly, the trend with regards to biologically related neutron experiments is upward.

The increasing number of biologically relevant experiments taking place is very much in line with the fact that many neutron facilities are interested in seeing biological problems elucidated with the various neutron scattering techniques available. Presently, biology is an educational outreach tool, that can connect with the public and policy makers in ways that many other sciences cannot. Experiments seen as a having some relevance to advances in medicine can be promoted within and beyond the facility. This has had the effect that new instruments devoted to biological sciences such as, the dedicated biological Advanced Neutron Diffractometer/Reflectometer (AND/R) at NIST, and a new 35 m small angle neutron scattering facility at ORNL, are coming online.

The succeeding chapters serve to illustrate the various techniques of neutron diffraction and spectroscopy, in detail. The importance of contrast variation that was introduced in this chapter will serve to demonstrate the broad usefulness that neutron diffraction has in biology.

Acknowledgments

The authors would like to thank V.A. Raghunathan (Raman Research Institute, India) for the many discussions, and M.J. Watson (National Research Council) for providing us with the illustrations used to assemble the various figures.

References

1. J. Chadwick, Nature **129**, 312 (1932)
2. J. Chadwick, Proc. Roy. Soc. A **136**, 692 (1932)
3. H. Dachs, Principles of neutron diffraction, in *Topics in Current Physics: Neutron Diffraction*, H. Dachs (Eds.) (Springer-Verlag, New York, Berlin, 1978) pp. 1–40
4. W.M. Lomer, G.G. Low, Introductory theory, in *Thermal Neutron Scattering*, P.A. Egelstaff (Eds.) (Academic Press, London, New York, 1965) pp. 1–52
5. B.P. Schoenborn, E. Pitcher, Neutron diffractometers for structural biology at spallation neutron sources, in *Neutron in Biology*, B.P. Schoenborn, R. B. Knott (Eds.) (Plenum Press, New York, 1996) pp. 433–444
6. D.L. Price, K. Sköld, Introduction to neutron scattering, in *Methods of Experimental Physics*, vol. 23 Part A Neutron Scattering, K. Sköld, D.L. Price (Eds.) (Academic Press, Orlando, 1986) pp. 1–97
7. *Institut Laue-Langevin 2002 Annual Report* (ILL, Grenoble, 2003)
8. *NIST Center for Neutron Research 2001 Annual Report* (NIST, Gaithersburg, 2001)
9. G.W. Lynn, M.V. Buchanan, P.D. Butler, L.J. Magid, G.D. Wignall, J. Appl. Cryst. **36**, 829 (2003)
10. R. Widreöe, Arch. Elektrotech. **21**, 387 (1928)

11. W.K.H. Panofsky, L.W. Alvarez, H. Bradner, H. Gordon, L.C. Marshall, F. Oppenheimer, C. Richman, R. Serber, C. Turner, J.R. Woodyard Science **106**, 506 (1947)
12. W.K.H. Panofsky, L.W. Alvarez, H. Bradner, J.V. Franck, H. Gordon, J.D. Gow, L.C. Marshall, F. Oppenheimer, C. Richman, J.R. Woodyard, Rev. Sci. Instrum. **26**, 111 (1955)
13. J. Als-Nielsen, D. McMorrow, *Elements of Modern X-Ray Physics* (John Wiley and Sons, England, 2001)
14. C.R. Cantor, P.R. Schimmel, *Biophysical Chemistry Part II: Techniques for the Study of Biological Structure and Function* (W.H. Freeman and Co., San Francisco, 1980)
15. M. Tomita, T. Hasegawa, T. Tsukihara, S. Miyajima, M. Nagao, M. Sato: J. Biochem. (Tokyo) **125**, 916 (1999)
16. T. Gutberlet, U. Heinemann, M. Steiner, Acta Cryst. **D57**, 349 (2001)
17. G.L. Squires, *Introduction to the Theory of Thermal Neutron Scattering* (Dover Publications, Mineola, New York, 1978)
18. G.D. Wignall, Small angle scattering characterization of polymers, in *Physical Properties of Polymers*, 3rd edn. J.E. Mark (Eds.) (Cambridge University Press, 2004) pp. 424–511
19. B. Jacrot, Rep. Prog. Phys. **39**, 911, (1976)
20. G. Zaccaï, Application of neutron diffraction to biological problems, in *Topics in Current Physics: Neutron Diffraction*, H. Dachs (Eds.) (Springer-Verlag, New York, Berlin, 1978) pp. 243–269
21. M.C. Weiner, S.H. White, Biophys. J. **59**, 174 (1991)
22. H. Durchschlag, P. Zipper: J. Appl. Cryst. **30**, 803 (1997)
23. *BENSC experimental reports 2002* (Hahn-Meitner-Institute, Berlin, 2003)
24. *Annual Report 2003 Rapport Annuel* (NRC-CNRC, Canada, 2003)
25. *Progress report on Neutron Scattering Research* (Japan Atomic Energy Research Institute, Tokai, 2004)

Part I

Elastic Techniques

2

Single Crystal Neutron Diffraction and Protein Crystallography

C.C. Wilson, D.A. Myles

2.1 Introduction

The neutron is a flexible probe for the study of condensed matter, having significant advantages over other forms of radiation in the study of the microscopic structure and dynamics of matter. Neutron scattering gives detailed information about the microscopic behavior of condensed matter, which has significantly affected our experimental and theoretical understanding of materials ranging from magnets and superconductors to chemical surfaces and interfaces and biological systems.

Neutron diffraction is the method of choice for many crystallographic experiments. The nature of the scattering of neutrons by atomic species is such that the technique offers a description of all atoms in a structure at approximately the same level of precision. This is due to the fact that neutrons are scattered by the nucleus rather than the electrons in an atom, and hence the scattering power does not have the strong dependence on Z found for many other scattering techniques such as X-ray or electron diffraction.

These properties give neutron diffraction the following features compared with other techniques:

- It is easier to sense light atoms, such as hydrogen, in the presence of heavier ones. For example, in the presence of relatively light carbon atoms, a hydrogen contributes only 1/36 (less than 0.03) of the X-ray scattering intensity from a carbon atom. The equivalent ratio for neutrons is around 0.32, meaning that hydrogen atoms are, roughly speaking, determined around 12 times more accurately with neutrons than X-rays in the presence of carbon atoms. This factor generally increases as the atomic number of the "heavy" atom increases, reaching 41 for H in the presence of oxygen (and almost 1,100 for hydrogen in the presence of lead).
- Neighboring elements in the periodic table generally have substantially different scattering cross-sections. For example manganese and iron have

$Z = 25$ and 26, respectively, giving a very small contrast for X-ray scattering while the respective neutron scattering lengths (−3.9 and 9.5 fm) are not only different in magnitude but also in sign, giving large contrast. For light elements in particular this is the only practical direct method of distinguishing neighboring elements.

- The dependence of the scattering on the nucleus allows isotopes of the same element to have substantially different scattering lengths for neutrons, thus allowing the technique of isotopic substitution to be used to yield structural and dynamical details. In the area of organic and biological molecular structures, the most relevant isotopic substitution is that of ^2H (deuterium, scattering length 6.67 fm) for ^1H (hydrogen, scattering length 3.74 fm). This also allows the use of contrast variation, where the scattering density of different parts of a molecule or of an H_2O–D_2O mixture is altered. This method is extremely powerful and has been a key to many successful applications of the technique of neutron scattering in chemistry and biology.
- The lack of a fall-off in scattering power as a function of scattering angle gives neutron diffraction the ability to study structures to very high resolution, although even for neutrons there will always be a fall-off in scattered intensity caused by thermal effects.
- Neutrons interact weakly with matter and are therefore nondestructive, even to complex or delicate materials – this is particularly relevant in the study of biological materials.

Correspondingly, neutrons are a bulk probe, allowing us to probe the interior of materials, not merely the surface layers probed by techniques such as X-rays, electron microscopy or optical methods. Neutrons also have a magnetic moment, allowing magnetic structure (the distribution of magnetic moments within a material) and magnetic dynamics (how these moments interact with each other) to be studied in a way not possible with other forms of radiation.

In the following, we briefly outline the techniques for single crystal neutron diffraction and summarize the application of these techniques in the study of molecular systems. A short review of applications in the field of structural biology is given and the history of single crystal neutron diffraction in this area is reviewed, along with some future directions. A more detailed account of single crystal neutron diffraction in the area of molecular systems has recently been published [1].

2.2 Single Crystal Neutron Diffractometers: Basic Principles

There are two main types of neutron source used for condensed matter studies: steady state (usually reactor) sources and pulsed (usually spallation) sources; the instrumentation used for single crystal diffraction at these sources is significantly different.

On a steady-state neutron source, traditional four-circle diffractometer techniques have traditionally been used with great success for chemical and small molecule single crystal diffraction, with a monochromatic beam and a single detector. Rotations of the crystal (and detector) are used to allow measurement of each reflection sequentially. There is also the potential for increasing the region of reciprocal space accessed in a single measurement by using an area detector. Alternatively it can be combined with a broad band (white) beam, and used for Laue or quasi-Laue diffraction, with a stationary crystal and detector. Since reflections in this technique are stimulated across a broad spectral bandwidth, the resultant data are typically scaled or normalized to account for the intensity distribution of the spectrum.

Structure factor data collected on a monochromatic steady-state source at present yields the ultimate in accuracy for neutron single crystal structure determination and the high time-averaged neutron flux make Laue diffraction at a steady-state source also extremely powerful (see below). The benefits of single crystal diffraction instrumentation on a steady state source can be summarized as follows:

- Well established diffractometry developed world-wide on four-circle X-ray diffractometers can be directly applied to a monochromatic neutron instrument. Diffractometer control software and step-scanning methods for intensity extraction can be directly transferred.
- In such a case, all reflections are observed with the same neutron wavelength, eliminating the need for wavelength dependent corrections. The constant wavelength nature of the data collection and the steady state of the source also removes the need for correcting data for the incident flux profile and leads to more straightforward error analysis. Large area detectors are also not essential, removing the systematic deviations caused by fluctuations of detector response.
- The time averaged flux at current high-flux steady-state sources is substantially higher than at present-day pulsed sources, allowing better counting statistics to be obtained in the same time, and allowing the study of smaller crystals or larger unit cells, particularly in the Laue technique – which is especially relevant for studies of macromolecular structures.

These factors tend to lead to more accurate structure factors, better internal agreement and ultimately to lower crystallographic R factors and somewhat more precise atomic parameters in most small molecule work. Constant wavelength single crystal diffraction is therefore the method of choice if the ultimate limit on precision is set by the instrument rather than by the sample. This is rarely the case for protein crystals, however, which are typically weakly diffracting, have large mosaicity, and are characterized by relatively large temperature factors that limit the ultimate extent of the data to quite modest resolutions (Bragg resolution $d_{\min} \sim 1.5$–$2.0\,\text{Å}$). Both monochromatic and Laue geometries have been successfully exploited for individual structure determination and the high-flux reactor sources are also currently favored for larger unit cells or smaller crystals.

As an aside, and picking up on one point raised above, the resolution of an experiment is an important data collection parameter. The word "resolution" in crystallography is open to misinterpretation as there are several definitions. Broadly speaking, however, in a single crystal experiment we speak of resolution as being the minimum d-spacing measured (d_{min}, often quoted by chemical crystallographers in the inverse form, maximum $\sin\theta/\lambda = 1/2d_{\mathrm{min}}$), while in a powder experiment resolution generally means the ability to separate adjacent peaks. For larger crystalline systems, including many large organometallic, supramolecular and biological systems, the scattering is less favorable for obtaining detailed "atomic resolution" pictures; there is normally high water content in many biological systems which often "blurs" the picture. Large molecules frequently exhibit a degree of disorder and at the very least are very flexible, which leads to high temperature factors and weaker scattering. For these reasons, even the best protein crystals, for example, rarely give diffraction patterns extending to d-spacings of less than 1.0 Å with X-rays or of less than 1.5 Å with neutrons. More typically neutron diffraction ends at around 2–3 Å resolution. Such data can still yield quasi-atomic resolution in the model of the molecule, when it is combined with a combination of prior knowledge, chemical and stereochemical arguments, model building, and constrained molecule or residue refinement packages.

For pulsed source instruments, the time-of-flight Laue diffraction (tofLD) technique is used; this method exploits the capability of a single crystal diffractometer on such a source to access large volumes of reciprocal space in a single measurement. This is due to the combination of the wavelength-sorting inherent in the time-of-flight (tof) technique with large area position-sensitive detectors (PSDs). tofLD thus samples a large three-dimensional volume of reciprocal space in a single measurement with a stationary crystal and detector.

The characteristics of the structure factor data collected on an instrument with a PSD on a pulsed source can have certain advantages for structural refinement:

- The collection of many Bragg reflections simultaneously in the detector allows the accurate determination of crystal cell and orientation from a single data frame (collected in one fixed crystal/detector geometry). It is also worthy of note that for some applications a single frame may be the only data required.
- The white nature of the incident beam enables the straightforward measurement of reflections at different wavelengths, which can be useful in the precise study of wavelength dependent effects such as extinction and absorption.
- The collection of data to potentially very high $\sin\theta/\lambda$ values (exploiting the high flux of useful epithermal neutrons from the undermoderated beams), can allow improved determination of heavily Q-dependent parameters to be obtained, for example anharmonic effects.

- The nature of the Laue method allows possibilities for the rapid collection of data sets, by removing the need to measure each reflection individually. This flexibility, long appreciated in synchrotron Laue methods for studying protein structures, has recently become recognized as a great strength of time-of-flight neutron Laue diffraction methods.

Time-of-flight single crystal diffraction is thus ideal for surveying reciprocal space, rapid determination of large numbers of reflections, and following structural changes using a subset of reflections. It also provides good accuracy and precision in standard structural refinements, while not matching the ultimate performance of a constant wavelength instrument in this area.

2.2.1 Development of Single Crystal Neutron Diffractometers

Historically, the main drawback of using neutron diffraction in protein structure determination has been the requirement for relatively large protein crystals of several cubic millimetres and the long data acquisition times of up to weeks per data set required to compensate the relatively low neutron flux that is available even at high-power neutron sources such as the reactor at the Institut Laue-Langevin (ILL) [2]. Recent progress in improving Laue diffraction (see below), new neutron optics, new detector technologies, and longer neutron wavelengths are now having dramatic impact on these problems, with possible huge gains in efficiency. Parallel improvements in modern molecular biology now allow fully (per)deuterated protein samples to be produced for neutron scattering that essentially eradicate the large – and ultimately limiting – hydrogen incoherent scattering background that has hampered such studies in the past. High quality neutron data can now be collected to near atomic resolution (~2.0 Å) for proteins of up to ~50 kDa molecular weight using crystals of volume ~0.1 mm^3. Spallation neutron sources with higher flux neutron beams and optimized signal-to-noise ratio for time-of-flight Laue methods promise further order of magnitude gains in performance that will revolutionize the field.

2.2.2 Achievements of Neutron Macromolecular Crystallography at Reactor Sources

Given the large size and weak scattering nature of biological macromolecules, coupled with the inherent difficulties of growing the large crystals that were required until recently ($\gg 1\,mm^3$), the major requirement for neutron protein crystallography is high incident neutron flux and high efficiency neutron detection. Neutron protein crystallography has therefore only been feasible at the brightest national and international neutron research facilities and, until recently, has been dominated by instruments at steady-state reactor sources. A major problem that has limited efficiency of biological neutron scattering has been reliant upon single crystal monochromatic diffractometers that collect reflections sequentially and often individually using counters or – at best – with relatively small area detectors. Since the number of reflections from even moderate protein crystal unit cell edges ($\sim 50-70$ Å) quickly exceed the tens of thousands, neutron protein crystallography has been restricted to all but the smallest protein systems. Recently, important advances have been made with the move toward large 2D area detectors that are able to capture much larger fractions – in some cases even all – of the large number of reflections that are stimulated simultaneously at each position of the crystal in monochromatic experiments. The emergence of white beam Laue techniques at steady-state reactor sources that deliver order of magnitude improvements in efficiency and data collection rates are now impacting significantly upon the field.

The number of instruments in the world that are suitable for macromolecular neutron crystallography is rather limited. In Europe, most research is concentrated on instruments at the ILL (Grenoble) where two classical single-crystal four-circle diffractometers (D19 and DB21) have been successfully used for collecting crystallographic data on biological systems. In addition, ILL recently commissioned a quasi-Laue diffractometer (LADI) equipped with a large ($>2\pi$) neutron image plate detector that is located on a cold neutron beam and dedicated to neutron protein crystallography (see below).

The D19 instrument is a thermal monochromatic neutron diffractometer that is optimized for samples with medium-sized unit cell edges of <40 Å and can operate at a number of wavelengths in the range 1.0–2.4 Å. The limited size of the area detector previously available on D19 has meant that the instrument has been used little for neutron crystallography from proteins, though it has found major applications in fibre diffraction analysis of complex biopolymers with small unit cell repeats, such as DNA and cellulose; the instrument is also extensively used for smaller macromolecules and in chemical crystallography. This instrument will benefit from a likely >20-fold improvement in capability for many experiments through provision of large array detectors in an upgrade which is currently in progress.

DB21 is a cold neutron diffractometer developed by ILL and EMBL-Grenoble for low-resolution (>10 Å) neutron protein crystallography of complex biological macromolecules (e.g., multimeric proteins or assemblies of proteins with nucleic acids, such as viruses or ribosomes). Experiments on this instrument typically use contrast-variation techniques to collect low resolution data sets at a series of H_2O/D_2O crystal solvent concentrations (or contrasts) that are varied to match and cancel the scattered signal from the individual components of complex systems. The technical challenges of collecting very low resolution data are overcome by using long-wavelength neutrons (7.5 Å) and the resolution of the instrument is adapted to large unit-cell edges of up to 1,000 Å. DB21 is thus specially designed to locate disordered components in large biological complexes (such as DNA/RNA in viruses and detergent structures in membrane proteins) that crystallize in very large unit cells (up to ~600 Å).

The most exciting recent development for neutron protein crystallography at ILL has been the quasi-Laue diffractometer LADI [3], which provides the advantages of rapid data collection using Laue geometry by the use of a cylindrical neutron image plate detector that surrounds the sample and provides $>2\pi$ Sr solid angle coverage (Fig. 2.1). In order to reduce the problems of both spatially and harmonically overlapped reflections, and to reduce the accumulation of the otherwise large background under the Laue diffraction pattern, LADI operates with a restricted broad-band wavelength range of around 1 Å (typically $3 < \lambda < 4$ Å). The combination of a broad band-pass quasi-Laue geometry with the novel 2π neutron-sensitive image-plate detector to record long-wavelength neutrons up to 4 Å provides 10–100-fold gains in data-collection rates compared with conventional neutron diffractometers (Fig. 2.1). The instrument is thus well suited to neutron macromolecular

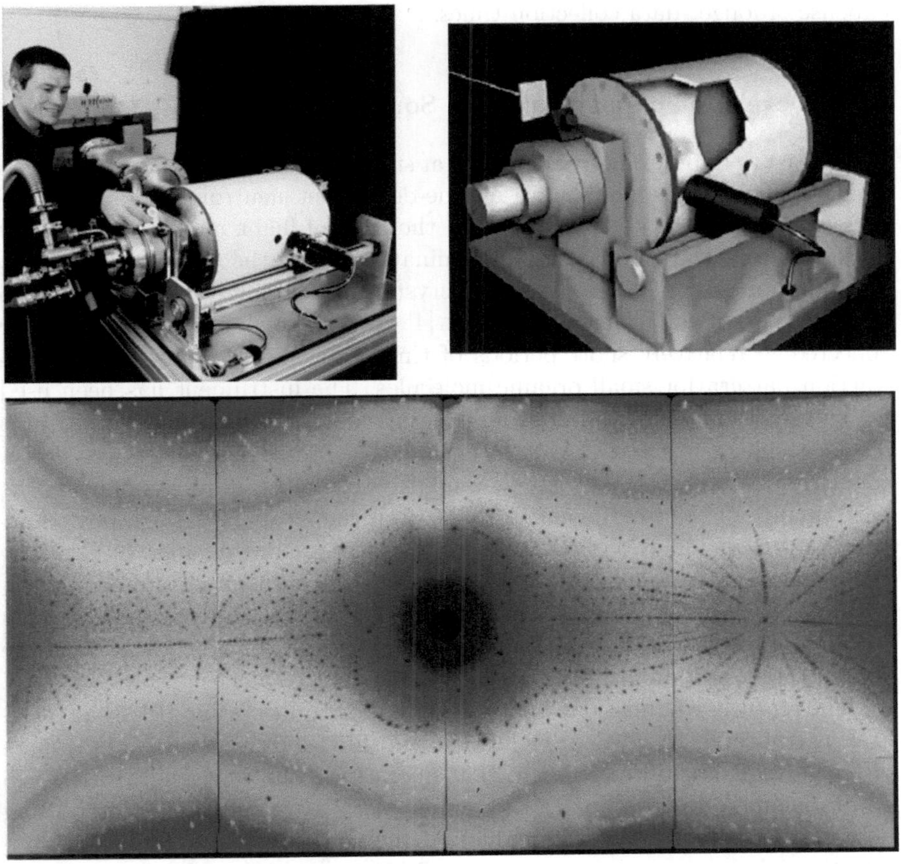

Fig. 2.1. Views of the LADI instrument at ILL, with its 2π image plate detector (*top*, also pictured, one of the authors (DAM)). Neutron Laue diffraction data collected on the LADI detector at ILL from a single crystal of sperm whale myoglobin (*bottom*)

crystallography and is used for single-crystal studies of small proteins up to 45–50 kDa at medium resolution (~2 Å), which is sufficient to locate individual H atoms of special interest, water structures, or other small molecules that can be marked by deuterium to be particularly visible.

At the Japanese neutron facility JAERI (Tokai-mura), two monochromatic thermal neutron protein diffractometers (BIX-3, BIX-4) are now in routine operation. These instruments are closely similar in design and exploit a cylindrical neutron-sensitive image plate design similar to that used on LADI to give $>2\pi$ solid angle of detection. Here, however, the neutron beam is monochromatized using a bent perfect Si crystal at wavelengths between 1.6 and 2.4 Å and the BIX instruments use a crystal-step scan method to produce high resolution (~1.5 Å) diffraction patterns. Although the data-collection efficiency is lower than that of quasi-Laue techniques, the accumulated background is significantly less and high resolution data can be collected, at the expense of longer data collection times.

2.2.3 Developments at Spallation Sources

In contrast to diffractometers operated at steady-state reactor neutron sources, spallation neutron sources have a time-dependent neutron flux and single-crystal diffractometry is performed in the time-of-flight mode, which allows background noise to be largely discriminated out by the counter electronics. This method is applied on the single-crystal diffractometer SXD at the ISIS Spallation Neutron Source in the UK [4] (Fig. 2.2), where data sets can be collected in relatively short periods of time, typically a day or less per diffraction pattern for small organic molecules. The instrument has been used for chemical crystallography, drug structure determination and for analysis of biopolymers such as DNA. A recent increase in the number of area detectors (from 2 to 11; Fig. 2.2) is now implemented, substantially enhancing the performance of the instrument in the area of chemical crystallography. However, the characteristics of SXD, and of the pioneer in this area, SCD at the IPNS source at Argonne National Laboratory, are not suited to macromolecular crystallography. A new time-of-flight single crystal diffractometer, which will be optimized for macromolecular crystallography, is now planned on the second target station of ISIS, now in construction.

The recently commissioned protein crystallography station (PCS) at LANSCE is dedicated to protein, membrane and fibre diffraction. PCS is a time-of-flight single-crystal instrument that operates in the wavelength range 1–5 Å, producing a neutron flux at the sample of 7×10^6 neutrons $s^{-1}cm^{-2}$ using a partially coupled, and thus enhanced flux, moderator. A large position-sensitive ^{3}He cylindrical detector covers 2,000 cm^2 with a spatial resolution of 1.3 mm FWHM and a counting rate $>10^6$ neutrons s^{-1}. To prevent spot overlap and improve the signal-to-noise ratio, a chopper system will eliminate the initial radiation pulse and provide a short-wavelength and long-wavelength

Fig. 2.2. The old, two PSD SXD at ISIS (left, being tended by one of the authors (CCW) on its last day of operation), and its upgraded replacement (right), with 11 detectors (six visible, the others beneath) offering 2π solid angle coverage

cutoff. The PCS instrument is the first purpose built neutron protein crystallography instrument at a spallation neutron source and first results are already encouraging [5, 6] (Fig. 2.3).

2.2.4 Forward Look for Instrumentation for Neutron Macromolecular Crystallography

The main problem with studying biological materials with neutrons is the fact that the structures are large and weakly scattering. A particularly exciting recent development has therefore been the increased exploitation of Laue methods of data collection from single crystal samples. The use of these broad bandpass techniques maximizes both the neutron flux at the sample and the number of reflections that are stimulated and recorded at the detector. Currently, the most promising neutron protein crystallography instrument is the quasi-Laue diffractometer LADI at the ILL, discussed above, which uses a cylindrical neutron image plate that surrounds the sample to give more than 2π solid angle coverage (Fig. 2.1). The combination of a broad band-pass Laue geometry and large $>2\pi$ Sr detector coverage has dramatically reduced data collection times by more than 100-fold compared to traditional diffractometers [7]. While capable of collecting data to 1.5 Å resolution in standard configuration (Fig. 2.3), the LADI instrument is limited to systems with unit cell edges of ~100 Å on edge, limited by the high density and spatial overlap of reflections at higher resolutions (<2.0 Å). A new and improved LADI instrument is now planned at the ILL which, when installed on its new cold neutron

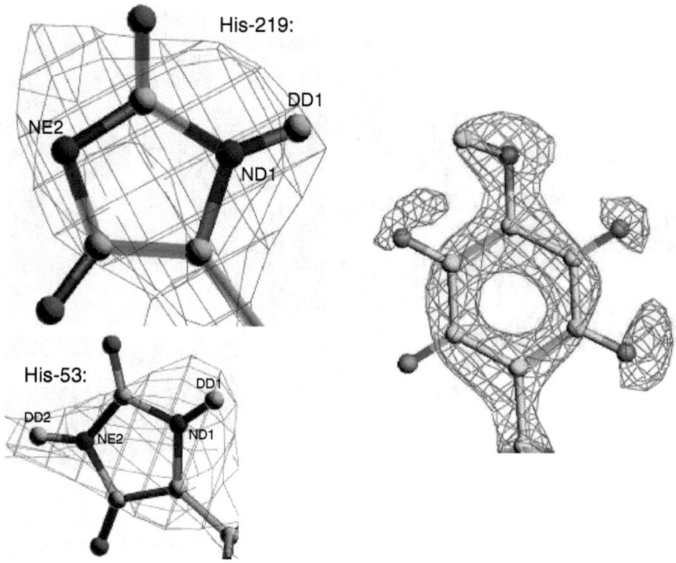

Fig. 2.3. Preliminary density maps from glucose isomerase, measured on PCS at LANSCE (*left*). The 1.6 Å $2F_o$-F_c positive nuclear density in blue and the negative nuclear density in red for the side chain of Tyrosine 10 in W3Y rubredoxin (Pf). An example map from neutron data collected at cryogenic temperature 15 K on LADI at the ILL (*right*)

beamline, promises to deliver a further order of magnitude improvement in performance for neuron protein crystallography.

In addition to its potentially revolutionary applications in biological crystallography, the LADI concept has obvious applications also in the study of smaller molecular systems, and in magnetism. For problems in chemical crystallography, the recently commissioned VIVALDI instrument, sited on a thermal neutron beam line with a wavelength range centered around 1.6 Å, is optimal. The image plate approach adopted on LADI is also exploited in the monochromatic Japanese BIX instruments for neutron protein crystallography [8]. Indeed, the BIX concept was the first to use the neutron image plate, developed largely by Niimura [9].

There is also a tremendous opportunity for the exploitation of neutron time-of-flight diffraction using single crystal samples. The upgraded SXD instrument recently installed at ISIS has detectors covering over 50% of the solid angle, with a total of 11 PSDs (Fig. 2.2). The detectors are based on the fibre-optically encoded, 3 mm resolution ZnS scintillator detectors previously used on SXD. Future implementations of related instruments, on beamline and moderator choices with better characteristics for larger unit cell structures, will offer the prospect of studying macromolecular structures; as mentioned above, such an instrument is planned for the ISIS Second Target Station, TS2 [10].

The next generation of high power spallation neutron sources, such as the SNS, being constructed at Oak Ridge National Laboratory in the USA [11] and the spallation source being constructed as part of the J-PARC facility in Tokai, Japan [12] offer new opportunities for neutron protein crystallography. At SNS, a dedicated Macromolecular Neutron Diffractometer (MaNDi) has been designed and is optimized for data collection to 1.5 Å resolution from crystals of volume 0.1–1 mm^3 with unit cell parameters of 150 Å. In addition, the MaNDi instrument will allow neutron data to be collected to between 2.5 and 3 Å, on large macromolecular complexes with unit cell dimensions up to 250 to 300 Å. The MaNDi instrument will be sited on a decoupled moderator at the 60-Hz SNS source, making optimal use of the SNS design to provide best signal-to-noise and highest possible resolution for large unit cell systems. The ability to measure high resolution neutron diffraction data sets within 1–7 days from such large and complex systems promises to greatly extend the range and number of macromolecular systems that are accessible to neutron protein crystallography. It is anticipated that the MaNDi instrument will therefore impact significantly on many areas of structural biology, including enzymology, protein dynamics, drug design, and the study of membrane proteins.

2.2.5 Improvements in Sources

It can be seen from the preceding discussion that the recent successful efforts of instrument designers have led to large improvements in the provision of instrumentation appropriate for the study of macromolecular systems, and it is clear that this will continue. However, a further step function increase in the capabilities of single crystal neutron diffraction will be greatly facilitated when the neutron sources themselves are improved. The consequent increases in flux will lead to the study of larger molecules, smaller crystals, and improved variable temperature, and even kinetic studies. For reactors, the prospect for improvement is somewhat limited, given that the power density in the reactor core limits the potential flux increase to around five times that of the present ILL at best. However, it is simple to construct moderators for cold neutrons in reactors and it is also easy to select, guide and detect the longer wavelength neutrons required for biology; there is much prospect for further exploitation of reactor instrumentation in macromolecular single crystal neutron studies.

For pulsed sources, still relatively in their infancy, there is more scope for improvement, already alluded to above. On both the accelerator and target station side, advancing technology allied with increasing experience of operating such sources promise a rich future. There are many moderator options for optimizing wavelength, flux, and pulse width characteristics for instruments on such sources; these issues are being tackled for the first time with the PCS protein crystallography beamline recently constructed at the Los Alamos pulsed source, built on a partially coupled moderator to optimize its high flux performance for biological studies. The ISIS second target station (under construction) also promises a source more optimized for large molecule structural studies than the existing high resolution target station.

On the accelerator side, as mentioned above there are higher intensity sources in construction in the USA (1.4 MW) and in Japan (600 kW). For still higher flux, the present design study for the ESS, the European Spallation Source, aims to provide a 5 MW source with a peak flux of some 30 times that of the present ISIS, with a time averaged flux equivalent to that of the ILL [13]. The use of time-sorted white beams from these sources means that full advantage can be taken of this flux increase and combining this with appropriately optimized moderators should allow for the provision of extremely powerful instrumentation for future single crystal studies, including macromolecular crystallography [14].

Plans are well advanced for instruments optimized for macromolecular crystallography at all of these sources, including a next generation LADI at ILL, the BIX instruments in Japan, MaNDi at SNS and LMX/Proteus at ISIS TS2.

2.3 Information from Neutron Crystallography

Neutron diffraction is the method of choice for many crystallographic experiments. Among other characteristics, the nature of the scattering of neutrons by atomic species is such that the method offers a description of all atoms in a structure at approximately the same level of precision. This "equivalence" of atoms is due to the fact that neutrons are scattered by the nucleus rather than the electrons in an atom, and hence the scattering power does not have the strong dependence on Z found for many other scattering techniques such as X-ray or electron diffraction. Of particular relevance to the discussion here is the ability of neutron diffraction to detect hydrogen (and deuterium) in chemical and biological structures, where these light elements have approximately equal contributions to the diffraction as do the other atoms in the material.

2.3.1 Neutron Crystallography of Molecular Materials

Neutron scattering has played a major role in developing an understanding of how structure affects the properties of crystalline materials, in areas of relevance to much of modern structural chemistry [15]. Areas accessible to single crystal and powder neutron diffraction include organic materials, pharmaceuticals, small biological macromolecules, zeolites, polymer electrolytes, battery materials, catalysts, superconductors, time-resolved and in situ studies, and chemical magnetism [16]. Neutron diffraction experiments are often carried out under extreme conditions of sample environment such as high and low temperature, under controlled atmospheres, high pressure and in chemical reaction cells. The combination of X-rays and neutrons is powerful in many studies, including the characterization of host–guest interactions in, for example, zeolites, and in determination of charge distributions in crystal structures.

Neutron diffraction is unparalleled in its ability to locate hydrogen atoms and refine their positions and thermal parameters. Hydrogen atoms can be located in metal clusters (e.g., hydride ligands) far more reliably than by any other method. Much of the structural work on hydrogen bonded systems (e.g., amino acids, nucleic acid components, carbohydrates, cyclodextrins) has used neutron diffraction. In addition, determination of the hydrogen anisotropic displacement parameters in short O–O hydrogen bonds allows, for example, the deduction of the shape of the potential well in which the atom sits. Neutron single crystal diffraction has an important role in defining the patterns of "weak" intermolecular interactions in complex molecular and supramolecular structures, as these often crucially involve hydrogen atoms. This leads directly to a strong impact in the expanding area of molecular and crystal engineering. Neutron diffraction also gives complementary information to X-ray diffraction for charge density studies. In X–N studies the neutron parameters fix the nuclear positions and the X-ray data determine the electron density involved in bonding and nonbonding interactions.

2.3.2 Neutron Crystallography in Structural Biology

In addition to the major impact the technique has had in chemical crystallography, single crystal neutron diffraction has made a significant and important contribution in the determination of biologically important structures. There are a number of examples where single crystal studies of proteins have had a profound influence on our understanding of how the protein might function. An excellent summary of the application of single crystal neutron diffraction in the biological area was given by Knott and Schoenborn [17] and in a more recent review by Tsyba and Bau [18].

In the field of structural biology, the relation between structure and function has been established since the early days of protein structure determination. A full understanding of this relationship depends on an appreciation of the detailed molecular interactions involved. These interactions occur through mechanisms such as hydrogen bonding, charge transfer, and other nonbonded interactions, and many of these are governed by the location of hydrogen atoms. Accurate neutron diffraction studies can define to high precision the geometry of an active site, and the role which this may play in important interactions, for example with drug molecules. Hydrogen atoms inevitably decorate much of the outer regions of both protein molecules and interacting small molecules. Many important protein functions can thus depend on the presence or absence of just one hydrogen atom and it is clearly important to locate these accurately.

It is this crucial aspect of the role of hydrogen atoms in biological function that has led to the continued pursuit of routine neutron protein crystallography through many years of effort and in spite of the intrinsic difficulty of the experiments. There are other areas where single crystal neutron diffraction has a unique contribution to make in this field:

- The ability of neutron scattering to distinguish clearly between nitrogen, carbon, and oxygen is important, for example in determining the orientation of histidine, apargine, and glutamine.
- The very large scattering length difference ("contrast") between hydrogen and deuterium can be exploited to allow the determination of exchangeable hydrogen atoms, yielding information on protein dynamics and on solvent accessibility.
- In a traditionally powerful area of application of neutron single crystal studies, analysis of thermal motions of hydrogen-containing groups in the structure can give information on the physics underlying the structure.
- Protein–solvent interactions are critical to life. The ordered solvent structure around protein molecules has also been elucidated by high resolution neutron single crystal diffraction studies. Neutron studies can reveal both the position and the orientation of the water molecules by locating not only oxygen atom positions but also the hydrogen (or deuterium) atoms as well.

2.3.3 Sample and Data Requirements for Single Crystal Neutron Diffraction

The requirement for rather large protein single crystals is one of the main drawbacks of single crystal neutron diffraction compared with X-ray methods. With the relatively low flux of neutron sources and the rather weak scattering of most materials, usually crystals of several cubic millimeters are required to allow collection of a good data set in a reasonable data collection time. The limit is usually regarded as being around 1 mm^3, given that data collection times of greater than 10–14 days per data set are usually impractical. This limit is now being lowered at facilities such as the LADI instrument at ILL and there are significant efforts being made by source and instrument designers to allow it to be further reduced. The major limitation on the signal-to-noise ratio in neutron diffraction from biological (as well as all other hydrogenous materials) is that hydrogen has a large incoherent neutron scattering factor of 80 barns that produces a high level background that significantly reduces the signal to noise ratio of the diffraction data. The corresponding value for the deuterium isotope is $\sim$2 barns; isotopic substitution of deuterium for hydrogen therefore results in huge reductions in the incoherent scattering background and order of magnitude improvement in signal to noise. It is now possible to prepare such isotopically H/D substituted or labeled proteins and other macromolecules in the laboratory. The ability to clone and over-express target proteins of interest for biochemical and biophysical analysis in host microbial systems is now considered routine in laboratories world-wide. These similar microbial systems can be adapted to growth in heavy water (D_2O) solutions and when fed with deuterated carbon sources, are able to produce functional protein molecules in which all hydrogen atoms in the structure have been replaced by the deuterium [19]. The reduced incoherent scattering background from crystals of such deuterated proteins results in order of magnitude improvements in signal to noise. This is a critical advantage for neutron protein crystallography that, together with instrument developments, promises to deliver further 100-fold reductions in the size of sample than can be used on future instruments, making larger and more complex systems amenable to neutron protein structure determination.

2.4 Brief Review of the Use of Neutron Diffraction in the Study of Biological Structures

Much of biological structure and function is mediated by hydrogen bonding interactions and proper account of the hydrogen atom structure of biological systems brings understanding of biological process at the most fundamental level. Neutron protein crystallography can play a key role in structural molecular biology by locating hydrogen atoms and water positions in proteins that cannot be seen by X-ray analysis alone. While NMR and X-ray techniques have unrivalled capacity for high-throughput structure determination, neutron diffraction, and other specialized techniques, make key and unique contributions to the field. Hydrogen bonding interactions mediate most of biological structure and function and the location of even a single hydrogen atom can help determine and define the mechanism and pathway of complex biological processes. However, hydrogen atoms can only be seen (if at all) by X-rays if protein crystals are sufficiently well ordered to provide data to atomic resolutions (<1.1 Å). While the position of many hydrogen atoms can be reliably inferred from the chemical groups to which they are bound, the positions of other more labile – and perhaps more interesting – atoms cannot and must be determined by other techniques.

A number of structural studies have been carried out using neutron diffraction, focusing very much on structures where the X-ray work has left some ambiguity over an important aspect of hydrogen atom location or solvent structure. In most cases, the neutron data have provided an indication of some structural feature which was left undefined by the X-ray data. Often neutron difference Fourier maps based on the known X-ray structure can reveal incorrect assignment of atoms or misorientation of important groups. For example, positions of oxygen and nitrogen atoms in asparagine and glutamine side-chain amides were distinguished and switched based on information from neutron data. Joint X-ray and neutron refinements were used to study the structure of ribonuclease A [20]. Neutron data extending to around 2 Å resolution were collected from a 30 mm^3 sample at the NBS reactor at NIST. The neutron structure refinement showed the proper rotation of key amino groups, including the orientation of the four histidine side chains, the Arg39 side chain was also completely reoriented, and the catalytically important Lys41 was completely rebuilt based on the neutron Fourier maps. The new configuration of this group increased the likelihood that it was involved in the activity of this molecule. The solvent structure was also significantly modified in the joint refinement. High resolution studies have also allowed high levels of detail to be obtained in smaller, and therefore well ordered, protein structures such as lysozyme [21,22] and crambin [23].

However, such studies have always been performed in the limiting context of the need for very large crystals, which has severely limited the number and range of structures amenable for study. Even at the most intense neutron sources, such as the 58 MW high flux reactor at ILL, conventional instruments such as D19 that is prominent in neutron chemical crystallography, have been restricted to small unit cell biological systems, such as vitamin B12 [25,26], lysozyme [21] and to fibre diffraction of biological polymers, hampered by the limited capacity of available instrumentation (see Forsyth et al. this volume). The new generation of protein crystallography instruments, such as LADI at ILL, the BIX instruments at JAERI and PCS at LANSCE, have delivered order of magnitude gains in performance that make feasible studies of larger biological complexes and smaller crystals than was previously possible. The further order of magnitude improvements in performance provided by the protein crystallography instruments planned at the 1.4 MW SNS and the 0.6 MW J-SNS facilities will be decisive in opening new fields of research.

2.4.1 Location of Hydrogen Atoms

The transfer and exchange of hydrogen atoms in biological systems is of fundamental importance in biology. The determination of the hydration and protonation state of proteins is thus a major focus in neutron protein crystallography, particularly where the geometry of these cannot reliably be predicted on the basis of the X-ray structure. A fine early example of this was given in the study of vitamin B_{12} [24]. The continuing interest in vitamin B_{12} coenzyme, both for interest in itself and as a model system for large molecule studies, has led to repeated high resolution neutron single crystal studies being carried out at both room and low temperature. Refinement against room temperature data showed that there was significant reorientation of part of the structure compared with the earlier X-ray study, due to a redistribution of hydrogen bonds around the phosphate group [25]. There is a complex hydrogen bonding network, made still more complex by the presence of disorder in one of the side chains of the corrin ring, characterized by the neutron diffraction studies. The low temperature study [26], carried out at 15 K, showed a substantial reduction in the static disorder present.

The location of a single hydrogen atom can also have profound implications for the biological mechanism of a large molecule, for example in the oxygen carrying mechanism of haemoglobin. The local geometry around the heme group was also the motivation for the early neutron diffraction studies on myoglobin [27]. Subsequent work on this system allowed an important distinction to be made between the binding modes of CO and O_2 molecules, with the finding that the imidazole group of His-64 was not protonated on CO binding [28], while it was found to be so in the situation of O_2 binding [29].

Another case in which the presence or absence of a single hydrogen atom has implications for the function of a protein is in the mechanism of action of the serine proteases [30]. Specifically, in the case of trypsin the location of a single proton, either on Histidine-57 or Asparagine-102, has implications for the catalytic activity of this protein. The neutron single crystal diffraction, collected to 2.2 Å resolution at Brookhaven on a 1.5 mm^3 crystal which had been soaked in D_2O to replace most of the exchangeable H for D, showed that the proton was attached to His-57, showing this to be the chemical base in the hydrolysis reaction [31]. This meant that mechanisms for the action of the serine proteases which had Asp-102 as the base could be eliminated.

In recent key studies, H/D locations determined at medium resolution (~2 Å) by neutron protein crystallography have provided additional information that could not be determined from atomic resolution (<1.2 Å) synchrotron X-ray data alone. For example, the 2 Å neutron structure of an aspartic protease, endothiapepsin, a transition state analogue complex, directly revealed the key hydrogen positions at the catalytic site of the protein (Fig. 2.4). The data provide convincing evidence that Asp-215 is protonated

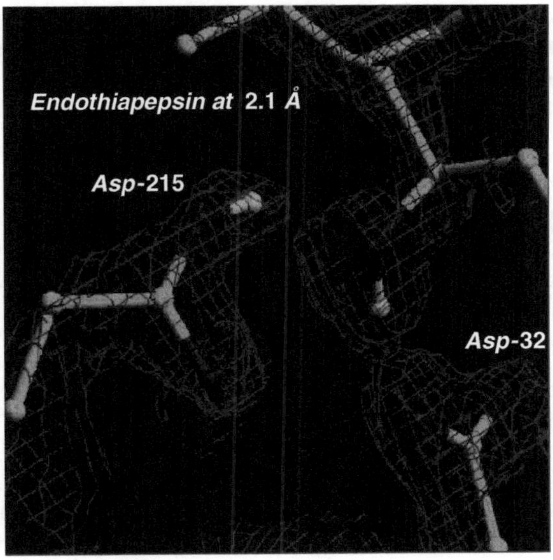

Fig. 2.4. The 2 Å neutron structure of Endothiapepsin directly revealed the key hydrogen positions at the catalytic site of the protein

and that Asp-32 is the negatively charged residue in the transition state complex. This has an important bearing on mechanistic proposals for this class of proteinase, resolving the long-standing controversy over the catalytic mechanism in this important family of enzymes [32]. In myoglobin, where hydrogen atom positions could not be visualized from X-ray data at <1.15 Å [33], the neutron structure of the perdeuterated protein determined at 2.0 Å resolution enabled deuterium atom positions of interest to be determined [29]. In particular, this work allowed the determination of the protonation sites as positive peaks near the N^{ε} or N^{δ} atoms of all 12 histidines in Mb at pH 6.2.

As for smaller organic structures, the hydrogen atoms in terminal methyl groups in proteins undergo torsional or rotational motion. The orientations of these "rotor" methyl groups are difficult to determine, but neutron diffraction has had success in doing so. One example is in the structure of trypsin, where the examination of neutron scattering densities as a function of torsion angle for the rotation of these clearly revealed the correct orientation and hence the location of the hydrogen atoms [34].

2.4.2 Solvent Structure

Neutron diffraction is also ideally suited to the study of solvent structure in biological systems. Typically 40–60% of an average protein crystal actually consists of water. The relative scattering power of solvent water is larger for neutron diffraction than X-ray diffraction, especially so in the case of heavy (D_2O) water and in high resolution studies, so that neutron analysis also gives the opportunity of distinguishing the proton/deuterium positions in the solvent structure. This has led, for example, to a detailed description of the hydration structure in vitamin B_{12} [35]. There has also been success in studying the solvent structure in larger molecule systems, especially in the case where water molecules are well ordered, such as when they are directly bound to the protein surface [36]. The interaction of other solvents with proteins can also be investigated, for example an analysis of the interaction of dimethyl sulphoxide (DMSO) with lysozyme [21] showed that the DMSO molecules interact with the protein surface both through hydrogen bonds and bonding through solvent methyl groups to the hydrophobic parts of the protein molecule, without significantly changing the protein configuration. Such studies offer the possibility of resolving the precise nature of the solvent–protein and solvent–solvent interactions present. This potential has recently begun to be more fully explored by combining high resolution studies with advanced modeling of solvent structure [37]. In a recent technical advance, it has been shown that by collecting neutron data at cryo temperatures, the dynamic disorder within a protein crystal is reduced, the definition of the nuclear density is improved, and a comparison between the 15 and 293 K neutron structures shows that overall, twice as many bound waters (as D_2O) are identified at 15 K than at 293 K [38].

2.4.3 Hydrogen Exchange

While NMR and radioactive labeling are the most common tools for monitoring hydrogen exchange in proteins, the high contrast between H and D makes neutron diffraction a realistic alternative [39]. In particular, neutron diffraction provides a powerful method of monitoring the propensity for exchange of amide protons in even large protein structures. The method does not of course provide kinetic data on hydrogen exchange, but instead provides snapshots of the structure. Such snapshots, however, are sufficient to indicate the degree of exchange and hence to highlight the less flexible regions of the protein molecule. For example, in a series of experiments conducted at the BIX-3 instrument at JAERI on the small and unusually thermostable protein rubredoxin from *P. furiosus* (an organism that grows optimally at 373 K), comparison of the hydrogen-bonding patterns of partially deuterated wild-type and triple mutants proteins provided insight into the H/D-exchange pattern of the N–H amide bonds of the protein backbone and information on the mechanism of unfolding [40].

2.4.4 Low Resolution Studies

In the presence of disorder, or where extremely large structural components are to be resolved, lower resolution data are of value because they reflect measurements on a scale (of order 10 Å) where the scattering density of the various components is roughly constant. Such lower resolution studies can be carried out with neutrons of wavelength of the order of 7–8 Å, thus benefitting from the higher reflectivity for scattering of such neutrons and allowing smaller crystals to be used. In the case of very large biological molecular complexes, or in cases where a major component of the structure is disordered, such studies can provide information on the location of individual components in large and complex systems, such as lipid, detergent, carbohydrate, or nucleic acids, which can be essential to construct a biologically meaningful picture of the structure as a whole.

Low resolution studies, usually coupled with contrast variation, have been used to obtain information on membrane-bound proteins [41], and to see lipids associated with protein complexes and assemblies [42], and in one-dimensional diffraction studies of membrane structures themselves [43]. In a recent example of work on membrane protein structure the detergent structure present in crystals of the peripheral light-harvesting complex of the purple bacteria Rhodopseudomonas acidophila has been determined at a maximal resolution of 12 Å by neutron crystallography (Fig. 2.5) [44].

2.4.5 Other Biologically Relevant Molecules

Neutrons can also be used to study other biologically relevant structures, such as pharmaceuticals and other biologically active molecules. These are in

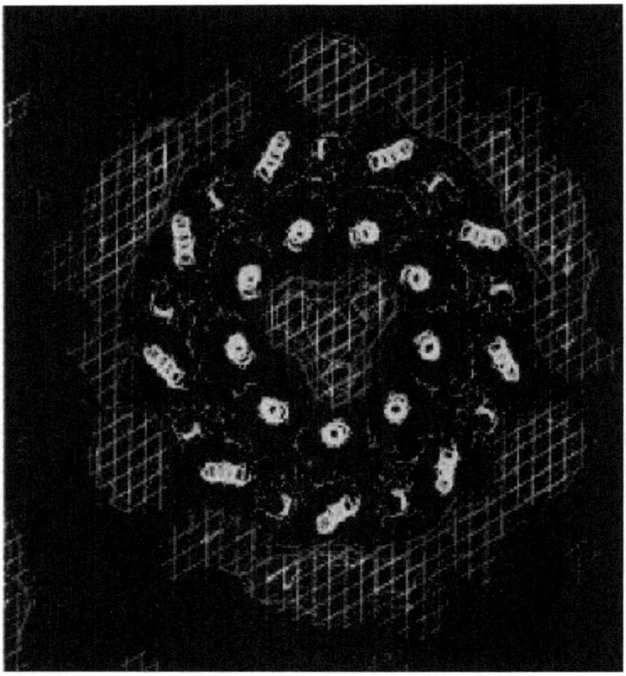

Fig. 2.5. Detergent structure present in crystals of the peripheral light-harvesting complex of the purple bacteria *Rhodopseudomoas acidophila* strain 10050 determined by neutron crystallography at 12 Å resolution

general "small" molecules, where the main interest is in understanding their conformation and interactions, aiming to project these properties into understanding their interactions with macromolecules and hence their function.

One of the largest such molecules subjected to high resolution single crystal neutron diffraction is cyclosporin A, an immunosuppresant drug with wide clinical application [45]. Stable refinements of this structure were obtained from data collected on H3A at Brookhaven on a 20 mm^3 sample, in spite of a low data to parameter ratio of just 2.3. In addition to defining fully the hydrogen atom geometry in this large organic molecule ($C_{62}H_{111}N3_{11}O_{12}{\cdot}H_2O$; 199 atoms in the asymmetric unit), the neutron study revealed the presence of a bound, ordered water molecule – an ordered solvent interaction.

In general, neutron single crystal diffraction is of enormous value in the study of pharmaceuticals, where many drug molecules crystallise with unit cells in the accessible cell range up to $\sim 10^4$ Å^3. Detailed neutron data can be vital to the understanding of molecular conformation, especially with regard to the often very small energy differences between active and inactive polymorphs. Neutrons also sample the bulk of such materials, again vital in the study of polymorphism in relation to production processes.

2.5 Recent Developments and Future Prospects

The recent developments in studying biological structures at high resolution with neutrons have focused around the use of neutron image plates along with modified Laue methods [46]. These developments, including the Japanese BIX instrument and LADI at the ILL, are discussed above and will be discussed in more detail by other contributors to this volume, as well as the potential for high impact of these methods in the field of biomolecular crystallography.

References

1. C.C. Wilson, *Single Crystal Neutron Diffraction from Molecular Materials* (World Scientific, Singapore, 2000)
2. http://www.ill.fr/
3. C. Wilkinson, M.S. Lehmann, Nucl. Inst. Methods **A310**, 411–415 (1991)
4. http://www.isis.rl.ac.uk/
5. X.M. Li, P. Langan, R. Bau, I. Tsyba, F.E. Jenney, M.W.W. Adams, B.P. Schoenborn, Acta Cryst. **D60**, 200–202 (2004)
6. B.L. Hanson, P. Langan, A.K. Katz, X.M. Li, J.M. Harp, J.P. Glusker, B.P. Schoenborn, G.J. Bunick, Acta Cryst. **D60**, 241–249 (2004)
7. N. Niimura, Y. Minezaki, T. Nonaka, J.C. Castagna, F. Cipriani, P. Hoghoj, M.S. Lehmann, C. Wilkinson, Nature Struct. Biol. **4**, 909–914 (1997); J.R. Helliwell, Nature Struct. Biol. **4**, 874–876 (1997)
8. S. Fujiwara, Y. Karasawa, I. Tanaka, Y. Minezaki, Y. Yonezawa, N. Niimura, Physica B **241**, 207–209 (1997)
9. Recent developments are discussed in S. Tazaki, K. Neriishi, K. Takahashi, M. Etoh, Y. Karasawa, S. Kumazawa, N. Niimura, Nucl. Inst. Methods **A424**, 20–25 (1999)
10. http://www.isis.rl.ac.uk/TargetStation2/
11. http://www.sns.gov/
12. http://www.j-parc.jp/
13. J. Kjems, A.D. Taylor, J.L. Finney, H. Lengeler, U. Steigenberger, ESS: A Next Generation Neutron Source for Europe, Volume I: The European Spallation Source; Volume II: The scientific case (ESS Council, Roskilde, 1997)
14. W. Jauch, M.S. Lehmann, L. Sjölin, C. Wilkinson, C.C. Wilson, ILL Internal Report, ILL97/JA19T (Institut Laue-Langevin, Grenoble, 1997)
15. C.C. Wilson, in *Implications of Molecular and Materials Structure for New Technologies*, ed. by J.A.K. Howard, F.H. Allen, G.P. Shields. NATO Science Series E: vol 360 (Kluwer, Dordrecht, 1999), pp. 11–21
16. C.C. Wilson, (1998) http://www.rsc.org/pdf/forwardlook/neutrongrp.pdf.
17. R.B. Knott, B.P. Schoenborn, in *Neutrons in Biology*, ed. by B.P. Schoenborn, R.B. Knott (Plenum, New York, 1996), pp. 1–15
18. I. Tsyba, R. Bau, Chemtracts **15**, 233–257 (2002)
19. F. Shu, V. Ramakrishnan, B.P. Schoenborn, Proc. Natl. Acad. Sci. USA **97**, 3872–3877 (2000)
20. A. Wlodawer, L. Sjölin, Biochemistry **22**, 2720–2728 (1983)

21. G.A. Bentley, E.D. Duee, S.A. Mason, A.C.J. Nunes, Chim. Phys. **76**, 817–821 (1979); G.A. Bentley, M. Delepierre, C.M. Dobson, R.E. Wedin, S.A. Mason, F.M.J. Poulsen, J. Mol. Biol. **170**, 243–247 (1983); S.A. Mason, G.A. Bentley, G.J. McIntyre, in *Neutrons in Biology*, ed. by B.P. Schoenborn (Plenum, New York, 1984), pp. 323–334
22. M.S. Lehmann, R.F.D. Stansfield, Biochemistry **28**, 7028–7033 (1989)
23. M.M. Teeter, A.A. Kossiakoff, in *Neutrons in Biology*, ed. by B.P. Schoenborn (Plenum, New York, 1984), pp. 335–348
24. H.F.J. Savage, P.F. Lindley, J.L. Finney, P.A. Timmins, Acta Cryst. **B43**, 280–295 (1987)
25. J.P. Bouquiere, J.L. Finney, M.S. Lehmann, P.F. Lindley, H.F.J. Savage, Acta Cryst. **B49**, 79–89 (1993)
26. F.M. Moore, B.T.M. Willis, D. Crowfoot-Hodgkin, Nature **214**, 130–133 (1967)
27. B.P. Schoenborn, Nature **224**, 143–146 (1969)
28. J.C. Hanson, B.P. Schoenborn, J. Mol. Biol. **153**, 117–146 (1981)
29. S.E.V. Phillips, B.P. Schoenborn, Nature **292**, 81–82 (1981)
30. A.A. Kossiakoff, S.A. Spencer, Biochemistry **20**, 6462–6474 (1981)
31. A.A. Kossiakoff, S.A. Spencer, Nature **288**, 414–416 (1980)
32. L. Coates, P.T. Erskine, S.P. Wood, D.A. Myles, J.B. Cooper, Biochemistry **40**, 13149–13157 (2001)
33. J. Vojtechovsky, K. Chu, J. Berendzen, R.M. Sweet, I. Schlichting, Biophys. J. **77**, 2153–2174 (1999); A.E. Miele, L. Federici, G. Sciara, F. Draghi, M. Brunori, B. Vallone, Acta Cryst. **D59**, 982–988 (2003)
34. A.A. Kossiakoff, S. Shteyn, Nature **311**, 582–583 (1984)
35. H.F.J. Savage, Biophys. J. **50**, 947–965 (1986); H.F.J. Savage, Biophys. J. **50**, 967–980 (1986)
36. S.E.V. Phillips, B.P. Schoenborn, Nature **292**, 81–82 (1981); N.V. Raghavan, B.P. Schoenborn, in *Neutrons in Biology*, ed. by B.P. Schoenborn, (Plenum, New York, 1984), pp. 247–259; M.M. Teeter, Proc. Natl. Acad. Sci. USA **81**, 6014–6018 (1984); H.F.J. Savage, A. Wlodawer, Methods Enzymol. **127**, 162–183 (1986)
37. B.P. Schoenborn, J. Mol. Biol. **201**, 741–749 (1988); X. Cheng, B.P. Schoenborn, Acta Cryst. **B46**, 195–208 (1990); A.A. Kossiakoff, M.D. Sintchak, J. Shpungin, L.G. Presta, Proteins: Struct., Funct. and Gen., **12**, 223 (1992)
38. M.P. Blakeley, M. Cianci, J.R. Helliwell, P.J. Rizkallah, Chem. Soc. Rev., **33**, 548–557 (2004)
39. A.A. Kossiakoff, Nature **296**, 713–721 (1983)
40. T. Chatake, K. Kurihara, I. Tanaka, I. Tsyba, R. Bau, F.E. Jenney Jr., M.W. Adams, N. Niimura, Acta Cryst. **D60**, 1364–1373 (2004)
41. M. Roth, A. Lewit-Bentley, H. Michel, J. Deisenhofer, R. Huber, D. Oesterhelt, Nature **340**, 659–662 (1989)
42. P.A. Timmins, B. Poliks, L.J. Banaszak, Science, **257**, 652–655 (1992)
43. G. Zaccai, J.K. Blasie, B.P. Schoenborn, Proc. Natl. Acad. Sci. USA, **72**, 376–380 (1975); J.F. Pardon, D.L. Worcester, J.C. Wooley, K. Tatchell, K.E. van Holde, B.M. Richards, Nucleic Acids Res. **2**, 2163–2176 (1975); D.L. Worcester, N.P. Franks, J. Mol. Biol. **199**, 359–378 (1976)
44. S.M. Prince, T.D. Howard, D.A.A. Myles, C. Wilkinson, M.Z. Papiz, A.A. Freer, R.J. Cogdell, N.W. Isaacs, J. Mol. Biol. **326**, 307–315 (2003)
45. R.B. Knott, J. Schefer, B.P. Schoenborn, Acta Cryst. **C46**, 1528–1533 (1990)
46. J.R. Helliwell, Nature Struct. Biol. **4**, 874–876 (1997)

3

Neutron Protein Crystallography: Hydrogen and Hydration in Proteins

N. Niimura

3.1 Introduction

The three-dimensional structure determination of biological macromolecules such as proteins and nucleic acids by X-ray crystallography has improved our understanding of many of the mysteries involved in life processes. At the same time, these results have clearly suggested that hydrogen and water molecules around proteins and nucleic acids play a very important role in many physiological functions. However, since it is very hard to determine positions of hydrogen atoms in protein molecules using X-rays, a detailed discussion of protonation and hydration sites is often very speculative upon so far. In contrast, neutron diffraction provides an experimental method of locating hydrogen atoms much more precise. Despite this quality, the examples of protein structure determination by neutrons are relatively low since the requested sample size of the protein crystals is larger and it takes a considerable amount of time to collect a sufficient number of Bragg reflections.

The recent development of a neutron imaging plate (NIP) became a breakthrough in the application of neutron protein crystallography (NPC) [1–3]. Its first application of NIP was the structure determination of tetragonal hen-egg-white lysozyme using the quasi-Laue diffractometer, LADI at the Institute Laue-Langevin (ILL) in Grenoble [4]. In the Japan Atomic Energy Research Institute (JAERI), several high-resolution neutron diffractometers (BIX-type diffractometers) dedicated to biological macromolecules have been constructed, which exploit NIP using monochromatized neutron beam [5–9]. Detailed descriptions on NIP and the BIX-type diffractometers are given in the contribution by Wilson et al. in this volume and in the original papers, respectively.

The general subject of NPC has been already reviewed by several authors [10–17]. Their articles are recommended and helpful in understanding the historical background of this approach. In this chapter, several topics of NPC relevant to hydrogen positions and hydration in proteins, obtained using BIX-type diffractometers will be presented. The proteins that will be treated in the present chapter are myoglobin (Mb) [18], wild type rubredoxin (Rb-w) [19], a mutant form of rubredoxin (Rb-m) [20], hen-egg-white lysozyme (HEWL) at pH 4.9 [21] and cubic porcine insulin [22].

3.2 Complementarity of Neutrons and X-rays

The distinctive features of neutrons are summarized as follows: (i) Since the neutron scattering lengths densities of hydrogen and deuterium are comparable to those of other elements, they are easily observed by neutrons. The X-ray atomic scattering factor of hydrogen is much less and hydrogen is hard to be observed by X-rays. (ii) Since a proton (H^+) has no electrons, it can not be seen by X-rays, on the contrary it can be observed by neutrons. Using heavy water (D_2O) in neutron diffractometry for the crystallization of proteins H^+ can be replaced by D^+ and observed by neutrons. Mobile hydrogen atoms can be distinguished if they are replaced by deuterium as the neutron scattering lengths densities of hydrogen and deuterium are different. Table 3.1 shows the neutron scattering lengths densities and X-ray atomic scattering factors of some of the elements which constitute proteins.

3.2.1 Refinement of Hydrogen Positions

Since the neutron scattering lengths densities of hydrogen and deuterium are comparable to those of other elements, in neutron protein crystallography they are not only identified but also their positions can be refined like other elements, such as carbon, nitrogen and so on. Several examples are shown in Fig. 3.1. Figure 3.1a shows a $2|F_o| - |F_c|$ nuclear density map of Phe48 of wild type rubredoxin at 1.5 Å resolution [19]. The neutron scattering lengths density of the hydrogen atoms is negative, while deuterium, carbon, nitrogen, and oxygen atoms all have positive neutron scattering lengths densities. In the map shown, the hydrogen atoms bound to carbon atoms are clearly visible. Their negative neutron scattering length density is clearly separated from the

Table 3.1. Neutron scattering lengths and X-ray atomic scattering factors

atom	neutron b_{coh} (10^{-12} cm)	X-ray $f_{X\text{-ray}}$ (10^{-12} cm)
D^+	0.67	0
H	−0.37	0.28
D	0.67	0.28
C	0.67	1.69
N	0.94	1.97
O	0.58	2.25
S	0.29	4.48

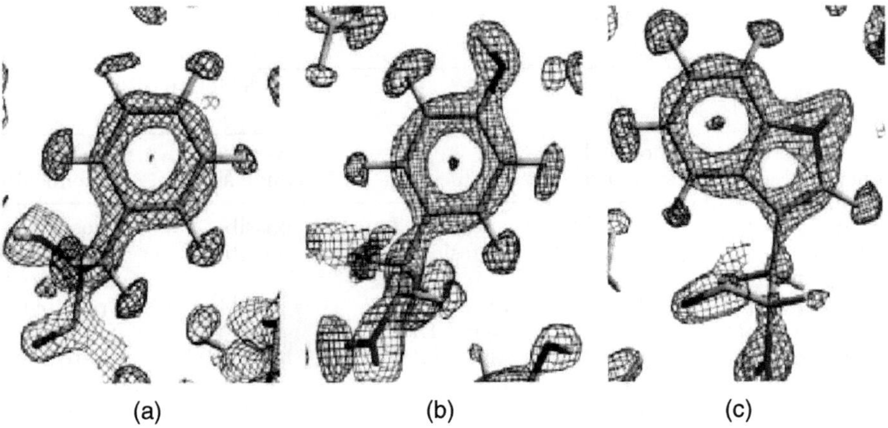

Fig. 3.1. $2|F_o| - |F_c|$ nuclear density maps around. (**a**) Phe48, (**b**) Tyr12, (**c**) Trp36

positive densities of the carbon atoms. Figure 3.1b shows a similar nuclear density map for Tyr12 of wild type rubredoxin at 1.5 Å resolution [19]. The hydrogen atoms produce a large amount of incoherent scattering, which results in an undesirably high level of background radiation in the neutron diffraction experiment. This effect can be partially overcome either by growing crystals from, or by soaking the crystals in, D_2O solutions. This treatment leads to the replacement of hydrogen atoms bound to nitrogen and oxygen (exchangeable hydrogens) by deuterium, as well as of the hydrogen atoms of the solvent molecules in the crystal, without modification of the overall structure of the macromolecule. In Fig. 3.1b, the density contours of the hydrogen atom of the O–H bond in Tyr12 have a positive value, thus it can be concluded that it has been replaced by a deuterium atom. Figure 3.1c shows the $2|F_o| - |F_c|$ nuclear density of Trp36 of wild type rubredoxin at 1.5 Å resolution [19]. It is seen that the N–H bond of the indole ring has a positive density value, i.e., it has become an N–D bond.

3.2.2 Hydrogen Atoms Which Cannot be Predicted Stereochemically

The positions of hydrogen atoms covalently bound to carbon atoms can be calculated stereochemically based on the coordinates of carbon and nitrogen atoms determined by high resolution X-ray crystal structure analysis. However, the positions of some hydrogen atoms covalently bonded to carbon atoms are difficult to be calculated stereochemically, and if hydrogen atoms bound to oxygen, nitrogen, and sulfur atoms become protons, they are impossible to be identified and refined because protons have no electrons scattering X-rays. These hydrogen atoms and protons are summarized in Table 3.2. Neutrons can identify and refine these hydrogen positions.

Table 3.2. Hydrogen atoms (protons) in proteins

functional group	chemical structure	residues	detection of hydrogen positions: X-ray analysis	detection of hydrogen positions: stereo-chemically
aromatic ring	Φ-H, Φ-D	Phe, Tyr, Trp, His	possible	possible
alkyl group (except methyl)	$-CH-$, $-CH_2-$	all residues	possible	possible
peptide group and –ND group	–ND–	all residues except His	possible	possible
methyl group	$-CH_3$	Ala, Ile, Leu, Met, Thr, Val	hard	hard
protonated amino group	$-ND_3$	N-terminus, Lys	hard	hard
hydroxyl group	–OD	Ser, Thr, Tyr	hard	impossible
protonated carboxyl group	–COOD	C-terminus, Asp, Glu	impossible/hard	impossible
amino group	$-ND_2$	Arg, Asn, Gln	hard	impossible
sulfhydryl group	–SD	Cys	hard	impossible

The hydrogen atoms in methyl groups can sometimes be significantly off their predicted positions because of free rotation around the C–C bonds. Consequently, if their precise positions are required they should be determined by neutron diffraction experiments. Figure 3.2 shows examples of some methyl hydrogen atoms in wild type rubredoxin determined from neutron diffraction data [19].

In X-ray protein crystallography sometimes it is very difficult to distinguish the nitrogen and oxygen atoms in Asn and Gln. In neutron protein crystallography such a difficulty does not occur on replacement of hydrogen of the amino groups by deuterium atoms ($-ND_2$) which are identified very easily. Figure 3.3 shows the $2|F_o| - |F_c|$ Fourier map of Asn21 in the mutant form of rubredoxin [20].

The protonation and deprotonation states of the two nitrogen atoms (N_π, N_τ) in the imidazole ring of histidine are often very important pieces of information that need to be known in order to fully understand the function of certain enzymes, as well as the metal complexation behavior of certain proteins. This information can be obtained from neutron diffraction. Figures 3.4a, b show the $2|F_o| - |F_c|$ nuclear density maps of the His5 and His10 residues, respectively, of the B-chain of cubic porcine insulin at 1.6 Å resolution [22]. The protein is a hetero-dimer, composed of an A-chain and a B-chain. For His5 of the B-chain, N_π is protonated and N_τ is deprotonated. In contrast, for His10, both N_π and N_τ are protonated. This means that His5 is electronically neutral while His10 is positively charged.

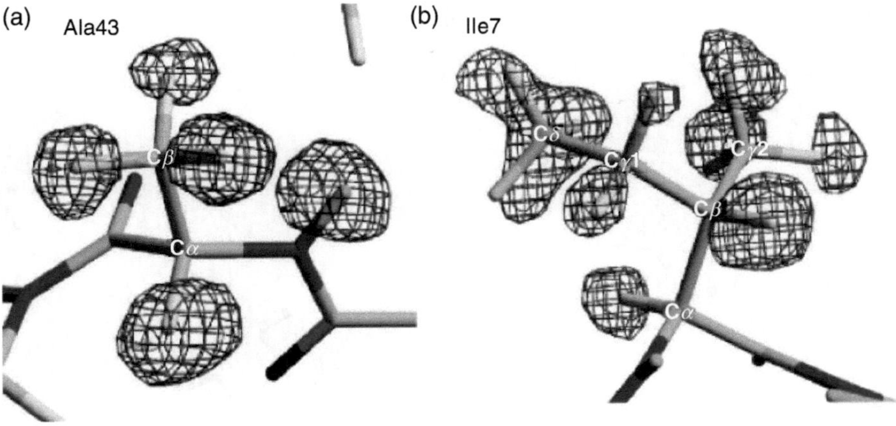

Fig. 3.2. $|F_o| - |F_c|$ omit map of the hydrogen atoms around the residues Ala43 (**a**), Ile7 (**b**)

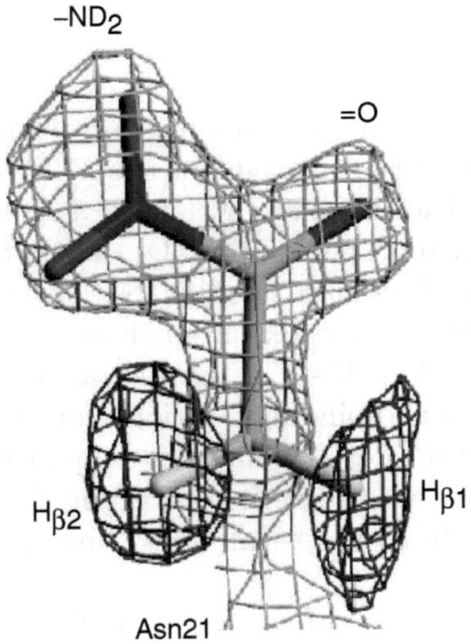

Fig. 3.3. The $2|F_o| - |F_c|$ Fourier map of Asn21 in the mutant form of rubredoxin

As mentioned earlier, polar hydrogen atoms, like those of N–H and O–H bonds, can be exchanged by deuterium if the protein crystal is soaked in a D_2O buffer. In contrast, hydrogen atoms bound to carbon are normally not exchangeable. An exception is the hydrogen atom bonded to the $C_{\varepsilon 1}$ carbon atom of histidine. The $C_{\varepsilon 1}$–H of the imidazole group is the most acidic C–H bond found in amino acids [23]. Therefore this hydrogen atom is in principle exchangeable, depending on its environment. Direct experimental evidence for this behavior was found in metmyoglobin. Figure 3.5 shows the nuclear density map for His97 in this protein [18].

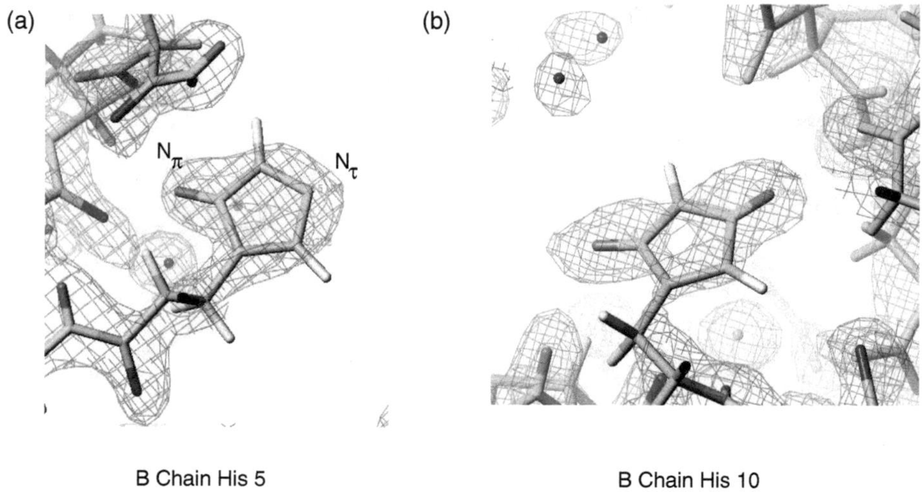

Fig. 3.4. $2|F_o| - |F_c|$ nuclear density map of (**a**) His5 and (**b**) His10 in the B-chain of cubic porcine insulin

The $H_{\varepsilon1}$-atom clearly shows positive neutron density contours near $C_{\varepsilon1}$, indicating an exchange of the H atoms of the C–H bond to deuterium. An occupancy refinement yields the ratio 80% D/20% H. The nitrogen atom $N_{\varepsilon2}$-atom is also deuterated (occupancy: 65% D). The alternative conformation, obtained by rotating the imidazole ring by 180° around the C_β–C_γ axis, cannot explain this finding since the $H_{\delta2}$-atom shows a full hydrogen occupancy (negative density in Fig. 3.5). His97 is located on the so-called proximal side of the heme plane (the ligand binding position is on the other site of the heme plane, the so-called distal side). To our knowledge, this is the first time neutron diffraction has been used to verify the acidic character of the H 1 atom of His97 in myoglobin. In hen-egg-white lysozyme, a similar conclusion was reported recently [24].

In the proposed mechanism of the reaction of lysozyme with oligosaccharides, consideration was given to the fact that the enzyme activity is maximal at pH 5 and is less active at pH 7. It is postulated that at pH 5, the carboxylate group of Glu35 is protonated, and it is this proton that is transferred to the oxygen atom on the bound substrate (sugar) during the hydrolysis process. During the reaction, another acidic residue Asp52, remains in its dissociated state [34]. In order to elucidate the role of hydrogen atoms in this reaction, neutron diffraction experiments of hen-egg-white lysozyme, crystals of which have been grown at different pH's (specifically, 4.9 [21] and 7.0 [4]), have been carried out. The detailed procedures of these neutron structure analyzes are given in [4, 21]. The results shown in Fig. 3.6a, b show the $2|F_o| - |F_c|$ nuclear density map around the carboxylate group of Glu35 at pH 4.9 (Fig. 3.6a) and pH 7.0 (Fig. 3.6b). As indicated by an arrow in Fig. 3.6a,

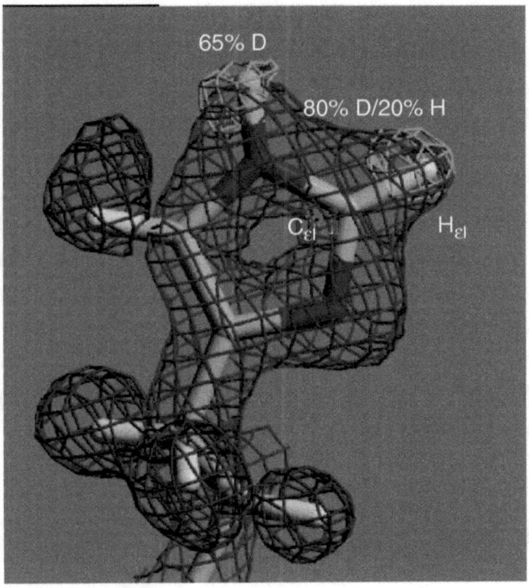

Fig. 3.5. Nuclear density maps of His97 in myoglobin: $2|F_o| - |F_c|$ map (positive) (dark gray); $|F_o| - |F_c|$ omit map (positive) (light gray); $|F_o| - |F_c|$ omit map (negative) (gray). All H(D)-atoms were omitted for the calculation of the F_c and j_c for the omit-map

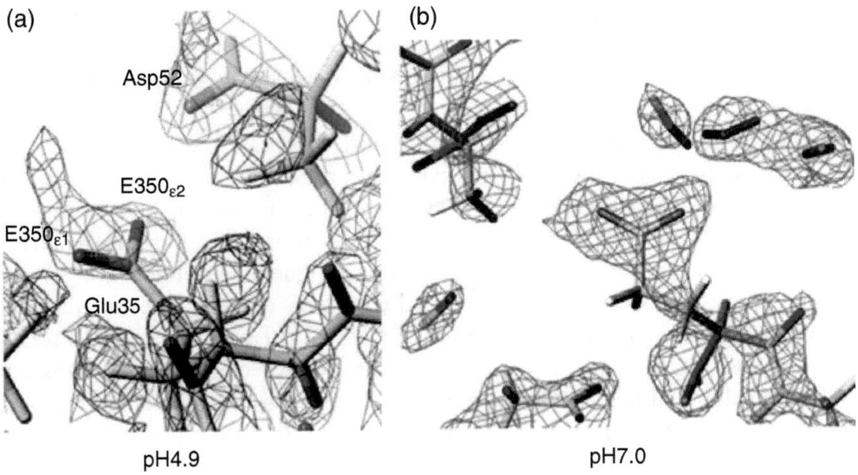

Fig. 3.6. $2|F_o| - |F_c|$ nuclear density map around the carboxylate group of Glu35 at pH 4.9 (**a**), and Fig. 3.15(**b**) at pH 7.0

neutron density was observed to extend from the position of the O atom of the carboxyl group labeled E35$O_{\varepsilon 1}$, suggesting that this carboxyl oxygen atom is a protonated atom. On the other hand, in Fig. 3.6b, it is seen that around this oxygen (E35$O_{\varepsilon 1}$) there is a water molecule at pH 7.0, but no indication of hydrogen (deuterium) atoms. The fact that this catalytic site is deprotonated explains why lysozyme has significantly reduced activity at pH 7.0. The results indicate that a water molecule around the carboxyl oxygen atom at pH 7.0 is kicked out as a result of the protonation of this oxygen atom at pH 4.9, and suggests that it is in fact this enzymatically-active proton which is subsequently transferred to the oxygen atom of the substrate (sugar) during the hydrolysis process. Mason et al. have carried out a neutron diffraction study of lysozyme at pH 4.2 using a triclinic crystal. They report a protonated carboxylate group of Glu35 [26].

3.3 Hydrogen Bonding

3.3.1 Weak and Strong Hydrogen Bonding

Hydrogen bonds play important roles in countless biological processes. The interaction energy of the hydrogen bond is intermediate between those involving covalent and van der Waals forces. Hydrogen bonds are directional and form several kinds of networks in biological macromolecules. However, since it is not easy to determine the positions of all the hydrogen atoms in protein molecules using X-rays or NMR alone, detailed discussions of hydrogen bonds, X–H–Y (in which X and Y are the hydrogen donor and acceptor, respectively), have often been limited, because of the absence of detailed positional information of H atoms.

Along with other investigators, Baker and Hubbard have extensively discussed H-bonds in globular proteins, using hydrogen atom positions predicted from atomic coordinates derived from high-resolution protein X-ray data [27]. Hydrogen atoms were added to the various protein models at their calculated positions, but only those that could be unambiguously defined by the protein geometry. As a matter of fact, no hydrogens were placed on amino or hydroxyl groups, such as those in Ser, Thr, Tyr, or Lys side chains. Thus, in their conclusions, the authors stressed the necessity of high-resolution neutron diffraction studies. Recently performed high-resolution neutron results meet their suggestion. Figure 3.7 shows one example, taken from the study of a mutant form of rubredoxin. In this diagram, the H-bonds, X–H(D)–Y have been plotted with the H(D) atom at the origin, the X–H(D) bond defining the horizontal axis, and the Y atom distributed in the (x, y) plane [20]. The resulting figure is consistent with the concept of the weak and strong H-bonds as proposed by Desiraju and Steiner [28]:

$$\text{Strong H-bonds: } 1.5\,\text{Å} < d[\text{H–Y}] < 2.2\,\text{Å},\ 130° < \text{angle}[\text{X–H–Y}] < 180°$$

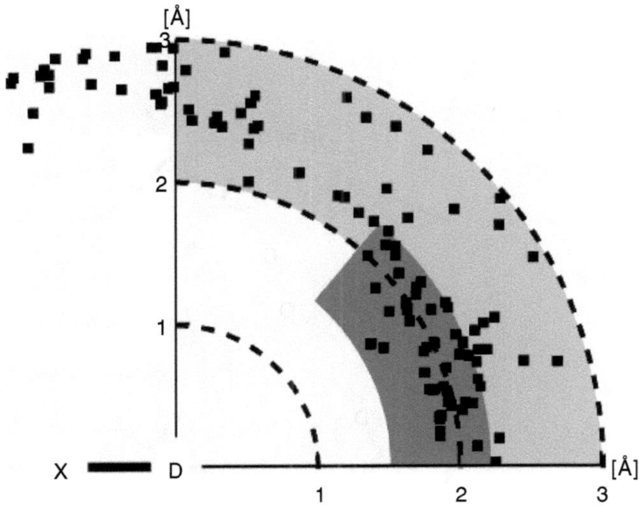

Fig. 3.7. The distribution of Y atoms in X–H(D)–Y hydrogen bonds, with the position of the H(D) atom fixed at the origin. The component of H(D)–Y along the X–H(D) direction is plotted along the horizontal axis, and the component of H(D)–Y perpendicular to X–H(D) is plotted in the vertical direction. The region of strong hydrogen bonds is indicated by the gray area, while weak hydrogen bonds are shown in the light gray area

Weak H-bonds: 2.2 Å $< d$[H–Y] $<$ 3.0 Å, 90° $<$ angle[X–H–Y] $<$ 180°

In Fig. 3.7, strong and weak H-bonds are indicated by gray and light gray areas, respectively. It can be seen that high-resolution neutron protein crystallography has allowed to examine H-bonds in more detail, and that numerous "weak H-bonds" in a protein structure can be identified with this method.

3.3.2 Bifurcated Hydrogen Bonds

In our studies of several small proteins, all the hydrogen bonds between the C=O and N–H groups in the helices of the proteins have been surveyed including their hydrogen atom positions. Figure 3.8 shows one example of how a conventional hydrogen bond in a α-helix of myoglobin is seen in a neutron diffraction experiment. The hydrogen positions as well as carbon, nitrogen and oxygen positions have been refined. This figure clearly indicates that the location of the experimental H positions with neutron data is usually unambiguous. However, we have found several exceptions to the conventional picture of H-bonds in α-helices.

When the hydrogen bonds in the α-helices were refined individually, several bifurcated hydrogen bonds were found. The occurrence of bifurcated hydrogen bonds in the α-helices of proteins has been proposed earlier, based on an

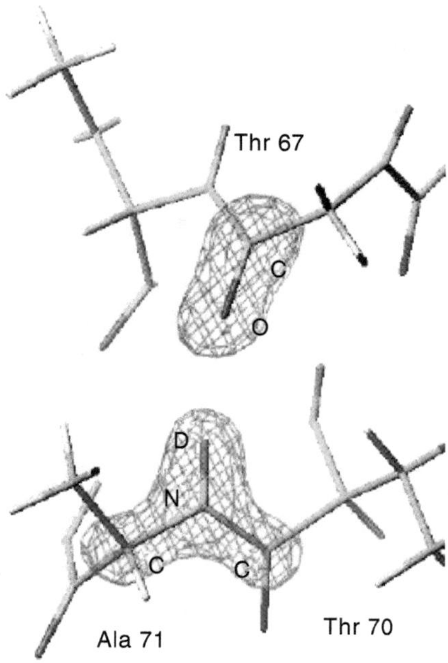

Fig. 3.8. Hydrogen bond in a α-helix. $|F_o| - |F_c|$ omit nuclear density map. The marked atoms were omitted for the calculation of F_c for the omit-map

analysis of calculated hydrogen atom positions from atomic coordinates derived from high-resolution X-ray data [27,29]. However, it is somewhat risky to discuss the detailed structure of bifurcated hydrogen bonds based solely on those predictions. In high-resolution neutron protein crystallography, the H atoms of the polypeptide backbone can be identified and refined unambiguously. In the case of myoglobin a positional refinement with loosened restraints for the planarity of the peptide plane was performed, i.e., the O–C–N–H group was allowed to deviate from a planar *trans*-configuration. The result is that the O–C–N–H torsion angle showed deviations up to 15° from planarity with an average value equal to 179.2° and a standard deviation of 6.3°. These values are in good agreement with those from ultra-high-resolution X-ray structure determinations [30,31].

3.4 H/D Exchange

In a recently-completed analysis of wild type rubredoxin [19], the protein solution used was subjected to a H_2O/D_2O exchange prior to the growth of the crystals. Out of a total of 74 hydrogen/deuterium atoms at potentially exchangeable sites, 24 atoms did not have significant positive (deuterium) peaks at the expected positions. Of those, 11 atoms were bound to nitrogen

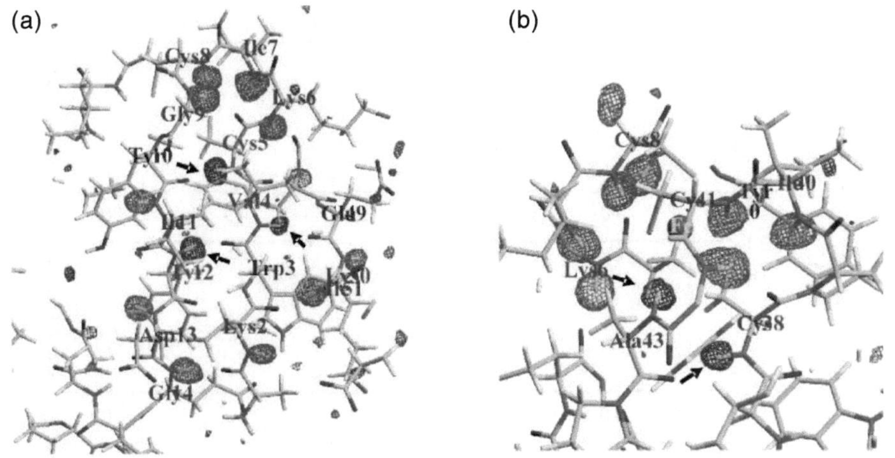

Fig. 3.9. $|F_o| - |F_c|$ omit maps in the β-sheet region (**a**) and near the iron–sulfur cluster. (**b**) Negative densities are marked with arrows

atoms of the main chain, implying that those positions are not fully accessible to the H/D exchange process. Moreover, five of them have prominent negative (hydrogen) peaks. Figure 3.9a, b show $|F_o| - |F_c|$ omit maps around those H atoms, calculated without contributions from any H and D atoms bonded to the main chain N atoms. Of the five residues whose backbone N–H groups did not exchange with D, Val4, Cys5, and Tyr12 are located at the central-sheet of the protein, while Cys38 and Ala43 are in the region of the unique FeS_4 redox site. Those atoms all form hydrogen bonds with neighboring oxygen atoms of the main chain. In Fig. 3.9a, b, negative densities are at those H atoms positions even though most other H/D positions have positive densities. Those H atoms (negative densities) did not engage in H_2O/D_2O exchange in D_2O presumably because of the poor solvent accessibility to those positions.

In order to obtain quantitative information about the distribution of hydrogen/deuterium populations in the rubredoxin molecule, the occupancies of the H and D atoms bonded to the main chain N atoms were refined. In this least-squares refinement, a D atom and an H atom were constrained to be in the same position, and their B-factors were set equal to the values of the N atoms to which they were bonded. During this refinement, the sum of the two occupancies of the H and D atoms was not constrained to be 1, but after the refinement these occupancies were recalculated and reset to give a sum of 1.

The result of this population refinement is shown in Fig. 3.10. The five atoms mentioned above have quite small (or nearly zero) values of the H/D exchange ratio. Comparing this result with the distribution of B-factors and accessible surface area (ASA) of the main chain atoms, it is seen that the H/D atoms having small H/D exchange ratios also have small B-factor values and small ASAs (Fig. 3.10, middle and bottom). These results not only show

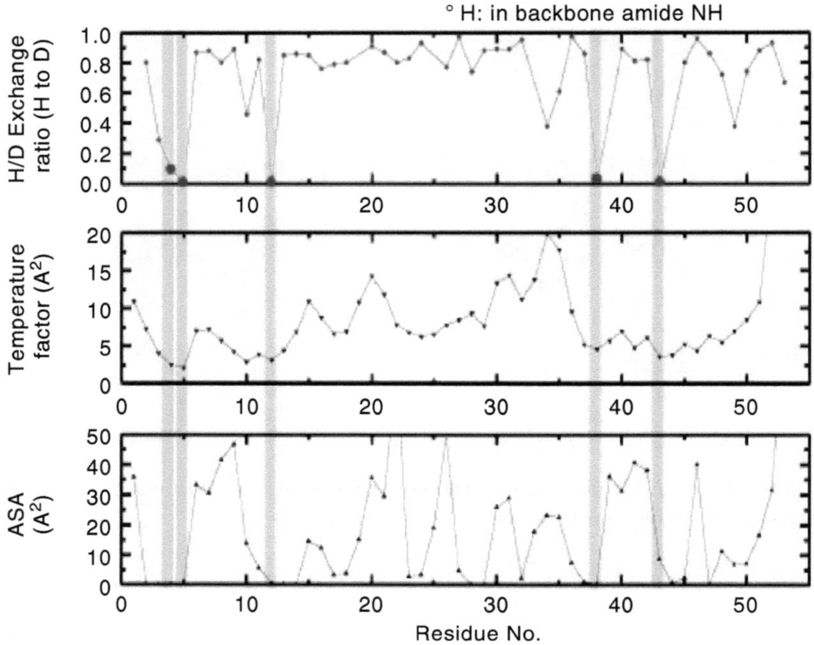

Fig. 3.10. H/D exchange ratio (*top*), *B*-factor (*middle*) and accessible surface area (ASA) (*bottom*) of main chain

that those atoms are located in the interior of the protein molecule, but also suggest that the regions around those atoms have such a rigid structure so that solvent molecules are unable to contact them. The same H/D exchange analyzes have been carried out on mutant form of rubredoxin and myoglobin and very similar results have been observed [18, 20].

It is interesting to compare the H/D exchange of wild type rubredoxin obtained by neutron protein crystallography with the one obtained by NMR [32]. Generally speaking, the trend is that an amide hydrogen bond, which has a fast H/D NMR exchange rate will show a high H/D exchange ratio in the neutron diffraction experiment, and conversely a slow rate of NMR exchange also corresponds to a low ratio in the neutron diffraction. However, a few exceptions can be found. Although a certain N–H bond has a slow H/D exchange rate according to NMR data, its H/D exchange ratio from neutron diffraction is very high and it was found that a conformational change around the amide hydrogen has occurred in the crystallization process. Figures 3.11a, b show such an example near residue Ile7 of wild type rubredoxin. In the solution structure of Ile7 determined by NMR (Fig. 3.11b), the amide hydrogen atom is completely surrounded by the sidechain of Ile7 and shielded from the water [31], H/D exchange under such conditions should be slow. On the contrary, in the present neutron diffraction crystal structure (Fig. 3.11a), C_δ of Ile7 is

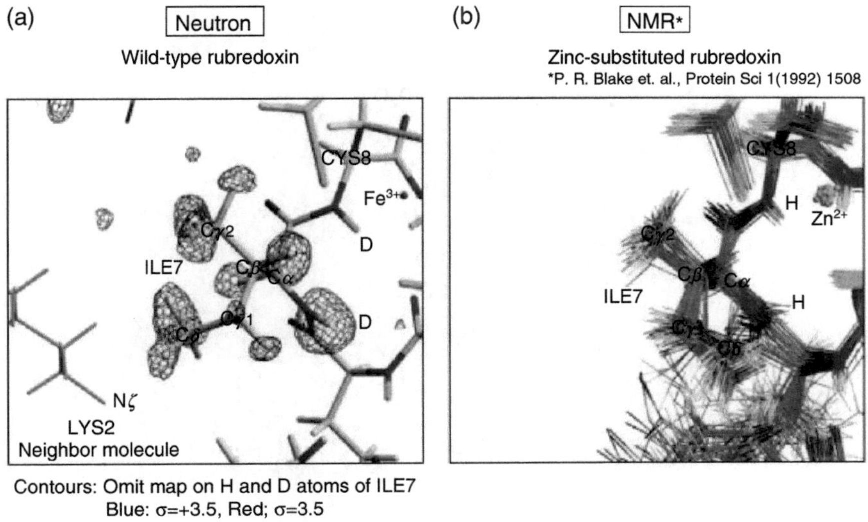

Fig. 3.11. (**a**) The omit map on H and D atoms of Ile7 of wild-type rubredoxin determined by neutron diffraction. (**b**) The portion of Ile7 of wild type rubredoxin (zinc-substituted) determined by NMR

bent to the outside of the proteins, the amide hydrogen atom is exposed to water and the H/D exchange ratio should be high.

3.5 Hydration in Proteins

3.5.1 Experimental Observation of Hydration Molecules

The hydration structure of myoglobin has been studied by Schoenborn et al. and the hydration layer structure such as radial distribution function of water around protein atoms were obtained [34–36]. In our recent study of myoglobin, the hydration structure of individual water molecules was presented [37]. Figure 3.12 displays one region of the hydration structure around myoglobin. It shows that all of the hydration water molecules are completely isolated. It is also remarkable that a sulfate group can be clearly distinguished in this map. In some cases, hydrogen (deuterium) atoms in water molecules can be clearly identified in triangular (boomerang) shaped peaks and the formation of the hydrogen bonds between two water molecules can be recognized as well. At the same time it is interesting to note that, near the two triangular-shaped contours, a spherically shaped water molecule can be found (Fig. 3.12). Moreover, water molecules with other shapes, such as ellipsoidal (stick-shaped) ones, have been found in other places. The interpretation of these shapes will be discussed in Sect. 3.5.2 [37].

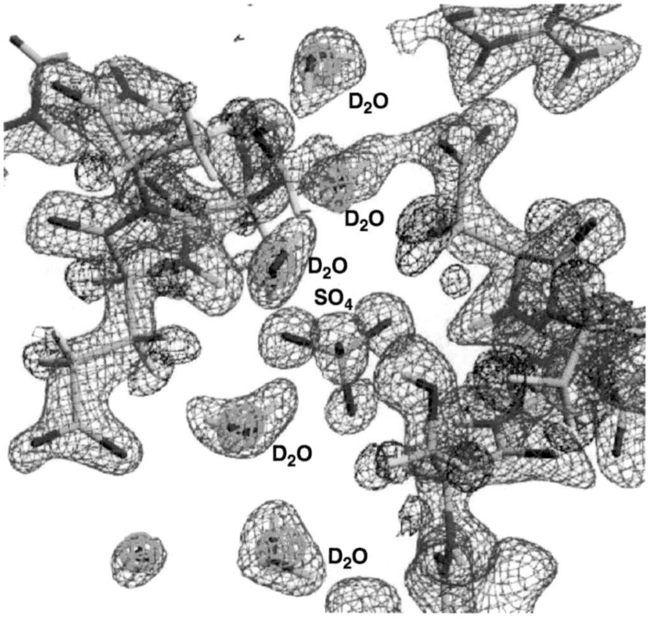

Fig. 3.12. Protein–protein contact region in the case of myoglobin. $2|F_o| - |F_c|$ nuclear density map contoured at positive and negative values. The $2|F_o|-|F_c|$ X-ray electron density map for the water molecules (D_2O) is superimposed. The triangular-shaped neutron contours correspond to D_2O molecules

3.5.2 Classification of Hydration

We have categorized observed water molecules into the following classes based on their appearance in Fourier maps: (i) triangular shape, (ii) ellipsoidal stick shape, and (iii) spherical shape. Moreover the second category, ellipsoidal stick shapes can be further sub-classified as (iia) short and (iib) long. We found that this classification conveniently reflects the degree of disorder and/or dynamic behavior of a water molecule. A typical example of the (i) triangular shape is shown in Fig. 3.13a-1,2, in which the contours indicate $2|F_o| - |F_c|$ maps calculated from neutron and X-ray data, respectively. The oxygen positions observed by X-ray and neutron scattering coincide within experimental error. In this case, the two deuterium atoms and the oxygen atom of the water molecule are H-bonded to nearby O/N and deuterium atoms, respectively. Thus, it can be seen that the orientation of this water molecule is well-defined. In fact, triangular shaped contours correspond to the most highly-ordered water molecules in our maps.

A typical example of a short ellipsoidal stick shape (iia) is shown in Fig. 3.13b-1,2. The oxygen position observed by X-rays is located at one end of the neutron Fourier peak, and only one deuterium atom could be observed.

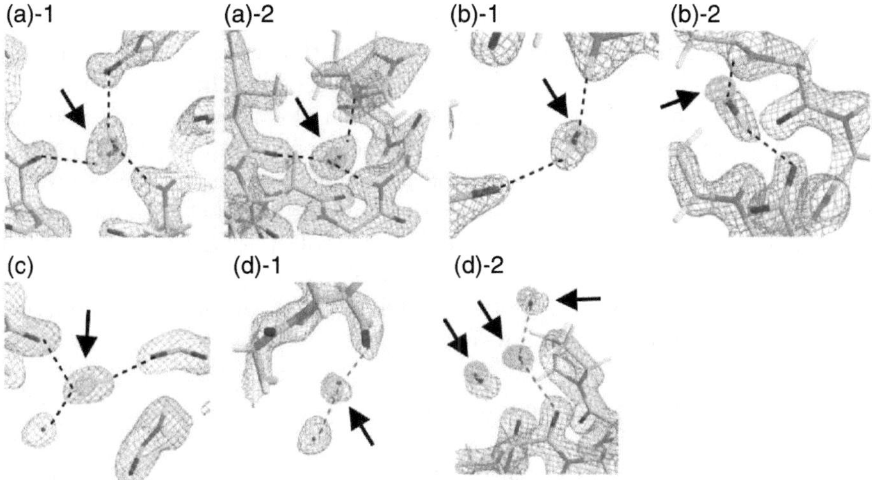

Fig. 3.13. $2|F_o| - |F_c|$ nuclear density maps of water molecules of hydration for myoglobin and the rubredoxin mutant observed by neutron protein crystallography. Examples shown are those of those of: (**a**) triangular shape, (**b**) short ellipsoidal shape, (**c**) long ellipsoidal shape and (**d**) spherical shape. In these maps, the contours correspond to neutron peaks, while the superimposed contours correspond to oxygen peaks from X-ray data (marked by *arrows*). Observed (located) atoms from the neutron data are shown as stick diagrams. Note that in Fig. 3.13a all atoms of the central D_2O molecule are visible, whereas in the other diagrams only some of the solvent atoms have been located: O, D (Fig. 3.13b), two D (Fig. 3.13c)and O only (Fig. 3.13d)

The observed D and O atoms are H-bonded to neighboring O/N and D atoms, respectively, but the other deuterium atom was not identified because of the molecular rotation (or packing disorder) around the fixed O–D bond. Thus, short ellipsoidal stick shaped peaks are interpreted to represent water molecules rotationally disordered around an O–D bond. A typical example of the long ellipsoidal stick-shaped peak (iib) is shown in Fig. 3.13c. The O position observed by X-rays (but not by neutrons) is located in the middle of the neutron Fourier peak, and the two D atoms are clearly observed in the neutron map. The entire appearance is that of an elongated stick. In this case, the two D atoms are H-bonded to neighboring O and/or N atoms, but the O atoms of the D_2O molecule cannot be identified because of the molecular rotation or packing disorder around the D–D axis. Finally, a typical example of the spherical-shaped peak (iii) is shown in Fig. 3.13d-1,2. Only the center of gravity of this type of water molecule can be defined because its orientation is totally disordered. A spherical peak in a neutron Fourier map always means that the whole water molecule is freely rotating, even if X-ray results (which only show the O atom) reveal no hint of this disorder.

Although the above classification has been carried out based on the appearance of peaks in Fourier maps, it was found that the shapes are strongly correlated with the existence of hydrogen bonds, which fix the positions of atoms of water molecules. Most of the triangular-shaped water molecules are fixed at three atoms (D, O, D), while ellipsoidal ones are fixed at two atoms (D, D or D, O). In contrast, some spherical shaped water molecules are not fixed by any observed H-bonds. The average number of "anchor points" of triangular, ellipsoidal and spherical-shaped water molecules are 2.3, 1.3, and 0.3, respectively. In the three proteins of myoglobin, wild type and mutant form of rubredoxin, the average populations of triangular, ellipsoidal, and spherical shapes are 29%, 16%, and 55%, respectively [37].

3.5.3 Dynamic Behavior of Hydration

The dynamic behavior of water molecules becomes clearer when the B-factors obtained by neutron and X-ray experiments are plotted against each other as shown in Fig. 3.14, in which the B-factors obtained from the neutron analysis are the averaged values from three atoms (D, O, and D), while those from the X-ray analysis are those of the O atoms only. The B-factors of oxygen atoms obtained by X-rays are in the range from 13 to 45 Å^2. It is observed that the

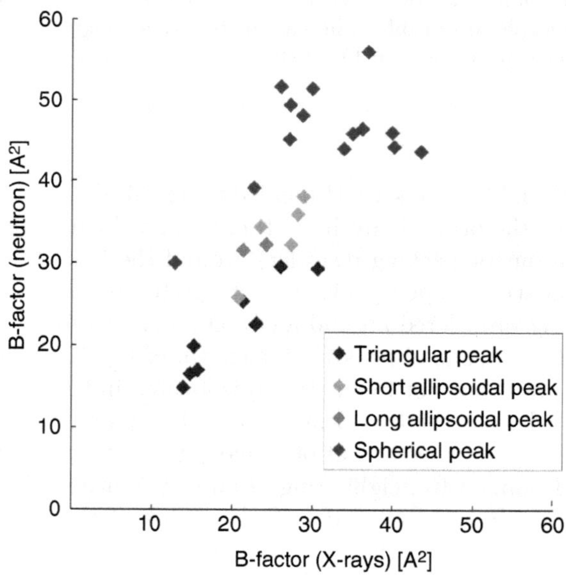

Fig. 3.14. Correlation between the B-factors of hydration water molecules obtained from neutron and X-ray scattering data for the rubredoxin mutant. Neutron B-factors were obtained using the average scattering lengths of D, O, and D atoms from every water molecule, regardless of shape; while in the case of X-ray B-factors, only those from O atoms were included

small, intermediate, and large B-factors from the X-ray analysis correspond to water molecules having the triangular, ellipsoidal, and spherical shapes, respectively. The spherical peak in the neutron Fourier map always means that the whole water molecule is freely rotating, even though the X-ray results (which show only the O atom) reveal no hint of this disorder.

The construction of a data base of hydrogen and hydration in proteins is now under way. The positional coordinates of all hydrogen atoms and hydration water molecules determined by neutron protein crystallography are stored according to the usual PDB format. The main function of the hydrogen hydration data base (HHDB) is (i) to extract the structural information relevant to hydrogen atoms, such as the stereochemical atomic configuration in the vicinity of a selected hydrogen atom, (ii) the search of all the H-bonds between main chains, the main chain and the side chain, and side chains and (iii) the statistical classification of H-bonds. During the analysis of H-bonds by the use of HHDB, very unfamiliar types of H-bonds have been discovered as shown in Fig. 3.15. Figure 3.15a shows that the nitrogen atom of amide (Lys78) in the main chain is an acceptor of H atoms of the neighbor amide (Lys79) (the H bond length H–N is 2.18 Å), and Fig. 3.15b shows that the H-bond is formed between the amide (Lys2) N–H and O=C in Lys2 (the H-bond length H–O is 2.50 Å).

3.6 Crystallization

One fundamental problem in neutron crystallography is the difficultly in obtaining large single crystals. It is really true that neutron protein crystallography necessitates the use of large protein crystals, the volume of which should be larger than 1 mm^3 currently. Usually such a large single crystal is difficult

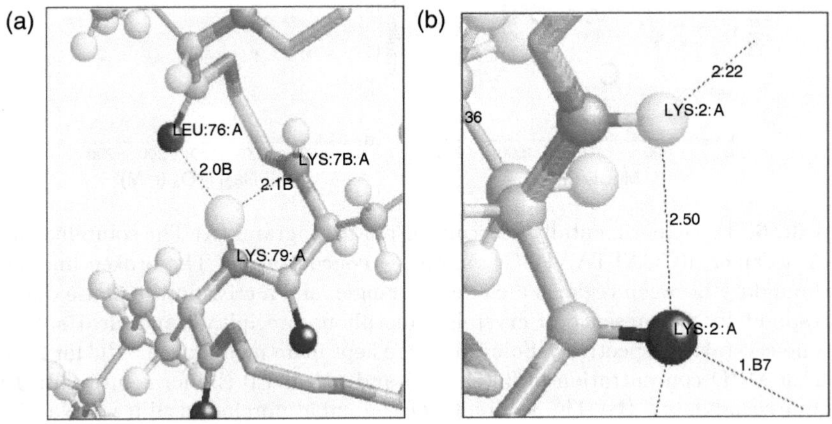

Fig. 3.15. Very unfamiliar types of H-bonds found in (**a**) myoglobin and (**b**) mutant form of rubredoxin by operating the hydrogen and hydration data base (HHDB)

to grow. However, we have found that one rational way to find the proper conditions to grow large single crystals is to establish the complete crystallization phase diagram, which includes determining the solubility curve [38]. Generally speaking, a large single crystal can be grown under supersaturated conditions close to the solubility boundary. As a matter of fact, the large single crystals of cubic porcine insulin, human lysozyme and a DNA oligomer that have been used in our studies have been grown using this method. The phase diagrams of the DNA oligomer [38] and cubic porcine insulin [22] are shown in Fig. 3.16. The corresponding crystals which are obtained on the basis of these phase diagrams are shown in Fig. 3.17, respectively. This method is applicable to grow not only large single crystals, but also crystals of high quality, which is essential for high-resolution crystallographic studies.

3.7 Conclusions and Future Prospects

A neutron protein crystallography experiment is still a time-consuming experiment at current. For example, when a single crystal of 1 mm^3 in volume, the unit cell of the lattice of which is less than 100 Å, is available, it takes 3 or 4 weeks to collect a 1.5 Å resolution data set. If the neutron intensity at the sample position would be increased by a factor of 100, the above mentioned restrictions (size of a single crystal, unit cell size of the lattice constant, data collection time) of neutron protein crystallography would become much more relaxed.

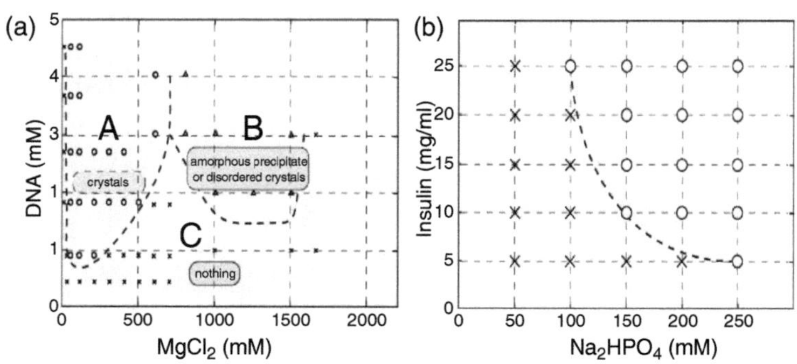

Fig. 3.16. The experimentally determined phase diagram. (**a**) The solubility of the DNA decamer d(CCATTAATGG) vs. $MgCl_2$ concentration. The broken lines show the boundary between regions: a circle, a triangle, and a cross in the phase diagram correspond to the presence of crystals, amorphous precipitate and clear solutions (i.e., no crystals), respectively. Solutions were kept in an incubator at 6°C for 20 days with an MPD concentration of 30% (v/v) and pH of 7.0 (buffer solution of 0.1 M sodium cacodylate). (**b**) The solubility of the cubic porcine insulin vs. Na_2HPO_4 concentration. A circle and a cross mean the presence and absence of cubic porcine insulin crystals in the crystallization. Solutions were kept in an incubator at 25°C for 7 days with 0.01 M Na_3-EDTA buffer solution

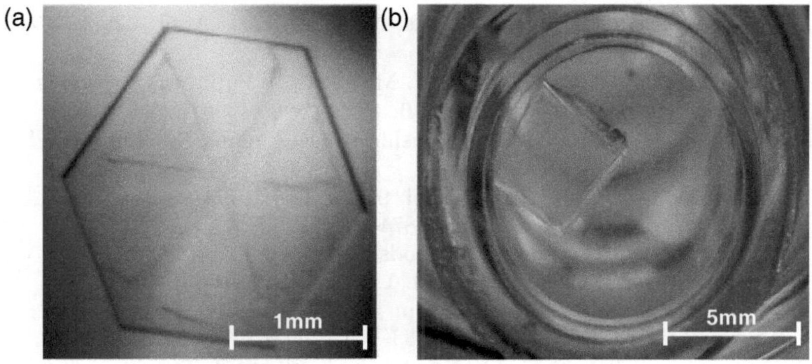

Fig. 3.17. The large crystals of the DNA decamer (**a**) and cubic porcine insulin (**b**) obtained on the basis of the phase diagrams in Fig. 3.16a, b, respectively

The J-PARC project in Japan for a 1 MW spallation neutron source and the SNS in the USA for a 2 MW spallation neutron source, which both now are under construction, will become capable to meet the above requirements in neutron intensity. In both projects, J-PARC and SNS, the construction of dedicated neutron diffractometers for protein crystallography (named BIX-P1 and MaNDi, respectively) is scheduled. At these new instruments, the neutron intensity at the sample position will become 50–100 times higher than at the current BIX-type diffractometers.

Acknowledgments

The studies presented were carried out as a part of a "Development of New Structural Biology Including Hydrogen and Hydration" Project, funded by the Organized Research Combination System (ORCS) and promoted by the Ministry of Education, Culture, Sports, Science and Technology of Japan.

References

1. N. Niimura, Y. Karasawa, I. Tanaka, J. Miyahara, K. Takahashi, H. Saito, S. Koizumi, M. Hidaka, Nucl. Instrum. Methods. A **349**, 521–525 (1994)
2. Y.K. Haga, S. Kumazawa, N. Niimura, J. Appl. Cryst. **32**, 878–882 (1999)
3. Y.K. Haga, K. Neriishi, K. Takahashi, N. Niimura, Nucl. Instrum. Methods Phys. Res. A **487**, 504–510 (2002)
4. N. Niimura, Y. Minezaki, T. Nonaka, J-C. Castanga, F. Cipriani, P. Hoghoj, M.S. Lehmann, C. Wilkinson, Nat. Struct. Biol. **4**, 909–914 (1997)
5. N. Niimura, I. Tanaka, Y. Minezaki, Y. Karasawa, I. Tanaka, K. Miki, M. Sato, M. Hidaka, Minakawa, Y. Morii, Physica B **213–214**, 786–789 (1995)
6. S. Fujiwara, Y. Karasawa, I. Tanaka, Y. Minezaki, Y. Yonezawa, N. Niimur, Physica B **241–243**, 207–209 (1998)

7. I. Tanaka, K. Kurihara, T. Chatake, N. Niimura, J. Appl. Cryst. **35**, 34–40 (2002)
8. I. Tanaka, K. Kurihara, Y. Haga, Y. Minezaki, S. Fujiwara, S. Kumazawa, N. Niimura, J. Phys. & Chem. Solids **60**, 1623–1626 (1999)
9. K. Kurihara, I. Tanaka, M. Refai-Muslih, A. Ostermann, N. Niimura, J. Synchr. Rad. **11**, 68–71 (2004)
10. N. Niimura, Curr. Opin. Struct. Biol. **9**, 602–608 (1999)
11. J.R. Helliwell, Nat. Struct. Biol. **4**, 874–876 (1997)
12. N.V. Raghavan, A. Woldawer, Methods Exp. Phys., Part C **23**, 335–365 (1987)
13. B.P. Schoenborn, Methods Enzymol. **114**, 510–529 (1985)
14. A.A. Kossiakoff, Annu. Rev. Biochem. **54**, 1195–1227 (1985)
15. I. Tsyba, R. Bau, Chemtracts **15**, 233–257 (2002)
16. N. Niimura, T. Chatake, A. Ostermann, K. Kurihara, I. Tanaka, Z. Kristallogr. **218**, 96–107 (2003)
17. N. Niimura, T. Chatake, K. Kurihara, M. Maeda, Cell Biophys. Biochem. **40**, 351–370 (2004)
18. A. Ostermann, I. Tanaka, N. Engler, N. Niimura, F.E. Parak Biohpys. Chem. **95**, 183–193 (2002)
19. K. Kurihara, I. Tanaka, T. Chatake, M.W.W. Adams, F.E. Jenny Jr., N. Moiseeva, R. Bau, N. Niimura, Proc. Natl. Acad. Sci. USA, **101**, 11215–11210 (2004)
20. T. Chatake, K. Kurihara, I. Tanaka, I. Tsyba, R. Bau, F.E. Jenny Jr., M.W.W. Adams, N. Niimura, Acta Cryst. D **60**, 1364–1373 (2004)
21. M. Maeda, S. Fujiwara, Y. Yonezawa, N. Niimura, J. Phys. Soc. Jpn. Suppl. A **70**, 403–405 (2001)
22. M. Maeda, T. Chatake, I. Tanaka, A. Ostermann, N. Niimura, J. Synchr. Rad. **11**, 41–44 (2004)
23. H. Matsuo, M. Oe, F. Sakiyama, K. Narita, J. Biochem. **72**, 1057–1060 (1972)
24. C. Bon, M.S. Lehmann, C. Wilkinson, Acta Cryst. D **55**, 978–987 (1999)
25. W. Gu, B.P. Schoenborn, Proteins **22**, 20–26 (1995)
26. S. Mason, A.G. Bentley, G.J. McIntyre, Deuterium exchange in lysozyme at 1.4A resolution, in: *Neutrons in Biology*, ed. by B.P. Schoenborn. (Plenum Press, New York, London, 1984), pp. 323–334
27. E.N. Baker, R.E. Hubbard, Prog. Biophys. Molec. Biol. **44**, 97–179 (1984)
28. G.R. Desiraju, T. Steiner, *The Weak Hydrogen Bond* (Oxford University Press, New York, 1999)
29. R. Preissner, U. Egner, W. Saenger, FEBS Lett. **288**, 192–196 (1991)
30. S. Longhi, M. Czjzek, V. Lamzin, A. Nicolas, C. Cambillau, J. Mol. Biol. **268**, 779–799 (1997)
31. T. Sandalova, G. Schneider, H. Kack, Y. Lindqvist, Acta Cryst. D **55**, 610–624 (1999)
32. P.R. Blake, J.B. Park, Z.H. Zhou, D.R. Hare, M.W.W. Adams, M.F. Summers, Protein Sci. **1**, 508–21 (1992)
33. W. Gu, A.E. Garcia, B.P. Schoenborn, Basic Life Science **64**, 289–298 (1996)
34. B.V. Daniels, B.P. Schoenborn, Z.R. Korszun, Basic Life Science **64**, 325–331 (1996)
35. T. Chatake, A. Ostermann, K. Kurihara, F.G. Parak, N. Niimura, Proteins: Struct. Funct. Genet. **50**, 516–523 (2003)
36. S. Arai, T. Chatake, Y. Minezaki, N. Niimura, Acta Cryst. D **58**, 151–153 (2002)

4

Neutron Protein Crystallography: Technical Aspects and Some Case Studies at Current Capabilities and Beyond

M. Blakeley, A.J.K. Gilboa, J. Habash, J.R. Helliwell, D. Myles, J. Raftery

4.1 Introduction

The present major driving forces of life science at the molecular and cellular scale are functional genomics and proteomics. Information on the specific functions of many more if not all proteins encoded in human and other genomes is seen as desirable. Major obstacles to these aims are the vast complexity of the individual proteins and the even more delicate interaction of different proteins and other biomolecules to form (transient) functional complexes.

Neutrons have a unique role to play in determining the structure and dynamics of biological macromolecules and their complexes. The similar scattering magnitude from hydrogen, deuterium, carbon, nitrogen and oxygen means that the effect of atomic vibration in lowering the visibility of these atoms in Fourier maps is no worse for the hydrogen/deuterium atoms. Moreover, the negative scattering length of hydrogen allows the well-known H/D contrast variation method to be applied. Also there is not a radiation damage problem using neutrons as the diffraction probe, unlike X-rays, which readily allows room temperature neutron data collection. Clearly, even with these advantages, the low flux of existing neutron facilities means that neutron protein crystallography (nPX) is not going to be a high throughput technique due to long measuring runs (e.g., currently typically between 1 and 4 weeks for a data set). However, the proteomics programmes of research are going to make many more candidate proteins accessible for nPX studies. Thus there is a renewed and growing interest in nPX studies today. The importance rests on knowing the details of the hydrogen and water substructure, which are involved in all the molecular processes of life virtually. This experimental structural information is mostly incomplete when studied by X-rays alone. Also many enzyme reactions involve hydrogen. So there is great potential for wide application of nPX if the technical capability can be found.

There are two major hurdles for wide application of neutron protein crystallography; first the size of crystals routinely available vs. the sizes required, and second a molecular weight ceiling of typically 40 kDa. Neutron source and apparatus developments could make an important impact here, e.g., enhancement plans for LADI at the ILL in Grenoble, a proposed LMX instrument on ISIS2 and the new (MAcromolecular Neutron Diffraction Instrument) MANDI planned for the new 1.4 MW USA source SNS. More routine use of full deuteration of the protein through microbiological expression of proteins for bacteria grown on deuterated media is possible and will make a major impact. For example it was recently shown [1] that a 1.7 Å neutron study of fully deuterated myoglobin was more effective than a 1.5 Å X-ray study in finding even the relatively static hydrogens (as deuteriums). At ILL in Grenoble, a new European funded perdeuteration Laboratory has recently come on line for a focus for protein production of fully deuterated proteins, which will improve signal-to-noise by an order of magnitude as well as opening-up new contrast variation experiments.

4.2 Data Collection Perspectives

Synergies between neutron and synchrotron radiation (SR) Laue crystallography, namely a commonality of knowledge of Laue geometry irrespective of radiation type, opened up a new path in neutron protein crystallography data collection [2–4]. The gain in speed over monochromatic neutron techniques has been notable with the advantage also allowing smaller crystals and bigger unit cells to be investigated. However, this is at the expense of signal-to-noise in the diffraction pattern. Nevertheless very high resolution studies (1.5 Å) on small proteins have been undertaken in times of 10–14 days. Narrow bandpass Laue and/or fully deuterated protein significantly improves S/N. On LADI the geometric limit of resolution is ~1.4 Å and the molecular weight ceiling is around 40 kDa in practice due to spot overlap congestion for the fixed radius and the crystal cross-sections commonly in use (up to 3 mm). In the limit, monochromatic techniques can also remain attractive at neutron reactor sources [5]. This is generally at the expense of longer data collection times (typically 30–60 days per monochromatic data set), even with large area coverage neutron image plates, and is restricted to rather small protein unit cells.

Another technical frontier involves the size of crystal that can be studied with current neutron image plate (IP) diffractometers. It is not always true that very big protein crystals cannot be grown and apparatus should be developed which would allow sample cross-sections of between 5 mm and 1 cm to be harnessed [6]. The crux of the new apparatus would be to allow for much bigger neutron diffraction spot sizes than hitherto imagined. Also the new MANDI planned for the new USA source SNS and, when funded, the proposed 5 MW European Spallation Source (ESS) would offer further gains.

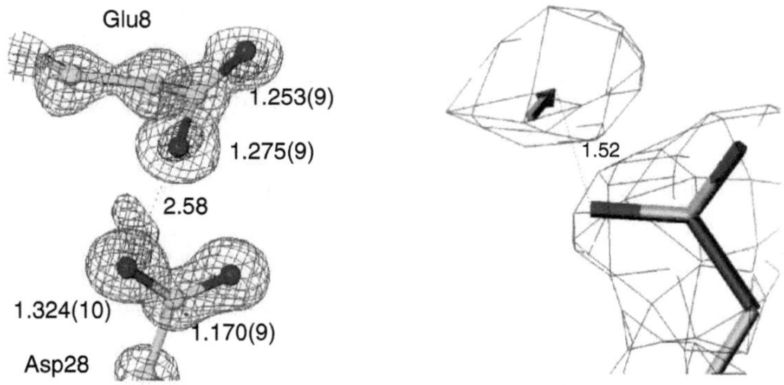

Fig. 4.1. Finding protein carboxy side chain hydrogens via (*left*) X-ray derived electron density maps and precise bond distances (standard uncertainty values in brackets) and (*right*) via neutron-derived nuclear density maps

4.3 Realizing a Complete Structure: The Complementary Roles of X-ray and Neutron Protein Crystallography

Whilst the determination of hydrogens in proteins is now feasible with ultra-high/atomic resolution protein SR X-ray crystallography (Fig. 4.1) mobility of hydrogens can kill their diffraction signal. Since neutron protein crystallography determination of deuteriums at around 2 Å or better resolution matches that at 1.0 Å by SR X-rays, then more mobile hydrogens are determinable by the neutron approach [7] (Table 4.1). Indeed the bound solvent is a whole category of deuterium atoms which are more efficiently sought by neutron techniques [7].

Atomic resolution is usually taken to be crystal structure studies where the X-ray diffraction data is still reasonable[1] at 1.2 Å and where, thus, the electron density shows resolved atoms. However, as emphasized in [8] with concanavalin A studied at 0.94 Å, the X-ray data to parameter ratio at atomic resolution (1.2 Å) is "only" around 2. Also, although the model structure dictionary restraints add data making an overall X-ray + restraints data to parameter ratio of $\sim$3, some of these restraints are not appropriate, e.g., for carboxyl side chains where the nonprotonated dictionary assumption is not correct. At resolutions better than 0.95 Å however, the X-ray data number has grown sufficient to allow the X-ray data to dominate the dictionary restraints

[1]The criteria for the data being reasonable ie observed to any given resolution usually include criteria like the resolution where $F/\sigma(F)$ crosses a value of 2 plus the completeness of data should be >50%. Sometimes the resolution where the $R_{\mathrm{merge}}(I)$ rises above 20% is also used. These are useful practical data quality indicators after all the efforts to make the best data collection experiments (choice of crystal, exposure time, temperature, etc.)

Table 4.1. Bound water comparison for the 2.4 Å neutron room-temperature structure and 0.94 Å X-ray cryo-structure [7,8]

	total no. of waters	common waters (within 1 Å)	D_2O	D–O	H_2O	H–O
neutron structure (room temp.)	148		62	20		
X-ray structure (cryotemp.)	319	88			15	35

where necessary in the structure, i.e., such as carboxy side chains. The emphasis at such resolutions, called ultrahigh for want of a better term, involves checking of bond distances and their standard uncertainties to define e.g., the presence or absence of a hydrogen, rather than the electron density map shape used at atomic resolution. At better resolutions than 0.95 Å, the X-ray data to parameter ratio improves further obviously. At much better resolutions still (~0.7 Å), the valence electron density becomes discernible and the emphasis switches back to the density, or rather the electron density along with bond distances. In these "smaller-molecule-crystallography accuracy" situations there is a difference from small molecule crystallography practice in that it is in fact not possible to remove the dictionary restraints because individual atoms or groups of atoms (such as on loops on the protein) do not diffract to the edges of the pattern and the dictionary structure restraints are essential to stop those parts of the structure "falling apart". It is in these more mobile parts of the protein structure that the role of neutron protein crystallography in completing the protein structure protonation details is clearly needed.

It is also a fair question to ask just how many protein crystal structures can be studied at 0.95 Å X-ray diffraction resolution or better. Only time and experience will reveal the answer to that along with answering the similar question of how many protein crystal structures will become available per annum as neutron sources and instruments for nPX improve and as more fully deuterated proteins are crystallized.

4.4 Cryo-Neutron Protein Crystallography

Due to the inherent low flux of neutron sources, large crystals are at present essential for high-resolution neutron crystallographic studies. Since neutrons do not cause radiation damage there has not been a pressing need to combine cryoprotection of crystals with a large volume. However, there is the advantage to be gained from cryo-crystallography of the reduction of atomic mobility that would lead to reduction in B-values and enhanced density definition. Furthermore collecting data at cryo-temperatures can also reduce background scattering (diffuse scattering), and therefore can aid the signal-to-noise ratio and hence the resolution limit. Cryo-crystallography also opens up the possibility of freeze trapping of intermediates in biological reactions triggered in a protein crystal. So, the idea of combining neutron and cryo-crystallography advantages together is of considerable interest.

There are, however, several problems that can occur by cooling the crystal. Cooling can create disorder in the molecule or unit cell, which reduces data quality as quantified by atomic and overall molecule B-factors, and so degrades the resolution limit and therefore reduces accuracy of electron or nuclear density maps. The whole crystal usually also becomes more mosaic and the monochromatic rocking width or Laue pattern spot size of reflections increases. Thus reflections can sink into the background more quickly than if they are "sharp". Again this manifests as a reduction of resolution limit in effect but is generally less severe than deterioration of the protein B-factors just referred to. In extreme cases, cooling the crystal can cause complete cracking and dislocations, making data collection impossible. The imperfections in a crystal can be described by a simple mosaic-block model, and that describes three characteristic parameters. These parameters are the size s of the mosaic blocks, the angular spread ω of the blocks and the variation in cell dimensions δa between blocks. A common effect observed by cooling macromolecular crystals is an increase in mosaicity. Small crystals at room temperature generally have small mosaicities ($<0.02°$), with this value being increased by the characteristics of the X-ray beam. When cooled, protein crystals generally have mosaicities of $0.2°$ or more. It seems to have been generally assumed that freezing large crystals is not possible. It has been shown that it is possible to freeze and collect high resolution X-ray (1.65 Å on a rotating anode, Fig. 4.2a) and neutron (2.5 Å on LADI) data from large concanavalin A protein crystals ($\sim$5 and $\sim$2 mm^3) as examples [9]. These data have allowed a combined "X+n" protein structure analysis to be undertaken [10] as performed previously with room temperature X-ray and neutron data sets [7]. Figure 4.2b, c compare an LADI image for concanavalin A recorded at 15 K with one at room temperature and where the high quality spot shape at both temperatures is evident. These results demonstrate the potential of protein cryo-crystallography with neutrons thus combining the advantages of the neutron and cryo-approaches for studying the structural details of bound water hydrogens (as deuteriums) and of protonation states of amino acid side carboxyl side chains. Perhaps most exciting of all this opens up the possibility of time-resolved neutron freeze-trap protein crystallography. Reviews of X-ray time-resolved protein crystallography are given in [9] and [11].

4.5 Current Technique, Source, and Apparatus Developments

At the neutron reactor source in Grenoble, the most powerful in the world, there is a coordinated millennium instrument refurbishment going on. This includes upgrades to the biological crystallography relevant instruments, namely D19 (a monochromatic neutron diffractometer with enlarged area

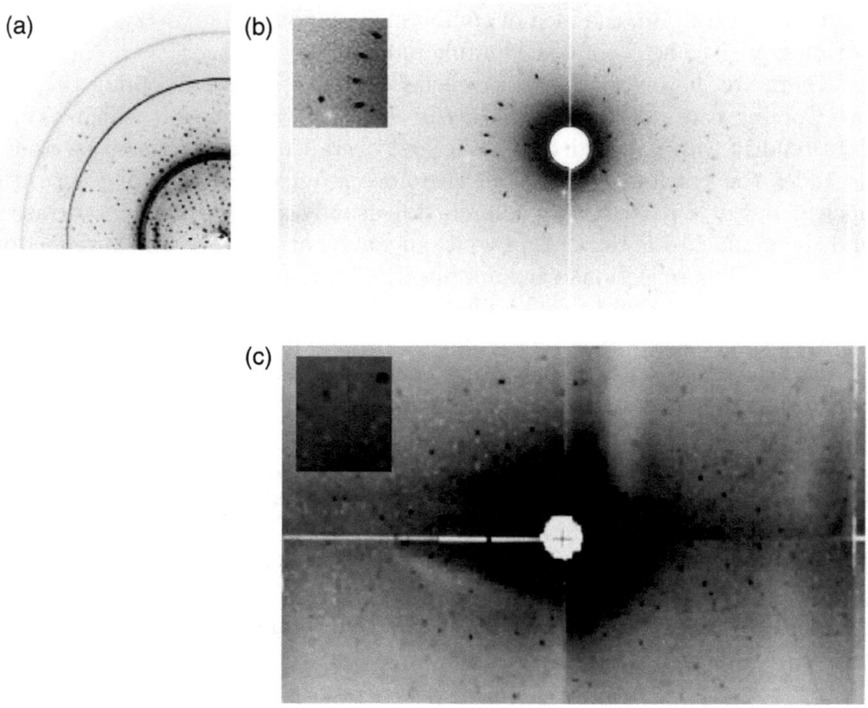

Fig. 4.2. (**a**) X-ray 0.05° rotation diffraction image from a frozen crystal of concanavalin A of volume ∼2 mm^3 showi ng diffraction to 1.65 Å resolution. Approximately 80% of this crystal diffracted to high resolution like this. Neutron LADI diffraction data was recordable to 2.5 Å resolution from this identical crystal. (**b**) LADI image recorded from a concanavalin A crystal at 15 K. (**c**) Likewise from another concanavalin A crystal but at room temperature. Both are accompanied by insets showing an enlarged view of individual spots. From [10]

detector coverage (thus improving data collection efficiency), and the LAue DIffractometer (LADI) which will have a new neutron image plate reader, new guide-optics and a higher flux location yielding gains of ∼10 or more in sensitivity + flux (thus reducing data collection time or allowing smaller crystal samples to be used or larger molecular weight proteins to be investigated or combinations thereof).

A vital next step beckons using the time-of-flight Laue approach, feasible at a spallation neutron source. This type of source employs accelerated proton pulses, which strike a target such as mercury. Neutrons are ejected (spallated) out of the target also in pulses and over a broad continuum of wavelengths. Laue spots containing overlapping Bragg reflections can be resolved by the time of arrival at the detector, and likewise noise pile up reduced, delivering a better signal to noise than can be realized using Laue techniques at steady state (reactor) sources. This approach has now recently been successfully exploited in protein crystallography at the Los Alamos Neutron Scattering Centre (LANSCE) [12]. A time-of-flight Laue protein crystallography facility with neutrons is planned at the SNS which will operate at 2 MW. The proposed ESS at 5 MW has been estimated [13] to reach the same average flux (in fact 0.5) as the ILL Grenoble 58 MW reactor neutron beam flux but with the addition of the benefit of the time-of-flight neutron measuring approach, which can "gate out" the noise falling onto a reflection spot intensity in ordinary Laue exposures i.e., which is without time-of-flight gating.

4.6 Plans for the ESS and nPX

The ESS Project Team studies recently allowed a review and assessment of future potentialities for nPX, quoting from [14]: "In high throughput structural biology research, the best sample size is rarely above $100 \times 100 \times 100\,\mu$m. It is essential for neutron protein crystallography to find source, instrument and sample (deuteration) combinations to face this challenge. There is also a barrier to the application of high resolution neutron structural study posed by molecular weight, which determines the unit cell volume, of large biological complexes. Such weakly scattering crystals cannot be studied currently. If, however, we combine the ESS source and instrument improvements, and improved knowledge of the protein preparation and crystallogenesis for the growth of large crystals, the unit cell size capability could reach $(250\,Å)^3$." ... "In the first case (then), a brighter neutron source, well-focussed beams and smaller-pixel detectors will be in the design (of the first nPX instrument). To meet the second challenge, one has to continue to harness the expertise of the crystallogenesis community to produce big crystals. Thus larger beams and bigger-pixel detectors are needed. This is a different (nPX) instrument. Also we should harness longer wavelengths to enhance the scattering-efficiency-with-wavelength-effect as well. A methane moderator tailored to wavelengths 1.5–5 Å should be investigated." These extracts show the scope imagined for nPX with such a new source and the nPX instrument solutions envisaged. The ESS is seen as a long-term goal for Europe, meaning that it was only recently still "not yet approved" for funding. As an example of the technical challenge currently Fig. 4.3 shows the protein crystal sample sizes and unit cell volumes marked for the cases of references [6,7,15] with [7] within current capability, [6] at the current limit and [15] well beyond current capabilities. Fig. 4.4 shows the LADI diffraction pattern for [6] with 50 kDa in the crystal asymmetric unit of concanavalin A with glucose bound ($I2_13$ a = 167.8 Å).

4.7 Conclusions and Future Prospects

The role of structural data in drug discovery in the pharmaceutical industry will increase when it is much more routine that hydrogen atoms and bound water deuterium atoms positional information can be incorporated.

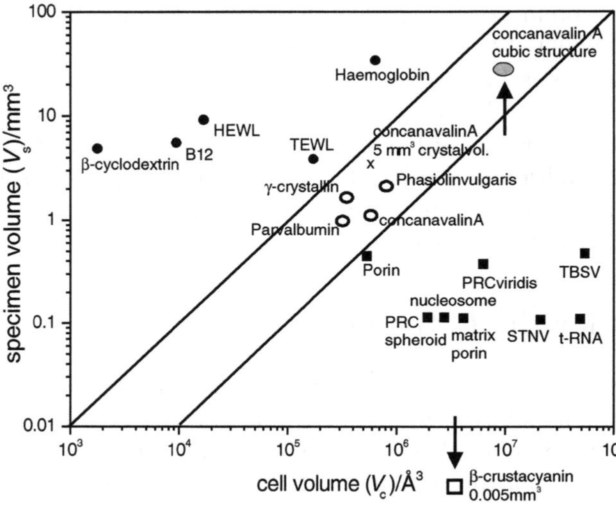

Fig. 4.3. Scatter plot of the examples of protein crystals investigated at ILL on the monochromatic neutron instrument D19 (*filled circles*) and DB21 (*filled squares*). The line through the origin shows the empirical limit and so the open circles show where LADI is capable. The three case studies of this paper are highlighted as marked. DB21 offers data collection at very low resolution (e.g., the TBSV experiment marked involved data collected to 16 Åresolution) in order to map out crystal packing details whereas LADI is used for high resolution neutron protein crystallography i.e. 1.5 to 2.5 Åresolution (as was D19 before it)

Thus the discovery of new pharmaceuticals and of enhanced efficiency compounds would then accelerate. Also, because neutrons are nondestructive, unlike X-rays, room temperature structure and vibration data can be provided, and which is the relevant temperature physiologically. Our most recent work however also shows that nPX data can be measured down to 15 K with benefits in the clarity of the nuclear density seen. Thus future approaches appear to be able to include structures being determined and refined with nPX from 15 K up to room temperature including freeze trapped protein crystal structures. Most recently Hazemann et al. [18] report their use of a radically smaller crystal volume of 0.15 mm^3 using fully deuterated aldose reductase protein in a 2.2 Å high resolution analysis. This was achieved due to a stronger crystal scattering efficiency for neutrons and also eliminating the hydrogen incoherent scatter. Clearly this is an exciting development. A conceptual design report (CDR) for the new MANDI SNS diffractometer for protein crystallography is now in press [19], and includes an estimate of a few days only to measure a complete high resolution data set from a fully deuterated protein crystal of 0.125 mm^3 total volume, and also the CDR for the new JPARC protein crystal diffractometer in Japan is recently published [20]. These two CDRs continue a fascinating new wave of instrumentation for this field of research.

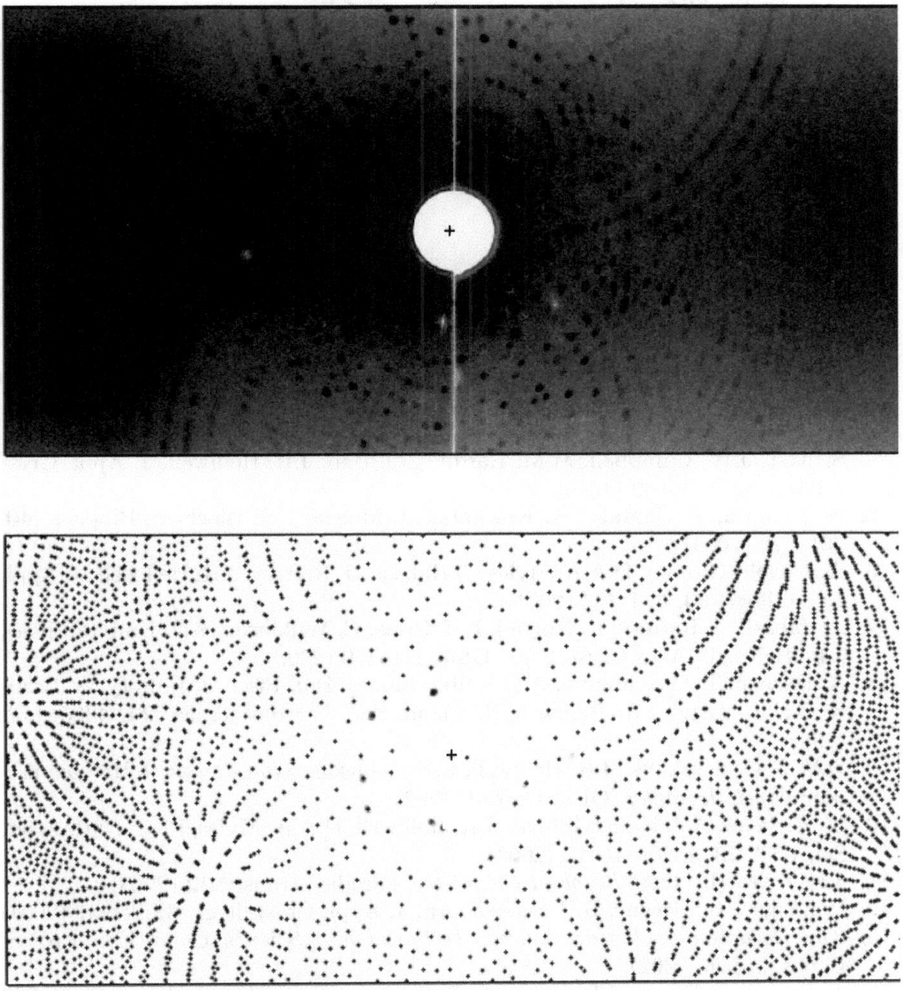

Fig. 4.4. The large molecular weight frontier. LADI diffraction recorded from a crystal of cubic concanavalin A where there is 50 kDa in the asymmetric unit of the crystal (from [6]). The diffraction extends to 3.5 Å thus new sources, as well as perdeuteration opportunities will extend the technical capability to reach high-resolution structure analyzes even in such cases. The actual LADI pattern is shown (*top*) and the predicted Laue pattern (*bottom*) using the Daresbury Laue software analysis package [16] modified for neutrons and a cylindrical detector geometry [17]

Acknowledgements

JRH is very grateful to his past and present colleagues and students in the University of Manchester Structural Chemistry Research Laboratory, which is devoted to biological and chemical crystallography and includes extensive crystallographic methods development. The research grant support of BBSRC, The Wellcome Trust and The Leverhulme Trust and of EPSRC,

BBSRC, British Council, the UK/Israel Fund and the Institut Laue Langevin for PhD studentship support and the EU (for Host Institute, Network and Marie Curie Training Centre awards) are all also gratefully acknowledged. The CHESS SR Facility Cornell, USA, ESRF Grenoble and SRS Daresbury and the Institut Laue Langevin in Grenoble are thanked for SR and neutron beamtime, respectively, for the studies described herein.

References

1. F. Shu, V. Ramakrishnan, B.P. Schoenborn, PNAS (USA) **97**, 3872–3877 (2001)
2. D.W.J. Cruickshank, J.R. Helliwell, K. Moffat, Acta Crytallogr. **A43**, 656–674 (1987)
3. J.R. Helliwell, C. Wilkinson, X-ray and neutron Laue diffraction, in *Neutron and Synchrotron Radiation for Condensed Matter Studies: Applications to Soft Condensed Matter and Biology* (Springer Verlag, Berlin, New York, 1994)
4. S. Arzt, J.W. Campbell, M.M. Harding, Q. Hao, J.R. Helliwell, J. Appl. Crystallogr. **32**, 554–562 (1999)
5. N. Niimura, T. Chatake, K. Kurihara, M. Maeda, Cell Biochem. Biophys. **40**, 351–370 (2004)
6. A.J. Kalb (Gilboa), D.A.A. Myles, J. Habash, J. Raftery, J.R. Helliwell, J. Appl. Crystallogr. **34**, 454–457 (2001)
7. J. Habash, J. Raftery, R. Nuttall, H.J. Price, C. Wilkinson, A.J. Kalb (Gilboa), J.R. Helliwell, Acta Crystallogr. **D56**, 541–550 (2000)
8. A. Deacon, T. Gleichmann, A.J. Kalb (Gilboa), H.J. Price, J. Raftery, G. Bradbrook, J. Yariv, J.R. Helliwell, J. Chem. Soc. Faraday Trans. **24**, 4305–4312 (1997)
9. D.W.J. Cruickshank, J.R. Helliwell, L.N. Johnson, *Time-Resolved Macromolecular Crystallography* (OUP, Oxford, 1992)
10. M. Blakeley, A.J. Kalb (Gilboa), J.R. Helliwell, D.A.A. Myles, Proc. Natl. Acad. Sci. USA **23**, 16405–16410 (2004)
11. J.R. Helliwell, *Time-Resolved Chemistry*, Faraday Trans. **122** (2003)
12. P. Langan, G. Greene, B.P. Schoenborn, J. Appl. Crystallogr. **37**, 24–31 (2004)
13. *The European Spallation Source: Technical and Science Case* (ESS Project Books, Jülich, 2002)
14. T. Bayerl, O. Byron, J.R. Helliwell, D. Svergun, J.-C. Thierry, J. Zaccai, in *The European Spallation Source: Science Case* (ESS Project Books, Jülich, 2002)
15. M. Cianci, P.J. Rizkallah, A. Olczak, J. Raftery, N.E. Chayen, J.R. Helliwell, PNAS (USA) **99**, 9795–9800 (2002)
16. J.R. Helliwell, J. Habash, D.W.J. Cruickshank, M.M. Harding, T.J. Greenhough, J.W. Campbell, I.J. Clifton, M. Elder, P.A. Machin, M.Z. Papiz, S. Zurek, J. Appl. Crystallogr. **22**, 483–497 (1989)
17. J.W. Campbell, Q. Hao, M.M. Harding, N.D. Nguti, C. Wilkinson, J. Appl. Crystallogr. **31**, 496 (1998)
18. I. Hazemann, M.T. Dauvergne, M.P. Blakeley, F. Meilleur, M. Haertlein, A. Van Dorsselaer, A. Mitschler, D.A.A. Myles, A.D. Podjarny, Acta Crystollogr. D **61**, 1413–1417 (2005).
19. A.J. Schultz, P. Thiyagarajan, J.P. Hodges, C. Rehm, D.A.A. Myles, P. Langan, A.D. Mesecar, J. Appl. Cryst., in press.
20. I. Tanaka, T. Ozeki, T. Ohara, K. Kurihara, N. Niimura, J. Neutr. Res. **13** (2005), 49–54.

5

Detergent Binding in Membrane Protein Crystals by Neutron Crystallography

P. Timmins

5.1 Introduction

The structures of membranes and membrane proteins are taking on ever increasing importance since the observation that they represent perhaps >30% of the proteins encoded by the human genome but much less than 1% of the structures determined to close to atomic resolution by X-ray crystallography. The difficulties encountered in X-ray crystallographic structure determination have been so serious that the first structure of a membrane protein to be solved at even medium resolution was by electron diffraction. This was the famous case of the purple membrane from *Halobacterium salinarum* where the structure of the bacteriorhodopsin embedded in its natural membrane was solved. Due to its unique crystallinity in the natural membrane, it is still the only whole membrane structure to have been solved.

The first membrane proteins to be crystallized were the photo-reaction center from *Rhodopseudomonas viridis* and OmpF porin from *E. coli* in the early 1980s. Since then a number of other membrane proteins have been crystallized and their structures solved by X-ray crystallography but these remain much less than 1% of the total of soluble proteins. A particularity of practically all membrane protein structures is that they are crystallized from detergent solubilized proteins and the detergent becomes an integral part of the crystal. Although the protein itself is well ordered and its structure can be obtained at high resolution, the detergent used to solubilize the protein is fluid and disordered and hence invisible in X-ray maps. Neutron crystallography, although up to now unable to locate high resolution features due to insufficiently large crystals, is ideally suited to locate the detergent phase and hence provide information on crystal packing and also on detergent–protein interactions analogous to lipid–protein interactions in the real membrane.

5.2 Advantages of Neutrons

Membranes can be studied either in their native state where the principal components are protein and lipid or in a solubilized state where the components are protein and lipid or protein and detergent. Each of these components has a characteristic scattering length density for neutrons and each may in principle be deuterium labelled allowing that scattering length density to be varied. Replacement of water by heavy water allows a kind of isomorphous replacement which facilitates phasing in one dimensional diffraction.

In fact the whole concept of contrast variation originated from experiments by Bragg and Perutz [1] to try and determine the overall shape of haemoglobin using salt solutions to modify the electron density of the crystal solvent with respect to the protein. The natural contrast between protein and water is very low for X-rays; the electron density of RNAse for example is 0.432 eÅ^{-3} and for pure water 0.335 eÅ^{-3}. The addition of salt to water raises the electron density even closer to that of protein. Small molecules such as sucrose or indeed high salt concentrations may be used to vary the contrast but the chemical effects of such additives are often such as to destabilize the crystal or even the complex itself or alter the conformations of the component molecules. The exchange of deuterium for hydrogen is a much less perturbative change. In addition it is not trivial to collect low resolution X-ray diffraction data from crystals of biological macromolecules although a number of workers have built specialized beam-lines to do this [2,3].

Manipulation of contrast in neutron scattering is particularly straightforward and has been heavily exploited for many years in neutron small angle scattering. The presence of a well-defined solvent (water) phase in crystals of macromolecular complexes means that the same principles may be applied in neutron crystallography – hence the term small-angle neutron crystallography. Figure 5.1 shows the neutron scattering length density for a number of chemically distinct components of biological molecules. These are calculated from the atomic composition of average proteins, nucleic acids etc. and do not vary greatly between for example different proteins. The values shown for detergents and lipids may vary considerably, however, depending particularly on the balance between hydrophilic head and hydrophobic tail.

At low resolution the scattering is due to the contrast between solvent and macromolecule, i.e., the difference in scattering length density. Thus for example proteins have a zero scattering length density difference with respect to water in a solution containing 40% D_2O/60% H_2O. This concentration at which a particle has zero contrast is known as the *isopicnic point*. It should be noted that the concept of zero contrast applies only to the average scattering length density and hence only to scattering at $Q = 0$. At all other Q-values there will be some contributions from internal scattering length density fluctuations within the molecule such that there are always some parts of the molecule that have a positive or negative contrast. Hence the Bragg scattering is a minimum at the *isopicnic point* but is not zero. A striking point of Fig. 5.1 is that the scattering length densities of H_2O and D_2O encompass those of all other natural components of biological macromolecules. The only exception to this is the case of fully deuterated protein or nucleic acid which have a scattering length density greater than that of pure D_2O.

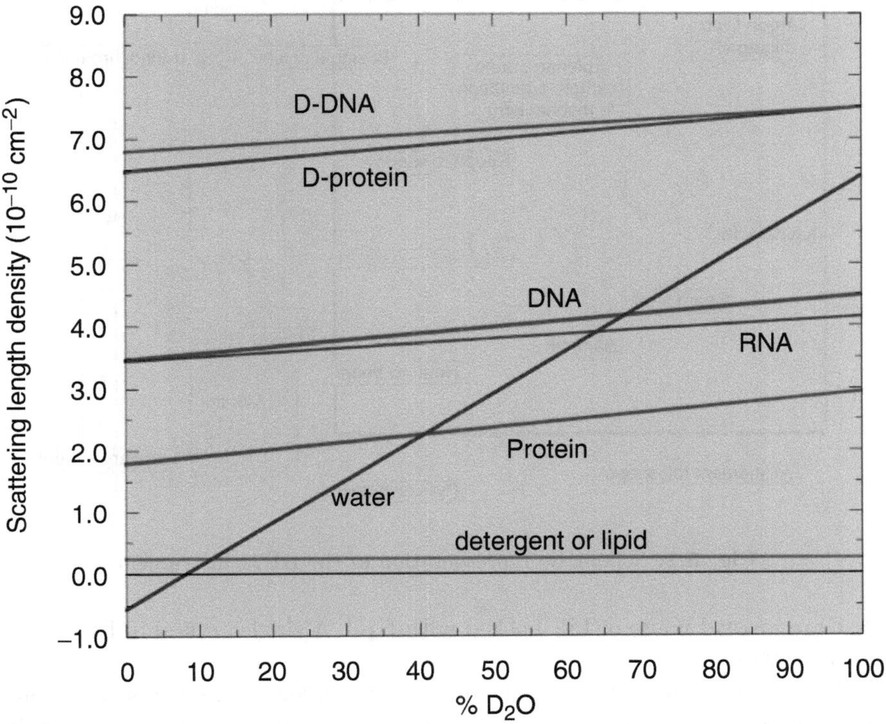

Fig. 5.1. Neutron scattering length densities of biological macromolecules as a function of the deuterium content of the solvent water

5.3 Instrumentation and Data Reduction

One of the difficulties encountered in low resolution protein crystallography is the problem of measuring data at very low resolution – including for example the first order of a 700 Å unit cell. Hence, cold neutrons and H/D contrast variation can be combined to measure neutron diffraction data in a rather simple way. A dedicated diffractometer was constructed at ILL some years ago as collaboration between ILL and EMBL. This instrument, DB21, is a four-circle diffractometer with multidetector and is described in some detail in Roth et al. [4, 5]. A schematic representation of the instrument is shown in Fig. 5.2

The beam is monochromated by a potassium intercalated graphite crystal giving a neutron wavelength of 7.56 Å and a wavelength spread of 2% (FWHM) or a pyrolytic graphite crystal giving a wavelength of 4.6 Å. The beam is collimated using LiF pinholes and a graphite collimator after passing a series of graphite filters for eliminating $\lambda/2$, $\lambda/3$, and $\lambda/4$ contamination. The detector is of the Anger camera type allowing a resolution of 1.75×1.53 mm. Due to the γ-sensitivity of this detector no cadmium is used in beam defining apertures

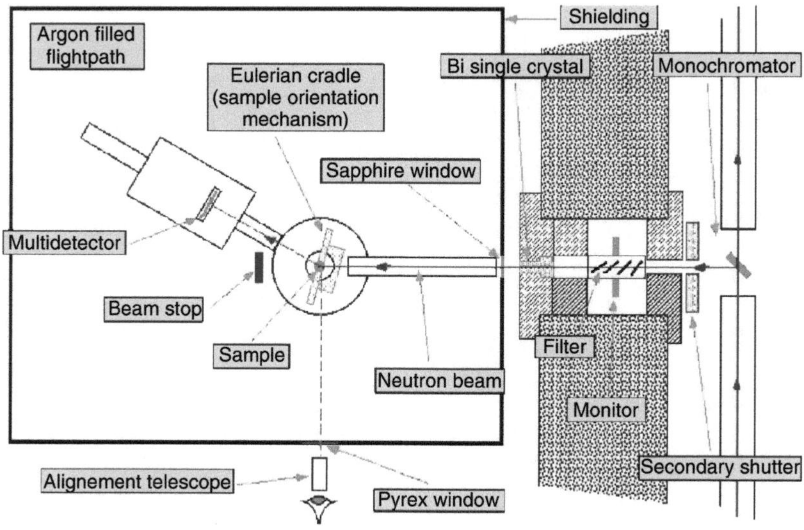

Fig. 5.2. Schematic representation of the DB21 instrument

that are instead made of LiF backed with B_4C. A single crystal of Bi removes in-beam γ-radiation.

Data acquisition is by ϕ- scans or ω-scans. In low symmetry space groups reorientation of the crystal may be required to obtain a full data step due to the rather restricted reciprocal space coverage of the current detector. Determination of the crystal orientation may be difficult at low resolution and many of the standard autoindexing algorithms have failed. A program based on the graphics package 'O' has been written to allow manual rotation of the known reciprocal cell into the reciprocal space diffraction pattern in order to determine the crystal orientation [6]. Data reduction is carried out using a modified version of XDS [7]. Another problem that arises is the scaling of data measured from crystals of different H_2O/D_2O content. This is done by using the parabolic relationship between intensities or the linear relationship between structure amplitudes in centric zones [8] as described in Eq. 5.2. This of course requires consistent indexing which in certain enantiomorphic space groups may be a problem. For example in space group P312 it is impossible to distinguish between *hkl* and *khl* reflections. In high resolution data this problem can be resolved when the final structure is known but in the neutron low resolution case it is absolutely necessary to carry out a correct scaling. In practise it may be necessary to perform the scaling using all possible combinations of index and to select that having the best scaling for all reflections.

5.3.1 The Crystallographic Phase Problem

The variation of the crystallographic structure factor as a function of contrast can be expressed as:

$$\boldsymbol{F}(h, X) = \boldsymbol{F}(h, 0) + X\boldsymbol{F}(h)_{\mathrm{HD}}, \tag{5.1}$$

where h is the reciprocal lattice point, X is the mole fraction of $[D_2O]/[D_2O]+[H_2O]$ in the crystal, $\boldsymbol{F}(h)_{\mathrm{HD}}$ is the vector difference between the structure factor in H_2O and that in D_2O. Multiplying by the complex conjugate we obtain the diffracted intensity:

$$I(h, X) = F(h, 0)^2 + 2X \cos\phi F(h, 0) F_{\mathrm{HD}}(h) + X^2 F_{\mathrm{HD}}^2(h), \tag{5.2}$$

where ϕ is the phase angle between $\boldsymbol{F}(h, 0)$ and $\boldsymbol{F}(h)_{\mathrm{HD}}$.

This relationship has several important consequences for low resolution crystallography including, as mentioned above, the possibility of scaling together data from different contrasts and the interpolation of missing data [8]. In terms of structure solution it is of fundamental importance as it means that the phase difference, ϕ between any two contrasts, of a reflection h, may be determined except for the sign $\pm$, if the amplitudes at 3 contrasts are known. Therefore if the structure is known at any one contrast then the phases at that contrast may be calculated and then determined at any other contrast except for knowledge of the sign. In the particular case of centrosymmetric reflections where $\phi = 0$ or π then there is of course no ambiguity and the phase may be calculated at any contrast [9].

In most studies carried out to date the structure of one component of the macromolecular complex has been determined by X-rays or could be modelled from other information. Hence structure factors calculated from the known part of the structure at a contrast where the other component is visible provide starting phases for the determination of the structure at any contrast. This is illustrated in Fig. 5.3 which demonstrates the vector relationships between structure factors at four different contrasts. The two triangles bounded by F_0, F_{HD}, and F_{D} are the two possible relationships which can be constructed through knowledge of the structure factor amplitudes alone following Eq. 5.1, and corresponding to the two possible signs of ϕ. This particular figure illustrates the case of, for example, a protein/detergent complex where data would be measured at 40% D_2O where the protein is invisible, 10% D_2O where the detergent is invisible and three other contrasts, 0%, 70%, and 100% D_2O. In this case we imagine that the protein structure is known and that the detergent structure is to be determined. We may therefore calculate the phase of the structure factor in 10% D_2O and thus determine the orientation of the phase triangle with just the ambiguity of sign corresponding to the two triangles shown. This is very closely analogous to the situation of single isomorphous replacement [10, 11] in X-ray protein crystallography. Once this (ambiguous) phase has been determined then the ambiguity may be resolved and an approach to the true phase may be made using density modification based on constraints such as the invariability of the known part of the structure, solvent flattening or noncrystallographic symmetry averaging [12, 13].

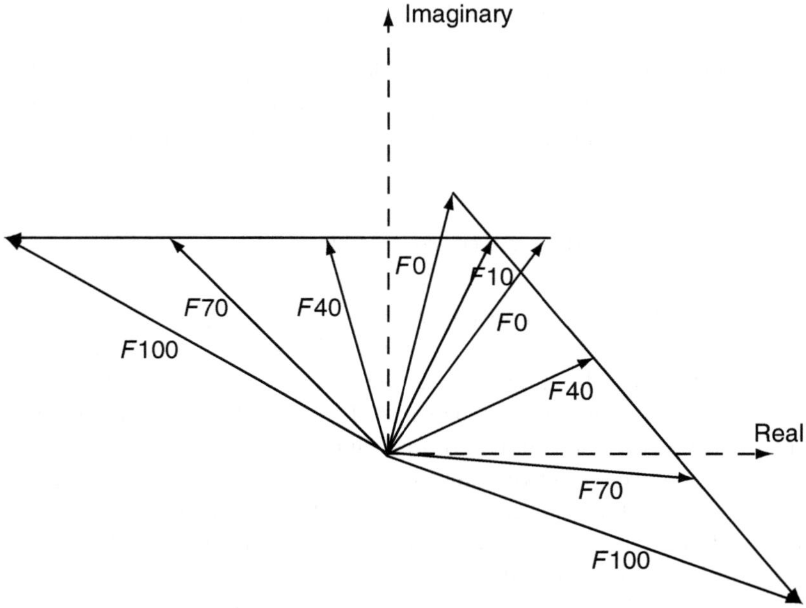

Fig. 5.3. Vector diagram illustrating the relationship between structure amplitudes and phases at different contrasts. The two vector triangles are oriented arbitrarily with the F_{HD} vector parallel to the real axis

5.4 Comparison of Protein Detergent Interactions in Several Membrane Protein Crystals

Because of the very different scattering lengths of protein and detergent and in particular the very high contrast obtained between detergent and D_2O, complexes of protein and detergent make ideal objects for study by neutron contrast variation. X-ray crystallographic studies have to date succeeded in resolving the structure of a number of membrane proteins and in some cases individual tightly bound detergent or residual lipid molecules have been observed but in no case has this method been able to visualize the solubilizing belt of detergent. Using the known X-ray structure and neutron diffraction data measured at a number of H_2O/D_2O component contrasts it has been possible to calculate neutron scattering length density maps which show only the detergent. The key factor here is the detergent content of the cell. The volume in which the protein is located is known from the X-ray structure but generally speaking the amount of detergent in the crystals is unknown. It is therefore necessary to estimate this and to perform solvent flattening and density modification as a function of the detergent content as a variable parameter [13].

5.4.1 Reaction Centers and Light Harvesting Complexes

The first membrane protein/detergent structures to be studied by low resolution neutron crystallography were the photosynthetic reaction centers of *Rhodopseudomonas viridis* [14] and *Rhodobacter sphaeroides* [13]. In the reaction center from *R. viridis* only one molecule of the detergent *N*,*N*'-dimethyldodecylamine-*N*-oxide (LDAO) could be visualized in the high resolution X-ray maps. The neutron diffraction results show the reaction center to be surrounded by a detergent belt some 25–30 Å thick. This corresponds to roughly twice the length of an extended LDAO molecule. The aliphatic chain of the detergent molecule observed in the X-ray maps falls within the neutron density although its polar head is outside. The lack of density corresponding to the head may be due to its small size compared with the resolution of the data and to its rather low scattering density arising from disorder and hydration. As Roth et al. point out it should also be noted that in the case of the reaction center the detergent does not necessarily mimic the biological membrane as in the real membrane the reaction center is surrounded by and makes contact with several light harvesting molecules.

The photoreaction center from *R. sphaeroides* crystallizes in the presence of *n*-octyl-β-glucoside (β-OG) and the small amphiphile heptane-1,2,3-triol (HP). Michel [15] was the first to use small amphiphiles in membrane protein crystallization and suggested that they had the effect of reducing the size of detergent micelles and thereby facilitating crystallization of the membrane protein/detergent complex. This effect was confirmed in neutron small angle scattering experiments by Timmins et al. [16] which showed that HP was indeed included in micelles at least of LDAO and decreased their radius of gyration. Gast et al. [17] also showed by turbidity measurements that addition of 5% HP decreased by a factor of two the amount of LDAO bound to the reaction center of *R. viridis.* The most striking observation in the neutron diffraction results on the *R. sphaeroides* reaction center was that the detergent ring formed by the β-OG was almost identical and size and shape to that formed by the LDAO around the *R. viridis* reaction center. Given the similarity in total length of β-OG and LDAO it would appear that detergent size/geometry plays a key role in the crystallization of membrane proteins.

As mentioned above the reaction centers are usually found in the membrane closely associated with several copies of light harvesting proteins and therefore the protein–detergent interactions observed in the crystal may not always be representative of protein lipid interactions in the membrane. Recently a study has been published on the light harvesting complex LH2 from the photosynthetic purple bacterium *Rhodopseudomonas acidophila*, a nonameric complex with a central hole which in vivo contains membrane lipids [18]. The complex was crystallized from solutions containing β-OG and experiments were performed using both hydrogenated and tail-deuterated detergent. As well as demonstrating the presence of a detergent ring as with other membrane proteins the experiments also showed the central hole to be occupied by two detergent/amphiphile micelles which must have displaced the original lipid during purification.

5.4.2 Porins

Another class of membrane proteins to be studied were the outer membrane proteins known as porins. In contrast to the bacterial photosynthetic reaction centers these proteins form multistranded β-barrel structures which are deeply embedded in the membrane with relatively small protruding loops. The OmpF porin from *E. coli* was in fact, along with the photosynthetic reaction center from *R. viridis*, the first membrane protein to be crystallized in the early 1980s [19, 20]. Due however to technical crystallographic problems its structure was not solved until 1994 when a trigonal form was crystallized and served as model for a molecular replacement solution [21]. The neutron crystallographic analysis of the tetragonal form [22] showed very clearly the detergent belt bound to the hydrophobic surface of the protein and suggested how the protein is anchored in the membrane. Figure 5.4 shows how the detergent binding surface is clearly delimited by two lines of aromatic amino acids with tyrosine residues directed towards the headgroups and phenyl alanines towards the hydrophobic acyl chains. This had been surmised from the X-ray structure but the absence of any density in the electron density maps did not allow this to be demonstrated. Another important observation to come from the neutron diffraction analysis concerned the packing of the molecules

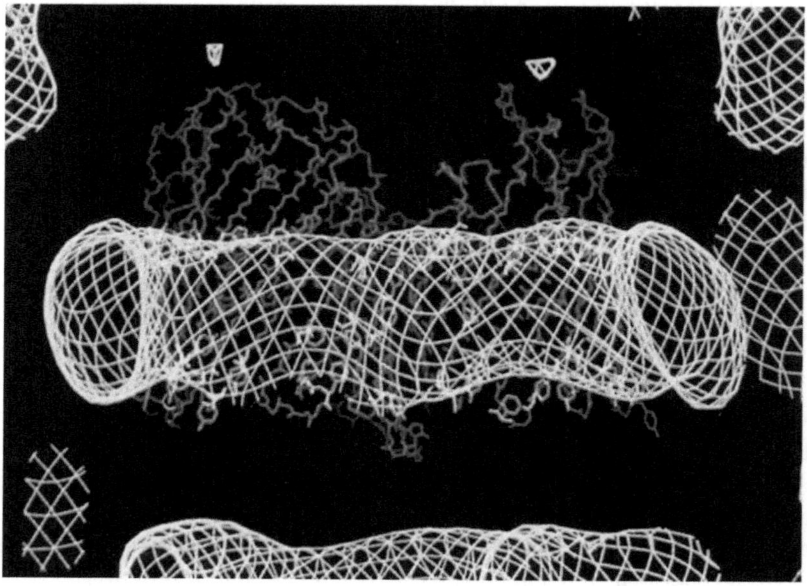

Fig. 5.4. Detergent binding in tetragonal crystals of OmpF porin. Note the binding surface delimited by aromatic residues

in the crystal. The X-ray structure shows the OmpF trimers to pack as two interpenetrating lattices with no protein–protein contacts between molecules in the separate lattices. The question then arises as to how such a structure is stabilized. The answer was provided by the neutron data which showed the two lattices to come into contact through the detergent belts of adjacent molecules. The crystal is therefore stabilized by protein–protein, protein–detergent, and detergent–detergent contacts.

The trigonal form of OmpF is very different from this [23]. The OmpF trimers here form columns of molecules running in opposite directions with protein–protein contacts being responsible for the interactions within each column. The columns of molecules are then held together by fusion of the detergent rings surrounding the hydrophobic protein surfaces. No distinct detergent belts remain but protein trimers interact directly through hydrophobic interactions or mediated by patches of detergent (Fig. 5.5). X-ray contrast variation using xenon has also been used to investigate detergent binding in these crystals [24]. Although the detergent containing volumes determined by neutrons and X-rays partly overlap there are significant differences. This is due most probably to the xenon penetrating only the most hydrophobic parts of the detergent (the tails) as well as into the surface of the protein [24].

Another outer membrane protein to have been studied is the outer membrane phospholipase A (OMPLA) from *E. coli*. The structure of this protein is a 12 stranded β-barrel and its active form is a dimer [25]. Here the neutron data showed a detergent structure distinct from any other of the known

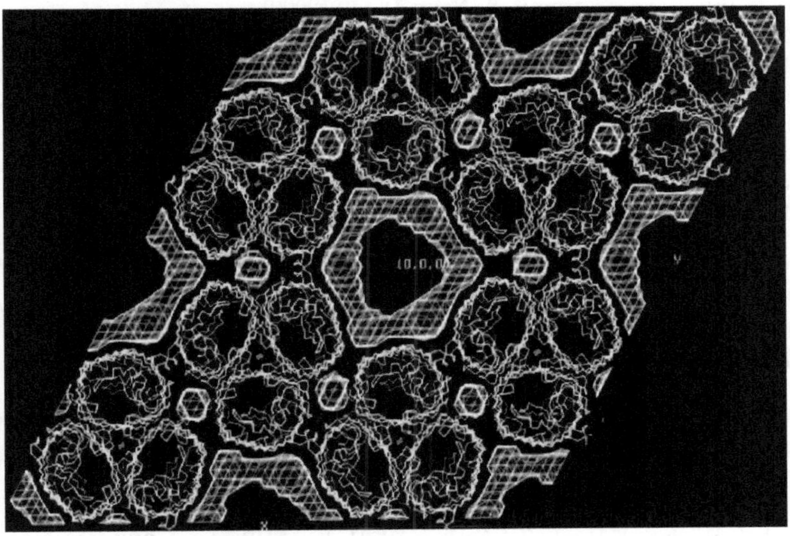

Fig. 5.5. Neutron scattering density map showing the detergent distribution (hashed regions) in trigonal crystals of OmpF porin from *E. coli.* Note the hydrophobic protein–protein contacts

structures with a continuous detergent phases throughout the crystal, somewhat reminiscent of lipid cubic phases [26]. An attempt was also made in this case to obtain the structure from X-ray diffraction of Xe soaked crystals but the results were not consistent with the neutron maps.

5.5 Conclusions

Neutron low resolution crystallography is a powerful technique for visualizing disordered regions in crystals of biomolecular complexes and in particular the localization of detergent in crystals of membrane proteins. The technique of course relies on the availability of crystals but these do not necessarily have to be very large as required for high resolution crystallography $-0.1\,\mathrm{mm}^3$ or less is sufficient. The membrane protein structures studied to date have demonstrated that not only are protein–protein interactions important in the formation of crystals but that in some cases protein–detergent and even detergent–detergent interactions can be crucial.

References

1. W.L. Bragg, M.F. Perutz, Acta Cryst. **5**, 277 (1952)
2. W. Shepard, R. Kahn, M. Ramin, R. Fourme, Acta Cryst. **D56**, 1288–1303 (2000)
3. H. Tsuruta, Synchrotron Rad. News **13**, 10 (2000)
4. M. Roth, A. Lewit-Bentley, Acta Cryst. **A38**, 670 (1982)
5. M. Roth, A. Lewit-Bentley, G.A. Bentley, in *Position Sensitive Detection of Thermal Neutrons*, P. Convert, J.P. Forsyth (Eds.) (Academic Press, New York, 1983)
6. S. Penel, P. Legrand, J. Appl. Cryst. **30**, 206 (1997)
7. W. Kabsch, J. Appl. Cryst. **26**, 795 (1993)
8. M. Roth, A. Lewit-Bentley, G.A. Bentley, J. Appl. Cryst. **17**, 77 (1984)
9. M. Roth, Acta Cryst. **A42**, 230 (1986)
10. M. Roth, Acta Cryst. **A43**, 780 (1987)
11. T.L. Blundell, L.N. Johnson, *Protein Crystallography* (Academic Press, London, 1976)
12. M. Roth, in *Crystallographic Computing 5*, D. Moras, A.D. Podjarny, J.C. Thierry (Eds.) (Oxford University Press, Oxford, 1992)
13. M. Roth, B. Arnoux, A. Ducruix, F. Reiss-Husson, Biochemistry **30**, 9403 (1991)
14. M. Roth, A. Lewit-Bentley, H. Michel, J. Deisenhofer, R. Huber, D.Oesterhelt, Nature **340**, 659 (1989)
15. H. Michel, J. Mol. Biol. **158**, 567 (1982)
16. P. A. Timmins, J. Hauk, T. Wacker, W. Welte, FEBS Lett. **280**, 115 (1991)
17. P. Gast, Hemelrijk, A.J. Hoff, FEBS Lett. **337**, 39 (1994)
18. S.M. Prince, T.D. Howard, D.A.A. Myles, C. Wilkinson, M.Z. Papiz, A.A. Freer, R.J. Cogdell, N.W. Isaacs, J. Mol. Biol. **326**, 307 (2003)
19. R.M. Garavito, J.P. Rosenbusch, J. Cell. Biol. **86**, 327 (1980)

20. R.M. Garavito, J.A. Jenkins, J.M. Neuhaus, A.P. Pugsley, J.P. Rosenbusch, Ann. Microbiol. (Paris) **133A**, 37 (1982)
21. S.W. Cowan, T. Schirmer, G. Rummel, M. Steiert, R. Ghosh, R.A. Pauptit, J.N. Jansonius, J.P. Rosenbusch, Nature **358**, 727 (1992)
22. E. Pebay-Peyroula, R.M. Garavito, J.P. Rosenbusch, M. Zulauf, P.A. Timmins, Structure **3**, 1051 (1995)
23. S. Penel, E. Pebay-Peyroula, J.Rosenbusch, G. Rummel, T. Schirmer, P.A. Timmins, Biochimie **80**, 543 (1998)
24. O. Sauer, M. Roth, T. Schirmer, G. Rummel, C. Kratky, Acta Cryst. **D58**, 60 (2002)
25. H.J. Snijder, I. Ubarretxena-Belandia, M. Blaauw, K.H. Kalk, H. Verheij, M.R. Egmond, N. Dekker, B.W. Dijkstra, Nature **401**, 717 (1999)
26. H.J. Snijder, P.A. Timmins, K.H. Kalk, B.W. Dijkstra, J. Struct. Biol. **141**, 122 (2003)

6

High-Angle Neutron Fiber Diffraction in the Study of Biological Systems

V.T. Forsyth, I.M. Parrot

6.1 Introduction

Fiber diffraction has made a critical impact in structural biology. It has broad application to the study of a wide range of biological and synthetic polymers, and continues to provide key information about the structure of these molecules. The purpose of this chapter is to illustrate the general scope of the method and in particular to demonstrate the impact of *neutron* fiber diffraction methods for the study of biological systems. Because in most neutron fiber diffraction work is carried out in combination with X-ray diffraction studies, this review will focus on the unique complementarity provided by the two approaches to structure analysis. Two specific examples have been chosen to illustrate the power of a combined X-ray and neutron approach by high-angle fiber diffraction – DNA, and cellulose. Both are examples where the involvement of neutron studies has added completely unique information.

The expanding interest in structural studies of biological and industrial polymers is opening up important new opportunities for neutron fiber diffraction. This is illustrated by current trends in X-ray fiber diffraction methods where there is increasing emphasis on the exploitation of the high fluxes available for time-resolved studies and for microdiffraction work, and the key aspects of hydration and hydrogen bonding, both of which are largely inaccessible to X-ray fiber diffraction methods, but which nonetheless having a vital bearing on biological function or physical properties of biological and synthetic polymers. Many of the assumptions made on the basis of initial X-ray fiber diffraction analyzes, even for relatively simple systems, are being reevaluated through the use of innovative techniques at modern X-ray and neutron sources. This is likely to become more important in the future as the new instruments and facilities currently being constructed make their impact.

6.2 Fibers and Fiber Diffraction

An aligned fiber is characterized as containing many regions in which the polymer molecules are arranged with their long axes parallel to the fiber axis. These are typically separated by less ordered amorphous regions. The molecular alignment within such a fiber is associated with varying degrees of order in the side-by-side packing of the chains [1–3]. To a large extent this ordering will determine the general appearance of the observed diffraction pattern which may vary from continuous molecular transform to the other extreme where the pattern is dominated by sharp Bragg reflections. Because there is no preferred orientation of microcrystallites about the fiber axis, an X-ray fiber diffraction pattern from a crystalline fiber has some of the character of that which would be obtained from a single crystal if it was rotated by 360° about one of its principal axes during data collection (see Fig. 6.1). The resolution attained in fiber diffraction experiments varies widely but can in favorable circumstances extend to atomic resolution. While there is some loss of information imposed by cylindrical averaging in fiber diffraction this has to be offset against the fact that in

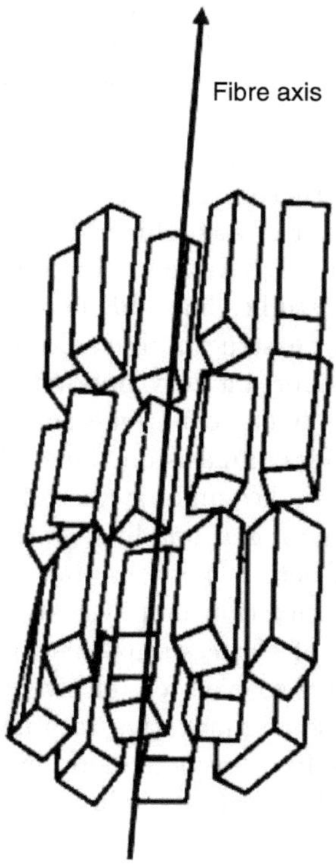

Fig. 6.1. Diagram to illustrate the random orientation of crystallites about the fiber axis in a crystalline fiber

most situations comparable measurements are not possible from single crystal systems.

In many biological structures such as muscle and connective tissue and also in synthetic industrial polymers, the fibrous state is functionally important and fiber diffraction provides the most powerful technique to establish both molecular structure and the higher order interactions of these molecules. It is also true to say that fiber diffraction offers one of the few quantitative methods which are well suited to the study of conformational pathways in polymer molecules. The fibrous state, with its typical mix of highly ordered and amorphous regions, can accommodate changes in molecular conformation and packing much more readily than can single crystals. This capability allows a systematic investigation of the dependence of molecular conformation on important biological parameters such as water and ionic contents in a fiber. In particular, conditions can be chosen which are much closer to the *in vivo* environment than is typically achieved in solution or in single crystal studies. In many biological systems changes in the water content of a fiber can be controlled by varying the relative humidity of the fiber environment, and it is possible to use synchrotron X-ray sources to carry out time-resolved studies of stereochemical pathways followed in structural transitions, that occur as a function of hydration. While this type of work provides quite novel information through the capacity to carry out time-resolved [4, 5] and even microdiffraction studies [6], X-ray diffraction is often a poor probe of water structure. It is fortunate therefore that neutron diffraction can be used, particularly in conjunction with hydrogen/deuterium isotopic replacement, to study the location of water or hydrogen in biopolymer molecules.

6.3 Neutron Fiber Diffraction: General Issues

The advantages of neutron diffraction are well known. One of the most important of these is the fact that, in contrast to the situation with X-rays, hydrogen atoms have a coherent scattering length that is comparable to the other atoms commonly found in biological systems (Fig. 6.2).

Furthermore, it is relatively easy to exchange hydrogen by its more strongly scattering isotope, deuterium, as illustrated in Fig. 6.3, and to use this isotopic replacement in determining the location of hydrogen/water in these systems. Although it has been argued (for the single crystal case) that this type of information can be obtained from high resolution (*i.e.*, atomic) X-ray diffraction studies, the real situation is more complex. Not only is it quite rare (especially in fibrous systems) for biological samples to diffract to atomic resolution, but there is evidence that even when they do, hydrogen atoms with large thermal displacement parameters are not well located. Neutron methods identify such atoms, which may be of biological interest, very clearly [7], even when the available resolution is relatively low.

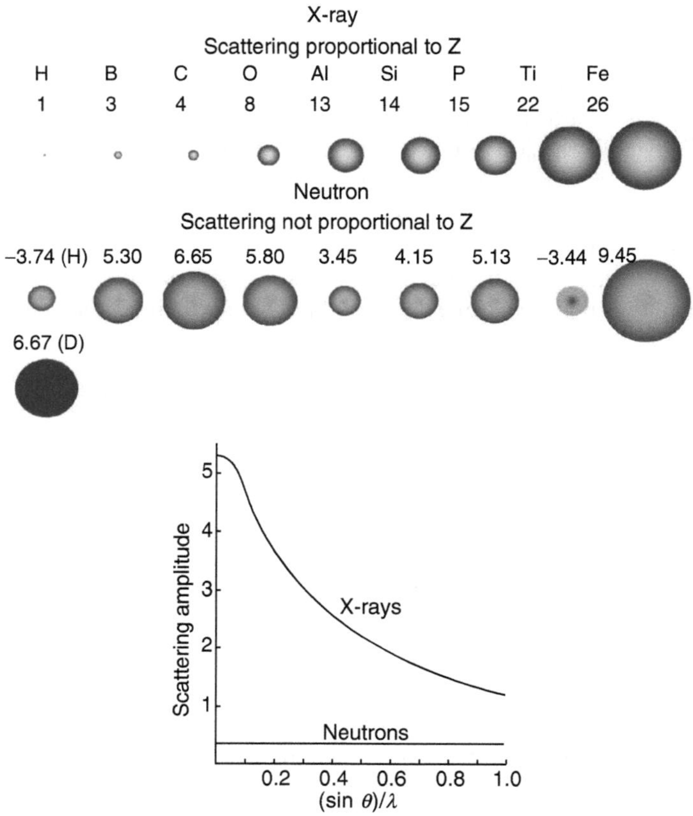

Fig. 6.2. *Top*: diagram giving a representation of the neutron scattering lengths, including some atoms of relevance to biological systems, and emphasizing the difference in neutron scattering lengths of H and D. *Bottom*: plot showing the variation of atomic form factor for X-ray and neutron scattering as a function of scattering angle

Another important aspect of neutron scattering arises from the fact that the atomic nucleus from which neutrons are scattered is essentially point-like in size compared to the wavelength of the incident beam. As a result, in contrast to the situation for X-rays, there is effectively no variation in the atomic form factor for neutrons over the relevant range of scattering angle. This means that in circumstances where the observed diffraction is not dominated by either thermal or spatial disorder, there exists the possibility to record neutron diffraction data at higher resolution than would be possible with X-rays. This point is well illustrated in Fig. 6.4, which compares X-ray and neutron fiber diffraction patterns recorded from a non-biological fiber, poly(*p*-phenylene terephthalamide) (PPTA) more commonly known as the commercial fibres Kevlar or Twaron.

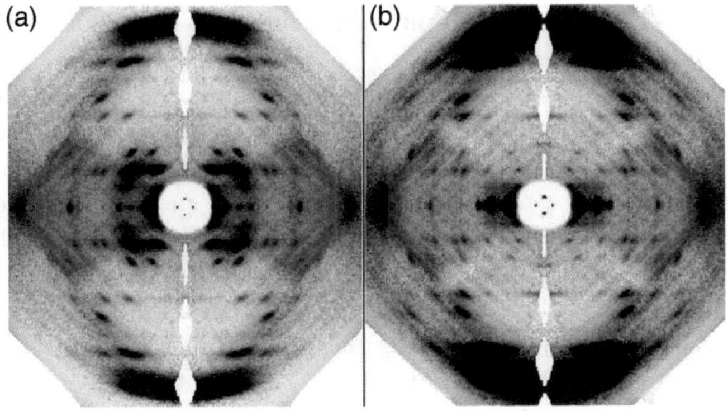

Fig. 6.3. Neutron fiber diffraction patterns recorded from DNA. (**a**) the pattern recorded when the DNA is hydrated with D_2O; (**b**) the pattern recorded when the DNA is hydrated with H_2O

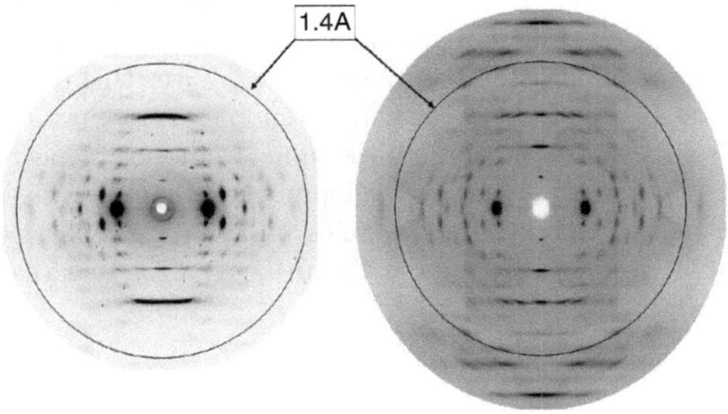

Fig. 6.4. X-ray and neutron fiber diffraction patterns recorded from a synthetic crystalline system, illustrating a falloff in the overall intensity distribution at high angles in the case of the X-ray pattern. In contrast, scattering of similar average intensity is observed well beyond this point in the neutron diffraction pattern

Because of the lower brilliance of the neutron beams, there are major issues to be faced for neutron diffraction studies of biological systems. By far the most important is the requirement for large sample size. Although this is a major obstacle in crystallography, there are various approaches that can be adopted in tackling this problem for fiber studies. The first high-angle studies of DNA at the ILL [8–10] were carried out using individual fibers (each about 100 μm in diameter) aligned in a parallel array. In later work, sheet samples made using the wet spinning method of Rupprecht [11] were used to obtain the required sample volume both for fiber diffraction and for neutron and X-ray spectroscopy [12,13]. For the work on cellulose described in the following sections of this chapter, Nishiyama and colleagues have used a shear flow method to deposit aligned microcrystals in a gel-like film which is then dried to produce large high-quality samples of various forms of cellulose that diffract to nearly atomic resolution.

For crystallographic and fiber studies in particular, the issue of sample size is further exacerbated by incoherent scattering from hydrogen in the solvent and in the macromolecule itself. It is relatively straightforward to replace H_2O solvent by D_2O for these studies; this results in a huge decrease in the hydrogen incoherent scattering "background", as illustrated in Fig. 6.5.

However, there is still a large problem arising from the hydrogen that is covalently bound to carbon atoms in the structure. This places limitations on sample size, data collection times, on the interpretation of the data, and indeed on the quality of the final analysis. However, the presence of hydrogen incoherent scattering and the associated limitations in neutron scattering can be effectively eliminated by sample perdeuteration. Although the benefits of sample deuteration have been widely appreciated for some time, it has never been easy for individual biologists to deuterate their systems in a routine way that allows best use to be made of valuable central facility resources; the expertise is rather and individual requirements vary quite considerably. As described in the concluding sections of this chapter this problem has been identified at the ILL and the EMBL in Grenoble [14] where these facilities, with advice from their user communities and a number of peer-review committees, took the view that major developments for modern instrumentation should be paralleled with strong emphasis on sample preparation issues. Of particular importance in this area is the utilization of selective and sample deuteration.

6.4 Facilities for Neutron Fiber Diffraction

Neutron high-angle fiber diffraction methods were developed at the ILL on the D19 diffractometer, and most of the work that has been published subsequently has used this facility. D19 was originally optimized for chemical crystallography where it was used for structure determination for systems having relatively large unit cell dimensions [15–19]. A key aspect of the instrument was the use of an area detector rather than smaller "single" detectors commonly used in crystallographic studies of small molecule systems. Such a detector is essential for fiber diffraction work, which requires not only that the intensities of the Bragg reflections are well measured, but also that the variation of the background over the entire pattern is measured as accurately as possible. Figure 6.6 shows a photograph of the D19 diffractometer alongside a sample dataset.

At the time that this photograph was taken the D19 detector was a long thin banana-shaped gas-filled device with an angular aperture of 4° in its equatorial plane and 64° in its vertical plane. The requirements for data collection, data reduction and analysis are different from those commonly used in single-crystal studies. The strategy for data collection has therefore always

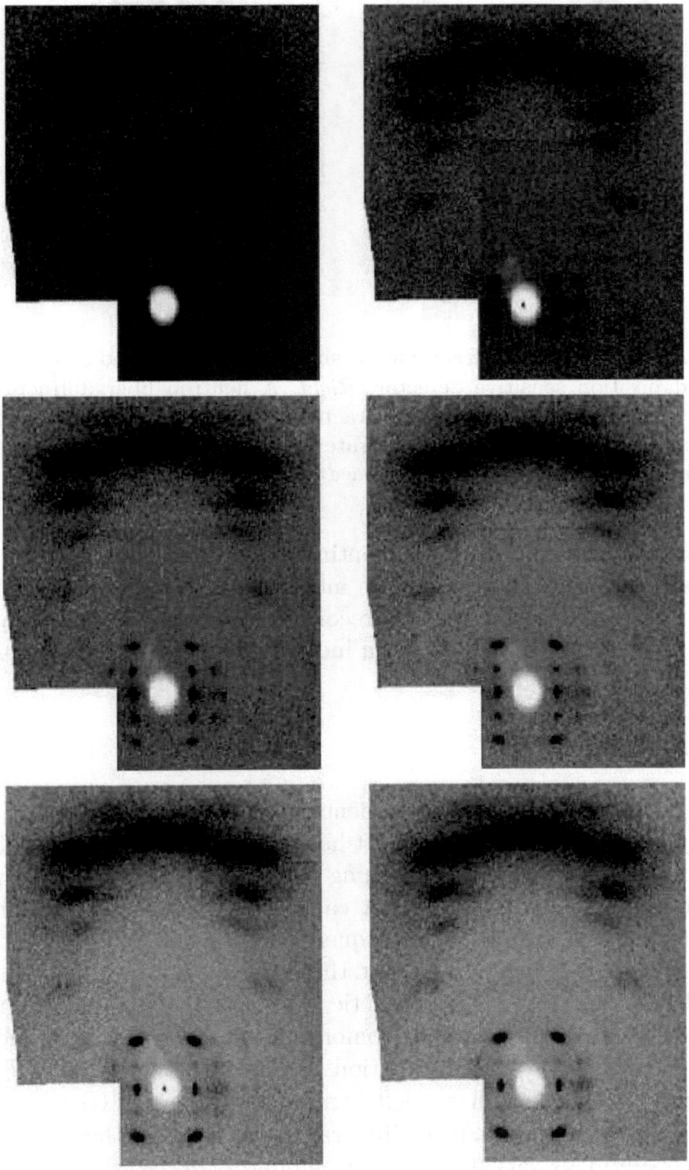

Fig. 6.5. High-angle neutron fiber diffraction patterns recorded from a sample of A-DNA as the water in the sample is exchanged by D_2O (*top left* pure H_2O to *bottom right* pure D_2O). This set of pictures illustrates the consequences of hydrogen incoherent scattering and the major benefits of deuterium exchange

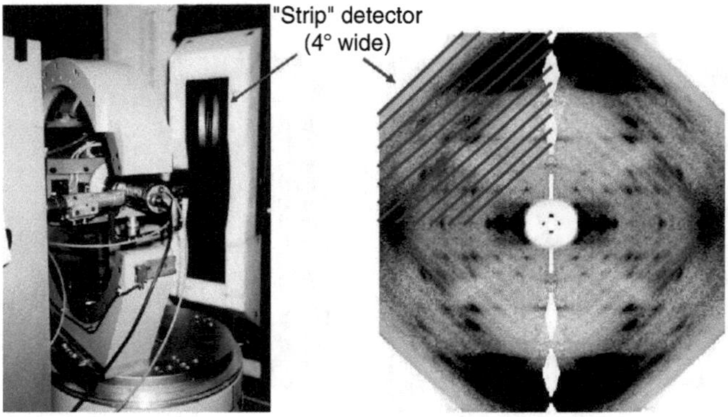

Fig. 6.6. *Left*: The D19 diffractometer, showing Eulerian cradle, sample environment, and position sensitive detector. *Right*: A neutron fiber diffraction pattern recorded from DNA using the D19 diffractometer. Individual detector acquisitions are indicated on the pattern to illustrate how data from the diffractometer are mapped into a continuous reciprocal space image

been to construct a complete and continuous diffraction pattern by mapping individually recorded data segments into reciprocal space. Figure 6.6 shows a neutron fiber diffraction pattern recorded from a DNA sample and shows graphically how this is mapped from individual single detector acquisitions.

6.5 Nucleic Acids

DNA in its natural habit is a filamentous molecule, with a high degree of regularity over a large length scale. It has been known for decades that water is of critical importance in maintaining this regularity and furthermore that variation of hydration around DNA causes major conformation changes in the double helix. Natural, "mixed-sequence" DNA can be drawn into aligned fibers and has been shown to adopt three main conformations, which have been called A, B, and C. For synthetic DNA polymers containing repetitive base-pair sequences, two further conformations called D-DNA and Z-DNA can be observed. The Z conformation is the only one of these five major DNA conformations that has a left-handed helical sense. Given appropriate conditions of ionic strength, reversible transitions between these structures can be induced simply by varying the relative humidity of the sample environment. This structural polymorphism in DNA is summarized in Figs. 6.7 and 6.8.

The biological significance of these DNA structures is not fully understood. However, it would be surprising if the ability of DNA to adopt these markedly different structures were not exploited in biological function. Further support for the biological significance of these structures comes from studies that confirm the presence of the B and the Z forms of DNA *in vivo* and that change to the Z form causes a number of functionally significant effects [20–24]. A variety

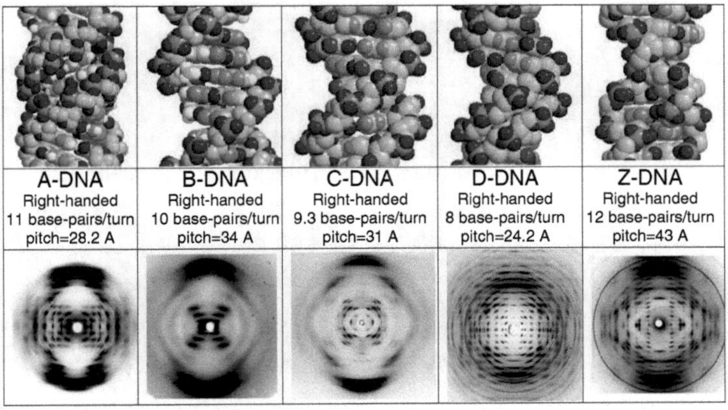

Fig. 6.7. The five major conformations of DNA, their helix parameters and their X-ray fiber diffraction patterns

NATURAL DNA	ALTERNATING A-T	ALTERNATING G-C	ALTERNATING G-m^5C	HOMOPOLYMER G-C	HOMOPOLYMER A-T
	A-T	G-C	G – m^5C	G-C	A-T
	T-A	C-G	m^5C - G	G-C	A-T
	A-T	G-C	G – m^5C	G-C	A-T
"Random"	T-A	C-G	m^5C - G	G-C	A-T
sequence	A-T	G-C	G – m^5C	G-C	A-T
	T-A	C-G	m^5C - G	G-C	A-T
	A-T	G-C	G – m^5C	G-C	A-T
	T-A	C-G	m^5C - G	G-C	A-T
	A-T	G-C	G – m^5C	G-C	A-T
	T-A	C-G	m^5C - G	G-C	A-T
C↔A↔B	D↔B	Z↔B	A→Z	A	B

Fig. 6.8. Summary of the structures adopted by various DNA sequences. The middle section of the diagram shows the sequence repeat in the DNA. The lower section shows the structures adopted and the transitions that occur

of proteins that bind to B, Z, A, and D type conformations have been identified [25–28], and there is a substantial amount of work that implicates a number of repetitive DNA sequences in regulatory processes. Whatever the biological importance of these structures, there is clearly great interest in understanding structural aspects relating to their stability as well as those that mediate transitions between them. In this respect, X-ray and neutron fiber methods have genuinely complementary roles. Modern synchrotron X-ray sources have sufficient flux to allow the study of DNA structural transitions in real time. As a good examples of this Fig. 6.9 shows the D↔B transition in synthetic DNA having a regular alternating adenine-thymine (A-T) repeat. The transition, first described by Mahendrasingam et al. [29] and Forsyth et al. [30] occurs through a stereochemical pathway in which the pitch of the DNA changes from the 24 Å characteristic of the D conformation to the

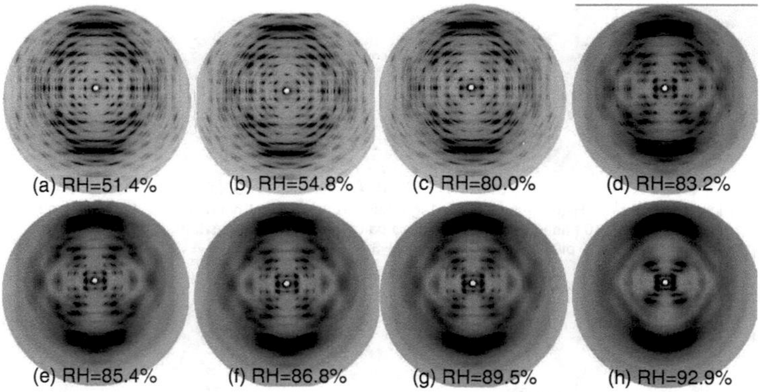

Fig. 6.9. Selected X-ray fiber diffraction patterns recorded during the water driven transition between the D and the B conformations of DNA. The data were recorded at the Daresbury SRS

34 Å characteristic of B-DNA. Figure 6.10 shows a further example involving the A–Z transition in an alternating G-C sequence in which the cytosine residues are methylated at the 5 position. More recent work on the transition between the A and the B forms is in progress [31].

The examples given above illustrate two aspects of DNA structure that may be important in biological function. First they illustrate cooperativity in DNA polymorphism – both transitions occur through a process during which substantial regions of the samples change in a highly cooperative way. Second, both transitions further emphasize the critical role that water plays in determining DNA structure and therefore the importance of establishing the location of water around each DNA structure and also during structural changes. X-ray diffraction studies have provided high-resolution information on hydration around the A, B, and Z conformations in oligonucleotide single crystals, with particular emphasis on local sequence dependence variation. In contrast, neutron fiber diffraction provides information on hydration at lower resolution (typically about 3 Å) but which relates to the regularity and cooperative properties of the long polymer molecule. The two methods therefore provide highly complementary information.

The first neutron diffraction experiments on DNA fibers were carried out on D-DNA [8, 9] at the Institut Laue Langevin in Grenoble, France. These experiments demonstrated the power of isotopic replacement of light by heavy water and allowed Fourier synthesis methods to be used to image the location of water around the D form of the DNA double helix (Fig. 6.11). Of particular significance in this work was the presence of water in the minor groove and its relationship to stabilizing cations that had been located by X-ray diffraction studies of isomorphous derivatives of D-DNA.

Similar work to that described for the D conformation of DNA has been carried out on the A and the B conformations [10, 32–34]. In the first study

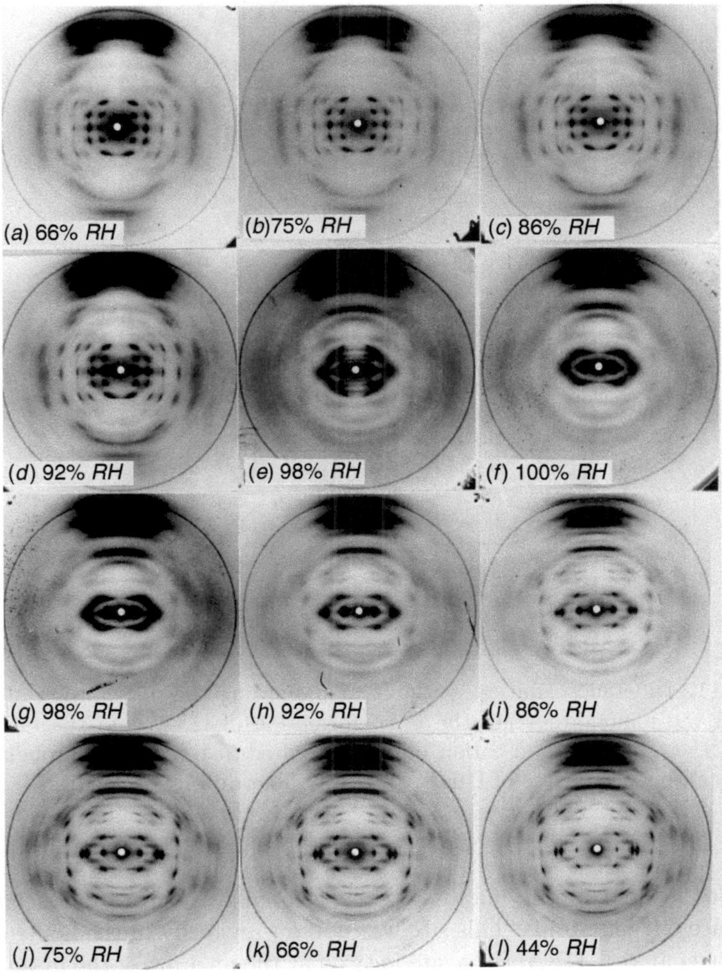

Fig. 6.10. Diffraction patterns recorded during the transition between the A and the Z conformations of the DNA polymer poly [d(G-m^{5}C)]

of A-DNA, hydrogenated material was used. The Fourier maps obtained showed a number of features that had been observed in previous X-ray diffraction studies of oligonucleotide single crystal – most notably water molecules located between successive phosphate oxygen atoms along the major groove (see Fig. 6.12).

In later studies, more detail was provided by Shotton et al. [35] who used perdeuterated DNA to study hydration in A-DNA. The DNA was obtained from *E. coli* cells grown in D_2O with a deuterated carbon source. Figure 6.13 summarizes the main results from this study. Four main sites were located:

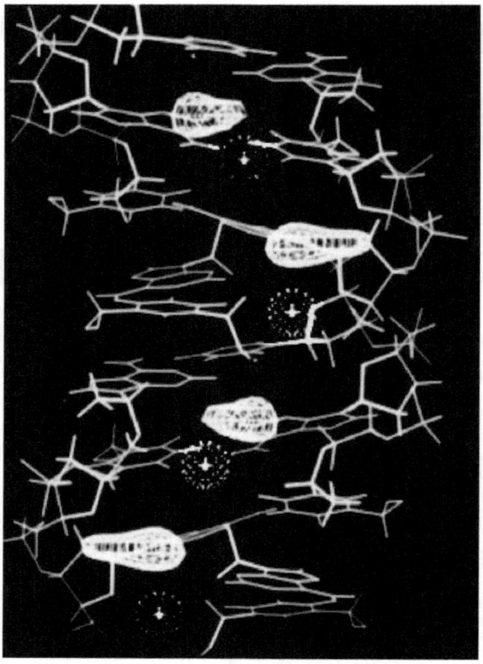

Fig. 6.11. Difference Fourier synthesis map showing the distribution of ordered water in the minor groove of the D conformation of DNA, and the relationship between these features and cation positions located in the groove through the use of heavy atoms derivatives of D-DNA in X-ray fiber diffraction studies

(a) a site located in the major groove at equal distances from neighboring phosphate oxygen atoms. This site was noted previously in the neutron study of hydrogenated A-DNA; (b) a site located at the opening of the major groove at equal distances from phosphates on either strand; (c) a site also located at the center of the major groove, but at a smaller radius; (d) a feature running down the "hollow" center of the molecule within possible hydrogen bonding distance to base edge atoms. This feature could not be completely interpreted as a result of the sequence averaging of the base pairs.

The structure of A-DNA is such that the base-pairs are displaced some 5 Å from the helix axis, creating a molecule which has a clear central hole when viewed along the length of the molecule. The central column of water described above and seen in Fig. 6.13d is located within this hole. Its apparent continuity and close relationship with the DNA was described by Shotton et al. [32]. It is clear that this column of hydration would have to be broken during a transition to the B form, where the base-pairs are located centrally on the helix axis. Fuller et al. [36] have suggested that the column may be significant in understanding the hydration-driven transition between the A

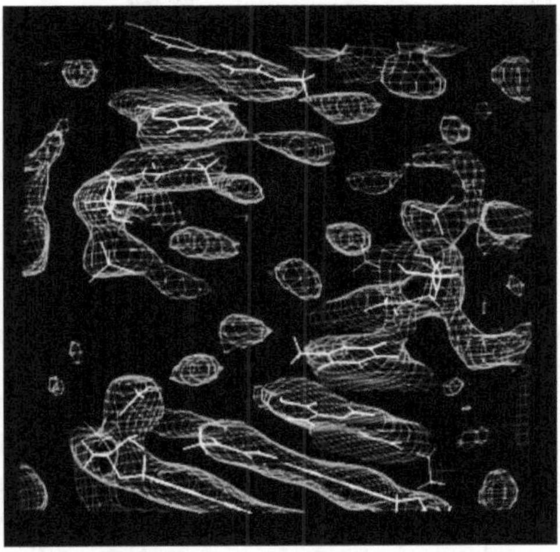

Fig. 6.12. A $2F_o - F_c$ Fourier map of the A conformation of DNA showing a string of water molecules along the major groove, bridging phosphate oxygen atoms

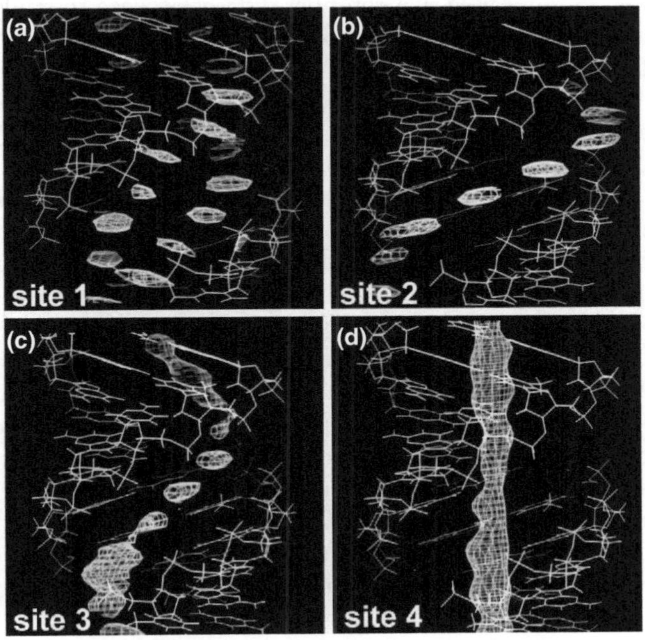

Fig. 6.13. The results of neutron fiber diffraction analysis showing the four ordered water sites in A-DNA [35]

and the B forms of the double helix, and that its continuity in A-DNA and disruption in B-DNA may be related to the cooperativity observed in the transition.

6.6 Cellulose

Cellulose exists in a number of different forms depending on its origin and on the way in which it is treated. During the last few years a number of studies have been published in which X-ray fiber diffraction data have been combined with neutron fiber diffraction data collected on instrument D19 at the ILL to yield novel information that in all likelihood would never have been derived using X-ray diffraction alone. The significance of these studies has recently been summarized by Jarvis [37], who while noting the basic simplicity of cellulose, also emphasizes the importance of hitherto unsolved questions relating to the side-by-side packing of the molecules in their different forms – packing effects where hydrogen-bonding is of key importance. In all of the recent fiber diffraction work carried out, X-ray data, typically to atomic resolution, has been recorded and used to determine the location of the "non-hydrogen" atoms in the cellulose structure. However, even at this resolution these workers were not able to determine hydrogen atom positions from their electron density maps [38], and used neutron fiber diffraction data to study hydrogen bonding interactions in these structures. The first study using this approach was that of cellulose II, where Langan et al. [39] were able to discriminate between two competing models and to unambiguously determine the hydrogen bonding network in this structure. Figure 6.14 shows a $2F_o - F_c$ Fourier synthesis map in which the observed amplitudes were measured from cellulose II samples in which the OH groups had been replaced by OD through a process of mercerization with NaOD.

In later studies, the same group of workers carried out detailed analyses of the cellulose I_β [40], and cellulose I_α [41] using high resolution X-ray and neutron fiber diffraction data. The neutron analyzes were critical in establishing the nature of the hydrogen bonding in each structure and in attempting to understand the conversion between the two forms. Figure 6.15a shows the neutron fiber diffraction patterns recorded both from a normal hydrogenated sample of cellulose I_β (yielding structure factor amplitudes F_h) and from an analogous sample in which the OH groups had all been replaced by OD (yielding structure factor amplitudes F_d). Figure 6.15b shows a difference Fourier map computed from these data using coefficients $(F_d - F_h)$, with phases calculated from the best X-ray refined model for the carbon and oxygen atom positions.

These results, combined with the synchrotron X-ray results, define the structure of all atoms including hydrogen, at atomic resolution. They highlight the well-defined nature of the intramolecular O3,. . .,O5 hydrogen bonding, and show that the hydrogen bonding involving the O2 and O6 atoms is disordered, imparting some stability to the sheet structure of cellulose I_β.

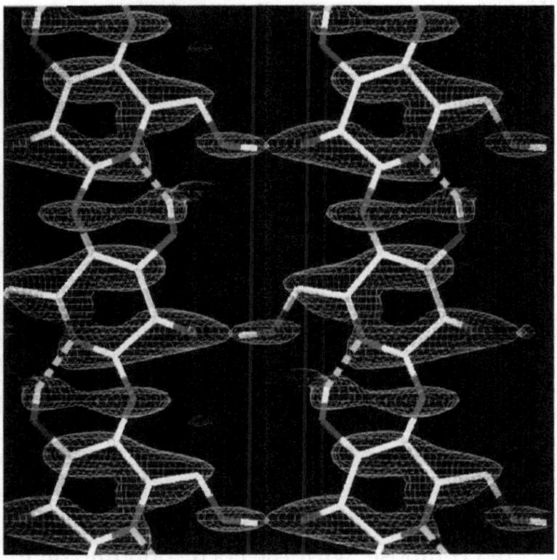

Fig. 6.14. $2F_o - F_c$ Fourier synthesis map in which observed amplitudes (F_o) were extracted from neutron data recorded from deuterated cellulose (mercerized with NaOD). This density map illustrates the presence of a network of hydrogen bonds that is different from that inferred by previous X-ray diffraction work (see Langan et al. [39])

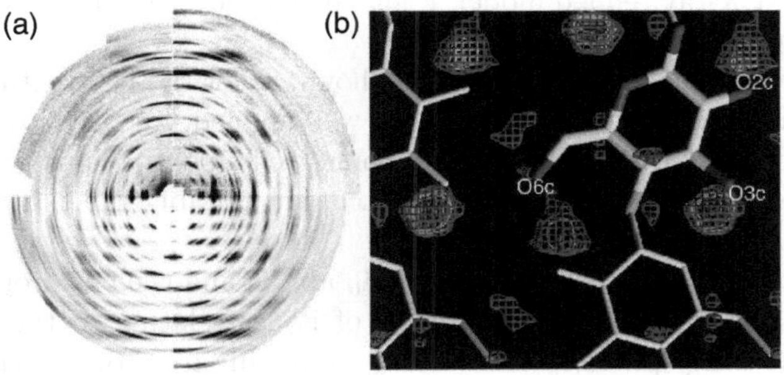

Fig. 6.15. (**a**) High-angle neutron fiber diffraction patterns recorded from cellulose I_β using the D19 diffractometer. The image is split into four quadrants - the left ones relating to the hydrogenated cellulose and the right ones relating to the deuterated cellulose. In each case the corresponding quadrants on the bottom show simulations of the fitted intensities. The meridian (fiber axis) is vertical. (**b**) Section of the $(F_d - F_h)$ difference Fourier map, showing the position of the deuterium atoms associated with the O3, O2, and O6 (see Figs. 6.4 and 6.5 in Nishiyama et al. [40])

Neutron results from the analogous study of cellulose I_α are shown in Fig. 6.16. Again, two highly aligned and crystalline samples were produced, one hydrogenated, and one deuterated. The neutron data were again recorded using the D19 diffractometer at the ILL (Fig. 6.16a). The analysis followed

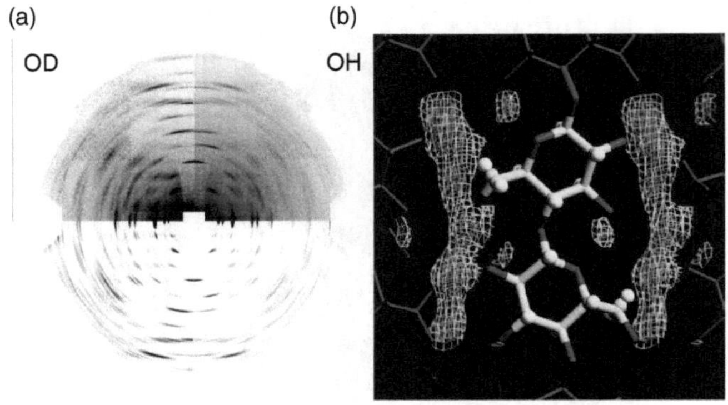

Fig. 6.16. (**a**) High-angle neutron fiber diffraction patterns recorded from cellulose I_α using the D19 diffractometer. As with Fig. 6.15a, the image is split into four quadrants, the left ones relating to the hydrogenated cellulose and the right ones relating to the deuterated cellulose, with the corresponding quadrants on the bottom show simulations of the fitted intensities. (**b**) Section through the $(F_d - F_h)$ difference Fourier map, showing the position of the deuterium atoms associated with the O3, O2, and O6 (see Figs. 6.4 and 6.5in Nishiyama et al. [41])

essentially the same procedure as described for cellulose I_α, with difference Fourier maps calculated using the coefficients $(F_d - F_h)$ and phases derived from the best X-ray refined model. Figure 6.16a shows an $(F_d - F_h)$ difference Fourier map for the I_α structure.

The neutron analysis of cellulose I_α shows, as for the I_β structure, that the O3,. . .,O5 hydrogen bond is single and well-defined, and that the hydrogen atoms associated with the O2 and O6 atoms are found in a number of partially occupied positions, with occupancies that are notably different in the two forms.

These definitive models for cellulose I_α and I_β and their hydrogen bonding networks have shed light on a number of issues relating to the stability of cellulose and likely pathway followed in converting from one to the other. In each structure the hydrogen bonding within the individual chains has been confirmed, and new information on the way in which inter-chain interactions occur has been revealed. As noted by Jarvis [37], these studies also describe a packing configuration that implies the involvement of an ordered arrangement of weak C – H, . . . ,O hydrogen bonds between sheets of cellulose chains. Along with hydrophobic interactions between sheets, this may explain the nature of sheet stacking in cellulose. This type of combined X-ray and neutron approach has also recently been applied to cellulose III_I [42].

6.7 Conclusions and Future Prospects

The work reviewed in this chapter illustrates the genuine complementarity of X-ray and neutron diffraction methods in the study of biological systems.

While X-ray fiber diffraction is well suited to definitive structural studies of filamentous molecules, isomorphous replacement studies, and to the study of conformational changes in polymer systems, it is usually very difficult to use X-ray data to determine the location of ordered water or hydrogen atoms. Early work [43–45] was successful in using Fourier synthesis methods to investigate features of the B-DNA double helix, but were not very successful in locating water positions around the molecule. Even though DNA samples of this type rarely diffract to better than 3 Å resolution, neutron studies have allowed a clear picture of the distribution of ordered water to be obtained, particularly in the A and the D conformations. The recent work that has been carried out on cellulose further emphasizes these points. Despite the availability of X-ray data at atomic resolution, neutron analyses were required to determine the intrachain, interchain, and intersheet hydrogen-bonding interactions responsible for the stability and properties of different forms of cellulose.

All of the neutron work described has been carried out on the D19 diffractometer at the ILL, using a thin banana-shaped detector that has an angular aperture of $4° \times 64°$. Despite being located on a thermal beam at one of the best neutron sources in the world, the diffracted neutrons have been sadly under-exploited in the past. The reason for this has been the limited size of the detector. The approach used to collect datasets was, as described earlier, to devise a sequence of detector and sample movements that allowed a contiguous diffraction pattern to be constructed in reciprocal space from a large number of individual detector measurements (see Fig. 6.6). At any given instant in time during such an experiment, less than 5% of the diffracted neutrons were detected. This has had huge implications for sample size, data collection times and indeed the overall scope of the technique. In particular, such experiments have until now been restricted to the study of polycrystalline fiber samples in which the average diffracted intensity per pixel at the detector is considerably higher than it is for samples that give continuous (layer line) diffraction.

The D19 diffractometer is about to undergo a major upgrade, as shown in Fig. 6.17. Further details of this project and its significance for chemical crystallography and fiber diffraction is given by [46] and [47]. The new instrument will mean that it will be possible to use samples that have hitherto been far too small to study using neutrons. It will also mean that for the first time detailed measurements will be possible of continuous diffraction from polymer molecules. This type of diffraction predominates in diffraction studies of many filamentous viruses such as Pf1 [48], TMV [49], a range of plant viruses [50], as well as drug-DNA and protein-DNA complexes, and it is also a key aspect of changes in ordering that occur during structural transitions [51]. For the work on Pf1 filamentous phage, Mitsch [52] and Langan [53] have demonstrated the feasibility of neutron fiber diffraction studies on D19 but were not able to record adequate data as a result of the limited detector size. It should be noted that this is exactly the type of work where the upgraded D19 will have a major impact.

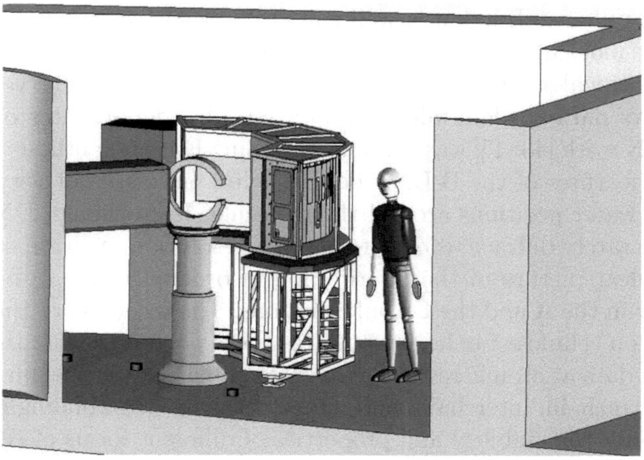

Fig. 6.17. Cartoon showing the new D19 diffractometer at the ILL. The centerpiece of this project is a large area detector that together with upgrades of other aspects of the diffractometer will provide an efficiency gain of ~25 on the original instrument

Examples of other work in progress that will benefit from this development are studies of chitin aimed at understanding the involvement of water in the structure, and studies of spider dragline silk that focus on the relationship between the structures of the crystalline and amorphous components of the system [54]. Numerous opportunities also exist for the investigation of key problems in the study of amyloid fibers, where despite considerable progress in the study of amyloid structures in a variety of systems [55–58], many structural questions remain. Other developments in instrumentation suitable for neutron fiber diffraction are under way at the Los Alamos LANCSE facility, where the PCS instrument has been built [59, 60], at JAERI which accommodates the BIX monochromatic diffractometers [61], at the FRM-II facility in Munich, at the Rutherford-Appleton ISIS facility in the UK, and at Oak Ridge National Laboratory in Tennessee.

Many of these studies will benefit from and exploit the availability of purpose-designed facilities for the selective and nonselective deuteration of biopolymer systems. Such a facility has been established jointly by the ILL and the EMBL in Grenoble [14] and is now operational with an active and expanding user-driven and in-house research programme. This facility, which will form a key part of the new Partnership for Structural Biology (PSB) between the ILL, the ESRF, the EMBL, and the IBS central facility operators in Grenoble, will have an important impact on the scope of future neutron fiber diffraction work on a wide range of important biological polymer systems, either through the provision of samples for which hydrogen incoherent scattering is effectively eliminated, or samples in which specific parts of the structure have been labelled. A similar laboratory has since been set up at the LANCSE facility at Los Alamos and one is also planned for the Spallation Neutron Source (SNS) in Tennessee. The ability to label biological macromolecules in this way has obvious advantages. For crystallographic work and high-angle fiber diffraction studies the elimination of hydrogen incoherent scattering has a major impact in terms of the required sample size as well as data collection times and data interpretation. It has been estimated that for a neutron crystallographic study of a typical protein molecule one can expect a gain of a factor of 10 in signal to noise if the protein is perdeuterated. Experiments on perdeuterated myoglobin [62] and on perdeuterated DNA [32] clearly demonstrate these advantages. Selective deuteration is also extremely powerful, as has been demonstrated by Gardner et al. [63] in studies of synthetic polymers. At lower resolution contrast variation can be used in studies where specific parts of multicomponent systems are deuterated while the rest of the structure remains hydrogenated and can be matched out [64].

Acknowledgments

We acknowledge S. Mason and J. Archer at the Institut Laue Langevin, P. Langan at LANSCE, Los Alamos and V. Urban and C. Riekel at the European Synchrotron Radiation Source for valuable discussion. We thank Y. Nishiyama for providing pictures taken from their work on cellulose. We also wish to acknowledge B. Guerard and other members of the ILL Detector Group for all of their efforts in the design and construction of new detector facilities that will soon be installed at the ILL. We acknowledge support from EPSRC under grants GR/R99393/01 and GR/R47950/01 and from the EU under contracts HPRI-2001-50065 and RII3-CT-2003-505925. I.P. acknowledges support from EPSRC and ILL for the provision of a studentship held at Keele University.

References

1. B.K. Vainshtein, *Diffraction of X-rays by Chain Molecules* (Elsevier, Amsterdam 1966)
2. R.D.B. Fraser, T.P. MacRae, in *Conformation in Fibrous Proteins* (Academic Press, US 1973)
3. R.P. Millane, W. Stroud, Int. J. Biol. Macromol. **13**, 202–208 (1991)
4. T. Narayanan, O. Diat, P. Boesecke, Nucl. Instrum. Methods Phys. Res. A **467–468**, 1005–1009 (2001)
5. V. Urban, P. Panine, C. Ponchut, P. Boesecke, T. Narayanan, J. Appl. Cryst. **36**, 809–811 (2003)
6. C. Riekel, Rep. Prog. Phys. **63**, 233–262 (2000)
7. I. Hazemann, M.T. Dauvergne, M.P. Blakeley, F. Meilleur, M. Haertlein, P. Timmins, A. Van Dorsselaer, A. Mitschler, D.A.A. Myles, A. Podjarny, Acta Cryst. D61, 1413–1417 (2005)
8. W. Fuller, V.T. Forsyth, A. Mahendrasingam, W.J. Pigram, R.J. Greenall, P. Langan, K. Bellamy, Y. Al-Hayalee, S.A. Mason, Physica B **156/157**, 468 (1989)
9. V.T. Forsyth, A. Mahendrasingam, W.J. Pigram, R.J. Greenall, K. Bellamy, W. Fuller, S.A. Mason, Int. J. Biol. Macromole. **11**, 236 (1989)

10. P. Langan, V.T. Forsyth, A. Mahendrasingam, W. J. Pigram, S.A. Mason, W. Fuller, J. Biomol. Struct. Dyn. **10**, 489 (1992)
11. A. Rupprecht, Biotechnol. Bioeng. **12**, 93 (1970)
12. H. Grimm, H. Stiller, C.F. Majkrzak, A. Rupprecht, U. Dahlborg, (1987), Phys. Rev. Lett. **59**, 1780–1783 (1987)
13. H. Grimm, M. Krisch, A. Mermet, V.T. Forsyth, A. Rupprecht, Phys. Rev. E, submitted
14. V.T. Forsyth, D.A.A. Myles, P.A. Timmins, M. Hartlein, Possibilities for the exploitation of biological deuteration in Neutron scattering, in *Opportunities for Neutron Scattering in the 3rd Millennium*, J. Dianoux (Eds.) (Institut Laue Langevin, Grenoble, 2001), pp 47–54
15. P. Langan, M. Lehmann, C. Wilkinson, G. Jogl, C. Kratky, Acta Cryst. D **55**, 51–59 (1999)
16. B. Arhens, M.G. Davidson, V.T. Forsyth, M. Mahon, A.L. Johnson, S.A. Mason, R.D. Price, P.R. Raithby, J. Am. Chem. Soc. **123**, 9164–9170 (2001)
17. C.K. Broder, M.G. Davidson, V.T. Forsyth, J.A.K. Howard, S. Lamb, S.A. Mason, Cryst. Growth Des. **2**, 163–169 (2002)
18. G. Vives, S.A. Mason, P.D. Prince, P. Junk, J.W. Steed, Cryst. Growth – Des. **3**, 699 (2003)
19. S. Detti, V.T. Forsyth, R. Roulet, R. Ros, A. Tassan, K.J. Schenk, Z. Kristallogr. **219**, 47–53 (2004)
20. A. Jaworski, W.-T. Hsieh, J.A. Blaho, J.E. Larson, R.D. Wells, Science **238**, 773–777 (1987)
21. L.J. Peck, J.C. Wang, Cell **40**, 129–137 (1985)
22. S. Wolfl, Biochim. Biophys. Acta **1264**, 294–302 (1995)
23. V. Muller, M. Takeya, S. Brendel, B. Wittig, A. Rich, Proc. Natl. Acad. Sci. USA **93**, 780–784 (1996)
24. S. Wolfl, C. Martinez, A. Rich, J.A. Majzoub, Proc. Natl. Acad. Sci. USA **93**, 3664–3668 (1996)
25. A.G. Herbert, K. Lowenhaupt, J.R. Spitzner, A. Rich, Proc. Natl. Acad. Sci. USA **92**, 7550–7554 (1995)
26. T. Schwartz, M.A. Rould, K. Lowenhaupt, A. Herbert, A. Rich, Science **284**, 1841–1845 (1999)
27. M.A. Schumacher, B.K. Hurlburt, R.G. Brennan, Nature **409**, 215–219 (2001)
28. N. Mizuno, G. Voordouw, K. Miki, A. Sarai, Y. Higuchi, Structure **11**, 1133–1140 (2003)
29. A. Mahendrasingam, V.T. Forsyth, R. Hussain, R.J. Greenall, W.J. Pigram, W. Fuller, Science **233**, 195–197 (1986)
30. V.T. Forsyth, R.J. Greenall, R. Hussain, A. Mahendrasingam, C. Nave, W.J. Pigram, W. Fuller, Biochem. Soc. Trans. **14**, 553–557 (1986)
31. I.M. Parrot, V. Urban, K.H. Gardner, V.T. Forsyth, Nucl. Instrum. Methods B, 238, 7–15 (2005)
32. M.W. Shotton, L.H. Pope, V.T. Forsyth, P. Langan, R.C. Denny, U. Giesen, M.-T. Dauvergne, W. Fuller, Biophys. Chem. **69**, 85–96 (1997)
33. L.H. Pope, M.W. Shotton, V.T. Forsyth, P. Langan, R.C. Denny, U. Giesen, M.-T. Dauvergne, W. Fuller, Physica B **241–243** 1156–1158, (1998)
34. V.T. Forsyth, M.W. Shotton, H. Ye, C. Boote, P. Langan, L.H. Pope, R.C. Denny, Fiber Diffr. Rev. **7**, (1998) 17–24.
35. M.W. Shotton, L.H. Pope, V.T. Forsyth, R.C. Denny, J. Archer, P. Langan, H. Ye, C. Boote, J. Appl. Cryst. **31**, 758–776 (1998)

36. W. Fuller, V.T. Forsyth, A. Mahendrasingam, Phil. Trans. Roy. Soc. **359**, 1237–1248 (2004)
37. M. Jarvis, Nature **426**, 611–612 (2003)
38. P. Langan, Crystallogr. Rev., 11, 125–147 (2005)
39. P. Langan, Y. Nishiyama, H. Chanzy, J. Am. Chem. Soc. **121**, 9940–9946 (1999)
40. Y. Nishiyama, H. Chanzy, P. Langan, J. Am. Chem. Soc. **124**, 9074–9082 (2002)
41. Y. Nishiyama, J. Sugiyama, H. Chanzy, P. Langan, J. Am. Chem. Soc. **125**, 14300–14306 (2003)
42. M. Wada, H. Chanzy, Y. Nishiyama, P. Langan, Macromolecules **37**, 8548–8555 (2004)
43. D.A. Marvin, PhD Thesis, University of London (1960)
44. S. Arnott, M.H.F. Wilkins, L.D. Hamilton, R. Langridge, J. Mol. Biol. **11**, 391–402 (1965)
45. D.A. Marvin, M. Spencer, M.H.F. Wilkins, L.D. Hamilton, Acta Cryst. **20**, 663–669 (1961)
46. V.T. Forsyth, S.A. Mason, J.A.K. Howard, M.G. Davidson, W. Fuller, D.A.A. Myles, Neutron News **12**, 10–15 (2001)
47. J.C. Buffet, J.F. Clergeau, R.G. Cooper, J. Darpentigny, A. De Laulany, C. Fermon, S. Fetal, F. Fraga, B. Gurard , R. Kampmann, A. Kastenmueller, G. Manzin, F. Meilleur, F. Millier, N. Rhodes, L. Rosta, E. Schooneveld, G.C. Smith, H. Takahashi, P. Van Esch, K. Zeitelhack, Nucl. Instrum. Methods B 554, 392–405 (2005)
48. L.C. Welsh, M.F. Symmons, J.M. Sturtevant, D.A. Marvin, R.N. Perham, J. Mol. Biol. **283**, 155–177 (1998)
49. K. Namba, G. Stubbs, Science **231**, 1401–1406 (1986)
50. G. Stubbs, Rep. Prog. Phys. **64**, 1389–1425 (2001)
51. M.W. Shotton, L.H. Pope, V.T. Forsyth, P. Langan, H. Grimm, A. Rupprecht, R.C. Denny, W. Fuller, Physica B, **241–243**, 1166–1168 (1998)
52. C. Mitsch, PhD Thesis, Cambridge University (1996)
53. P. Langan, Physica B **234–236**, 213 (1997)
54. D. Sapede, T. Seydel, V.T. Forsyth, M. Koza, R. Schweins, F. Vollrath, C. Riekel, Macromolecules 38, 8447–8453 (2005)
55. K. Papanikolopoulou, G. Schoen, V. Forge, V.T. Forsyth, C. Riekel, J.-F. Hernandez, R.W.H. Ruigrok, A. Mitraki, J. Biol. Chem. **280**, 2481–2490 (2005)
56. H. Retsos, K. Papanikolopoulou, C. Filipini, C. Riekel, K.H. Gardner, V.T. Forsyth, A. Mitraki, in preparation
57. S.O. Makin, E. Atkins, P. Sikorski, J. Johansson, L.C. Serpell, Proc. Nat. Acad. Sci. **102**, 315–320 (2005)
58. U. Baxa, N. Cheng, D.C. Winkler, T.K. Chiu, D.R. Davies, D. Sharma, H. Inouye, D.A. Kirschner, R.B. Wickner, A.C. Steven, J. Struct. Biol. **150**, 170–179 (2005)
59. B.P. Schoenborn, P. Langan, J. Synch. Rad. **11**, 80 (2004)
60. P. Langan, G. Greene, B.P. Schoenborn, J. Appl. Cryst. **37**, 24 (2004)
61. I. Tanaka, K. Kurihara, T. Chatake, N. Niimura, J. Appl. Cryst. **35**, 34–40 (2002)
62. F. Shu, V. Ramakrishnan, B.P. Schoenborn, Proc. Natl. Acad. Sci. (USA) **97**, 3872–3977 (2000)
63. K.H. Gardner, A. English, V.T. Forsyth, Macromolecules **37**, 9654–9656 (2004)
64. W.A. King, D.B. Stone, P.A. Timmins, T. Narayanan, A.A. von Brasch, R.A. Mendelson, P.M. Curmi, J. Mol. Biol. **345**, 797–815 (2005)

7

Neutron Scattering from Biomaterials in Complex Sample Environments

J. Katsaras, T.A. Harroun, M.P. Nieh, M. Chakrapani, M.J. Watson, V.A. Raghunathan

7.1 Introduction

The study of materials under difficult environmental conditions (such as high magnetic fields, high pressures, shear, and 100% relative humidity) is by no means straight forward and requires specialized equipment. These conditions may at first seem nonbiological, except for those organisms adapted to extreme environments, but a deeper understanding of biologically relevant materials has been gained from such studies.

In many cases, these experiments are made easier by the fact that neutrons interact weakly, thus nondestructively, with many commonly available materials, like aluminum and its alloys, suitable for the construction of sample cells. Their relatively low cost and useful physical characteristics mean that complex sample environments can readily be accessed with neutrons.

Lipid bilayers in water are perhaps the biologically relevant system most studied under various experimental conditions. The complex phase behavior they exhibit is of general interest to material science, as well as biology. Lipids have been subjected to extremes of temperature and pressure; have undergone detailed hydration studies; and have been aligned under shear and externally applied magnetic fields. The intrinsic properties of neutrons along with the ease of designing and constructing neutron sample environments have enabled us to probe each of these conditions.

In this chapter, we will elucidate, with a variety of recent examples, the power of neutron scattering as a tool to study biologically relevant materials in complex sample environments.

7.2 Alignment in a Magnetic Field

In order to obtain structural details on the atomic scale, the use of a single crystal sample is usually a prerequisite. However, obtaining single crystals of a desired sample is not always possible as many molecules (e.g., deoxyribose nucleic acid (DNA)) do not lend themselves to crystallization. In many such cases, however, the use of aligned samples makes it possible to determine certain structural features of the system which can provide sufficient information to construct realistic models. Examples of aligned systems providing unique structural information are: DNA [1], plant viruses such as, tobacco mosaic virus (TMV) [2, 3] and papaya mosaic virus (PMV) [4], and various lipid bilayers [5, 6], to name a few.

Over the years, various strategies have been devised to orient samples that have proven either difficult or impossible to crystallize. One such strategy is to align biomolecules in an externally applied magnetic field, **B**.

The effect of externally applied magnetic fields on biological systems has been the subject of many studies. In the 1930s, Pauling and Coryell [7] first reported the paramagnetic susceptibility of deoxyhemoglobin and the diamagnetic susceptibility of oxyhemoglobin. More recently, Higashi et al. [8] studied the orientation of erythrocytes in magnetic fields up to 8 T (tesla) and found them to orient with their disk plane parallel to **B**. Similar behavior was observed with erythrocytes at 4 T [9].

Besides red blood cells, fibrinogen, a plasma protein, is polymerized and aligned in magnetic fields [9]. Maret et al. [10], showed that fragments of high-molecular weight native DNA partially align perpendicular to **B** and that bases possessing diamagnetic anisotropy are responsible for this alignment. Moreover, Brandes and Kearns [11] demonstrated that liquid crystalline phases of DNA align with the long molecular axes perpendicular to **B**. Other biological systems that have been aligned in magnetic fields are nematic phases of TMV [12] and membrane complexes such as retinal rods [13] and purple membranes of *Halobacterium halobium* [14].

With regards to living organisms, frog embryos in a 1 T field [15] exhibited no morphological differences from unexposed controls, suggesting that magnetic fields have little or no effect with normal embryonic development. However, a recent study on hemolymph samples from adult bees that had undergone pupal development and emergence in a 7 T field, contained a lower percentage of glucose than controls implying that trehalase enzyme activity is depressed in high magnetic fields [16].

7.2.1 Magnetic Alignment of Lipid Bilayers

It is generally known that lipid membranes orient with their bilayer normals perpendicular to **B** [17] as shown in (Fig. 7.1a). This is a result of the overall negative diamagnetic anisotropy exhibited by the lipid hydrocarbon chains and their high internal order. Magnetically oriented lipid bilayered micelles, or so-called "bicelles" [18–20], possess great potential as biomimetic substrates in aligning membrane associated peptides and proteins for in-depth structural and dynamic studies. They are composed of a combination of short-chain and long-chain phosphatidylcholines (PCs) such as, dihexanoyl PC (DHPC) and dimyristoyl PC (DMPC), respectively. It is believed that the function of the

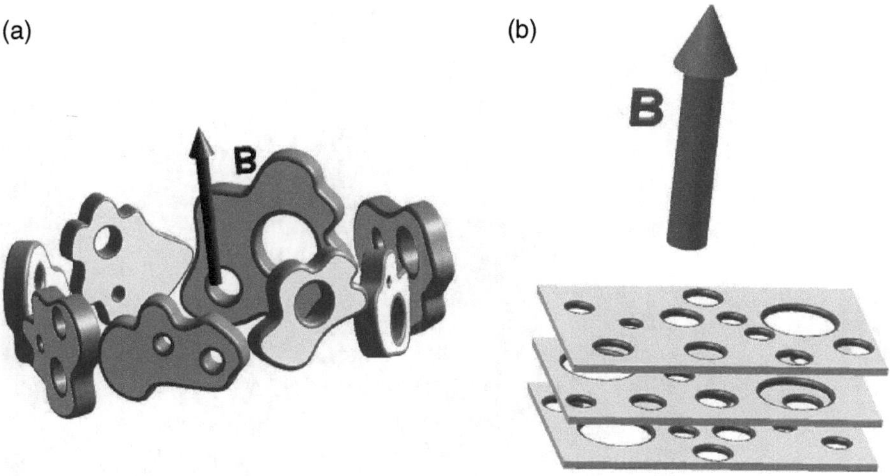

Fig. 7.1. Cartoon of (**a**) non-doped DMPC/DHPC mixture (DMPC:DHPC 3.2:1) in the presence of 2.6 T applied magnetic field, **B**, and a temperature of 315±1 K. Extended bilayered micelles or "finite" lamellar sheets align with their bilayer normals perpendicular to **B**. (**b**) The same system as in (a) but doped with Tm^{3+} ions. In this case, the extended lamellar sheets have their bilayer normals aligned parallel to **B**. In both the doped and nondoped cases, the bilayers are believed to be perforated [28, 29]. The long-chain DMPC molecules form the bilayer while the short-chain DHPC molecules partition, primarily at the edges of the perforations and the micelles

short-chain lipid is to coat the edges of the relatively small (diameter ~10–100 nm) bilayered micelle, thus protecting the hydrophobic DMPC chains from coming into contact with water. The size of the bicelles is also dependent on the molar ratio of the two lipid species [19, 21].

In a magnetic field, the orientation of DMPC/DHPC mixtures is such that the average bilayer normal, **n**, is perpendicular to **B** (Fig. 7.1a). In 1996, Prosser et al. [22] doped mixtures of DMPC/DHPC with paramagnetic ions, such as Tm^{3+}, and found that the orientation of the system altered such that **n** was now parallel to **B** (Fig. 7.1b). Compared to nondoped bicelles, the orientation of the lanthanide (e.g., Eu^{3+}, Er^{3+}, Tm^{3+}, and Yb^{3+}) doped bicelles resulted in better resolved NMR spectra. Moreover, the alignment of the nondoped bicelles restricts, due to inhomogeneous broadening of the NMR lines, the size of the membrane associated peptides that can be studied. This limitation is not there in the case of the doped bicelles [23]. Although the DMPC/DHPC bicelle mixture was reconstituted with a number of membrane-associated peptides and proteins [20, 24–26], the morphology of this magnetically alignable substrate was debatable.

In a series of publications, the structures of the lanthanide-doped, DMPG-doped (dimyristoyl phosphatidylglycerol), and nondoped DMPC/DHPC systems (3.2:1, DMPC:DHPC) were reported as a function of temperature

Fig. 7.2. N5 triple-axis spectrometer with M2 superconducting magnet/cryostat located at the NRU reactor (Chalk River, Canada). 2.37 Å wavelength neutrons were selected using the (002) reflection of a pyrolytic graphite monochromator. The M2 magnet/cryostat is a somewhat unique instrument in that it produces a horizontal, rather than a vertical magnetic field

and total lipid concentration [27–29]. Using the C5 and N5 triple-axis spectrometers (Fig. 7.2) located at the National Research Universal (NRU) reactor (Chalk River, Canada) and 2.37 Å neutrons, the samples were subjected to a 2.6 T horizontal magnetic field (Fig. 7.2), as in the NMR experiment.

At a temperature of 315±1 K, the nondoped DMPC/DHPC mixture suspended in 77 wt% D_2O formed a nematic phase, characterized by a single broad peak centered at Q ~0.05 Å^{-1} ($Q = 2\pi/d$, where d is the lamellar repeat spacing) (Fig. 7.3a), and resulting from bilayered micelles or small bilayer sheets possessing long-range orientational order but lacking positional order [27]. In this phase, the system's bilayer normals are perpendicular to the magnetic field (Fig. 7.1a) and analogous to a lipid/detergent system studied by X-ray scattering [30]. Upon addition of Tm^{3+} ions (DMPC:Tm^{3+}, 7.5:1), the system underwent a nematic → smectic transition, as exemplified by the

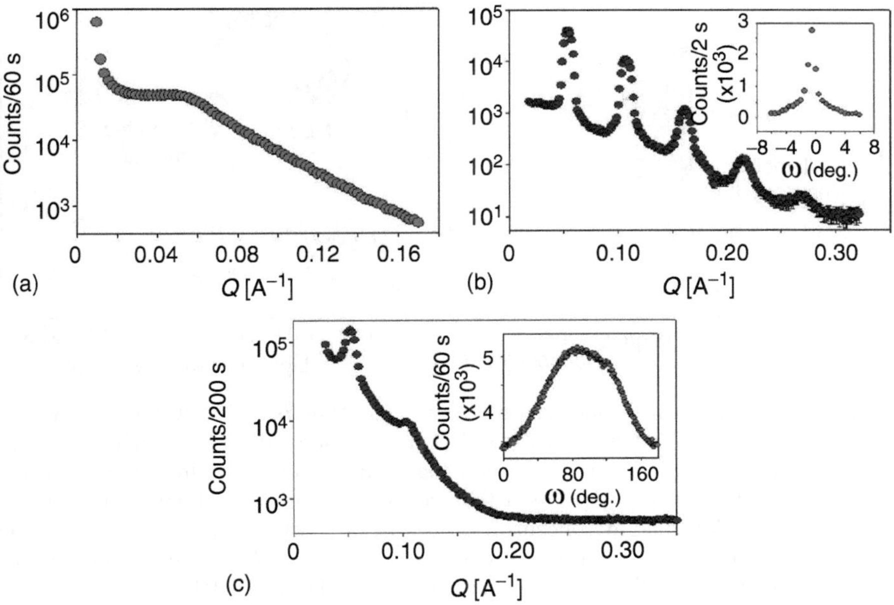

Fig. 7.3. Scan in Q of (**a**) DMPC/DHPC system in the absence of Tm^{3+}, at a T of 315 K and a 2.6 T field. The broad peak centered at $\sim 0.05\,\text{Å}^{-1}$ is indicative of a nematic phase (1D ordering, see Fig. 7.1). (**b**) The addition of Tm^{3+} ions results in a smectic phase (2D order) with well-defined Bragg reflections. The inset to the figure shows that the phase is highly aligned, within a degree, or so, of the applied magnetic field. (**c**) Removal of the magnetic field results in a less ordered smectic phase, as indicated by the rocking curve (inset), with the lamellar spacing remaining unaltered. For further details the reader is referred to [27]

appearance of well-defined Bragg reflections (Fig. 7.3b), and indicative of a system possessing a well-defined interbilayer spacing, d, of 116 Å. Moreover, the system was shown to be highly aligned with the rocking curve having an FWHM of $\leq 1°$ (inset to Fig. 7.3b). In the absence of an applied magnetic field, the orientation of the system is, for the most part, lost (Fig. 7.3c and inset) while the phase remained unaltered. In summary, the doping of the DMPC/DHPC mixture with Tm^{3+} ions resulted in the system undergoing a nematic $\rightarrow$ smectic transition while the magnetic field imparted an alignment to the system [27]. The above-mentioned study was later refined, in the absence of a magnetic field, using small-angle neutron scattering (SANS) [28,29] and whose partial phase diagrams are presented below (Fig. 7.4).

7.2.2 Neutron Scattering in a Magnetic Field: Other Examples

In 1989, Hayter et al. [31], reported on SANS measurements of ferrofluids containing TMV and tobacco rattle virus (TRV). In this case, the nonmagnetic

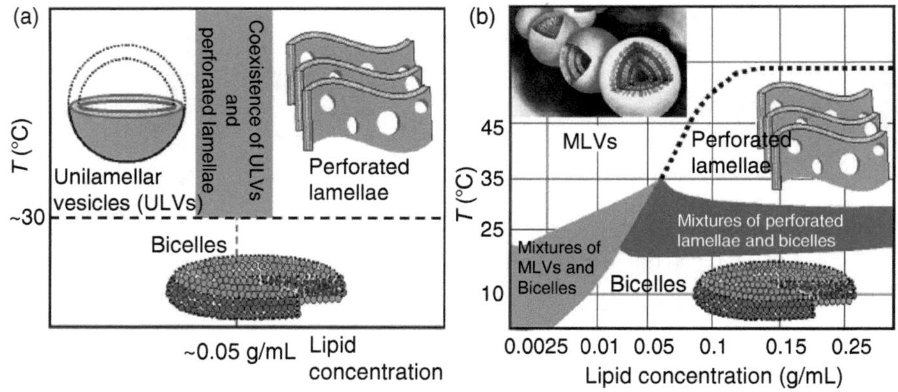

Fig. 7.4. Partial phase diagrams of (**a**) the Tm^{3+}-doped DMPC/DHPC system at a ratio of 3.2:1 (DMPC:DHPC) and (**b**)the non-doped DMPC/DHPC system. In the Tm^{3+}-doped system two morphologies are observed at high temperatures (T): Unilamellar vescicles (ULVs) at lipid concentrations approximately $0.01\,g\,ml^{-1}$ and perforated lamellae at concentrations $0.05\,g\,ml^{-1}$ wt%. For T below 15°C, the mixture exhibits an isotropic phase composed of bilayered micelles. Compared to Tm^{3+}-doped DMPC/DHPC mixtures, the nondoped DMPC/DHPC system exhibits a much more complex phase behaviour, and the appearance of multilamellar vesicles (MLVs) instead of ULVs seen previously in the Tm^{3+}-doped system. The SANS data used to determine the various morphologies were collected at the National Institute of Standards and Technology (NIST, Gaithersburg, USA) using the NG-7 30 m instrument

viruses were aligned by the magnetic ferrofluid in a modest external field. Using this colloidal dispersion the contrast between the dispersed particles and the ferrofluid carrier was altered giving rise to information with regards to some structural features of these systems. Since most biological materials possess neither sufficiently anisotropic magnetic properties to align in a magnetic field nor morphological characteristics to respond to alignment *via* shear, ferro-dispersed suspensions offer a method of aligning colloidal particles in suspension. In addition, their ability to align in low concentrations is particularly important when it comes to samples which are not readily available in large quantities.

Groot et al. [32] reported on SANS studies carried out using Na-DNA fragments at concentrations between 190 and $285\,mg\,ml^{-1}$. Applying a magnetic field either perpendicular or parallel to the incident neutron beam they were able to deduce the cholesteric or chiral nematic structure of the liquid crystalline solutions. When **B** was applied in a direction parallel to the incident neutron beam the small-angle scattering was found to be isotropic. This is not surprising as the incident beam was parallel to the pitch of the cholesteric phase. On the other hand, when the direction of **B** was changed to be perpendicular to the incident neutron beam, the resultant scattering was anisotropic. It should be noted that the average direction of DNA molecules is perpendicular to the magnetic field.

Kiselev et al. [33], determined the orientation of pure DMPC MLVs below and close to the main gel–liquid crystalline transition, T_M, and of DMPC/$C_{12}E_8$ (dodecyl-octaethyleneoxide) mixed micelles in magnetic fields from 1 to 4 T. It was determined that spherical DMPC vesicles deform to an ellipsoidal shape at $\mathbf{B} = 2\,\mathrm{T}$ while the mixed micelles of DMPC/$C_{12}E_8$ forms a Gaussian-coil, composed of rod-like micelles, irrespective of the magnetic field strength. In the case of liquid crystalline DMPC vesicles, the degree of deformation was more pronounced than gel phase DMPC vesicles.

Mucins are polyelectrolytes whose rigidity can be altered as a function of pH. For stomach mucins, molecular weights of between 2×10^5 and 1.6×10^7 Da have been reported with their structure related to the function that they perform, namely to protect the stomach epithelium from its surrounding environment. They supposedly do so by forming dense viscoelastic gels at low pH (e.g., pH 2) [101] and the side chain interdigitation is crucial in the network's formation [102]. A recent study by Waigh et al. [103] showed that in the absence of a magnetic field these side chains form a polydomain nematic phase, while a monodomain phase is induced when a 1.48 T magnetic field is applied. The magnetic field was found to orient the molecules with their long axis pointing in the direction of the field. Moreover, the field was used to study the nature of entanglement couplings between the side chains.

7.3 High Pressure Studies

The potential of pressure in biological systems as a thermodynamic variable remains largely unexplored even though pressures experienced by many aquatic organisms is in the range of ~50 MPa, or greater. At these pressures, there are most likely, significant effects on macromolecular structure and function.

Pressure has the effect of reversibly denaturing proteins and can therefore be used as a means of studying protein folding and protein interactions [34,35]. In the recent past, high pressure has emerged as a method to stabilize folding intermediates [34]. The molecular basis of protein–RNA and protein–DNA recognition is intricately related to the thermodynamics of the system. Recent studies have shown that pressure can inactivate viruses while preserving their immunogenic properties [36, 37].

One of the least developed areas using pressure is high-pressure protein crystallography. Kundrot and Richards [38] carried out the first high pressure X-ray crystallographic study using hen egg-white lysozyme at a pressure of 100 MPa using a dead end-bored beryllium rod [39]. A similar device was used to study sperm whale myoglobin at 150 MPa [40]. More importantly, Urayama et al. [40] developed a technique whereby the pressurized crystal is cooled, "freezing-in" pressure-induced collective movements and eliminating a pressure cell during data collection. Studies on myoglobin [41, 42], lysozyme [39,43,44] and staphylococcal nuclease [45] show that protein crystals are robust and can withstand substantial amounts of pressure.

An area of ongoing interest is the effect of hydrostatic pressure on lipid phase behavior and dynamics. The response of lipid bilayers to pressure can provide some insight into the effect of other perturbations at ambient pressure. Pressure dependent structure and phase behavior of lipid systems has been studied over the years by Winter and co-workers using a combination of X-ray and neutron scattering [46–49].

7.3.1 Hydrostatic Pressure and Aligned Lipid Bilayers

The main gel–liquid crystalline transition (T_M) in lipid bilayers has attracted a great deal of attention in the last few decades. In the case of phosphatidylcholine lipids such as DMPC, one outstanding issue is with regards to the structural changes occurring in the vicinity of the main transition. On decreasing temperature, the lamellar repeat spacing, d, of liquid crystalline DMPC bilayers increases nonlinearly. This nonlinear increase in lamellar repeat spacing, or "anomalous swelling," in the vicinity of T_M, has previously been reported by various groups studying PC bilayers [50–59]. The commonly accepted view is that this anomalous swelling is a pretransitional effect.

One possibility, put forth by Nagle in 1973, is that a critical transition gets intercepted by the first-order main transition [60]. Another point of view is that due to some intrinsic bilayer property, the main transition itself is weakly first-order [61]. Recently, Pabst et al. [62] demonstrated that the majority of the anomalous swelling is the result of increasing interbilayer water, and a sudden decrease of the bilayer bending rigidity, K_c. Of importance is that the functional form of K_c follows a power law dependence near T_M.

In 1986, Lipowsky and Leibler [63] predicted the critical unbinding (i.e., loss of periodicity) of a membrane stack, due to steric repulsion, independent of the anomalous swelling phenomenon occurring in lipid bilayers. One reason that leads to membranes unbinding, is a reduction in K_c causing bilayers to undulate and repel each other [63]. It therefore seems that one can relate thermal unbinding and anomalous swelling, both the result of a decrease in K_c, leading to a temperature dependence of the lamellar periodicity, given by $d \approx (T-T_c)^{-\psi}$, where T_c is the unbinding temperature. The critical exponent, ψ, is predicted to be unity.

If the functional form of K_c with respect to temperature is reflected in the functional form of the anomalous swelling, then pressure can be used to interrogate the region in the vicinity of T_M. Pressure also allows one to study the behavior of short chain lipids whose T_M is below 0°C.

Compared to isotropic or "powder" samples the use of aligned samples is highly desirable as the signal from these samples is anisotropic and usually easier to decipher. In the case of X-ray or neutron scattering an oriented sample allows for the differentiation of the inter-bilayer (lamellar repeat spacing) and intra-bilayer (hydrocarbon chain correlations) organization [64]. Also, due to the fact that the signal is not spread-out over 2π, much less sample is required to obtain a good signal to noise ratio.

Watson et al. [65] recently constructed a sample cell suitable for neutron scattering from aligned lipid multibilayers and capable of exerting hydrostatic pressures up to 370 MPa over a temperature range of between −10 and 100°C (Fig. 7.5a). The advantage of this cell compared to other high-pressure neutron cells [66, 67] is that it allows for the study of samples whose quantities are limited and in conjunction with a 2D detector the in-plane and out-of-plane correlations can easily be obtained both as a function of temperature and pressure.

Aluminum was chosen as the material to construct the cell as it is practically transparent to neutrons. At ambient temperatures Al is reasonably corrosion resistant. However, the same cannot be said at elevated temperatures. In order to retard the corrosion process the sample block was hard anodized (Fig. 7.5b). Although the measures taken did reduce the amount of

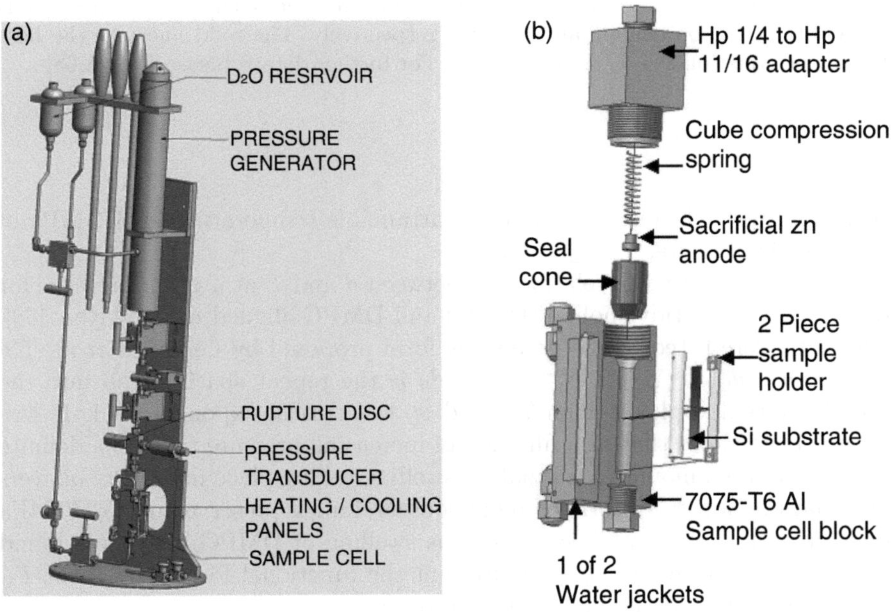

Fig. 7.5. (**a**) Pressurized sample cell assembly rated for hydrostatic pressures up to 370 MPa and suitable for neutron diffraction of aligned biomimetic systems. (**b**) Neutron sample cell assembly constructed from 7075-T6 Al alloy. The sample cell was hard anodized to reduce corrosion and fitted with helicoils, on both ends, to reduce stretching of the threads. A Zn sacrificial anode was used to further retard the corrosion process much evident at elevated temperatures. For further details please refer to [65]

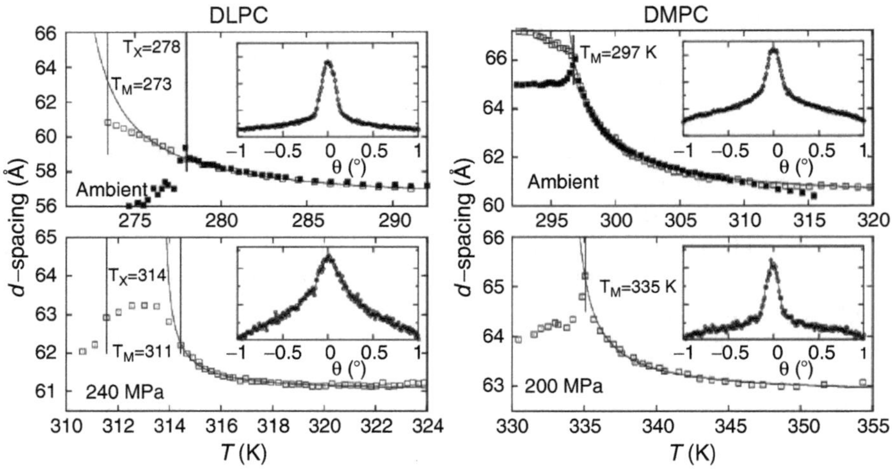

Fig. 7.6. Lamellar repeat spacings, d, as a function of temperature and a given hydrostatic pressure for fully hydrated dilauroyl phosphatidylcholne (DLPC) and DMPC multibilayer stacks. The insets to the figures depict so-called "rocking curves," a direct measure of the samples alignment. The open and closed symbols were obtained upon cooling and heating, respectively. The solid lines are the best fits to the equation $d - d_0 \propto (T - T^{\star})^{-\psi}$. For further details please see [65, 68]

corrosion, nevertheless the maximum attainable temperature at 370 MPa of hydrostatic pressure, was ≤60°C.

Figure 7.6 shows the relationship between d and T at a given pressure for dilauroyl phosphatidylcholine (DLPC) and DMPC aligned multibilayers [68]. The data were fitted to the power law form proposed by Lemmich et al. [53] namely $d - d_0 \propto (T - T^{\star})^{-\psi}$ where d_0 is the repeat spacing well into the liquid crystalline phase (high T), and ψ, the critical exponent, is 1. It was interesting to note that as a function of increasing pressure there is a definite decrease in the amount of anomalous swelling taking place in DMPC bilayers and that the power law form of anomalous swelling is preserved up to 240 MPa of hydrostatic pressure. The anomalous swelling of DMPC bilayers is found to decrease with increasing pressure, but the functional form of K_c near T_M is preserved even at the highest pressure used.

An important result from these studies was that in DLPC bilayers complete unbinding may take place at hydrostatic pressures in excess of 290 MPa [68]. Presently, we have been unable to carry-out the requisite experiments to test this prediction as our sample cell has proven, due to corrosion, incapable of attaining the necessary hydrostatic pressures. However, we are in the process of designing and constructing a new cell made out of copper/beryllium.

7.3.2 High Pressure Neutron Scattering Experiments: Other Examples

Czeslik et al. [46] studied the lateral organization of the binary lipid mixture, DMPC/DSPC (distearoyl phosphatidylcholine) at hydrostatic pressures up to 100 MPa. What was observed was an increase of 22°C/100 MPa of applied pressure of the two phase coexistence region. They also noted the existence of fractal-like membrane morphologies within the gel–liquid crystalline coexistence region and not the kind of phase separation that one would anticipate on the basis of the thermodynamic equilibrium phase diagram. Compared to ambient pressure, the fractal exponent of coexistence mixture changed slightly at 100 MPa.

Worcester and Hammouda [69] studied, as a function of temperature and pressure, the behavior of PC lipids with C20 (diarachidoyl, DAPC) and C22 (dibehenoyl, DBPC) hydrocarbon chains. Worcester and Hammouda observed that DBPC formed interdigitated bilayers at pressures <60 MPa while DAPC formed a similar phase at 60 MPa of pressure showing that the minimum pressure for interdigitation changes systematically with the length of the hydrocarbon chains. Other disaturated PCs, such as DPPC and DSPC (distearoyl phosphatidylcholine) have also been observed to form such interdigitated phases [70].

Doster and Gebhardt [71] reported on the dynamics and stability of myoglobin. As a function of pressure, the evolution of the protein–solvent bonds and the unfolding transition were observed. The pressure-induced unfolding of the protein took place above 300 MPa with $\approx$40% of the protein's helical structures being preserved in the unfolded state. Doster and Gebhardt concluded that pressure enhanced protein–solvent interactions may be a factor in destabilizing the native state of the protein.

Loupiac et al. [72] reported on horse azidometmyoglobin (MbN_3) at pressure up to 300 MPa. As a function of pressure the protein's radius of gyration remained unaltered up to 300 MPa. From the second virial coefficient of the protein solution the authors determined that the protein–protein repulsive forces, although diminished, were never overcome even at 300 MPa while the specific volume of MbN_3, compared to atmospheric pressure, decreased by 5.4% at 300 MPa.

Köhling et al. [73] studied the phase behavior of dioctyl sulfosuccinate sodium (AOT)-n-octane–water mesophases as a function of pressure (0.01–300 MPa). The incorporation of the water-soluble enzyme α-chymotrypsin with the surfactant mixtures resulted in significant changes to the structure and phase behavior of the various surfactant mesophases with the observed changes enhanced with increasing pressure. The application of pressure resulted in fluid lamellar and bicontinuous surfactant phases. Ultimately, the changes in α-chymotrypsin activity, as a function of pressure, were attributed to changes in the surfactant mesophase structure and not to any changes in tertiary or secondary protein structure.

7.4 Shear Flow Induced Structures in Biologically Relevant Materials

Some of the earliest reports of the use of shear flow to study soft materials were by Scheraga and Backus [74], and Ackerson and Clark [75]. Since then, the use of shear has allowed the observation of shear-induced structural transformations in a wide variety of soft materials [76]. Shear-induced transformations in complex fluids include: micellar elongation and alignment [77], isotropic to nematic transitions [78] and the formation of multilamellar vesicles [79–81]. In the case of biologically relevant materials shear has been used to crystallize various fats (e.g., milk fat, cocoa butter) [82], study the aggregation of casein micelles in undiluted skim milk [83], measure the extent and rate of adhesion of leukemia cells [84], and the alignment of lecithin reverse micelles [85], to name a few. In all of the above-mentioned studies, shearing devices of different geometries have been developed to induce the necessary shear.

7.4.1 Shear Cells Suitable for Neutron Scattering

Over the years, a variety of shear cells have been developed for the study of shear-induced structures using X-ray [86–90] and neutron [91–98] scattering techniques. Shear gradients $>10^3$ s^{-1} needed to study colloidal particles and micellar solutions are readily achievable by either Poiseuille or Couette type cells (Fig. 7.7). Generally, Couette flow is preferable because the cell diameter (d) is much smaller than the gap width (r) resulting in a constant gradient across the gap, whereas the characteristic flow in a Poiseuille cell has a parabolic velocity profile [99].

The first widely used Couette type cell suitable for neutron scattering was constructed by Lindner and Oberthur at the Institut Laue-Langevin (ILL). In the Couette geometry the sample is sheared between two concentric cylinders, usually made out of polished quartz. The inner cylinder, the stator, is stationary while the outer one rotates (rotor). The difference in velocity between the outer and inner cylinders divided by the gap separating them, gives rise to the average applied shear experienced by the sample. Although the basic Couette design has remained relatively unaltered since its inception, nevertheless in the last couple of years improvements to the basic design have been made. One such improvement has been made by Porcar et al. [98], whereby they have designed a cell capable of operating at shear rates up to 15,000 s^{-1} without liquid losses due to evaporation. The cell, like many others of its type, is temperature controlled and capable of accepting sample volumes as low as 7 ml.

A shear cell suitable for the study of liquid–solid interfaces by neutron reflectometry and SANS was designed a decade ago by Baker et al. [94]. The shear rates were altered by changing, over three orders of magnitude, the volume flow through the cell under laminar flow conditions. Recently, a new type of shear cell designed for the study of interfaces was described by Kuhl

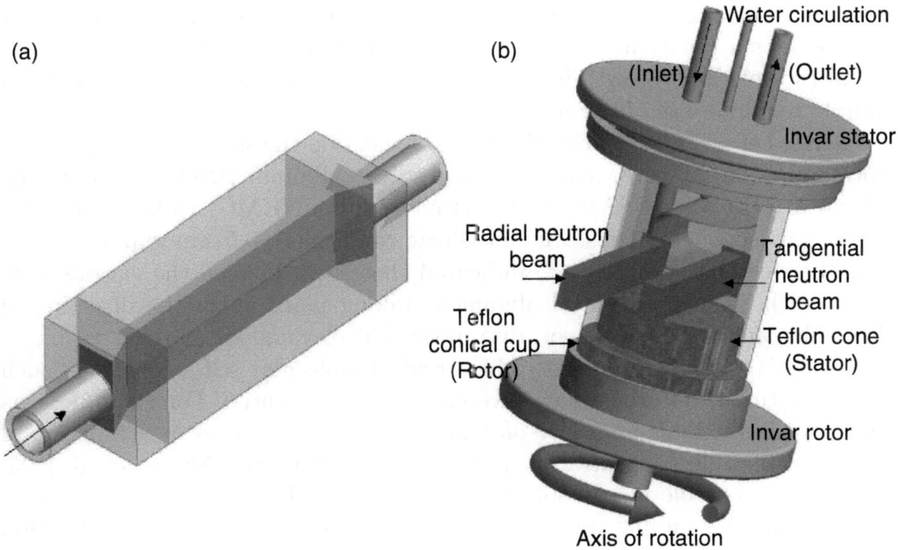

Fig. 7.7. (**a**) Poiseuille flow cell made out of quartz. (**b**) A typical concentric cylinder Couette type shear cell. Couette flow results in a constant gradient across the gap, whereas the characteristic in a Poiseuille cell is that shear rate tends to zero toward the center of the flow cell. Both the Poiseuille and Couette type shear cells are capable of being interrogated in the radial and tangential directions. For further information the reader is referred to [95] and [98]

et al. [97]. This shear cell, suitable for neutron reflectometry, has the ability to control surface separation (i.e., gap) and alignment under applied loads. The gap size is variable from millimeters to <100 nm and capable of exerting steady shear rates from 0.001 to 20 s^{-1}. The difference between the two above-mentioned reflectometry shear cells is that the one by Kuhl et al. [97] achieves shear by the lateral motion of the lower substrate relative to the stationary upper substrate. Throughout the shearing process the substrates maintain a defined gap separation. The difference between the Baker et al. [94] and Kuhl et al. [97] shear cells is that for the latter case, the shear is occurring at the substrate interface rather than the solvent flow/sample interface as in the case of the cell by Baker et al.

7.4.2 Shear Studies of Biologically Relevant Systems

Shear cells have traditionally been used to examine polymeric systems, however, over the years there have been examples of studies investigating biologically relevant materials. Schurtenberger et al. [85] studied the alignment of lecithin/isooctane solutions using a Couette type shear cell and SANS. They obtained, as a function of shear rate, direct evidence of water-induced cylindrical (anisotropic) growth in reverse micelles in a 1 mm gap. The amount of sample required was only 8 ml.

Renard et al. [100] studied the effect of shear on the structure of a protein–polysaccharide mixture, namely bovine serum albumin (BSA)/hydroxyethyl cellulose (HEC) or BSA/carboxymethyl cellulose (CMC). SANS measurements carried out under static and shear conditions (0.5 mm gap and shear rates between 0.1 and 100 s^{-1}) indicated that shear aligned the various mixtures, with some preferential alignment taking place along the direction of flow. This anisotropy, however, disappeared at elevated shear rates.

There is a growing interest in hierarchical molecular self-assembly as such nanostructured materials may have commercial potential. For example, certain peptides exhibit a variety of supramolecular structures as a function of increased peptide concentration in water [104]. Recently, Mawer et al. [105] studied the possible mesoscopic structures responsible for the nonlinear rheology of self assembling peptide fibrils and fibrillar networks. As a function of shear rate (0–500 s^{-1}), the orientation of the nematic director in the fluid and gel phases was studied using SANS. In the velocity direction (radial), self assembled fibril structures consisted of 8–10 single β-sheet tapes (single molecule thick) which upon gelation increased to between 10 and 12 tapes. At moderate shear rates, SANS data was found to be consistent with that of an oriented nematic gel network formed of semiflexible fibrils, while at high shear rates the linkages between the fibrils broke leading to a reduction in sample viscosity.

7.5 Comparison of a Neutron and X-ray Sample Environment

Under any circumstance, the study of materials in difficult environments is not trivial. However, because of their penetrating power (interact weakly) with many commonly available materials, particularly aluminum and its alloys, neutrons have a distinct advantage over X-rays in construction simplicity and cost. Besides aluminum, other commonly used materials for sample cell environments are vanadium and Ti_{66}:Zr_{34} commonly used as a null scattering alloy. As mentioned previously, Cu–Be alloy and Maraging steel are suitable for high pressure studies, while for high temperatures sapphire and Inconel have been used [106]. All of these materials have almost no transparency to X-rays. Here we present an example of a neutron and X-ray sample cell capable of fully hydrating aligned lipid multibilayer stacks.

7.5.1 100% Relative Humidity Sample Cells

In elucidating structure, there are advantages of studying aligned lipid multibilayer stacks as opposed to isotropic multilamellar vesicles. The problem was that when the lipid bilayers aligned on a solid support were hydrated in a 100% relative humidity (RH) environment, the lamellar repeat spacing, d, was found to be consistently smaller that the same MLV material immersed in bulk water [107–109]. This posed a serious problem as in equilibrium, the chemical potential of water vapor at 100% RH and that of bulk water, are the same. Since these results are paradoxical, this discrepancy between samples hydrated from 100% RH and bulk water came to be known as the vapor pressure paradox (VPP) [110]. Moreover, in 1997 a theory was published to explain the underlying mechanism of VPP [111].

The theory by Podgornik and Parsegian [111] stated that lipid bilayers aligned on rigid supports experience a global suppression of bilayer fluctuations, not just at the sample interfaces, as a result of the rigid substrate and the lipid/water vapour interface. This reduction in bilayer fluctuations results in smaller entropic repulsion pressures and concomitantly, reduced d. A somewhat less elegant explanation was that all of the data contributing to the VPP were obtained from experiments utilizing sample cells that were incapable of attaining 100% RH.

To elucidate this discrepancy between theory and experiment, a sample environment suitable for neutron diffraction was designed with the following characteristics: (a) Reduce temperature gradients. (b) Minimize the volume around the sample. (c) Have an "evaporative surface" in close proximity to the sample. A sample cell, similar the one in Fig. 7.8a, was designed and built at Chalk River Laboratories (Canada). The neutron diffraction results conclusively demonstrated that VPP was an artifact due to poorly designed sample cells over a period of three decades [112].

The concepts of the 100% RH neutron cell (Fig. 7.8a) were transferred to a sample cell suitable for X-ray diffraction (Fig. 7.8b) [64]. Comparing the two cells (Fig. 7.8), one can easily come to the conclusion that the X-ray cell is a much more complicated device. This was necessary as X-rays are generally not highly penetrating and require special, nonabsorbing "window" materials. These windows possess different thermal properties than the other materials used in constructing the sample cell, leading to the possibility of thermal gradients and the reality of RHs $<$100%. Nevertheless, the X-ray sample cell, shown in Fig. 7.8b, was able to achieve the requisite humidities and yielded results indistinguishable from those obtained from neutrons scattering experiments. However, the costs of design, construction, and implementation of the X-ray cell were $\approx$20 times that of the neutron sample environment.

7.6 Conclusions

It is the hope of the authors that this brief review has provided the reader with comprehensive information to the various sample environments, suitable for biologically relevant studies, and presently used by the various neutron scattering laboratories worldwide. It should be said that there are few, if

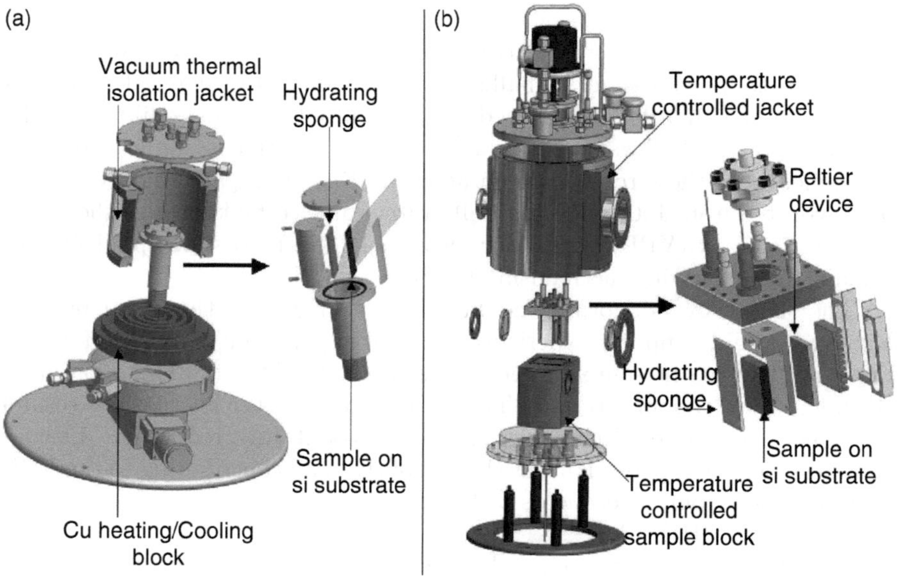

Fig. 7.8. Comparison of (**a**) 100% relative humidity (RH) cell suitable for neutron scattering and (**b**) similar cell suitable for X-ray diffraction. Because X-rays are easily absorbed, choosing the materials to construct various parts of the sample cell is not trivial and results, in comparison to the neutron sample cell, in a rather complicated design with concomitant costs. Moreover, and unlike the X-ray sample cell, the one for suitable neutrons can be filled with liquid water. For further information with regards to these two samples environments the reader is referred to [64, 112, 113]

any, sample environments that exist for neutron scattering that cannot be replicated for use with X-rays. Generally speaking, however, because the interaction of neutrons with many commonly used materials is weak, the design of a particular sample environment, compared to the one for use with X-rays, is simplified.

Up to now there have not been an abundance of biologically relevant studies that have used the sample environments described in the present review. However, the hope is that the benefits presently experienced by the colloidal and polymer communities will become evident to those studying biomimetic materials especially, the use of shear to align systems as shear cells are ubiquitous in neutron scattering laboratories.

The use of hydrostatic pressure is another potential growth area as the interest in protein unfolding is ever increasing. Future samples cells capable of routinely exerting 500–600 MPa of hydrostatic pressure are not out of the question.

References

1. J.D. Watson, F.H.C. Crick, Nature **171**, 737 (1953)
2. J.D. Bernal, I. Fankuchen, J. Ge. Physiol. **28**, 111 (1941)

3. J. Gregory, K.C. Holmes, J. Mol. Biol. **13**, 796 (1965)
4. P. Tollin, J.B. Bancroft, J.F. Richardson, N.C. Payne, T.J. Beveridge, Virology **98**, 108 (1979)
5. J. Katsaras, V.A. Raghunathan, Phys. Rev. Lett. **74**, 2022 (1995)
6. V.A. Raghunathan, J. Katsaras: Phys. Rev. Lett. **74**, 4456 (1995)
7. L. Pauling, C. Coryell, Proc. Natl. Acad of Sci. U.S.A. **22**, 210 (1936)
8. T. Higashi, A. Yamagishi, T. Takeuchi, N. Kawaguchi, S. Sagawa, S. Onishi, M. Date, Blood **82**, 1328 (1993)
9. A. Yamagishi, J. Mag. Mag. Mat. **90**, 43 (1990)
10. G. Maret, M. V. Schickfus, A. Mayer, K. Dransfeld, Phys. Rev. Lett. **35**, 397 (1975)
11. R. Brandes, D. Kearns, Biochemistry **25**, 5890 (1986)
12. X. Ao, X. When, R.B. Meyer, Physica A **176**, 63 (1991)
13. M. Chabre, Proc. Natl. Acad. Sci. USA **75**, 5471 (1978)
14. B. Lewis, L.C. Rosenblatt, R.G. Griffin, J. Courtemanche, Biophys. J. **47**, 143 (1985)
15. S. Ueno, K. Harada, K. Shiokawa, IEEE Trans. Magn. **MAG-20**, 1663 (1984)
16. J. Kefuss, K. M'Diaye, M. Bounias, J. Vanpoucke, J. Ecochard, Bioelectromagnetics **20**, 117 (1999)
17. G. Maret, K. Dransfeld, D.M. MacKay, Strong and ultrastrong magnetic fields and their applications, in: *Topics in Applied Physics*, vol. 57, ed. by, F. Herlach (Springer, New York, 1985), pp. 143–204
18. P. Ram, J.H. Prestegard, Biochim. Biophys. Acta **940**, 289 (1988)
19. C.R. Sanders, J.P. Schwonek, Biochemistry **31**, 8898 (1992)
20. C.R. Sanders, B.J. Hare, K.P. Howard, J.H. Prestegard, Prog. NMR Spectrosc. **26**, 421 (1994)
21. R.R. Vold, R.S. Prosser, J. Magn. Reson. B **113**, 267 (1996)
22. R.S. Prosser, S.A. Hunt, J.A. DiNatale, R.R. Vold, J. Am. Chem. Soc. **118**, 269 (1996)
23. R.S. Prosser, V.B. Volkov, I.V. Shiyanovskaya, Biophys. J. **75**, 2163 (1998)
24. C.R. Sanders, G.C. Landis, Biochemistry **34**, 4030 (1995)
25. G.C. Sanders, R.S. Prosser, Ways & Means **6**, 1227 (1998)
26. J. Struppe, E.A. Komives, S.S. Taylor, R.R. Vold, Biochemistry **37**, 15523 (1998)
27. J. Katsaras, R.L. Donaberger, I.P. Swainson, D.C. Tennant, Z. Tun, R. R. Vold, R. S. Prosser, Phys. Rev. Lett. **78**, 899 (1997)
28. M.-P. Nieh, C.J. Glinka, S. Krueger, R.S. Prosser, J. Katsaras, Langmuir **17**, 2629 (2001)
29. M.-P. Nieh, C.J. Glinka, S. Krueger, R.S. Prosser, J. Katsaras, Biophys. J. **82**, 2487 (2002)
30. B.J. Hare, J.H. Prestegard, D.M. Engleman, Biophys. J. **69**, 1891 (1995)
31. J.B. Hayter, R. Pynn, S. Charles, A.T. Skjeltorp, J. Trewhella, G. Stubbs, P. Timmins, Phys. Rev. Lett. **62**, 1667 (1989)
32. L.C.A. Groot, M.E. Kuil, J.C. Leyte, J.R.C. van der Maarel, Liq. Cryst. **17**, 263 (1994)
33. M.A. Kiselev, M. Janich, P. Lesieur, A. Hoell, J. Oberdisse, G. Pepy, A.M. Kisselev, L.V. Gapienko, T. Gutberlet, V. L. Aksenov, Appl. Phys. A [Suppl.] **74**, S1239 (2002)
34. J.L. Silva, D. Foguel, C.A. Royer, Trends Biochem. Sci. **26**, 612 (2001)
35. C. Balny, P. Masson, K. Heremans, Biochim. Biophys. Acta **1595**, 3 (2002)

36. L.P. Gaspar, A.C. Silva, A.M. Gomes, A.P.D. Ano Bom, W.D. Schwarcz, J. Mestecky, M.J. Novak, D. Foguel, J.L. Silva, J. Biol. Chem. **277**, 8433 (2002)
37. A.M.O. Gomes, A.S. Pinheiro, C.F.S. Bonafe, J.L. Silva, Biochemistry **42**, 5540 (2003)
38. C.E. Kundrot, F.M. Richards, J. Mol. Biol. **193**, 157 (1987)
39. C.E. Kundrot, F.M. Richards, J. Appl. Cryst. **19**, 208 (1986)
40. P. Urayama, G.N. Phillips Jr., S.M. Gruner, Structure **10**, 51 (2002)
41. U.F. Thomanek, F. Parak, R.L. M¨ossbauer, H. Formanek, P. Scwager, W. Hoppe, Acta Crystallogr. **A29**, 263 (1973)
42. R.F. Tilton Jr., G.A. Petsko, Biochemistry **27**, 6574 (1988)
43. A. Katrusiak, Z. Dauter, Acta Crystallogr. **D52**, 607 (1996)
44. R. Fourme, R. Kahn, M. Mezouar, E. Girard, C. Hoerentrup, T. Prangè, I. Ascone, J. Synchrotron Radiat. **8**, 1149 (2001)
45. F. Österberg, Induced changes in the diffuse X-ray scattering background from protein crystals. PhD Thesis (Princeton University, New Jersey, 1996)
46. C. Czeslik, J. Erbes, R. Winter, Europhys. Lett. **37**, 577 (1997)
47. R. Winter, J. Erbes, C. Czeslik, A. Gabke, J. Phys.: Condens. Matter **10**, 11499 (1998)
48. R. Winter, C. Czeslik, Z. Kristallogr. **215**, 454 (2000)
49. R. Winter, Biochim. Biophys. Acta **1595**, 160 (2002)
50. S. Kirchner, G. Cevc, Europhys. Lett. **23**, 229 (1993)
51. T. Hønger, K. Mortensen, J.H. Ipsen, J. Lemmich, R. Bauer, O.G. Mouritsen, Phys. Rev. Lett. **72**, 3911 (1994)
52. R. Zhang, W. Sun, S. Tristram-Nagle, R.L. Headrick, R.M. Suter, J.F. Nagle, Phys. Rev. Lett. **74**, 2832 (1995)
53. J. Lemmich, K. Mortensen, J.H. Ipsen, T. Hønger, R. Bauer, O.G. Mouritsen, Phys. Rev. Lett. **75**, 3958 (1995)
54. F.Y. Chen, W.C. Hung, H.W. Huang, Phys. Rev. Lett. **79**, 4026 (1997)
55. J.F. Nagle, H.I. Petrache, N. Gouliev, S. Tristram-Nagle, Y. Liu, R.M. Suter, K. Gawrisch, Phys. Rev. E **58**, 7769 (1998)
56. F. Richter, L. Finegold, G. Rapp, Phys. Rev. E **59**, 3483 (1999)
57. S.S. Korreman, D. Posselt, Eur. Phys. J. **1**, 87 (2000)
58. P.C. Mason, J.F. Nagle, R.M. Epand, J. Katsaras, Phys. Rev. E **63**, 030902 (2001)
59. S.S. Korreman, D. Posselt, Eur. Biophys. J. **30**, 121 (2001)
60. J.F. Nagle, Proc. Natl. Acad. Sci. USA **70**, 3443 (1973)
61. D.P. Kharakoz, E.A. Shlyapnikova, Phys. Chem. B **104**, 10368, (2000)
62. G. Pabst, J. Katsaras, V. Raghunathan, M. Rappolt, Langmuir **19**, 1716 (2003)
63. R. Lipowsky, S. Leibler, Phys. Rev. Lett. **56**, 2541 (1986)
64. J. Katsaras, M.J. Watson, Rev. Sci. Instrum. **71**, 1737 (2000)
65. M.J. Watson, M.-P. Nieh, T.A. Harroun, J. Katsaras, Rev. Sci. Instrum. **74**, 2778 (2003)
66. R. Winter, W.-C. Pilgrim, Ber. Bunsenges. Phys. Chem. **93**, 708 (1989)
67. G.W. Neilson, D.I. Page, W.S. Howells, J. Phys. D **12**, 901 (1979)
68. T.A. Harroun, M.-P. Nieh, M.J. Watson, V.A. Raghunathan, G. Pabst, M.R. Morrow, J. Katsaras; Phys. Rev. E **69**, 031906 (2004)
69. D. Worcester, B. Hammouda, Physica B **241–243**, 1175 (1998)
70. L.F. Braganza, D.L. Worcester, Biochemistry **25**, 2591 (1986)
71. W. Doster, R. Gebhardt, Chem. Phys. **292**, 383 (2003)

72. C. Loupiac, M. Bonetti, S. Pin, P. Calmettes, Eur. J. Biochem. **269**, 4731 (2002)
73. R. Köhling, J. Woenckhaus, N.L. Klyachko, R. Winter, Langmuir **18**, 8626 (2002)
74. H.A. Scheraga, J.K. Backus, J. Am. Chem. Soc. **73**, 5108 (1981)
75. B.J. Ackerson, N.A. Clark, Phys. Rev. A **30**, 906 (1984)
76. P.D. Butler, Curr. Opin. Colloid Interface Sci. **4**, 214 (1999)
77. P.D. Butler, L.J. Magid, W.A. Hamilton, J.B. Hayter, B. Hammouda, P.J. Kreke, J. Phys. Chem. **100**, 443 (1996)
78. E. Cappelaere, J.F. Berret, J.P. Decruppe, R. Cressely, P. Lindner, Phys. Rev. E **56**, 1869 (1997)
79. O. Diat, D. Roux, F. Nallet, J. Phys. II **3**, 1427 (1993)
80. J. Zipfel, P. Lindner, M. Tsianou, P. Alexandridis, W. Richtering, Langmuir **15**, 2599 (1999)
81. J. Zipfel, J. Berghausen, P. Lindner, P. Richtering, J. Phys. Chem. B **103**, 2841 (1999)
82. G. Mazzanti, S.E. Guthrie, E.B. Sirota, A.G. Marangoni, S.H.J. Idziak, Cryst. Growth Des. **3**, 721 (2003)
83. D. Lehner, P. Worning, G. Fritz, L. Øgendal, R. Bauer, O. Glatter, J. Colloid Interface Sci. **213**, 445 (1999)
84. D.G. Swift, R.G. Posner, D.A. Hammer, Biophys. J. **75**, 2597 (1998)
85. P. Schurtenberger, L.J. Magid, J. Penfold, R. Heenan, Langmuir **6**, 1800 (1990)
86. R.J. Plano, C.R. Safinya, E.B. Sirota, L.J. Wenzel, Rev. Sci. Instrum. **64**, 1309 (1993)
87. J.A. Pople, I.W. Hamley, G.P. Diakun, Rev. Sci. Instrum. **69**, 3015 (1998)
88. Ch. Münch, J. Kalus, Rev. Sci. Instrum. **70**, 187 (1999)
89. P. Baroni, C. Pujolle-Robic, L. Noires, Rev. Sci. Instrum. **72**, 2686 (2001)
90. M. Kisilak, H. Anderson, N.S. Babcock, M.R. Stetzer, S.H.J. Idziak, E.B. Sirota, Rev. Sci. Instrum. **72**, 4305 (2001)
91. P. Lindner, R.C. Oberthur, Rev. Phys. Appl. **19**, 759 (1984)
92. G.C. Straty, H.J.M. Hanley, C.J. Glinka, J. Stat. Phys. **62**, 1015 (1991)
93. L. Kalus, G. Neubauer, U. Schmelzer, Rev. Sci. Instrum. **61**, 3384 (1990)
94. S.M. Baker, G. Smith, R. Pynn, P. Butler, J. Hayter, W. Hamilton, Rev. Sci. Instrum. **65**, 412 (1994)
95. V.M. Cloke, J.S. Higgins, C.L. Phoon, S.M. Richardson, S.M. King, R. Done, T.E. Cooper, Rev. Sci. Instrum. **67**, 3158 (1996)
96. G.C. Straty, C.D. Muzny, P.D. Butler, M.Y. Lin, T.M. Slawecki, C.J. Glinka, H.J.M. Hanley, Nucl. Instrum. Methods Phys. Res. A **408**, 511 (1998)
97. T.L. Kuhl, G.S. Smith, J.N. Israelachvili, J. Majewski, W. Hamilton, Rev. Sci. Instrum. **72**, 1715 (2001)
98. L. Porcar, W.A. Hamilton, P.D. Butler, G.G. Warr, Rev. Sci. Instrum. **73**, 2345 (2002)
99. P.G. Cummins, E. Staples, B. Millen, J. Penfold, Meas. Sci. Technol. **1**, 179 (1990)
100. D. Renard, F. Boué, J. Lefebvre, Physica B **234–236**, 289 (1997)
101. X. Cao, R. Bansil, K.R. Bhaskar, B.S. Turner, T.J. LaMont T. Niu, N.H. Afdhal, Biophys. J. **76**, 1250 (1999)
102. L.A. Sellers, A. Allen, E.R. Morris, S.B. Ross-Murphy, Carbohydr. Res. **178**, 93 (1988)
103. T.A. Waigh, A. Papagiannopoulos, A. Voice, R. Bansil, A.P. Unwin, C.D. Dewhurst, B. Turner, N. Afdhal, Langmuir **18**, 7188 (2002)

104. A. Aggeli, I.A. Nyrkova, M. Bell, R. Harding, L. Carrick, T.C.B. McLeish, A. N. Semenov, N. Boden, Proc. Natl. Acad. Sci. USA **98**, 11857 (2001)
105. P.J. Mawer, T.A. Waigh, R. Harding, T.C.B. McLeish, S.M. King, M. Bell, N. Boden, Langmuir **19**, 4940 (2003)
106. I.F. Bailey, Z. Kristallogr. **218**, 84 (2003)
107. J. Torbet, M.H.F. Wilkins, J. Theor. Biol. **62**, 447 (1976)
108. N.P. Franks, W.R. Lieb, J. Mol. Biol. **133**, 469 (1979)
109. J. Katsaras, J. Phys. Chem. **99**, 4141 (1995)
110. R.P. Rand, V.A. Parsegian, Biochim. Biophys. Acta **988**, 351 (1989)
111. R. Podgornik, V.A. Parsegian, Biophys. J. **72**, 942 (1997)
112. J. Katsaras, Biophys. J. **75**, 2157 (1998)
113. G. Pabst, J. Katsaras, V. A. Raghunathan, Phys. Rev. Lett. **88**, 128101 (2002)

8

Small-Angle Neutron Scattering from Biological Molecules

J.K. Krueger, G.D. Wignall

8.1 Introduction

The technique of small-angle neutron scattering (SANS) has developed into a powerful tool for the study of biological macromolecules in solution. SANS is applicable over a wide range of length scales (~10–1000 Å) and yields information about size, shape, and domain orientations, conformational changes and/or flexibility, as well as molecular associations in solution – all of which can be key to discerning molecular interactions that are essential for intracellular function. Using SANS, the contrast of individual components within a macromolecular complex can be manipulated systematically through either isotopic labeling (deuteration) and/or an appropriate choice of solvent. Thus, SANS is a unique technique offering the only available method for extracting structural information on each component within a composite of interacting biomolecules under near physiological conditions. In combination with advances in molecular biology techniques (and the increasing number of high-resolution data on component structures), substantial improvements in neutron sources, and instrumentation at leading neutron scattering institutions have broadened the potential impact of SANS in modern structural molecular biology.

This article will discuss instrumentation, theory, and practical aspects with biological examples to highlight the power of the SANS technique, and will attempt to explain the physics of scattering with the minimum of unnecessary detail and mathematical rigor. The aim is to aid potential users who have a general scientific background, but no specialist knowledge of scattering to apply the technique to provide new information in areas of their own particular research interests.

8.1.1 Why Neutron Scattering is Appropriate and Comparison with Other Low-Q Scattering Techniques

Neutron scattering had its origin in 1932, the year that marked the discovery of the neutron by Chadwick [1]. Initially, the technique of neutron scattering was used mainly for the study of "hard" crystalline materials. Such studies continue to give important structural information [2], though during the last two decades, the technique has been used increasingly by scientists from other disciplines, including those with "soft" matter applications, such as organic polymers and biological macromolecules. Many of these systems contain copious amounts of hydrogen, either in their molecular constitution or when suspended in aqueous or H^1-containing solvents, thus they are particularly suitable for applying deuterium-labeling techniques.

Scattering techniques have been employed for decades to provide information on the spatial arrangements of macromolecules and for those that crystallize, most high-resolution structures have been determined by X-ray diffraction via Bragg's law

$$\lambda = 2d \sin \phi/2, \tag{8.1}$$

where d is the distance between crystallographic planes, λ is the wavelength and ϕ is the angle of scatter. In the literature, 2θ is widely used as an alternative symbol for ϕ, the angle of scatter. For elastic scattering, where the energies of the incident and scattered radiation (neutrons, X-rays etc.) are the same, the intensity, $I(Q)$, is measured as a function of the momentum transfer, Q

$$Q = 4\pi\lambda^{-1} \sin \phi/2. \tag{8.2}$$

Combining equations (1) and (2) gives

$$d = 2\pi/Q. \tag{8.3}$$

Although Bragg's law (Eq. 8.1) does not apply to noncrystalline materials, a Fourier or inverse relationship between the structure in real space (r) and the scattering in Q-space, means that Eq. 8.3 may be applied to first order for all types of scattering. Thus, data at small-Q probe longer distances, and in order to study the length scales ($\sim 10^1$–10^3 Å) that are important for biomolecules, we need to work at low Q-values (10^{-3}– 10^{-1} Å^{-1}) and collect data at small angles ($2\theta = \phi < 15°$), using long wavelength ($5 < \lambda < 20$ Å) or "cold" (i.e., low energy) neutrons. These measurements are conventionally referred to as small angle scattering, although it is the Q-range which determines the size of objects studied, and radiations with other wavelengths (e.g., light, X-rays) can obviously provide complementary information in different angular ranges. For example, light scattering (LS), with wavelengths in the range 2–6×10^3 Å, probes a much smaller Q-range ($\sim 10^{-6} < Q < 10^{-3}$ Å^{-1}) than small-angle X-ray scattering (SAXS), even though the angular range can be quite large (up to $2\theta = \phi \sim 160°$). Hence, LS measurements probe distance scales, via Eq. 8.3, up to $\sim$ 10 μm and the technique has been used extensively since the 1940s to determine the molecular weights and global dimensions of macromolecules and aggregates, for example in dilute solution [3].

X-ray scattering using Cu K_α ($\lambda = 1.54$ Å) probes the interatomic length scales of ~10^1–10^3 Å over a larger Q-range (~$10^{-3} < Q < 10^{-1}$ Å^{-1}) than LS. Neither SAXS nor LS can be applied to the condensed state or to concentrated solutions, due to the difficulties of separating the intermolecular and intramolecular contributions to the structure. However, SANS eliminates this limitation and has been widely used to study macromolecules in overlapping, "crowded" environments. By deuterium-labeling a fraction of the molecules, it is possible to measure the individual molecular dimensions [4–6] and such "high-concentration labeling" methods are important for synthetic polymers in the condensed state. However, they are not particularly relevant to biological samples, which are often very dilute solutions (typically in the micromolar range), so biopolymers are usually studied in the limit of zero concentration. In principle, SANS measurements in dilute solution offer the same information as SAXS, which permit the elucidation of molecular dimensions via the electron density contrast between a macromolecule and solvent. However, even in this limit, SANS has distinct advantages. Greater signal to noise may be obtained with the neutron technique since it is less sensitive to dust particles than is LS. Also, contrast for LS depends on the refractive index difference, while that for X-rays depends on the electron density difference. These differences between organic biomolecules and their contaminants are not great, so in general the signal-to-noise is less than for neutrons, where strong contrast may be achieved by means of deuterium labeling.

For LS, the scattering patterns are dependent on the polarization directions, yet chemical bonding has little effect on SAXS or SANS where there is negligible influence of the differences between the directions of radiation polarization and molecular orientation [7]. Hence, polarization effects can be neglected in SAXS and SANS experiments.

Neutrons demonstrate convincingly the wave–particle duality of matter and scattering experiments exploit both aspects of neutron behavior. A typical "cold" neutron with incident particle velocity, v_o, of 750 m/s, has a wavelength of 5.3 Å (via the de-Broglie relation, $\lambda = h/mv_o$ where h is Planck's constant and m is the neutron mass), which is the same order as the nearest neighbor spacing between biological macromolecules. The neutron lifetime is 885.7 ± 0.8 s [8], and as the maximum length of neutron scattering instruments is ~ 10^2 m for SANS (and even less for other instruments), the time of flight during an experiment is typically $\ll 1$ s.

The kinetic energy, E_0, is given by $E_0 = mv_o^2/2 = 0.003$ eV or 4.7×10^{-15} ergs [9], which are much lower than X-ray photons (~10 keV). If no energy change takes place in the scattering process, the energies of the incident and scattered beam are equal and the scattering is termed elastic. If energy is transferred, there is a finite difference ($\Delta E \neq 0$) between the energies of the incident (E_0) and scattered (E_1) beams, which may be regarded as a Doppler shift in the scattered wavelength due to thermal motion of the nucleus, and the process is termed inelastic. If ΔE is small compared to the incident energy ($E \ll E_0$), the scattering is termed quasi-elastic.

Most neutron scattering measurements on biomolecules have involved scattering at small values of the momentum transfer ($Q \rightarrow 0$) and as mentioned above, this type of measurement is conventionally referred to as small-angle (rather than small Q) neutron scattering though the terms are equivalent for the long wavelengths or "cold" neutrons ($\lambda > 5\,\text{Å}$). It may be shown [10] that for such long wavelengths, $Q \rightarrow 0$ implies $E_1 \rightarrow E_0$ and the scattering is predominantly elastic, as any neutron scattered with a large energy transfer, E, could not satisfy both energy and momentum conservation at small-Q.

For most applications discussed in this chapter, neutron and X-ray scattering are examples of predominantly elastic scattering, where the incident and scattered radiation have the same energy or wavelength. Such experiments give information on the time-averaged structure of the system. There has been work involving inelastic processes, where there is a change of energy on scattering and the incident and scattered radiation have different wavelengths. This technique gives valuable information on dynamics [11], though this methodology is beyond the scope of this article, which will seek to illustrate how SANS has complemented and expanded the information from other scattering techniques, while emphasizing the analogies and differences between neutron and photon scattering.

8.1.2 Complementary Aspects of Light, Small-Angle Neutron and X-ray Scattering for Solution Studies

Light scattering is useful for studying the state of association or conformation of biological macromolecules in solution [12]. Both static (elastic) and dynamic (quasi-elastic) light scattering techniques are generally easy to perform and can be done on solutions with relatively low concentrations of analyte. The static light scattering (SLS) experiment monitors the total light scattering intensity averaged over time and can provide information on the "apparent" molecular weight (M_{app}) and the radius of gyration (R_{g}) of the macromolecule in solution.

Dynamic light scattering (DLS) experiments monitor fluctuations in the intensity of light scattered by small volume elements in solution, which is directly related to the Brownian motion of the solutes thereby providing information on the hydrodynamic radius, R_{H}, which also can be related to an apparent molecular weight. In either case, light scattering techniques can be used as an initial probe of biopolymer (protein, DNA) conformations to monitor aggregation or conformational changes in varying solution environments.

Light scattering can be utilized to effectively screen for previously defined conformational changes in macromolecules. For example, Vogel and co-workers [13] used DLS to assess the influence of different synthetic peptides, each comprising the calmodulin (CaM)-binding domains from various CaM-binding proteins on the structure of apo-CaM (calcium-free) and Ca^{2+}-bound CaM [13]. The large scale conformational changes in Ca^{2+}-CaM upon binding a few archetypal peptides had previously been well-characterized by small-angle X-ray and neutron scattering [14], NMR [15] and X-ray crystallography [16]. Thus, interpretation of the light scattering results, as reflected by an increase in the R_H, proved to be useful as a rapid screen for peptides that induced the previously characterized conformational changes in CaM.

Similarly, light scattering experiments may detect major conformational changes in proteins that can be used as a guide to design experiments to further characterize those changes using small-angle X-ray and/or neutron scattering. More detailed structural information may be obtained from analysis of the pair-distance distribution function, $P(r)$. The $P(r)$ represents the frequency of vectors connecting small-volume elements within the entire volume of the scattering particle and is calculated from an inverse Fourier transformation (FT) of the scattering data (see below).

Small-angle scattering (SAS) experiments yield information on the overall shapes and electron (or nuclear) density distribution within macromolecules in solution (for a review see [17–19]). In addition, the different neutron scattering properties of the isotopes of hydrogen, combined with the ability to uniformly label biological macromolecules with deuterium, allow one to characterize the conformations and relative dispositions of the individual components of an assembly of biomolecules. There are several examples now of neutron scattering experiments that have used solvent matching or the more rigorous methods of contrast variation to provide a unique view of the interactions within molecular complexes where one component within the complex is deuterated. The results from these experiments have provided insight into many dynamic processes, such as enzyme activation, as well as into the highly regulated and coordinated interactions of complex systems such as muscle. Several of these experiments will be discussed at the end of this paper or elsewhere in this volume. High-resolution techniques such as X-ray crystallography have the capability of providing detailed structural views of these protein:protein complexes, however, SAS techniques are applied in solution and can give insights into systems in which inherent flexibility may cause problems for crystallization.

8.2 Elements of Neutron Scattering Theory

8.2.1 Coherent and Incoherent Cross-Sections

As pointed out in the contribution by Harroun et al. in this volume, the neutron scattering cross-section (σ_s) splits into coherent (σ_c) and incoherent (σ_i) components. Only the former contains information on interference effects arising from spatial correlations of the nuclei in the system, i.e., the structure of the sample. The incoherent cross-section contains no information on interference effects and forms an isotropic (flat) background which must be subtracted off in SANS structural investigations. The incoherent component of the scattering does however contain information on the motion of single atoms (particularly hydrogen) which may be investigated via energy analysis of the scattered beam.

While most of the atoms encountered in neutron scattering from biopolymers are mainly coherent scatterers (e.g., carbon, oxygen), there is one important exception. In the case of hydrogen (^{1}H), the spin-up and spin-down scattering lengths have opposite sign ($b^+ = 1.080 \times 10^{-12}$ cm; $b^- = -4.737 \times 10^{-12}$ cm), and

$$\sigma_c = 1.76 \times 10^{-24}\,\text{cm}^2; \quad \sigma_i = 79.7 \times 10^{-24}\,\text{cm}^2. \tag{8.4}$$

For photons (SAXS or LS), there is no strict analog of incoherent scattering of neutrons due to nonzero spin in the scattering nucleus. Compton scattering, which occurs for X-rays, is similar in that it contains no information on interference effects, i.e., the structure of the sample, and forms a background to the coherent signal. However, to a good approximation, this background goes to zero in the limit $Q \to 0$ and is usually neglected in SAXS studies. Table 1.1 in the contribution by Harroun et al. in this volume gives the cross-sections and scattering lengths for atoms commonly encountered in synthetic and natural materials. These cross-sections refer to bound protons and neglect inelastic effects arising from interchange of energy with the neutron. For coherent scattering, which is a collective effect arising from the interference of scattered waves over a large correlation volume, this approximation is reasonable, especially at low Q where recoil effects are small. However, for incoherent scattering, which depends on the uncorrelated motion of individual atoms, inelastic effects become increasingly important for long wavelength neutrons with the result that the ^{1}H-incoherent cross-section is a function of both the incident energy and sample temperature [20]. Thus, the transmission of H_2O is a function of both these variables, and the ^{1}H-incoherent cross-section ($\sigma_i = 79.7 \times 10^{-24}$ cm^2), almost never applies to real biopolymer systems.

Figure 8.1 dramatically shows that the total (σ) scattering cross-section (scattering plus absorption) of a light water molecule (H_2O) is a strong function of the energy of the incident neutrons (E_0). For high E_0, the energies associated with the vibrational and translational motion of water molecules are negligible and the cross-section approaches the sum of the free atom cross-sections for the hydrogen and oxygen atoms ($\sigma_{\text{FREE}} \sim 44 \times 10^{-24}$ cm^2 or 44 barns). However, as $E_0 \to 0$, the scattering does not plateau at the bound atom cross-section (~167 barns) and varies continuously with energy. This variation is due mainly to inelastic processes affecting the incoherent scattering which is the main component of the cross-section. Also, because of inelastic effects due to torsion, rotation, and vibration, the effective ^{1}H incoherent cross-section is also a function of the particular chemical group (methyl, hydroxyl, etc.) in which the proton is situated [21]. The total ^{1}H atom cross-section is dominated by the incoherent component (σ_{inc}), and hence is a strong function of λ (Fig. 8.1) and only approaches ~80 barns at $\lambda \sim 4.5$ Å [9].

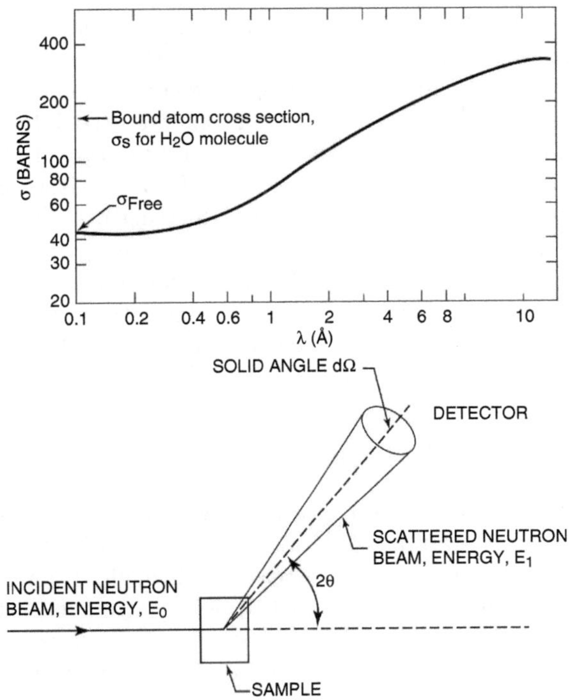

Fig. 8.1. Total cross-section for water molecule (H_2O) vs. neutron wavelength at $T = 293\,\mathrm{K}$ according to Brookhaven National Laboratory Tables (*top*) The basic scattering experiment (*bottom*)

There is a large difference in the coherent scattering length between deuterium ($b_D = 0.667 \times 10^{-12}$ cm) and hydrogen ($b_H = -0.374 \times 10^{-12}$ cm) and the latter value is actually negative. This arises from a change of phase of the scattered wave and results in a marked difference in scattering power (contrast) between molecules labeled with deuterium or hydrogen, or suspended in light–heavy water. The basic experiment [9] consists of an incident neutron beam (wavelength, λ), which is scattered by an assembly of nuclei through an angle $\phi = 2\theta$ into a solid angle $\mathrm{d}\Omega$ (Fig. 8.1), and for both SANS and SAXS it is assumed that any change in energy on scattering is small compared to the incident energy, E_0. The coherent component of the scattering contains information on the correlations between different nuclei [9] and hence reflects the relative spatial arrangement of atoms in the system (e.g., the structure). Thus, the angular or Q-dependence of the scattering is related inversely, via a Fourier transform, to the spatial variation of the structure Eq. 8.3.

In principle, the incoherent cross-section contains information on the correlations between the *same* nucleus and hence gives information on the time dependence of the position of an individual atom (e.g., vibration, diffusion, etc.). However, extracting such information would require an energy analysis of the scattered beam, which has not hitherto been performed for the vast majority of SANS experiments. These are conventionally undertaken by integrating the scattered neutrons over all energies, so information on the time dependence of the structure is not normally obtained in practice, and the incoherent component of the cross-section forms an isotropic (flat) background which must be subtracted off in SANS structural investigations. This signal arises from nuclei with nonzero spin (e.g., hydrogen) and due to multiple scattering, effects, this background is a function of the sample dimensions, transmission, etc. and thus cannot be expressed as a true cross-section [22,23]. However, it is usually smaller than the coherent signal and may be subtracted off to good accuracy by empirical methods [24].

8.2.2 Scattering Length Density

In general, radiation incident on a medium whose scattering power is independent of position is scattered only into the forward direction ($\phi = 2\theta = 0$). For every volume element (S) which scatters through an angle $\phi = 2\theta > 0$, there is another volume element (S') which scatters exactly (180°) out of phase, (see Fig. 8.2 where $PS - S'S = \lambda/2$). Therefore, all scattering cancels unless the scattering power is different at S and S', i.e., fluctuates from point-to-point in the sample. X-rays and light photons interact with electrons in the sample and hence are scattered by fluctuations in electron density. Neutrons do not interact with electrons (apart from unpaired spins, where the interaction arises from the magnetic moment of such elements as rare earths, transition metal, etc.). In general, biopolymers and solvents do not contain such elements, so the only interaction is via nuclear scattering. Because each nucleus has a different scattering

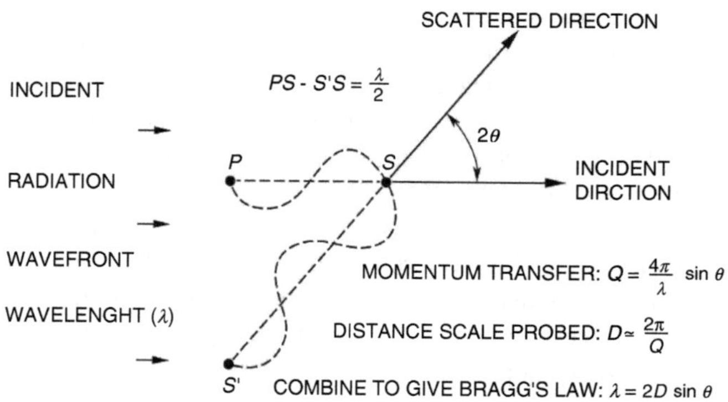

Fig. 8.2. For every point S which scatters radiation through an angle $2\theta > 0$, there is another point S', which scatters radiation exactly 180° out of phase. Therefore, all scattering cancels unless the scattering power is different at S and S', i.e., fluctuates from point to point in the sample

amplitude, the scattering length density (SLD) is defined as the sum of coherent scattering lengths over all atoms lying in a given volume, V, divided by V [9, 23]. Table 1.2 in the contribution by Harroun et al. this volume gives representative values of the scattering lengths and volumes of some common amino acids and proteins. The SLD is the ratio of these quantities and is typically in the range 1.4–5.4 × 10^{10} cm^{-2}, though the actual volumes (and hence the associated SLDs) are not universal constants and differ according to conditions (e.g., solvent, salt concentration, environment, etc.). For a discussion of such effects, see D. Svergun et al. [29]. For X-rays or light, the (photon) SLD is the electron density multiplied by the Thompson scattering factor of one electron, $r_T = 0.282 \times 10^{-12}$ cm [23, 26].

8.2.3 Contrast Variation

Contrast variation methods have found wide application in structural biology, where they can be used to distinguish the scattering due to individual components within a macromolecular complex; provided the components have different neutron scattering densities. In the case of a multiprotein complex, selective deuteration of an individual protein component provides an approach to selectively altering that component's neutron scattering density. Subsequently, by changing the deuterium level in the solvent, the neutron scattering contrast of each component is varied. Under certain conditions where the mean solvent density matches that of one of the components, the solvent matched component becomes "invisible" in the neutron experiment and any measured scattering intensity is due primarily to the non solvent-matched component. The concepts underlying solvent matching are most easily demonstrated by an equivalent experiment with visible light, illustrated in Fig. 8.3, made by D.M. Engelman (Yale University). Both tubes contain two Pyrex beads embedded in (borosilicate) glass wool, which has a different refractive index to the beads. When light shines on the tube at right, both the beads and glass wool scatter light, but only the glass wool can be seen because it dominates the scattering. In order to observe the beads, the tube on the left has been filled with a solvent which has the same refractive index as the glass wool. Thus, the electron density and hence the scattering power of the glass wool has been matched with that of the solvent, thus eliminating this component of the scattering and making the wool transparent to light.

This principle can be used in SANS experiments via isotopic solvent mixtures (e.g., H_2O–D_2O), as light water (H_2O) has an SLD of -0.562×10^{10} cm^{-2} (or -0.00562×10^{-12} cm Å^{-3}) while that of heavy water (D_2O) is 6.404×10^{10} cm^{-2} (or 0.06404×10^{-12} cm Å^{-3}), so the SLD of a mixture is a linear function of the percentage of D_2O and is zero for 8 vol% D_2O. Because of proton exchange, the SLD of a biological unit (e.g., protein) will vary even if it is not specifically deuterated at nonexchangeable position, by immersion in a solvent containing D_2O, and is therefore a function of the H_2O–D_2O ratio.

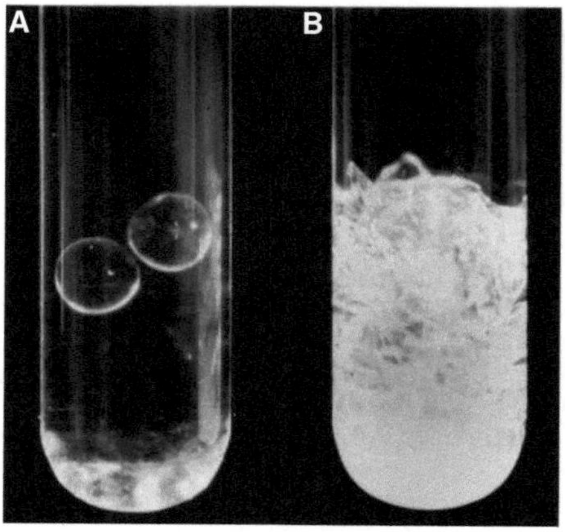

Fig. 8.3. Two tubes containing Pyrex beads in glass wool and solvent: (**A**) Refractive index of solvent matches that of glass wool. (**B**) Refractive index of solvent is different to that of glass wool or Pyrex beads and scattering from the glass wool dominates (reproduced with permission of D.M. Engelman)

Solvent matching thus provides a means for extracting structural information on the individual components within a complex. The effectiveness of the solvent matching experiment depends upon having uniform density components such that the internal density fluctuations can be ignored, as well as on very precise matching of the component and the solvent densities, which can be tricky.

A more robust approach to utilizing contrast variation methods with neutron scattering for extracting structural information from macromolecular complexes is to measure a "contrast series" in which the solvent deuteration level is systematically varied over the widest range possible. For a complex of two components with different mean neutron SLD, the total scattering can be written as:

$$I(Q, \Delta\rho_{\rm A}, \Delta\rho_{\rm B}) = \Delta\rho_{\rm A}^2 I_{\rm A}(Q) + \Delta\rho_{\rm A}\Delta\rho_{\rm B} I_{\rm AB}(Q) + \Delta\rho_{\rm B}^2 I_{\rm B}(Q). \qquad (8.5)$$

The subscripts A and B refer to each component and $\Delta\rho_{\rm X} = \rho_{\rm X} - \rho_{\rm S}$ where $\rho_{\rm X}$ is the mean SLD for the individual components (A or B) and $\rho_{\rm S}$ is the mean SLD for the solvent. Equation 8.5 assumes the difference between mean scattering densities for the individual components is much greater than any internal density fluctuations within each component. The three terms in Eq. 8.5 correspond to the three *basic scattering functions*. $I_{\rm A}(Q)$ and $I_{\rm B}(Q)$ represent the scattering of components A and B, respectively, while $I_{\rm AB}(Q)$ is the cross-term which is due to interparticle scattering thus its inverse Fourier transform (IFT) provides information about vector distances between the two scattering particles and the first moment of this transform gives the separation of the centers-of-mass. A set of neutron scattering measurements with different D_2O:H_2O ratios in the solvent gives a set of equations in the form of Eq. 8.5

which can be solved to give the three basic scattering functions from which one can derive the structural parameters for each component as well as information on their relative dispositions. Contrast variation using neutrons for studies of biological molecules was demonstrated by Ibel and Stuhrmann [27] and there are number of excellent reviews on the topic [18, 28–31]. Specific examples on the use of both solvent matching as well as contrast variation to obtain unique information will be presented for SANS structural studies of protein/protein complexes at the end of this paper.

8.3 Practical Aspects of SANS Experiments and Data Analysis

8.3.1 SANS Instrumentation

The first instrument [32, 33] suitable for the study of biopolymers was built in the early 1970s at the FRJ2 reactor at the Forschungszentrum Jülich, Germany, and pioneered the use of long wavelength neutrons and large overall instrument length (> 20 m). It was also the first to boost the flux of the long wavelength (λ > 5 Å) or "cold neutron" component of the Maxwellian spectrum by moderating the neutrons to a lower temperature by means of a cold source containing a small volume of liquid hydrogen at T ~20 K. This gives flux gains of over an order of magnitude at λ ~10 Å, and it was on this instrument that the initial SANS experiments on biopolymers were performed. The D11 facility, built on the High Flux Reactor (HFR) at the Institut Laue-Langevin (ILL), Grenoble, France, incorporated many of the features of the FRJ2 instrument, including a cold source and long (~80 m) dimensions [34]. The FRJ2 and HFR facilities have both been subsequently upgraded [35, 36] and expanded to be among the most productive SANS facilities worldwide. At the time of writing, the most effective SANS facilities for biological studies in the US are at the National Institute for Standards and Technology (NIST) [37].

Currently, over 30 SANS instruments are now in operation or under construction worldwide, most of which are reactor based, and a schematic diagram is shown in Fig. 8.4. Fission neutrons are produced in the core, which is surrounded by a moderator (e.g., D_2O, H_2O) and reflector (e.g., Be, graphite) which reduce the neutron energy. Because of the λ^{-4} factor which enters into the calculation [33] of the scattering power for a given resolution ($\Delta Q/Q$), it is highly advantageous to use long wavelengths and to increase the flux in this region. This may be accomplished by further moderating the neutrons to a lower temperature by means of a cold source containing a small volume of liquid or superfluid hydrogen, placed near the end of the beam tube. The neutrons are transported from the source to the instrument by neutron guide tubes. These are often coated by natural Ni or isotopic ^{58}Ni, and operated by total internal reflection to transport the neutron beam from the cold source to

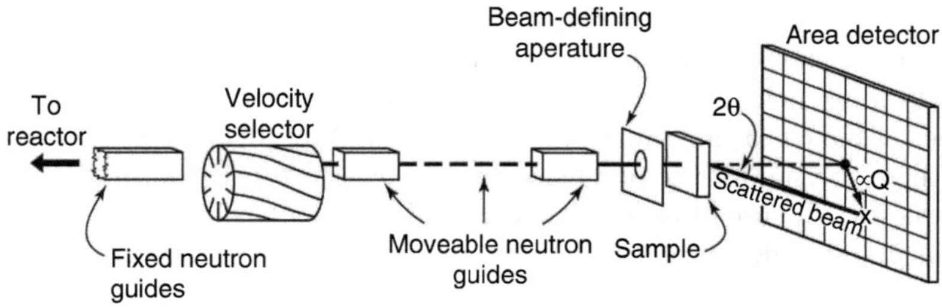

Fig. 8.4. Schematic of a reactor-based SANS facility

the sample, in a manner analogous to the way that light may be transported by fiber optics. The guide system (Fig. 8.4) provides a gap for the insertion of a velocity selector to define the wavelength ($5 < \lambda < 30$ Å) and bandwidth ($\Delta\lambda/\lambda \sim$ 5–35%) of the neutron beam. In addition to fixed neutron guides, most instruments have translatable guide sections and apertures that may be moved in and out of the neutron beam to define the incident beam collimation. This is followed by an accessible section (1 – 2 m) at the sample position to accommodate temperature-controlled sample changers, flow cells, etc. Thus, when all the moveable guides are removed from the beam, the source slit is typically ∼10 – 20 m from the sample, and this distance is reduced to 1 – 2 m, when all the guides are translated into the beam to increase the sample flux. An area detector (typically a $64 \times 64\,\mathrm{cm}^2$ or $100 \times 100\,\mathrm{cm}^2$ proportional counter) is often positioned via a motor-driven carrier mounted on rails [32–38] in the post-sample flight-tube, ∼1 – 20 m in length. Like the incident neutron guides, this is normally evacuated to reduce air scatter, which would otherwise be strong, given overall instrument lengths ∼20–40 m.

The majority of area detectors are multiwire proportional counters [34,39], with active areas up to $1\,\mathrm{m}^2$, and an element (cell) size $\sim 0.5 - 1\,\mathrm{cm}^2$, which is chosen to be of the same order as the sample size to equalize the various contributions to the instrumental resolution [33,34]. In general, the detector response function, $R(Q)$, is Gaussian with a full width at half maximum ∼0.5 – 1 cm and the spatial variation of the detector efficiency (ϵ) is usually measured via an incoherent scatterer (e.g., light water), which has an angle-independent intensity in the Q-range measured [23,40].

Reactor sources also produce appreciable background (e.g., fast neutrons, γ-rays). By introducing some curvature into the guides, it is possible to separate out this component, which is not reflected as efficiently as cold ($\lambda \sim 5$ – 30 Å) neutrons. Alternatively, the beam may be deflected by supermirrors [41,43] and such mirrors may be designed to reflect up to 3 – 4 times the critical angle for internal reflection that can be achieved by natural Ni guide coatings ($\theta_c = 0.1\lambda$ (Å)).

The size of the beam at the sample is defined by slits (irises) made of neutron absorbing materials (e.g., Li^6, cadmium, boron), for which the ratio of scattering to absorption is virtually zero. This has the result that neutron beams can be very well collimated [10,32] and the ratio of parasitic scattering to the main beam intensity is very small (typically $<10^{-5}$ within $\sim$1 mm from the beam stop). For X-rays on the other hand, materials which have high absorption (to define an SAXS beam) also have high scattering power, as both parameters are a strong function of the atomic number, and parasitic scattering is usually higher for SAXS.

At the time of writing there are over 40 neutron sources around the world operating as user facilities [42]. Of these sources, 36 are reactor facilities, the majority of them are commissioned more than 30-years ago and consequently now have increasingly finite lifetimes. A forward survey [43] estimated that over the next two decades, the installed capacity of neutron beams for research could decrease substantially. Fortunately, the expected decline in the availability of reactor-based SANS instruments has been offset by two competing trends. First, several new reactors are under construction worldwide [44], along with upgrades to existing sources (e.g., at the ILL in the mid-1990s, NIST during 1995–2002 and Oak Ridge National Laboratory during 2000–2006). In addition, a range of accelerator-based SANS instruments have been developed over the past 15-years, and in particular, a "next generation" Spallation Neutron Source is under construction at Oak Ridge [45] and at Tokai-mura [46]. Similarly, the second target station for the ISIS-pulsed facility [47], currently under construction, and a proposed [48] European Spallation Source will do much to assure the availability of SANS facilities in the future.

The spallation process involves bombarding a heavy metal (e.g., Ta, W, or Hg) target with high-energy protons, thus placing those nuclei into a highly excited state. These lose energy by "evaporating" nuclei, and in the case of a tungsten target, each proton results in the production of $\sim$15 neutrons. The protons are usually accelerated in pulses and so neutron production also occurs in pulses, which allows the use of time-of-flight (TOF) techniques. Shorter wavelength neutrons travel faster and arrive at a detector earlier than longer wavelength neutrons, so there is thus no need to employ a velocity selector to monochromate the incident beam. Another benefit of the TOF approach is that any given point on a detector corresponds to several different Q values, determined by the wavelength of the neutrons arriving there. Hence, a greater the range of Q values can be measured with any given configuration of [49] instrument. Pulsed-source SANS instruments therefore have a greater dynamic range in Q than reactor-source instruments, though the latter can be increased by moving the detector "off axis" [37,50].

As the main applications of the SANS technique have been undertaken on reactor sources, these instruments have been optimized over the past several decades, and the flux of instruments planned on new or upgraded reactor sources will either be less than or equal to the current state of the art instruments (e.g., the D22 instrument at the ILL [50]). However, this is not the case for pulsed facilities, which have not yet begun to reach their full potential so we can still expect order of magnitude gains over the current facilities. Thus, it seems likely that pulsed sources will make a greater contribution to SANS studies of biopolymers in future than they have in the past.

8.3.2 The Importance of Absolute Calibration and Having Well-Characterized Samples

This section will emphasize the importance of placing intensity data on an absolute scale, typically in the form of a differential scattering cross-section $\mathrm{d}\Sigma/\mathrm{d}\Omega(\boldsymbol{Q})$, in units of cm^{-1}. While the use of absolute units is not essential for the measurement of the spatial dimensions (e.g., determining the radius of gyration, R_g, of a molecule or particle), it forms a valuable diagnostic tool for the detection of artifacts to which scattering techniques are often vulnerable [23]. Because the cross-section varies as the sixth power of the dimensions [51], it is a sensitive indicator of whether an appropriate structural model has been chosen. For example, SANS studies of colloidal solutions may be modeled by core–shell spherical micelles as a function of a set of parameters describing the particle structure and interactions [52]. On an arbitrary intensity scale, Hayter and Penfold have pointed out that it is possible to produce excellent fits of the particle shape, which may be in error by as much as 3–4 *orders of magnitude* in intensity [53]. Thus, absolute calibration allows such artifacts to be recognized, and the model parameters may be restricted to those, which reproduce the observed cross-section.

In view of the maturity of the SANS technique, it is surprising that data are still published in arbitrary units which are functions of the time scale of the experiment and/or the sample dimensions (e.g., thickness). Conversion to an absolute scale may be accomplished by multiplying by a calibration constant and the absolute cross-section $[\mathrm{d}\Sigma/\mathrm{d}\Omega(\boldsymbol{Q})]$, is defined [54] as the ratio of the number of neutrons (neutrons s^{-1}) scattered per second into unit solid angle divided by the incident neutron flux (neutrons $\mathrm{cm}^{-2}\,\mathrm{s}^{-1}$) and thus has the dimensions of area (cm^2). On normalizing to unit sample volume, $\mathrm{d}\Sigma/\mathrm{d}\Omega(\boldsymbol{Q})$ has units of cm^{-1}. For all systems discussed in this chapter, the scattering is azimuthally symmetric about the incident beam, i.e., $\mathrm{d}\Sigma/\mathrm{d}\Omega(\boldsymbol{Q})$ is a function only of the magnitude of the scattering vector $|\boldsymbol{Q}| = 4\pi\lambda^{-1}\sin\theta$. Thus, the relationship between the cross-section and the measured intensity or count rate $I(Q)$ (counts s^{-1}) in a detector element with area, Δa, and counting efficiency, ϵ, situated normal to the scattered beam at a distance, r, from the sample, is given by

$$\mathrm{d}\Sigma/\mathrm{d}\Omega(Q) = \frac{I(Q)r^2}{\epsilon I_0 \Delta a A t T}, \tag{8.6}$$

where I_0 is the intensity (counts $\mathrm{s}^{-1}\,\mathrm{cm}^{-2}$) on a sample of area A, thickness t, and irradiated volume $= At$. The measured transmission T is given by $T = \mathrm{e}^{-\mu t}$ where μ is the linear attenuation coefficient and accounts for the attenuation of the beam on passing through the sample. For SANS it is assumed that the attenuation factor is the same for all scattered neutrons and this approximation is reasonable for $\phi = 2\theta < 10°$. Similarly, Eq. 8.6 assumes that the solid angle subtended by a detector element is independent of 2θ and this approximation again holds for small angles where cos 2θ is close to unity. Since the time dimension cancels in both the numerator and denominator of Eq. 8.6, absolute calibration reduces to measuring the constant $K_{\mathrm{N}} = \epsilon I_0 \Delta a$, which may be determined by comparison with a standard of known cross-section, run in the same scattering geometry for the same time. If an incident beam intensity monitor is employed, as is normally the case, comparisons are made for the same number of monitor counts, i.e., the same number of incident neutrons. Various calibration measurements have been used to measure the calibration constant, both for SANS [22, 55, 56] and SAXS [57], including direct measurement of the beam flux, calibration via a predominantly incoherently scattering material (e.g., vanadium or water) and various other standards.

Specific factors that must be considered in SANS calibrations have been discussed [9] and in particular, multiple scattering and sample preparation are important when using vanadium, which has virtually no coherent cross-section. One disadvantage of this standard is that the cross-section is low and also isotropic, so the run times for calibration are relatively long. Due to limited beam-time allocations, arising from the high demand for SANS facilities, users are naturally reluctant to devote a significant fraction of their instrument time for calibration runs. For this reason, it has been a matter of policy at many SANS facilities, to provide strongly scattering precalibrated samples to allow users to perform absolute scaling with brief calibration runs, which do not detract significantly from the available beam time. The scattering from light water (H_2O) is predominantly incoherent and because the absorption cross-section is small, this system has the advantage of much higher intrinsic scattering for calibration purposes [55, 58], and hence has lower sensitivity to statistical errors and artifacts than vanadium. One disadvantage is that, for 1–2 mm samples, the multiple scattering is much higher (>30% than for vanadium (~10%) and cannot be calculated to the same degree of accuracy [22], because an appreciable fraction of neutrons are scattered inelastically. Such effects are very difficult to model [59–61] and moreover, the detector efficiency is a function of the wavelength and this introduces sample-dependent and instrument-dependent factors, depending on how a given detector responds to the inelastically scattered neutrons [59]. The use of Eq. 8.6 would lead to apparent cross-sections, which are functions of wavelength and are also detector dependent. Also, because of the strong multiple scattering, the intensity for water or protonated (^{1}H-labeled) polymer samples is not proportional to the product, tT, as in Eq. 8.6, and hence it is not possible to define a true cross-section which is a material (intensive) property, independent of the sample dimensions. The scattering is a nonlinear function of the thickness, though such samples may still be used for calibration, provided the thickness is minimized (~1 mm) and they are calibrated against primary standards for a given instrument to take advantage of the intrinsically high signal-to-noise ratio for light-water samples [9, 55, 56]. However, even with the higher cross-section of water or protonated biopolymers, such isotropic scatterers cannot be used at low Q values (long sample-detector distances, r) as the intensity falls as $(1/r^2)$, and standardization requires a measurement at a low sample-detector distance, followed by scaling to the r-value of the measurement via the inverse square law [23].

The spatial variation of the detector efficiency (ϵ) is measured via an incoherent scatterer such as water or a protonated polymer (e.g., polymethyl-methacrylate) and despite the fact that multiple scattering in such materials is not fully understood, the data are independent of angle to a good approximation [23, 40]. Thus, the variation in the measured signal is proportional to the detector efficiency, and may be used in the data analysis software to correct for this effect on a cell-by-cell basis. Second-order corrections representing departures from isotropic scattering and unequal path lengths through different regions of the active gas are usually wavelength-dependent, instrument-dependent and, even detector-dependent, and Lindner and co-workers have discussed how such adjustments may be customized for a particular facility [58].

Extraction of structural information on individual macromolecules or macromolecular complexes in solution from scattering data requires samples that are rigorously aggregation free. The zero angle or forward scattering, $I(0)$, is directly proportional to the molecular weight squared of the scattering particle and hence is extremely sensitive to aggregation [57]. For X-ray experiments, most proteins in aqueous solution have essentially the same contrast. Therefore, by using a standard protein of known concentration that is also known to be monodisperse in solution, its $I(0)$ can be used to calculate very precise concentration values. Alternatively, if the concentrations are known, the $I(0)$ of the standard can be used to check for sample aggregation. The relationship used for these types of analyzes, for proteins of molecular weight M_{X} and in solution at a concentration c_{X}, given in milligrams/milliliter, is:

$$I(0)_{\mathrm{X}}/M_{\mathrm{X}}c_{\mathrm{X}} = I(0)_{\mathrm{STD}}/M_{\mathrm{STD}}c_{\mathrm{STD}}. \tag{8.7}$$

For neutron scattering experiments, the best method for determining that samples are not aggregated is by comparing the measured $I(0)$ values at the each of the contrasts with the expected $I(0)$ values based on the calculated contrasts and this is only possible if the $I(0)$ values have been put on an absolute scale using an appropriate calibration standard [23, 57].

8.3.3 Instrumental Resolution

Experimentally measured scattering data differ from the actual (theoretical) cross-sections because of departures from point geometry in a real instrument [62–66]. In general, instrumental resolution effects are smaller for SANS than for SAXS. This is because most SANS experiments are performed in point geometry whereas a significant proportion of X-ray experiments have used long slit sources (e.g., Kratky cameras), where smearing effects are larger, particularly at small angles [67,68]. Less attention has been paid to resolution effects in SANS experiments, largely because the corrections are in general smaller for point geometry. However, the corrections are not always negligible, particularly for sharply varying scattering patterns and large scattering dimensions.

In a pinhole SANS instrument (Fig. 8.4), there are essentially three contributions to the smearing of an ideal curve: (i) the finite angular divergence of the beam, $\Delta\theta/\theta = \Delta\phi/\phi$, (ii) the finite resolution of the detector, $R(Q)$ and (iii) the polychromatic nature of the beam, $\Delta\lambda/\lambda$. As mentioned above, for all systems discussed in this chapter, the scattering is azimuthally symmetric about the incident beam, i.e., $\mathrm{d}\Sigma/\mathrm{d}\Omega(\boldsymbol{Q})$ is a function only of the magnitude of the scattering vector $|\boldsymbol{Q}| = 4\pi\lambda^{-1}\sin\theta$. In this case, once the instrumental parameters are well characterized, it is possible by numerical techniques not only to smear a given ideal scattering curve, but also to desmear an observed pattern by means of an indirect Fourier transform to obtain the actual Q dependence [62–66].

A dramatic example of smearing effects is illustrated by SANS data from Kilham rat virus (KRV), which has a core–shell structure, and was modeled by a hollow-shell form factor [65, 69]. KRV may be prepared either "empty" or "full" of nucleic acid and a scattering from the former is shown in Fig. 8.5, which compares the actual measured scattering pattern in a pinhole SANS instrument with the simulated curve in the absence of instrumental resolution

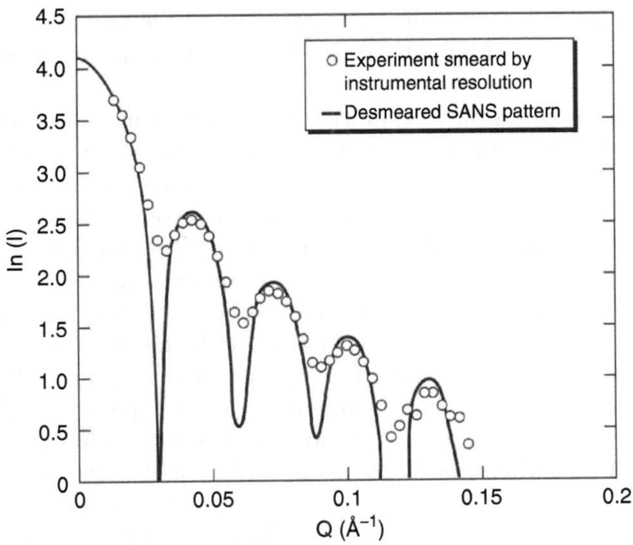

Fig. 8.5. Smeared and desmeared SANS data from KRV empty capsids in D_2O

effects using IFT methods [69]. Although SANS data are routinely analyzed without reference to smearing effects, it is clear from Fig. 8.5 that this omission can sometimes lead to gross errors.

Another model system which has sharp features and is ideal for investigating instrumental resolution effects is monodisperse protonated poly(methyl methacrylate) (PMMA-H) latex particles, suspended in an H_2O–D_2O solvent [67, 70, 71]. The scattering from a homogeneous sphere is a Bessel function [72], with sharp maxima and minima which are averaged by instrumental smearing effects to produce a smoothly varying curve. Desmearing via IFT methods, using an algorithm due to Moore [63], leads to a sharply varying desmeared curve and a particle diameter calculated from the positions of the maxima and minima of $D = 990$ Å. This may be compared with a value of $D = 992$ Å calculated from the particle radius of gyration ($R_g = 384$ Å) derived from the desmearing procedure. The desmearing algorithm [63] used a set of transformed size functions whereas Glatter [64, 66] has employed a set of cubic splines. The latter procedure leads to $D = 988$ Å [73] and thus both routines give dimensions and extrapolated $Q = 0$ intensities which agree within 1%. These data have also been analyzed via an algorithm employing analytical expressions for the wavelength and angular smearing in a pinhole SANS camera [74]. This simplification allows a rapid on-line least-squares desmearing analysis to be performed which leads to $D = 996$ Å, in good agreement with the above determinations. Similar agreement has been achieved for hollow (core–shell) polymer latex scattering [70, 71]. As mentioned in Section 8.3.2, IFT methods also give rise to a length distribution function, $P(r)$, which represents the frequency of vectors connecting volume elements within the scattering particle and goes to zero at a value corresponding to the maximum dimension of the particle. $P(r)$ is more readily interpreted in terms of structural information than the scattering profile and is sensitive to the overall shape and to the relationships between domains or repeating structures.

Where the assumption of azimuthal symmetry cannot be made, the above smearing and desmearing procedures are not applicable, and alternative procedures based on Monte Carlo (MC) techniques have been developed, which simulate the experimental smearing of a given theoretical scattering pattern that can be expressed analytically or numerically [62]. This procedure permits the estimation of resolution effects even in anisotropic systems, but cannot facilitate the desmearing of the observed pattern. Taken together, MC and IFT methods permit a realistic evaluation of the circumstances where resolution effects warrant correction. Both procedures have been illustrated via a range of results of experiments performed on a typical pinhole SANS facility [62], where it was shown that for experiments with scattering dimensions <200 Å smearing effects are small ($<5\%$) and that dimensions up to ~ 1000 Å may be resolved after proper evaluation of resolution effects. Smearing effects may be reduced by decreasing the wavelength range ($\Delta\lambda/\lambda$) or the angular spread ($\Delta\theta/\theta$), though the measured intensity is a strong function of the resolution and Schelten has pointed out that a reduction of a factor of two in $\Delta Q/Q$ will reduce the scattered intensity by over three orders of magnitude [33].

8.3.4 Other Experimental Considerations and Potential Artifacts

For sample containment, there are several materials (e.g., quartz, single-crystal Si), which have very little absorption or scattering for neutrons. For SAXS on the other hand, materials which have high absorption (to define a SAXS beam) also have high scattering power, as both parameters are a strong function of the atomic number, and parasitic scattering is usually higher for SAXS. Thus, the high penetrating power of neutrons makes it relatively easy to contain samples with a minimum of instrumental backgrounds.

For singly scattered neutrons, the intensity $I(Q)$ is proportional to the sample thickness (t) and transmission ($T = \mathrm{e}^{-\mu t}$) and is maximized for $\mu t = 1$, where μ is the linear attenuation coefficient. Thus, the optimum sample thickness is ~1–2 mm for H_2O and ~1 cm for D_2O. Measurements in the intermediate-angle scattering range (~$0.1 < Q < 0.6\,\text{Å}^{-1}$) are particularly sensitive to the incoherent background, which can be of the same order of magnitude as the coherent signal. This is because the coherent scattering falls rapidly with angle (e.g., as Q^{-2} for Gaussian coils or as Q^{-4} in the Porod regime [23]). The coherent intensity of singly scattered neutrons, $I(Q)$ is proportional (Eq. 8.6) to the sample thickness (t), transmission (T) and sample area (A). Thus, measurements on samples with different dimensions (t, A) and transmission (T) may be normalized to the same volume to give a (coherent) cross-section which is an intensive (material) property, independent of the sample dimensions. This is based on the assumption that neutrons are scattered only once before being detected and this has been shown to be a reasonable approximation for coherent SANS from polymeric [77] and other materials [76], with cross-sections $\mathrm{d}\Sigma/\mathrm{d}\Omega(0)$ typically $<10^3\,\mathrm{cm}^{-1}$, which includes most biological materials. For samples with higher cross-sections that exhibit substantial coherent–coherent multiple scattering, a common way to recognize and minimize this artifact is to measure the cross-section as a function of the sample thickness and to extrapolate to $t = 0$.

For incoherent scattering, 1–2 mm samples containing hydrogen (H_2O, protonated polymers, etc.) give rise to appreciable multiple scattering [22,23]. The difficulties in estimating an incoherent background to subtract from a given "sample" and thus isolate the residual coherent cross-section are illustrated in [77] where the apparent cross-section produced of protonated PMMA-H blanks, after normalizing via Eq. 8.6 was shown to vary by >50% over a typical range (~0.2–1.2 mm) of sample thicknesses. Similarly, the scattering of light water contains appreciable multiple scattering [22, 23, 55, 56, 58–61], which is not proportional to the thickness or transmission, and cannot be normalized to a true cross-section which is independent of the sample dimensions. Moreover, as explained above, the bound-atom cross-section cannot be used to calculate the background, because the hydrogen incoherent cross-section ($\sigma_{\text{inc}} = 79.7 \times 10^{-24}\,\text{cm}^2$), although widely quoted in the literature, almost never applies to real biological or aqueous-based systems. However, the incoherent scattering is independent of Q to a good approximation, and empirical methods have been developed to subtract this background [24].

8.3.5 Data Analysis: Extracting Structural and Shape Parameters from SANS Data and $P(r)$ Analysis

Several comprehensive reviews and books have been published describing the application of SAS to biological systems [66, 78, 79] and current examples are given in this volume (e.g., see contribution by S. Krueger et al.). In combination with advances in molecular biology techniques that facilitate production of large amounts of pure protein using bacterial expression systems, substantial improvements in neutron sources and instrumentation [35–37] have broadened the impact of SAS in modern structural molecular biology. The absolute cross-section is proportional to the scattered intensity Eq. 8.6 and an initial analysis of the data may be performed to determine the scattering molecule's R_g, along with values of the forward cross-section $\text{d}\Sigma/\text{d}\Omega(0)$, in a model-independent way [51] via the Guinier approximation:

$$\text{d}\Sigma/\text{d}\Omega(Q) = \text{d}\Sigma/\text{d}\Omega(0)\text{e}^{-Q^2R_\text{g}^2/3}. \tag{8.8}$$

A plot of $\ln \text{d}\Sigma/\text{d}\Omega(Q)$ versus Q^2 gives a straight line with a slope of $-R_\text{g}^2/3$ and an extrapolated intercept $\ln \text{d}\Sigma/\text{d}\Omega(Q)$ in the region where $QR_\text{g} \leq 1$ (the precise upper limit of QR_g for which the Guinier approximation is valid is dependent on particle shape).

For a dilute solution of monodisperse, identical particles the scattered intensity $I(Q)$ (which is proportional to the absolute cross-section $\text{d}\Sigma/\text{d}\Omega)(Q)$ is related to the distribution of interatomic distances $P(r)$ in the scattering particle by a Fourier transformation:

$$I(Q) = 4\pi \int P(r)[\sin(Qr)/Qr]\text{d}r. \tag{8.9}$$

Eq. 8.9 assumes that the electron density of the particle is homogeneous which means that $P(r)$ is a continuous function of r. The $\sin(Qr)/Qr$ term comes from a spatial average of all particle orientations and assumes that they are random. The inverse relationship of Eq. 8.9:

$$P(r) = \frac{r^2}{2\pi^2} \int I(Q)Q^2[\sin(Qr)/Qr]\text{d}Q, \tag{8.10}$$

can be used to derive the $P(r)$ function from the experimental scattering profile. Several algorithms exist for calculating the $P(r)$ from the scattering cross-section or intensity [63,64,80], that have also been used to model instrumental resolution effects (see above).

$P(r)$ gives a real-space representation of the structure and thus, is more readily interpreted in terms of structural information than the scattering profile, $\mathrm{d}\Sigma/\mathrm{d}\Omega(Q)$. $P(r)$ is sensitive to the overall shape of the scattering particle and to the relationships between domains or repeating structures. Several specific pieces of structural information can be extracted from the $P(r)$ analysis: (i) The $P(r)$ goes to zero at a value corresponding to the maximum dimension of the particle, D_{max}. (ii) The zeroth moment of $P(r)$ gives the forward or zero-angle scattering, $I(0)$, which as mentioned earlier, is proportional to the square of the molecular weight of the scattering particle. $I(0)$ is therefore a very sensitive test for monodispersity in a protein solution of known concentration. Alternatively, it can be a sensitive indicator of macromolecular association and polymerization. Additionally, (iii) a value for R_{g} can be calculated using the entire angular range of the scattering profile by calculating the second moment of the $P(r)$ distribution Eq. 8.11.

$$R_{\mathrm{g}}^2 = \int P(r) r^2 \mathrm{d}^3 \frac{r}{2} \int P(r) \mathrm{d}^3 r. \tag{8.11}$$

Shape information in the form of electron (or nuclear) density distribution is contained within all SAS data. Extraction of that shape information has become more and more sophisticated in the past decade and one should be familiar with the various approaches and limitations so as to avoid the penchant of over-interpreting their data. A significant limitation of this approach to the interpretation of solution scattering data arises from the fact that the molecules are randomly oriented and hence there is an inherent spherical averaging. Three-dimensional data is being extracted from a one-dimensional data set. Nonetheless, given enough constraints, for example from a complete set of contrast data and/or additional structural information from complementary biophysical techniques, the resultant shape models can be highly informative.

One common approach to extracting shape information has been to begin with a general shape assumption, usually based on known shape information for the system under study. For example, most enzyme structures are globular and to a good approximation uniformly packed with atoms so a reasonable shape assumption would be an ellipsoid of uniform density. To model the neutron scattering data collected on this enzyme, one would start with an ellipsoidal structure randomly filled with points of uniform neutron scattering length density. When more than one geometric shape is used to build a model structure, each can contain points of a different, uniform SLD. After allowing the geometric parameters of the structure to vary and calculating a distance distribution for each new set of parameters, one then searches for the parameters that result in a model that best fits the experimental scattering data. There are several programs that have used this approach, each newer version of which continues to build in sophistication and degree of user friendliness [81–84]. In general each of these programs begin by generating scattering points, via a MC method, to fall within a given volume (e.g., sphere, ellipsoid, cylinder, etc.). To simulate a uniform SLD within the given sub volume, the total number of points is proportional to that volume.

Other types of shape restoration from SAS data hold the promise of providing more detailed structural information than the geometric modeling approach described above because there are no assumptions about the basic shape. A number of shape restoration methods have become available in recent times; e.g., using spherical harmonics [85–92] or aggregates of spheres [93–97]. Many current methods use a larger number of degrees of freedom to reconstruct the shape of a scattering object in more detail. Again, though, one must bear in mind that the problem of shape restoration for solution scattering data is particularly complex because the rotationally isotropic nature of the samples results in a one-dimensional (1D) scattering intensity profile. For this reason, the uniqueness of a three-dimensional (3D) structure associated with a 1D scattering profile cannot be guaranteed and multiple shapes that fit the data equally well can result from shape restoration methods. One shape restoration approach that addresses the issue of multiple solutions is GA_STRUCT [98]. The method for calculation of SAS intensity differs from the Debye formula for calculating the SAS intensity of a collection of non-overlapping spheres [99] because the spheres used by GA_STRUCT are allowed to overlap, there by eliminating the internal gaps in the particle volume and providing a truly uniform interior density. The scattered intensity profile is calculated using an MC approach implemented previously [100] that first calculates $P(r)$. Then, $I(Q)$ is calculated by the Fourier transform defined in Eq. 8.9. Several independent runs of the minimization process are automatically performed to generate a family of structures. This family is then characterized for similarity, and a consensus envelope is produced from the set of structures that represents the most common structural features of the family. GA_STRUCT characterizes the reproducibility of the shape restoration and provides an "average" shape, called the consensus envelope. The consensus envelope is not necessarily the "best fit" model to the scattering data, it simply represents those features most frequently emerging in the population of best fit models. An evaluation of how well the consensus envelope represents this family is made by reviewing the individual members of the family.

The following section describes a series of neutron scattering experiments that were performed, over the past decade, on a biological complex between the protein calmodulin and the skeletal muscle isoform of myosin light chain kinase (or, in some cases, a smaller peptide representing the portion of the kinase that contains a CaM-binding sequence). The structural information that was acquired from small angle scattering data has revealed new insights and understanding on the calcium-dependent regulation of muscle contraction. An underlying theme behind these pioneering experiments is that there is a continuous enhancement and confidence in the interpretation of the scattering data as the analysis and modeling methodologies improved.

Additionally, instrumentation improvements and the cold-source upgrade at neutron facilities were absolutely essential to their success.

8.4 SANS Application: Investigating Conformational Changes of Myosin Light Chain Kinase

8.4.1 Solvent Matching of a Specifically Deuterated CaM Bound to a Short Peptide sequence

Calmodulin is the major intracellular receptor for Ca^{2+}, and is responsible for the Ca^{2+}-dependent regulation of a wide variety of cellular processes via interactions with a diverse array of target enzymes including a number of kinases. The Ca^{2+}/calmodulin (CaM)-dependent activation of myosin light chain kinase (MLCK) is a model for CaM-kinase interactions that has been investigated extensively. All isoforms of MLCK include a conserved catalytic core homologous to that of other protein kinases, followed immediately by a carboxyl-terminal regulatory segment consisting of both an auto-inhibitory sequence and a CaM-binding sequence [101]. In its inhibited conformation, the regulatory segment of MLCK maintains numerous contacts with the catalytic core, thus preventing substrate binding and its subsequent phosphorylation [102–104]. CaM has an unusual dumbbell-shaped structure with two globular lobes connected by an extended helix, each having two Ca^{2+}-binding, aka. "EF hand" motifs [105]. A ribbon representation of the peptide backbone structure of $4Ca^{2+}$-CaM from its crystal structure is shown in Fig. 8.6a. When Ca^{2+} binds to calmodulin, hydrophobic clefts on each globular lobe that are important in target enzyme recognition and binding are exposed (reviewed in [106, 107]).

Small-angle X-ray and neutron scattering [14] were the first experiments to demonstrate that CaM undergoes a dramatic conformational collapse upon binding a 25 amino acid peptide with a sequence homologous to the CaM-binding region from MLCK. Figure 8.6c shows the $P(r)$ that was determined from a SANS "solvent-matching" experiment on perdeuterated CaM and a nondeuterated MLCK-I peptide in a buffer containing 37% D_2O. Deuterated calmodulin has a neutron SLD that is greater than that of 100% D_2O, while the nondeuterated peptide has a SLD approximately equal to that of the buffer in 37% D_2O. Thus, in the 37% D_2O buffer, deuterated CaM is strongly contrasted against the solvent but nondeuterated MLCK-I peptide has the same mean SLD as the solvent and hence does not contribute to the scattering. The maximum linear dimension of CaM in the complex (where the $P(r)$ goes to zero) is approximately 50 Å versus 68 Å for CaM without the peptide present, which could only be achieved if the two globular domains of CaM come into close contact. This observed collapse of CaM was proposed to be achieved via flexibility in the interconnecting helix region that

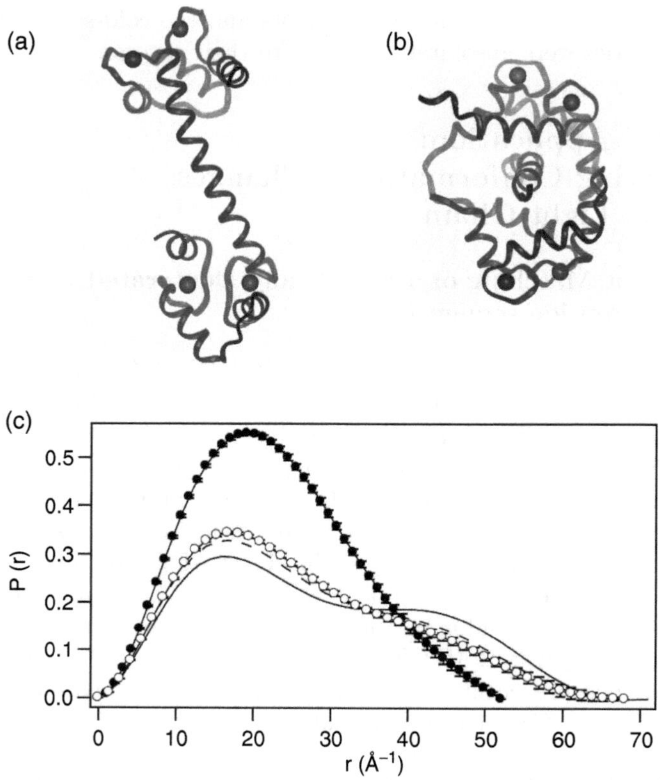

Fig. 8.6. (**a**) Ribbon representation of the backbone structure of CaM in the crystal structure [16] and (**b**) in its complex with the peptide MLCK-I derived from the NMR data [15]. (**c**) $P(r)$ functions, each scaled to the square of the molecular weight, calculated from the crystal structure of $4Ca^{2+}$/CaM (*solid line*) and measured using solution scattering from CaM (*dashed line*), $4Ca^{2+}$/CaM (*open circles*), and the solvent-matched $4Ca^{2+}$/CaM from the neutron scattering experiment on perdeuterated CaM bound to the MLCK-I peptide (*closed circles*)

allows the two lobes of the dumbbell-shaped CaM to come into close contact, encompassing the peptide as the hydrophobic clefts in the globular lobes of CaM interact with hydrophobic residues in the helical target peptide. Later, this collapse was confirmed and further detailed by higher resolution studies using NMR [108] (see Fig. 8.6b) and X-ray crystallography [16] on complexes of CaM with isolated peptides based on CaM-binding sequences from smooth and skeletal muscle MLCKs.

8.4.2 Contrast Variation of Deuterated CaM Bound to MLCK enzyme

It has been proposed that the regulatory segment of MLCK, which includes both autoinhibitory and CaM-recognition sequences, folds back on the catalytic core to inhibit kinase activity [104]. This idea is consistent with the crystal structure of the autoinhibited form of CaM-dependent protein kinase I [109]. In addition, selected-site mutagenesis studies collectively show that the autoinhibitory sequence of MLCK forms an extensive network of contacts with the surface of the catalytic core [102, 103, 110]. The effect of CaM-binding to MLCK had been proposed to involve release of autoinhibition of the kinase via some sort of movement of the autoinhibitory sequence [111, 112]. Neutron scattering studies with contrast variation provided the first direct structural evidence in support of the autoinhibitory hypothesis for MLCK activation [113].

While there has been an abundance of structural data on calmodulin–peptide complexes, until the neutron scattering contrast variation studies mentioned herein, there was very little structural data on CaM complexed with a functional enzyme. Specifically, in the case of the CaM–MLCK interactions, this situation led to speculation about whether the MLCK C-terminal regulatory region could be released from its interactions with the surface of the catalytic core such that the CaM-binding sequence would be sterically unrestricted and able to form the tight interaction with the conformationally collapsed CaM as was observed for the CaM–peptide structures. SANS contrast variation experiment on the complex formed between deuterium-labeled CaM bound to a catalytically active MLCK revealed the surprising answer. The basic scattering functions for the individual components of each complex were extracted from the contrast series yielding the R_g and $P(r)$ distributions for the CaM and MLCK components as well as the distances between the centers of mass of the two components in each complex. The results showed that indeed CaM undergoes an unhindered conformational collapse upon binding MLCK that is very similar to that observed with the isolated CaM-binding peptides. An MC integration modeling procedure, BIOMOD [114], was used to systematically test against the scattering data all possible two-ellipsoid uniform-density models for the complex within the set constrained by the known structural parameters. Figure 8.7 (left) shows the resultant two-ellipsoid model of the scattering data that led to an autoinhibitory hypothesis for MLCK activation. It was clear from the model that CaM binding to the enzyme must induce a significant movement of the kinase's CaM-binding and autoinhibitory sequences away from the surface of the catalytic core. Major factors that were critical to the success of this contrast variation experiments include: (i) working at low concentrations ($\sim 1 - 2\,\mathrm{mg\,ml}$) to avoid time-dependent aggregation of the complex, (ii) collecting a complete contrast series to extract basic scattering functions as the concentration of the CaM component was so low that the 40% D_2O solvent-matched contrast was very weak, and thus, (iii) the higher intensity of the neutron beam as a result of the cold source upgrade at NIST [37].

Neutron contrast series data were collected for deuterated CaM bound to MLCK in the presence of substrates (a nonhydrolyzable analog of adenosine triphosphate, AMPPNP, and a peptide substrate that includes the

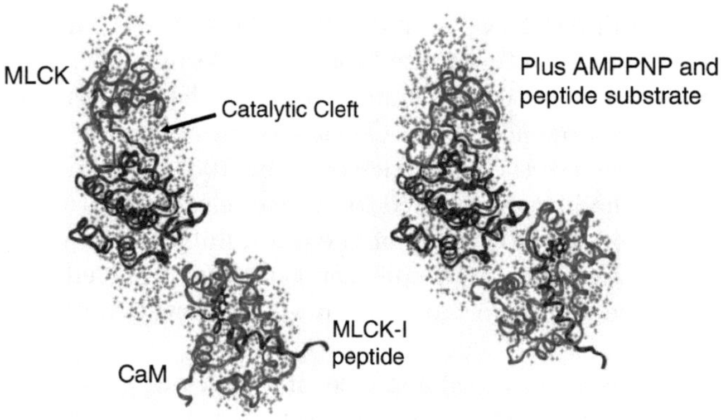

Fig. 8.7. Two ellipsoid models derived from the neutron scattering data for the $4Ca^{2+}$–CaM–MLCK complexes with (right, [115]) and without (left, [113]) substrates. The conserved portion of the inhibited kinase catalytic core [117] and the NMR structure of CaM complexed with MLCK-I peptide [15] are fit within the dimensions of the larger and the smaller ellipsoids, respectively. The upper and lower lobes of the catalytic core, with the catalytic cleft (labeled) between them, are represented as gray and black ribbon drawings. CaM is represented as a gray ribbon drawing, with its bound MLCK-I peptide in black, and a CPK representation of its hydrophobic Trp residue near the N-terminal end. This Trp residue is key to recognition and binding by the C-terminal CaM domain. This figure is adapted from [115]

phosphorylation sequence for myosin regulatory light chain) [115]. Comparison of the R_g values determined for the complex with bound substrates (31.6 ± 1.2 Å) to that without substrates present (34 ± 0.7 Å) indicated that there are significant structural differences. As would be expected, R_g and $P(r)$ analysis of the basic scattering functions determined for the CaM component was similar for the two complexes (18.1 ± 1.5 vs. 17.3 ± 0.4 Å). The R_g determined from the basic scattering function for the MLCK component decreases by almost 3 Å upon binding substrates indicating a compaction of MLCK that is also reflected in the $P(r)$ analysis. This observed compaction of MLCK upon substrate binding is similar to that arising from the closure of the catalytic cleft in cAMP-dependent protein kinase upon binding pseudosubstrate. In addition, the distances between the centers-of-mass of the two components in each complex were determined from the basic scattering function cross-terms to be 57 ± 9 and 49 ± 10 Å.

A newer version of the MC integration modelling procedure, SASMODEL [82], was used to systematically test uniform-density two-ellipsoid models for the CaM–MLCK complexes against all of the scattering data. This two-ellipsoid modelling exercise was ideal for this system because there was good evidence that both CaM and MLCK had compact globular structures. Figure 8.7 shows how the known high-resolution structure of CaM complexed with the 20 residue MLCK-I helical peptide [116] and the conserved catalytic core of the kinase (based on the cAMP-dependent protein kinase structure [117]) fit within the ellipsoid shapes derived from the scattering data. The empty spaces in the ellipsoid representing MLCK most likely are occupied by N-terminal and C-terminal sequence segments whose structures have not yet been determined. The center-of-mass separation between the two ellipsoids in the models are 57 Å for the minus substrate complex and 45 Å for the plus substrates complex. These values are consistent with those values determined from analysis of the basic scattering function of the cross-terms.

The models show that CaM binds to the kinase such that there must be a significant movement of the CaM-binding and autoinhibitory sequences away form the surface of the catalytic core. Upon binding substrates there is a movement of CaM approximately 12 Å closer to the catalytic cleft. Additionally, the models suggests that there is a reorientation of CaM with respect to the kinase that results in interactions between the N-terminal sequence of CaM and the kinase that were not observed in the complex without substrates.

The neutron scattering and contrast variation data presented a structural view which follows sequentially the conformational transitions in the CaM-dependent activation of MLCK in solution. These studies provided important structural data that defined the mechanistic steps in the release of autoinhibition of MLCK by CaM, as well as subsequent substrate binding and activation.

8.4.3 Mechanism of the CaM-Activation Step: SAXS/SANS Studies of a (Deuterated) Mutant CAM

Neutron scattering and contrast variation experiments on the CaM-dependent activation of MLCK in solution have elucidated the sequential conformational transitions involved in the CaM-dependent activation mechanism of MLCK. Structural data determined from solution scattering experiments in combination with high resolution structural data on the individual components have defined the mechanistic steps responsible for the release of autoinhibition of MLCK by CaM, as well as subsequent substrate binding and activation. Additionally, a $2Ca^{2+}$ intermediate had been proposed based on spectroscopic studies [118,119] and this was further supported by small-angle X-ray scattering [120]. The purpose of such an intermediate could be to restrain the CaM from diffusing away in rapidly cycling functions such as muscle contraction and relaxation. Since the Ca^{2+} affinities of CaM are strongly affected by its different target binding sequences, it has been further suggested that CaM-binding sequences in different enzymes may serve the purpose of "tuning" the calcium affinities of the Ca^{2+}-binding sites so as to optimize for the formation of such intermediates when needed. CaM's N-terminal lobe would possess the regulatory function, alternately binding and releasing the autoinhibitory sequence of MLCK in response to the Ca^{2+} signal.

The modeling program GA_STRUC was used to generate low-resolution models for three complexes; $2Ca^{2+}$–CaM/MLCK, $4Ca^{2+}$–CaM/MLCK, and

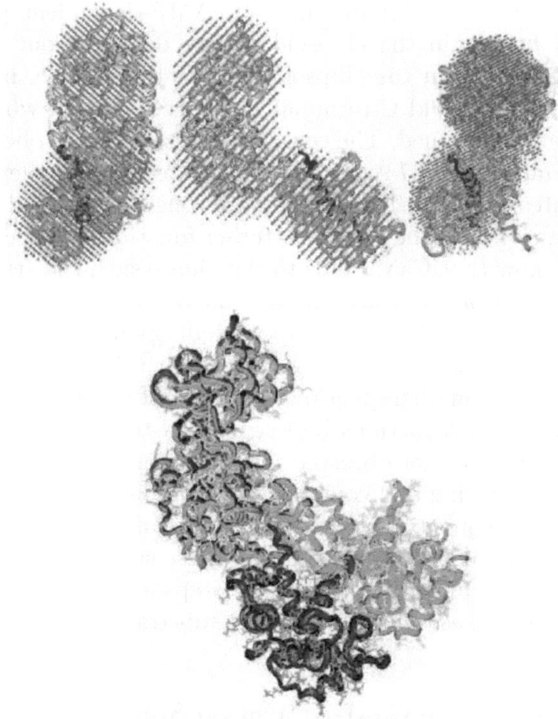

Fig. 8.8. Results of shape restoration for $2Ca^{2+}$–CaM/MLCK SAXS data [115]. Three orthogonal views of consensus envelope from GA_STRUCT (*top*). The conserved catalytic core for protein kinases in the open cleft conformation [117, 121] and collapsed CaM (2BBK [108]) structures have been docked by hand into the consensus envelope. The high-resolution structures resulting from the docking of the extended cPKA structure and collapsed CaM structure (2BBK) are shown in two views (the $2Ca^{2+}$–CaM/MLCK (grey) and $4Ca^{2+}$–CaM/MLCK (black) (*bottom*). The two complexes are overlaid such that the cPKA structures are coincident. The resulting view shows how far the CaM translocates away from the catalytic cleft of skMLCK when all four Ca^{2+} binding sites are occupied

$4Ca^{2+}$–CaM/MLCK with bound substrate. These models were used in conjunction with high-resolution structures of the protein components to better understand the interactions between them (Fig. 8.8 [98]). In the case of the $2Ca^{2+}$–CaM/MLCK, the consensus envelope is consistent with CaM in a fully collapsed state with its two globular lobes in close contact with each other while the catalytic cleft of the kinase is open. The consensus envelope for the $4Ca^{2+}$–CaM/MLCK indicates that the collapsed CaM has swung further away from the open catalytic cleft of the MLCK compared to the $2Ca^{2+}$ complex, and further that substrate binding to this complex results in closure of the kinase catalytic cleft, in agreement with previous neutron scattering results. Most importantly, the GA_STRUCT models indicate that activation of MLCK by CaM can only occur once CaM is fully translocated away from the catalytic cleft, which is presumably linked to full release of the pseudo-substrate/inhibitory sequence and this step is completed only when all four calcium binding sites are loaded.

All amino acid residues in CaM make up the two globular Ca^{2+}-binding domains with the exception of residues 76–81 found in the central helix and residues 1–8 ($A^1DQLTEEQ^8$) at the N-terminus, hereinafter referred to as the N-terminal leader sequence. It has been shown that deletion of the N-terminal leader sequence results in a CaM mutant (DNCaM) capable of recognizing and binding to MLCK yet incapable of activating this kinase [122].

SANS contrast variation data on a specifically deuterated DNCaM mutant bound to MLCK has been collected. It is anticipated that analysis of this data will provide a more detailed atomic description of the binding events between CaM and MLCK prior to the kinase activation step, providing a molecular description of the regulatory mechanism for an archetypal calmodulin-mediated Ca^{2+} response in the cell. A high-resolution model of the DNCaM–MLCK complex has to be built from available atomic-resolution structures of CaM and the catalytic core of MLCK within the confines of the molecular envelope shapes, boundaries and relative dispositions. A docking procedure can then be used to develop "best fit" models of the complex similar to the procedure report recently by Tung et al., [123] that was used to develop a structural model of the catalytic subunit-regulatory subunit dimeric complex of the cAMP-dependent protein kinase (Fig. 8.9). The detailed "atomic" modeling presented in this article is an example of how the shape constraints provided by SANS can be combined with high-resolution crystal or NMR structure information and further constrained by other biophysical measurements. Higher detail models built from high-resolution crystal and NMR data of various substructures within the molecular dimensions determined from SAS and constrained by interresidue distances determined, for example, from chemical cross-linking and peptide mapping or FRET distances or mutagenesis data, is poised to become an important tool for an integrated structural approach to visualizing the protein:protein interactions that are essential for intracellular function. Approaches to modeling protein complexes with hybrid experimental data will become increasingly important as crystallographers continue to rapidly grow the structural data base with domain and subunit structures and we begin to turn our attention to understanding how these substructures function in the complex interactions within the cell.

8.5 Conclusions and Outlook

Neutron scattering with contrast variation can provide unique views of the interactions within molecular complexes involved in dynamic processes such as enzyme activation as well as in the highly regulated and coordinated interactions of complex systems such as muscle. The ability to selectively label components in an assembly and extract information about their conformations within that assembly can be quite powerful. SAS techniques are applied in solution and hence can give insights into systems in which inherent flexibility may cause problems for crystallization. Importantly, the techniques can be applied to systems over a wide range of sizes, from 10 to 1000s of Å. Neutron scattering does suffer the limitation that it requires access to large, expensive facilities of which there are a limited number. Thus, the techniques should only be applied when they can contribute unique information on an important problem. Understanding the molecular mechanisms underlying motility in biological systems and its regulation is one such problem. The complexity and the dynamic nature of motile function make it a very compelling system for study using neutrons.

As mentioned in the contribution by Harroun et al. in this volume, biology can be an educational outreach tool, that can connect with the public and policy makers in ways that many physics experiments cannot, particularly if they have some relevance to advances in medicine. This has had the effect that new instruments devoted to biological sciences such as the dedicated biological Advanced Neutron Diffractometer/Reflectometer (AND/R) at NIST are coming on line. In addition, a new 35 m SANS facility at ORNL [42] is being constructed as part of a Center for Structural and Molecular Biology (CSMB).

Finally, it may be worth re-emphasizing a point made initially in the context of SANS studies of synthetic polymers [124]: "The greatest limitation for SANS experimentalists is the securing of suitable samples. To take full advan-

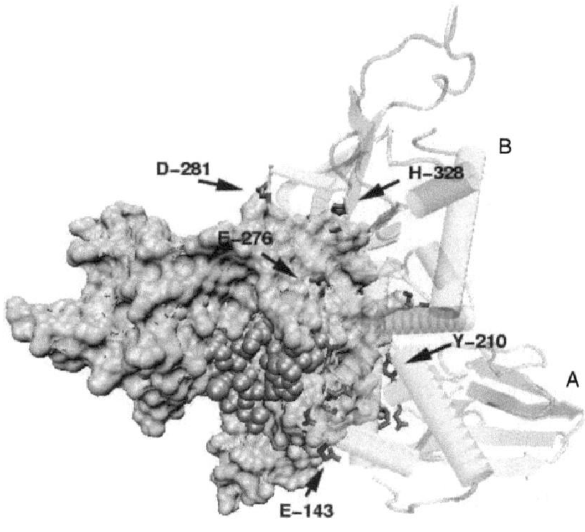

Fig. 8.9. Best fit model for the R–C heterodimer of the cAMP-dependent protein kinase shows the dimer poised for dissociation. C (catalytic) subunit is shown as surface representation and the R (regulatory) subunit is shown as a cartoon representation of the backbone structure with some residues identified to be at interface, in ball-and-stick, labeled

tage of the power of SANS, samples should be selectively deuterated in designated places." Similarly, deuteration, partial or full, of biological molecules such as proteins, nucleic acids, lipids, sugars, is essential to exploit fully the techniques of neutron scattering and to highlight and analyze selected parts of macromolecular structures in situ. The commitment of a small fraction of the planned large investments in instrumentation to an in vivo labeling program will dramatically increase the overall impact and productivity of future research on biopolymers. The ILL in collaboration with European Molecular Biology Laboratory, has established a laboratory for the deuteration of biological molecules [125]. Similarly, as part of its strategy for the expansion of neutron scattering in the life sciences, the CSMB is planning a Deuterium Labeling Facility at ORNL. The provision of deuterated macromolecules will greatly enhance both the quality and quantity of experiments that can be done using neutron scattering, and in many cases will make feasible new and more sophisticated experiments than can presently be performed.

Acknowledgments

The work at Oak Ridge was supported by the Division of Materials Science, U.S. Department of Energy under contract DE-AC05-00OR22725 with the Oak Ridge National Laboratory, managed by UT-Battelle, LLC. The work at University of North Carolina Charlotte was supported by the National Science Foundation CAREER award MCB-0237676. The authors would like to thank D.M. Engelman (Yale University) who provided Fig. 8.3, C.-S. Tung who provided Fig. 8.9 and G. Zaccai for helpful advice in understanding the factors, which affect the partial specific volumes of biological molecules. They also wish to acknowledge their many co-workers for permission to include data from their joint publications, particularly J.M. O'Reilly, V. Ramakrishnan, J. Trewhella, and W.T. Heller.

References

1. J. Chadwick, Nature **129**, 312 (1932)
2. T.E. Mason, A.D. Taylor, Mat. Res. Soc. Bull. **24**, 14 (1999)
3. P. Lindner, T. Zemb, *Neutron, X-ray and Light Scattering* (Elsevier Publishers, New York, 1991)
4. C.E. Williams et al., J Polym. Sci. Pt. C **17**, 379 (1979)
5. A.Z. Akcasu et al., J. Polym. Sci. [B] **18**, 863 (1980)
6. J.S. King et al., Macromolecules **18**, 709 (1985)
7. J.S. Higgins, R.S. Stein, J. Appl. Cryst. **11**, 346 (1978)
8. D.E. Groom, et al., (Particle Data Group), Eur. Phys. J. [C] **15**, 1 (2000) and 2001 partial update for edition 2002 (URL, http,//pdg.lbl.gov)
9. G.D. Wignall, in *Encyclopedia of Polymer Science and Engineering*, vol. 10, 2nd edn. (Wiley, New York, 1987), pp. 112–184

10. W. Schmatz, T. Springer, J. Schelten, K. Ibel, J. Appl. Cryst. **7**, 96 (1974)
11. J.S. Higgins, H. Benoit, *Polymers and Neutron Scattering* (Clarendon Press, 1994)
12. R.M. Murphy, Curr. Opin. Biotechnol. **8**, 25 (1997)
13. A.L. Papish, L.W. Tari, H.J. Vogel, Biophys. J. **83**, 1455 (2002)
14. D.B. Heidorn et al., Biochemistry **28**, 6757 (1989)
15. M. Ikura, G. Barbato, C.B. Klee, A. Bax, Cell Calcium **13**, 391 (1992)
16. W.E. Meador, A.R. Means, F.A. Quiocho, Science **257**, 1251 (1992)
17. J. Trewhella, J.K. Krueger, in *Methods of Molecular Biology*; vol. 173, H.J. Vogel (eds.) (Humana Press, 2001), pp. 137–160
18. J. Trewhella et al., Sci. Prog. **81**, (1998)
19. J, Trewhella, Curr. Opin. Struct. Biol. **7**, 702 (1997)
20. A. Maconnachie, Polymer **25**, 1068 (1984)
21. L.D. Coyne, W.L. Wu, Polymer Commun. **30**, 312 (1989)
22. G.D. Wignall, F.S. Bates, J. Appl. Cryst. **20**, 28 (1987)
23. G.D. Wignall, in *Physical Properties of Polymers*, 3rd ed., ed. by J.E. Mark (Cambridge University Press, 2004), pp. 424–511
24. W.S. Dubner, J.M. Schultz, G.D. Wignall, J. Appl. Cryst. **23**, 469 (1990)
25. D. Svergun et al., Proc. Natl. Acad. Sci. (USA) **95**, 2267 (1998)
26. G.D. Wignall, in *Polymer Properties Handbook*, ed. by J.E. Mark (Cambridge University Press, 1996), pp. 299–310
27. K. Ibel, H.B. Stuhrmann, J. Mol. Biol. **93**, 255 (1975)
28. P.B. Moore, J. Appl. Cryst. **14**, 237 (1981)
29. G. Zaccai, B. Jacrot, Ann. Rev. Biophys. Bioeng. **12**, 139 (1983)
30. D.L. Worcester, J. Appl. Cryst. **21**, 669 (1988)
31. H.B. Stuhrmann, Zeit. fur Krist. **178**, 208 (1987)
32. J. Schelten, Kerntechnik **14**, 86 (1972)
33. J. Schelten, in *Scattering Techniques Applied to Supramolecular and Nonequilibrium Systems*, vol. **73**, ed. by S.H. Chen, B. Chu, R. Nossal (Plenum Press, 1981), pp. 75–85
34. K. Ibel, J. Appl. Cryst. **9**, 296 (1976)
35. Neutronenstreuexperimente am FRJ2 in Jülich (English and German texts are available from the Forschungszentrum, Jülich, 1997)
36. P. Lindner, R.P. May, P.A. Timmins, Physica B **180**, 967 (1992)
37. C.J. Glinka et al., J. Appl. Cryst. **31**, 430 (1998)
38. W.C. Koehler, Physica B and C **137**, 320 (1986)
39. R.K. Abele, G.W. Allin, W.T. Clay, C.E. Fowler, M.K. Kopp, IEEE Transact. Nuc. Sci. **28**, 811 (1981)
40. R.E. Ghosh, A.R. Rennie, J. Appl. Cryst. **32**, 1157 (1999)
41. G.W. Lynn et al., J. Appl. Cryst. **36**, 829 (2003)
42. J.B. Hayter, H. Mook, J. Appl. Cryst. **22**, 35 (1989)
43. http://www.oecd.org/dsti/sti/s_t/ms/prod/scattering.htm: *A Twenty Years Look Forward at Neutron Scattering Facilities* (1998)
44. G.D. Wignall et al., in *Scattering Methods for the Investigation of Polymers*, ed. by J. Kahovec (Wiley-VCH, Weinheim, 2002), pp. 185–200
45. http://www.sns.gov/
46. http://j-parc.jp/
47. http://www.isis.rl.ac.uk/targetstation2/
48. http://www.ess-europe.de/ess_js/index.html

49. P. Thiyagarajan et al., J. Appl. Cryst. **30**, 280 (1997)
50. http://www.ill.fr
51. A. Guinier, G. Fournet, *Small-Angle Scattering of X-rays* (John Wiley, New York, 1955)
52. L.J. Magid, Colloids and Surfaces **19**, 129 (1986)
53. J.B. Hayter, J. Penfold, Coll. Pol. Sci. **261**, 1022 (1983)
54. V.F. Turchin, in *Slow Neutrons* (Sivan Press, Jerusalem, 1965), p. 16
55. B. Jacrot, G. Zaccai, Biopolymers **20**, 2413 (1981)
56. P. Lindner, J. Appl. Cryst. **33**, 807 (2000)
57. W.R. Krigbaum, F.R. Kugler, Biochemistry **9**, 1216 (1970)
58. P. Lindner, F. Leclercq, P. Damay, Physica B **291**, 152 (2000)
59. R.P. May, K. Ibel, J. Haas, J. Appl. Cryst. **15**, 15 (1982)
60. J.R.D. Copley, J. Appl. Cryst. **21**, 639 (1988)
61. W. Boyer, J.S. King, J. Appl. Cryst. **21**, 818 (1988)
62. G.D. Wignall, D.K. Christen, V. Ramakrishnan, J. Appl. Cryst. **21**, 438 (1988)
63. P.B. Moore, J. Appl. Cryst. **13**, 168 (1980)
64. O. Glatter, J. Appl. Cryst. **10**, 415 (1977)
65. V. Ramakrishnan, J. Appl. Cryst. **18**, 42 (1985)
66. O. Glatter, O. Kratky, *Small-Angle X-ray Scattering* (Academic Press, New York, 1982)
67. G.D. Wignall, J. Appl. Cryst. **24**, 479 (1991)
68. P.W. Schmidt, J. Appl. Cryst. **3**, 257 (1970)
69. C.R. Wobbe, S. Mitra, V. Ramakrishnan, Biochemistry **23**, 6565 (1984)
70. G.D. Wignall et al., J. Mol. Cryst. Liqu. Cryst. **180A**, 25 (1990)
71. L. Fisher et al., J. Coll. Interface Sci. **123**, 24 (1988)
72. L. Rayleigh, Proc. R. Soc. London Ser. A **84**, 24 (1911)
73. O. Glatter, personal communication
74. J.S. Pederson, D. Posselt, K. Mortensen, J. Appl. Cryst. **23**, 321 (1990)
75. P.S. Goyal, J.S. King, G.C. Summerfield, Polymer **24**, 131 (1983)
76. J. Schelten, W. Schmatz, J. Appl. Cryst. **13**, 385 (1980)
77. J.M. O'Reilly, D.M. Teegarden, G.D. Wignall, Macromolecules **18**, 2747 (1985)
78. P.B. Moore, Methods Exper Phys **2**, 337 (1982)
79. L.A. Feigin, D.I. Svergun, *Structure Analysis by Small-Angle X-ray Scattering* (Plenum Press, New York and London, 1987)
80. D.I. Svergun, J. Appl. Cryst. **26**, 258 (1993)
81. G.A. Olah, J. Trewhella, Biophys. J. **66**, A311 (1994)
82. J.K. Zhao et al., J. Biol. Chem. **273**, 30448 (1998)
83. http://www.sans.chem.umbc.edu/
84. http://www.embl-hamburg.de/ExternalInfo/Research/Sax/
85. F. Spinozzi, F. Carsughi, P. Mariani, J. Chem. Phys. **109**, 10148 (1998)
86. H.B. Stuhrmann, Acta Cryst. **A26**, 297 (1970)
87. J.G. Grossmann et al., Biochemistry **32**, 7360 (1993)
88. D.I. Svergun, M.H.J. Koch, I.N. Serdyuk, J. Mol. Biol. **240**, 66 (1994)
89. D.I. Svergun et al., Proc. Natl Acad. Sci. USA **91**, 11826 (1994)
90. D.I. Svergun et al., Acta Cryst. **A 52**, 419 (1996)
91. D.I. Svergun et al., J. Appl. Cryst. **30**, 798 (1997)
92. D.I. Svergun, H. B. Stuhrmann, Acta Cryst. **A 47**, 736 (1991)
93. P. Chacon et al., Biophys. J. **74**, 2760 (1998)
94. P. Chacon et al., J. Mol. Biol. **299**, 1289 (2000)

95. D.I. Svergun, Biophys. J. **76**, 2879 (1999)
96. D.I. Svergun, M.V. Petoukhov, M.H.J. Koch, Biophys. J. **80**, 2946 (2001)
97. D. Walther, F.E. Cohen, S. Doniach, J. Appl. Cryst. **33**, 350 (2000)
98. W.T. Heller, J.K. Krueger, J. Trewhella, Biochemistry **42**, 10579 (2003)
99. P. Debye, Ann. Phys. **46**, 809 (1915)
100. D.B. Heidorn, J. Trewhella, Biochemistry **27**, 909 (1988)
101. B.E. Kemp et al., Trends Biochem. Sci. **19**, 440 (1994)
102. J.K. Krueger, R.C. Padre, J.T. Stull, J. Biol. Chem. **270**, 16848 (1995)
103. P.J. Gallagher et al., J. Biol. Chem. **268**, 26578 (1993)
104. B.E. Kemp, R.B. Pearson, Biochim. Biophys. Acta **1094**, 67 (1991)
105. Y.S. Babu, C.E. Bugg, W.J. Cook, J. Mol. Biol. **204**, 191 (1988)
106. G.M. Clore et al., Curr. Opin. Struct. Biol. **3**, 838 (1993)
107. M. Ikura, Trends Biochem. Sci. **21**, 14 (1996)
108. M. Ikura et al., Science **256**, 632 (1992)
109. J.Goldberg, A.C. Nairn, J. Kuriyan, Cell **84**, 875 (1996)
110. D.P. Fitzsimons et al., J. Biol. Chem. **267**, 23903 (1992)
111. P.J. Kennelly et al., J. Biol. Chem. **262**, 11958 (1987)
112. B.E. Kemp et al., J. Biol. Chem. **262**, 2542 (1987)
113. J.K. Krueger et al., Biochemistry **36**, 6017 (1997)
114. G.A. Olah, J. Trewhella, Biochemistry **33**, 12800 (1994)
115. J.K. Krueger et al., Biochemistry **37**, 13997 (1998)
116. M. Ikura, L.E. Kay, G. Barbato, S. Spera, A. Bax, FASEB J. **6**, A403 (1992)
117. D.R. Knighton et al., Science **258**, 130 (1992)
118. O.B. Peersen, T.S. Madsen, J.J. Falke, Protein Sci. **6**, 794 (1997)
119. P.M. Bayley, W.A. Findlay, S.R. Martin, Protein Sci. **5**, 1215 (1996)
120. J.K. Krueger et al., Biochemistry **37**, 17810 (1998)
121. G.A. Olah et al., Biochemistry **32**, 3649 (1993)
122. A. Persechini, K.J. Gansz, R.J. Paresi, Biochemistry **35**, 224 (1996)
123. C.S. Tung, D.A. Walsh, J. Trewhella, J. Biol. Chem. **277**, 12423 (2002)
124. R.S. Stein, in *Neutron Scattering in the Nineties* (IAEA, Vienna, 1985), p. 335
125. http://www.ill.fr/Yellowbook/deuteration

9

Small Angle Neutron Scattering from Proteins, Nucleic Acids, and Viruses

S. Krueger, U.A. Perez-Salas, S.K. Gregurick, D. Kuzmanovic

9.1 Introduction

This chapter will focus on SANS applications to complex biological macromolecules such as proteins, nucleic acids, viruses, micelles, and vesicles. Because of its sensitivity to the biologically important light elements such as H, C, N, and O, SANS can provide unique information on the structure and function of biological macromolecules. Recent advances in biochemistry, crystallography and structural NMR have made it possible to prepare greater quantities of deuterium-labeled proteins and to determine an ever-increasing number of high-resolution structures. Thus, SANS has also come into wider use as a complementary tool for comparing the structures in crystal and solution phases and for elucidating the unresolved regions in a crystal structure. Since the measurements are performed in solution, SANS gives unique structural information under conditions that more closely mimic the molecule's natural environment, and thus can provide critical insights in a number of bioengineering areas.

The SANS experimental method has been described previously. Detailed information on SANS from biological macromolecules can be found in this book and in several review articles [1–3]. In the case of complex systems such as viruses, nucleic acids, and proteins, it is often far easier to obtain data than to interpret what the data mean. One simple, model-independent analysis of the scattered intensity, $I(Q)$, that is normally performed is the Guinier approximation [4], given by,

$$I(Q) = I(0) \exp\left[-\frac{Q R_{\mathrm{g}}{}^{2}}{3}\right], \tag{9.1}$$

where R_{g} is the radius of gyration, $I(0)$ is the forward scattered intensity and $Q = 4\pi \sin(\theta)/\lambda$, where λ is the neutron wavelength and 2θ is the scattering angle. This approximation is only valid in the region where $QR_{\mathrm{g}} \approx 1$. A real-space representation of the data can be obtained from the distance distribution function, $P(r)$, which is related to $I(Q)$ by

$$I(Q) = 4\pi V_o \int_0^{D_{\max}} P(r) \frac{\sin(Qr)}{Qr} \mathrm{d}r, \tag{9.2}$$

where V_o is the volume of the scatterer. The integral is carried out to a value $D_{\max}$, defined as the maximum distance beyond which there is no significant scattering mass of the biological sample. A number of indirect Fourier transformation methods exist [5–7] for calculating $P(r)$ from $I(Q)$. Typically, data are analyzed by first using Eq. 9.1 on the low-Q portions of the data to obtain initial values for R_g and $I(0)$. An indirect Fourier transformation method, which makes use of all of the data, rather than a limited data set at small Q values, is then used to determine $P(r)$, R_g, $I(0)$, and $D_{\max}$. $D_{\max}$ is chosen to obtain the best fit to the $I(Q)$ vs. Q data with a R_g value that agrees well with that found using Eq. 9.1. While $P(r)$ can help reveal the shape of the scatterer, further structural analysis requires comparison to model structures.

9.1.1 Modeling SANS Data

High Resolution Starting Structure is Available

When the X-ray crystal or NMR structure of the biological macromolecule is known, it is possible to calculate a model SANS intensity and R_g which allows for a direct comparison with the experimental data. One widely used program, CRYSON, calculates SANS intensities using spherical harmonics [8]. Because SANS is a low-resolution technique, atomic resolution is lost. However, if each residue in the protein were to be treated as a sphere of appropriate scattering length density and size, related to the particular amino acid sequence, then it is possible to gain back some structural resolution. Appropriate size beads are simply strung along the backbone to represent the protein. Then a scattering curve is calculated by using a Monte Carlo simulation [9]. Currently, such a program, XTAL2SAS, is being developed in collaboration with NIST and UMBC. XTAL2SAS is based on the original work of Glenn Olah, which relied on the method reported by Heidorn and Trewhella [10] to calculate a scattering profile from protein crystal structures. In the original program, a protein crystal structure (PDB file) is read into the program and each C_α carbon is taken as the center of a scattering sphere. The radius, molecular weight, volume, and neutron scattering length density (SLD) of each scattering sphere is dependent on the residue type [11]. In order to simulate the scattering profile, $I(Q)$ vs. Q, the calculation of the distance distribution function, $P(r)$, is first performed. This is generated by a Monte Carlo simulation of the scattering experiment whereby the spheres are randomly filled with points of known SLD dependent upon the residue type. $P(r)$ is then calculated by summing all possible distances between all possible pairs of points in the total structure, weighted according to the neutron SLD for each point. R_g and $I(Q)$ are then calculated from an integration of the $P(r)$ function as shown in Eq. 9.2.

However, if the solution structure deviates from the crystal structure, then the calculation of a reasonable model to fit the experimental scattering data is extremely difficult, as was illustrated for the case of the conformational change in cAMP Receptor Protein (CRP). Recent studies found that when CRP is complexed with cAMP and DNA, it undergoes a rather drastic conformational change, as evident by the large increase in the protein radius of gyration and a shift in the neutron scattering curve [12]. To model this conformational change, a method of a constrained walk along well-defined conformational coordinates was developed. This was the first application of such a procedure to SANS and it enabled the calculation of a best fit structural model to the experimental data [12].

A solvent accessible surface area (SASA) like approach to treat the effects of protein hydration, as determined by SANS, has also been developed as part of the XTAL2SAS program. A sphere of radius 6.5 Å is traced along the protein surface. A determination is made as to whether each surface residue is polar or nonpolar. For each polar surface residue, up to 5.0 Å of bound water is fit into the volume of the corresponding probe sphere. The scattering length density inside this hydration layer is that of bound water [13]. The scattering profile, $I(Q)$ vs. Q, is then calculated as above by first determining the distance distribution function, $P(r)$, by a Monte Carlo simulation of the scattering experiment. The scattered intensity is then calculated by Eq. 9.2.

High Resolution Starting Structure is Not Available

When a high resolution structure is not available, a low resolution structure can be built from one or more simple geometric shapes. The solid geometric structure is then randomly filled with points of uniform neutron SLD, each representing the average SLD of the molecule of interest. When more than one geometric shape is used to build a model structure, each can contain points of a different, uniform, SLD. By optimizing the geometric parameters of the structure to best fit experimental scattering data, a low resolution model is obtained. This is the basis behind the LORES program [14]. The first part of the program, involving the generation of a geometric model, relies on the same procedure as described in [9] for the generation of a set of scattering points within a given sub-volume. However, this original work has been extended to include many different shapes and to include an optimization procedure to determine the best fit geometric shape to inputted experimental data. The scattering points are generated, via a Monte Carlo method, to fall within a given volume (e.g., sphere, ellipsoid, cylinder, etc.). To simulate a uniform SLD within the given sub-volume, the total number of points is proportional to that volume. This method will ensure a uniform distribution of random points within a structure. It was found in the original work that the number of Monte Carlo points must be at least 1,000 in order to obtain a distribution that is indeed uniform [9].

Once a candidate structure is selected, the starting parameters and a given range for each parameter must also be input. During the course of the geometry optimization new parameters are generated randomly, subject to this chosen range. The scattered intensity, $I(Q)$ vs. Q, is calculated and compared with the experimental scattering profile. This is accomplished by calculating the distance distribution function, $P(r)$, by making a histogram representation of all possible distances between all possible pairs of scattering points within the given structure, weighted according to the neutron SLD for each point. A radius of gyration, R_g is also calculated. The Monte Carlo optimization algorithm strives to minimize the χ^2 distribution, in a least squares manner. For each model, a regression coefficient, R^2, is also calculated. The Monte Carlo optimization will minimize the χ^2 value and maximize the R^2 value simultaneously. Values of R_g and volume can be input as additional optimization parameters. The program output consists of a family of possible models (in PDB format) as well as scattering profiles to best fit the data.

Last, an on-line, user friendly web based software package is being developed for the molecular modeling of small angle scattering data of biological macromolecules. This interface is composed of a front end with html-like documents. The back-end interfaces to XTAL2SAS, LORES, and other useful programs via a series of PERL wrappers, one for each program of interest. A prototype website is also available at [15].

9.1.2 Contrast Variation

Often advanced modeling techniques will be used in addition to the contrast variation technique, in which the isotopic substitution of D for H is routinely used to change the scattering length density of the macromolecule or solvent, in order to separate the scattering from the individual components in a multicomponent complex and model them independently [16]. Thus, the conformation of a particular component bound in the complex can be directly compared to that of its counterpart free in solution. For a two-component system, the scattering from the two components can be written as:

$$I(Q) = \Delta\rho_1^2 I_1(Q) + \Delta\rho_1 \Delta\rho_2 I_{12}(Q) + \Delta\rho_2^2 I_2(Q), \tag{9.3}$$

where $I(Q)$ is the measured scattered intensity of the complex and the contrast, $\Delta\rho = (\rho - \rho_s)$, is the difference between the mean scattering length density of the molecule, ρ, and that of the solvent, ρ_s. $I_1(Q)$ and $I_2(Q)$ are the scattered intensities of components 1 and 2, bound in the complex. $I_{12}(Q)$ is the cross term between the two components. The Q value at which the cross term first reaches zero can be used to approximate the separation of the centers of masses of the two components, $D = 2\pi/Q$. If scattered intensities of the complex are measured in solvents with different H_2O/D_2O ratios, then a set of simultaneous equations can be solved in order to determine the unknowns, $I_1(Q)$, $I_2(Q)$ and $I_{12}(Q)$. Here, the measured scattered intensities, $I(Q)$, as well as the contrasts, $\Delta\rho_1$ and $\Delta\rho_2$ are the known quantities. Specific examples of the use of contrast variation to obtain unique information will be presented on SANS structural studies of protein/protein and protein/RNA complexes.

9.1.3 Experimental Examples

Three specific examples of using SANS to study the structure of biological systems are presented here. The systems are quite diverse, RNA, to protein/protein and protein/RNA complexes. In each case, different tools are used to model the structures measured by SANS. Although the RNA system is seemingly the simplest, its measured $I(Q)$ curves cannot be fit to simple model shapes. Rather, a high resolution model structure is compared to the data using the CRYSON program [8]. For the protein/protein complex and protein/RNA complex, or phage, the contrast variation technique was used to separate the scattering from the two components. Then, the components and the complexes were modeled separately using both the LORES and XTAL2SAS programs.

All SANS measurements shown in the following examples were performed on the 30-m SANS instruments at the NIST Center for Neutron Research in Gaithersburg, MD [17]. Typical neutron wavelengths, (λ), were 5 or 6 Å, with a wavelength spread, $\Delta\lambda/\lambda$ of 0.15. Raw counts were normalized to a common monitor count and corrected for empty cell counts, ambient room background counts and nonuniform detector response. Data were placed on an absolute scale by normalizing the scattered intensity to the incident beam flux. The two-dimensional data were then radially averaged to produce $I(Q)$ vs. Q curves. The one-dimensional scattered intensities from the samples were then corrected for buffer scattering and incoherent scattering from hydrogen in the samples. Guinier radii were found using Eq. 9.1 and the GNOM program [6] was used to calculate $P(r)$.

9.2 Nucleic Acids: RNA Folding

9.2.1 Compaction of a Bacterial Group I Ribozyme

Like proteins, certain RNA molecules fold into unique three-dimensional structures that are essential for their biological activity. Ribozymes, RNA fragments that have enzymatic activity, are an example of this class of molecule. Typically, a precursor RNA (pre-RNA) fragment contains two coding exons separated by a noncoding intron (the ribozyme). The ribozyme must fold into a unique conformation in order to join the two exons together that form a full coding sequence, and then remove itself by self-splicing. The mechanism by which the folded structures form from the unfolded or denatured, state has become the subject of intense investigation, [32–34, for example]. In contrast to proteins, where hydrophobic interactions drive the collapse of the polypeptide chain, RNA folding requires counterions to neutralize the electrostatic repulsion between phosphate groups. The collapse of RNA chains to intermediate (non-native) structures in the presence of counterions is of fundamental importance because this determines the probability of forming biologically active structures in a short time.

Theoretical and experimental studies of DNA and RNA show that counterion condensation around nucleic acids reduces the effective phosphate charge by 75–90% [35, and references therein]. Theoretical models of polyelectrolytes suggest that counterion condensation initially produces an ensemble of compact forms that contain both native and non-native interactions that slowly diffuse to the native state [33].

The presence of collapsed intermediates in RNA folding has been detected by biochemical [36, 37] and small angle X-ray scattering (SAXS) experiments [38, 39] and has demonstrated that counterions induce compact structures at concentrations below what is required to stabilize the native structure. Furthermore, in accordance with theoretical predictions [33], structural studies have shown that the initial collapse can occur in 1–10 ms [40][and references therein], which is a much shorter time than required to form the native RNA.

An important question is the extent to which the native interactions stabilize these compact folding intermediates. To address this question, SANS was used to measure changes in the global dimensions of a 195-nucleotide ribozyme of the *Azoarcus* bacterium [41] (Fig. 9.1) that is responsible for forming the RNA sequence that matches to the amino acid isoleucine ($tRNA^{ile}$). The collapse transition detected by SANS was compared with two conformational phase transitions previously defined by biochemical probes of RNA structure [41]: a transition from unfolded (U) RNA to a more ordered intermediate (I_C) at low counterion concentrations that involve the assembly of helices in the core of the ribozyme, and a second transition from I_C to the native tertiary structure (N) in higher Mg^{2+} concentrations that coincides with the

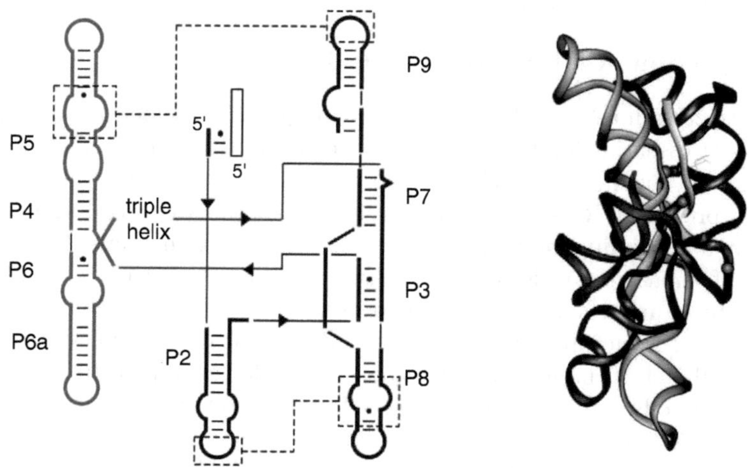

Fig. 9.1. The model structure of the *Azoarcus* group I ribozyme. The secondary and tertiary structure [41, and references therein] was modeled from comparative sequence analysis. Base-paired (P) regions in the ribozyme are indicated

appearance of catalytic activity [37]. A variation of the latter is a transition from I_C to I_F in higher Na^+ concentrations where the resulting structure is folded but inactive [42].

The *Azoarcus* ribozyme was transcribed in vitro, gel separated and purified following standard protocols [41, 42, and references therein]. RNA solution (2 mg/ml) was made in H_2O buffer containing 0–20 mM $MgCl_2$ or 0–2 M NaCl and maintained at 32°C during the SANS measurements [43]. This was the highest RNA concentration attainable for which the scattering profile showed no evidence of particle–particle interactions, particularly when in buffer alone. The distance distribution functions, $P(r)$, for the *Azoarcus* ribozyme in solution were compared to two standard analytic $P(r)$ models: the Gaussian polymer chain and the sphere [43]. In addition, the $P(r)$ corresponding to the sample with highest Mg^{2+} concentration (20 mM) was compared to the $P(r)$ computed from the 3D atomic model of the *Azoarcus* ribozyme [8], as shown in Fig. 9.1.

Change in RNA Conformation

SANS curves for the ribozyme in increasing concentration of $MgCl_2$ (0–20 mM)and NaCl (0–2 M) are shown in Fig. 9.2a, b respectively. In both panels, the scattering curves are observed to fall into two distinct classes. The change in the scattering at low Q suggests that, for the lower counterion concentrations, the particles have a relatively larger R_g than for the higher counterion concentrations. Comparing Fig. 9.2a, b it is clear that for the Mg^{2+} titration series the transition between the two types of scattering curves occurs abruptly between 1.6 and 1.7 mM $MgCl_2$, whereas the transition for the Na^+ titration series is more gradual as the salt concentration varies between 0 and 450 mM NaCl. The fact that higher concentrations of Na^+ are required to condense the RNA is consistent with the monovalent counterions being less efficient at charge neutralization [37, 42].

$P(r)$ functions, determined from the scattering curves shown in Fig. 9.2 are shown in Fig. 9.3. It is evident from Fig. 9.3 that the two classes of $P(r)$ functions relate to two distinct particle shapes: an extended shape at low counterion concentrations and a significantly more compact state at higher counterion concentrations. The variation in the maximum extension of the particles, D_{max}, for both the extended and the compact shapes was approximately 7%. R_g, which can be computed from $P(r)$ [43], decreased from an average of 53 ± 1 Å below 1.6 mM $MgCl_2$ to 31.5 ± 0.5 Å above 1.7 mM $MgCl_2$ and to 33.4 ± 0.2 Å in 2 M NaCl.

To evaluate the nature of the unfolded state, the $P(r)$ functions at low counterion concentrations were compared to a standard Gaussian chain model (random coil) with an equivalent R_g. This is shown in Fig. 9.4, where it is clear that the mass of the RNA measured in H_2O buffer with no added salts is distributed over shorter distances than predicted by the random coil model. The experimental data also show that the value of $P(r)$ is greater over distances of 90–130 Å than what would be expected for a random coil model. This

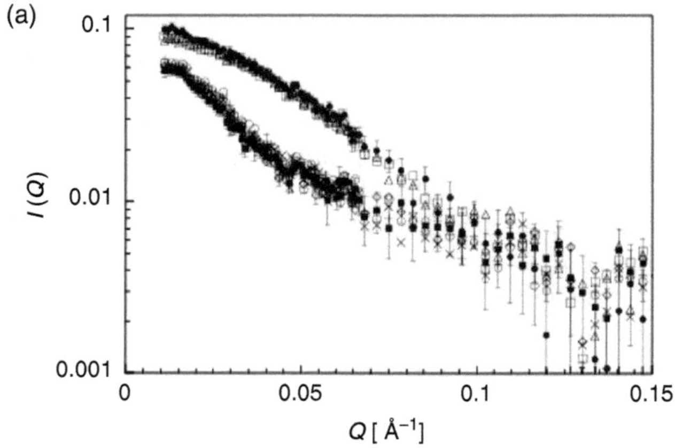

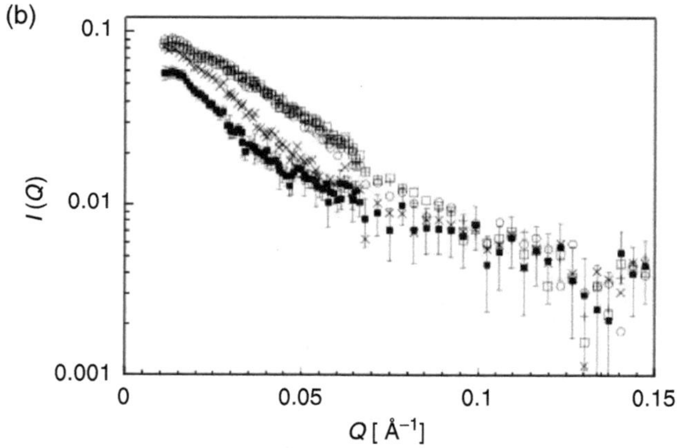

Fig. 9.2. (**a**) Mg^{2+} concentration dependence for RNA in H_2O buffer with no added salts (*filled square*), 1 mM Mg^{2+} (*diamond*), 1.3 mM Mg^{2+} (x), 1.5 mM Mg^{2+} (*circle*), 1.6 mM Mg^{2+} (+), 1.7 mM Mg^{2+} (*square*), 4 mM Mg^{2+} (*triangle*), 20 mM Mg^{2+} (*filled circle*). (**b**) Na^+ concentration dependence for RNA in H_2O buffer with no added salts (*filled square*), 100 mM Na^+ (x), 450 mM Na^+ (*circle*), 750 mM Na^+ (+), 2 M Na^+ (*square*)

suggests that RNA is more rigid than a Gaussian chain and this local stiffness is presumably due to double helical segments in the unfolded RNA.

Comparing the corresponding $P(r)$ curve for the 20 mM Mg^{2+} sample, where the ribozyme is in its native conformation, to the predicted real space density correlation function for the 3D model of the ribozyme shown in Fig. 9.1, $P(r)_{3D}$, and to the real space density correlation function for a sphere, $P(r)_{\text{sphere}}$, it is clear that the experimental $P(r)$ curve for the 20 mM Mg^{2+} sample has a greater resemblance to $P(r)_{3D}$ than to $P(r)_{\text{sphere}}$, especially for $r < R_g$. This is reinforced by the fact that the computed scattering curve from the 3D model of the ribozyme, $I(Q)_{3D}$, is similar to the SANS data for

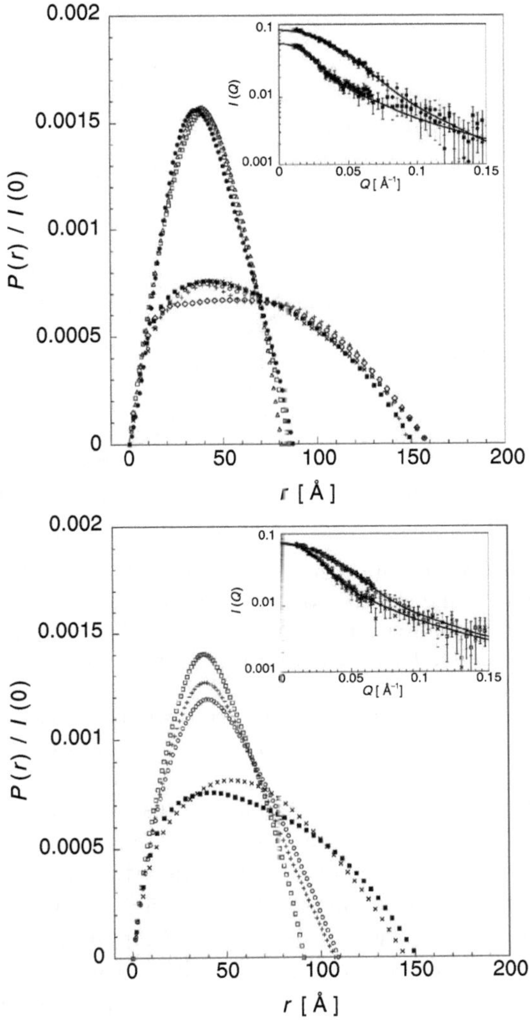

Fig. 9.3. $P(r)$ distributions were obtained from the SANS data in Fig. 9.2 according to Eq. 9.2 and scaled by $I_{\mathrm{calc}}(0)^{-1}$. Symbols are as in Fig. 9.2. (**a**) Mg^{2+} titration. (**b**) Na^{+} titration. Insets: Scattering curves computed from $P(r)$ distribution functions are compared with neutron scattering data from samples containing 0 and 20 mM $MgCl_2$ or 0 and 2 M NaCl, respectively

the 20 mM Mg^{2+} sample (inset in Fig. 9.4), except that the 3D model yields a smaller R_g of 30 Å compared to the experimental R_g of 31.5 ± 0.5 Å. This difference cannot be attributed to experimental error. The difference in the most probable value of r between the $P(r)$ curve for the 20 mM Mg^{2+} sample and $P(r)_{3D}$ could be due to either conformational fluctuations in the native state or errors in the model, which is based on comparative sequence analysis [41] .

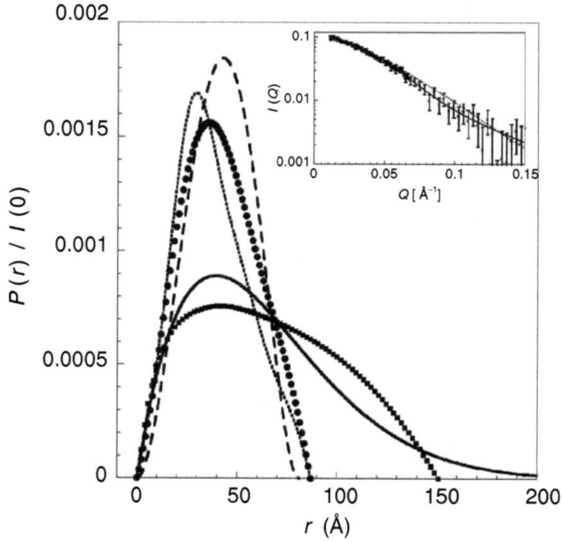

Fig. 9.4. $P(r)$ functions obtained from SANS data for RNA in H_2O buffer with no added salt (*filled squares*); 20 mM Mg^{2+} (*filled circles*). The curves represent $P_{\mathrm{randomcoil}}(r)$ for a random coil (*solid line*), $R_g = 53$ Å; $P_{\mathrm{sphere}}(r)$ for a uniform sphere (*long dashed line*), $R_g = 31.5$ Å; $P_{\mathrm{3D}}(r)$ for the 3D atomic model (*dashed line*), $R_g = 30$ Å. Inset: SANS data for RNA in H_2O buffer plus 20 mM Mg^{2+}. The continuous curves correspond to $I_{\mathrm{calc}}(Q)$ computed from the experimental $P(r)$ (*solid line*) and $I_{\mathrm{3D}}(Q)$, computed from the 3D model (*dotted line*)

Differences in the size of the compact states formed in Mg^{2+} and Na^+ are small, with a deviation in R_g at the largest salt concentrations of about 2 Å (Fig. 9.3). A slightly less compact shape is attained in 2 M NaCl ($R_g = 33.4 \pm 0.2$ Å) than in 20 mM $MgCl_2$ ($R_g = 31.5 \pm 0.5$ Å). If saturation was not reached even at 2 M NaCl, it is possible that at higher concentrations the difference in R_g becomes smaller. The similarity of R_g values obtained in Mg^{2+} and Na^+ is consistent with biochemical results showing that the ribozyme forms many tertiary interactions in monovalent salts, lacking only a few within the active site [42].

9.2.2 RNA Compaction and Helical Assembly

Counterion-mediated Collapse

Multivalent cations drive the compaction of RNA chains more efficiently than monovalent cations [37, and references therein]. In the case of the *Azoarcus* ribozyme, a 100-fold lower concentration of Mg^{2+} than Na^+ is required to induce compaction. Because electrostatic repulsion of the phosphates is a major force opposing RNA folding, the R_g of an approximately spherical folded RNA is expected to correlate with its residual net charge after counterion condensation. The *Azoarcus* ribozyme forms a compact structure in the presence of counterions when approximately 90% of the phosphate charge is neutralized, which is qualitatively consistent with previous work [44, 45]. The idea that the collapse transition of the RNA is not driven by site-specific coordination of metal ions is supported by the fact that the net charge per phosphate is roughly equal in Na^+ and Mg^{2+}.

Collapse Correlates with Helix Assembly

Because of nearly complete neutralization of the backbone charges due to nonspecific counterion condensation, an important question is whether the metal ion induced decrease in R_g of the ribozyme correlates with the degree of native structure. Two macroscopic conformational transitions in the *Azoarcus* ribozyme occur with increasing Mg^{2+} concentration [41]. Under conditions with no added salts, only the P2, P4, P5, and P6a stem-loops are detected by protection of guanine nucleotides from RNase T1 digestion, and the RNA appears largely unfolded (U). At moderate concentrations of monovalent or divalent salts, the double helices in the core of the ribozyme (I_C) are stabilized, including the P3/P7 pseudoknot and a triple helix that mediates interactions between the P4–P6 and P3–P9 domains (Fig. 9.1). Higher Mg^{2+} concentrations are required to form the native tertiary structure (N) and for catalytic activity [41]. Similar transitions are observed with other monovalent and divalent counterions, except that the resulting structure is folded but inactive (I_F) [42].

To determine which of these transitions (U $\rightarrow$ I_C or, correspondingly, I_C $\rightarrow$ N in Mg^{2+} or $I_C \rightarrow I_F$ in Na^+) correlate with compaction of the RNA, the secondary structure of the ribozyme was probed by partial digestion with RNase T1 and splicing assays under the conditions of the SANS experiments. In partial RNase T1 digestion assays, RNase T1 reacts with solvent accessible guanine (G) residues of 5′-^{32}P-labeled ribozyme. Separation of the products, done through a standard sequencing gel, maps the specific G nucleotides along the RNA sequence affected by RNase T1 digestion [41, 42, and references therein]. Self-splicing assays indicate that the amount of catalytic activity in ^{32}P-labeled pre-tRNA. Using a standard size exclusion gel, catalytic activity is quantified by the amount of spliced product [41,43, and references therein].

Addition of counterions resulted in the protection of G nucleotides in the core of the ribozyme. The midpoint of base pairing in the core is close to the counterion concentration at which the collapse of the RNA was observed (1.7 mM Mg^{2+} and 450 mM Na^+, respectively). By contrast, a splicing assay showed that fivefold higher Mg^{2+} concentrations were required for self-splicing activity under these conditions, the midpoint of the transition to the native structure being 4.5 mM Mg^{2+}, with maximal activity above 20 mM Mg^{2+}. That the assembly of helices in the ribozyme core occurs at low Mg^{2+} concentrations and precedes tertiary folding is suggested in previous and recent experiments [43, and references therein]. Even after this fivefold increase in Mg^{2+} concentration, no further compaction of the ribozyme is observed by

SANS, indicating that changes in the size and shape of the ribozyme during the transition from the intermediates to the native state are smaller than the 1 Å error of these measurements.

Taken together, these experiments provide strong support for the idea that counterions induce the collapse of polynucleotide chains, but that collapse alone is not sufficient to produce the native state. That the collapse transition of the *Azoarcus* ribozyme produces an I_C state with near-native R_g values suggests that Mg^{2+} leads to a specific collapse, in agreement with biochemical assays [41]. In contrast, other RNAs, both smaller and larger than *Azoarcus*, have been found to form intermediates that are about 5–15% less compact than their native structures [38, 39]. More expanded intermediates could reflect the increased presence of non-native interactions or greater structural dynamics of the folding intermediates compared with the collapsed intermediates formed by the *Azoarcus* ribozyme.

This example shows how SANS, in combination with biochemical assays and structural modeling techniques, can provide insight into the collapse of RNA molecules. The overall structural information provided by SANS can be directly related to biochemical activity if the biochemical assays are performed under the same conditions as the SANS experiments. Such analyzes can also be applied to protein folding problems.

9.3 Protein Complexes: Multisubunit Proteins and Viruses

9.3.1 Conformation of a Polypeptide Substrate in Model GroEL/GroES Chaperonin Complexes

The role of molecular chaperones in mediating and controlling intracellular as well as in vitro protein folding has broad implications for biotechnology. There is now considerable insight into the possible mechanisms whereby chaperone proteins recognize, stabilize, and release non-native polypeptide chains in a manner whereby they are able to productively refold. However, there are a number of important, fundamental gaps in the understanding of chaperone action, that remain to be resolved. Among the growing list of chaperone families, which are thought to play essential roles in a variety of fundamental cellular processes, are the chaperonins GroEL and GroES, which have been intensively studied. Knowledge of how misfolded protein substrates physically interact with GroEL should provide vital clues necessary to unravel the process by which GroEL mediates the proper folding of a wide variety of unfolded and misfolded protein substrates. The ultimate goal is to determine the mechanism by which GroEL transforms its substrate proteins and then releases them in a form able to refold to their native conformation.

One of the key issues in establishing a molecular mechanism for GroEL is to describe in structural terms the conformations of polypeptide substrates when bound to various chaperonin complexes. Any mechanism for chaperonin action will require the answer to several questions. For example, does a non-native polypeptide substrate unfold further upon binding to GroEL? On the other hand, when a chain is released from GroEL in the presence of the co-chaperonin GroES, does it adopt a more folded, or unfolded conformation? These questions are difficult to resolve with naturally occurring proteins since they refold so readily when released from chaperonin complexes. One approach to address these issues, however, is to utilize a family of mutationally altered protein substrates that are unable to adopt their native conformation. A number of such protein systems are readily available, one of which is a non-native subtilisin variant (PJ9) that is unable to refold when released [47].

Wild-type and Single-ring GroEL

In order to evaluate any changes in polypeptide conformation in association with models of chaperonin complexes, it is necessary to describe a model for the solution conformation of the single ring GroEL variant (srGroEL) used in this study. Data were obtained from srGroEL in both H_2O and D_2O buffers and compared to data obtained from wild-type GroEL under the same conditions. Analysis of these data were enhanced by parallel studies with a mutational variant of GroEL for which 16 C-terminal residues have been deleted [48] as well as by using information obtained from previous SANS studies on GroEL and chaperonin complexes [49, 50]. These additional data sets enabled an assignment of the crystallographically disordered C-terminal domain of GroEL, and were also helpful in modeling studies of the solution structure of the single ring chaperonin.

The scattering curve for srGroEL in D_2O buffer is presented in Fig. 9.5 along with the corresponding curve for the wild type GroEL. The solid line in Fig. 9.5 represents the scattered intensity calculated, using XTAL2SAS, from just one of the rings of the double-ring crystal structure [51] with the added SANS-derived model for the disordered C-terminal domain. The best fit to the data was obtained when the disordered C-terminal domain is positioned along the inner wall of the GroEL ring, a possibility also suggested in [50], and when no adjustments are made to the location of the flexible apical domains. Equally good fits to the data can still be obtained if the apical domains are allowed to rotate up to 10° in a similar manner to that described in [49]. Thus, the solution structure of srGroEL can be described well using one ring of the double ring crystal structure and by positioning the disordered C-terminal domain along the inner wall of the ring.

GroEL Complexes with Substrate

Single-ring GroEL mutants have been shown to assist in the refolding of non-native polypeptide chains [52, 53] and they form unusually stable complexes with GroES upon the addition of nucleotides. This attribute was exploited to

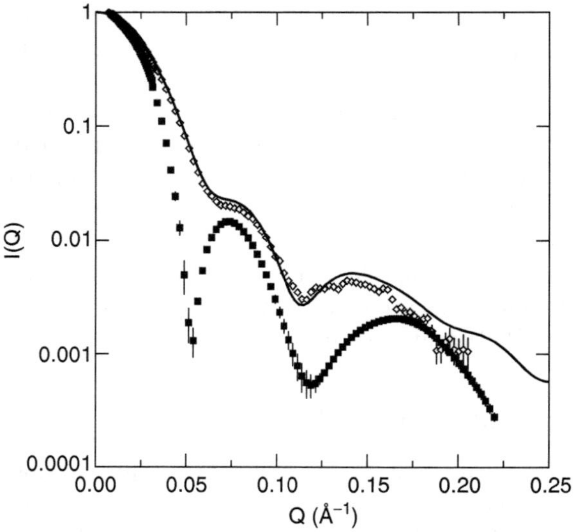

Fig. 9.5. SANS data of the srGroEL variant (*diamond*), compared with that of wild-type GroEL (*filled squares*). The solid line represents the scattered intensity calculated from one ring of the wild-type GroEL crystal structure, including the SANS-derived model for the disordered C-terminal residues

obtain a low-resolution structure for a non-native variant of the serine protease subtilisin polypeptide (PJ9) when bound to GroEL. The subtilisin was 86% deuterated (dPJ9) so that its SLD contrasted sufficiently with the chaperonin, allowing the contrast variation technique to be used to separate the scattering from the two components bound in the complex. The srGroEL mutant assured that dPJ9 and GroES were each bound in a 1:1 stoichiometry with the single ring of GroEL, providing an advantage over previous SANS experiments [49] which included mixed stoichiometries of GroEL. For comparison, a complex between dPJ9 and wild-type, double-ring GroEL was also studied. Care was taken to ensure that two dPJ9 molecules were bound to each GroEL molecule, in order to maintain the 1:1 stoichiometry between dPJ9 and each ring of GroEL.

SANS contrast variation data for the wild-type GroEL/dPJ9 and single ring srGroEL/dPJ9 complexes are shown in Fig. 9.6. Measurements were made in 0%, 20%, 70%, 85%, and 100% D_2O buffers in each case [54]. The extensive data set in Fig. 9.6 allows the scattering from each of the components in the complex to be separated using Eq. 9.3. Here, $I_1(Q)$ and $I_2(Q)$, refer to the GroEL (or srGroEL) component and the dPJ9 component of the GroEL/dPJ9 complexes, respectively. The cross-term, $I_{12}(Q)$, represents the interference function between the GroEL and dPJ9 components. Fig. 9.7 shows the scattered intensities for the wild-type GroEL and dPJ9 components, as determined from the contrast variation data. The peak in the dPJ9 curve at

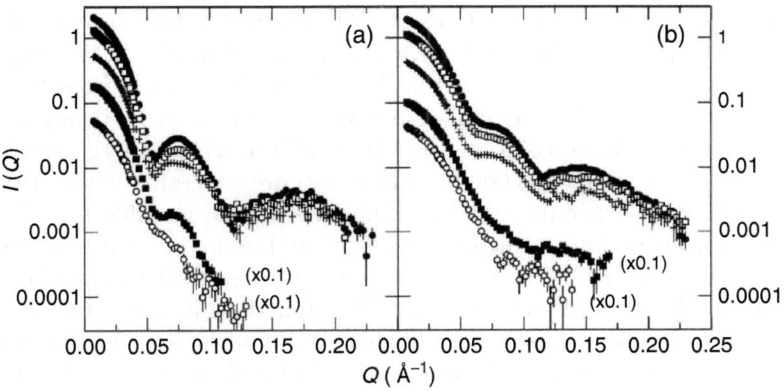

Fig. 9.6. Contrast variation data from (**a**) GroEL/dPJ9 and (**b**) srGroEL/dPJ9 complexes measured in 100% (*filled circle*), 85% (*square*), 70% (+), 20% (*circle*), and 0% (*filled square*) D_2O. The data in 20% and 0% D_2O solution are shifted by the factor 0.1, as indicated, for clarity

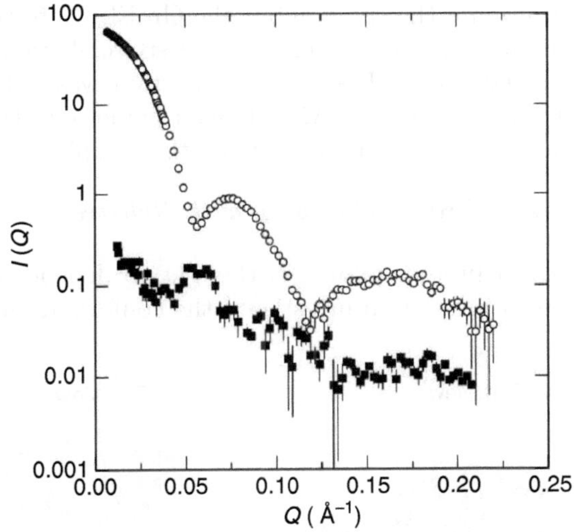

Fig. 9.7. Scattered intensities $I_{EL}(Q)$ (*circle*) and $I_{\mathrm{dPJ9}}(Q)$ (*filled square*), the GroEL and dPJ9 components, as bound in the complex, respectively

$Q \approx 0.05\,\text{Å}^{-1}$ is due to the interaction between the two dPJ9 molecules at each end of the GroEL/dPJ9 complex. A comparison of the scattered intensity for bound GroEL and that measured free in solution (Fig. 9.5) indicates that little or no change in GroEL conformation occurs upon binding the substrate polypeptide. The separation of the centers of masses of the two dPJ9 molecules in the complex is approximately 125 Å, determined from the location of the peak in $I_{\mathrm{dPJ9}}(Q)$, using $D = 2\pi/Q_{\mathrm{peak}}$.

A similar analysis for the srGroEL/dPJ9 contrast variation series of data revealed that the R_g value for dPJ9 bound in the srGroEL/dPJ9 complex is 19.0 Å. However, its maximum extent is 55 Å, as determined from the $P(r)$ function, suggesting that the molecule is very asymmetric. Modeling was accomplished by a Monte Carlo method (LORES) in which a large number of models (>10,000, and >100,000 in some cases) are generated within the constraints of the SANS data and the minimum volume possible for dPJ9, as calculated from its known molecular weight and assuming a partial specific volume of 0.73 $cm^3\,g^{-1}$. The models were then tested to determine how well the calculated scattered intensities and $P(r)$ functions fit the data. After initially testing simpler ellipsoidal and cylindrical models, and finding them to be a poor fit to the data, mushroom models for the dPJ9 were explored. Such a model lends itself well to the geometry of the srGroEL molecule and is also suggested for rhodanese in [49].

Fig. 9.8 presents two views of a complete model for the srGroEL/dPJ9 complex. The significant result is that the dPJ9 component has an asymmetric shape and that part of the polypeptide must beyond the cavity inside the srGroEL ring and up into the space above the GroEL. Note that the bottom portion of the dPJ9 mushroom penetrates the srGroEL cavity and the top portion sits above the cavity. The srGroEL portion was obtained from the crystal structure with the added SANS-derived model for the disordered C-terminal domain, which is seen at the bottom the complex in the side view.

Single-ring GroEL/GroES Complex with Substrate

Because physiological protein folding is thought to depend on both GroEL and GroES, it was of interest to investigate the conformational changes in a

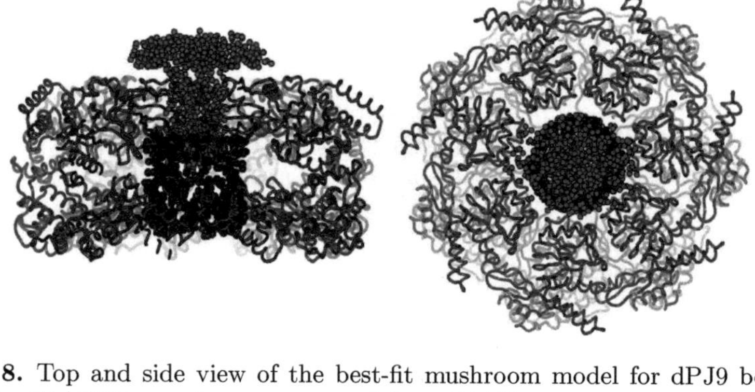

Fig. 9.8. Top and side view of the best-fit mushroom model for dPJ9 bound in the srGroEL/dPJ9 complex, constructed from SANS contrast variation and crystallography data. The srGroEL is represented by the ribbon structure, the dPJ9 is represented by the light spheres and the disordered C-terminal residues are represented by the dark spheres

polypeptide substrate within an active chaperonin complex. In this way, it would be possible to detect any changes in the conformation of the substrate under conditions that either promote, or do not promote, protein folding. However, because chaperonin complexes are in a dynamic equilibrium linked to the ATPase activity of GroEL, a simpler model of these components was needed for the relatively long times needed to collect adequate neutron scattering data. Thus, the single-ring GroEL variant was used to trap dPJ9 within GroEL by GroES upon the addition of ADP (adenosine diphosphate) or ATP (adenosine triphosphate).

Unlike ATP, ADP does not cause dissociation of dPJ9 from GroEL. If dPJ9 is no longer covalently bonded to GroEL, would its location in the GroEL/GroES complex change? To answer this question, two contrast variation series of measurements were performed on the srGroEL/GroES/dPJ9 complex [54]. One set of measurements was obtained with ADP present and the other with ATP present. Measurements were made in 0%, 20%, 70%, 85%, and 100% D_2O buffers in each case. The scattering from each of the components in the complex were then separated into $I_{\mathrm{ELES}}(Q)$ and $I_{\mathrm{dPJ9}}(Q)$, for the srGroEL/GroES component and the dPJ9 component of the GroEL/GroES/dPJ9 complex, respectively, using Eq. 9.3.

Using this method, the location and approximate shape of dPJ9 in the srGroEL/GroES/PJ9 + ADP complex was determined as modeled in Fig. 9.9. The significant result is that the dPJ9 component retains its asymmetric shape and, again, part of the polypeptide must extend beyond the cavity inside the srGroEL ring and up into the space surrounded by GroES. The significant difference in the srGroEL/GroES/dPJ9 complex formed from ATP is that the shape of the bound dPJ9 molecule changes from an asymmetric shape such as that shown in Fig. 9.9 to a more symmetric shape. Figure 9.10 shows the distance distribution functions for bound dPJ9 in both the

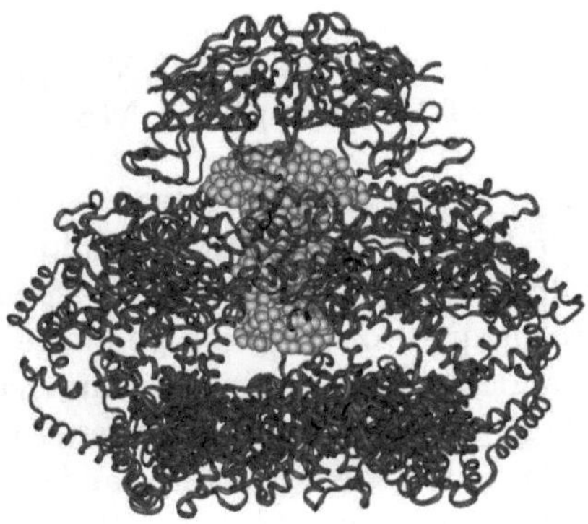

Fig. 9.9. Side view of a model for the srGroEL/GroES/dPJ9+ADP complex constructed from SANS contrast variation and crystallography data. The srGroEL/GroES complex is represented by the ribbon structure and the dPJ9 is represented by the light spheres. The disordered C-terminal residues are not shown

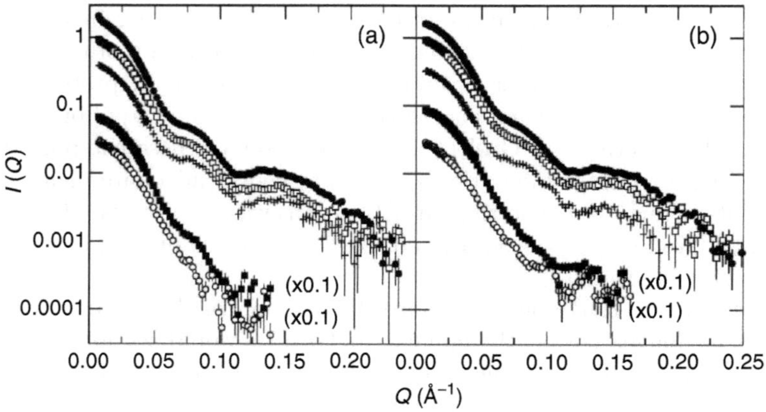

Fig. 9.10. Normalized distance distribution functions, $P(r)$ vs. r, for dPJ9 bound to srGroEL (*solid line*), srGroEL/GroES + ADP (*dotted line*) and srGroEL/GroES + ATP (*dashed line*)

srGroEL/GroES/dPJ9 + ADP and srGroEL/GroES/dPJ9 + ATP complexes. The most probable distance increases from approximately 22 to 30 Å , with a similar increase in the radius of gyration, R_g, from 19.0±0.5 Å to 21.0±0.5 Å. The shape in the presence of ATP is clearly more symmetric, as indicated by the greater symmetry of the distance distribution function. This suggests that dPJ9 is transformed into a more expanded form in the ATP complex. This conformational change either was not supported by the complex formed from ADP or was insufficient to generate a lasting change in shape in that case, and dPJ9 instead relaxed back to a form close to its original conformation. This important observation reflects the relative ability of ATP to promote refolding of protein substrates relative to ADP.

9.3.2 Spatial Distribution and Molecular Weight of the Protein and RNA Components of Bacteriophage MS2

The MS2 bacteriophage is a model organism for a number of important areas of research including viral replication, infection, and assembly [56]. Recently, noninfectious, genetically modified forms of the MS2 phage that contain varying amounts of RNA (compared to the wild-type phage) have been developed for use as biological standards [57, 58]. These commercially available recombinant particles, Armored RNAs, are used as reference material in research assays for the HIV, Ebola, Borna, Hepatitis A, C, and G, Dengue, Enterovirus, West Nile, and Norwalk viruses, among others [59].

The clinical use of these particles as biological standards in public health screening of humans and livestock has been hampered by the lack of rapid quantitative methods to analyze the physical properties of this family of particles. As these particles are not found in nature, they cannot be scientifically characterized by traditional methods. Specifically, these MS2-like biomarkers, because of their small size and the necessity that they be noninfectious, cannot be rapidly or reliably counted. As a result, this new generation of biological reference material cannot be cheaply characterized for general use in public health laboratories. This is solely due to the fact that their physical properties in solution cannot be quantified or confirmed. Thus, there is a need for instrumentation that can count biological particles, about which nothing is known, and that also can provide structural information about their properties in solution.

The creation of these new forms of MS2 has made it increasingly important to both understand the relationship of the indigenous MS2 RNA to its protein shell and to measure the molecular weight (Mw) of the wild-type RNA molecule in vivo under biological conditions. For practical purposes, the analysis of biological materials by SANS is almost exclusively used for structural analysis of molecules in combination with a variety of other techniques and not for the characterization of unknown viruses [3]. This is due in part because of the technical challenges associated with accurately determining particle concentration. Typically, particle number is measured by optical density (OD) in milligrams per milliliter using conventional spectrophotometry. Optical density measurements are possible only if the molar absorption coefficient of the sample is known. The molar absorption coefficient is a constant unique to the sample under study and assumes that the molecular weight of the sample is known [60]. Thus, characterization of an unknown virus or phage can only be accomplished by combining SANS with a novel virus counting instrument, the Integrated Virus Detection System (IVDS). The use of the IVDS instrument for virus counting is required because there is no other method to rapidly count small (<100 nm) biological particles with unknown properties in solution in the absence of viral infectivity or information about the particle Mw.

Purified MS2 phage was isolated by cesium chloride equilibrium gradient using a protocol similar to that described by Sambrook and Russell [61]. The measured density of the MS2 particles was 1.38 ± 0.01 g cm^{-3}, which is the same density value reported by [62]. Samples for SANS measurements were made in buffers containing 0%, 10%, 65%, 85%, and 100% D_2O. The purity of the samples was confirmed by SDS polyacrylamide gel electrophoresis. The samples were dialyzed in the appropriate buffers for 2 h at room temperature, with two changes of buffer, then transferred to sample holders.

Since MS2 can be approximated very well by a spherical shell at the resolution level of the SANS measurements, the data were also fit to a core-shell sphere model [4] in order to obtain the radius of the protein shell and RNA core. The neutron scattering length density of the RNA core was an additional fitting parameter that allowed the amount of water, versus RNA, in the core to be calculated. The scattered intensities from the MS2 protein/RNA complex were decomposed into the scattering from their components, $I_{\mathrm{PROT}}(\mathrm{Q})$ and $I_{\mathrm{RNA}}(Q)$ using Eq. 9.3. The Mw values of the protein and RNA components of MS2 were calculated in a similar manner as described in [63]. It is important to note that $I(0)$ must be on an absolute scale, usually in cm^{-1}, in order to obtain accurate Mw values.

Number density determinations were made using two methods: (1) the concentration was measured by optical density (OD) using a conventional spectrophotometer and then the number density was estimated using this information and (2) the number density was obtained directly using the Integrated Virus Detection System (IVDS) which is a particle counting method [64].

Molecular Weight of the Protein and RNA Components

To obtain the Mw of the protein and RNA components of MS2, the number density, n, must be known. Both OD and IVDS methods were used to obtain this information, and the resultant number densities agreed very well. IVDS analysis thus makes possible a novel use of SANS as a tool for the identification and physical characterization of unknown viruses or phage. The Mw of the MS2 RNA and coat protein components, calculated from the number density information [63], are $1.0 \times 10^6 \pm 0.2 \times 10^6\,\mathrm{g\,mol^{-1}}$ and $2.5 \times 10^6 \pm 0.3 \times 10^6\,\mathrm{g\,mol^{-1}}$, respectively. Mw_{PROT} agrees well with expected value from the 2.8 Å resolution crystal structure [65]. The total Mw of the MS2 bacteriophage is $3.5\times10^6\pm0.5\times10^6\,\mathrm{g\,mol^{-1}}$, in good agreement with total Mw measurements using other techniques [63, and references therein]. These results show that the use of SANS in combination with IVDS makes possible quantitative physical characterization of viruses and phage.

Spatial Distribution of the Protein and RNA Components

The SANS data for a typical contrast variation series of measurements are shown, on an absolute scale, in Fig 9.11. At the resolution of the SANS measurements, the shape of an MS2 particle can be approximated very well by a spherical shell, with inner radius, $R1$, outer radius, $R2$, and shell thickness, $t = R2 - R1$. A sample model fit, made with and without correcting the model for instrumental resolution effects, is shown for the 100% D_2O data in Fig. 9.12. In all cases, the outer radius of the shell, $R2$, consistently falls between the values of 134 and 144 Å. The (core) inner radius, $R1$, falls between 110 Å and 118 Å, except for the 10% D_2O buffer sample, which consistently shows a much smaller $R1$ value for both experiments. The RNA in the core scatters strongly in comparison with the protein shell under these solvent conditions. Thus, at this contrast, the lower value for $R1$ could be an indication that the RNA is actually packed compactly and does not completely fill the core region, with the remainder of the core being mostly solvent. The amount of water in the core was calculated from the fitted scattering length density of the core region for the data at each contrast [63]. The average fraction of water in the core region was found to be 0.81 ± 0.04. If the values obtained in 10% D_2O are excluded, the average parameters obtained from the core-shell model fit are $R1 = 115 \pm 1$ Å, $R2 = 136 \pm 1$ Å, $t = 21 \pm 1$ Å.

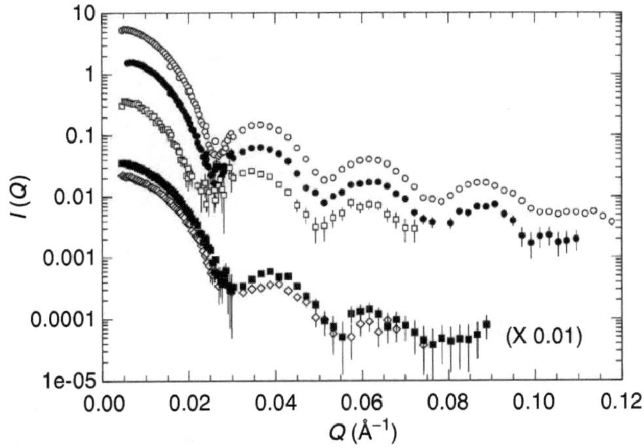

Fig. 9.11. MS2 contrast variation series of scattered intensity curves from samples in 100% (*circle*) 85% (*filled circle*), 65% (*square*), 10% (*diamond*), and 0% D_2O (*filled square*). The scattered intensity curves for 10 % D_2O and 0% D_2O have been multiplied by 0.01, for clarity

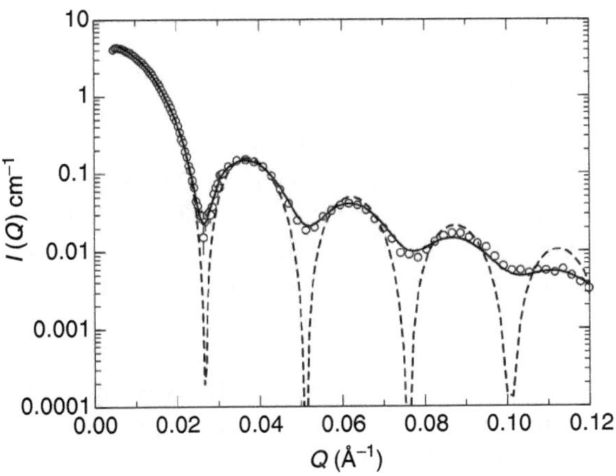

Fig. 9.12. A sample core-shell model fit for MS2, with (*solid line*) and without (*dashed line*) correcting the model for instrumental resolution effects, for the 100% D_2O data (*circle*)

Distance distribution functions, $P(r)$, were obtained from the data and are plotted in Fig. 9.13. The $P(r)$ functions are normalized so that the peak value is equal to 1.0 in each case. The maximum distance, $D_{\max}$, in all cases was found to be 300 Å, which is larger than $2 \times R2$. By definition, $D_{\max}$ is the distance at which $P(r)$ goes to zero. Thus, $D_{\max}$ suggests a sharp boundary between the particle and its surroundings. Since the shape of the MS2 coat protein region is actually icosahedral, this boundary is not sharp and the

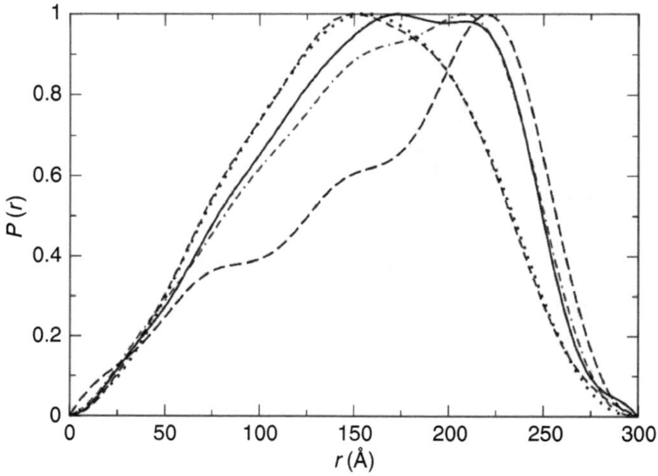

Fig. 9.13. Distance distribution functions, $P(r)$ vs. r, from the MS2 data for samples in 100% (*solid line*), 85% (*dot-dashed line*), 65% (*long dashed line*), 10% (*dashed line*), and 0% D_2O (*dotted line*)

$P(r)$ functions suggest that the particle does actually extend beyond $2 \times R2$. However, the number of probable distances beyond $2 \times R2$ drops sharply.

The $P(r)$ function for the 65% D_2O sample is consistent with that of a hollow spherical shell. In this case, the peak of the distance distribution is at 200 Å, consistent with the fact that the most probable distances are occurring beyond $2 \times R1$. On the other hand, the peak of the distance distribution function occurs at values smaller than $2 \times R1$ for the data obtained at the other contrasts. The RNA component is contributing more to the total scattering in 0% and 10% D_2O and this is reflected as a shift in the peak in $P(r)$ to smaller r values. In 85% and 100% D_2O, the RNA component is contributing to the total scattering, but the scattering from the protein component is much stronger. Thus, the peak in $P(r)$ falls in between the 65% D_2O case and the 0% and 10% D_2O cases.

The scattered intensities from the RNA and protein components were separated from each other using Eq. 9.3. The resultant $P(r)$ functions are shown in Fig. 9.14. While D_{max} for the protein shell remains at 300 Å, D_{max} for the RNA core was found to be 165 Å. Thus, the RNA component appears to be confined mostly within a radius of ≈83 Å. The peak of the RNA $P(r)$ distribution is also around this value. These results agree very well with the $R1$ values from the core-shell model fits for the samples measured in 10% D_2O. Recall, that the 10% D_2O solvent condition is where the RNA scattering is the strongest relative to that of the protein. Indirect genetic and biochemical results hint that a variety of mechanisms may act in concert to fold and compact the MS2 genomic RNA. This body of in vitro experiments suggests strongly

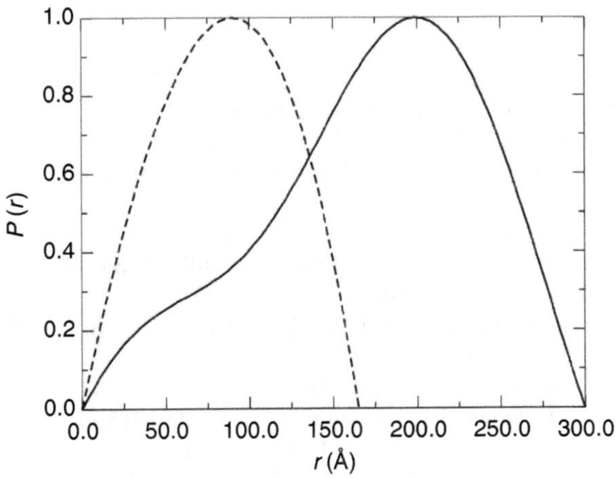

Fig. 9.14. Distance distribution functions, $P(r)$ vs. r, for the protein (*solid line*) and RNA (*dashed line*) components of the MS2 particles

that the MS2 RNA is tightly compacted and that the degree of packing is important for transcriptional regulation and genomic integrity. This work is the first study to directly measure the spatial distribution of the MS2 genomic RNA under indigenous conditions and to confirm that it is indeed compact in vivo.

Acknowledgments

The work presented in this chapter is the result of close collaborations and we would like to take this opportunity to explicitly thank our collaborators. In particular, UPS and SK would like to thank Prof. Sarah Woodson, Dr. Pranshanth Rangan, Mr. Robert Moss, Prof. Deverajan Thirumalai and Prof. Robert Briber for the group effort on the RNA work. SKG and SK would like to thank Prof. Edward Eisenstein and Mr. James Zondlo for their kind collaboration on the chaperonin project and Ms. Jing Zhou for her work in developing the LORES program. DK and SK would like to thank Drs. Charles Wick and Ilya Elashvili for collaborating with them on the MS2 bacteriophage work. Partial support for the RNA compaction research came from grants from the National Institutes of Health. UPS was supported by the National Research Council/National Institute of Standards and Technology Postdoctoral Associateship Program. SKG thanks the DOE/Sloan Foundation for a postdoctoral fellowship in computational molecular biology that helped support her work. DK was supported by the National Research Council/National Institute of Standards and Technology Postdoctoral Associateship Program. The NIST SANS facilities are partially supported by the National Science Foundation under Agreement Nos. DMR-9423101 and DMR-9986442.

References

1. P.A. Timmins, G. Zaccai, Eur. Biophys. J. **15**, 257–268 (1988)
2. J. Trewhella, Cur. Opin. Struct. Biol. **7**, 702–708 (1997)
3. S. Krueger, Physica B **241**, (1998) 1131–1137
4. A. Guinier, G. Fournet, *Small-Angle Scattering of X-rays* (John Wiley and Sons, 1955)
5. P.B. Moore, J. Appl. Cryst. **13**, 168–175 (1980)
6. A.V. Semenyuk, D.I. Svergun, J. Appl. Crystallogr. **24**, 537–540 (1991)
7. O. Glatter, J. Appl. Cryst. **10**, 415–421 (1977)
8. D.I. Svergun, S. Richard, M.H.J. Koch, Z. Sayers, S. Kuprin, G. Zaccai, Proc. Natl Acad. Sci. USA **95**, 2267–2272 (1998)
9. S. Hansen, J. Appl. Cryst. **3**, 334–346 (1990)
10. D.B. Heidorn, J. Trewhella, Biochemistry **27**, 909–915 (1988)
11. B. Jacrot, Rep. Prog. Phys. **39**, 911–953 (1976)
12. S. Krueger, S. Gregurick, Y. Shi, S. Wang, B.D. Wladowski, F.P. Schwarz, Biochemistry **42**, 1958–1968 (2003)
13. S.J. Perkins, Eur. J. Biochem. **157**, 169–180 (1986)
14. J. Zhou, private communication
15. http://www.sans.chem.umbc.edu/
16. H.B. Stuhrmann, A. Miller, J. Appl. Cryst. **11**, 325–345 (1978)
17. C.J. Glinka, J.G. Barker, B. Hammouda, S. Krueger, J.J. Moyer, W.J. Orts, J. Appl. Cryst. **31**, 430–445 (1998)
18. N.E. Gabriel, M.F. Roberts, Biochemistry **23**, 4011–4015 (1984)
19. P. Schurtenberger, N. Mazer, W. Koenzig, J. Phys. Chem. **89**, 1042–1049 (1985)
20. B. Carion-Taravella, J. Chopineau, M. Ollivon, S. Lesieur, Langmuir **14**, 3767–3777 (1998)
21. M. Johnsson, K. Edwards, Langmuir, **16**, 8632–8642 (2000)
22. M. Ollivon, S. Lesieur, C. Grabrielle-Madelmont, M. Paternotre, Biochim. Biophys. Acta **1508**, 34–50 (2000)
23. C.R. Sanders II, J.P. Schwonek, Biochemistry, **31**, 8898–8905 (1992)
24. C.R. Sanders II, B.J. Hare, K.P. Howard, J.H. Prestegard, Prog. NMR Spectro. **26**, 421–444 (1994)
25. C.R. Sanders II, J.H. Prestegard, Biophys. J. **58**, 447–460 (1990)
26. J. Chung, J.H. Prestegard, J. Phys. Chem. **97**, 9837–9843 (1993)
27. M.-P. Nieh, C.J. Glinka, S. Krueger, R.S. Prosser, J. Katsaras, Langmuir **17**, 2629–2638 (2001)
28. M.-P. Nieh, C.J. Glinka, S. Krueger, R.S. Prosser, J. Katsaras, Biophys. J. **82**, 2487–2498 (2002)
29. G. Gregoriadis, Trends Biotechol. **13**, 527–537 (1995)
30. M.-P. Nieh, T.A. Harroun, V.A. Rughunathan, C.J. Glinka, J. Katsaras, Phys. Rev. Lett. **91**, 158105 (2003)
31. J.B. Hayter, J. Penfold, Mol. Phys. **42**, 109–118 (1981)
32. P. Brion, E. Westhof, Ann. Rev. Biophys. Biomol. Struct. **26**, 113–137 (1997)
33. D. Thirumalai, N. Lee, S.A. Woodson, D.K. Klimov, Ann. Rev. Phys. Chem. **52**, 751–762 (2001)
34. T.R. Sosnick, T. Pan, Curr. Opin. Struct. Biol. **13**, 309–316 (2003)
35. V.K. Misra, D.E. Draper, Biopolymers **48**, 113–135 (1998)
36. K.L. Buchmueller, A.E. Webb, D.A. Richardson, K.M. Weeks, Nat. Struct. Biol. **7**, 362–366 (2000)

37. S.L. Heilman-Miller, D. Thirumalai, S.A. Woodson, J. Mol. Biol. **306**, 1157–1166 (2001)
38. R. Russell, I.S. Millett, S. Doniach, D. Herschlag, Nat. Struct. Biol. **7**, 367–370 (2000)
39. X. Fang, K. Littrell, X.-J. Yang, S.J. Henderson, S. Siefert, P. Thiyagarajan, T. Pan, T.R. Sosnick, Biochemistry **39**, 11107–11113 (2000)
40. R. Russell, I.S. Millett, M.W. Tate, L.W. Kwok, B. Nakatani, S.M. Gruner, S.G. Mochrie, V. Pande, S. Doniach, D. Herschlag, L. Pollack, Proc. Natl. Acad. Sci. USA, **99**, 4266–4271 (2002)
41. P. Rangan, B. Masquida, E. Westhof, S.A. Woodson, Proc. Natl. Acad. Sci. USA, **100**, 11574–1579 (2003)
42. P. Rangan, S.A. Woodson, J. Mol. Biol. **329**, 229–238 (2003)
43. U.A. Perez-Salas, P. Rangan, S. Krueger, R.M. Briber, D. Thirumalai, S.A. Woodson, Biochemistry **46**, 1746–1753 (2004)
44. V.K. Misra, D.E. Draper, J. Mol. Biol. **294**, 1135–1147 (1999)
45. V.K. Misra, D.E. Draper, J. Mol. Biol. **317**, 507–521 (2002)
46. D. Thirumalai, B.Y. Ha, Phys. Rev. A **46**, R3012–R3015 (1992)
47. Z. Lin, E. Eisenstein, Proc. Natl. Acad. Sci. USA, **93**, 1977–1981 (1996)
48. N.F. McLennan, A.S. Girshovich, N.M. Lissin, Y. Charters, M. Masters, Mol. Microbiol. **7**, 49–58 (1993)
49. P. Thiyagarajan, S.J. Henderson, A. Joachimiak, Structure **4**, 79–88 (1996)
50. R. Stegmann, E. Manakova, M. Rossle, H. Heumann, S.E. Nieba-Axmann, A. Pluckthun, T. Hermann, R.P. May, A. Wiedenmann, J. Struct. Biol. **121**, 30–40 (1998)
51. K. Braig, Z. Otwinowski, R. Hegde, D.C. Boisvert, A. Joachimiak, A.L. Horwich, P.B. Sigler, Nature **371**, 578–586 (1994)
52. I.S. Weissman, Chem. Biol. **2**, 255–260 (1995)
53. S.G. Burston, J.S. Weissman, G.W. Farr, W.A. Fenton, A.L. Horwich, Nature **383**, 96–99 (1996)
54. S. Krueger, S.K. Gregurick, J. Zondlo, E. Eisenstein, J. Struct. Biol., **141**, 240–258 (2003)
55. J.F. Hunt, A.J. Weaver, S.J. Landry, L. Gierasch, J. Deisenhofer, Nature **379**, 37–45 (1996)
56. P.G. Stockley, N.J. Stonehouse, K. Valegard, Int. J. Biochemistry **26**, 1249–1260 (1994)
57. B.L. Pasloske, C.R. Walkerpeach, R.D. Obermoeller, M. Winkler, D.B. DuBois, J. Clinical Microbiol. **36**, 3590–3594 (1998)
58. P.G. Stockley, R.A. Mastico, Methods Enzymol. **326**, 551–569 (2000)
59. http://www.ambiondiagnostics.com/products/armored_rna.html/
60. D. Eisenberg, *Physical Chemistry with Applications to the Life Sciences* (Benjamin/Cummings, 1979)
61. J. Sambrook, D.W. Russell, *Molecular Cloning a Laboratory Manual* (Cold Spring Harbor Press, 2001)
62. J.H. Strauss, R.L. Sinsheimer, J. Mol. Biol. **7**, 43–54 (1963)
63. D.A. Kuzmanovic, I. Elashvili, C. Wick, C. O'Connell, S. Krueger, Structure **11**, 1339–1348 (2003)
64. C.H. Wick, P.E. McCubbin, Toxicol. Methods, **9**, 245–252 (1999)
65. R. Golmohammadi, K. Valegard, K. Fridborg, L. Liljas, J. Mol. Biol. **234**, 620–639 (1993)

10

Structure and Kinetics of Proteins Observed by Small Angle Neutron Scattering

M.W. Roessle, R.P. May

10.1 Introduction

Proteins are the machines of life. These molecules, composed of several hundred to thousands of atoms, are involved in all the processes and reactions inside a living organism. Depending on their functions and tasks, proteins can be classified into two major groups: First the structural proteins, which are responsible for the formation of passive overall structures such as hair fibres or the cytoskeleton of the cell. The second group are the function-related proteins, which facilitate the cell metabolism and regulation. Each protein has at least one three-dimensional structure in which it is stable and active under biological conditions. However, proteins are no rigid bodies, and a stable protein structure can be transformed to other stable states [1,2]. The transition between the distinct protein conformations is induced thermally or by specific binding of ligands or substrates, and several structural intermediates can be adopted. It is evident that transient intermediates can be more easily identified if one observes a signal that is related to structure, and rate constants can be derived very simply. Simultaneously, direct information about the protein structure during the reaction cycle can be obtained.

10.2 Solution Scattering

Protein structures can be analyzed at high resolution with several methods. X-ray crystallography and nuclear magnetic resonance (NMR) are able to resolve protein structures at the level of atomic sizes (1–3 Å), whereas electron microscopy (EM) provides a resolution of about 10–20 Å. A wealth of structural and functional information of proteins has been obtained by these methods.

Small angle scattering, a method that allows one to study macromolecular scattering in solution, is an attractive alternative, with in principle no restrictions such as size of the protein or its ability to form single crystals [3,4]. The price to be paid is limited resolution (similar to EM) and the observation of structures that are averaged over all orientations. On the other hand, there is practically no limitation to the composition of the scattering solution and the conditions applied to it, e.g., pressure, temperature, etc.

Since the zero-angle scattering is proportional to the square of the molecular mass of a molecule in solution at a fixed concentration, this parameter allows one to follow the formation of a complex of a macromolecule with another one; the decrease in concentration of the free second molecule (which is proportional to its concentration) diminishes the scattering less than the gain due to square term mentioned before.

The zero-angle scattering cannot be observed directly, due to the direct beam intensity that usually needs to be hidden behind a beamstop, but it can be calculated by extrapolation using the Guinier or Zimm approximations that are valid in a small angular range, in which also the radius of gyration R_G of the scattering particles can be obtained. This radius of gyration is equivalent to the radius of inertia in mechanics and is a measure for the elongation of the particles.

Scattering intensities are plotted vs. the momentum transfer Q rather than the scattering angle 2Θ. Q is defined as $Q = (4\pi/\lambda)\sin\Theta$, where λ is is the neutron wavelength. The Guinier and Zimm approximations are valid in a range of $Q < 1/R_G$.

Beyond this range, the shape of the scattering curves becomes specific for the shape and composition of a particular molecule. For example the particular symmetry of a sphere can be recognized by characteristic minima and maxima of the scattering curve.

In many cases, however, the scattering properties of a molecule are better understood if one transforms the scattering curve $I(Q)$ by a sine Fourier transformation into a pair distance distribution function $p(r)$ according to

$$p(r) = \frac{1}{2\pi^2}\int_0^\infty I(Q)\,Qr\,\sin Qr\,\mathrm{d}Q. \tag{10.1}$$

Since this transform is infinite with respect to Q, an indirect Fourier transform (IFT) method has been proposed by Glatter [5]. IFT uses a least-squares fit of the amplitudes c_i of a set of N Fourier-transformed equidistant B-spline functions ϕ_i to the measured scattering curves for obtaining the smoothest $p(r)$ function that is compatible with the scattering data.

$$I(Q) = 4\pi\int_0^{D_{\max}} p(r)\,\frac{\sin Qr}{Qr}\,\mathrm{dQ}, \text{ where } p(r) = \sum_{i=1}^{N} c_i\phi_i(r). \tag{10.2}$$

As can be seen in Eq. 10.2, the integral over r is limited to $D_{\max}$, i.e., the system must consist of noninteracting particles with a maximal dimension $D_{\max}$ at low concentration. Since in general the number of splines used for the fit exceeds the number of free parameters, a regularization procedure is required that will not be discussed here.

10.2.1 Specific Aspects of Neutron Scattering

Due to the particle and wave duality, neutrons can be used like X-rays as a probe for the internal structure of matter. Contrary to X-rays which are sensitive to the electric field of the atoms, neutrons are deviated by interactions with the nuclei. The strength of this interaction is different for every single isotope. In particular, neutrons "see" natural abundance hydrogens (^{1}H or H) completely different from deuterons (^{2}H or D; hydrogens that contain an additional neutron in their nucleus). Due to the difference in mass of hydrogen and deuterium, "heavy water" can be enriched by physical methods. Chemically, hydrogen and deuterium can often be hardly distinguished. This opens the way for the contrast variation method.

In the first instance, this means that the different components of biological material, namely proteins, lipids, sugars, and nucleic acids, can be distinguished by neutrons varying the percentage of D_2O in a mixture of H_2O and D_2O. This is due to the fact that the scattering amplitude, also called scattering length, of H is negative and about half as big as that of D or O (oxygen). Lipids, proteins, sugars, and nucleic acids contain different and decreasing amounts of hydrogen. If one defines a scattering length density as the sum of all scattering lengths in a volume, divided by that volume, one can observe that the different scattering lengths can be matched by specific H_2O/D_2O mixtures, about 0% D_2O for the fatty acid chains, 40% for proteins, about 50% for sugar, and 70% for nucleic acids.

Second, one can grow microorganisms relatively easily, in particular bacteria and yeast, in heavy-water containing media, using either deuterated rich carbon sources or minimal media forcing the microorgnisms to introduce the deuterium by metabolic pathways, and thus produce perdeuterated or partially deuterated proteins, etc. [6, 7]. Perdeuterated proteins even exceed the scattering length of D_2O, partially deuterated protein can be tailored such that they are matched by a chosen level of D_2O. In Fig. 10.1 the different contrast $\Delta\rho\,(= \rho_{\text{protein}} - \rho_{\text{solution}})$ of H-proteins and D-proteins is shown.

Finally, the interaction of neutrons with most matter is very weak, allowing one to use bulky sample environment without major effect on the scattering. The weak interaction also means that biological molecules do not get damaged by irradiation with long-wavelength "cold" neutrons, contrary to what happens with X-rays or electrons.

10.3 Time-Resolved Experiments: Dynamics vs. Steady State

In general, the high resolution techniques mentioned in Sect. 10.2 are able to characterize stable conformations of proteins, for instance the initial conformation before a reaction starts and the final conformation after the reaction

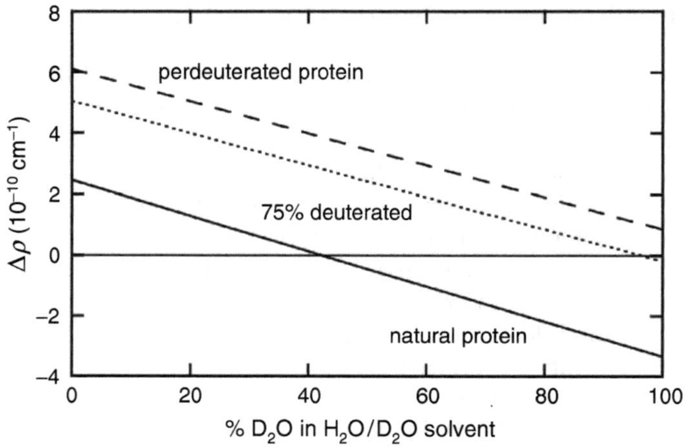

Fig. 10.1. Contrast $\Delta\rho$ of protonated, perdeuterated and 75% deuterated protein as a function of the D_2O content in the buffer solution. The scattering length density of protonated (natural) protein is matched by a solvent containing about 40% D_2O, that of 75% deuterated protein by nearly pure D_2O, while the perdeuterated protein remains always "visible" for neutrons

is finished. But time-resolved methods have to be used in order to trace the reaction between the two conformations. Table 10.1 shows that the conformational changes of proteins can occur on a wide range of time scales and spatial extents. The investigation of these structural changes requires suitable time-resolved methods such as time-resolved crystallography, NMR, or cryo-EM. However, several experimental parameters such as the size of the protein, time scale of the reaction and the triggering of the reaction [8, 9] restrict the use of these techniques. For instance, time-resolved Laue crystallography monitors conformational changes in the nanosecond time range with atomic resolution [10, 11], but is restricted to smaller proteins. Evidently, the crystal lattice itself must be maintained during the observation of the reaction, a condition that is not easily fulfilled, but also the conditions required for obtaining protein crystals may be prohibitive for studying a given reaction. Electron microscopy allows one to investigate larger molecules [12,13], but the solution must be frozen at discrete reaction steps [14]. Finally, NMR proofs time-dependent structural changes in detail, but it is also restricted to small proteins [15].

10.3.1 Protein Motions and Kinetics

The protein can perform work if energy is released by these processes. These protein dynamics occur on different time scales with different spatial extents. The different parameters of protein movements are listed in Table 10.1.

Table 10.1. The spatial extent and the time scales of movements which occur in proteins (adapted from S. Cusack in [16])

motion	spatial extent (Å)	characteristic time scale (s)
vibrations of bonded atoms	2–5	10^{-14}–10^{-13}
elastic vibration of globular regions	10–20	10^{-12}–10^{-11}
rotation of side chains at surface	5–10	10^{-11}–10^{-10}
relative motions of globular regions	10–20	10^{-10}–10^{-7}
allosteric transitions	5–50	10^{-4}–10^{0}
protein folding		10^{-5}–10^{1}

A special group of function-related proteins are the *motor proteins*. In contrast to enzymes, which catalyze chemical reactions, working or motor proteins are able to provide mechanical energy. These molecules bind nucleoside triphosphate (NTP) and convert it into nucleoside diphosphate (NDP) by hydrolysis of one phosphate. The released energy is used for processive or continuous motions [17, 18]. During these processive motions mechanical work can be performed. The more general term working protein is used if the protein can provide mechanical work; the motion itself is rather stepwise than continuous [19].

If the transition between the stable protein states requires a cascading process with several structural intermediates, a regulatory process has to be introduced. The simplest regulation is obtained by binding ligands that induce the transition between the inactive and the active conformation to the protein. The ligand acts as an activator for the transition inactive → active conformation, or it can act as an inhibitor for the conversion from the active state to the inactive state. This regulation process is possible for monomeric proteins with one ligand binding site. If the protein is assembled from several identical subunits, a more complicated process takes place. One speaks of an allosteric regulation process, if every subunit of a multimeric protein can bind the ligand, but the whole protein multimer is only functional if all the subunits have bound the ligand. This allosteric regulation is often characterized by special kinetics of the ligand binding [20, 21].

10.3.2 Cooperative Control of Protein Activity

As mentioned above, allosteric regulation is found in multimeric protein assemblies. If these proteins provide more than one ligand binding site, but all binding sites have to bind the ligands before the whole protein is converted into the active state, a regulatory scheme is needed. A so-called allosteric regulation scheme was proposed 1965 by Monod et al. [22]. Its main feature is the enhancement of the affinity for ligand binding after the first ligand is bound. This behavior is called ***cooperative*** binding of ligands. The opposite of this behavior, where the binding of a ligand inhibits the binding of other ligands, is called negative or ***anticooperativity***. In large protein assemblies, consisting of several subunits, the allosteric regulation can cause large conformational changes that may even involve the whole protein [23]. In addition, cooperative and anticooperative behavior can occur at the same protein using the same ligand. In these reactions the cooperative binding favors subunits to bind the ligand (which are forced in the tense T state), while it inhibits ligand binding on the other subunits (which stay in the relaxed R state) [23]. The biological advantage of allosteric regulation is that no additional control mechanism for the ligand is needed, because structural changes of the protein upon ligand binding control the reaction. The transition between the T and the R state can occur concerted, sequentially or by a combination of both. Since the analysis of cooperative and allosteric protein activity gives insights into the reaction mechanism and its regulation scheme, the investigation of these processes is of major interest in biophysics and biochemistry.

A typical representative of a cooperatively driven reaction is the chaperonin GroEL from *Escherichia coli.* Chaperonins assist protein folding and facilitate the repair of misfolded proteins as a part of the shock response of bacteria. For performing this task proteins of the chaperonin family are able to bind denatured proteins as substrates after recognizing hydrophobic patterns on the protein surface [24–26]. GroEL is a multimeric protein that consists of 14 identical subunits with a sevenfold symmetry revealing three different subdomains. The entire GroEL is build of two rings of seven subunits, that are stacked together back to back forming the typcial cylindrical shape of class II chaperonins. The regulatory scheme of this large protein assembly (800 kDa) and its 14 ATP binding sides exhibits an remarkable interplay between cooperative and anticooperative nucleotide binding. Only one of the GroEL rings can bind nucleotides, whereas the binding of the same nucleotide type is inhibited at the second ring. This behavior controls the substrate binding as well as the binding of the smaller co-chaperonin GroES to one end of the GroEL cylinder. Figure 10.2 shows the crystal structure and a low resolution model of the GroEL–GroES complex based on SANS data.

Cooperative protein kinetics is often investigated by biochemical steady-state methods. For instance, the steady-state formation rate of molecules in the T state can be measured as a function of ligand concentration using radioactive or fluorescence markers. From this analysis the Hill coefficient, a parameter for cooperative ligand binding can be obtained. Systems far from equilibrium, where the assumption of steady state is no longer valid, need to be studied using transient-state kinetics. For the analysis of a transient reaction mechanism, time-resolved techniques have to be used.

10.4 Protein Kinetic Analysis by Neutron Scattering Experiments

For kinetic studies of protein–protein interactions the reaction can be triggered by rapidly mixing the reaction partners [27–30], or by rapidly changing

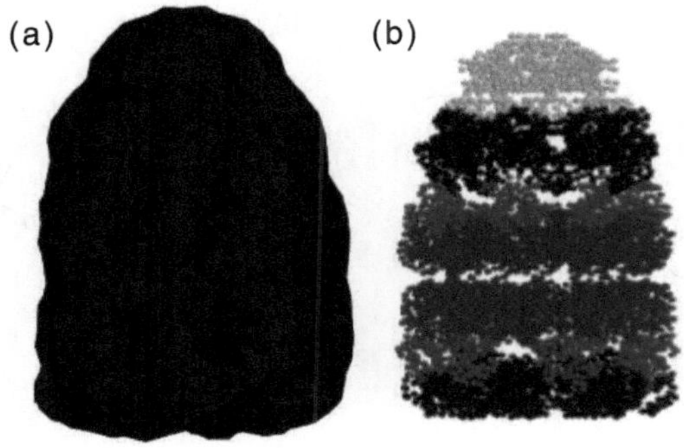

Fig. 10.2. Structures of the GroEL in complex with the co-chaperonin GroES (bound on one side of the GroEL cylinder), (**a**) Envelope reconstruction based on SANS data [7], (**b**) X-ray crystallographic data (Protein Data Bank entry: 1AON)

environmental conditions such as pH or ionic strength [31]. Beyond this approach, reaction intermediates can be stalled by introducing ligand substitutes that block the further reaction pathway. In special cases the reaction can be slowed down by cooling or by using viscous solutes, and the reaction intermediates can be investigated.

A typical approach for the substitution of a ligand is the exchange of adenosine triphosphate (ATP) by a nonhydrolysable equivalent (see Fig. 10.3). Such substitutes bind in many cases as specifically to the protein and can facilitate a structural change in the same way as ATP. However, the subsequent secession of the triphosphate during the hydrolysis reaction is inhibited. Thus, these kinetically stalled structures are reaction intermediates just before the hydrolysis reaction takes place.

10.4.1 Trapping of Reaction Intermediates: The (αβ)-Thermosome

The trapping approach was used for the investigation of a chaperonin from the archaebacterium *Thermoplasma acidophilum*. The thermosome is a member of the chaperonin family (see Section 10.3.2) It consists of two different subunits (α and β) that are alternating in an eightfold ring. Two of these rings bind back-to-back forming the typical double donut structure of the chaperonins. The ADP-bound form facilitates the binding of the misfolded substrate protein. ADP is replaced by ATP and the subsequent hydrolysis of ATP to ADP triggers the refolding process. The thermosome exists in at least two different conformations; an open cylindrical and a closed ball-like structure were observed by protein crystallography [32] and cryo-electron microscopy [33, 34], respectively (see Fig. 10.4). The two structures are considered as reaction intermediates in the reaction cycle of the chaperonin. The open conformation corresponds to the substrate acceptor state and the closed state to one of maybe several refolding intermediates.

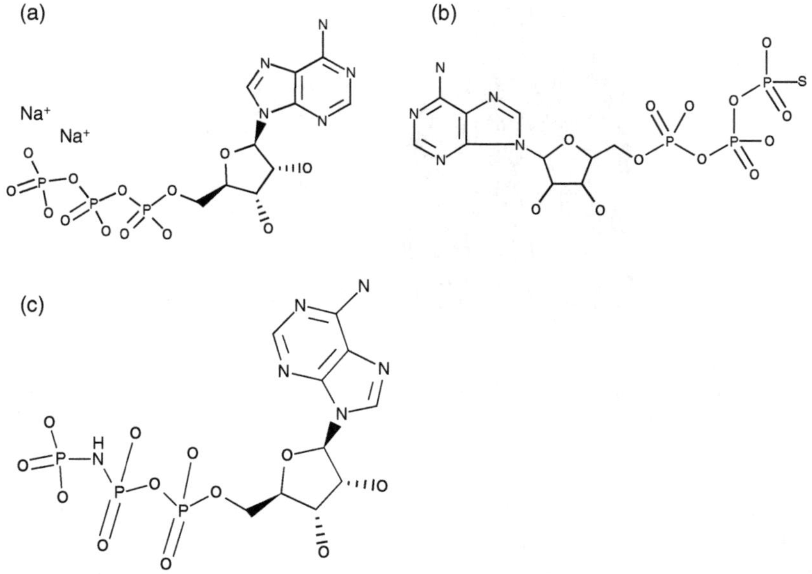

Fig. 10.3. (**a**) ATP and substitutes: (**b**) ATP-γS, (**c**) AMP-PNP

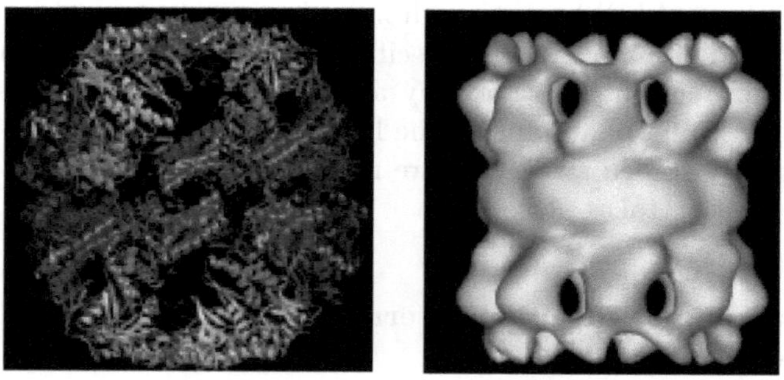

Fig. 10.4. Structures of the thermosome as derived by protein crystallography and cryo-electron microscopy

The energy gained in the ATP hydrolysis reaction is mainly used for triggering the chaperonin into a new reaction intermediate where the refolding of the substrate protein takes place. The SANS investigation of the reaction intermediates was facilitated by using different ligand substitutes and conditions. From other chaperonins such as the GroEL/GroES system from *E. coli* it is known that the structure of the reaction intermediates depends on the nucleotide bound to the protein. In contrast to the GroE system the thermosome does not possess a co-chaperonin. Thus, the necessary closing of the refolding vessel must be accomplished by the different domains of the thermosome. The SANS approach with trapping the reaction intermediates can be used in this case for stepping through the reaction.

The ATP-binding structure was observed by mimicking ATP with its nonhydrolysable analogue AMP-PNP, whereas the subsequent hydrolysis step was investigated by adding ADP together with an excess of phosphate to the thermosome. The phosphate-ADP condition drives the protein back to the state shortly after the hydrolysis reaction where the phosphate is no longer covalently bound, but still not ejected from the binding site. In order to probe the temperature dependence of the nucleotide-driven structural changes the experiments were performed by scanning different temperatures and nucleotides. Since the thermosome is found in an archea, which can stand high temperatures, a temperature range up to 50°C was chosen. These extreme experimental parameters are easily accessible to small-angle scattering.

The scattering curves corresponding to the closed and open states of the thermosome are shown in Fig. 10.5. The difference of the two states can be clearly distinguished by SANS. The experiments were performed on protonated thermosome in a buffer containing D_2O in order to enhance the scattering signal-to-noise ratio. As can be seen in Fig. 10.1, the contrast of natural protein is highest in D_2O, and at the same time, the incoherent scattering from D_2O is significantly lower than from H_2O.

The results listed in Table 10.2 show the transition of the open thermosome conformation to the closed state upon the ATP hydrolysis. Only at 50°C, the

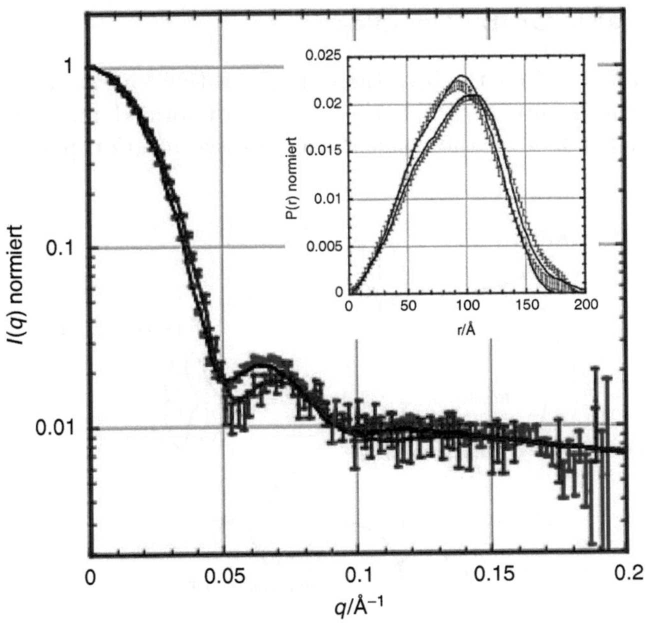

Fig. 10.5. Experimental data for the open and closed thermosome structure

Table 10.2. Conformation of the thermosome under different nucleotide conditions at 50°C

nucleotide	conformation
AMP-PNP	closed
ADP-AlF	open
ADP-Pi	closed
ADP	open
Pi (control)	open

thermosome a closed thermosome-ATP structure was observed, i.e., the thermosome behaved like GroE, while at lower temperatures, thermosome-ATP remained in an open conformation. It would be very difficult to observe this important finding with another method than solution scattering. Based on the structural data of this archebacterial chaperonin the individual reaction intermediates of the complete reaction cycle were identified [34, 35]. The proposed reaction cycle is shown the next Fig. 10.6.

The comparison of these results together with biochemical data and structures found by cryo-EM allowed to round off the understanding of the thermosome ATPase activity cycle, and the ADP-Pi state was found to be the rate-limiting step of the reaction.

10.4.2 Quasi-static Analysis of Reaction Kinetics–The Symmetric GroES–GroEL–GroES Complex

A standard method for the investigation of reaction kinetics are steady-state titration experiments in which the amount of formed reaction product is measured depending on the concentration of one reaction partner [36]. This

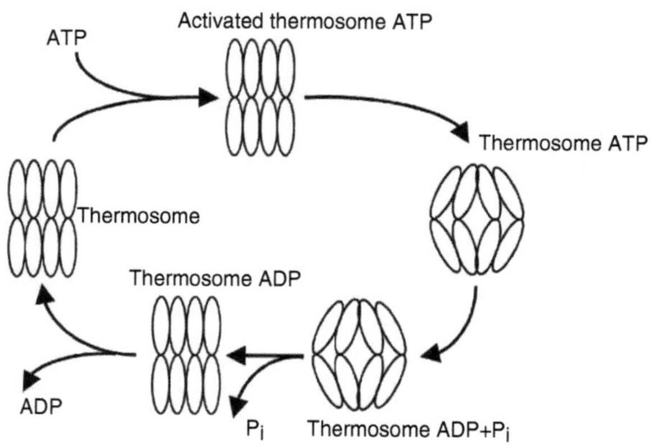

Fig. 10.6. Proposed reaction cycle for the thermosome upon nucleotide binding and hydrolysis [34]

approach can also be used in SANS, simply by measuring the overall scattering power, which is proportional to the molecular mass of the scattering particle (see Sect. 10.2).

However, much more information can be obtained by taking advantage of the different contrast of protonated and deuterated proteins. Let us look at the formation of the symmetric chaperonin complex GroEL–GroEL–GroES, a reaction intermediate that is found in the presence of high concentrations of ATP [37]. Although the physiological relevance of the symmetric complex for substrate folding is not proven, it is interesting for the study of the cooperative interactions of the GroE/nucleotide system

For the steady-state titration analysis, ATP was substituted by its non-hydrolysable analogue AMP-PNP. In order to override the anticooperativity of the GroEL the concentration of ANP-PNP was in a 100-fold excess with respect to GroEL. Under these conditions the second GroEL ring is switched into the GroES acceptor state. Protonated GroEL was matched by 40% D_2O, and the concentration of perdeuterated GroES was increased stepwise. Figure 10.7 shows the $p(r)$ functions derived from the scattering data.

As described above, only the deuterated GroES protein contributes to the scattering signal. Thus, the first maximum in the $p(r)$ function is due to the increasing amount of GroES; it stems from all distances within single, bound or unbound, GroES molecules. Besides this dominant maximum a smaller

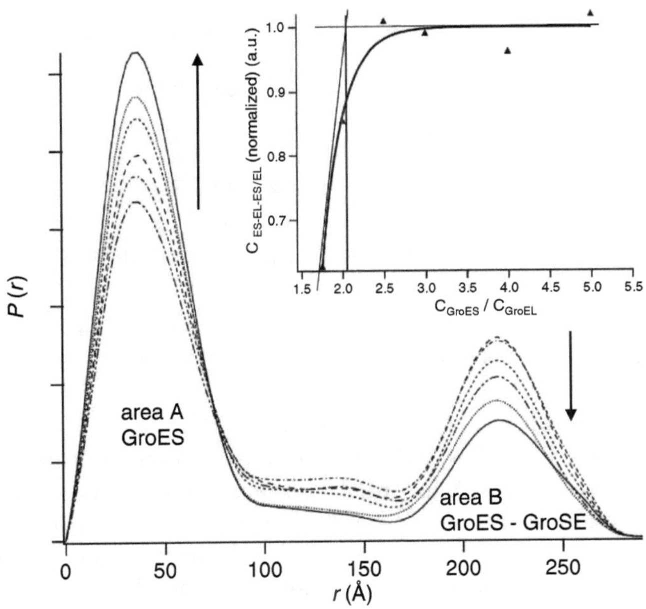

Fig. 10.7. Formation of the symmetrical GroES–GroEL–GroES complex. With increasing GroES concentration the amount of symmetric complex increases

peak at 210 Å indicates the binding of a second GroES to the opposite side of the GroEL cylinder; it is due to all distances linking a volume element in one GroES molecule to a volume element of another GroES at a fixed position at the opposite end of a GroEL molecule. This signal can be used to derive kinetic data for the formation of the symmetric GroES–GroEL–GroES complex. Since the integral of the $p(r)$ function is directly related to the molecular mass, the integral of the peak at 210 Å reflects the amount of formed symmetrical complex. The integral of the first maximum can be used for normalization. The data can be fitted by solving the equation for a reaction over an activated complex:

$$\text{GroES} + \text{GroEL} \Longrightarrow \text{GroES–GroEL} \Longrightarrow \text{GroES–GroEL–GroES}$$

In order to solve the differential equation of this reaction one can assume the following conditions:

- In the presence of a high amount of AMP-PNP every GroEL has bound at least one GroES
- For c_{EL}:$c_{\text{ES}} \leq 1$ all GroES is bound to GroEL in the asymmetric or symmetric complex

Under these assumptions the time-dependent concentrations of the symmetric complex are determined by the two differential equations [36]:

$$c_{\text{EL}} = c_{\text{asym.comp}} + c_{\text{sym.comp}}, \tag{10.3}$$

$$c_{\text{ES}} = c_{\text{asym. comp}} + 2 \cdot c_{\text{sym. comp}} + c_{\text{free ES}}. \tag{10.4}$$

By defining a dissociation constant K_{D}

$$K_{\text{D}} = \frac{k_{-1}}{k_1} = \frac{c_{\text{sym. comp}}\, c_{\text{ES}}}{c_{\text{asym. comp}}}, \tag{10.5}$$

the equation has to be solved for $c_{\text{asym. comp}}$, the parameter derived from the analysis of the peak integral:

$$c_{\text{asym. comp}} = \tfrac{1}{2}c_{\text{ES}} + \tfrac{1}{2}K_{\text{D}} - \sqrt{c_{\text{EL}} + c_{\text{EL}}c_{\text{ES}} + \tfrac{1}{4}c_{\text{ES}}^2 + \tfrac{1}{2}c_{\text{ES}}K_{\text{D}} + \tfrac{1}{4}K_{\text{D}}}. \tag{10.6}$$

The data fit is shown in Fig. 10.4 (right-hand side).

A value of 2×10^{-7} M is derived for K_{D}, corresponding to a rather stable binding of the second GroES to GroEL. The analysis of the tangential behavior of the fit function allows one to determine the binding ratio of GroEL and GroES. The crossing point at a ratio of 1:2 indicates that no excess of GroES is necessary for the formation of the symmetric complex. It is interesting to note that during the expression of the chaperonins in the cell GroEL and GroES are expressed exactly in this ratio.

10.4.3 Chasing Experiments (Slow Kinetics)

An elegant way of employing the contrast matching technique is materialized in so-called "chasing" experiments. GP31 a protein found in the bacteriophage T4 [38], is able to substitute itself for GroES in bacteria and to bind in its place to one end of the GroEL cylinder. Despite a low sequence homology the two high-resolution crystal structures available for GP31 and GroES are rather similar. Like the co-chaperonin GroES, GP31 consists of seven identical subunits arranged in a ring. GP31 has a larger molecular weight (12 kDa; GroES 9 kDa), and the elongated flexible loops of GP31 could help to enlarge the refolding cavity after binding to the GroEL cylinder. The replacement of GroES by GP31 is maybe necessary for the folding of GP23, an important protein for the assembly of the T4 phage head that folds only in the presence of GP31 into its active conformation. In Fig. 10.8 a comparison of GroES and GP31 exhibits the different structures derived by high-resolution protein crystallography.

Chasing can be observed when an excess of partially deuterated GroES is added to GroEL that is prebound with GroES or GP31 in a buffer solution containing an excess of ADP and that matches the scattering-length density of the partially deuterated GroES, rendering the latter "invisible" (see Fig. 10.9). Varying the relative concentration of GroES and GP31 allows one to determine their binding constants, since the scattering intensity of the sample decreases with time, as the apparent scattering mass of the H-GroEL–D-GroES complex is lower than that of the initial H-GroEL–H-GroES or H-GroEL–H-GP31 [41].

The data can be fitted with a double-exponential revealing two different dissociation constants (see Fig. 10.9). For the fast reaction a value of $1.3 \times 10^{-3}\,s^{-1}$ for the chasing of protonated GroES by deuterated GroES (control) and $4.9 \times 10^{-4}\,s^{-1}$ for the corresponding GP31 experiment. The second constants were $2.3 \times 10^{-5}\,s^{-1}$ for the chasing of GP31 by GroES and $3.9 \times 10^{-5}\,s^{-1}$ for the control. The double exponential behavior indicates two different reaction mechanisms. First, in the fast reaction the real chasing of

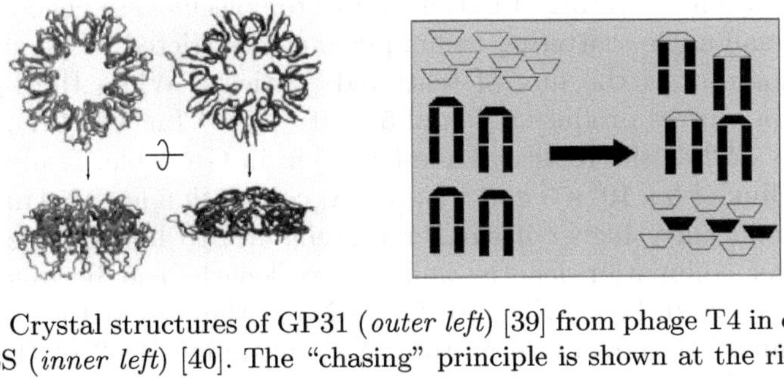

Fig. 10.8. Crystal structures of GP31 (*outer left*) [39] from phage T4 in comparison with GroES (*inner left*) [40]. The "chasing" principle is shown at the right picture (see text)

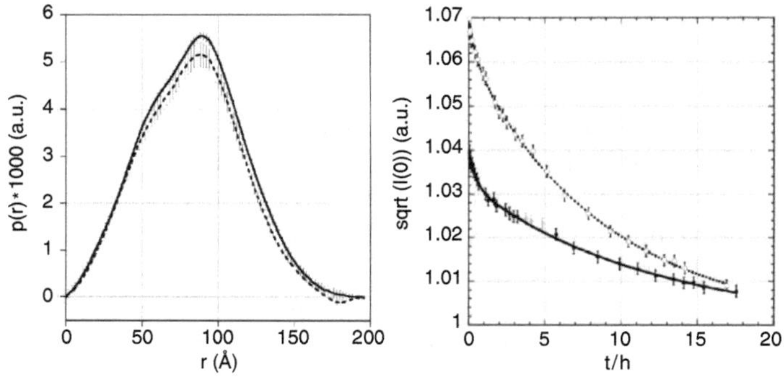

Fig. 10.9. Pair-distance distribution function of protonated GroEL and protonated GP31 with partially deuterated GroES in 80% D_2O (*left*), and data fitted with double-exponential (*right*)

bound GroES or GP31 by invisible GroES takes place. The already chased GroES or GP31 starts to compete with the invisible GroES. This competition slows down the apparent dissociation constant and can be described by a second exponential decay function. The different dissociation constants in the fast process indicate a higher stability of the GroEL–GP31 complex than the native GroEL–GroES assembly. GP31 is necessary for the folding of GP23, a special protein in the phage cycle one can argue that the higher stability of the GroEL–GP31 complex indicates an enhanced folding rate of the GP23 in the phage-infected bacteria. These higher stability and binding affinity are facilitated by the different binding sites of GP31 and GroES to the interface of GroEL.

10.4.4 Time Resolved Small-Angle Neutron Scattering

Since the advent of third-generation synchrotron sources (e.g., ESRF, APS, Spring 8) time resolved small-angle scattering (TR-SAS) was used successfully for solution scattering of biological macromolecules. In general, the time resolved small-angle scattering technique is not restricted to X-rays, but the limiting quantity is the flux of scattered particles. While third-generation synchrotron sources produce a flux of $5 \times 10^{12}\,\mathrm{ph\,s^{-1}}$ (at $\lambda = 1\,\text{Å}$), the SANS instrument D22 at the Institut Laue-Langevin in Grenoble, France, provides a neutron flux of $3 \times 10^7\,\mathrm{s^{-1}\,cm^{-2}}$ ($\lambda = 8\,\text{Å}$, collimation length 4 m). The flux can be significantly higher if smaller proteins are to be examined, because then shorter collimation lengths and/or wavelengths can be used. Another difference between TR-SAXS and TR-SANS is the size of the neutron and X-ray beams. The beam size at the SAXS beamline ID02 of the ESRF is $0.6 \times 0.2\,\mathrm{mm^2}$ while the beam size of D22 is $55 \times 40\,\mathrm{mm^2}$. For the optimal use of these neutron beam parameters a flat cell has to be employed, and the experiment has to be repeated several times.

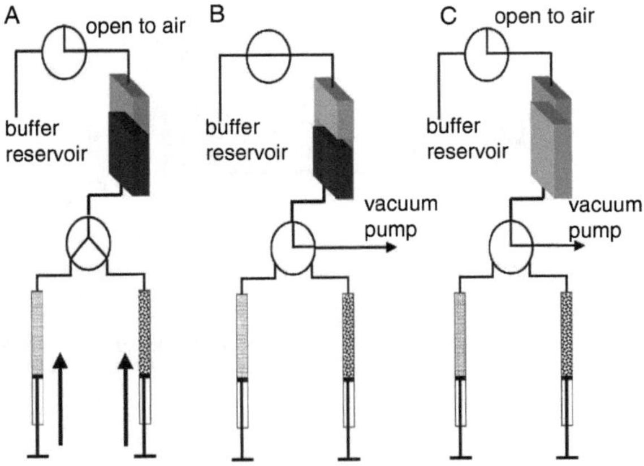

Fig. 10.10. Schematic design of a stopped flow apparatus for time resolved small angle neutron scattering: (**A**) mixing and filling step; (**B**) rinsing step with buffer/washing solution; (**C**) emptying of the cell for the next experiment

These special geometrical constrains of the TR-SANS approach have to be considered for the experimental setup. In Fig. 10.10 a schematic design of a stopped-flow apparatus is shown. It takes into account that – unlike the thin flow-through capillaries for X-ray scattering – the flat cell has to be emptied and rinsed after every experiment by separate steps. For proper rinsing two valves can be employed that allow first to empty the cell with a vacuum pump and second to switch to a washing step.

The rinsing and emptying steps ensure the same starting conditions for every new mixing experiment and do not use sample volume for emptying as used in TR-SAXS experiments where due to the use of thin flow-through capillaries the remaining sample is pulled out by next sample volume. However, in order to minimize the consumption of protein the exposure times of one time frame were restricted to a minimum of 1 s, and averaging over ten independent experiments led to sufficient statistics of the scattering data (see Fig. 10.11).

Frequently, the large difference in the scattering-length density between protonated and deuterated protein is used such that the scattering contribution of the protonated protein moiety in a complex is made equal to zero (at zero angle), while the deuterated protein is still (well) visible. Another option in TR-SANS is to enhance the signal for the protein–protein interaction by performing scattering experiments of H-protein and fully deuterated protein in 99% D_2O buffer.

In this experiment, the protonated protein has a strong negative scattering contrast while the deuterated protein has a positive contribution [30]. The interference cross-term of the scattering in this case is also negative, so that the radius of gyration of the H-GroES–D-GroEL complex is smaller than that of a homogeneous (protonated or deuterated) GroEL–GroES complex.

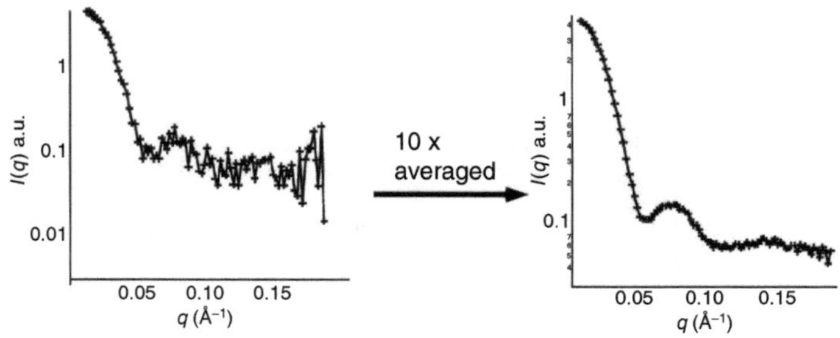

Fig. 10.11. Neutron scattering intensities of a single experiment and the averaging over 10 independent experiments

The decrease of the R_g value with time can be used to follow the binding reaction of the two proteins. This stratagem is safer than following an increase of intensity that can also be due to aggregation. In Fig. 10.12 the time dependence of radius of gyration for the binding of H-GroES to D-GroEL in the presence of a high concentration (30 mM) of the nonhydrolysable ATP analogue AMP-PNP is shown. Under these conditions the symmetric GroES–GroEL–GroES complex is formed. The time progression can be fitted to a double exponential function, indicating the successive binding of the second GroES molecule to GroEL.

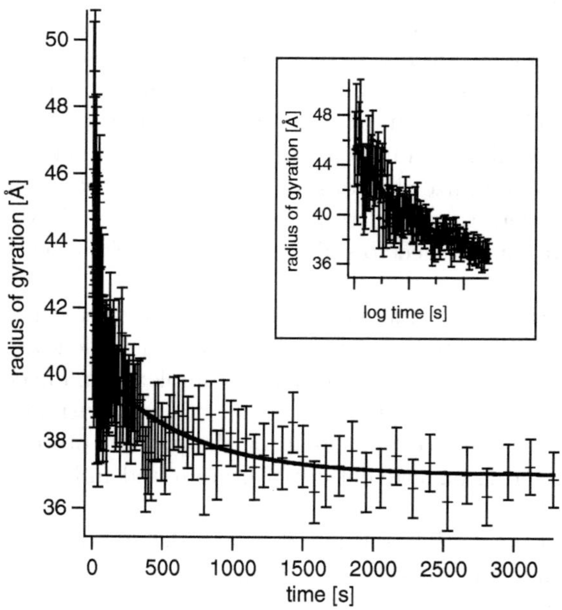

Fig. 10.12. Time dependence of the radius of gyration for the formation of the symmetric GroEL–$(GroES)_2$ complex [30]

10.5 Conclusions and Outlook

Small-angle scattering is a convenient and powerful tool for the investigation of protein–protein and protein–nucleic acid interactions. The lower flux of neutron sources compared to the high brilliance of third-generation X-ray sources is partially compensated by the higher contrast (= scattering power) of proteins in D_2O. The main advantages of ("cold") neutrons compared to X-rays are the absence of radiation damage and the benefits of contrast variation. Solvent contrast-variation and specific deuteration allow one to dissect biological macromolecules into partial structures. Novel approaches for the analysis of small-angle scattering data using these techniques are promising a rapid structure determination in the low-resolution regime [42]. Time-resolved neutron small-angle scattering is emerging as a tool for unique information concerning the structure of products and intermediates in the reaction of proteins. Since the derived kinetic data can be directly pinned down to specific conformational rearrangement processes, time-resolved small angle scattering permits the structural analysis of single reaction steps in protein interactions. Complete sets of such kinetic data would be an important step towards systems biology. The recent development of high-count-rate neutron detectors and of neutron-focusing optics will permit to measure the structures and kinetics of smaller molecules. The advent and further improvement of new pulsed neutron sources with higher fluxes will in future allow one to study more rapid reactions of macromolecules in solution and/or to use less of the sometimes very precious material.

References

1. H. Frauenfelder, S.G. Sligar, P.G. Wolynes, Science **254**, 1598–1603 (1991)
2. A.E. Garcia, J.A. Krumhansl, H. Frauenfelder, Proteins **29**, 153–160 (1997)
3. D.I. Svergun, M.H.J. Koch, Rep. Progr. Phys. **66**, 1735 (2003)
4. D.I. Svergun, M.H.J. Koch, Curr. Opin. Struct. Biol. **12**, 654 (2002)
5. O. Glatter, J. Appl. Cryst. **10**, 415–421 (1997)
6. H. Lederer, R.P. May, J.K. Kjems, W. Schaefer, H.L. Crespi, H. Heumann, Eur. J. Biochem. **156**, 655–659 (1986)
7. R. Stegmann, E. Manakova, M. Roessle, H. Heumann, S.E. Nieba-Axmann, A. Pluckthun, T. Hermann, R.P. May, A. Wiedenmann, J. Struct. Biol. **121**, 30–40 (1998)
8. K. Moffat, Acta Cryst. A **54**, (1998) 833–841
9. I. Schlichting, Accounts Chem. Res. **33**, 532–538 (2000)
10. B. Perman, V. Srajer, Z. Ren, T. Teng, C. Pradervand, T. Ursby, D. Bourgeois, F. Schotte, M. Wulff, R. Kort, K. Hellingwerf, K. Moffat, Science **279**, 1946–1950 (1997)
11. B. Perman, S. Anderson, M. Schmidt, K. Moffat, Cell. Mol. Biol. **46**, 895–913 (2000)
12. M.H. Chestnut, D.P. Siegel, J.L. Burns, Y. Talmon, Microscopy Res. Tech. **20**, 95–101, (1992)

13. E.M. Mandelkow, E. Mandelkow, R.A. Milligan, J. Cell Biol. **114**, 977–991 (1991)
14. H.R. Saibil, Nat. Struct. Biol. **7**, 711–714 (2000)
15. A.K. Bhuyan, J.B. Udgaonkar, Proteins **32**, 241–247 (1998)
16. S. Cusack, *Neutron and Synchroton Radiation for Condensed Matter Studies*, Les editions de physique, vol. III (Springer Verlag, 1994)
17. S.A. Endow, Nat. Cell Biol. **1**, 163–167 (1999)
18. R.D. Vale, R.A. Milligan, Science **288**, 95–100 (2000)
19. K.Jr. Kinosita, FASEB J. **13**, 201–208 (1999)
20. A. Mattevi, M. Rizzi, M. Bolognesi, Curr. Opin. Struct. Biol. **6**, 824–829 (1996)
21. K.E. Neet, Methods Enzymol. **249**, 519–567 (1995)
22. J. Monod, J. Wyman, J.P Changeux, J. Mol. Biol. **12**, 88–118 (1965)
23. M.F. Perutz, Quart. Rev. Biophys. **22**, 139–237 (1989)
24. R.J. Ellis, Nature **328**, 378–379 (1987)
25. T. Langer, G. Pfeifer, J. Martin, W. Baumeister, F.U. Hartl, EMBO **11**, 4757 (1992)
26. M. Klumpp, W. Baumeister, FEBS Lett. **430**, 73–77 (1998)
27. M. Roessle, E. Manakova, I. Lauer, T. Nawroth, J. Holzinger, T. Narayanan, S. Bernstorff, H. Amenitsch, H. Heumann, J. Appl. Cryst. **33**, 548–551 (2000)
28. A. Neidhart, T. Nawroth, M. Huetsch, K Dose, FEBS Lett. **280**, 179–182 (1991)
29. M. Roessle, E. Manakova, I. Lauer, T. Nawroth, R. Gebhardt, T. Narayanan, H. Heumann, ESRF Newslett. **33**, 10–11 (1999)
30. M. Roessle, E. Manakova, J. Holzinger, K. Vanatalu, R.P. May, H. Heumann, Physica B **276–278**, 532–533 (2000)
31. Th. Nawroth, M. Rusp, R.P. May, Physica B **350**, 635–638 (2004)
32. L. Ditzel, J. Lowe, D. Stock, K.O. Stetter, H. Huber, R. Huber, S. Steinbacher, Cell **93**, 125–138 (1998)
33. I. Gutsche, J. Holzinger, M. Roessle, H. Heumann, W. Baumeister, R.P. May, Curr. Biol. **10**, 405 (2000)
34. I. Gutsche, O. Mihalache, W. Baumeister, J. Mol. Biol. **300**, 187 (2000)
35. I. Gutsche, O. Mihalache, R. Hegerl, D. Typke, W. Baumeister, FEBS Lett. **477**, 278 (2000)
36. H. Gutfreund, *Kinetics for the Life Science* (Cambridge University Press, University of Bristol, 1995)
37. O. Llorca, S. Marco, J.L. Carrascosa, J.M. Valpuesta, FEBS Lett. **405**, 195–199 (1997)
38. A. Richardson, S.M. van der Vies, F. Keppel, A. Taher, S.J. Landry, C. Georgopoulos, J. Biol. Chem. **274**, 52–58 (1999)
39. J.F. Hunt, S.M. van der Vies, L. Henry, J. Deisenhofer, Cell **90**, 361–371 (1997)
40. J.F. Hunt, A.J. Weaver, S.J. Landry, L. Gierasch, J. Deisenhofer, Nature **379**, 37–45 (1996)
41. J. Holzinger, *Untersuchung der Reaktionszyklen von Chaperoninen aus Escherichia coli und Thermoplasma acidophilum mit Hilfe der Neutronenkleinwinkelstreuung.* (PhD Thesis, LMU, München, 2002)
42. M.B. Kozin, D.I. Svergun, J. Appl. Cryst. **34**, 33 (2001)

11

Complex Biological Structures: Collagen and Bone

P. Fratzl, O. Paris

11.1 Introduction

Biological materials such as wood, bone, or tendon are hierarchically structured and extremely complex. Properties have been shown to depend on a large extend on all size levels. For a better understanding of structure–function relations it is, therefore, necessary to get a better insight into their structure at all levels. The molecular and supramolecular levels are, in principle, accessible to electron microscopic or scattering techniques. However, the progress was slow mostly because of the difficulty to disentangle the contributions of the various size levels. One of the advantages of neutron and X-ray diffraction (or scattering) is the fact that information can be collected at the nanometer level in mostly intact macroscopic specimens. This advantage has been used in recent years to advance considerably the understanding of structure–function relations in collagen and bone. There are also some specific aspects where the use of (synchrotron radiation) X-rays was superior and others were the application of neutrons proved advantageous. These aspects will be worked out in more detail in the following sections. Roughly speaking, the high brilliance of synchrotron radiation is the key for position-resolved and/or time-resolved experiments. Position resolution is essential if one aims at exploring quantitatively the hierarchical nature of the structure (scanning X-ray diffraction or scanning small-angle scattering). Time resolution allows to focus on mechanisms of deformation, for example, because the material structure can be monitored during the deformation. Neutrons, on the other hand, profit from the fact that the scattering length densities are not directly linked to electron densities. In particular, for organic/inorganic mixtures such as bone, information can be collected on the structure of the organic component, even in fully mineralized tissue. With X-rays on the contrary, the huge signal from the inorganic phase can completely cover the scattering contribution from the organic component. Another strength of neutron scattering in this context is the possibility of contrast variation with D_2O/H_2O mixtures, which allows to highlight specific components in the structure, such as covalent cross-links in collagen for example. Inelastic (or quasielastic) neutron scattering can also give information on protein dynamics in such systems or can be used as a spectroscopic tool to identify phase compositions. Of course, there are further applications of neutrons not related to scattering, such as neutron activation analysis of bone to study the calcium–phosphorus ratio [1]. In this chapter, we will restrict ourselves to the applications of scattering (or diffraction) of neutrons and X-rays to study the complex biological materials, collagen and bone. Progress in this field has been closely linked also to the combined use of X-rays and neutrons which clearly emphasizes the complementarities of these two scattering approaches.

11.2 Collagenous Connective Tissue

11.2.1 Structure and Dynamics by Neutron Scattering

Collagen is among the most abundant proteins in vertebrates and a major constituent in many hierarchically structured biological tissues. It is found in tendon, bone, cartilage, cornea, skin, blood vessels, and other connective tissue and its function is mainly mechanical. A typical example is the tendon consisting mostly of collagen type I shown schematically in Fig. 11.1. Collagen molecules are triple helices with a length of about 300 nm. Two chains in this triplet (called type 1) are associated with a third chain (called type 2), which is similar but not identical. Molecules are assembled into fibrils by an axial staggering with a periodicity of 67 nm (Fig. 11.1d). Fibrils typically have a diameter of a few hundred nanometers. They are decorated with proteoglycans, which form a matrix between fibrils (Fig. 11.1b,c). Fibrils are assembled into fascicles and finally, into a tendon (Fig. 11.1a). While the axial stagger of the molecules in the fibrils (Fig. 11.1d) is well established, the full three-dimensional arrangement is still a matter of debate. There is a predominant liquidlike order with some degree of long-range molecular ordering on a quasi-hexagonal lattice in the cross-section of fibrils [2,3]. The outstanding mechanical properties of tendons are due to the optimization of their structure (see Fig. 11.1) on many level of hierarchy. One of the challenges is to work out the respective influence of these different levels. Only a few covalent cross-links are connecting each of the collagen molecules to its neighbors [4]. Nevertheless, these cross-links are crucial for the mechanical functioning of the collagenous tissue. Indeed, when the cross-link formation is inhibited, the strength of the tissue reduces dramatically and the mechanical behavior resembles more a viscous fluid than a solid [3]. One of the strengths of neutron scattering is that – using the methodology of contrast variation by D_2O/H_2O substitutions – the position of cross-links may be studied within the collagen tissue. Using neutron scattering, Wess and co-workers determined two positions of natural cross-links within the axial period of the collagen fibrils structure [5,6]. In addition, pathological cross-linking by glycation may occur in diseases such as diabetes.

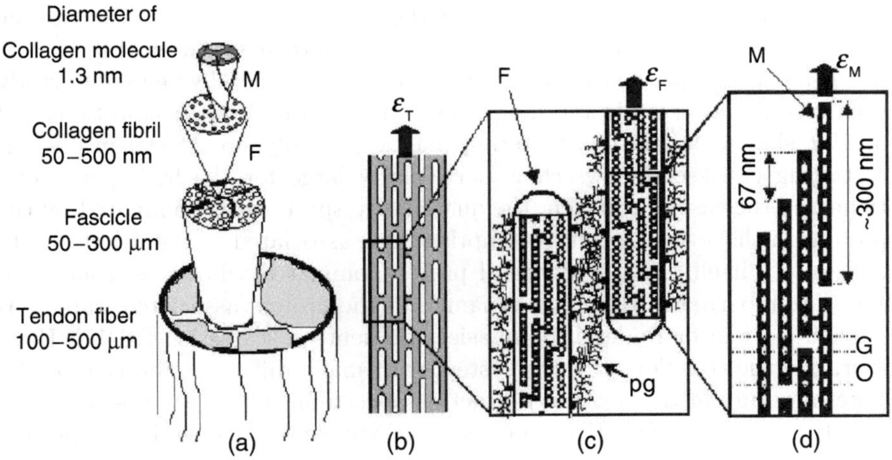

Fig. 11.1. (**a**) Simplified tendon structure (from [3]). Tendon is made of a number of parallel fascicles containing collagen fibrils (marked F), which are assemblies of parallel molecules (marked M). (**b**) The tendon fascicle can be viewed as a composite of collagen fibrils (having a thickness of several hundred nanometres and a length in the order of 10 μm) in a proteoglycan-rich matrix. (**c**) Proteoglycans are coating the fibril surfaces and connecting adjacent fibrils (**F**). (**d**) Triple-helical collagen molecules (M) are packed within fibrils in a staggered way with an axial spacing of $D = 67\,\text{nm}$. Since the length of the molecules (300 nm) is not an integer multiple of the staggering period, there is a succession of gap (**G**) and overlap (**O**) zones. The lateral spacing of the molecules is around 1.5 nm. The full three-dimensional arrangement is not yet fully clarified, but contains both elements of crystalline order and of disorder [2]

X-ray diffraction reveals that the electron density distribution along the period of the collagen fibrils changes in a systematic way with glycation [7]. Using neutron scattering and contrast variation, the glycation could be studied in even greater detail [8], giving additional information on the density and position of the cross-links along the period of the collagen fibril.

Collagen is not only the major constituent of tendon, but also of the cornea. Cornea is a tissue where collagen fibrils of uniform diameter are regularly arranged within a matrix of proteoglycans. In contrast to tendon, the fiber direction is not uniform, but the cornea consists of a succession of superposed layers with different fiber orientation each. The collagen fibrils in each individual layer are parallel, but not touching each other laterally as in tendon. They are kept at a significant distance from each other (probably by a thick proteoglycan layer [9]), which is essential for the transparency of the tissue (for reviews, see [10, 11]). While most structural information about cornea stems from electron microscopy and X-ray scattering, some details of the collagen structure in cornea could be better addressed by neutron rather than

X-ray scattering [12,13]. More recently, the attachment of chlorine ions to the cornea was studied by neutron scattering and contrast variation [14].

Inelastic and quasielastic neutron scattering can generally be used to study the dynamics of crystals and molecules giving information on vibrations and on diffusion. Quasielastic scattering makes generally use of the incoherent scattering cross-section which is particularly large for the hydrogen atom. Hence, in the case of collagen, the quasielastic spectrum is dominated by the protons in the structure which comprise those associated to water but also to the protein itself. First evidence of proton-dominated vibrations from collagen were reported in [15]. The dynamics of the protein associated water was studied later in more detail by quasielastic neutron scattering [16, 17]. However, due the complexity of the system with many different vibration modes superimposing in the neutron diffraction spectrum, it was not possible until now to reach a quantitative description of vibrational modes in the protein itself.

11.2.2 Elastic and Visco-elastic Behavior of Collagen from In situ Mechanical Experiments with Synchrotron Radiation

The outstanding mechanical properties of collagenous tissues are directly related to their hierarchical structure (Fig. 11.1). For instance, the stress–strain curve of tendon collagen shows a region of low stiffness at small strains (toe region), followed by an upward bending (stiffening) of the curve (heel region) and finally a linear region with high elastic modulus at large strains [18]. It is generally agreed that these different regions in the stress–strain curve are due to structural changes at the different hierarchical levels (Fig. 11.1), and X-ray diffraction using synchrotron radiation and simultaneous tensile testing has revealed different strains at different levels in the same tendon [19–24]. The toe region can be attributed to the removal of a macroscopic crimp with a period of roughly 100 µm, being visible also in the light microscope [25]. In the heel region, an entropic mechanism has been deduced from in situ synchrotron radiation studies of the equatorial scattering, which explains the increasing stiffness by straightening out molecular kinks in the gap region of collagen fibrils [20]. The linear region is thought to be due to the stretching of the collagen molecules themselves, as well as the cross-links between the molecules (Fig. 11.1), implying also a side by side gliding of the molecules [21–27]. The unique quality of such in situ experiments lies in the possibility to measure strains at different hierarchical levels such as the macroscopic strain of the whole tendon (ϵT), the strain of the collagen fibrils (ϵF) and the strain of the collagen molecules (ϵM). The finding that $\epsilon M < \epsilon F < \epsilon T$ underlines the importance of the cross-links as well as the influence of the proteoglycan matrix, which contributes presumably to the total tendon extension through a shear deformation. This could very recently be demonstrated by systematically investigating the strain rate dependence from tendons with normal and cross-link deficient collagen and a simple visco-elastic model was set up to qualitatively describe the observed effects [24]. In this respect, we also see a great potential for in situ neutron diffraction as a complementary tool to in situ X-ray studies. In situ tensile testing in different D_2O/H_2O solutions could further help to clarify the role of cross-links for the mechanical properties of tendon collagen.

11.3 Bone and other Calcified Tissue

11.3.1 Structure of Mineralized Collagen: Contributions from Neutron Scattering

Bone consists mainly of collagen type I and mineral particles deposited inside this organic matrix. The mineral is carbonated hydroxyapatite (dahlite) in the form of extremely small particles (about 2–4 nm thickness [28, 29]) containing many crystal defects. The basic building block of bone is the mineralized collagen fibril as shown in Fig. 11.2. The collagen fibrils are similar to those in tendon (except for a different cross-link pattern [4]) and show a succession of gap and overlap zones along the axial direction of the fibril (see Fig. 11.2). It is believed that the mineral particles nucleate inside the gap zones [30] and then grow also into the overlap regions.

The earliest evidence for this picture of the mineralized collagen fibril came from neutron scattering investigations [31–33]. As shown in Fig. 11.2, the fact that mineral is located primarily in the gap regions changes the electron density distribution along the fibril axis from a situation where the electron density is lower in the gap (U) to one where it is higher (M). In both cases, one expects Bragg reflections along the meridional direction and they are, indeed, observed for both mineralized and un-mineralized collagen by small-angle X-ray diffraction (see, e.g., [34]). When neutron are used, however, the scattering length density of the mineral is not totally dominating the signal from the protein and – using contrast variation principles – it could actually be proven that mineral particles are deposited with a periodicity corresponding to the gap-overlap succession [31, 33]. This general picture was later confirmed by transmission electron microscopy and three-dimensional reconstruction [28].

In addition to the axial structure, neutron scattering has also been operational to determine the meridional packing of collagen molecules in fully mineralized tissues [35–41]. In fact, the precise packing of collagen molecules in the fibrils has not yet been fully clarified [2, 3]. Most probably, it consists of regions of three-dimensional order [42] and others with more liquid-crystalline character. A possible arrangement of molecules reconciling these aspects of order and disorder [43] is shown in Fig. 11.3. Obviously, there is a typical spacing between neighboring molecules which gives rise to a broad peak in the X-ray or neutron diffraction pattern (Fig. 11.4). In this respect, it is very advantageous to use neutrons, when fully mineralized collagen tissues are to be investigated,

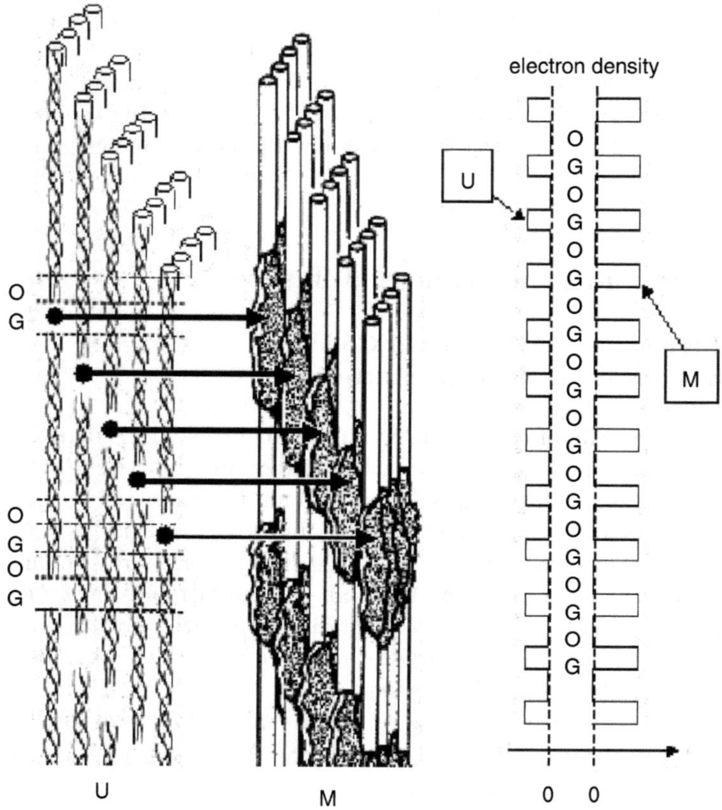

Fig. 11.2. Sketch of the location of mineral particles inside a collagen fibril. The un-mineralized fibril (**U**) is similar to the collagen fibrils found in tendon (except the covalent cross-links between the molecules which might be different [4]), see Fig. 11.1. Due to the axial stagger of the molecules there is a succession of gap (**G**) and overlap (**O**) zones. Mineral is thought to nucleate in the gaps and the plate-like crystals then spread into the collagen matrix. As a consequence, mineral particles are disposed according to the succession of gap and overlap zones inside the mineralized fibril (**M**). The drawing of mineral particles in collagen was adapted from Landis [30]. As a consequence, the axial trace of projected electron densities (*right*) is roughly periodic for both the un-mineralized (**U**) and the mineralized (**M**) tendon: For U the density is smaller in the gap regions and for M it is larger

since the (comparatively weak) signal from the collagen molecules can still be seen on top of the small-angle scattering from the mineral. When X-rays are used, the large electron density from the mineral can completely dominate the signal. The intermolecular spacing is in the order of 1.1 nm for fully dry fibrils. In un-mineralized collagen, the intermolecular distance can grow up to 1.6 nm and the diffraction peak shifts accordingly (see Fig. 11.4). In mineralized tis-

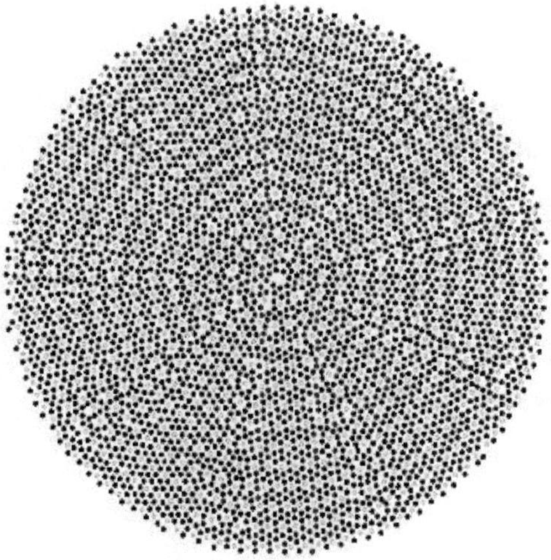

Fig. 11.3. Model for the cross-section through a collagen fibril according to [43]. Each dot corresponds to a collagen molecule. The lighter dots are molecules with the same axial stagger position. The darker spots are molecules with any of the other four axial stagger positions. The structure is dominated by short-range ordering (as in a two-dimensional liquid) with some elements of long-range ordering, possibly such as in grains of a polycrystal

sues, neutron scattering has shown that the swelling is limited by the amount of mineral in the fibril (see Fig. 11.5). This kind of data was essential to prove that the mineralization process of the bone matrix corresponds essentially to a replacement of water by mineral [39,40]. Therefore, the largest possible mineral contents seems to be defined by the original water content of the tissue (Fig. 11.5). Moreover, these data also show that the collagen molecules are not enclosed individually inside a mineral "wall". They are being compressed by extra-fibrillar or interfibrillar mineral, perhaps in the way sketched in Fig. 11.4 (line IV).

Quite recently, there is a renewed interest in the actual composition of the bone mineral, which is close to, but not quite hydroxyapatite ($Ca_{10}(PO_4)_6OH$). It is well known that the mineral is deficient in phosphate which is partially substituted by carbonate. The availability of very large neutron-scattering intensities from the nuclear-spin incoherence in hydrogen has permitted the determination of atomic motion of hydrogen in synthetic hydroxyapatite and in de-proteinated isolated apatite crystals of bovine and rat hone without the interference of vibrational modes from other structural units [44]. From such studies it was claimed initially that virtually all OH groups are also substituted by carbonate in bone mineral [44]. More recent data suggest, however, that this substitution is not complete and that at least 50% of the OH are present in bone mineral [45, 46].

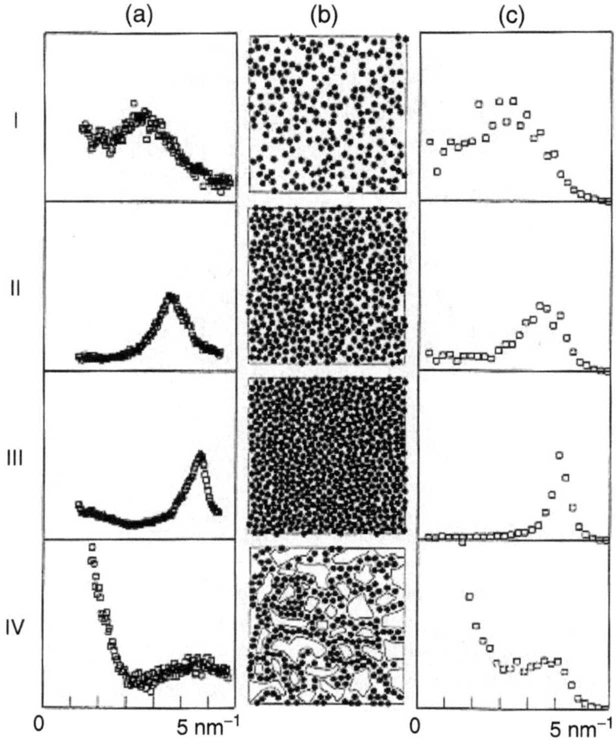

Fig. 11.4. (**a**) Meridional X-ray diffraction peak as a function of the length of the scattering vector q from turkey leg tendon with different degrees of hydration from fully wet (I) to dry (III). The peak is seen to shift to larger q-values indicating that molecules come closer with drying. (**b**) shows a simulated arrangement of molecules with a calculated diffraction pattern as shown in (**c**). The last line (IV) is a proposed arrangement for mineral particles in bone. The data (IVa) are from the mineralized part of the turkey leg tendon (figure taken from [34])

11.3.2 Investigating the Hierarchical Structure of Bone

Bone is a hierarchical tissue which adapts its detailed structure to the functional needs [47]. As an example, Fig. 11.6 shows several hierarchical levels in the shaft of a typical mammalian long bone (femur, tibia, etc.). The outer shells of such bones are made of compact bone which has typically a lamellar character (Fig. 11.6b). The lamellae consist of layers of parallel mineralized collagen fibrils (Fig. 11.6c). The typical diameter of collagen fibrils is in the range of a few hundred nanometers and they are reinforced by platelets with a thickness of 2–4 nm. These fibrils are arranged in parallel to form layers, whose orientations turn in a plywood-like manner [48, 49]. Since the mineral particles are crystalline, their Bragg reflections can be used to determine the crystallographic orientation distribution of the mineral particles in the entire

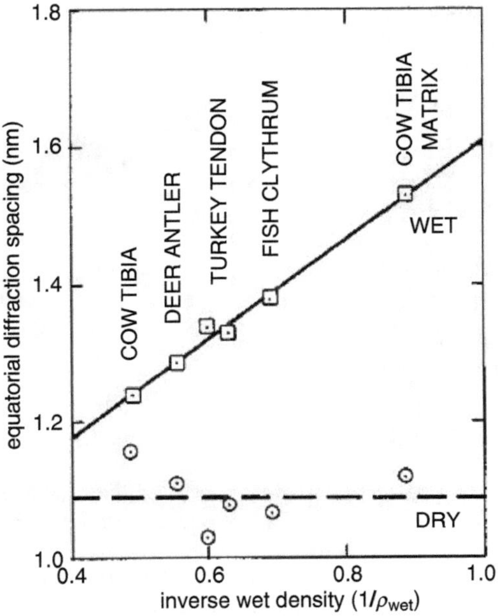

Fig. 11.5. Equatorial diffraction spacing (derived from the position of the maximum of a diffraction signal such as in Fig. 11.4a) as a function of mineral content (expressed in terms of the macroscopic density) for wet and dry tissue (from [37]). The molecular spacing in dry tissue is always around 1.1 nm irrespective of the mineral content. The higher the mineral content, however, the smaller the intermolecular spacing in wet tissue

tissue. Such texture investigations have been carried out by X-ray diffraction [50–52], and also by neutron scattering [53–56]. In the case of long bones, it was generally observed that the crystals are oriented with their longest dimension parallel to the long axis of the bone (which corresponds to the crystallographic *c*-axis of the hydroxyapatite). Moreover, this orientation gets more pronounced with age, which points to a continuous optimization of the bone's nanostructure to withstand bending loads. Neutrons were useful in particular for anatomical studies with the aim to correlate the local mineral orientation to the stress patterns produced by specific loading situations in different types of bones [56]. In contrast to diffraction using neutrons or X-rays – including also small-angle diffraction to study for instance the axial periodicity of collagen – small-angle scattering (SAS) is usually understood as diffuse scattering, being sensitive to weakly ordered structures in the range from about 1 nm to about 100 nm [57–59]. Hence, structural parameters from the mineral particles in the collagen matrix of bone may be derived from SAS measurements. Small-angle X-ray scattering (SAXS) has been established and applied systematically in recent years to study size, shape, and orientation of mineral particles in bone [51, 59–66] and other mineralized tissue [29, 67–69],

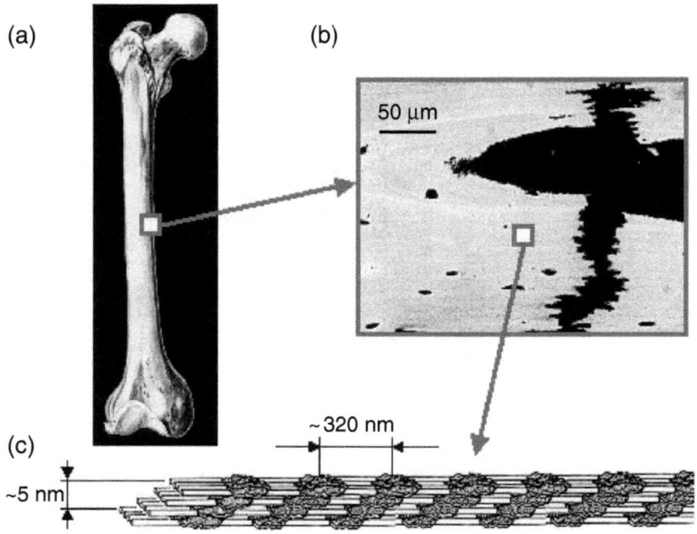

Fig. 11.6. Hierarchical structure of compact bone in mammalian long bones. It consists typically of mineralized collagen fibrils (**c**, see also Fig. 11.2) which are assembled into a lamellar structure (**b**). The fibril orientation is typically rotating inside the lamellae in a plywood-like fashion [48, 49]. The typical size of mineral particles is a few nanometers, the diameter of fibrils a few hundred nanometers, the thickness of lamellae a few micrometers and the thickness of the cortical bone shell a few millimeters

and in the meantime it is a well-recognized analytical tool to study for instance medical questions related to bone diseases [70–77].

One of the consequences of the hierarchical architecture of bone (Figs. 11.6 and 11.7) is the fact that nanometer sized features such as the size and orientation of mineral particles may vary systematically throughout the tissue on a larger (typically micrometer-sized) scale. This introduces the technical difficulty that the nanometer-scale structures need, in principle, be characterized in a position-resolved way.

As an example, three distinct hierarchical levels are shown in Fig. 11.7 for human cancellous bone, i.e., the macroscopic anatomy of the vertebral body (mm), the size of single trabeculae (~200 μm) and the scale of single lamellae within the trabeculae (~20 μm). A very important impact of SAXS in recent years was indeed the possibility to use a beam size adapted to the hierarchy level of interest, allowing a position-resolved analysis by scanning of bone sections across the X-ray beam [59, 62, 64]. The principle of a scanning SAXS or scanning XRD experiments is very simple and is depicted schematically in Fig. 11.8. The specimen can be moved in two directions (x and y) perpendicular to the incoming X-ray beam. Scanning the sample through the beam while successively recording scattering patterns yields a two-dimensional image of

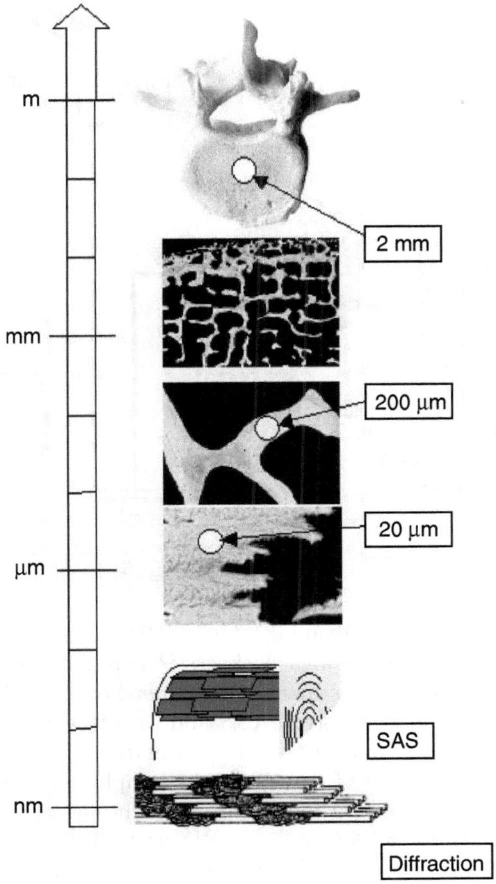

Fig. 11.7. Hierarchical structure of mammalian cancellous bone. As compared to compact bone (Fig. 11.6), cancellous bone such as vertebra exhibits one more hierarchical level, namely a foam-like structure with struts of 100 – 300 µm – the so called trabeculae – pointing preferentially into the direction of principal stresses. The structures at the nanometer scale can be investigated by small-angle scattering (SAS) and diffraction techniques. Combination with scanning techniques using a beam adapted to the hierarchical level of interest (shown for three examples by the white circles) permits to image the nanostructure with a resolution corresponding to the beam size

these patterns. The lateral size of the beam – together with the specimen thickness, which should be adapted to the beam size – defines the spatial resolution of the scanning image. The scattering patterns are usually recorded in transmission geometry using a high resolution area detector as depicted in Fig. 11.8, allowing in some cases to record even both, the XRD and the SAXS signals simultaneously [78,79]. The evaluation of the scattering patterns

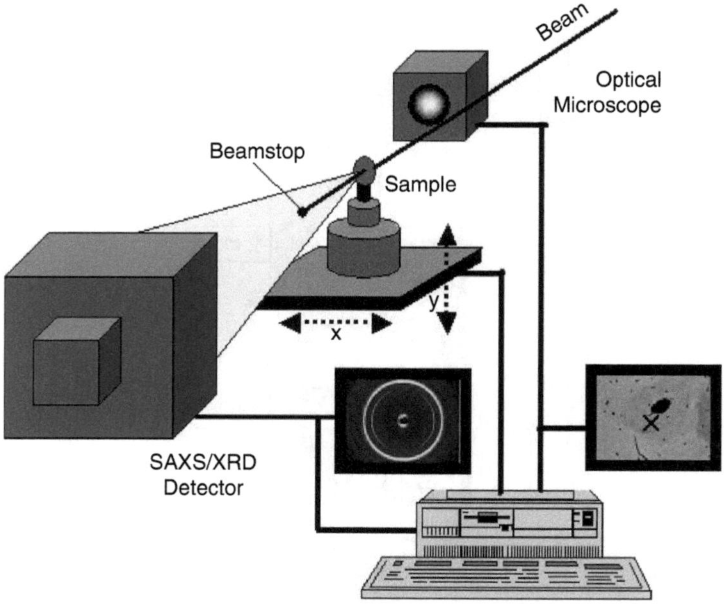

Fig. 11.8. Schematic picture of a scanning SAXS/XRD setup. An X-ray microbeam of lateral size D penetrates a sample with thickness D, defining an interaction volume of D^3. The sample is scanned in two dimensions perpendicular to the primary beam with a step width of D, and successive SAXS and/or XRD patterns are collected for any scanning step using a high resolution area detector. To keep measurement time reasonable, scanning regions of interest at the sample can be selected with the help of an optical microscope. Data reduction to obtain one or several nanostructural parameters from every SAXS and/or XRD pattern, and matching of these parameters with real space images of the specimen (e.g., a light microscopy image) is presently done offline after the experiment. In principle, fully automated data reduction, matching, and visualization may be possible online for certain application and will be a challenge for future developments

provides structural parameters from the nanometer level at every position of the scan. Such parameters may then be superimposed on a micrograph of the specimen, for instance on an X-ray scanning absorption image which has been taken prior to the scattering experiment [62,64,80]. Other options are to use an online, off-axis microscope such as in Fig. 11.8, or an off-line, on-axis microscope at a calibrated position with respect to the X-ray beam [79]. A scanning resolution in the order of about 200 μm can be achieved using laboratory X-ray sources [62], but the most promising development is due to the high brilliance of third generation synchrotron radiation sources. X-ray microbeams of some micrometers diameter are nowadays available routinely at many dedicated microfocus beamlines (for instance at the beamline ID13 at the European Synchrotron Radiation Facility, ESRF, in Grenoble [79]), and current developments permit to reach the 100 nm regime in the near future. By building up scanning devices in combination with X-ray (or neutron) scattering it is thus possible, in principle, to cover the size-ranges from atomic dimensions to about 100 nm (by XRD and SAXS in reciprocal space) and above 100 nm (by moving the specimen stepwise across the beam in real space) in a single experiment. Moreover, the capability of imaging a specimen in the micrometer range opens the possibility of combining several experimental approaches. For instance, the combination of scanning SAXS with light microscopy and scanning electron microscopy (SEM) [63, 65], with FTIR-Microspectroscopy [81], and even with local imaging of mechanical properties using nano-indentation [67] has been demonstrated for the analysis of mineral particles in calcified tissue.

Scanning SAXS with 200 μm Resolution: Case Example of Cancellous bone

As shown in Fig. 11.7, the trabeculae in the spongiosa of human cancellous bone have a thickness of a few hundred micrometers. Thus, the mineral crystals in individual trabeculae can be investigated by scanning SAXS using a laboratory X-ray source with a beam diameter of about 200 μm [62]. Figure 11.9 shows a typical result for a 200 μm thick section of human vertebra embedded in polymethylmethacrylate. Figure 11.9a shows a scanning X-ray radiography taken by scanning the sample through the beam and measuring the transmitted intensity through the specimen at each position. The trabeculae appear as darker regions (higher absorption) in the radiography. Scattering patterns from selected areas on the sample are shown in Fig. 11.9b, where the scattered intensity is presented in a pseudo-gray scale, dark corresponding to low and bright to high intensity. Two of the scattering patterns taken at positions within the trabeculae have an elliptical shape, the third pattern (top left of Fig. 11.9b) was recorded in a region outside the trabeculae and shows isotropic background scattering from the resin.

An important nanostructural parameter derived from the SAXS patterns (Fig. 11.9b) is the ratio between the total volume and the total surface of mineral per unit volume of bone matrix, respectively [61]. It can easily be shown that for constant composition (around 50 vol% mineral content) this so called T-parameter is a measure for the smallest particle dimension, i.e., the thickness of the mineral platelets in bone. Figure 11.9c shows the values of the T-parameter obtained at various positions in the vertebra section. Smaller values of T at some outer trabecular regions can be attributed to younger bone formed by remodeling processes [63]. Two further parameters, which can directly be deduced from the orientation and the eccentricity of the elliptically shaped SAXS patterns, are the main orientation of the plate shaped mineral particles and their degree of alignment, respectively [61]. Particle orientation and degree of alignment are indicated by the direction and the length of the bars in Fig. 11.9d, respectively. It is seen that the long axis of the mineral

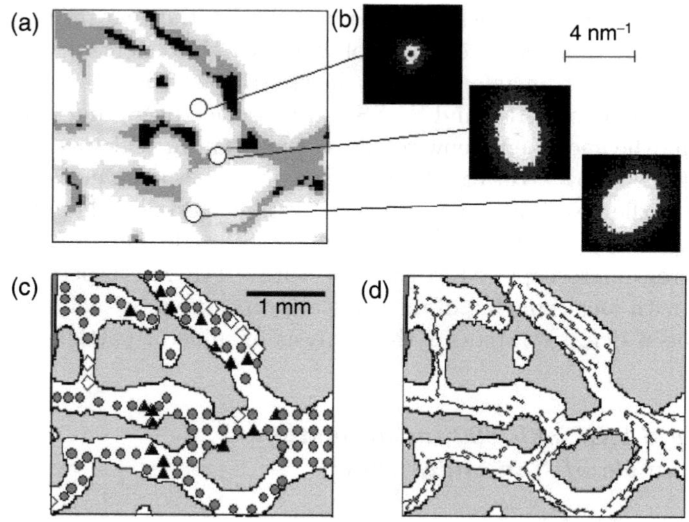

Fig. 11.9. Example of scanning SAXS results from trabecular bone using an X-ray beam of 200 µm size (from [62]). (**a**) shows scanning absorption image with the trabeculae in dark (high absorption) embedded in resin (bright = low absorption). SAXS patterns are shown in (**b**) for selected regions (indicated by circles) of the trabecular structure and in the resin. The length scale for the scattering vector q is also indicated and the scattered intensity is presented in a logarithmic pseudo-gray scale (dark = low intensity, bright = high intensity). (**c**) shows the contours of the trabeculae with the "average thickness" T of the plate-like mineral particles superimposed. T is calculated from the integrated intensity and the Porod constant P of the azimuthally averaged SAXS patterns, $T = 4\tilde{I}/\pi P = 4\Phi(1-\Phi)/\sigma$, where Φ and σ are the total volume and surface of mineral per unit volume of bone matrix, respectively. *Triangles*: $T > 3.5$ nm, *circles*: $3.5\,\text{nm} < T < 3.0\,\text{nm}$, *diamonds*: $T < 3.0$ nm. (**d**) shows the orientation (direction of the bars) and the degree of alignment ρ (length of the bars) of the mineral particles with respect to the trabecular structure. The degree of alignment is a parameter varying between 0 and 1, where 0 corresponds to random orientation and 1 to a perfectly parallel alignment of the mineral platelets. The longest bars in (**d**) correspond to $\rho \approx 0.5$

particles points preferentially into the direction of the trabeculae and such a behavior is found typical for human cancellous bone [63, 65].

Knowing that the trabecular orientation corresponds also to the principal stress directions, the optimization of the material bone with regard to its mechanical function at all levels is impressively demonstrated by the results in Fig. 11.9. A number of fundamental [63, 65, 66, 81], and medical questions [74, 75, 77] have in turn been studied on calcified tissue with the laboratory equipment providing 200 µm scanning resolution. Very recently, the experiments on human cancellous bone have been extended to local 3D SAXS measurements within single trabeculae by applying an additional sample rotation [51]. This allows one to construct so called SAXS pole-figures, which quantify the 3D distribution of the local average habit plane of mineral platelets with respect to an external coordinate system (e.g., the trabeculae direction). By combining this SAXS pole-figure analysis with classical pole-figure analysis using wide angle X-ray diffraction from the bone mineral on exactly the same specimen positions, complementary information about the crystallographic texture (XRD) and the morphological texture (SAXS) could be gained [51].

Towards One Micrometer Position Resolution: Lamellar Bone and Other Calcified Tissue

While for the investigation of the trabecular structure of bone a spatial resolution of about 200 μm is sufficient, many other structural features require a smaller beam size and consequently, synchrotron radiation comes into play. A first dedicated synchrotron setup for scanning SAXS of bone was designed to have a standard beam size of 20 μm by using simple pinhole collimation [80]. Even though this was a rather inefficient way to create a microbeam, a flux of about 108 photons per second could be reached at the synchrotron radiation source ELETTRA, permitting SAXS scans on bone and other calcified tissues within some hours by carefully selecting small regions of interest [80]. An example for scanning SAXS with 20 μm resolution is depicted in Fig. 11.10, showing an overlay of two scanning images from human cortical bone: in the background a backscattered electron image, and in the foreground an image of SAXS patterns from the same specimen positions exactly matched to the BE image [69, 80]. The different gray levels in the BE image correspond to different local mineral density (brighter gray level means higher mineral density) [82]. It is obvious that the mineral density is clearly lower in the osteons, which are the regions surrounding the elliptical holes [83]). The 2D image of SAXS patterns is overlaid to part the backscattered electron image, with the contours of the osteons indicated for better visualization. The SAXS intensity is clearly lower in the regions of the osteons, in qualitative agreement with the BEI images. The elliptical shape of the iso-intensity gray levels in the SAXS patterns indicates a preferred orientation of the mineral particles also in cortical bone, and their changing orientation suggests local changes of the morphological texture. It could indeed be shown that the particles are aligned in a cylindrical manner around the osteons [64]. Other examples where a position resolution of 20 μm was used to study mineralized tissue concern the investigation of the interface region between bone and different types of mineralized cartilage [69] or the region near the edge of the mineralization front in mineralized turkey leg tendon [68]. Beside scanning SAXS, some first position resolved X-ray diffraction experiments on calcified tissue with highest position resolution around – or even below – 1 μm have recently been reported from the European Synchrotron Radiation Facility (ESRF) in Grenoble, such as for instance local investigations of archeological bone samples [84] or studies of reconstructed bone near prosthetic surfaces [85]. It has also been

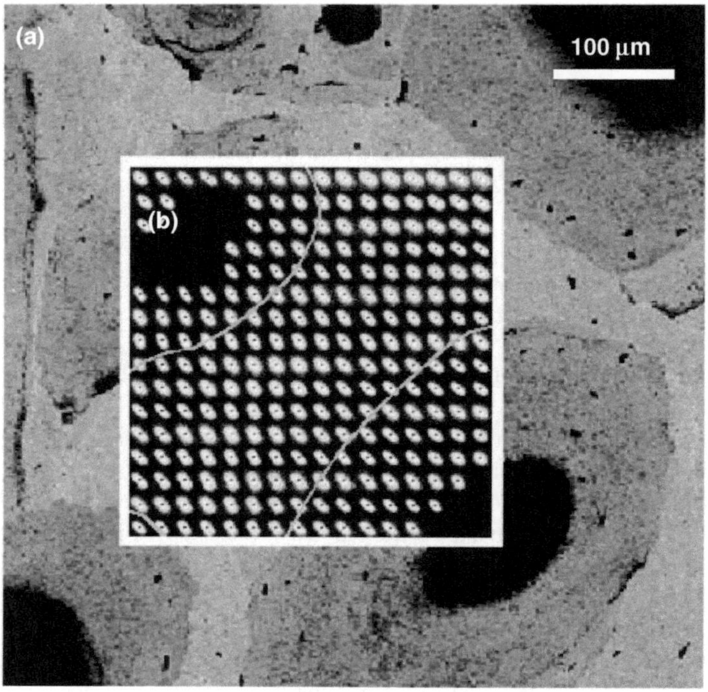

Fig. 11.10. Scanning SAXS imaging of cortical bone. The background image (**a**) shows a backscattered electron image (BEI) from a 20 µm thick section of cortical human bone (cross-section from a long bone such as in Fig. 11.6). The gray-levels in the BE image can be used to quantify the mineral density locally (qBEI) [82]. The overlaid image (**b**) consists of 2D SAXS patterns measured at exactly the corresponding specimen positions with 20 µm beam size, and matched to the BE image. For clarity, the hollow channels and the borders of the osteons are indicated by black and gray colors in the image (**b**), respectively

demonstrated that local texture analysis in bone tissue is possible down to a few micrometers spatial resolution [86, 87]. In combination with scanning SAXS/XRD, this will open fascinating new possibilities to investigate the complex 3D-structure of lamellar bone at the level of single lamellae.

Scope for Position Resolved Neutron Scattering

The success and the rapid development of X-ray microbeams is a direct consequence of the extremely high brilliance of wiggler and undulator sources on the one hand, and the rapid development of X-ray optics on the other hand (see e.g., [79]). While the brilliance of neutron sources is generally low and will not increase considerably in the near future, all the focusing principles used for X-rays may in principle be applied with neutrons as well. When comparing different focusing devices, a quantitative description of the efficiency of a certain optical system to produce a small beam is given by the so called gain factor. This value describes essentially the ratio between the integrated flux in the focus of the optical system and the flux one would obtain by preparing a beam with the same size and divergence properties using conventional slit collimation. Several neutron optical devices were tested in recent years with similar gains as compared to X-ray optics, allowing thus to prepare sufficiently intense neutron beams with a diameter below 100 μm. For instance, a neutron beam of about 80 μm diameter with a gain of 25 was reported using tapered capillaries [88] and similar gains were obtained by using compound refractive lenses [89]. Moreover, neutrons can be deflected in magnetic fields due to their magnetic moment. This has led to the construction of magnetic neutron lenses and focusing of a neutron beam with a gain of 35 using a permanent sextupol magnet has been reported [90]. Finally, even the feasibility of a sub-micron sized neutron (line) focus has been demonstrated recently by using a thin film wave-guide [91]. All these developments suggest that neutron beams with diameter in the order of 50–100 μm may well be feasible in the future with sufficiently high flux and low divergence for neutron diffraction and for SANS experiments. We propose that such neutron "microbeams" could be used to establish scanning SANS as a complementary tool to scanning SAXS on a length scale of about 50–100 μm. In particular the sensitivity of neutrons to the organic part of mineralized tissue would make position resolved neutron scattering on bone an attractive technique. In cancellous bone for instance, the detailed structure and arrangement of the collagen fibrils at the level of single trabeculae could be studied by this means with neutrons and complemented with the detailed structure of the mineral as obtained from X-rays.

References

1. J.L. Irigaray, H. Oudadesse, V. Brun, Biomaterials **22**, 629–640 (2001)
2. D.J S. Hulmes, J. Struct. Biol. **137**, 2–10 (2002)
3. P. Fratzl, Curr. Opin. Coll. Interf. Sci. **8**, 32–39 (2003)
4. L. Knott, A.J. Bailey, Bone **22**, 181–187 (1998)
5. T.J. Wess, A. Miller, J.P. Bradshaw, J. Mol. Biol. **213**, 1–5 (1990)
6. T.J. Wess, L. Wess, A. Miller, Alcohol Alcohol **29**, 403–409 (1994)
7. J.C. Hadley, K.M. Meek, N.S. Malik, Glycoconjugate J. **15**, 835–840 (1998)
8. T.J. Wess, L. Wess, A. Miller et al, J. Mol. Biol. **230**, 1297–1303 (1993)
9. P. Fratzl, A. Daxer, Biophys. J. **64**, 1210–1214 (1993)
10. D.M. Maurice, J. Physiology-London **136**, 263 (1957)
11. K.M. Meek, D.W. Leonard, Biophys. J. **64**, 273–280 (1993)
12. G.F. Elliott, J.M. Goodfellow, A.E. Woolgar et al, J. Appl. Cryst. **11**, 496–496 (1978)
13. G.F. Elliott, Z. Sayers, P.A. Timmins, J. Mol. Biol. **155**, 389–393 (1982)
14. J.W. Regini, P.A. Timmins, G.F. Elliott et al, Biochim. Biophys. Acta **1620**, 54–58 (2003)
15. H.D. Middendorf, R.L. Hayward, S.F. Parker et al, Biophys. J. **69**, 660–673 (1995)

16. H.D. Middendorf, A. Traore, L. Foucat et al, Physica B **276**, 518–519 (2000)
17. H.D. Middendorf, U.N. Wanderlingh, R.L. Hayward et al, Physica A **304**, 266–270 (2002)
18. J.F.V. Vincent, in: *Structural Biomaterials* (Princeton University Press, Princeton, 1990)
19. N. Sasaki, S. Odajima, J. Biomech. **29**, 1131–1136 (1996)
20. K. Misof, G. Rapp, P. Fratzl, Biophys. J. **72**, 1376–1381 (1997)
21. P. Fratzl, K. Misof, I. Zizak et al, J. Struct. Biol. **122**, 119–122 (1998)
22. N. Sasaki, N. Shukunami, N. Matsushima et al, J. Biomech. **32**, 285–292 (1999)
23. V. Ottani, M. Raspanti, A. Ruggeri, Micron **32**, 251–260 (2001)
24. R. Puxkandl, I. Zizak, O. Paris et al, Phil. Trans. R. Soc. Lond. B **357**, 191–197 (2002)
25. F.H. Silver, Y.P. Kato, M. Ohno et al, J. Long-Term Effects Med. Implants **2**, 165–195 (1992)
26. W. Folkhard, E. Mosler, W. Geercken et al, Int. J. Biol. Macromol. **9**, 169–175 (1987)
27. E. Mosler, W. Folkhard, E. Knorzer et al, J. Mol. Biol. **182**, 589–596 (1985)
28. W.J. Landis, K.J. Hodgens, J. Arena et al, Microscopy Res. Tech. **33**, 192–202 (1996)
29. P. Fratzl, M. Groschner, G. Vogl et al, J. Bone Miner. Res. **7**, 329–334 (1992)
30. W.J. Landis, J.J. Librizzi, M.G. Dunn et al, J. Bone. Miner. Res. **10**, 859–867 (1995)
31. D.J.S. Hulmes, A. Miller, S.W. White et al, Int. J. Biol. Macromol. **2**, 338–346 (1980)
32. J.W. White, A. Miller, K. Ibel, J. Chem. Soc. Faraday Trans. II **72**, 435 (1976)
33. S.W. White, D.J.S. Hulmes, A. Miller et al, Nature **266**, 421–425 (1977)
34. P. Fratzl, N. Fratzl-Zelman, K. Klaushofer, Biophys. J. **64**, 260–266 (1993)
35. L.C. Bonar, S. Lees, H.A. Mook, J. Mol. Biol. **181**, 265–270 (1985)
36. D.D. Lee, M.J. Glimcher, J. Mol. Biol. **217**, 487–501 (1991)
37. S. Lees, L.C. Bonar, H.A. Mook, Int. J. Biol. Macromol. **6**, 321–326 (1984)
38. S. Lees, D. Hanson, E. Page et al, J. Bone Miner. Res. **9**, 1377–1389 (1994)
39. S. Lees, D.W.L. Hukins, Bone Mineral **17**, 59–63 (1992)
40. S. Lees, H.A. Mook, Calcif. Tissue Int. **39**, 291–292 (1986)
41. J.M.S. Skakle, R.M. Aspden, J. Appl. Cryst. **35**, 506–508 (2002)
42. J. Orgel, A. Miller, T.C. Irving et al, Structure **9**, 1061–1069 (2001)
43. D.J.S. Hulmes, T.J. Wess, D.J. Prockop et al, Biophys. J. **68**, 1661–1670 (1995)
44. C.K. Loong, C. Rey, L.T. Kuhn et al, Bone **26**, 599–602 (2000)
45. M.G. Taylor, S.F. Parker, P.C.H. Mitchell, J. Mol. Struct. **651**, 123–126 (2003)
46. M.G. Taylor, S.F. Parker, K. Simkiss et al, Phys. Chem. Chem. Phys. **3**, 1514–1517 (2001)
47. S. Weiner, H.D. Wagner, Ann. Rev. Mater. Sci. **28**, 271–298 (1998)
48. S. Weiner, T. Arad, I. Sabanay et al, Bone **20**, 509–514 (1997)
49. A. Bigi, M. Burghammer, R. Falconi et al, J. Struct. Biol. **136**, 137–143 (2001)
50. A. Ascenzi, E. Bonucci, P. Generali et al, Calcif. Tissue Int. **29**, 101–105 (1979)
51. D. Jaschouz, O. Paris, P. Roschger et al, J. Appl. Cryst. **36**, 494–498 (2003)
52. N. Sasaki, Y. Sudoh, Calcif. Tissue Int. **60**, 361–367 (1997)
53. G.E. Bacon, P.J. Bacon, R.K. Griffiths, J. Appl. Cryst. **10**, 124–126 (1997)
54. G.E. Bacon, P.J. Bacon, R.K. Griffiths, J. Anatomy **139**, 265–273 (1984)
55. R.K. Griffiths, G.E. Bacon, P.J. Bacon, J. Anatomy **124**, 253–253 (1977)

56. G.E. Bacon, Neutron Anatomy, in: *Neutrons in Biology*, ed. by B.P. Schoenborn, R. Knott (Plenum Press, New York, 1996), pp. 17–27
57. A. Guinier, G. Fournet, in: *Small-angle Scattering of X-rays* (Wiley, New York, 1955)
58. O. Glatter, O. Kratky, in: *Small-angle X-ray scattering* (Academic Press, New York, 1983)
59. P. Fratzl, J. Appl. Cryst. **36**, 397–404 (2003)
60. P. Fratzl, N. Fratzl-Zelman, K. Klaushofer et al, Calcif. Tissue Int. **48**, 407–413 (1991)
61. P. Fratzl, S. Schreiber, K. Klaushofer, Conn. Tissue Res. **35**, 9–16 (1996)
62. P. Fratzl, H.F. Jakob, S. Rinnerthaler et al, J. Appl. Cryst. **30**, 765–769 (1997)
63. S. Rinnerthaler, P. Roschger, H.F. Jakob et al, Calcif. Tissue Int. **64**, 422–429 (1999)
64. O. Paris, I. Zizak, H. Lichtenegger et al, Cell Mol. Biol. **46**, 993–1004 (2000)
65. P. Roschger, B.M. Grabner, S. Rinnerthaler et al, J. Struct. Biol. **136**, 126–136 (2001)
66. P. Roschger, H.S. Gupta, A. Berzanovich et al, Bone **32**, 316–323 (2003)
67. W. Tesch, N. Eidelman, P. Roschger et al, Calcif. Tissue Int. **69**, 147–157 (2001)
68. H.S. Gupta, P. Roschger, I. Zizak et al, Calcif. Tissue Int. **72**, 567–576 (2003)
69. I. Zizak, P. Roschger, O. Paris et al, J. Struct. Biol. **141**, 208–217 (2003)
70. P. Fratzl, P. Roschger, J. Eschberger et al, J. Bone Miner. Res. **9**, 1541–1549 (1994)
71. P. Fratzl, O. Paris, K. Klaushofer et al, J. Clin. Invest. **97**, 396–402 (1996)
72. P. Fratzl, S. Schreiber, P. Roschger et al, J. Bone Miner. Res. **11**, 248–253 (1996)
73. P. Roschger, P. Fratzl, K. Klaushofer et al, Bone **20**, 393–397 (1997)
74. B. Grabner, W. J. Landis, P. Roschger et al, Bone **29**, 453–457 (2001)
75. P. Roschger, S. Rinnerthaler, J. Yates et al, Bone **29**, 185–191 (2001)
76. B.M. Misof, P. Roschger, W. Tesch et al, Calcif. Tissue Int. **73**, 251–257 (2003)
77. W. Tesch, T. Vandenbos, P. Roschger et al, J. Bone Miner. Res. **18**, 117–125 (2003)
78. O. Paris, D. Loidl, H. Peterlik et al, J. Appl. Cryst. **33**, 695–699 (2000)
79. C. Riekel, Rep. Prog. Phys. **63**, 233–262 (2000)
80. I. Zizak, O. Paris, P. Roschger et al, J. Appl. Cryst. **33**, 820–823 (2000)
81. N.P. Camacho, S. Rinnerthaler, E.P. Paschalis et al, Bone **25**, 287–293 (1999)
82. P. Roschger, P. Fratzl, J. Eschberger et al, Bone **23**, 319–326 (1998)
83. A. Boyde, S.J. Jones, Microscopy Res. Tech. **33**, 92–120 (1996)
84. T.J. Wess, M. Drakopoulos, A. Snigirev et al, Archaeometry **43**, 117–129 (2001)
85. A. Cedola, V. Stanic, M. Burghammer et al, J. Phys. IV **104**, 329–332 (2003)
86. H.R. Wenk, F. Heidelbach, Bone **24**, 361–369 (1999)
87. F. Heidelbach, C. Riekel, H.R. Wenk, J. Appl. Cryst. **32**, 841–849 (1999)
88. V.A. Sharov, Q.-F. Xiao, I. Ponomarev et al, Rev. Sci. Instrum. **71**, 3247–3253 (2000)
89. H.R. Beguiristain, I.S. Anderson, C.D. Dewhurst et al, Appl. Phys. Lett. **81**, 4290–4292 (2002)
90. H.M. Shimizu, Y. Suda, T. Oku et al, Nuclear Instrum. Methods Phys. Res. A **430**, 423–434 (1999)
91. F. Pfeiffer, V. Leiner, P. Hoghoj et al, Phys. Rev. Lett. **88**, 055507-1-4 (2002)

12

Structural Investigations of Membranes in Biology by Neutron Reflectometry

C.F. Majkrzak, N.F. Berk, S. Krueger, U.A. Perez–Salas

12.1 Introduction

Membranes are an essential part of every living cell. Determining the nanometer scale structure of these partitions is of interest for the understanding of important cellular processes on a molecular level, including, for example, transport mechanisms into and out of the cell interior and the functioning of protein sensors embedded in the membrane [1].

In "real" space, probes such as atomic force and electron microscopies, at present, can provide localized images of a material *surface* with nanometer scale resolution. However, *scattering* techniques employing neutrons and X-rays have proven to be especially well-suited for "viewing", in comparable detail, the distribution of matter *beneath* the surface. The reasons for this subsurface sensitivity are manifold, but principally are a consequence of the wave nature of the radiation, the relative strengths of interaction (between photon and atomic electrons or between neutron and nucleus), and the ability to accurately measure and analyze the diffraction pattern that the material density distribution of the film gives rise to.

The sensitivity of diffraction as a probe of membrane structure is considerably enhanced if a homogeneous specimen of the film can be constrained to lie on a flat surface (either a solid or liquid substrate). A membrane so confined is, effectively, a quasi two-dimensional scattering object. Treating the neutron as a plane wave, having a wavevector k proportional to its momentum, the coherent (in phase), elastic (no energy transfer) reflection of that neutron from a flat film can be then separated into two distinct types, specular and non-specular. Specular scattering refers to the condition in which the glancing angle θ between the reflected neutron wavevector and the surface is equal to that of the incident wave. In this case the momentum transfer is exactly perpendicular to the surface. Analysis of the specular reflectivity reveals the depth profile of the film's density along the surface normal. If there are no variations in the composition or material density within the plane of the film, then *only* specular scattering can occur. In a case where in-plane fluctuations of the density are present, the specular component of the reflected intensity is caused by a film density that is averaged, at a given depth, over the in-plane area for which the neutron plane wavefront is coherent (typically of the order of microns). In addition, however, in-plane fluctuations produce non-specular scattering wherein the momentum transfer has a component parallel to the surface. Non-specular scattering data thus contains information about in-plane structure. Non-specular reflectometry has great potential for the study of biofilms, for example, in determining the sizes and distribution of various entities, such as cholesterol "rafts", within the plane of a membrane. However, research in this area is not yet as developed as that involving specular reflectometry. The reasons for this involve a number of technical difficulties, including the preparation of specimens of sufficient size and homogeneity and the theoretical interpretation of the non-specular scattering, particularly at wavevector transfers where the Born approximation (discussed below) is not valid. The present chapter is concerned primarily with specular reflection. The interested reader is referred elsewhere for discussions of non-specular scattering, e.g., [2, 3].

The resolution of a material distribution in real space deduced from diffraction data is, ultimately, inversely proportional to the range in wavevector transfer Q over which the reflected intensity is measured. The wavevector transfer for the reflected beam is $k_\mathrm{f}-k_\mathrm{i} = -2k\sin(\theta) = -4\pi\sin(\theta)/\lambda = -2k_{0z}$, where k_{0z} is the component of the incident wavevector normal to the film. We will always define $Q = 2k_{0z}$, so that the reflected wavevector transfer is $-Q$. For instance, the spatial resolution in the compositional depth profile obtained from analysis of the specular reflectivity (defined as the reflected intensity divided by the incident intensity) measured out to a wavevector transfer of $0.7\,\text{Å}^{-1}$ corresponds to a spatial resolution of the order of 0.5 nm. Given the strength of available neutron sources, sample areas of several square millimeters or more are therefore necessary in practice to obtain sufficiently accurate data. Reflectivity experiments (i.e., "reflectometry") conducted over a range of wavevector transfers similar to that just given are to be distinguished from diffraction studies performed at higher wavevector transfers corresponding to interatomic scale resolution.

The purpose of this contribution is to provide an overview of the experimental and theoretical methods now employed in the study of membrane structures by specular neutron reflectometry. Nonetheless, a reasonable amount of detail is included here so that the researcher new to the technique can better judge what structural information is obtainable from reflectometry and can assess what actually is required to prepare a suitable specimen, perform measurements, and subsequently analyze the data. Reviews of current research in which neutron reflectometry has been applied to the study of biological or biomimetic membrane structures, including systems with embedded proteins and peptides, are given in other contributions of this volume by Lu, Gutberlet et al., Salditt et al., and elsewhere [4]. However, a representative example of a neutron reflectometry study of a lipid bilayer membrane into which the peptide melittin has been introduced is included here, to further illustrate the technique.

12.2 Theory

The theory of neutron reflectivity and diffraction is well-established [5–10], although there have been relatively recent developments in methods for phase determination and inversion (see, for example, the review [11] and references therein). This section summarizes key features of the theory pertinent to the study of thin films and membranes.

12.2.1 The Exact ("Dynamical") Solution

The specular reflectivity from a flat surface or film effectively reduces to a one-dimensional wave mechanics problem (see Merzbacher [13], for example)

$$-\frac{\partial^2\psi(z)}{\partial z^2} + 4\pi\rho(z)\psi(z) = k_{0z}^2\psi(z)\,, \tag{12.1}$$

where k_{0z} is the wavevector of the neutron in vacuum and $\rho(z)$ is the the scattering length density (SLD) "profile," which describes the neutron interaction with the film and its surrounding media everywhere along the z-axis, normal to the film surface. For neutrons with wavelengths of the order of several Ångstroms, the SLD at any "point" is the compositionally weighted average of the coherent neutron scattering lengths in a volume element having linear dimensions on the order of the neutron wavelength, divided by the volume of the element. Scattering length densities thus have dimensions of inverse area. Scattering lengths are the fundamental measure of the neutron–nucleus interaction and vary from one isotope to another in an essentially random but fixed manner. Coherent, in this context, refers to the component of the interaction that enables neutrons scattered by nuclei at different points in the film to interfere, much as ripples on a pond. It is such interference which makes the scattering dependent on spatial structure.

We have assumed the ideal situation for specular scattering where the SLD varies only along the surface normal. In general the SLD $\rho(x, y, z)$ can vary in all three directions in the film, so that the true reflection problem is inherently three-dimensional. Thus, the SLD $\rho(z)$ appearing in Eq. 12.1 is defined as

$$\rho(z) = \lim_{S\to\infty}\frac{1}{S}\iint_S \rho(x,y,z)\,\mathrm{d}S \equiv \langle\rho(x,y,z)\rangle_{xy}\,, \tag{12.2}$$

where S denotes the surface area of the film. In-plane variations of the SLD give rise to non-specular scattering. In the most extreme case, the specular reflection caused by $\rho(z)$ and the non-specular reflection caused by $\Delta\rho(x, y, z)$

interfere with one another, so that the resultant reflectivity cannot be expressed as two distinct contributions. However, in many cases of interest, either the non-specular component is negligible or the two contributions are separable. When lateral variations of the SLD are random, the measured reflectivity represents a "thermodynamic" average of the reflectivity over a suitable ensemble of such configurations. For cases where $\rho(x,y,z)$ is "self–averaging," i.e., where $\langle\rho(x,y,z)\rangle_{xy} = \langle\rho(x,y,z)\rangle_{\text{therm}}$, it can be shown that the specular reflection determined by $\rho(z)$ and the non-specular reflection induced by the residual $\Delta\rho(x,y,z)$ are decoupled from one another [14]. Even then, however, we need to know how to separate them. The specular reflectivity is defined as the ratio of the specularly reflected intensity to the incident intensity; the non-specular intensity affects both. When the instrument is configured to collect the specular signal (i.e., on the *specular ridge*), some fraction of the non-specular intensity is also counted and, therefore, must be subtracted. At the same time, non-specular intensity in other directions diminishes incident intensity which would otherwise cause specular reflection.

For both computational and analytic purposes, an SLD profile $\rho(z)$ of any shape can be accurately represented, for measurements up to a finite maximum $Q = Q_{\text{max}}$, by a piecewise continuous subdivision, $\rho_{\text{pwc}}(z)$, into a sufficient number of rectangular slices, or "bins"' of widths $\Delta z \ll \pi/Q_{\text{max}}$, where the SLD within each slice is taken to be constant, as depicted schematically in Fig. 12.1. The fundamental quantity describing the specular reflection of the neutron by the membrane is the spectrum (as a function of Q) of the reflection amplitude r, a complex number, $r = |r|\,\mathrm{e}^{\mathrm{i}\phi}$ of modulus $|r|$ and a phase ϕ. Similarly, the transmitted wave is characterized by a transmission amplitude t, but it turns out that all of the relevant information is contained in the

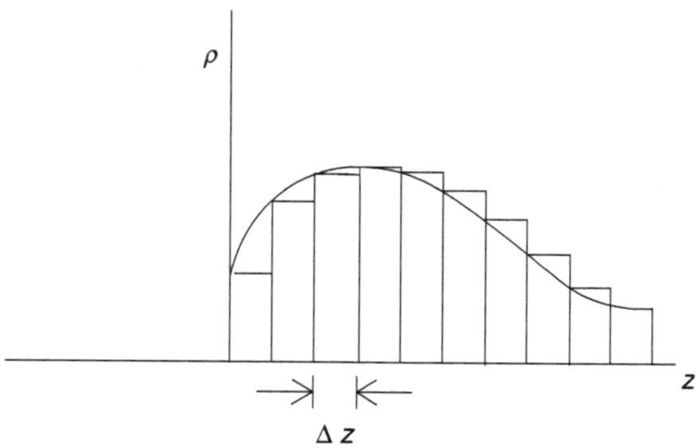

Fig. 12.1. Scattering length density depth profile, along the surface normal, of arbitrary shape represented by rectangular bins or slices over each of which the density is taken to be constant

spectrum of r. To set up equations which describe the relationship between r, t, and ρ, we first make the piecewise continuous rendering of $\rho(z)$ explicit with

$$\rho_{\text{pwc}}(z) = \begin{cases} \rho_j & \text{if } (j-1)\Delta z \le z < j\Delta z, \\ 0 & \text{otherwise,} \end{cases} \tag{12.3}$$

where $j = 1, \cdots, N$. Thus $\rho_{\text{pwc}}(z)$ is a "histogram" of N bins of uniform width $\Delta z = L/N$, where $\rho_{\text{pwc}}(z) = \rho_j$ in the jth bin. Next, we partition the z-axis (along the film normal) into three contiguous regions: region I, the "fronting," where $z < 0$; region II, the "film of interest," where $0 \le z \le L$; and region III, the "backing," where $z > L$. The fronting is defined as the region containing the incident and reflected beams, while the backing is the region of the transmitted beam, regardless of how the film is mechanically supported. In region I, $\rho(z) = \rho^{\mathbf{I}}$, and in region III, $\rho(z) = \rho^{\mathbf{III}}$, where ρ^{I} and ρ^{III} are known constants (typically, the SLD values for air or vacuum, silicon, sapphire, and mixtures of water and heavy water, as appropriate to the experiment). With each of the regions of constant SLD, viz., I and III and in the slabs comprising ρ_{pwc} in II, we can associate a wavevector component along the z-axis

$$k_z^{\text{I,II,III}} = k_{0z}\sqrt{1 - 4\pi\rho^{\text{I,II,III}}/k_{0z}^2} \equiv n_z^{\text{I,II,III}} k_{0z} \,. \tag{12.4}$$

(From now on we will suppress the "z" subscript on k.) Note that in region II, where $\rho = \rho_{\text{pwc}}$ has values ρ_j, k_z^{II} has the corresponding values k_j^{II}. In regions I and III, the physical solutions of Eq. 12.1 have the simple plane wave forms

$$\psi^{\text{I,III}}(z) = \begin{cases} \mathrm{e}^{\mathrm{i}k^{\text{I}}z} + r\mathrm{e}^{-\mathrm{i}k^{\text{I}}z} & \text{for } z < 0, \\ t\mathrm{e}^{\mathrm{i}k^{\text{III}}z} & \text{for } z > L. \end{cases} \tag{12.5}$$

These solutions (and their derivatives) are "transferred" across region II by the matrix equation [11, 15],

$$\begin{pmatrix} t \\ in^{\text{III}}t \end{pmatrix} \mathrm{e}^{\mathrm{i}k^{\text{III}}L} = M \begin{pmatrix} 1+r \\ in^{\text{I}}(1-r) \end{pmatrix}, \tag{12.6}$$

where the *transfer matrix* $M = \begin{pmatrix} A & B \\ C & D \end{pmatrix}$ is a 2×2 real-valued matrix having unit determinant, $AD - BC = 1$. For ρ_{pwc}, this is the matrix product

$$M = M_N M_{N-1}, \cdots, M_j, \cdots, M_2 M_1 \,, \tag{12.7}$$

where M_j is the transfer matrix for the jth bin,

$$M_j = \begin{pmatrix} \cos(k_j^{\text{II}}\Delta z) & \sin(k_j^{\text{II}}\Delta z)/n_j^{\text{II}} \\ -n_j^{\text{II}}\sin(k_j^{\text{II}}\Delta z) & \cos(k_j^{\text{II}}\Delta z) \end{pmatrix}. \tag{12.8}$$

In general Eq. 12.7 can represent any useful decomposition of $\rho(z)$ into N contiguous, non-overlapping segments.

Equation 12.6 stands for two simultaneous linear equations, which are straightforwardly solved for r and t as a function of the matrix elements $A, B, C,$ and D as functions of k_{0z}. For the case of a "free" film, i.e., a film in contact with vacuum fronting and backing, the result for the reflection amplitude is

$$r = \frac{B + C + \mathrm{i}(D - A)}{B - C + \mathrm{i}(D + A)} = \frac{B^2 + D^2 - A^2 - C^2 - 2\mathrm{i}(AB + CD)}{A^2 + B^2 + C^2 + D^2 + 2}, \tag{12.9}$$

while the reflectivity, $|r|^2 = r^* r$, is most simply represented by

$$2\frac{1 + |r|^2}{1 - |r|^2} = A^2 + B^2 + C^2 + D^2 . \tag{12.10}$$

While it is straightforward to compute the reflectivity for a given model SLD profile, the so-called "direct problem," deducing ρ from reflectivity data, the "inverse problem," is much more problematic and inherently ambiguous because of the "lost" phase angle ϕ. Indeed, we see from Eqs. 12.9 and 12.10 that full knowledge of r needs three combinations of A, B, C, and D, viz., $A^2 + C^2$, $B^2 + D^2$, and $AB + CD$ (because AD − BC = 1,these are not completely independent); while knowledge of $|r|^2$ implies only the sum of squares combination, $A^2 + B^2 + C^2 + D^2$. In practice, the determination of an SLD profile from reflectivity data employs fitting schemes based on either model-dependent or model-independent methods (see [17, 18], for example). Figure 12.2 shows SLD profiles for a pair of model thin film structures, having thicknesses and SLD values typical of those of interest to us here (Fig. 12.2a), and the corresponding specular neutron reflectivities (Fig. 12.2b), which are nearly identical and thus demonstrate the importance of phase information – or its absence. Even though it might not be possible to deduce from the reflectivity alone which of two or more SLD profiles is the veridical one, i.e., the one that actually produced the data, it can be concluded whether or not a given model SLD profile is at least consistent with the measured reflectivity. Furthermore, a priori knowledge of the SLD in part of the film or the adjacent substrate can be used to recover, in effect, some of the phase information: this can also be accomplished by controlled manipulation of the SLD in certain sections of the film, i.e., by exchanging hydrogen for deuterium [19]. Such partial phase information can significantly reduce the number of acceptable solutions.

Methods have been developed to recover phase information through the use of various reference structures – either adjacent films or surrounding media [20–25]. For some of these, the reflection amplitude for the unknown part of the film can be obtained "locally" (i.e., independently at any k) and exactly [22, 24]. It has been shown, that the reflection amplitude and the SLD profile are in one-to-one correspondence for a large class of film potentials. This means that the given profile produces a unique spectrum of r and that a given r, if known for all Q, produces a unique SLD profile, when using the appropriate mathematical tools to retrieve it [26]. Figure 12.3 shows the real part

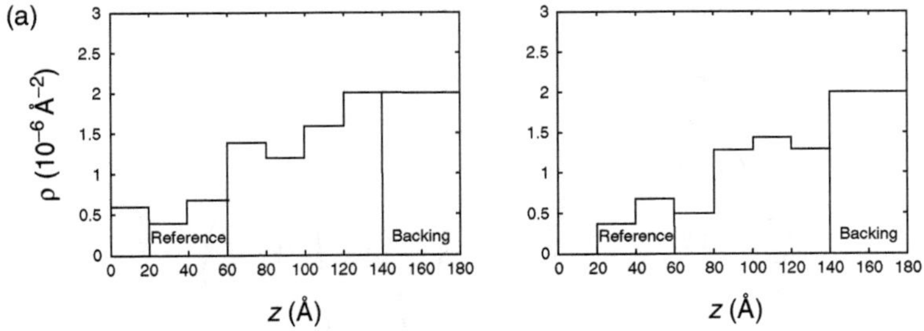

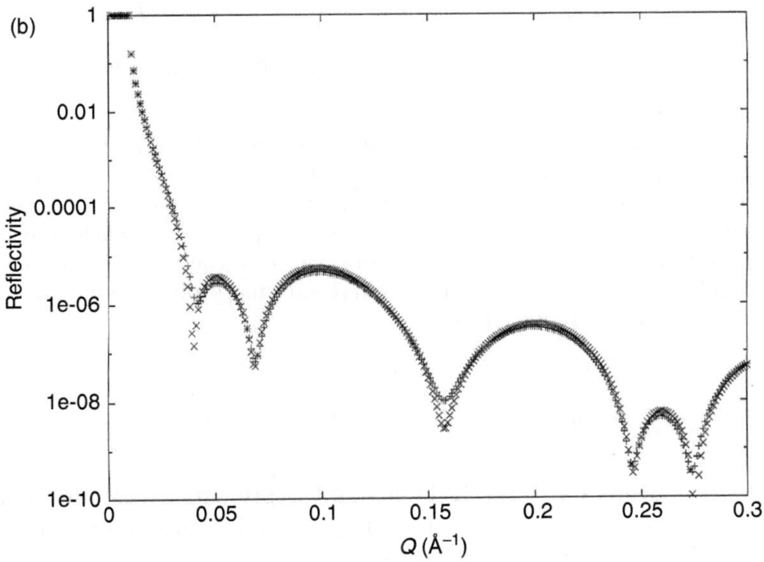

Fig. 12.2. *Top*: **(a)** Model SLD (neutron) profiles similar to two of the profiles considered for X-ray reflection (Fig. 12.3 of [44]). Both profiles share a common "reference" or known segment between $z = 20$ Å and $z = 60$ Å. *Bottom*: **(b)** Corresponding neutron reflectivity curves calculated for the two composite SLD profiles. The two curves are practically indistinguishable from one another (after Fig. 12.10 of [16])

of r (multiplied by Q^2) for each of the two model SLD profiles of Fig. 12.2a. In stark contrast to the two corresponding reflectivity curves of Fig. 12.2b, there is a marked, clearly distinguishable difference. An actual example which demonstrates the phase inversion technique is given in Sect. 12.5. In practice, the solution of the inverse problem is limited by the finite range of Q over which it is possible to measure the reflectivity, but ambiguities introduced by data truncation are systematic and to a limited extent, treatable [12, 27].

The reflection amplitude has a number of useful theoretical representations. If we know the solution ψ of Eq. 12.1 in region II, then an alternative, and quite general, expression for the free film r can also be derived [28] using the wave equation in Eq. 12.1,

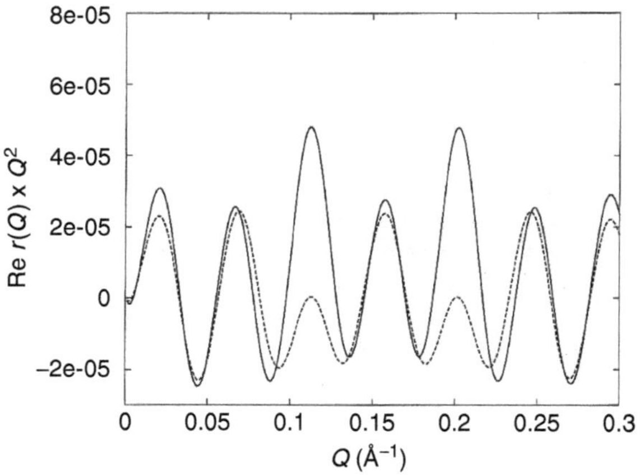

Fig. 12.3. $Q^2 \mathrm{Re}\, r(Q)$ for the (reversed) film structures of Fig. 12.2a (not including the backing but incorporating the known or reference sections of the films). These $\mathrm{Re}\, r(Q)$ correspond to what would be retrieved, for example, by phase-sensitive reflectivity experiments (for each of the two SLD profiles) in which the backing SLD was varied according to the methods discussed in the text. In contrast to the situation illustrated in Fig. 12.2b, these curves are markedly different over a wide range of Q. (after Fig. 12.9 of [16])

$$r = \frac{4\pi}{2\mathrm{i}k_{0z}} \int_{-\infty}^{\infty} \psi(z)\rho^{\mathrm{II}}(z)\mathrm{e}^{\mathrm{i}k_{0z}z}\,\mathrm{d}z\,. \tag{12.11}$$

Because ψ depends on r, Eq. 12.11 actually represents an implicit equation for r, but it does provide a useful starting point for formal analysis and for some practical approximation schemes.

12.2.2 The Born Approximation

In general, as seen from Eq. 12.11, the weaker the potential and the higher the wavevector transfer, the smaller the reflectivity becomes. For reflectivities of the order of a few percent or less, the neutron wave function within the scattering medium is not significantly distorted from its free space, plane wave form. In this case, $\psi(z)$ in Eq. 12.11 can be approximated by the incident wave function, leading to the Born approximation (BA) or so-called "kinematic" result,

$$r^{\mathrm{BA}}(Q) = \frac{4\pi}{\mathrm{i}Q} \int_{-\infty}^{\infty} \rho^{\mathrm{II}}(z)\mathrm{e}^{\mathrm{i}Qz}\,\mathrm{d}z\,. \tag{12.12}$$

Thus, $Qr^{\mathrm{BA}}(Q)$ and $\rho^{\mathrm{II}}(z)$ are related by Fourier transformation. The factor of Q^{-1} multiplying the integral in Eq. 12.12 does not result from the BA; it is the same factor appearing in Eq. 12.11, the general expression, and is inherent in the effective one-dimensionality of the specular reflection problem (i.e., the infinite-slab geometry of the 3D problem). The essential difference is that the exact $\psi(z, Q)$ approaches zero as Q goes to zero, unlike its plane wave approximation, thus keeping $r(Q)$ finite at $Q = 0$ in Eq. 12.11. Of course, it is to be expected that the BA will fail as $Q \to 0$, since, as $|r(Q)| \to 1$ at the origin, $\psi(z, Q)$ becomes poorly approximated in region II by the "undistorted" incident wave. Figure 12.4 shows a model SLD profile for a lipid bilayer similar to that deduced in a neutron reflectivity study of DOPC multilayers by Wiener and White [29]. In Fig. 12.4 the specular neutron reflectivity $|r|^2$ for the SLD profile is plotted as a function of Q, calculated using the exact theory and in the Born approximation for a freely standing single bilayer surrounded by vacuum. Also shown is the reflectivity for the same bilayer on a substrate (thick enough that it is effectively semi-infinite) as predicted by the exact theory. In the latter case the Born approximation would fail not only at the origin, but also in the neighborhood of the critical angle for total external or mirror reflection (below which the reflectivity is unity).

12.2.3 Multilayers

In certain cases it is advantageous to reflect from a repeating or multilayered assembly of membrane films instead of a single membrane unit. Since the

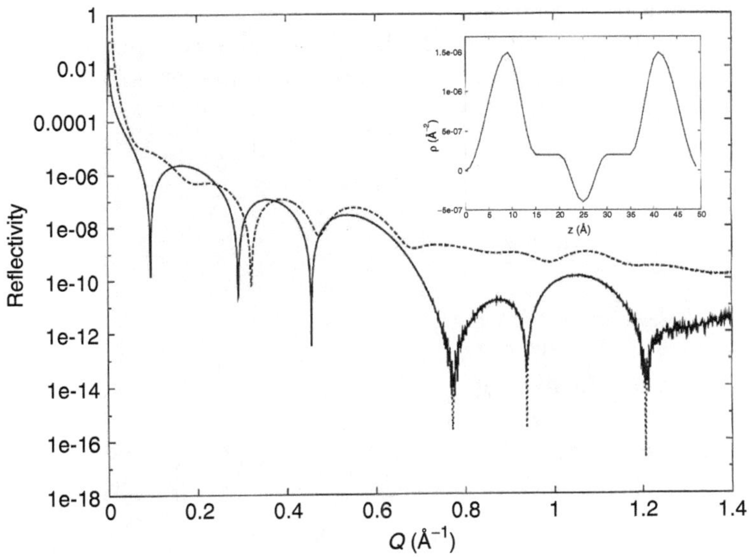

Fig. 12.4. Model SLD profile for a lipid bilayer as discussed in the text (*inset*). Specular reflectivity for the SLD profile calculated according to the exact theory as well as in the Born approximation, assuming the bilayer to be free standing. Also plotted is the reflectivity calculated according to the exact theory for the same bilayer film but on a semi-infinite substrate of Si. The reflectivities according to the exact theory and the BA are virtually indistinguishable on a logarithmic scale for the free standing films, except in the neighborhood of the origin. Only the reflectivity for the film on the substrate has a region of total external reflection (*long-dashed curve*)

reflection from such structures tends to be concentrated at higher Q-values than for single-layer thin films, the Born approximation can be particularly valuable in analyzing reflection from them. For a periodic multilayer structure (assuming ideally flat, parallel layers of uniform density and thickness), the reflection amplitude in the BA is given by ([28], for example)

$$r_{\mathrm{ML}}^{\mathrm{BA}}(Q) = \frac{4\pi}{\mathrm{i}Q}\left[\frac{\sin(MQD/2)}{\sin(QD/2)}\right]\mathrm{e}^{\mathrm{i}(M-1)DQ/2}\int_0^D \rho(z)\mathrm{e}^{\mathrm{i}Qz}\,\mathrm{d}z\,, \qquad (12.13)$$

for M repeats of a unit film (e.g., the bilayer) of thickness D. The integral over $\rho(z)$ is limited to the unit film. The effect of the M repeats appears only in the prefactor, where (using L'Hospital's rule) the ratio of sine functions in brackets acts as a concentrator of reflection about the *Nyquist lattice* points, $Q = Q_m$, as M increases, where $Q_m = 2\pi m/D$ for integer m. Thus, for M large (but not so large as to invalidate the BA), $r_{\mathrm{ML}}^{\mathrm{BA}}(Q)$ is strongly peaked on the Nyquist points, in the manner of Bragg peaks in crystallography. Figure 12.5 compares the reflectivities, calculated with the exact and kinematic formulas, for $M = 50$ bilayers having the SLD profile of Fig. 12.4 (and assuming no substrate). Note how the reflected intensity of the multilayer is localized about the Nyquist lattice, in contrast to being more uniformly distributed over the entire Q-range, as evident in Fig. 12.4. Clearly the kinematic theory can give a good account of multilayer reflection.

Now the SLD profile of Fig. 12.4 is centrosymmetric along the z-axis and can, consequently, be represented by a Fourier cosine series [30]

$$\rho(z) = A_0 + 2\sum_{m=1}^{\infty} A_m \cos(Q_m z)\,, \qquad (12.14)$$

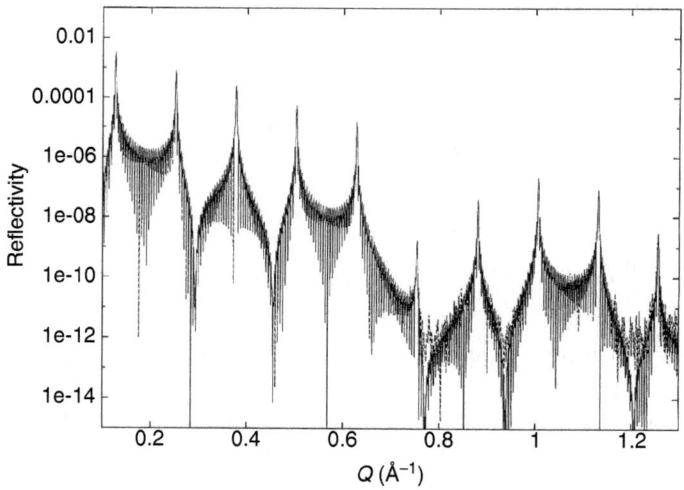

Fig. 12.5. Comparison of the reflectivities, calculated according to the exact and kinematic formulas, for $M = 50$ bilayers having the SLD profile of Fig. 12.4 (assuming no substrate)

where the A_m are real numbers. From the reflectivity curve plotted in Fig. 12.5 (corresponding to the bilayer of Fig. 12.4, with $M = 50$), the peaks up to the 10th order, inclusive, were used to determine the Fourier series of Eq. 12.14, truncated at $m = 10$. The resulting $\rho(z)$ is plotted in Fig. 12.6, along with the original SLD profile of Fig. 12.4 for comparison. The agreement displayed in Fig. 12.6 is qualitatively good. But even ten perfectly "measured" orders do not provide all the detail in the veridical $\rho(z)$.

12.2.4 Scale of Spatial Resolution

In assessing the value of SLD profiles inferred from reflection measurements, we must know how much spatial detail is meaningful to expect from the analysis; i.e., we can ask, what is the scale of spatial resolution – let us quantify this as a length l – of the resulting $\rho(z)$?

Let us say that we have perfect knowledge of the reflection amplitude $r(Q)$ up to a maximum value of $Q = Q_{\text{max}}$. (The experimental factors determining Q_{max} will be discussed in a later section.) The dominant factor limiting the spatial resolution ℓ of $\rho(z)$ inferred from this knowledge is the value of Q_{max}, according to $\ell = \pi/Q_{\text{max}}$ [11,12]. For our purposes, this holds that the number of spatial degrees of freedom N in $\rho(z)$ for a film of thickness L, when $r(Q)$ is known for $|Q| \leq Q_{\text{max}}$ is given by (the integer part of) $N = Q_{\text{max}}L/\pi$, referred to variously as the Nyquist or the Slepian number. Associating the corresponding scale of spatial resolution with $\ell = L/N$, one has $\ell = \pi/Q_{\text{max}}$ directly. In addition [12], ℓ also emerges explicitly from wavelet representations of $\rho(z)$, where ℓ is identified with the scale length of its most rapidly varying "detail," and N is the number of wavelets needed to fully describe $\rho(z)$ on this length scale. Indeed [11], as a special case, if we model $\rho(z)$ by N bins of uniform SLD

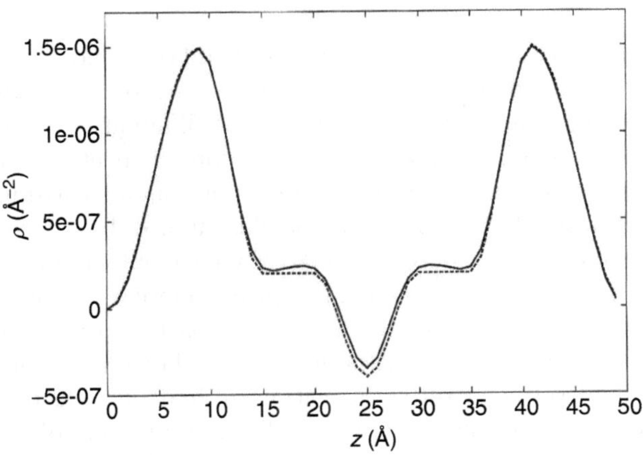

Fig. 12.6. SLD profile of Fig. 12.4 (*dashed curve*) compared to that obtained by the Fourier series analysis of the reflectivity curve (exact result) in Fig. 12.5 described in the text

and equal widths ℓ, as in $\rho_{\rm pwc}(z)$ of Eq. 12.3, then $r_{\rm BA}(Q)$ in Eq. 12.12 is exactly invertible for $\rho_{\rm pwc}(z)$ over the Q-range, $|Q| \leq Q_{\max}$, where $Q_{\max} = \pi/\ell$.

Now in general [12], let $r(Q)$ be perfectly known for $|Q| \leq Q_{\max}$, and call this conditional knowledge the function $r(Q|Q_{\max})$. Then the inversion of $r(Q|Q_{\max})$ by a fixed procedure – namely the one we would use for $Q_{\max} = \infty$, but setting $r(Q|Q_{\max}) = 0$ for $Q > Q_{\max}$ – determines a distorted or "smeared" version of the veridical $\rho(z)$, say $\rho(z|Q_{\max})$, which effectively parameterizes $\rho(z)$ on the spatial scale ℓ. The maximum Q of the measurement thus inherently limits the spatial resolution of $\rho(z)$ that can be reliably determined; the larger the value of $Q_{\max}$, the smaller the scale ℓ of detail we can know reliably.

What if our knowledge is limited to the reflectivity $|r(Q|Q_{\max})|^2$? If we consider the Born approximation, Eq. 12.12, as an adequate basis for analysis, then we may appeal to the well-known result that the Fourier transform of $Q^2|r^{\rm BA}(Q)|^2$ directly determines the auto-correlation function

$$\gamma(z) = \int_{-\infty}^{\infty} \rho(z - z')\rho(z')\,\mathrm{d}z' \,, \tag{12.15}$$

which describes the smearing of $\rho(z)$ by itself. This implies that for given $Q_{\max}$, if $\rho(z|Q_{\max})$ is resolved to scale ℓ, then $\gamma(z|Q_{\max})$ is resolved to scale 2ℓ. More carefully, if $\rho(z)$ is supported on an interval of length L, then $\gamma(z)$ has support of length $2L$. Thus applying the *same number* of spatial degrees of freedom to $\rho(z|Q_{\max})$ and to $\gamma(z|Q_{\max})$ leads to a scale of resolution $2\ell = 2\pi/Q_{\max}$ for the latter. The loss of phase information thus leads to a loss of spatial resolution for finite $Q_{\max}$.

12.3 Basic Experimental Methods

Neutron reflection can be done at both pulsed and continuous neutron sources. The only essential differences between them in regard to instrumental technique involve the means by which neutrons of different wavelengths are utilized and identified. With pulsed sources, the broad spectrum of wavelengths present in each pulse can be used because, for elastic scattering, the wavelength distribution of the beam can be determined by time-of-flight measurement. For continuous sources, a relatively narrow band of wavelengths, as defined by a crystal monochromator, is typically employed. The discussions to follow assume a continuous beam; for the most part, however, the experimental methodology described is applicable to pulsed beam reflectometers as well.

Another relatively general classification of neutron reflectometers can be made. For studying interfaces between solid and another solid, fluid, or gas, a sample can be oriented with its reflecting surface(s) vertical (and with the scattering plane, as defined by nominal incident and reflected wavevectors, horizontal). On the other hand, practical study of gas–fluid interfaces needs the liquid to be horizontal. The primary difference between these two types of reflectometers involves the mechanisms employed to direct the incident beam onto the sample and subsequently detect the reflected beam. For the sake of conceptual simplicity, we will assume the reflecting surface(s) of the sample to be vertical so that the nominal direction of the incident beam remains fixed relative to its source. Again, this choice does not limit, in any essential way, the relevance of the discussion to the one configuration.

12.3.1 Instrumental Configuration

Figure 12.7 is a schematic diagram of a typical neutron reflectometer, which is representative of the NG-1 reflectometer at the NIST Center for Neutron Research. The polarizing and spin flipping devices shown can be ignored for the present discussion, but are essential for magnetization depth profile measurements performed with polarized beams [32]. Within the core of the reactor, neutrons are produced by nuclear fission. The relatively high energies of these neutrons are subsequently moderated by collisions with heavy water at room temperature, resulting in a characteristic distribution of wavelengths with a peak in *elastically reflected intensity* (for $Q = 0.1\,\text{Å}^{-1}$, and $\Delta Q/Q = 0.05$) occurring at a wavelength about 1.5 Å. The energy distribution is further moderated by the liquid hydrogen "cold source," shown schematically in Fig. 12.7, shifting the peak in elastically reflected intensity to approximately 5 Å. The beam of cold neutrons is transported through an evacuated rectangular guide, the smooth, flat interior walls of

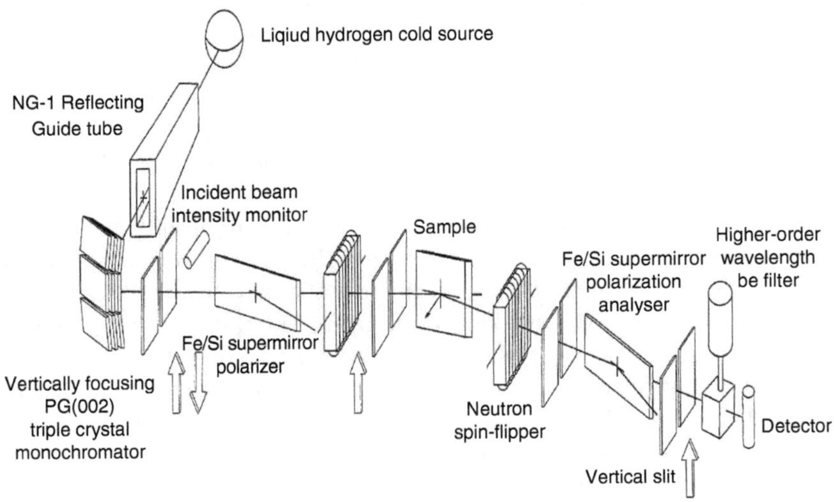

Fig. 12.7. Schematic of a typical neutron reflectometer (representative of the NG-1 polarized beam reflectometer at the NIST Center for Neutron Research; the polarizing and spin flipping devices are used in the determination of the vector magnetization depth profile in magnetic films and can be ignored for this presentation)

which are coated with a Ni film which gives a relatively large critical angle for total external or mirror reflection of about 0.1 deg Å^{-1} of incident wavelength.

This beam then impinges upon a pyrolytic graphite monochromating crystal array which Bragg reflects a vertically focussed beam onto the sample. The (002) atomic planes of the graphite crystal reflect a beam with a nominal wavelength $\lambda = 4.75$ Å for the chosen 90° scattering angle $2\theta_{\mathrm{M}}(\lambda = 2d \sin \theta_{\mathrm{M}})$, where d is the (002) atomic plane spacing, approximately 3.354 Å, and θ_{M} is the glancing angle of incidence measured from the crystal surface. The pyrolytic graphite consists of microcrystallites, which are essentially perfect single crystals of hexagonally arrayed carbon atoms, having dimensions of hundreds to thousands of Ångstroms, both along the (002) direction and perpendicular to it.

The beam reflected onto the sample by the monochromating crystal has a wavelength distribution determined mainly by the angular distribution of neutrons within the guide, the FWHM of the angular distribution of the graphite crystal's mosaic blocks, the interplanar spacing of the (002) graphite atomic planes, and the horizontal angular collimation of the beam (defined by the pair of vertical slits preceding the sample, as shown in Fig. 12.7.) A typical horizontal angular divergence is between 1 min and 10 min of arc, and because this is relatively small compared with that in the guide and the crystal mosaic angular spread, the wavelength resolution, $\Delta\lambda/\lambda$, principally depends on the latter two fixed quantities and is about 1 %.

As pictured in Fig. 12.7, each finger of a vertically focusing monochromator array is a stack of several graphite crystals, slightly inclined relative to one another, so as to create a broader (but effectively anisotropic) mosaic, thereby widening the wavelength band (to increase the intensity within the limits allowed by a given Q-resolution). At the NCNR NG-1 reflectometer, the 15 cm beam height in the guide is focused down to about 3 cm at the sample position so that the vertical angular divergence is approximately 2.5°. For specular reflectivity measurements, this relatively relaxed vertical angular divergence has a negligible effect on the Q-resolution.

Downstream of the sample position is a second pair of slits before the detector which are primarily used to suppress incoherent scattering background. For specular reflectivity measurements, the wavelength distribution and angular divergence of the incident beam, in conjunction with the glancing angle of incidence that the beam makes with the surface of a flat sample, determine a nominal value of Q and its associated resolution width. In addition, the slit just before the detector is needed to properly shape the instrumental resolution function in performing non-specular scattering scans perpendicular to the specular direction. In the latter case, the slit before the detector is significantly narrowed as the sample angle is rotated with the detector at a fixed scattering angle (this results in a trajectory in reciprocal space that is nearly orthogonal to the specular direction for sufficiently small angles).

For a sample that is distorted enough from perfect flatness, slits following the sample may block contributions to the reflected intensity from certain distorted regions of the sample surface from reaching the detector, thereby effectively improving the Q-resolution for specular reflection.

12.3.2 Instrumental Resolution and the Intrinsic Coherence Lengths of the Neutron

The theory of neutron reflection discussed in Section 12.2 assumed that each neutron in a beam of noninteracting, independent particles could be described by a single plane wave of infinite spatial extent. Generally, a neutron is more accurately described as a superposition of component plane waves, commonly known as a wave packet [13]. The wave packet description follows from localization of the neutron in space and imparts characteristic *coherence lengths*, both parallel and perpendicular to the direction of propagation defined by a nominal neutron wavevector **k**. These lengths – a measure of the combined uncertainties in position and momentum that must be associated with an individual neutron – determine the effective volume of the scattering medium with which the neutron wave packets *coherently* interact, and, consequently, they are important to interpret specular reflectivity data. However, because the neutrons constituting the incident beam originate from different, uncorrelated points within the source, there is an additional component of uncertainty in a distribution of nominal neutron wavevectors. This *incoherent* component can dominate the coherent, wave-packet-spread component, ultimately leading to the familiar "instrumental resolution" distribution in Q. In-depth, quantitative treatments of neutron coherence and instrumental resolution are given in several places [33–38]. Nonetheless, it is worthwhile here to consider these points further, at least qualitatively.

Incoherent vs. Coherent Effects

Figure 12.8 shows a liquid hydrogen moderator which acts as an incoherent source of neutrons for a specular reflectivity experiment to be performed downstream in a geometrically well-defined beam. Each of two neutrons, "A" and "B", radiates from a separate region of the liquid hydrogen cold source, as a result of an incoherent scattering event involving a single hydrogen nucleus. In general, all the coherent interactions of either neutron with objects along its path to the sample – e.g., the guide walls, a particular monochromator microcrystallite, and the pair of rectangular apertures preceding the sample – contribute to redefining the size and shape of the wave packet representing that neutron when it eventually encounters the sample. Nonetheless, let us assume that neutrons A and B are represented by wave packets of the same size and shape. Each neutron wave packet then possesses the same characteristic coherence lengths related to the uncertainties in position $(\Delta x, \Delta y, \Delta z)$

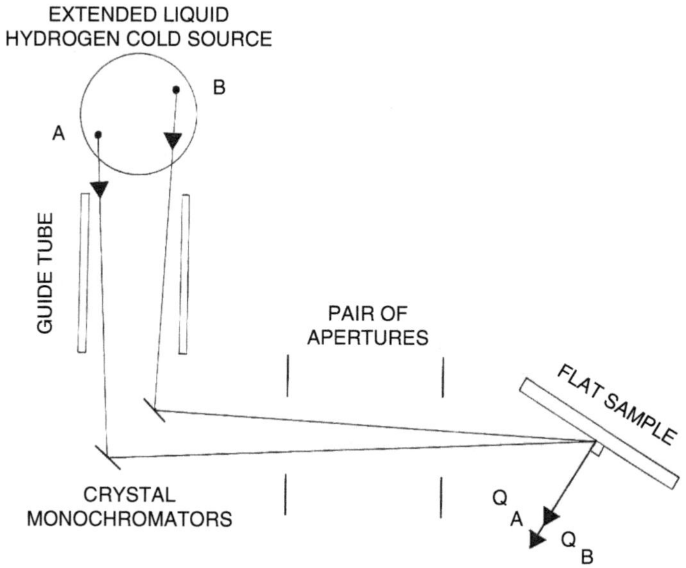

Fig. 12.8. Schematic representation of two independent, noninteracting neutrons, "A" and "B", emanating from different places in the cold source and passing through common instrumental optical elements en route to the sample. The size and shape of the wave packet describing neutron A is similar to that of B, but each packet has a different nominal wavevector direction. See the discussion in the text regarding coherent vs. incoherent components of the effective instrumental resolution

and wavevector $(\Delta k_x, \Delta k_y, \Delta k_z)$ of the neutron, that, as mentioned above, define a volume over which the neutron interacts with the sample. Any size and shape wave packet is an appropriately weighted superposition of plane waves [13]. The two neutrons travel in different directions, defined by their nominal wavevectors, so that each neutron is Bragg reflected from a separate monochromator crystal segment (but through the same pair of apertures) onto the sample. The latter fact means that the values of the normal component of the incident wavevector $k_{0z} = Q/2$ for the two neutrons differ from one another.

For specular reflection, we need only consider the z-axis normal to the plane of the film. In practice, at a continuous source, specular reflection measurements are performed with neutrons having nearly the same nominal k_0 incident at different glancing angles θ, so that the resulting range of $k_{0z} = k_0 \sin\theta$ values are obtained by changing the angle of incidence. Then, the relevant longitudinal coherence length effectively is the projection of the neutron wave packet coherence length along z. (See [39] for the case of normal incidence with ultra-cold neutrons.) The reflectivity, $|r|^2$, which results for a single plane wave incident on a perfectly flat and homogeneous Ni film 1,000 Å thick is plotted in Fig. 12.9. As is well-known, the oscillations evident in Fig. 12.9, the so-called "Kiessig fringes", are produced by interference

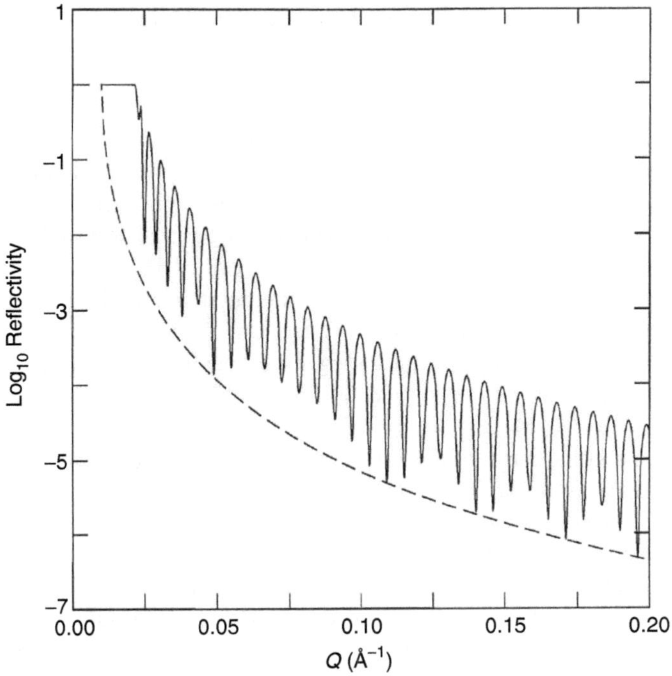

Fig. 12.9. Specular neutron reflectivity for a free-standing Ni film, 1,000 Å thick. The oscillations are a result of the interference which occurs in the simultaneous scattering of the wave from front and back surfaces of the film, as discussed in the text. The period of the oscillations is approximately $2\pi/L$

between parts of the wave that are scattered from front and back film surfaces; the period of the oscillations is approximately $2\pi/L$. Such interference requires that the incident plane wave interact with both interfaces "coherently," i.e., simultaneously.

For a one-dimensional incident wave packet with a finite characteristic coherence length along the z-axis Eq. 12.1 must be solved. We can describe the incident wave function as a wave packet $\psi_{\mathrm{WP}}(z)^{\mathrm{coh}}$ consisting of a superposition of plane waves, each component having a well-defined value of k_z, viz.,

$$\psi_{\mathrm{WP}}^{\mathrm{coh}}(z) = \int_{-\infty}^{\infty} \phi(k_z|k_{0z})\mathrm{e}^{\mathrm{i}k_z z}\,\mathrm{d}k_z\,, \tag{12.16}$$

where the normalized weighting $\phi(k_z|k_{0z})$ might, for instance, be represented by a Gaussian distribution centered on the nominal k_{0z}, viz.,

$$\phi(k_z) = \frac{2}{\Gamma_{\mathrm{coh}}}\sqrt{\frac{\ln 2}{\pi}}\mathrm{e}^{\frac{-4\ln 2}{\Gamma_{\mathrm{coh}}^2}(k_z-k_{0z})^2}\,, \tag{12.17}$$

where Γ_{coh} is the FWHM of the distribution. For such a Gaussian wave packet, the relationship between uncertainties Δk_z and Δz of wavevector and position, respectively, along z is given by the Heisenberg uncertainty relation

$$\Delta z \Delta k_z = \frac{1}{2} \,. \tag{12.18}$$

Although we will not explicitly solve the wave equation for the case of an incident wave packet here, let us call the result of that calculation for the reflection amplitude $r_{\mathrm{WP}}^{\mathrm{coh}}(k_{0z})$, where k_{0z} denotes the nominal k_z for the wave packet. To a first approximation, $r_{\mathrm{WP}}\mathrm{coh}$ is a superposition of "components" $r(k_z)$ with the same weighting $\phi(k_z|k_{0z})$ as in Eq. 12.16. The resulting specular reflectivity for the case where the incident neutron is described as a wave packet with a coherence length $L' << L$ along z does not display the pronounced Kiessig fringes appearing in Fig. 12.9 because the degree to which the neutron can coherently interact with front and back surfaces is significantly diminished.

Let us now consider a collection of neutrons which constitute an incident beam. Let us assume that every neutron in the beam is described by the same one-dimensional wave packet and coherence length, but that there now exists a distribution of different nominal wavevector magnitudes, k_{0z}, distinct from the distribution of k_z about a given k_{0z} in a wave packet. We can, for convenience, choose this distribution also to be Gaussian. However, the distribution of k_{0z} describes an incoherent association of our "test" neutrons A and B, in the sense that each neutron in the beam reflects from the film independently. Thus, the *measured reflectivity* $R_{\mathrm{M}}(Q_{\mathrm{M}})$, at nominal $Q = Q_{\mathrm{M}}$, for an incident beam of such neutrons is the average over $Q = 2k_{0z}$ of the wave packet reflectivities $|r_{\mathrm{WP}}^{\mathrm{coh}}(Q)|^2$; viz.,

$$R_{\mathrm{M}}(Q_{\mathrm{M}}) = \frac{2}{\Gamma_{\mathrm{inc}}} \sqrt{\frac{\ln 2}{\pi}} \int_{-\infty}^{\infty} |r_{\mathrm{WP}}^{\mathrm{coh}}(Q/2)|^2 \mathrm{e}^{-\frac{4 \ln 2}{\Gamma_{\mathrm{inc}}^2}(Q-Q_{\mathrm{M}})^2} \mathrm{d}Q \,, \tag{12.19}$$

where Γ_{inc} is the FWHM of the Q-distribution. The convolution in Eq. 12.19 is the "instrumental resolution" commonly employed – but usually with $r_{\mathrm{WP}}^{\mathrm{coh}}$ replaced by r for the ideal case of an incident plane wave – and also contributes to smearing the fringes in Fig. 12.9. Thus, instrumental resolution should be as tight as reasonably possible, especially where eventual knowledge of $r_{\mathrm{WP}}^{\mathrm{coh}}$ is the goal of the measurement.

Thus, the source of the neutrons and their interactions with instrumental components, which combine to define the size, shape and direction of each neutron wave packet, determine both coherent and incoherent distributions of possible wavevector components in the measurements. The coherent contribution characterizes the wave packets describing individual neutrons, our A or B, while the incoherent contribution emanates from the pathways taken by different neutrons, A and B. For example, diffraction by a sufficiently narrow slit aperture may significantly distort the nominal neutron plane wave, leading to a coherent distribution of wavevectors (common to A and B), while the mosaic structure of the monochromator induces an incoherent spread of wavevectors incident on the sample, distinguishing A from B.

12.3.3 In-plane Averaging

In-plane structure causes non-specular reflection, as previously mentioned, but even when this is weak enough to be ignored, the observed specular reflection will be influenced by lateral variations of the depth profile. The common assumption is that the laterally averaged scattering length density produces the specular "component" of reflection. That is, if the SLD profile is described everywhere in the film by $\rho(x, y, z)$, then the reflection amplitude $r(Q)$ is caused by its lateral average $\rho(z) = \langle \rho(x, y, z) \rangle_{xy}$, as introduced in Eq. 12.2 and the related discussion. This can be true, however, only to the extent that the neutron beam is laterally coherent over the surface of the film, so that such an average is meaningful.

In three dimensions, we can ascribe two coherence lengths to the incident neutron wave packet: a longitudinal coherence length $\ell_z^{\rm coh}$, which is the coherence length we had in mind in discussing the immediate implications of Eq. 12.18 in terms of film thickness; and a lateral or in-plane coherence length $\ell_{xy}^{\rm coh}$, which limits the size of the surface the incident wave packet "coherently sees." Note that the coherence lengths discussed here are the projections of the neutron wave packet coherence lengths, parallel and perpendicular to its nominal wavevector, projected onto the coordinate axes of the sample.

Now in Eq. 12.19 we implicitly assumed that the film was laterally homogeneous. More generally, however, the reflectivity $\left|r_{\rm WP}^{\rm coh}\right|^2$ appearing in the "incoherent" convolution integral must be replaced by a lateral average $\left\langle \left|r_{\rm WP}^{\rm coh}\right|^2 \right\rangle_{xy}$. For example, a sample characterized by partial coverage might comprise a film composed of two (fully and partially covered) components and a corresponding scale of inhomogeneity ℓ_{xy} equal to the larger of the dimensions associated with the fully and partially covered regions. In cases where $\ell_{xy}^{\rm coh} \gg \ell_{xy}$, the film appears to the neutron beam as homogeneous, and the specular reflectivity is caused by the corresponding lateral average, $\langle \rho(x, y, z) \rangle_{xy}$, as in Eq. 12.2. However, when $\ell_{xy}^{\rm coh} \ll \ell_{xy}$, the film appears, instead, as a collection of several types of films, each of which reflects the neutrons according to their "local" $\rho(z)$. Then the measured reflectivity is an areally weighted average of reflectivities from several different films. That is,

$$\left\langle \left|r_{\rm WP}^{\rm coh}\right|^2 \right\rangle \simeq \begin{cases} \left|r_{\rm WP}^{\rm coh}\right|^2 & \text{for } \ell_{xy}^{\rm coh} >> \ell_{xy}, \\ \sum_j w_j \left|r_{{\rm WP},j}^{\rm coh}\right|^2 & \text{for } \ell_{xy}^{\rm coh} << \ell_{xy}, \end{cases} \tag{12.20}$$

where w_j is the weighting for the jth type of in-plane component. In the second case, unlike the first, there is no single physically defined $\rho(z)$, so any attempt to analyze the reflectivity in terms of one such SLD profile must fail, unless $w_J \approx 1$ for some $j = J$.

Thus we must ask, how do we know which case of Eq. 12.20 is the correct one for a given experiment? In some cases a visual inspection of the sample may suffice to say; if we can literally see, i.e., with visible light, evidence for lateral homogeneities much greater than the neutron coherence length, then the much shorter wavelength neutron beam can "see" it too. However, a film may appear visibly homogeneous while still behaving as if $\ell_{xy}^{\rm coh} << \ell_{xy}$ and thus acting as an inhomogeneous collection of reflectors.

Now consider the specific case in which regions of two different SLDs are distributed within the plane of the film as shown in Fig. 12.10a, b. If the linear dimensions of the area of either SLD is much smaller than the in-plane projection of the neutron coherence length, as schematically represented by the straight line in Fig. 12.10a, then the neutron wave effectively averages over the SLDs of the two regions; i.e., the measured specular reflectivity is that for the areally weighted average, as plotted in Fig. 12.11a. However, if the linear dimensions of either SLD component are much larger than the in-plane coherence length, the measured reflectivity is the incoherent sum of two areally weighted reflectivities, as in Eq. 12.20, each corresponding to one region of SLD, as plotted in Fig. 12.11b (assuming equal weightings). This suggests that use of samples with known in-plane SLD distributions, such as might be fabricated by lithographic techniques, could be used to infer neutron coherence lengths independently, to some degree, of the incoherent instrumental resolution.

12.3.4 Q-Resolution for Specular Reflectivity, Assuming an Incoherent Beam

It is instructive and practical to consider the common situation where the wave packets are well approximated by ideal plane waves (wave packets having a very narrow distribution of wavevectors), so that resolution in fact is dominated by an incoherent distribution of mean wavevectors. The instrumental Q-resolution for specular reflection is then determined by applying the simple laws of geometrical optics for reflection and refraction to the reflecting guide, the mosaic crystal monochromator (for which Bragg's law is also imposed), the pair of slits preceding the sample, and the surface of the sample itself (since the flatness of the sample also affects the measured value of Q).

Figure 12.12 resolves the incident and reflected wavevectors, $\mathbf{k}_{\rm i}$ and $\mathbf{k}_{\rm f}$, respectively, into their rectangular components. From the diagram we can write

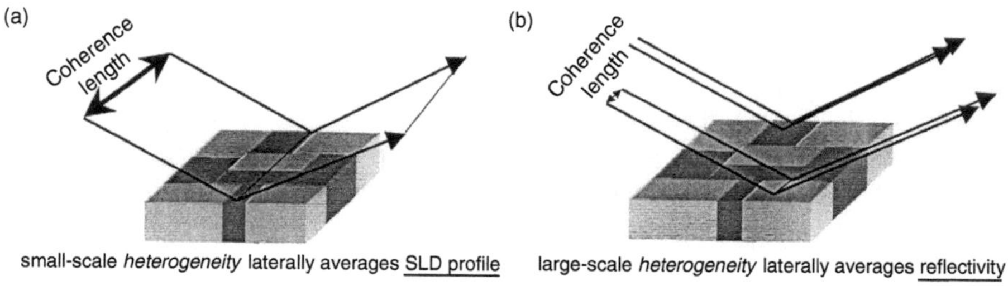

Fig. 12.10. Schematic representation of neutron coherence length and in-plane dimensions of homogeneous sample areas

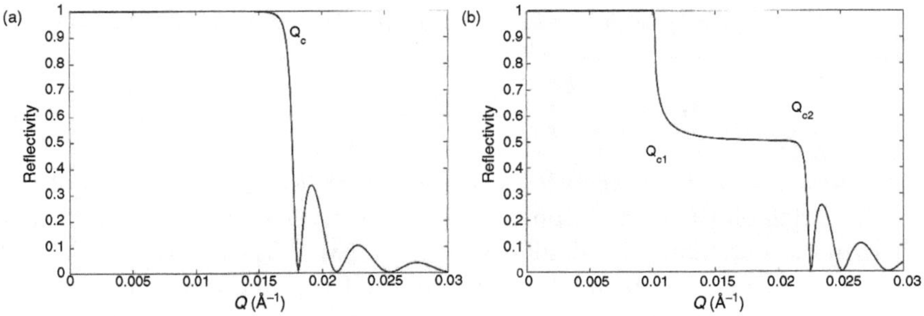

Fig. 12.11. Specular reflectivity: (**a**) corresponding to picture in Fig. 12.10a; and (**b**) corresponding to picture in Fig. 12.10b. It is assumed that the two different SLDs cover equal areas in both cases

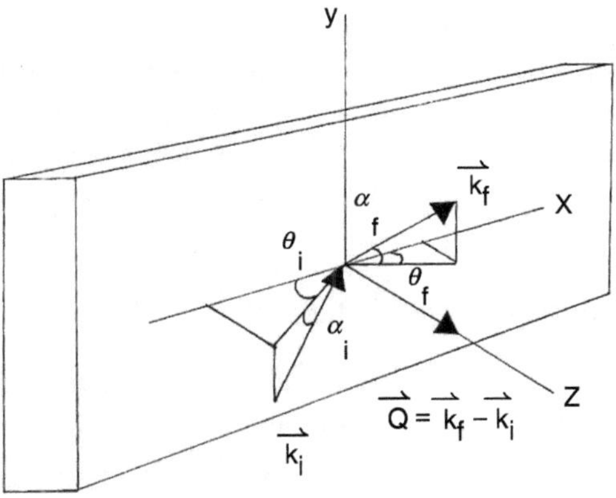

Fig. 12.12. Incident and reflected neutron wavevectors resolved into their respective rectangular components

$$
\begin{aligned}
k_{\mu x} &= k \cos \alpha_\mu \cos \theta_\mu , \\
k_{\mu y} &= k \sin \alpha_\mu , \\
k_{\mu z} &= k \cos \alpha_\mu \sin \theta_\mu ,
\end{aligned}
\qquad (12.21)
$$

where $\mu = \mathrm{i}, \mathrm{f}$ and $k = 2\pi/\lambda$. For specular reflection $\theta_\mathrm{i} = \theta_\mathrm{f}$, and, given that α typically is at most a few degrees, the expression for Q (i.e., $Q = -Q_z = k_\mathrm{i} - k_\mathrm{f}$, as defined in Sect. 12.1) reduces, to a good approximation, to the familiar

$$
Q = 2k \sin \theta . \qquad (12.22)
$$

In terms of wavelength λ and the grazing angle θ, the fractional uncertainty in Q then is

$$
\frac{\delta Q}{Q} = \frac{\delta \lambda}{\lambda} + \frac{\delta \theta}{\tan \theta} , \qquad (12.23)
$$

which, for the typically small angles in reflectivity experiments, is approximately

$$\frac{\delta Q}{Q} \approx \frac{\delta\lambda}{\lambda} + \frac{\delta\theta}{\theta}, \tag{12.24}$$

since $\tan\theta \approx \sin\theta \approx \theta$. As mentioned earlier, the angular divergences of the beam impinging on the monochromator crystal and the monochromator's mosaic distribution are normally significantly greater than the divergence defined by the slits which determine $\delta\theta$. In this case, the fractional wavelength uncertainty is nearly independent of θ, and the two contributions to the fractional uncertainty in Q can be taken to be independent, so that

$$\frac{\delta Q}{Q} \approx \sqrt{\left(\frac{\delta\lambda}{\lambda}\right)^2 + \left(\frac{\delta\theta}{\theta}\right)^2}. \tag{12.25}$$

Usually in specular reflectivity measurements, the slits preceding the sample are opened proportionally with θ, once the sample has fully intercepted the entire width of the incident beam, so that $\delta Q/Q$ (as well as the "footprint" of the beam on the sample) remain approximately constant with θ and Q (in the small angle approximation). A typical value of $\delta Q/Q$ is 0.025.

12.3.5 Measurement of the Reflectivity

To obtain the specular reflectivity, the reflected intensity is first measured as a function of wavevector Q – at a continuous source, by varying the incident angle θ at fixed wavelength and using Eq. 12.22 – up to a maximum value $Q_{\max}$ at which the signal to noise ratio S/N becomes prohibitively low. Background from incoherent scattering within the sample, substrate, or surrounding media, as well as from external sources must be measured and then subtracted from the measured reflected intensity. The resultant signal next must be divided by the incident beam intensity (which is also a function of θ if the slits are opened with increasing reflection angle). Corrections to the reflectivity spectrum also must be applied at values of Q below which the sample does not fully intercept the width of the incident beam (the so-called "footprint" correction). Finally, at least in principle, the instrumental resolution function needs to be deconvolved from the measured reflectivity data, when the resolution correction is significant enough to warrant it. However, in practice deconvolution is a mathematically problematic operation on finite data sets. In the following sections, some of the practical aspects of data reduction are discussed.

Sample Alignment

In order to obtain quantitatively accurate reflectivity data, proper sample alignment is essential. The procedure for accomplishing this is straightforward but can be complicated by substrates deformed from perfect flatness. The goal is to align the sample surface such that it is parallel to and bisects the width of the incident beam, viewed as a ribbon. (It is presumed that the centers of the beam and sample surface coincide.) A rough orientation of the sample can be obtained optically by translating the reflecting surface close to the center of the beam defined by the slit apertures and rotating the sample to be parallel with the beam. Any angular tilt of the sample away from vertical, about the horizontal axis of the incident beam, can be eliminated either using a laser beam reference or even a mechanical plumb line.

Then, the detector can be set at zero scattering angle (for specular reflection, the detector is always positioned at a scattering angle twice the reflection angle of the beam relative to the sample surface) with the pair of slits preceding the sample set so that the horizontal divergence is relatively tight, of the order of a minute of arc. The slit immediately following the sample can be set wide enough to accept the entire divergent width of the beam, but the last slit before the detector should be set to a width comparable to that of the first two slits in order to be sensitive to rotations of the sample. The sample is then translated across the incident beam in a scan in which the transmitted intensity is measured at each step. Once the translational position of the interface is located the sample is rotated in θ at this position with the detector still at zero scattering angle. The occurrence of a central peak corresponds to the position of the sample face being approximately parallel to the incident beam; regions of minimum intensity on either side correspond to the incident beam being reflected by the surface of the sample at a finite scattering angle (and, therefore, away from the detector which is at zero scattering angle). The two-step procedure just described can be repeated iteratively until convergence.

With the nominal zero of the sample angle θ defined, the sample can be rotated to a finite angle corresponding to a Q of 0.005 Å^{-1} ($\theta \approx 0.1°$ for $\lambda =$ 4.75 Å) with the detector at twice that angle (slit apertures unchanged). Now a sequence of three scans can be performed: a rotation of the sample in theta (about the vertical axis through the sample surface), referred to as a "rocking" curve; a translation of the sample through the incident beam; an angular tilt of the sample about a horizontal axis through the center of the sample surface. This sequence of scans is performed iteratively until convergence of the sample rotation (to a peak position in θ that occurs at half the scattering angle), translation, and tilt angle are each achieved. This process can also be carried out at negative reflection and scattering angles, which corresponds to the beam being incident from within the substrate (which is possible for the case of a single crystal Si substrate which is highly transparent to neutrons). Although no critical angle for total reflection may exist in going from the denser (Si) medium to air for certain films on the surface, the reflectivity is typically high enough.

A flat sample surface should result in a smoothly shaped rocking scan curve resembling a Gaussian with a FWHM close in value to the angular divergence defined by the pair of slits upstream. Any significant deviation from this (assuming that the tilt was properly optimized), especially manifest as asymmetric or multiple peak shapes, is indicative of a non-flat sample surface. As already discussed, a non-flat surface results in a broadened Q-resolution which must be accounted for. If the broadening is acceptable, in terms of resolution, precaution must still be taken that the slits downstream of the sample open sufficiently to fully accept the increased divergence of the specularly reflected beam on its path to the detector. This can be accomplished in a straightforward manner by measuring the reflected intensity at a given θ as a function of slit opening until a plateau is achieved. If a critical angle exists for the sample being examined, also a longitudinal scan (i.e., the specular θ–2θ scan) can be performed. If the sample is long enough, a plateau is reached below the critical angle, where the reflectivity is practically unity.

Geometrical Beam Footprint Correction

If the sample is not long enough to fully intercept the width of the incident beam, at lower reflection angles, then a decreased reflectivity is measured. If the sample has a critical angle, θ_c, above the point at which the surface intercepts the full width of the incident beam, then the correction below that point is trivial; the reflectivity simply is defined as unity for $0 < \theta \leq \theta_c$. However, if a critical angle is too small or nonexistent, then another sample of the same size, but with a critical angle that lies above the point of full interception, can be measured under identical conditions to obtain the proper geometrical scaling as a function of glancing angle. However, if the sample is not flat enough, an accurate footprint correction may not be achievable.

12.3.6 Sample Cell designs

Material Fronting Medium and Beam "Side" Entry

Taking advantage of the near transparency to neutrons of Si, sapphire, or quartz single crystals, the reflectivity of films deposited on such substrates can be measured with a beam incident upon the film from within the substrate. This makes it possible for a film of interest to be in contact with a neutron-attenuating aqueous reservoir or other fluid medium, as shown in the following subsection on cell design. In practice, incidence from within a substrate typically requires the beam to enter through a surface of the substrate perpendicular to the film, i.e., through a side of the crystal, as shown in Fig. 12.13. The beam incident from vacuum on the left enters the fronting medium (single crystalline Si, for example) through a face which is perpendicular to the plane of the film. The regions about the side boundary face and the film surface (schematically indicated in the figure by the rectangular perimeters in the figure) are assumed to be sufficiently separated that the neutron wave packet does not interact with both interfaces simultaneously.

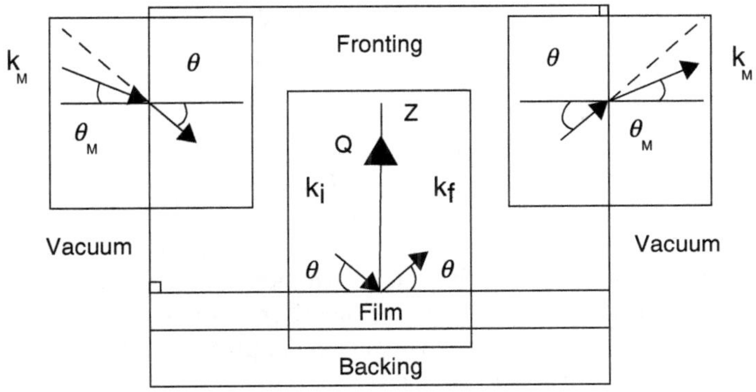

Fig. 12.13. Side-entry geometry typically employed in the case of a beam incident through material (non-vacuum fronting)

As described in Sect. 12.1, for vacuum fronting, the value of the wavevector transfer $2k_z$ in specular reflection, as measured in the laboratory, satisfies

$$2k_z = |\mathbf{k}_\mathrm{f} - \mathbf{k}_\mathrm{i}| = 2k_\mathrm{M} \sin\theta_\mathrm{M} = 2k_{0z} = Q\,. \tag{12.26}$$

Here the subscript M denotes quantities measured on the instrument in the laboratory, as indicated in Fig. 12.13 (since, for vacuum, there is no side interface to cross through). On the other hand, for nonvacuum fronting, a refractive bending occurs as the neutron crosses the side boundary, which, from Snell's law, is

$$\sin\theta_\mathrm{M} = n_\mathrm{f} \sin\theta, \tag{12.27}$$

where n_f is the refractive index of the fronting medium

$$n_\mathrm{f} = \sqrt{1 - \frac{4\pi\rho_\mathrm{f}}{k_\mathrm{M}^2}}\,, \tag{12.28}$$

where $\rho_\mathrm{f} = \rho^\mathrm{I}$. The index of refraction is not to be confused with n_z^I, defined in Eq. 12.4. Using Eq. 12.27, the value of k_z *inside* the fronting medium then is

$$k_z = k\sin\theta = k_\mathrm{M} n_\mathrm{f} \frac{\sin\theta_\mathrm{M}}{n_\mathrm{f}} = k_\mathrm{M}\sin\theta_\mathrm{M}\,. \tag{12.29}$$

The value of k_z in the fronting is the value $k_{\mathrm{M}z}$ measured on the instrument by measuring θ_M and by computing $k_\mathrm{M} = 2\pi/\lambda$. However, according to the 1D description in Eq. 12.1, for a given k_{0z}, the wavevector incident on the film is k^I of Eq. 12.4, as if side entry of the incident beam had not occurred. Therefore, to adapt Eq. 12.1 to side entry, its wavevector *parameter* k_{0z}, or the corresponding Q, must be identified in terms of the *measured* k_z, as given in Eq. 12.29. With Eqs. 12.4 and 12.29

$$k_z^{\mathrm{I}} = \sqrt{k_{0z}^2 - 4\pi\rho_{\mathrm{f}}} = \frac{2\pi \sin\theta_{\mathrm{M}}}{\lambda}\,, \tag{12.30}$$

so that, solving for k_{0z},

$$Q = 2k_{0z} = \sqrt{\left(\frac{4\pi \sin\theta_{\mathrm{M}}}{\lambda}\right)^2 + 16\pi\rho_{\mathrm{f}}}\,. \tag{12.31}$$

Therefore, for nonvacuum fronting and with side entry, in comparing the reflectivity measured at an angle θ_{M} to a reflectivity calculated for a model SLD profile, the value of Q at which the theoretical expression for $r(Q)$ (or $|r(Q)|^2$) must be computed is given by Eq. 12.31.

Sample Cell Designs with Liquid Reservoirs

In the study of biomimetic films, it is often required that the film be in contact with an aqueous reservoir. As already discussed, the high transparency of neutrons through single crystalline materials such as Si, Al_2O_3, and SiO_2 make it possible to construct fluid cells in which the single crystal serves both as substrate and fronting medium for the neutron beam. In principle, the design of a fluid cell is straightforward but, as is discussed in the following section, contributions to the background from the media surrounding the film can be the predominant factor which limits the maximum Q at which the reflectivity can be measured and consequently, the spatial resolution of the SLD depth profile.

Figure 12.14 shows face and end-on views of a liquid cell that has evolved as a standard piece of equipment for reflectivity measurements. The single crystal fronting and backing are assembled from 7.62 cm diameter discs of various thicknesses. Under the correct conditions, such a cell, in which the incident, transmitted, and reflected neutron beams in the vicinity of the sample are entirely within the single crystal media, typically allows maximum Q in the range $Q_{\max} \approx 0.3$ Å to $Q_{\max} \approx 0.4$ Å. The single crystal discs are normally polished on one side. The thickness of the reservoir next to the film is defined

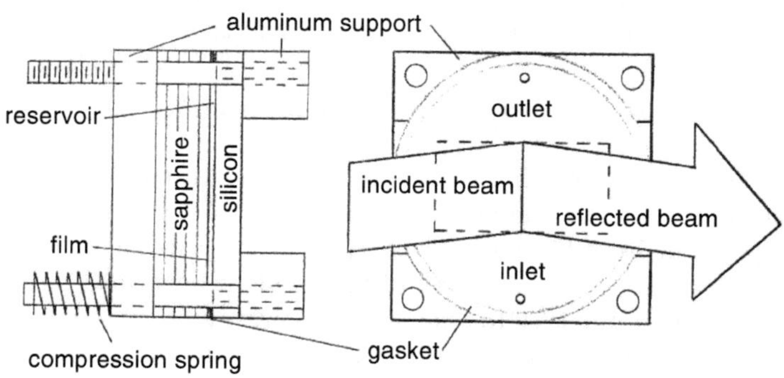

Fig. 12.14. Schematic views, face-on and end-on, of fluid reservoir cell used in neutron reflectivity measurements as described in detail in the text

by an annular gasket (e.g., nitrile or other similarly impervious material). This dimension can be as small as about 25 μm without any significant effect on the measured film reflectivity from the face of the backing crystal, but for reservoirs that are too thin, the possibility of coherent contributions from the face of the backing crystal needs to be considered. Fluid is introduced through a hole (e.g., ultrasonically drilled through the single crystal Si or Al_2O_3 and of diameter 1–2 mm) near the bottom of the backing disc; a similar hole diametrically opposed at the top of the disc serves as an outlet. A cylindrical "top hat" made of aluminum can be placed around the sample and the volume surrounding the cell filled with argon gas, which scatters neutrons significantly less than air. Brass or copper heating/cooling blocks can be attached to the aluminum cell frame at top and bottom. Temperature control (over a range from about $-10°C$ to $80°C$) can be maintained by a combination of fluid flow through the blocks and electrical resistance heater cartridges.

It is difficult to overemphasize the importance of using substrates that have been polished smooth and flat and of maintaining flatness in the compressed sandwich of the cell. A root mean square (RMS) roughness about 3–5 Å is obtainable and desirable since this ultimately limits the spatial resolution in the measured SLD depth profile. Flatness, on the other hand, as commonly used, is associated with in-plane areas comparable to or greater than the coherence length (of order micrometers); the normals to these areas should not deviate more than about 0.01° from the nominal direction. As discussed earlier, deviations from perfect flatness also degrade the effective instrumental Q-resolution for specular reflection measurements.

For a lipid bilayer on a Au film (thickness ≈100 Å) deposited on a 0.5 mm thick Si substrate and placed next to a D_2O reservoir of thickness ≈25 μm (in this case defined by a gasket and another 0.5 mm Si crystal as backing), specular neutron reflectivities have been measured for $Q_{max} = 0.73\,Å^{-1}$ [40].

If the sample film can be exposed to a humid atmosphere instead of an aqueous reservoir (e.g., water vapor in Ar), then it is advantageous for reduction of background to deposit the film of interest on a thin (e.g., 0.5 mm) single crystalline substrate. The humidity can be controlled either by saturated salt solutions or mechanical humidity generators.

12.3.7 Sources of Background

Normally, a single lipid bilayer membrane is itself a negligible source of incoherent scattering background. For a well-shielded instrument, external sources of background can also be relatively insignificant. The major contribution to the background in a specular reflectivity measurement most often originates in the media surrounding the film which is exposed to an incident beam that can be relatively intense at larger Q-values, where the slits are opened wide. For polycrystalline substrates, even though the wavelengths are often long enough that no Bragg scattering can occur (e.g., aluminum at $\lambda = 5.0\,Å$), small angle scattering from the crystal grains, as well as incoherent and inelastic scattering, can contribute. Single crystalline substrates can produce a significant amount of incoherent and inelastic scattering, as well, but are usually preferred to polycrystalline or amorphous (e.g., glass) materials. If the substrate contains a neutron absorber, e.g., boron in pyrex glass, the scattering that contributes to the background can be reduced, although the presence of significant absorption requires that the substrate be used only as a backing medium and that an imaginary component of the scattering length density for the substrate be taken into account in the analysis of the measured reflectivity. In any event, one way to judge the potential of a substrate for producing background, absent absorption, is to measure its transmission. Away from the critical angle, specular reflection falls rapidly with Q, at least as fast as Q^{-4} at large Q. Thus, at large Q, most of the beam should be transmitted through the backing with a transmission close to unity. Measuring a reduced value of the transmission, say, about 0.85 for a Si single crystal substrate 7.5 cm thick, implies that a substantial number of non-reflected neutrons are scattered elsewhere, some fraction of which enter the detector as background.

In addition to substrates, an aqueous reservoir adjacent to the sample film can also contribute a substantial amount of incoherent background, especially if it contains H_2O. Where possible, it is advantageous to use D_2O in place of ordinary water and to minimize the reservoir thickness. Note that single scattering of a neutron from a hydrogen nucleus is most often an incoherent event, resulting in an angularly isotropic distribution of scattered radiation.

Even when it is possible to use a thin single crystalline substrate, the air surrounding the sample which is intercepted by the incident and transmitted beams and simultaneously viewed by the detector can be a substantial source of background. This background can be eliminated by placing the sample in an evacuated chamber or by replacing the air with He or Ar gas, which scatter significantly less than nitrogen and oxygen.

Background Measurement

To measure the background at a given Q, the detector angle 2θ is set close to the specular condition but offset far enough to miss the specular signal. The amount of offset for given slit openings and beam width can be determined by performing a transverse scan along a direction perpendicular to the film normal (z-axis) and with the horizontal width of the aperture in front of the detector sufficiently tight; a rocking curve is normally a satisfactory approximation. Note, in particular, that non-specular reflection is not background, which is more or less isotropic, but is scattering from in-plane variations in SLD in the sample. As discussed earlier, the observation of significant non-specular scattering requires proper evaluation of the validity of the use of the one-dimensional specular scattering theory. Figure 12.15 is a plot of intensity vs. rocking angle θ at a fixed scattering angle for a metallic Ni/Ti multilayered sample having a relatively large number of interfaces with roughnesses

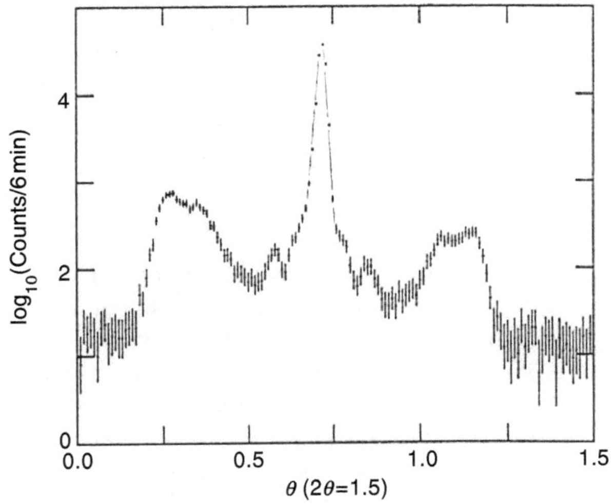

Fig. 12.15. Plot of intensity vs. rocking angle (θ) at a fixed scattering angle for a multilayer Ni/Ti sample with significant in-plane SLD variations, as described in the text

that could be correlated from one layer to another, thereby manifesting some degree of three-dimensional order.

Background Suppression

In our discussion above, we already mentioned that background can be suppressed by using thin single crystal substrates to support the film and to replace the surrounding air with vacuum, He, or Ar gas. It was also mentioned how the pair of slits downstream of the sample have no effect on the specular reflectivity measured from a flat sample. Instead, the slits following the sample act to discriminate the specular-reflected signal from scattering having a wider angular divergence. However, it can happen that the distance between the two slits which define the incident beam angular divergence is greater than that for the pair of slits which precede the detector. In such a case, opening the slits after the sample just enough to allow the full width of the specularly reflected beam through to the detector can result in a wider angular acceptance than that defined by the incident beam slits. Consequently, more of the isotropic incoherently scattered background is allowed into the detector. To remedy this, a set of parallel channels, called a "Soller" collimator, can be used to accept a wider beam at a narrower angular divergence more closely matched to that of the incident beam. Either a Soller collimator with reflecting partitions or one with nonreflecting, absorbing walls can be employed for the purpose.

Alternatively, a mosaic crystal with an appropriate angular distribution of mosaic blocks can be used to discriminate against a more widely divergent incoherent background. Unfortunately, a Soller collimator or analyzer crystal produces transmission and reflectivity losses for the specular signal, which typically range between 20% and 50%. Therefore, proper analysis of the signal to noise ratio, including efficiency, counting statistics, and error propagation considerations, is required for the proper use of these devices.

Signal Enhancement

To improve the signal to noise ratio, it can be important to boost the signal, as well as reducing the background. One method of signal enhancement that has proven useful in the study of single lipid bilayer systems is to deposit a Au layer, about 100 Å thick, onto a Si substrate. Then the film of interest is affixed to the gold layer: e.g., an alkane thiol layer followed by a phospholipid layer, with a D_2O reservoir as a backing [40]. Because the reflectivity can be calculated for a model of such a system, it can be relatively straightforward to determine a feasible combination of the film of interest and signal boosting layers, or surrounding media, that significantly increase the sensitivity of the measurement.

12.3.8 Multilayer Samples: Secondary Extinction and Mosaic

In discussing the theoretical treatment of multilayer reflectivity within the Born approximation, it is implicit that the reflectivity is sufficiently low that the reflectivity of a given reflection order is proportional to the square of the number of bilayers M, as we derived above. However, in practice, multilayer samples of lipid bilayers actually form structures similar to mosaic crystals, having an angular distribution of coherently scattering blocks, each consisting of a stack of bilayers. This angular distribution typically is centered about the mean surface normal of the substrate, with the normal of an individual block perpendicular to the plane of the lipid bilayers in that stack. As the incident beam penetrates such a sample, its intensity can be diminished by successive reflections from various stacks, so that a given reflection peak intensity no longer is proportional to M^2. This troublesome effect, called secondary extinction [5,29,41,42], introduces further complications into the multilayer analysis. It is necessary to recognize and take into account secondary extinction when it occurs so that an error is not made in determining relative reflection peak intensities. Recently, multibilayers of biofilm materials have been made with a well-defined, relatively small number of bilayer repeats which are appropriate for analysis using the dynamical theory outlined in Sect. 12.2.1 [43].

12.3.9 Data Collection Strategies for Time-Dependent Phenomena

In measuring specular reflectivity from thin film systems which may undergo structural changes with time, specular scans must be performed over a given range of Q in a time less than that required for any significant changes to occur. This can be directly determined by superimposing reflectivity plots for successive scans; successive runs can be added together to improve statistical accuracy once equilibrium has been achieved. Whether the film under study exhibits time-dependent behavior or not, it is prudent to perform rocking curves in between specular or other scans, such as background, to verify correct alignment of the sample.

12.4 Phase Determination Techniques

Earlier in the chapter we discussed the connection between the phase of the complex reflection amplitude and the uniqueness of SLD profiles. Here we continue discussion of phase-sensitive specular reflectometry techniques, outlining practical methods for determining the phase of reflection for a film of interest using reflectivity measurements of composite film structures, i.e., film sandwiches composed of the "unknown" film adjacent to a reference layer or to a known surrounding medium. These methods have been recently reviewed in depth [11].

12.4.1 Reference Films

Figure 12.16a illustrates the measurements which are performed to determine the SLD profile of a film, in this case a Cr/Au layer deposited on a Si substrate. The reference layer consists of a ferromagnetic Fe layer with a magnetization which is saturated in the plane of the film. For a polarized neutron in the "+" spin state (one of two possible spin eigenstates), the SLD of the Fe layer is a sum of two parts, one associated with the nuclear interaction and the other with the magnetic potential which exists between the magnetic moments of the neutron and the Fe atoms. In contrast, a neutron polarized in the "−" state sees a SLD which is the difference of the nuclear and magnetic components. By measuring two reflectivity data sets, one with a beam of neutrons in the "+" spin state and the other in the "−" state, plotted in Fig. 12.16a, the imaginary part of the reflection amplitude for the the Cr/Au film, can be determined uniquely, exactly, and independently at each Q [48]. The result is shown in Fig. 12.16b. The imaginary part of the reflection amplitude can then be inverted by a first principles calculation [26, 46, 47]. (More formally, either $\mathrm{Re}\, r(Q)$ or $\mathrm{Im}\, r(Q)$ suffices for most of the SLD profiles of interest to biology.) The result of inverting $\mathrm{Im}\, r(Q)$ of Fig. 12.16b is also shown in the figure. The SLD profile so obtained is unique, to the extent allowed by the finite wavevector range over which the original reflectivity data was collected. In solving for $\mathrm{Im}\, r(Q)$ of the unknown, two roots of a quadratic equation are obtained, only one of which is physical [25, 48]. The physical branch $\mathrm{Im}\, r(Q)$ can be determined, in principle, because $\mathrm{Im}\, r(Q)$ must be a continuous function of Q with known behavior at the origin, viz., $\mathrm{Im}\, r(Q) \leftarrow 0$ from negative values for an overall positive SLD. However, it can happen in practice that the separation of the two

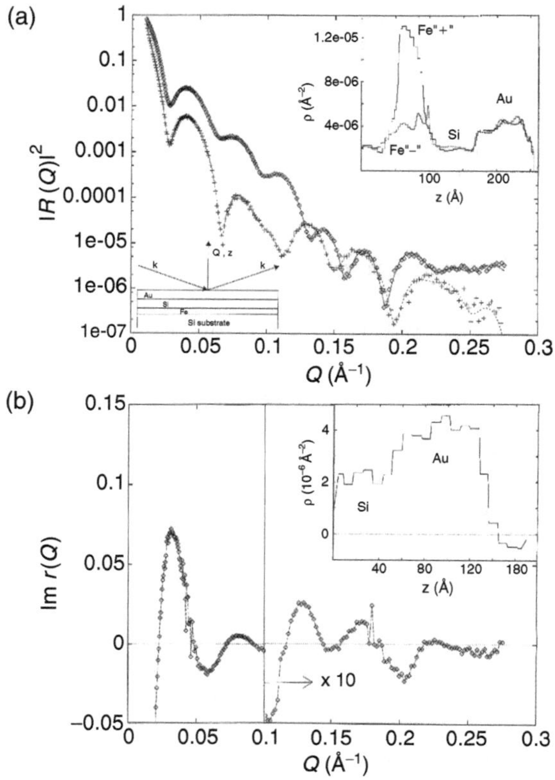

Fig. 12.16. Diagram illustrating the measurements which are performed to determine the SLD profile of an "unknown" film, in this case a Cr/Au layer deposited on a Si substrate (**a**). By measuring two reflectivity data sets the imaginary part of the reflection amplitude for the "unknown" film of interest, in this example the Cr/Au layers, can be determined uniquely at each value of Q; the result is shown in (**b**). The upper right corner inset of (**a**) shows the SLD profiles corresponding to independent fits of the reflectivities for the two composite film systems. The imaginary part of the reflection amplitude can then be inverted by a first principles calculation, as discussed in the text, the result of which is also shown in (**b**). (after Figs. 12.2 and 12.3 of [48])

branches is problematic, especially for noisy data. The use of three reference layers eliminates this problem – and in fact, was the first of the exact reference techniques for specular reflection [22, 23, 49] – but three references are difficult to achieve using a single magnetic layer. Furthermore, in any finite reference layer method for phase determination, the entire SLD density profile of each reference layer used must be known with an accuracy commensurate with the spatial resolution desired in the sample film profile. And of course, magnetic references, in particular, require the availability of polarized neutron beams.

12.4.2 Surround Variation

A reference method closely related to that employing different layers of finite thickness, as described above, involves varying the surrounding media, fronting or backing. This "variation on a theme" has the important advantage that only two constant SLD values, for either the fronting or backing, are required to obtain $\mathrm{Re}\, r(Q)$ corresponding to the sample film, independently at each Q, and without branch ambiguities [24], since, the resulting surround variation equations are linear. One approach that has been successfully employed involves depositing the sample film on two different substrates, e.g., Al_2O_3 and Si, simultaneously and under identical conditions [27]. Care must be taken to limit any differences between the two samples which could be present, such as the presence of a native oxide layer on the Si or a layer of different SLD on the Al_2O_3 due to the effects of surface polishing.

A less cumbersome approach employs a single sample and an adjacent liquid reservoir of variable SLD. Figure 12.17a contains a schematic of a surround variation method for phase determination in which the backing medium SLD can have (at least) two values, in this particular example that of D_2O and Si-(SLD) matched water – i.e., an H_2O and D_2O mixture with approximately-38% D_2O by volume. Figure 12.17 also shows the corresponding composite reflectivity curves for these two backing media adjacent to the Cr/Fe/Au/alkane thiol film indicated in the upper right hand corner of the figure. This is similar to the film structure of Fig. 12.16, except that the Fe "+" and "−" layers are now treated as part of the "unknown" film. Included in Fig. 12.17a is $\mathrm{Re}\, r(Q)$ for the unknown film, one in which the SLD of the saturated magnetization of the ferromagnetic Fe layer is that seen by a spin "+" state neutron beam. Last, Fig. 12.17b shows the SLD profile obtained by direct inversion of the $\mathrm{Re}\, r(Q)$ of Fig. 12.17a. For comparison, the SLD profile obtained for the "−" state neutron beam is also shown [45]. Note the consistency of the two results; the Au layer SLD is virtually identical in both sandwich structures, the one with the Fe "+" layer and the other with the Fe "−" film.

Given the importance and ubiquity of aqueous solutions in the study of biomembranes, the method of choice in phase-sensitive reflectivity measurements would very likely be variation of the backing medium using a suitable fluid, except for one crucial concern. If the fluid differentially penetrates the adjacent film of interest, then the reference measurement is destined to fail, since an essential premise of the technique is that the film of interest be invariant to the change in references. This restriction therefore precludes the use of variation by D_2O /H_2O substitution if water penetrates the membrane, which indeed is known to occur. This problem can be solved if an aqueous solution could be found in which a suitable solute is the agent of SLD variation without interfering with the film: possibly, for example, a sugar in D_2O , where sugar molecules – of variable concentration – do not penetrate or modify the film, whether or not the constant D_2O component is integral to the film. This would indeed be a "sweet solution" for surround variation in some problems.

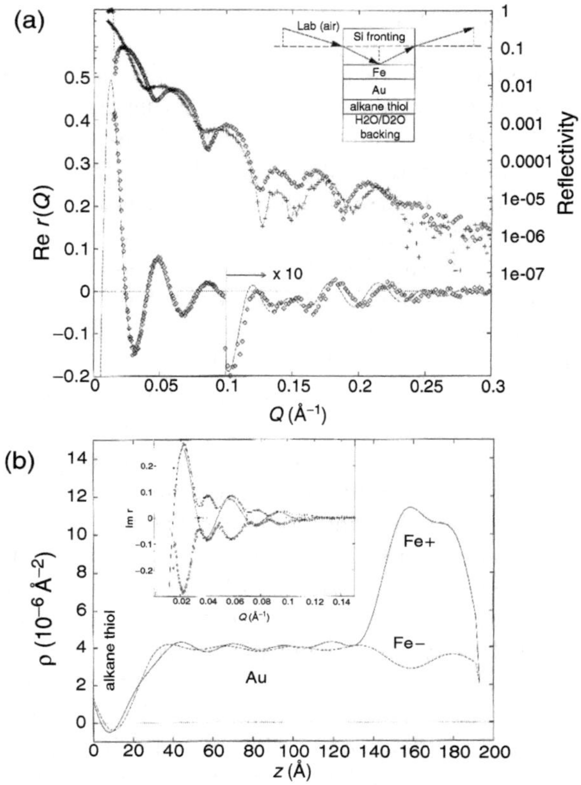

Fig. 12.17. Schematic representation of a surround variation method for phase determination in which the backing medium SLD can have (at least) two values, in this particular example that of D_2O and Si-SLD-matched water (**a**). (**a**) shows plots of the corresponding composite reflectivity curves for these two backing media SLD values adjacent to the Cr/Fe+ /Au/alkane-thiol film; note that this is similar to the film structure of Fig. 12.16 except that the Fe "+" layer is now treated as part of the "unknown" film. (**a**) also shows $\mathrm{Re}\, r(Q)$ for the "unknown" film, one in which the SLD of the saturated magnetization of the ferromagnetic Fe layer is that seen by a spin "+" state neutron beam. Last, (**b**) shows the SLD profile obtained by direct inversion of $\mathrm{Re}\, r(Q)$ of (**a**). For comparison, the SLD profile obtained for the "−" state neutron beam is also shown. (after Figs. 12.1 and 12.2 of [45])

12.4.3 Refinement

The formal inversion methods alluded to earlier begin with a Fourier transform of $\mathrm{Re}\, r(Q)$ and thus require this information at all values of Q for exact implementation. Thus, the resulting SLD profiles are distorted by unavoidable data truncation, the effect decreasing systematically with increasing Q_{max}. This means that the $\rho(z|Q_{\mathrm{max}})$ obtained by inverting $\mathrm{Re}\, r(Q|Q_{\mathrm{max}})$ will not exactly reproduce $\mathrm{Re}\, r(Q|Q_{\mathrm{max}})$ without additional refinement. Useful approaches to this problem [12, 18, 51] take $\rho(z|Q_{\rm max})$ as a starting point for model independent fitting procedures designed to accept only spatial detail consistent with the spatial resolution, $l = \pi/Q_{\rm max}$. The resulting refinement, say $\tilde{\rho}(z|Q_{\rm max})$, effectively represents the most that can be said about the veridical $\rho(z)$ at the given resolution.

Only the real part of the reflection amplitude, $\mathrm{Re}\, r(Q)$, is necessary to obtain the SLD profile by first-principles inversion for most films of interest, as mentioned in our discussion of surround variation in Sect. 12.4.2. Now the same information which gives $\mathrm{Re}\, r(Q)$ also predicts $\mathrm{Im}\, r(Q)$, but only up to a quadratic branch ambiguity, similar to that discussed in Sect. 12.4.1 for the technique using two finite references layers. This ambiguity is of no concern to obtaining $\rho(z)$, but the ancillary, if incomplete, knowledge of $\mathrm{Im}\, r(Q)$ that also results from surround variation happens to be a useful diagnostic of film quality, because of a seemingly arcane mathematical property of $r(Q)$. It turns out that for a perfect but arbitrary film of thickness L, the spectrum of $\mathrm{Im}\, r(Q)$ must possess a more-or-less uniform sequence of zeros near multiples of $Q = 2\pi/L$, suggestive of the Kiessig fringes seen in the reflectivity, as described in Sect. 12.3.2 [45]. On the other hand, $\mathrm{Re}\, r(Q)$ need display these zeros only if the film is perfectly centrosymmetric. So, in fact, the Kiessig fringes observed in $|r(Q)|^2$ normally are not the manifestation of zeros in $r(Q)$ but rather of $\mathrm{Im}\, r(Q)$ alone. In physical terms, these zeros are a property of coherent reflection from laterally homogeneous film and are readily detectable even in the presence of branch ambiguities. The absence of zeros, i.e., the presence of branch "splittings" in $\mathrm{Im}\, r(Q)$, thus is a strong indication that the film under study is defective in these terms.

For example, as discussed in Sect. 12.3.2, if a film is laterally inhomogeneous on a scale large compared to the neutron coherence length, then the measured specular reflectivity is an average of areally weighted reflectivities from the separate inhomogeneous components, as given in Eq. 12.20. In this case, there is no single SLD profile associated with the measured reflectivity, and any attempt to extract one, whether by inversion or fitting techniques, will produce unphysical results. The absence of $\mathrm{Im}\, r(Q)$ splittings, beyond those consistent with noise effects, is a good indication of acceptable film quality [27].

12.5 An Illustrative Example

To illustrate the application of neutron reflectometry to the study of biofilms, we consider the recent structural investigation of a hybrid bilayer membrane (HBM) and its interaction with melittin [40]. In this particular study, specular neutron reflectometry was used to probe the interactions of the peptide toxin, melittin, with supported bilayers of phospholipid (d54-dimyristoyl phosphatidycholine or dDMPC) and octadecanethiol ($HS(CH_2)_{17}CH_3$) or thiahexa(ethylene oxide) alkane ($HS(C_2H_4O)_6(CH_2)_{17}CH_3$ or THEO-C18) on gold. This supported lipid bilayer consisting of adjacent "leaflets" of alkanethiol and phospholipid forms a model biomimetic membrane. The primary objectives of the study were to locate the position and orientation of the melittin within the membrane and also to determine whether the ethylene oxide moieties are hydrated when the HBM is in contact with water. Sample preparation and other details of the experiments and analysis can be found in the original work [40].

Figure 12.18 shows the SLD profiles of the THEO-C18/dDMPC HBMs next to a D_2O reservoir with and without melittin, as obtained from model-independent fitting of the corresponding reflectivity data shown in the inset [40]. Note $Q_{\text{max}} \approx 0.73$ Å^{-1}, corresponding to a spatial resolution about 0.5 nm.

In order to verify that profiles so obtained were physically meaningful, phase-sensitive neutron reflectivity measurements were performed [27] on an almost identical pair of samples: self-assembled THEO-C18 on a Cr/Au metallic bilayer, predeposited on Si and Al_2O_3 single crystal substrates, followed by the dDMPC layer. In this case, the Si and Al_2O_3 substrates served as two different fronting media, with a common backing of Si SLD-matched water, for collection of the pair of composite reflectivity data sets shown in Fig. 12.19. $\text{Re}\, r(Q)$ for the common film sandwich determined from that reflectivity data by the surround variation solution is also shown in the figure, along with a schematic for the phase-sensitive reflectivity measurements [27]. Figure 12.19 shows the $\rho(z)$ obtained by first-principles inversion of the $\text{Re}\, r(Q)$ using the techniques of Sect. 12.4.2. This unique solution is compared to the prediction

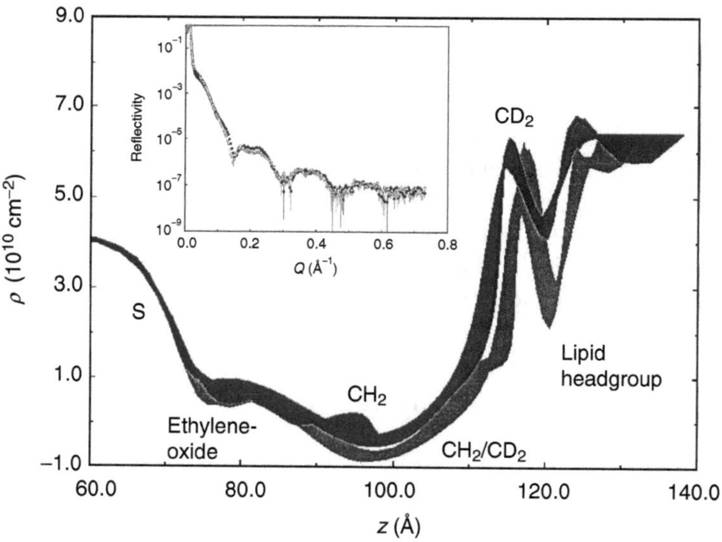

Fig. 12.18. SLD profiles of the THEO–C18/dDMPC HBMs described in the text next to a D_2O reservoir with and without melittin (*darker shaded thick curve*) as obtained from model-independent fitting of the corresponding reflectivity data (*filled symbols without melittin*) plotted in the inset [40]. (The Cr/Au metal layers, Cr, 20 Å thick, and Au, 65 Å thick, on Si, are not shown.) Note that $Q_{\text{max}} \approx 0.73$ Å^{-1}, corresponding to a spatial resolution about 0.5 nm

of a molecular dynamics simulation [31]. The close similarity of the SLD profiles of Fig. 12.18 (without melittin) and Fig. 12.19, gives confidence in the results.

The neutron reflectivity study described above indicates that melittin strongly perturbs the phospholipid headgroup region, but also affects the

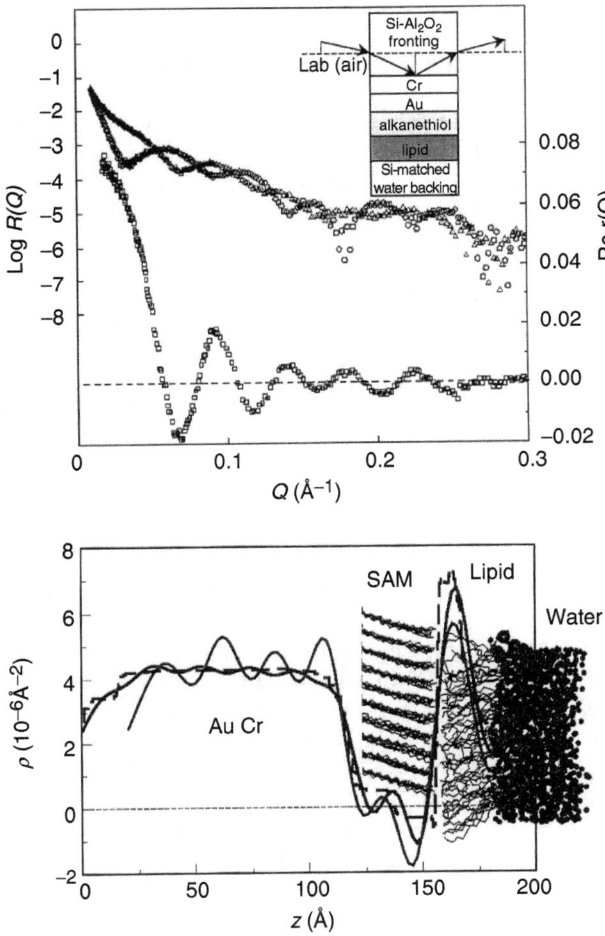

Fig. 12.19. Phase-sensitive neutron reflectivity measurements performed on a self-assembled THEO-C18 layer on a Cr/Au metallic bilayer, pre-deposited on Si and Al_2O_3 single crystal substrates, followed by a dDMPC layer (*top*). $\mathrm{Re}\, r(Q)$ for the common film sandwich determined from that reflectivity data is shown in the lower part, along with a schematic for the phase-sensitive reflectivity measurements in the upper right corner (details of the neutron reflectivity measurements and analysis are given in [27]). The SLD profile obtained by first-principles inversion (*solid curve* with pronounced oscillations in the Au/Cr region due in part to truncation of the data at $Q_{\max}$) of the $\mathrm{Re}\, r(Q)$ is shown below [27]. This unique solution is compared to the prediction of a molecular dynamics simulation (*other solid curve*) [31]

alkane chain region of the bilayer. Among other findings [40], these results demonstrate the utility of neutron reflectometry in determining subnanometer structural changes in biomimetic membranes caused by biologically relevant molecules.

References

1. B. Alberts, D. Bray, J. Lewis, M. Raff, K. Roberts, J.D. Watson, in *The Cell*, 3rd ed. (Garland Publishing, New York, 1994)
2. S.K. Sinha, E.B. Sirota, S. Garoff, H.B. Stanley, Phys. Rev. B **38**, 2297 (1988)
3. R. Pynn, Phys. Rev. B **45**, 602 (1992)
4. S. Krueger, Cur. Opin. Coll. & Interface Sci. **6**, 111 (2001)
5. G.E. Bacon, in *Neutron Diffraction*, 3rd ed. (Oxford University Press, London, 1975)
6. J. Penfold, R.K. Thomas, J. Phys.: Condens. Matter **2**, 1369 (1990)
7. T.P. Russell, Mater. Sci. Rep. **5**, 171 (1990)
8. *X-ray and Neutron Reflectivity: Principles and Applications*, ed. by J. Daillant, A. Gibaud (Springer, Berlin, 1999)
9. C.F. Majkrzak, Acta Physica Polonica A **96**, 81 (1999)
10. C.F. Majkrzak, J. Kwo, M. Hong, Y. Yafet, D. Gibbs, C.L. Chien, J. Bohr, Adv. Phys. **40**, 99 (1991)
11. C.F. Majkrzak, N.F. Berk, U. Perez-Salas, Langmuir **19**, 7796 (2003)
12. N.F. Berk, C.F. Majkrzak, Langmuir **19**, 7811 (2003)
13. E. Merzbacher, *Quantum Mechanics*, 2nd ed. (Wiley, New York, 1970)
14. N.F. Berk, C.F. Majkrzak, unpublished
15. M. Born, E. Wolf, in *Principles of Optics*, (Pergammon Press, Oxford, 1987), p. 51
16. C.F. Majkrzak, N.F. Berk, Physica B **336**, 27 (2003)
17. J.F. Ankner, C.F. Majkrzak, SPIE Proc. **1738**, 260
18. N.F. Berk, C.F. Majkrzak, Phys. Rev. B **51** 11296, (1995)
19. G. Fragneto, R.K. Thomas, A.R. Rennie, J. Penfold: Science **267**, 657 (1995)
20. W. Leslauer, J.K. Blasie, Acta Cryst. A **27**, 456 (1971)
21. M.K. Sanyal, S.K. Sinha, A. Gibaud, K.G. Huang, B.L. Carvalho, M. Rafailovich, J. Sokolov, X. Zhao, W. Zhao, Europhys. Lett. **21**, 691 (1993)
22. C.F. Majkrzak, N.F. Berk, Phys. Rev. B **52**, 10827 (1995)
23. V.O. de Haan, A.A. van Well, S. Adenwalla, G.P. Felcher, Phys. Rev. B **52**, 10831 (1995)
24. C.F. Majkrzak, N.F. Berk, Phys. Rev. B **58**, 15416 (1998) (Erratum Phys. Rev. B **60**, 16211 (1999))
25. R. Lipperheide, J. Kasper, H. Leeb, Physica B **248**, 366 (1998)
26. K. Chadan, P.C. Sabattier, *Inverse Problems in Quantum Scattering Theory* (Springer, New York, 1989)
27. C.F. Majkrzak, N.F. Berk, S. Krueger, J.A. Dura, M. Tarek, D. Tobias, V. Silin, C.W. Meuse, J. Woodward, A.L. Plant, Biophys. J. **79**, 3330 (2000)
28. C.F. Majkrzak, N.F. Berk, D. Gibbs, in *Magnetic Multilayers*, ed. by L.E. Bennett, R.E. Watson (World Scientific, Singapore, 1994), p. 299
29. M.C. Wiener, S.H. White, Biophys. J. **59**, 162 (1991)

30. J.K. Blasie, B.P. Schoenborn, G. Zaccai, in *Neutron Scattering for the Analysis of Biological Structures*, Brookhaven Symposia in Biology No. 27 (BNL50453) (NTIS, Springfield, VA, 1976), p. III-58
31. M. Tarek, K. Tu, M.L. Klein, D.J. Tobias, Biophys. J. **77**, 964 (1999)
32. C.F. Majkrzak, Physica B **221**, 342 (1996)
33. R. Gaehler, J. Felber, F. Mezei, R. Golub, Phys. Rev. A **58**, 280 (1998)
34. S.K. Sinha, M. Tolan, A. Gibaud, Phys. Rev. B **57**, 2740 (1998)
35. H.J. Bernstein, F.E. Low, Phys. Rev. Lett. **59**, 951 (1987)
36. R. Golub, S.K. Lamoreaux, Phys. Lett. A **162**, 122 (1992)
37. H. Rauch, S.A. Werner, *Neutron Interferometry* (Clarendon Press, Oxford, 2000)
38. A. Zeilinger, R. Gaehler, C.G. Shull, W. Treimer, in *AIP Conference. Proceedings of Neutron Scattering*, ed. by J. Faber (AIP, 1982), p. 93
39. A. Steyerl, K.A. Steinhauser, S.S. Malik, N. Achiwa, J. Phys. D, Appl. Phys. **18**, 9 (1985)
40. S. Krueger, C.W. Meuse, C.F. Majkrzak, J.A. Dura, N.F. Berk, M. Tarek, A.L. Plant, Langmuir **17**, 511 (2001)
41. M.J. Yeager, in *Neutron Scattering for the Analysis of Biological Structures*, Brookhaven Symposia in Biology No. 27 (BNL50453) (NTIS, Springfield, VA, 1976), p. VII-77
42. D.L. Worcester, in *Neutron Scattering for the Analysis of Biological Structures*, Brookhaven Symposia in Biology No. 27 (BNL50453) (NTIS, Springfield, VA, 1976), p. III-37
43. U. Mennicke, T. Salditt, Langmuir **18**, 8172 (2002)
44. K.-M. Zimmermann, M. Tolan, R. Weber, J. Stettner, A.K. Doerr, W. Press, Phys. Rev. B **62**, 10377 (2000)
45. C.F. Majkrzak, N.F. Berk, V. Silin, C.W. Meuse, Physica B **283**, 248 (2000)
46. P.E. Sacks, Wave Motion **18**, 21 (1993)
47. N.F. Berk, C.F. Majkrzak, J. Phys. Soc. Jpn **65**, Suppl. A 107, (1996)
48. C.F. Majkrzak, N.F. Berk, Physica B **267–268**, 168 (1999)
49. C.F. Majkrzak, N.F. Berk, J.A. Dura, S.K. Satija, A. Karim, J. Pedulla, R.D. Deslattes, Physica B **248**, 338 (1998)
50. J.M. Cowley, in *Diffraction Physics* (North Holland, Amsterdam, 1990), p. 131.
51. C.F. Majkrzak, N.F. Berk, Appl. Phys. A **74**, S67 (2002)
52. R.G. Nuzzo, F.A. Fusco, D.L. Allara, J. Am. Chem. Soc. **109**, 2358 (1987)

13

Protein Adsorption and Interactions at Interfaces

J.R. Lu

13.1 Introduction

Protein adsorption at interfaces is a complicated molecular process occurring in many technological applications [1,2]. The need to manipulate protein adsorption mainly comes from two strands of interest. There is a range of technological processes where protein adsorption is undesired. In bioseparation and purification, for example, protein deposition can block membrane pores, leading to the fast decline of permeate flux and the halt of the separation process. Protein adsorption is also a major source for cross-contamination of protein related diseases (e.g. prions) through reusable medical devices. Thus removal of surface-deposited blood proteins on reusable medical devices is a major challenge requiring extensive knowledge of the interfacial interaction between surface-bound blood protein and surfactant. In contrast to the undesired protein deposition, protein adsorption is strongly encouraged in many biomedical and biotechnological applications. Examples include biosensors such as the fertility test system working on the principle of surface immobilization of bioactive proteins and their specific recognition of hCG, a hormonal protein. In these cases, however, it is the specific protein recognition that is desired. Furthermore, the performance of the biosensors strongly hinges on the structural conformations of surface immobilized proteins. Protein interaction with biomaterials also implicates the success of the integration of cardiovascular implants and the progress of tissue engineering [3,4].

A number of physical techniques have been developed to reveal different aspects of information from adsorbed protein layers, concerning the adsorbed amount and structural conformations [5–8]. These include X-ray reflection, surface plasmon resonance spectroscopy (SPR), spectroscopic ellipsometry (SE), FTIR-ATR, and the more recently developed infrared-visible sum frequency generation (SFG) vibration spectroscopy [9]. Techniques such as small angle neutron scattering (SANS), circular dichorism (CD), and NMR are very appropriate for revealing useful information about proteins adsorbed on particulate surfaces [10–14]. The combined use of these techniques has provided a rich list of information about the vast amount of details relating to adsorbed protein layers. There are, however, some common drawbacks of existing techniques. For example, they are largely insensitive to a single protein layer and ineffective for performing in situ measurements under water. Furthermore, they cannot distinguish water from polypeptides inside the protein layer. These limitations can be well alleviated by neutron reflection.

In the past five years, we have explored the feasibility of application of neutron reflection in studying protein adsorption and the interfacial interactions between protein and surfactant [15–19]. These studies have been carried out at the air–water and solid–water interfaces. The neutron reflection work has revealed new structural features from protein layers adsorbed under different surface and solution conditions and the patterns of interactions between different proteins and surfactants. As protein adsorption and interactions between proteins and protein-binding species such as lipid, peptide, anesthetic are of strong interest both fundamentally and technologically, the exploratory experiments we have undertaken so far will have a profound impact to the future exploitation of this technique in a wide range of biointerfacial studies.

In this chapter, a number of recently studied protein systems will be used to demonstrate the typical structural information that can be revealed from neutron reflection. Lysozyme has been mainly used as a model protein because it has well defined globular structure and stability, and many studies have already been carried out to characterize its adsorption. The adsorption of proteins on the surface of water will be shown first to illustrate how structural information can be optimized from selective application of isotopic contrasts of the solvent. Protein adsorption at the solid–solution interface will then be shown to demonstrate how different surface chemistry affects the deformation and unfolding of protein molecules. Finally, examples of surfactant binding to proteins adsorbed at the solid–solution interface will be given to show neutron's strength in studying interfacial mixtures.

13.2 Neutron Reflection and Concept of Isotopic Contrast Variation

The technical advantage of neutron reflection is well illustrated in its ability in revealing useful structural information at the solid–solution interface. Under this condition, the deposited protein layer is buried; it is also heavily mixed with water and often less than 100 Å in the overall layer thickness. Although many optical techniques can perform the measurement, their sensitivity to the structure and composition of the protein layer is very limited. Because of the difficulties associated with the direct measurement under water, it is a widely adopted practice to try to estimate the adsorbed amount at the air–solid interface instead. The associated sample rinsing, drying, and treatments such as staining may alter the amount and biophysical state of the adsorbed protein. The results measured from the dry surface may have little correlation with the in situ structure and composition at the solid–solution interface.

Another important feature from neutron reflection is its ability to distinguish individual components across the interface through partial deuterium labeling, making it a unique tool for revealing the in situ structural profile of individual components.

In our previous studies of interfacial adsorption of surfactants and synthetic polymers, we demonstrated that neutron reflection was capable of detecting volume fraction distribution of any species present along the surface normal direction [15, 16]. When applied with selective isotopic substitution, the technique is sensitive to a given species across the interfacial layer with depth resolution at the level of 1–2 Å. Unlike synthetic polymers, proteins are not easily deuterated and this limits isotopic contrast variation to either the solvent or the support surface in the case of solid–solution interface. The concern arises as to how much useful information can be derived from neutron reflection under such circumstance.

An important structural feature of protein molecules is their secondary and tertiary structures. The key issue is whether neutron reflection can provide any, useful structural information that is indicative of conformational changes arising from either interfacial adsorption or from binding and complexation with another compound, e.g., coenzyme. The examples to be shown in the following will focus on addressing this issue. Lysozyme adsorption on the surface of water will be first shown to illustrate how the choice of isotopic contrast can lead to the optimal derivation of structural information. The measurement of lysozyme adsorption at the air–water interface allows us to address two very fundamental issues. First, we would like to know if the adsorption gives rise to deformation or unfolding given that air is a hydrophobic solvent. The adsorption of lysozyme onto the polarized surface may deteriorate the globular framework although lysozyme is very robust. Second, lysozyme has a rather hydrophobic outer surface with an uneven distribution of charged groups. It is interesting to know if the whole adsorbed molecule is staying afloat or immersed in the aqueous solution.

To address the first issue, we have selected an H_2O and D_2O mixture containing 8.1 vol% D_2O. This mixed water has a zero scattering length and is called null reflecting water (NRW). When neutron reflectivity is measured at the air–NRW interface, the only specular signal comes from the adsorbed protein layer [20, 21].

Figure 13.1 shows the reflectivity profiles measured at three different lysozyme concentrations, all at pH 7. It can be seen from Fig. 13.1 that the level of the reflectivity increases with increasing concentration, indicating a clear correlation with surface adsorbed amount (surface excess). While the profiles corresponding to the two low lysozyme concentrations are parallel, the third one decays faster indicating a much thicker layer formation at the highest concentration. Although the common approach to extract quantitative information from the measured reflectivity is to perform model fitting based on the optical matrix formula, a useful alternative is via kinematic approach and structural information is then obtained through model fitting to the partial structure factors [15, 16].

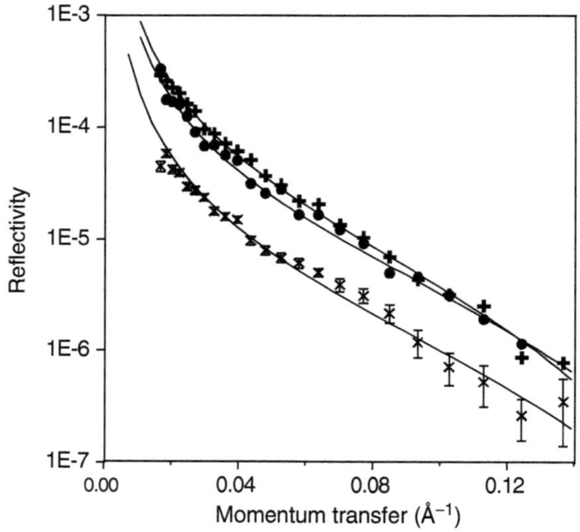

Fig. 13.1. Lysozyme adsorption on the surface of null reflecting water (NRW) at 10^{-3} (×), 0.1 (•) and 1 (+) g dm^{-3} at pH 7

In the kinematic approximation, neutron reflectivity (R) can be analytically related to area per molecule (A) and layer thickness (τ). If lysozyme adsorption forms a uniform layer on the surface of NRW, this relationship can be expressed as

$$h_{\mathrm{pp}}Q^2 = \frac{RQ^4}{16\pi^2 b_{\mathrm{p}}^2} = \frac{4}{A^2\tau^2}\sin^2\left(\frac{Q\tau}{2}\right), \tag{13.1}$$

where h_{pp} denotes the partial structure factor for protein and Q is the momentum transfer and is equal to $4\pi \sin\theta/\lambda$ (where θ is the beam incidence angle and λ is the wavelength). Since changes in A and τ affect the level and shape of reflectivity differently the model fitting to the measured reflectivity profiles leads to a reasonably reliable decoupling of the two parameters. The continuous lines shown in Fig. 13.1 represent the best uniform layer fits obtained from Eq. 13.1. The results show that at the lowest concentration the lysozyme layer is about 30 Å thick and is consistent with the formation of sideways-on monolayer and the lysozyme molecule is adsorbed with its short axis perpendicular to the surface. This is consistent with the area per molecule of 2800 Å^2 and is greater than the required minimum of 1350 Å^2. At the intermediate concentration the layer increases to 34 Å. This is accompanied by the decrease of A to 1300 Å^2 and is within the experimental error comparable to the limiting value for sideways-on adsorption. These changes together with the electrostatic repulsion within the layer suggest the possible tilting of the lysozyme toward headways-on adsorption, that is the molecules adsorbed with its long axis perpendicular to the surface. At the highest concentration of 1 g dm^{-3}, the layer is 47 Å thick and is comparable to the long axial length of the globular structure, indicating that the molecules adopt an entirely headways-on conformation. The area per molecule is 950 Å^2 and is close to the limiting value of 900 Å^2 required for headways-on adsorption. These results suggest a progressive transition of conformation of adsorbed lysozyme molecules with increasing bulk concentration.

A common concern in the data analysis of neutron reflectivity is the sensitivity of the model fit to the shape of the layer distribution. It is unlikely that the true protein layer distribution in this case follows the exact uniform layer model. However, the data may not be of sufficiently high quality to distinguish one shape from another. It is possible that the lysozyme layer resembles a Gaussian layer distribution more closely. Under the Gaussian model, Eq. 13.1 becomes

$$h_{\mathrm{pp}} = \frac{RQ^2}{16\pi^2 b_p^2} = \Gamma^2 \exp\left(-\frac{Q^2\sigma^2}{8}\right). \tag{13.2}$$

Figure 13.2 compares the best Gaussian model fit from Eq. 13.2 with that calculated from Eq. 13.1 based on the uniform layer model. Clearly, the Gaussian model represents the measured data better over the entire Q range. This is in contrast to the lower values calculated from the uniform layer model over the higher Q range. The better representation from the Gaussian model is reasonable given that the packing density in the middle of the lysozyme is likely to be higher than the two edges.

To address the second issue of the extent of immersion of the layer, reflectivity profiles were also measured in D_2O and a mixed D_2O and H_2O in the ratio of 1:1. Under these conditions, the reflectivity contains contributions from the protein layer, the water interface and the interference between the

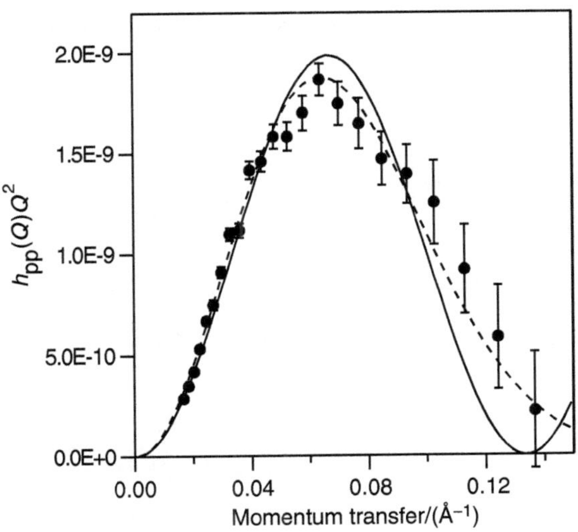

Fig. 13.2. Comparison of the best uniform layer model fit (*continuous line*) and the best Gaussian model fit (*broken line*)

two, thereby providing unique information about the relative location between the two distributions. It is worthwhile to note that under the 1:1 mixed water, the scattering length density of the water phase (ρ_w) is 2.9×10^{-6} Å^{-2} and is almost identical to that of lysozyme. Thus the portion of lysozyme layer immersed in water is indistinguishable from the water subphase. The measurement under this condition provides a unique sensitivity to the portion of the layer staying out of the water in the airside. This isotopic contrast together with the simultaneous measurements under NRW and D_2O offers a reliable determination of the extent of layer immersion and lysozyme conformation at a given solution condition. The outcome of this study is depicted in Fig. 13.3 where the shaded background indicates the surface of water. This study shows that the surface adsorption of lysozyme has not resulted in the breakdown of its globular framework and that in contrary to previous assumption the adsorbed layer is only partially immersed in water. Figure 13.3 also shows that the degree of its immersion is also dependent on surface conformation and packing. The shade area on the back of lysozyme molecules as indicated in Fig. 13.3 denotes the more hydrophobic region where less charge groups are present. Over the low surface concentration, it is plausible that this side stays out of water but under the high-surface packing this region is forced into water.

13.3 Adsorption of Other Proteins at the Air–Water Interface

The neutron reflection from lysozyme adsorption offered a useful experimental methodology for revealing molecular features relating to surface adsorption of

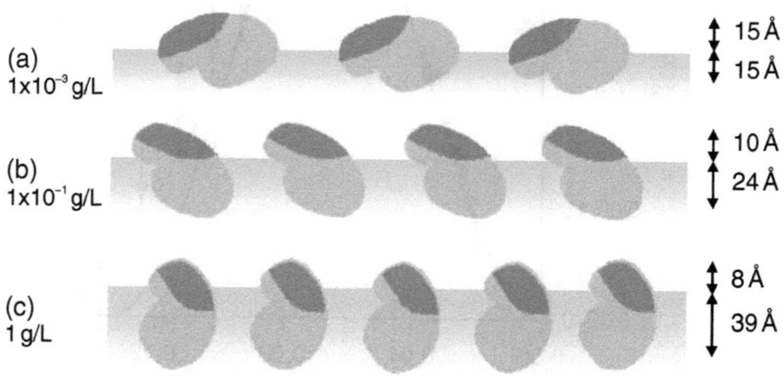

Fig. 13.3. Schematic representation of the change of orientation and packing of lysozyme molecules at the air–water interface with bulk concentration. Note that the darkend edges represent the surface of water and the shaded patch on the back of lysozyme donates the region where less charge groups are found

other proteins. We have carried out a systematic assessment of human serum albumin HSA and bovine serum albumin BSA adsorption and found that over a wide concentration range these proteins adopt sideways-on adsorption with their short axial length projected normal to the surface [22, 23]. Both molecules have a short axial length close to 40 Å. Below $1\,\mathrm{g\,dm^{-3}}$, the adsorbed layers are between 25 and 40 Å thick and are reasonably well represented by a uniform layer model, indicating a strong deformation upon surface adsorption. However, no indication of further structural deterioration leading to the structural characteristics of polypeptide adsorption was detected.

An interesting common feature of surface adsorption of proteins is their pH dependence. This feature is shown in Fig. 13.4 for lysozyme, BSA, and HSA. Maximal surface excess occurs at their respective isoelectric points (IP) at which the net charge within the protein molecule is zero [21–23]. This shows that the lateral charge repulsion plays an important role in governing the total amount of adsorption. As the solution pH is shifted away from the IP, the adsorbed amount goes down, but the rate of decrease varies with the size and stability of the protein. This clearly reflects the apparent "damping" effect arising from the structural flexibility of larger but less rigid proteins.

13.4 Adsorption at the Solid–Water Interface: The Effect of Surface Chemistry

The bare silicon oxide is hydrophilic with contact angle close to zero. As its weak negative charge over the normal pH range does not cause any major

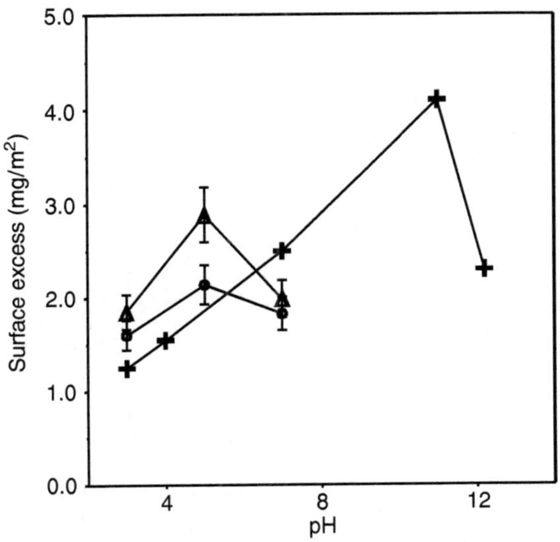

Fig. 13.4. The general trend of pH dependent adsorption HSA (o), BSA (Δ), and lysozyme (+) at a fixed protein concentration of $1\,\mathrm{g\,dm^{-3}}$

complication, the SiO_2 surface has been used as a standard to check the reproducibility of adsorption. These experiments rely on the polishing of optically flat silicon oxide surface with a layer of native oxide layer of 12–15 Å thick and at the same time to regenerate the surface hydrophilicity [24]. From these SiO_2 surfaces, we have been able to anchor self-assembled monolayers (SAM) bearing terminal CH_3, NH_2, COOH, and phosphorylcholine (PC) groups using silane chemistry. Using the high-structural resolution of neutron reflection, it was possible to control layer uniformity and density by fine-tuning the surface coating conditions. This approach of combining surface coating with stringent surface characterization allows to reproduce high-quality SAM layers with different terminal functions and is in contrast to the vast literature work in this area where organic monolayers are often poorly formed due to the lack of adequate characterization technique. In the contribution by Gutberlet and Lösche in this volume further approaches to establish and refine organic layer-coated interfaces for studying structure and interaction with proteins are considered.

Figure 13.5 shows the neutron reflectivity measured at the hydrophilic SiO_2–solution interface at 1 g dm^{-3} lysozyme and pH 7 under (a) D_2O, (b) the solution contrast matched to a scattering length density (ρ) of 4.0×10^{-6} Å^{-2} (CM4) and (c) contrast matched to silicon, with $\rho = 2.1 \times 10^{-6}$ Å^{-2} (CMSi) [25, 26]. The adsorbed layer is highlighted from different water contrasts and the combined measurements clearly improve structural sensitivity. Simultaneous fitting of all the reflectivity profiles gives a two-layer model with 30 Å each. The inner layer contains more protein than the outer layer and the total adsorbed amount is around 3.7 mg m^{-2}. Lysozyme has an approximate

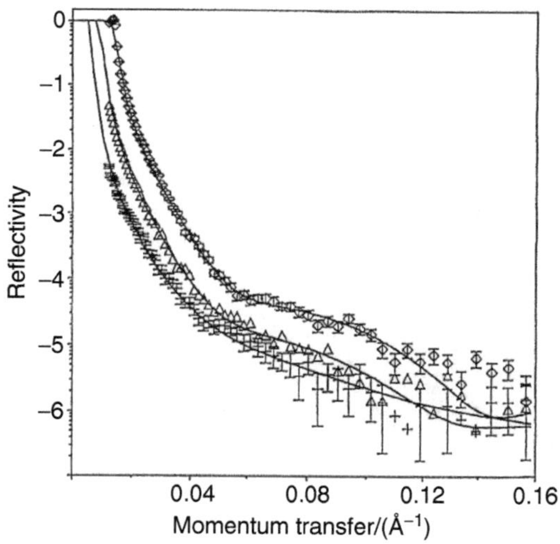

Fig. 13.5. A two-layer fit to the reflectivity profiles at pH 7 in the presence of 1 g dm^{-3} lysozyme: (**a**) (◇) D_2O, (**b**) (Δ) CM4 ($\rho = 4 \times 10^{-6}$ Å^{-2}), (**c**) (+) CMSi ($\rho = 2.1 \times 10^{-6}$ Å^{-2}). The continuous lines were calculated using surface-adsorbed-amount of 3.6 mg m^{-2}; $\tau_1 = \tau_2 = 30$ Å

dimension of $30 \times 30 \times 45\,Å^3$, and the formation of the two 30 Å layers at the interface suggests the adoption of two sideways-on molecular sublayers. Other information obtained from this work supports the retaining of globular entity of the protein after adsorption.

An important characteristic of the adsorbed protein layer is the extent of reversibility with respect to its solution concentration and pH. Hysteresis can be caused by many contacts between protein and the surface [27,28]. Although the fraction of segments in contact with the surface may be small and typically less than a few percent, the adsorption energy can easily be in excess of $100\,kJ\,mol^{-1}$ because of the large total number of contacts. Thus, irreversible adsorption is not necessarily associated with denaturation of the protein. Since lysozyme is robust in bulk solution over a wide pH range, we have measured the effect of pH on the adsorbed lysozyme structure [25,26]. The effects of pH on lysozyme adsorption at the hydrophilic SiO_2–water interface are schematically shown in Fig. 13.6. The surface coverage decreases with decreasing pH as a result of increased repulsion between the molecules inside the monolayer and this increased level of repulsion is also reflected in the reduced adsorption. The model depicts the formation of a sideways-on bilayer at pH 7 and at the higher bulk concentration, but only monolayer adsorption occurs at pH 4. This trend indicates a strong effect of electrostatic repulsion within the protein layer.

In comparison with the adsorption at the solid–solution interface, the adsorption at the air–water interface as described previously may be treated as a model for assessing the interfacial effect in the absence of the interference from the solid surface. We recall that at pH 7, lysozyme adsorption progressively changes from sideways-on to end-on conformation as the bulk protein concentration increases [20,21]. The changeover clearly enables the surface to accommodate more protein molecules. The end-on monolayer gives a thickness of 45 Å, but such a change is not observed at the silica–water interface at the high coverage at pH 7. It should be noted that neutron reflection is sensitive to the difference between 30 and 45 Å. The conformational difference appears

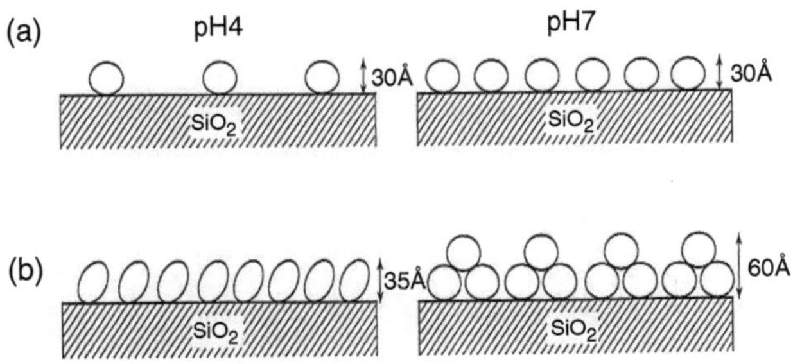

Fig. 13.6. Schematic diagram to illustrate the variation of surface coverage and structural conformation of lysozyme adsorbed at the silica–water interface with solution pH and bulk concentration. The lysozyme concentrations are (**a**) $0.03\,g\,dm^{-3}$ and (**b**) $1\,g\,dm^{-3}$

to lie in the interaction with the solid substrate, but the subtlety is more likely to arise from the fine balance between the lateral interaction within the protein layer and the interaction with the surface. At both interfaces, proteins retain their globular framework.

The effect of surface hydrophobicity was assessed using the SiO_2 surface chemically anchored with a dense and well-defined monolayer of octadecyltrichlorosilane (OTS). Partial labeling to the OTS layer (e.g., using $C_6H_{13}C_{12}D_{24}SiCl_3$) was found to help highlight the protein layer [29]. At the hydrophobed solid–solution interface, lysozyme adsorption was found to be irreversible to solution pH. In all cases, the interfacial layers were well represented by two main regions, a dense layer of 12–15 Å on the inner surface containing some 80% polypeptide and an outer diffuse layer of 50–80 Å containing some 20% polypeptide. The inhomogeneous distributions of the polypeptides, as shown schematically in Fig. 13.7, clearly indicate the unfolding of the globular framework of lysozyme caused by the strong interaction with the support substrate. Similar trend was also observed for BSA.

To assess the effect of surfaces with intermediate hydrophobicity, a SiO_2 surface grafted with $C_{15}OH$ was used to perform the same protein adsorption [30]. The advancing contact angle (θ_a) for the hydroxy surface was 53° and was intermediate between bare silicon oxide (ca. 0°) and OTS (ca. 110°). In many polymeric hydrogels, monomers with short alkyl chains bearing hydroxyl groups are readily incorporated to facilitate the water uptake, which is essential for many biomaterials applications such as contact lenses. It is thus useful to examine how this type of surface implicates protein adsorption. It was found that the amount of lysozyme adsorbed was substantially reduced in comparison with the bare SiO_2 and OTS surfaces. However, an interesting observation was the formation of a dense but rather uniform protein layer of some 20 Å around 1 g dm^{-3} lysozyme concentration, indicating

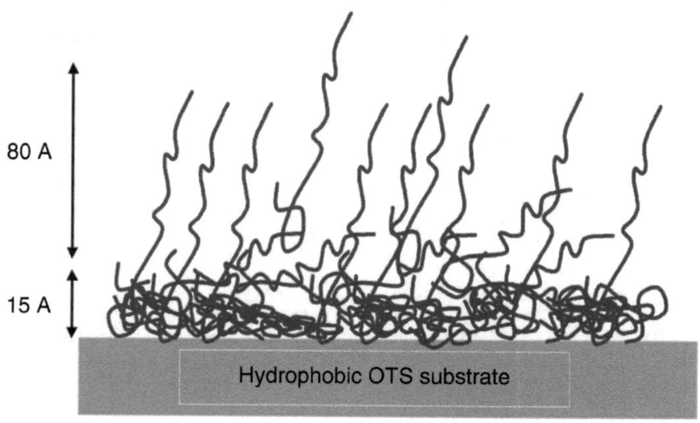

Fig. 13.7. Schematic diagram to illustrate the distribution of unfolded lysozyme at the hydrophobic OTS–water interface.

a substantial structural deformation. The structural deformation is consistent with the large extent of irreversible adsorption observed. These structural features may lead to clues as to why traditional hydroxy-based methacrylate polymers are not the best candidates as biomaterials for contact lenses and implant coatings.

In the course of studying protein adsorption on SAMs bearing usual terminal chemical functions, we also developed synthetic schemes for chemical attachment of SAMs bearing terminal phosphorylcholine (PC) groups. Adsorption at PC monolayer–solution interface [31] was pursued in parallel to the PC polymer films [32–34] carried out in collaboration with Biocompatibles UK Ltd. A particular version of PC molecules used for surface chemical grafting onto SiO_2 surface is shown in Fig. 13.8. The coated surface gave a typical advancing contact of some 30° and was found to be very effective at reducing protein adsorption. These SAM surfaces were found to be as effective as ultrathin PC polymer films at reducing protein adsorption. However, when compared with other more conventional polymeric materials, the reduction of nonspecific protein deposition by PC materials is more substantial. While these features can be attributed to the chemical nature of different materials, they set challenges for us to seek more appropriate explanations. It is relevant to note that while both PC and $C_{15}OH$ surfaces reduce protein adsorption substantially, the structural conformations of the adsorbed protein layers were different. The $C_{15}OH$–water interface induces strong deformation of lysozyme, but this is in contrast to the loose lysozyme distribution over 80–100 Å at the PC monolayer–water interface, indicating reduced contacts between the protein molecules and the PC surface. This feature is thought to arise from the high-surface hydration of PC groups, creating a hydration barrier. This structural characteristic is in agreement with the high degree of reversibility of adsorption on this surface.

Adsorption of other proteins such as BSA, HSA, and IgG were found to show similar behavior to lysozyme, e.g., the attainment of maximal adsorption at the solution pH close to their IPs [35,36]. However, as the size goes up and stability goes down, the influence of pH becomes less obvious. It was also

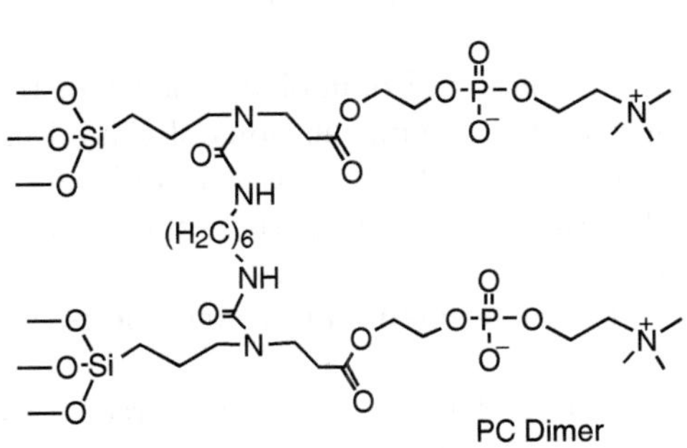

Fig. 13.8. Molecular structure of a PC dimer used for surface grafting onto SiO_2

observed that upon adsorption onto PC monolayer surface, the greater the protein size, the further reduced the adsorption. This trend is opposite when proteins are adsorbed on surfaces such as OTS and SiO_2. However, the exact nature of this behavior remains unclear. As the size increases, the extent of reversibility decreases, as expected. The experimental methodology developed here will add to the concerted effect in the understanding of the molecular mechanistic processes underlying surface biocompatibility [37–42].

Exchange of labile hydrogens in protein molecules with bulk D_2O is an issue that deserves some proper discussion. Uncertainty in the extent of incomplete exchanges would affect the surface excess although this has no effect on layer thickness. When proteins retain their globular structures, some labile hydrogens on the amino acid side chains and the backbones may be inhibited to exchange because of hydrogen bonding and hydrophobic encapsulation [43]. We have shown that for both lysozyme and BSA, the exchanges with bulk D_2O are complete within a few percent. We have further supported this verdict by comparing the protein surface excesses before and after structural unfolding induced by sodium dodecyl sulphate (SDS) [44,45]. Since H/D exchanges have been widely used as a probe for detecting structural perturbations, this experiment shows that neutron reflection has the potential for following the H/D exchanges with time when this effect is sufficiently slow (in the timescale of 10 min and beyond).

In summary, the main advance we have made in this part of study is the demonstration of simultaneous determination of in situ protein layer structure and composition at the solid–solution interface. This information together with the known three-dimensional structures of proteins allows reliable assessment of the extent of protein deformation and unfolding to be made. Although the deuteration of proteins is difficult, we have shown that by appropriate use of solvent isotopic contrasts, different parts of the interfacial layer can be highlighted and their structural distributions measured with sufficient resolution. We can summarize the main observations as follows:

- Protein molecules retain their globular frameworks at the hydrophilic solid–water interface but unfold completely at the hydrophobic solid–water interface. The amount of adsorption on these interfaces may however be comparable.
- Surfaces with intermediate hydrophobicity show substantially reduced adsorption, but the structural conformations of the adsorbed protein layers are different between the hydroxy ($-C_{15}H_{30}OH$) and phosphorylcholine (PC)-terminated surfaces, indicating the subtle effects of the nature of surface chemistry.
- Adsorption tends to reach maximum around the isoelectric point (IP) of the protein.
- The extent of structural deformation and degree of irreversibility of adsorption increases with the size of proteins.
- Labile hydrogens within globular proteins are completely exchanged when adsorbed at the interfaces.

13.5 Interaction Between Surfactant and Protein

In this section first the coadsorption of protein and surfactant at the air–water interface will be introduced, again because this surface can be used as model to examine the nature of the interaction without the interfering effect from the solid surface. An extensive work carried out previously [2] has indicated the effect of surfactant head groups on the mode of differing interactions. The representative data from the coadsorption of nonionic surfactant is shown. Although our previous work has shown that the binary mixture of lysozyme and nonionic $C_{12}E_5$ produced a typical model of competitive adsorption, structural deformation of lysozyme was revealed during the coadsorption at the surface [46]. The structural detail concerning lysozyme deformation was probed using hydrogenated $C_{12}E_5$ adsorbed from NRW. Under this isotopic contrast the reflectivity obtained arose from the lysozyme layer with little contribution from the hydrogenated $C_{12}E_5$.

When anionic SDS was used, completely different interfacial processes were observed [47]. Surface tension measurements indicated a rather complicated interfacial event. Just relying on surface tension alone, it would be impossible to unravel the key interfacial molecular processes and be able to outline the main picture. The surface tension shown in Fig. 13.9a is marked by a maximum around 1 mM SDS, followed by a break point well below the CMC of pure SDS. It is useful to comment that in the absence of the neutron data the apparent discontinuity at the low SDS concentration would be attributed to the onset of formation of micelles on the protein. The neutron reflection measurement, however, shows that over this region there is a steady increase of adsorbed amount of both lysozyme shown in Fig. 13.9b and SDS shown in Fig. 13.9c, with increasing SDS concentration. Thus the discontinuity is associated with the formation of highly surfaceactive SDS–lysozyme complexes and has nothing to do with the micellization on the protein. The almost constant surface tension over the SDS concentration between 0.05 and 0.5 mM conceals a more complex surface behavior, which is clearly indicated by the dramatic variation of the adsorbed amount of both SDS and lysozyme (Fig. 13.9b, c), as revealed by neutron reflection. These results together show that electrostatic interactions determine the low-surfactant concentration behavior and hydrophobic interactions prevail over the high surfactant concentration range. The combination of interactions over the crossover region can give rise to a range of quite different effects, including multilayer formation. The neutron data is able to reveal the role of the electrostatic interactions very clearly, not just by monitoring the composition of the surface but also by leading to the complex structure of the composite layer. The density profiles for the protein itself over the lowest SDS concentration region are approximately uniform and comparable with lysozyme on its own, indicating that no major structural deformation occurs. Thus, it appears that, in this case, the strong electrostatic interaction has a large effect on the thermodynamic behavior but is not strong enough to induce loss of tertiary structure. Further SDS

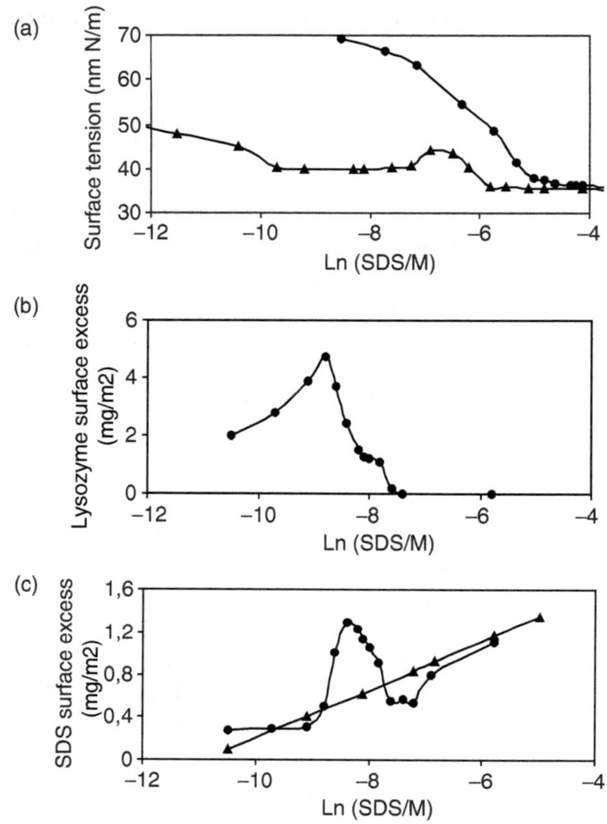

Fig. 13.9. (**a**) Surface tension for pure SDS (•) and SDS–lysozyme mixture (▲), (**b**) surface excess for lysozyme and (**c**) surface excess of SDS measured on the surface of binary mixture (•) as compared with its excess in pure form (▲)

addition into the "hydrophobic" region leads to the breakdown of the lysozyme framework as the hydrophobic interactions build up enough to dominate the electrostatic interactions.

Interactions at the solid–solution interface have mainly been done using the hydrophilic SiO_2, using preadsorbed protein layers in contact with pure surfactant solution. This experimental process mimics the cleaning of medical devices well. The results show that while nonionic surfactant such as $C_{12}E_5$ shows little tendency of association with preadsorbed proteins, both anionic and cationic surfactants interact strongly, resulting in different extent of protein removal. The high sensitivity of neutron reflectivity is well demonstrated from SDS binding to preadsorbed BSA, and this study was performed in conjunction with SDS labeling [44,45]. The modeling of reflectivities clearly shows that the interfacial mixtures are unevenly distributed, with SDS distributions skewed towards the bulk water. The studies also show that the removal of the adsorbed protein does not start until a critical SDS concentration is reached, that the critical concentration varies with pH and salt concentration and that binding of SDS induces structural unfolding of the preadsorbed protein.

SDS binding to the immobilized lysozyme was also studied to compare the effect of the nature of protein [48]. A significant difference was observed between the extent of surfactant bound to BSA and lysozyme. For SDS–BSA system, the amount of SDS bound to each gram of protein adsorbed at the interface (weight ratio) was 0.43, close to that found by Tanford et al. [49] for the binding carried out under similar bulk solution conditions. This is in contrast to the observed weight ratio of 0.1 for SDS–lysozyme. This difference is clearly attributable to the nature of the proteins, although Tanford et al. have shown that in bulk solution these differences, if any, are much smaller. These results together enforce the view that the interactions at the interface are very different from bulk solution. This statement is consistent with the fact that the surfactant–protein interaction is nonideal and that the interfacial structure and composition is not expected to be the same as in bulk solution.

Figure 13.10 depicts the pattern of protein removal from the hydrophilic SiO_2 surface using anionic SDS and cationic $C_{12}TAB$ (dodecyltrimethyl ammonium bromide) [50]. Clearly, when $C_{12}TAB$ is used, it can progressively remove lysozyme, but the extent of removal never reaches completion. Also, with the increasing lysozyme removal, the interfacial excess of $C_{12}TAB$ tends to increase, indicating the preferential binding of $C_{12}TAB$ to the interface. The results shown in Fig. 13.10 clearly show the effect of surfactant head groups and the underlying differences in the molecular processes of interfacial interactions.

These examples and other related preliminary work show that neutron reflection can provide better explanations of the molecular processes involved

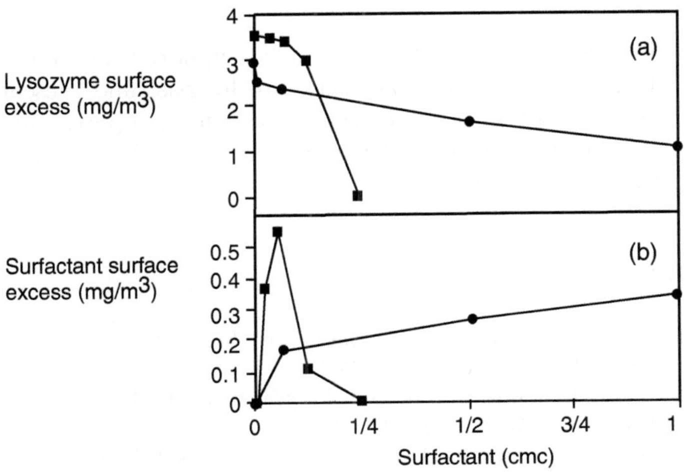

Fig. 13.10. Comparison of protein elution capabilities of SDS (■) and $C_{12}TAB$ (•) shown as the variation of surface excess of lysozyme (**a**) and surfactant (**b**) with bulk surfactant concentration

in protein–surfactant complexation. The complexity of the interactions both at the surface and in bulk solution is such that no reliable model can be established in the absence of a correct description of the complexation at the interfaces. This statement lends its support from further neutron reflection studies of surfactant complexation with protein at the hydrophobed solid–solution interface where the pattern of interactions is very different from the data obtained at the hydrophilic SiO_2–solution interface.

13.6 Future Prospects

Although extensive studies have been made on the interactions between proteins and surfactants in bulk solution, less is known about their behavior at the interfaces. It is important to realize that the nonideal behavior of protein adsorption and protein–surfactant interactions at the interfaces cannot be predicted by understanding their behavior from bulk solution because of interfacial effects and more importantly, because of our inability in predicting how and when a protein molecule deforms and unfolds when it arrives at a given interface. The current work provides a useful experimental methodology for further research into the understanding of molecular mechanistic processes related to surface-induced structural deformation and denaturation. The strong relevance of this research to a wide range of conventional and emerging technological applications will stimulate computational effort and theory development to attempt to corroborate the structural information obtained in neutron reflection with macroscopic behavior.

Acknowledgements

The author would like to thank the financial support from Biocompatibles UK Ltd, BBSRC, EPSRC. Thanks also go to his past and present research students, collaborators, mentors for their assistance, support, and direction. We thank the American Chemical Society (ACS) for the permission to use Figs. 13.5, 13.6, 13.9, and 13.10.

References

1. M. Malmsten, *Biopolymers at Interfaces, 2nd Ed, Revised and Expanded*, Surf. Sci. Ser. **110** (Dekker, New York, 2003)
2. T.A. Horbett, J.L. Brash, *Proteins at Interfaces II*, ACS Symposium Series **602** (Am. Chem. Soc., Washington DC, 1995)
3. K.B. McClary, T. Ugarova, D.W. Grainger, J. Biomed. Mater. Res. **50** (2000), 428
4. J. Andrade, *Surface and Interfacial Aspects of Biomedical Polymers*, Vol. 2 (Plenum, New York, 1985)

5. R.J. Green, R.A. Frazier, K.M. Shakesheff, M.C. Davies, C.J. Roberts, S.J.B. Tendler, Biomaterials **21** (2000), 1823
6. E. Ostuni, B.A. Grzybowski, M. Mrksich, C.S. Roberts, G.M. Whitesides, Langmuir **19** (2003), 1861
7. J.L. Keddie, Curr. Opin. Colloid Interface Sci. **6** (2001), 102
8. R.J. Green, I. Hopkinson I, R.A.L. Jones, Langmuir **15** (1999), 5102
9. J. Kim, G.A. Somorjai, J. Am. Chem. Soc. **125** (2003), 3150
10. T.J. Su, J.R. Lu, Z.F. Cui, R.K. Thomas, J. Membr. Sci. **173** (2000), 167
11. J.C. Marshall, T. Cosgrove, K. Jack, A. Howe, Langmuir **18** (2002), 9668
12. A. Kondo, F. Murakami, K. Higashitani, Biotech. Bioeng. **40** (1992), 889
13. M.H. Tian, W.K. Lee, M.K. Bothwell, J. McGuire, J. Colloid Interface Sci. **200** (1998), 146
14. P. Billsten, P.O. Freskgard, U. Carlsson, B.H. Jonsson, H. Elwing, FEBS Lett. **402** (1997), 67
15. J.R. Lu, E.M. Lee, R.K. Thomas, Acta Cryst. **A52** (1996), 11
16. J.R. Lu, R.K. Thomas, J. Chem. Soc., Faraday Trans. **94** (1998), 995
17. J.R. Lu, R.K. Thomas, in *Physical Chemistry of Biological Interfaces*, Eds. A. Baszkin, W. Norde, (Dekker, New York, 2000) p. 609–650
18. J.R. Lu, Annu. Rep. Prog. Chem. C **95** (1999), 3
19. J.R. Lu, S. Perumal, E. Powers, J. Kelly, J. Webster, J. Penfold, J. Am. Chem. Soc. **125** (2003), 3751
20. J.R. Lu, T.J. Su, R.K. Thomas, J. Penfold, J. Webster, J. Chem. Soc., Faraday Trans. **94** (1998), 3279
21. J.R. Lu, T.J. Su, B. Howlin, J. Phys. Chem. B **103** (1999), 5903
22. J.R. Lu, T.J. Su, J. Penfold, Langmuir **15** (1999), 6975
23. J.R. Lu, T.J. Su, R.K. Thomas, J. Colloid Interface Sci. **213** (1999), 426
24. J.B. Brzoska, N. Shahidzadeh, F. Rondelez, Nature **360** (1992), 719
25. T.J. Su, J.R. Lu, R.K. Thomas, Z.F. Cui, J. Penfold, Langmuir **14** (1998), 438
26. T.J. Su, J.R. Lu, R.K. Thomas, Z.F. Cui, J. Penfold, J. Colloid Interface Sci. **203** (1998), 419
27. C.A. Haynes, E. Sliwinski, W. Norde, J. Colloid Interface Sci. **164** (1994), 394
28. C.A. Haynes, W. Norde, J. Colloid Interface Sci. **169** (1995), 313
29. J.R. Lu, T.J. Su, R.K. Thomas, A.R. Rennie, R. Cubit, J. Colloid Interface Sci. **206** (1998), 212
30. T.J. Su, R.J. Green, Y. Wang, E.F. Murphy, J.R. Lu, R. Ivkov, S.K. Satija, Langmuir **16** (2000), 4999
31. J.R. Lu, E.F. Murphy, T.J. Su, A.L. Lewis, P.W. Stratford, S.K. Satija, Langmuir **17** (2001), 3382
32. E.F. Murphy, J.L. Keddie, J.R. Lu, J. Brewer, J. Russell, Biomaterials **20** (1999), 1501
33. E.F. Murphy, J.R. Lu, A.L. Lewis, J. Brewer, J. Russell, P. Stratford, Macromolecules **33** (2000), 4545
34. E.F. Murphy, J.R. Lu, J. Brewer, J. Russell, J. Penfold, Langmuir **15** (1999), 1313
35. T.J. Su, J.R. Lu, R.K. Thomas, Z.F. Cui, J. Phys. Chem. B **103** (1999), 3727
36. T.J. Su, J.R. Lu, R.K. Thomas, Z.F. Cui, J. Penfold, J. Phys. Chem. B **102** (1998), 8100
37. K.L. Prime, G.M. Whitesides, Science **252** (1994), 1164
38. C.M. Roth, A.M. Lenhoff, Langmuir **11** (1995), 3500

39. K.L. Prime, G.M. Whitesides, J. Am. Chem. Soc. **115** (1993), 10714
40. S. Herrwerth, W. Eck, S. Reinhardt, M. Grunze, J. Am. Soc. Chem. **125** (2003), 9359
41. Y. Iwasaki, A. Fujike, K. Kurita, K. Ishihara, N. Nakabayashi, J. Biomater. Sci. Polym. Edn. **8** (1996), 91
42. A.L. Lewis, Colloid Surf. B, Biointerfaces **18** (2000), 261
43. S.E. Radford, M. Buck, K.D. Topping, C.M. Dobson, P.A. Evans, Proteins: Struct. Func. Gene. **14** (1992), 237
44. J.R. Lu, T.J. Su, R.K. Thomas, J. Phys. Chem. B **102** (1998), 10307
45. J.R. Lu, T.J. Su, R.K. Thomas, J. Penfold, Langmuir **14** (1998), 6261
46. R.J. Green, T.J. Su, J.R. Lu, J. Webster, J. Penfold, Phys. Chem. Chem. Phys. **2** (2000), 5222
47. R.J. Green, T.J. Su, H. Joy, J.R. Lu, Langmuir **16** (2000), 5797
48. R.J. Green, T.J. Su, J.R. Lu, J. Penfold, J. Phys. Chem. B **105** (2001), 1594
49. C. Tanford, J. Mol. Biol. **67** (1972), 59
50. R.J. Green, T.J. Su, J.R. Lu, J.R.P. Webster, J. Phys. Chem. B **105** (2001), 9331

14

Complex Biomimetic Structures at Fluid Surfaces and Solid–Liquid Interfaces

T. Gutberlet, M. Lösche

14.1 Introduction

One major contributor to the recent success of nanotechnology is its profoundly interdisciplinary nature, involving supramolecular chemistry, biotechnology, bioinspired materials science, large-scale computational methods, and physical characterization techniques. One important role of physics in this context is the development of characterization methods that are sufficiently sensitive for the investigation of ever smaller sample sizes – down to monomolecular sensitivity – and ever smaller sample dimensions, as well as techniques that probe directly the relevant intermolecular and intramolecular interactions. A variety of surface and interface sensitive techniques, such as scanning force and fluorescence microscopy, FT-IRRAS, XPS, surface plasmon spectroscopy and surface-sensitive X-ray and neutron scattering have been developed in the past 15 years to reveal such information on nanoscopic systems. Planar lipid membrane mimics, such as floating monolayers on aqueous surfaces or immobilized bilayers on solid supports, have been established to correlate structural, functional and dynamic aspects of biomembrane models [1].

Neutron and X-ray scattering in nanotechnology share common grounds. The two techniques closely related to each other not only in terms of the underlying physics – neutrons scatter from the nuclei of a molecular species while X-rays scatter from their electron clouds – but also the formalisms that quantitate the optics of these processes are very similar. Neutrons and X-rays are also largely synergistic in their application to the characterization of interfacial structures [2,3]. Neutrons feature distinctive advantages over X-rays due to the possibility of isomorphic contrast variation and their penetration into condensed matter for the characterization of deeply buried interfaces [4,5]. Following the advent of third-generation synchrotron sources [6], however, in particular X-ray techniques have been an indispensable tool that has driven the development of nanoscience. Current perspectives of significant increases in neutron flux at sources under construction or in the planning stage [7,8] on the other hand, is expected to boost the application of neutrons in nanoscience considerably and will thus increase the overall capabilities of surface-sensitive scattering even further.

In this chapter, we review recent achievements in the molecular-level characterization of bioinspired interface and surface architectures using surface-sensitive neutron scattering and will briefly describe practical aspects of the application of reflectivity techniques. Then we will survey recent work with the aim of exemplifying the capabilities of neutrons in the elucidation of submolecular structures. For a deeper discussion of experimental and technical aspects of neutron scattering at interfaces, we refer to the contribution by Majkrzak and coauthors in this volume. For a review on aspects of protein adsorption and their interactions at interfaces, the reader is referred to the contribution by Lu, in this volume.

14.2 Surface-Sensitive Scattering

As reviewed elsewhere [9], surface-sensitivity in a scattering experiment, i.e., the discrimination of scattering from an interface-associated layer of (organic) material of nanometer thickness against the vast background of molecules contained in the adjacent bulk phase, is generally achieved by impinging the beam at grazing angles. Near the critical angle for total external reflection, this creates an evanescent wave which penetrates the bulk phase only by a few 10 nm [10]. This enables three classes of experiments: (i) specular and (ii) off-specular reflection, as well as (iii) grazing-incidence ("in-plane") diffraction (GID) (Fig. 14.1). Of these generic interface-sensitive scattering techniques, only specular reflection plays a role in the *neutron* scattering from organic monolayers, as the cross-sections for nonspecular scattering and for GID are orders of magnitude lower than that for specular scattering. The latter experiments are thus currently limited to synchrotron X-ray experiments (grazing-incidence X-ray diffraction GIXD). Specular reflection of neutrons at interfaces, however, is developing into a work horse in biologically oriented nanoscience.

14.2.1 Specular Reflectivity

As discussed by Majkrzak et al., in this volume, specular reflectivity reveals only one-dimensional (1D) information on the scattering length density (SLD) at an interface. Another limitation is that the method does not provide *atomic*-scale information on the system, comparable to, e.g., X-ray crystallography. Advanced structure-based data analysis techniques, however, provide a quasi-2D structural assessment of the investigated system, as delineated below. In addition, such techniques provide a framework for the consistent mapping of different contrasts onto each other. Thus, the simultaneous evaluation of neutron data taken from samples with different deuteration patterns or of neutron and X-ray data sets from identically prepared systems enhance greatly

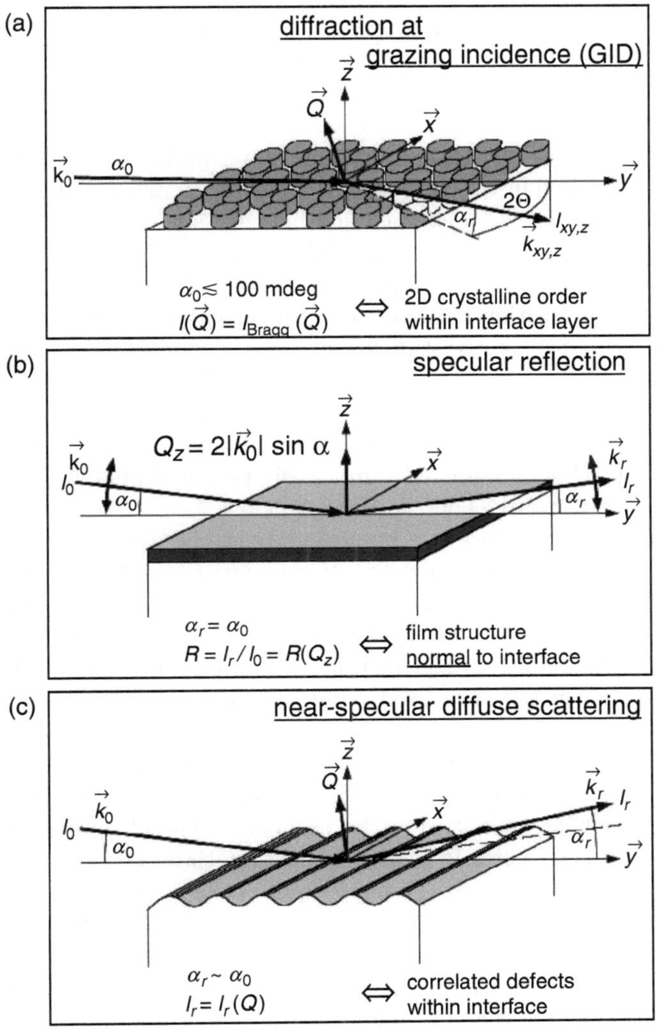

Fig. 14.1. Surface-sensitive scattering experiments at grazing incident angles [2]

the "virtual" resolution. One of the most striking advantages, reflectivity measurements do not require *crystalline* samples to obtain high resolution, and are thus capable of probing membraneous systems in their native, disordered state.

To achieve the highest possible resolution of a structure in reflectivity measurements, there are strict requirements on the sample with respect to the quality of the interface and to in-plane sample homogeneity. Clearly, to achieve subnanometer resolution on a sample film, the substrate has to be of ideal geometry down to the nanometer scale: interfaces with residual rms roughnesses in the Ångstrom regime are generally required. Moreover, a microscopically smooth but macroscopically curved sample surface also reduces resolution. This is one of the reasons why reflectivity measurements from systems at *fluid* surfaces have been relatively successful. Due to the action of gravity, a free water surface is "by definition" perfectly flat on the macroscopic length scale. On the microscopic length scale, thermally excited capillary waves, controlled by a competition between surface tension and thermal excitation, define the quality of the fluid substrate. For water with its high surface tension, the residual roughness is on the order of 3 Å – perfectly suited for high-quality reflection measurements. A second reason fluid surface monolayers have been frequently studied with neutron scattering is that the free fluid surface permits easy access and control of most relevant systems parameters, such as molecular density within the film and subphase chemistry and temperature, etc. Moreover, in situ manipulation, e.g., the injection of peptides or proteins underneath a previously prepared and characterized lipid film, is straight-forward as shown further below.

Another concern is in-plane homogeneity of the samples. Solid-substrate borne monolayer systems – so-called "*Langmuir–Blodgett*" films [11] – as well as lipid surface monolayers, particularly within a first-order phase transition [12], show frequently domain structures on the micrometer scale [13]. Since phase information is lost in the determination of the scattered *intensity*, the experimentally measured reflectivity cannot be directly inverted to obtain an SLD profile. The usual work-around is data modelling, i.e., to parameterize the SLD and optimize parameters by fitting to the experimental data. For this process to be tractable, in-plane heterogeneities have to be avoided, as they increase largely the number of model parameters that describe the microstructure. It is thus important that the sample preparation protocol ensures that films are homogeneous on the length scale of the in-plane coherence length of the probe beam. In angle-dispersive reflectivity measurements, this coherence length is determined by the geometry of the instrument and is typically largely anisotropic within the film plane – on the order of 10^5 Å in the projected direction of the beam onto the sample and on the order of 100 Å perpendicular to this direction. If one takes a geometric mean of these numbers as a criterion, then the sample needs to be homogeneous on the micrometer length scale. This quantification suggests that optical microscopy [14] is an appropriate technique to check routinely the suitability of samples for reflectivity characterization.

Given in-plane homogeneity of the sample and optimal substrate geometry, the ultimate limitation to resolution derives from the steep drop of the scattered intensity as a function of momentum transfer, Q_z (see e.g., Majkrzak et al., in this volume). The reflected intensity R drops as Q_z^4 for $Q_z \gg Q_z^c$, the critical momentum transfer of total reflection. This results in R being vanishingly small, on the order of 10^{-6}–10^{-8} once $Q_z \geq 0.3\,\text{Å}^{-1}$ for X-ray or neutron measurements of aqueous surface films. Clearly, X-ray probes, particularly at high-brilliance third-generation synchrotron sources, have great advantage over neutrons as it comes to determining this *one* scattered probe particle out of 10^6 incident probes! Ultimately, however, the limits of resolution cannot be indefinitely pushed to ever higher Q_z values by using ever-increasing primary beam intensities. Not only is beam damage a serious issue in X-ray scattering at synchrotron sources, but also both near-specular scattering from capillary waves [15] and incoherent scattering from the bulk phase – excited by the impinging beam which penetrates deeply into the bulk at high Q_z, and hence high incident angles – limit the detection of the minute number of specularly reflected probes in practical terms.

Typical maximum Q_z values amount to $\sim 0.8\,\text{Å}^{-1}$ in X-ray reflection and 0.3–0.6 Å^{-1} in neutron reflection measurements, largely depending on the isotopic nature of aqueous bulk phase and design of the sample cell. To reduce incoherent background subphase thickness needs to be kept at an absolute minimum [16, 17]. As the sample theorem would suggest, such $Q_z^{\max}$ values correspond to a "canonical" [18] resolution of $\Delta z = \pi/Q_z^{\max} \approx$ 5–10 Å. The practical consequences are illustrated in Fig. 14.2. Based on a *single* neutron reflectivity measurement that spans the range out to $Q_z \approx 0.25\,\text{Å}^{-1}$, it is not even possible to determine the thickness of the hydrophobic slab within a lipid surface monolayer to any reasonable certainty – even as the methylenes are fully deuterated.

14.2.2 Structure-Based Model Refinement

Fortunately, this situation has been remedied by the development of structure-based model refinement techniques [18–21]. "*Composition-space refinement*" [18] takes the parameterization from a level where one describes 1D SLD profiles to a level where one parameterizes the molecular structure of the interface architecture [18, 19]. This can be implemented in terms of the atomic content of the slabs in a box model [19] or, more directly, in terms of thermally broadened distributions of molecular subfragments. One realization of the latter approach is a "volume-restricted distribution function" (VRDF) parameterization [20, 21], in which the molecular fragments are subject to the condition that they just fill the available space. Thus, volumetric information on the molecular subfragments is essential and is usually derived from molecular dynamics simulations [22]. Since this approach determines the packing of molecular subfragments within the plane of the monolayer, the retrieved information is "quasi-2D" in character.

There are various benefits of the VRDF approach to modelling reflectivity data:

- At a resolution of $\pi/Q_z^{\max}$ better than ≈5 Å, the description of a monolayer even of the simplest phospholipid in terms of a *homogeneous* headgroup slab breaks down within the conventional box model: significant discrepancies between models and data that are frequently encountered at high Q_z [21]. VRDF models reconcile these discrepancies.

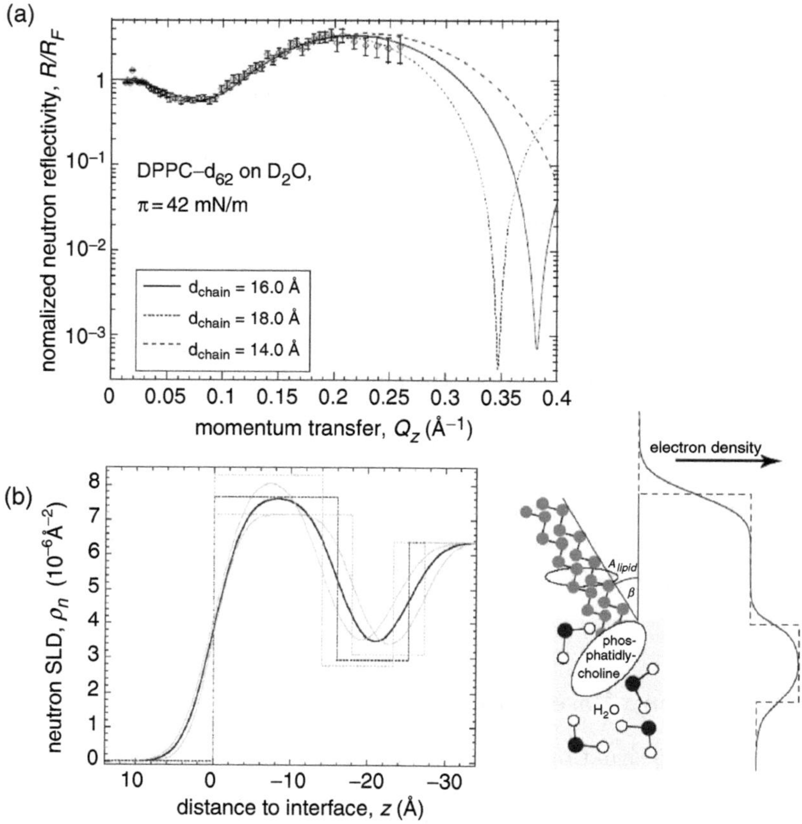

Fig. 14.2. Qualitative assessment of the information content of a reflection spectrum. Shown is a typical neutron reflectivity data set, R/R_{F} (chain-perdeuterated DPPC-d$_{62}$ on D$_2$O at $\pi = 42\,\mathrm{mN\,m^{-1}}$ [19]) (**a**) in comparison with various model reflectivities (**b**). A scheme of the molecular arrangement at the interface is shown to the right. In a *combination* of X-ray and neutron reflection measurements, the hydrophobic layer thickness has been determined to be $\mathrm{d_{hphob}} = 16.0\,\text{Å}$. The corresponding scattering length density (SLD) profile is shown as a bold line. To assess the information content of the *single* neutron reflection data set shown, detuning of d_{hphob} by $\pm 2\,\text{Å}$ and readjusting the corresponding SLD and surface roughness to fit the DPPC-d$_{62}$/D$_2$O data leads to models – shown as dashed lines – that cannot be discriminated within the resolution of the single experiment. However, a combination of X-ray and neutron scattering is very well capable of resolving the structure, even as the resolution of every single experiment is insufficient. Adapted from [21]

- The parameterization of the system chemistry permits a strict coupling of different data sets – obtained for various isotopic contrasts or neutron and X-ray measurements on isomorphic systems – thus boosting the *effective* resolution greatly.
- Finally, even if only one data set for a single contrast is available (e.g., one single high-resolution X-ray reflectivity data set), the chemically intuitive parameterization permits an intuitive molecular-level interpretation of the results – an advantage over multilayer box models which do not generally lend themselves to a straight-forward interpretation in terms of a chemical structure.

In Section 14.4, below results on various classes of supramolecular systems at aqueous surfaces and interfaces are collected to enable the reader to judge what information may be retrieved with surface-sensitive neutron scattering in various areas of bioinspired materials science.

14.3 Floating Lipid Monolayers: Structural Investigations and the Interaction of Peptides and Proteins with Lipid Interfaces

Amphiphilic molecules, such as phospholipids, self-assemble into free-floating monolayers ("*Langmuir* monolayers", LM) at the air–water interface due to their hydrophilic headgroup linked to hydrophobic acyl chains. As mentioned above, LMs are physicochemically well-controlled model systems, which provide unique opportunities for the investigation of structure and interaction of a biomembrane mimic with biomolecules, e.g., peptides and proteins [23]. Consequently, LMs have been early on studied in X-ray and neutron scattering experiments [24, 25] that have revealed their SLD distribution along the surface normal, indicative of their time-averaged thermally broadened structure.

The relatively low resolution in neutron reflectivity experiments is compensated for by isomorphic contrast variation via exchange of protons with deuterium in biomolecules, which exploits the large difference in neutron scattering length of the two isotopes (see also contribution by Lu, in this volume). Molecular subunits within a monolayer, embedded or adsorbed molecules – such as surfactants, peptides or proteins – or phase separation within the layer plane can be characterized by choosing appropriate contrasts in multiple scattering experiments. In pioneering work using neutron diffraction, this was utilized for the localization of methylene segments in the hydrophobic region and headgroup orientation of selectively deuterated phospholipid bilayers [26, 27].

14.3.1 Single Phospholipid LMs

The structure and organization of single phospholipid monolayers on aqueous subphase using different SLD contrasts have been investigated in several studies with neutron reflectometry [19,28–31]. The most convenient way to achieve contrast variation in LM systems is by preparation of the monolayer aqueous subphase with either H_2O or D_2O. The adsorption of protonated molecular species at the monolayer–D_2O interface decreases the SLD there substantially and is sensitively detected in a reflectivity experiment. Thus, the electrostatic coupling of poly-L-lysine to DMPC monolayers and to DMPC–dimyristoyl phosphatidylglycerol (DMPG) mixed monolayers was studied on D_2O subphase. Similarly, the penetration of the protein spectrin into the monolayer headgroup region was characterized [32] and the adsorption of F-actin filaments to cationic LMs of dimyristoyl trimethylammonium propane (DMTAP) was investigated [33].

Another convenient method to vary the contrast in LM systems is by using lipids with perdeuterated hydrophobic chains. The interaction of hisactophilin with the lipid interface was explored [34]. The protein, which has a myristic acyl chain anchor attached, binds tightly to negatively charged membranes. Changes in the SLD profile upon penetration of the protonated myristic acid anchor into the deuterated lipid matrix, distinguished adsorption of protein from penetration of protein into the lipid interface, thus discriminating electrostatically and hydrophobically driven lipid–protein interaction. With acyl-deficient hisactophilin only the formation of a protein layer adsorbed was observed beneath the charged phospholipid monolayer that retained its SLD at the value prior to protein exposure. For a fully synthetic system, the orientation of a palmitoylated diα-helical peptide with different deuterium-labeled positions beneath a LM of dilaurylphosphatidylethanolamine (DLPE) was examined [35] in combined X-ray and neutron reflectivity experiments. Similarly, the interaction of myoglobin with lipid monolayers was investigated with a combination of neutron and X-ray reflection and GIXD [36, 37]. Underneath mixed monolayers of DPPC/PSIDA, a diacylglycerol lipid with a metal-chelating imino diacetate headgroup, the formation of a densely packed monolayer of myoglobin was observed with thickness and density consistent with crystallographic data for myoglobin. With Cu^{2+} ions in the subphase, the protein adsorbed in a side-on orientation at low packing density but tilted to an end-on orientation at high packing density. With Ni^{2+} ions, on the other hand, myoglobin was observed to adsorb in an end-on orientation at all densities. These differences were attributed to a stronger interaction of the histidine moieties with the imino diacetate anchor mediated by Cu^{2+}.

A number of studies have been conducted to understand the structure and function of components lining the interface of the alveoli, known as lung surfactant [38]. The lipid layer consists mainly of a mixture of DPPC and the anionic lipid dipalmitoylphosphatidylglycerol (DPPG) and forms a matrix for four surfactant-associated proteins (SP-A, SP-B, SP-C, and SP-D).

X-ray reflectivity measurements on surfactant-associated protein monolayers have been performed in conjunction with pressure–area isotherms and modelling to suggest that the protein undergoes changes in its tertiary structure at the air–water interface under the influence of surface pressure variations, indicating the likely role of such changes in surfactant squeeze out as well as lipid exchange between the air–alveoli interface and the underlying subphase. Recent neutron reflectivity data on bovine SP-B monolayers on aqueous subphases are consistent with the exchange of a large number of labile protons as well as the inclusion of a significant amount of water, which is partly squeezed out of the protein monolayer at elevated surface pressures [38]. In a mixed SP-B and DPPC system, squeeze out of the protein and re-adsorption upon compression and re-expansion was observed, with most of the protein material predominantly associated with the interface. Only small quantities of lipid followed the protein on leaving the monolayer at compression as shown by comparison of SLD profiles with protonated and deuterated DPPC [39].

Bacterial S-layer proteins [40] crystallize at a wide range of interfaces and surfaces, including phospholipid membranes [41]. The microscopic interactions between recrystallized bacterial S-layers and floating phosphatidylethanolamine (PE) LMs have been analyzed using FT-IRRAS spectroscopy, X-ray reflectivity and grazing-incidence diffraction, and neutron reflectivity at air-water interfaces [42–44]. A slight increase of the lipid acyl chain order was observed in GIXD upon protein adsorption, indicative of an increase in local lipid density. Corefinement of X-ray and neutron reflectivity data suggested that protein interpenetrates the lipid monolayer only in the headgroup region. Since only a small amount of protein material is observed within the headgroups, it was inferred that structural intact protein motifs enter the PE headgroups at localized, and possibly repetitive, interaction points within the S-layer crystal lattice – rather than a laterally homogeneous interaction of the protein occurring with the lipid surface monolayer [45].

14.3.2 Functionalized Phospholipid LMs

Various studies utilized the strong specific binding of biotin to streptavidin to probe protein interactions with functionalized lipid monolayers on aqueous subphases [46–49]. Fluorescence microscopy using FITC-labeled streptavidin showed that the protein forms – presumably crystalline – domains and established preparation conditions where these domains covered the surface quantitatively. Neutron reflection experiments showed the formation of a monomolecular protein layer with an effective thickness of 44 ± 2 Å. Quantitative binding was observed already at ultralow biotin surface concentrations. A combination of X-ray and neutron scattering experiments subsequently revealed distortions of the LM by the tightly bound protein (Fig. 14.3) [47]. Introducing a spacer between the functionalized lipid surface anchors and the biotin moiety resulted in a hydration layer between streptavidin and the lipid monolayer [49], concomitant with a reduction of monolayer distortion.

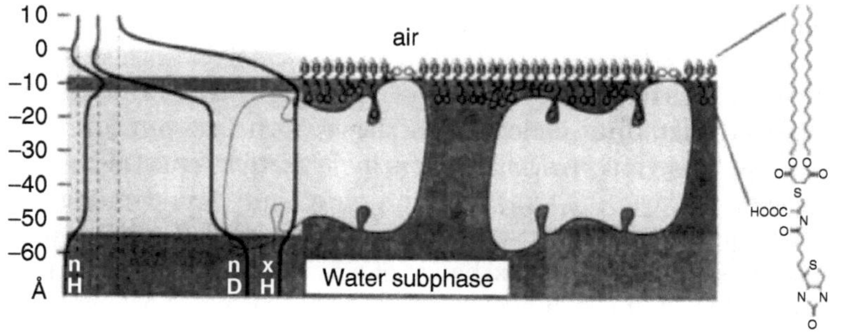

Fig. 14.3. Schematic representation of streptavidin interaction with a biotinylated monolayer. The lipid monolayer is distorted by the tightly bound protein. SLD profiles are shown on the left (n, neutrons; x, X-rays; H, H_2O; D, D_2O) [47]

The binding dynamics of the recombinant protein, lumazine synthase at biotinylated lipid monolayers were studied with neutron reflectometry [50]. The protein was biotinylated and coupled to streptavidin. The binding of these constructs to biotinylated LMs was monitored by following the neutron reflection. A model including a densely adsorbed protein monolayer and progressively more dilute protein layers was used to describe the experimental data.

Recently neutron reflectometry was applied to characterize the structure of ganglioside GM1-functionalized LMs upon binding of cholera toxin ($CT\alpha\beta_5$) and its β subunit ($CT\beta_5$) [51]. Structural parameters such as the density and thickness of the lipid layer, extension of the GM1 headgroup, orientation and position of the protein upon binding were inferred from the data. As shown in Fig. 14.4, the α subunit of the toxin was observed to be located underneath the pentameric ring of the B subunit and the $CT\beta_5$ is not penetrating into the lipid layer.

14.4 Lipopolymers

The grafting of water-soluble polymers to phospholipid membranes offers an opportunity for a targeted modification of the surface properties of, e.g., liposomes. Depending on the grafting density, the organization of polymer chains at interfaces may vary between a "mushroom"-like conformation and an extended, 'brush'-like configuration of the molecules ("polymer brush") [52]. Such polymer brushes might act as soft, flexible cushions between adsorbed biomolecules and solid interfaces in applications that depend on surface-modification with biopolymers.

The most prominent and best studied examples of such polymer brushes at lipid interfaces are comprised of polyethylene glycols (PEGs) as water soluble polymers that are chemically grafted to a phospholipid headgroup. Neutron

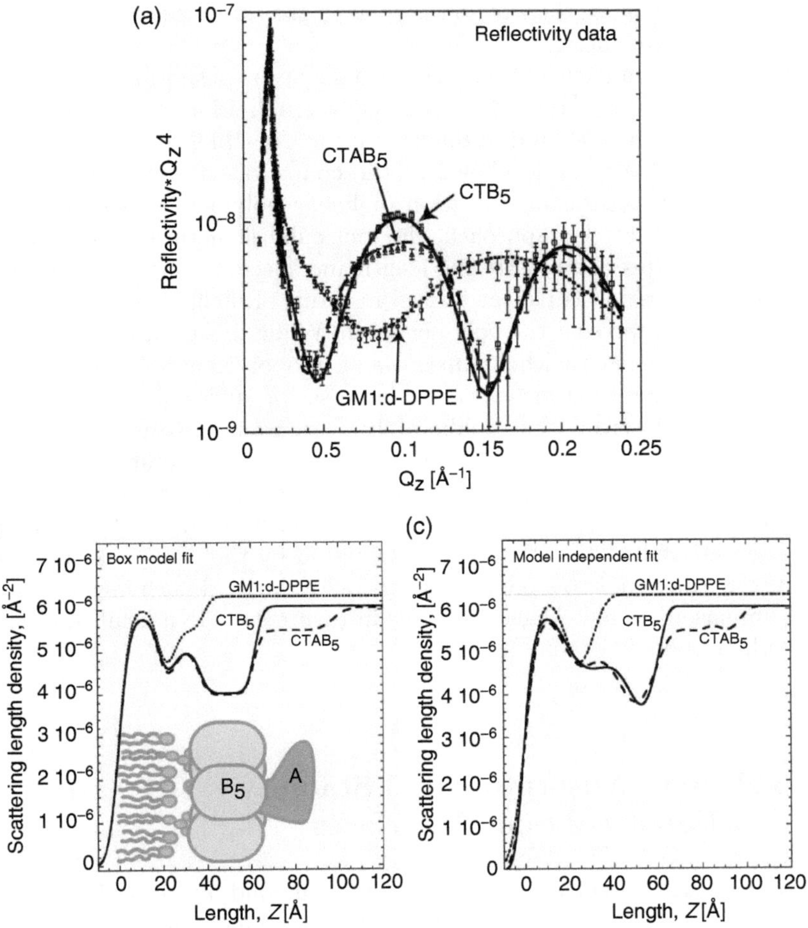

Fig. 14.4. Neutron reflectivity of GM1/d-DPPE monolayer, monolayer with bound $CT\beta_5$, and monolayer with bound $CT\alpha\beta_5$ (**a**). Lines indicate the fits corresponding to the SLD profiles from box-model fits in (**b**). The α subunit resides below the β pentamer, facing away from the lipid layer. An alternate set of SLD profiles from a model-independent spline fitting routine is shown in (**c**). The corresponding fits to the data were slightly better than the box-model fits [51]

reflectometry was applied to study the organization and structure of such PEGylated phospholipid monolayers in detail at the air–water interface [53]. Mixed with distearoylphosphatidylethanolamine (DSPE), a PEGylated lipid with 45 EG units exhibited the mushroom-to-brush transition upon monolayer compression, i.e., increasing polymer grafting density. The structure of the monolayer at the air–water interface was greatly perturbed by the presence of the bulky PEG–lipid headgroup resulting in a large increase of the apparent thickness of the headgroup region normal to the interface and a concomitant roughening of the interface.

Using X-ray and neutron reflectivity, monolayers of short chain poly(methyl oxazoline) (PMeOx) attached to a diacylglycerol lipid anchor have been studied as neat systems [54] and in binary mixtures with DMPC [55]. A new data evaluation method was developed in the course of that work to specifically evaluate the structural organization of linear polymers at surfaces and interfaces [56, 57]. In this approach, the molecular structure of the polymer is described in terms of its scattering length and specific volume per unit length, and its configuration is parameterized in terms of inclination angles between rigid rods that represent the polymer chain. While this is already overparameterizing the problem, a whole *ensemble* of lipopolymer molecules is encoded for the data evaluation in terms of a large set of angles and other structural parameters of interest, thus deliberately overparameterizing the model even more. Working with this model on a multitude of X-ray and neutron data, functionally relevant configurations may be identified by an evolution–strategy algorithm. In the case of PMeOx, a phase transition upon film compression has been attributed to a tensile stress exerted by the chains on their anchor points [58] that leads to a partial immersion of the lipid anchors into the aqueous subphase by a few Ångstrom. This in turn results in a local condensation of the hydrophobic chains.

14.5 Protein Adsorption and Stability at Functionalized Solid Interfaces

Adsorption and organization of biomolecules at solid interfaces bears implications to a broad range of areas, such as tissue engineering, biocompatibilization, and biosensorics [1]. Cell adhesion, cell–cell interaction, unfolding and denaturation of proteins are other important objectives related to the interaction of biomolecules with interfaces (for a comprehensive review see [59]). A number of aspects of protein adsorption to liquid and solid interfaces investigated with neutron reflectometry are already presented in the contribution by Lu, in this volume. Here we will discuss only investigations on protein interactions with surface-modified interfaces relevant for the construction of biosensors and biocompatible interfaces.

14.5.1 Hydrophobic Modified Interfaces

A well-established modification of native silicon–silicon oxide interfaces is silanization via e.g., octadecyltrichlorosilane (OTS) to hydrophobize the surface [52]. The structure and composition of bovine β-casein was investigated upon adsorption to OTS-modified silica surfaces in aqueous buffer [60, 61].

Adjacent to the surface, a dense protein layer was observed that protrudes into the solution with a structure of lower density as schematically shown in Fig. 14.5. A proteolytic enzyme, endoproteinase Asp-N, which cleaves the hydrophilic part of β-casein, affected only the outermost β-casein layer. Similar experiments were performed with β-lactoglobulin [62]. Effects of surfactants and of cations on β-casein adsorption to hydrophobized silica were also investigated [63, 64]. In comparison, protein adsorption to bare (i.e., unmodified), hydrophilic silica interfaces showed similar structural features, albeit at significantly slower adsorption rates [65]. The study thus demonstrates well that neutron reflection may be useful for studies of dynamics of interfacial phenomena in protein systems.

In a similar vein, the adsorption of lysozyme to OTS modified interfaces was observed to be irreversible [67]. The protein formed a densely packed, thin layer at the OTS surface with a diffuse, thicker proportion facing the bulk solution – a similar structure as observed with β-casein. As none of the dimensions of these structures corresponded to those of the globular protein in solution – unlike after adsorption at the hydrophilic silica–water interface – lysozyme is believed to denature at the OTS–water interface. The interfacial structure of lung surfactant has been studied at OTS-hydrophobized silica solid–liquid interfaces [68]. Results on lung surfactant from rabbit were interpreted within a two-layer model. An inner layer of ∼20 Å thickness contained about 50% solvent, consistent with a loosely packed phospholipid monolayer with substantial amounts of solvent and protein incorporated. The outer layer was ∼80 Å thick and incorporated ∼10% protein. Addition of the water-soluble phospholipase PLA_2 led to a densification of both layers. Whether these findings bear implications for a general understanding of surfactant function remains to be seen, since porcine lung surfactant showed somewhat different behavior. A phospholipid monolayer of higher density was observed, and the addition of PLA_2 exerted much less effect than in the rabbit system.

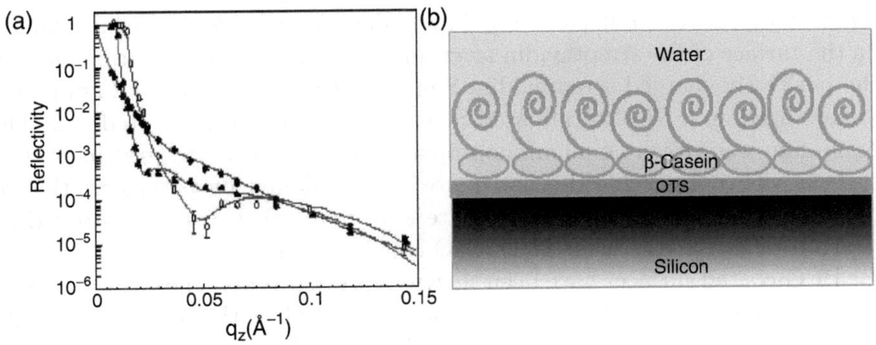

Fig. 14.5. (**a**) Neutron reflectivity and model fits measured for β-casein adsorbed at pH 7 on deuterated OTS in (○) D_2O, (•) H_2O, and (▲) water with an SLD of 4.5×10^{-6} Å^{-2} (details see [61]). (**b**) Schematic interpretation of these results [3]

14.5.2 Hydrophilic Modified Interfaces

Self-assembled monolayers (SAMs) provide a well-established method for the control of the surface-chemistry of solid substrates. Neutron reflectivity is among the standard techniques for the molecular-scale characterization of such interfaces. It has, for example, been used to characterize the adsorption of human serum albumin (HSA) onto silicon-supported NH_3^+-terminated SAMs [69]. Upon incubation with high protein concentration ($0.1\%\,\mathrm{wtV}^{-1}$), HSA formed a two-layer adsorption structure. Directly adjacent to the solid surface a ~40 Å thick protein layer was observed. A secondary layer extended an additional 40 Å into the solution.

In a slightly different context, chemical interface modification with SAMs also enables stable anchoring of biomolecules to polar or apolar surfaces [70]. For example, yeast cytochrome *c* (YCC) was covalently bound to the interface of a SAM with mixed $-CH_3$/–SH or –OH/–SH endgroups via disulfide linkage between a cysteine residue on the protein surface and the exposed thiol moieties of the SAM. If the SAM is prepared on top a stratified Fe/Si or Fe/Au/Si surface nanostructure, the resulting SLD striations allow for an *interferometric* approach to data inversion. This approach reduces the intrinsic limitation of the resolution that is due to the restrictions in attainable Q_z range. The potential of similar approaches for an unambiguous phase determination to calculate the SLD profile [71] is described in detail in the contribution by Majkrzak et al., in this volume. Measurements of the system in air with H_2O and D_2O hydrating the protein monolayer provided water distribution profiles [70] (Fig. 14.6). These profiles were consistent with corresponding electron density profiles determined previously via X-ray interferometry [72].

As discussed above, biotin–streptavidin technology is another standard that enables biospecific ligation at interfaces. In the context of functionalization of solid interfaces, streptavidin has been immobilized at an aminopropyl-terminated silicon surface. Biotinylated DNA strands have subsequently been grafted to this surface [73], and the evolution of the complex structure has been followed by neutron reflectometry. It was determined that DNA is collapsed on the surface of the streptavidin layer due to its electrostatic interaction with the positively charged streptavidin. Similarly, the formation of a lignin layer immobilized on a SAM-terminated Si surface has been investigated [74]. The lignin film was grown on an aminopropyl modified Si-wafer by grafting of this surface with polysaccharides and peroxidase, followed by polymerization at the solid–liquid interface. Intermediate and final structures were studied by neutron reflectometry using H_2O–D_2O contrast variation.

PEG-coated surfaces have been shown to prevent protein adsorption. Using neutron reflectivity, this was demonstrated for PEGs with a molecular weight of ≈5000 Da [75] as well as for short-chain SAMs of methoxy-tri(ethylene glycol) [76, 77]. A different way to passivate the hydrophilic silicon oxide surface against protein adsorption is by chemically anchoring an organic monolayer bearing terminal phosphorylcholine (PC) groups, as described in the

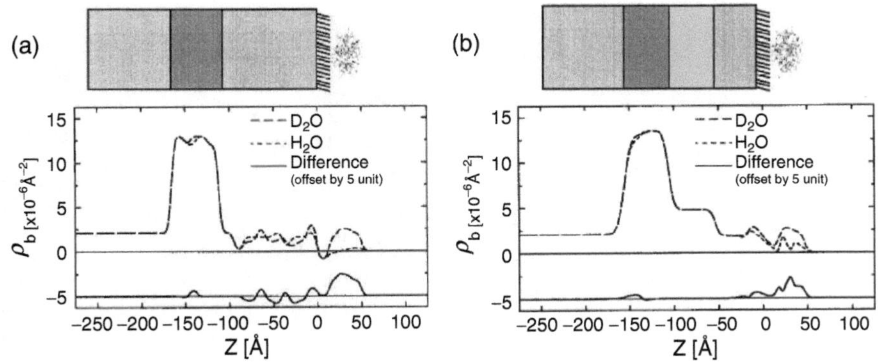

Fig. 14.6. Neutron SLD profiles of YCC on SAMs with probe spins parallel to the iron magnetization for partial hydration with D_2O and H_2O, and their difference profile for both a nonpolar (**a**) and a polar, (**b**) SAM. Schematic representations of the composite structures are shown on *top* [70]

contribution by Lu in this volume. The approach was shown to be effective in reducing protein adsorption of lysozyme, fibrinogen, and bovine serum albumin (BSA) [78]. Salt-induced protein resistance of poly(acrylic acid) (PAA) brushes against BSA has been demonstrated by neutron reflectometry [79]. BSA molecules penetrates deeply into the PAA brush at low sodium chloride concentrations, but did not interact with the PE cushion at >500 mM salt. The adsorption of enzyme staphylococcal nuclease (SNase) on negatively charged poly(styrene sulfonate) surface has been studied with neutron reflectometry [80] and it was observed that the degree of protein adsorption onto the charged surface depends largely on direct protein–protein interactions (Fig. 14.7).

14.6 Functionalized Lipid Interfaces and Supported Lipid Bilayers

14.6.1 Solid-Supported Phospholipid Bilayers

Phospholipid bilayers attached to solid supports are considered model systems that may be useful for the investigation of biological membranes of limited complexity, thus hoped to offer insight into underlying organization and interaction principles. Preparation of such systems can be performed either by spontaneous fusion of vesicles onto solid–liquid interfaces [81] or by Langmuir–Blodgett transfer techniques [82]. Neutron reflectometry is one of the work horses applied to probe such systems at the molecular level. For example, the structure of DMPC vesicles adsorbed on planar quartz surfaces [83] and that of DPPC on silicon single crystals [84] has been investigated. The formation of single phospholipid bilayers immobilized on solid surfaces was thus demonstrated and found to be separated from the surface by a water layer

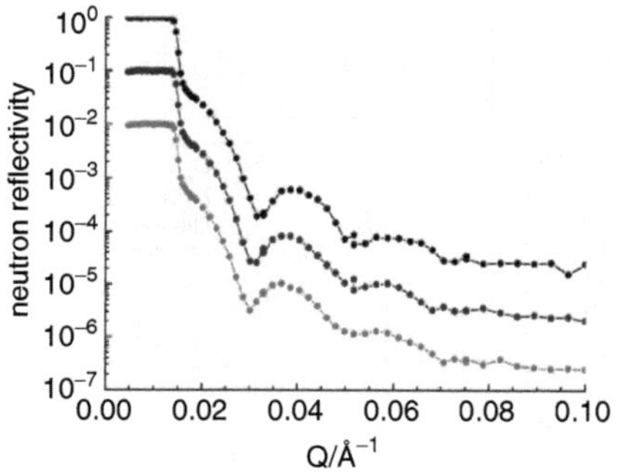

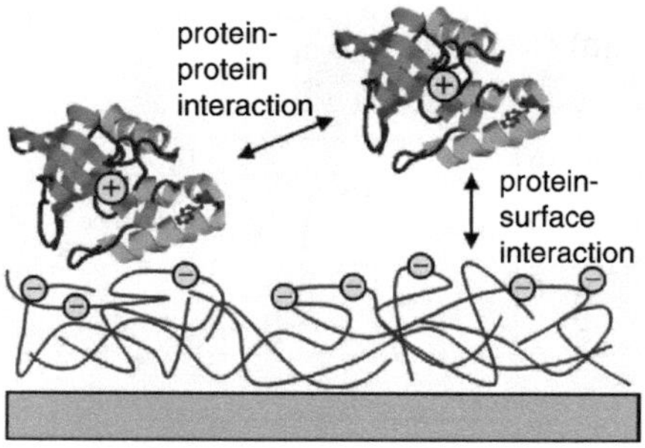

Fig. 14.7. *Left*: Neutron reflectivity curves of silica–water interface coated with a negatively charged polyelectrolyte. The upper curve refers to this interface without adsorbed protein, the other curves have been measured when staphylococcal nuclease was adsorbed (*middle*: 23°C, *lower*: 43°C). *Right*: Sketch of positively charged staphylococcal nuclease adsorbed on negatively charged polyelectrolyte surface. The degree of adsorption largely depends on protein–protein interactions, whereas attractive electrostatic interactions between the protein and the surface play a minor role [80].

of a few Ångstrom thickness. The process of phospholipid bilayer formation has been followed by time resolved measurements [85] (Fig. 14.8). Results were consistent with AFM and quartz microbalance (QMB) studies [86–88]. The interaction of biomolecules with such supported lipid bilayers has been exploited in various studies. For example, the penetration of the bacterial toxin, pneumolysin into mixed phospholipid membranes (10:10:1 molar ratio of PC:cholesterol:dicetyl phoshate) was investigated on silicon oxide interfaces [89].

PLA_2 interaction with phospholipid bilayers (DPPC, DOPC, or POPC co-adsorbed with dodecyl maltoside) at silica surfaces [90] has been studied [91]. The conventional model of PLA_2–membrane interaction assumes adsorption of the enzyme on the bilayer, followed by partial extraction of a substrate molecule into the enzyme, hydrolysis and release of the products [92]. In distinction, neutron reflectometry results suggest that the enzyme actually penetrates into the layer until the active site, located at the top of the enzyme, is at the same level with the lipid headgroups of the outer phospholipid layer. The penetration of the enzyme into the lipid layer would then imply that the rate determining step of the overall reaction is either the initial adsorption or the final desorption, rather than the actual hydrolysis.

In an attempt to better control the formation of lipid bilayers on solid support, a sequential deposition of phospholipid monolayers in a combination of Langmuir–Blodgett and Langmuir–Schaefer transfer techniques has been utilized [93]. Neutron reflection measurements allowed precise, nondestructive characterization of the structure, hydration and roughness of the deposited layers. Beyond the first bilayer, deposited by Langmuir–Blodgett transfer, a second, "free" and fully hydrated bilayer was formed and its physical

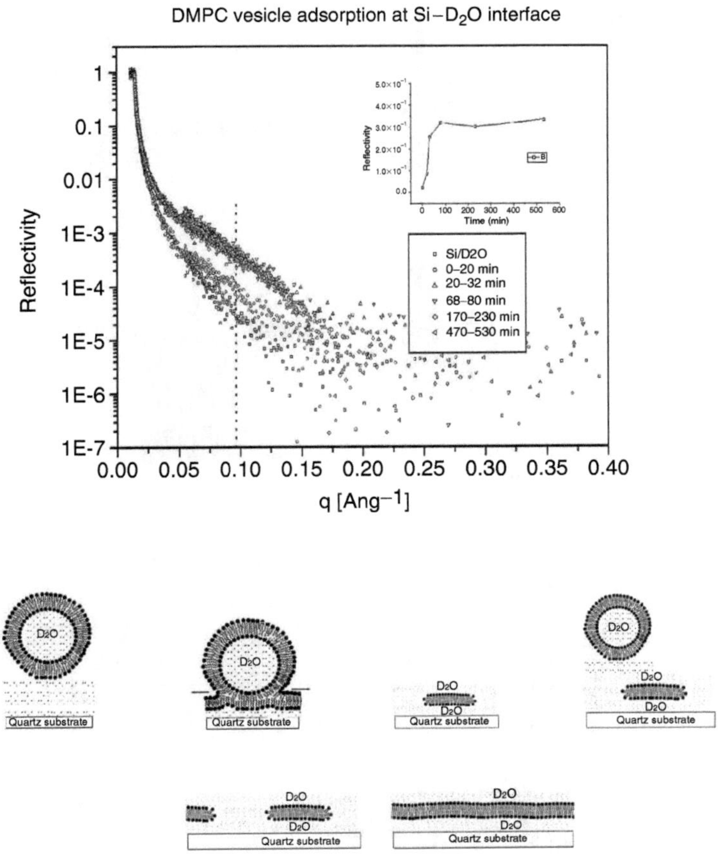

Fig. 14.8. *Top*: Neutron reflectivity demonstrating the adsorption of DMPC vesicles at an Si–SiO_2 surface in D_2O. The inset shows the change in reflected intensity at a fixed momentum transfer, $Q_z = 0.09\,\text{Å}^{-1}$. *Bottom*: Schematic depiction of the adsorption of uni-lamellar phospholipid vesicles onto a solid hydrophilic interface [85]

properties examined (Fig. 14.9) [94]. Utilizing this model system, the insertion of a model peptide that was determined to be an active transmembrane shuttle for drug delivery to cells demonstrated uniform peptide distribution in the interfacial lipid region [95]. In the presence of dipalmitoyl phosphatidylserine (DPPS), the peptide was detected mainly in the lipid headgroup region and increased bilayer roughness was observed in the neutron reflectometry measurements.

14.6.2 Hybrid Bilayer Membranes

A variation of the theme of mimicking biomembranes on solid interfaces are the so-called hybrid bilayer membranes (HBMs). A lipid monolayer is fused onto a chemically grafted SAM or polymer layer [16, 96–98]. The basic details of this approach are outlined in the contribution by Majkrzak et al., in this volume.

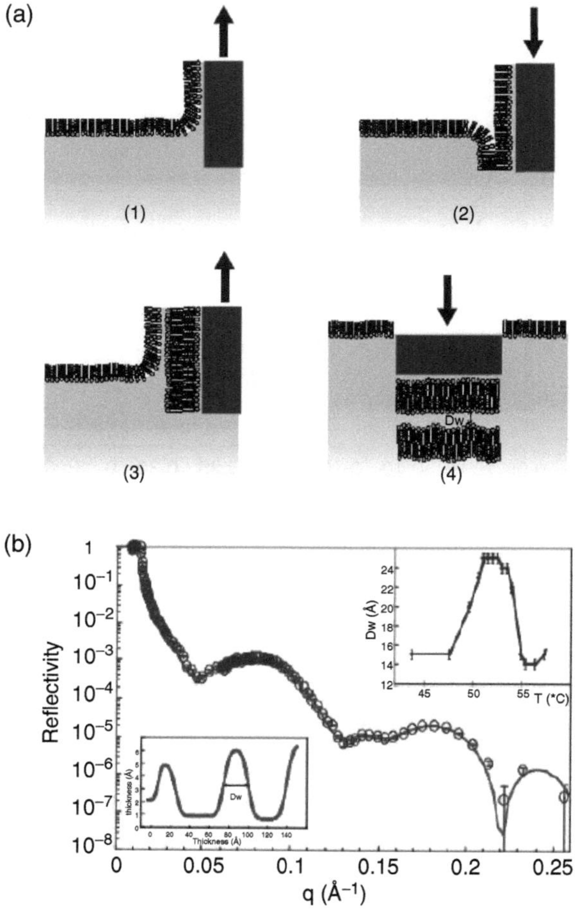

Fig. 14.9. (**a**) Cartoon demonstrating the four-step preparation of a double bilayer. Layers 1–3 are deposited by the Langmuir–Blodgett technique; layer 4 is deposited by the Langmuir–Schaefer method [93]. (**b**) Neutron reflectivity profile and fitted SLD profile (*bottom left inset*) for a DSPC double bilayer in D_2O at 59°C. *Upper inset*: variation of the inter-bilayer distance (D_w) with temperature [94]

By preparing a HBM on a gold-coated Si wafer with an ultrathin (only several micrometer) aqueous subphase in a solid–liquid sample cell – to reduce incoherent background – neutron reflectivity data with a momentum transfer of $0.7\,Å^{-1}$, spanning a dynamic range of $\sim 10^8$, have been recorded [17]. This work showed that the peptide melittin perturbs the phospholipid headgroup strongly and also affects the acyl chain region of the distal bilayer leaflet of the HBM [17]. Also the combination of an HBM and a subsequently transferred bilayer leads to the formation of a "free" phospholipid bilayer floating atop the HBM. Such systems have been studied with neutron reflectometry, and interpreted in the VRDF approach [97]. The free bilayer is organized according to the steric demands of the lipids rather than the influence of the substrate. The advantage of the free supported bilayer compared to more conventional supported systems is that the mobility of the phospholipid molecules makes them suitable for investigations of transmembrane processes. This was demonstrated by the incorporation of a deuterated peptide into the bilayer that afforded the observation of peptide orientational changes induced by small potential changes across the membrane structure [99].

Phospholipid monolayer formation on a hydrophobic polymer film (polystyrene, PS, spin-coated on a Si wafer) proceeds very fast after exposure of the surface to a vesicle suspension [98]. Contrast variation via D_2O–H_2O exchange and the use of deuterated compounds (DMPC-d_{54}, deuterated PS) made the approach highly sensitive to the adsorbed layer structure, even minor contaminations or structural changes caused by membrane active molecules–such as proteins, DNA, or detergents–are readily accessible. Using this approach, the interaction of the neurotoxic β-amyloid peptide Aβ (25–35) with a phospholipid monolayer obtained via fusion of small unilamellar vesicles onto a PS-coated substrate has been studied. It was shown that the peptide adsorbed rather than penetrated the supported lipid monolayer [100], in contrast to results of neutron diffraction experiments on phospholipid multilayer stacks, which indicated a deep penetration of Aβ amyloid into the lipid matrix [101].

14.6.3 Polymer-Supported Phospholipid Bilayers

Another problem in the context of membrane protein incorporation into biomembrane mimics is the nonphysiological interaction of the proteins with the solid support that would result in lateral immobilization or denaturation. The development of free bilayers or of bilayers that are tethered to the solid interface by means of polymer cushions has been proposed to solve this problem [1,102]. Promising systems that have been investigated for the preparation of polymer cushions have involved PEGs and polyethylenimines (PEIs). Also, polyelectrolyte (PE) multilayer films, where well-defined systems up to micrometer thickness may be obtained via sequential, layer-by-layer deposition of alternately charged PEs [104], have been employed. Their internal structure have been analyzed using neutron and X-ray reflectivity [48, 106].

The formation of DMPC bilayers on PEI-coated quartz substrates using a variety of deposition techniques has thus been studied by neutron reflection [103]. Using the layer-by-layer preparation technique, the attachment of individual phospholipid bilayers on poly(styrene sulfonate), PSS, and poly(allylamine hydrochloride), PAH, has been demonstrated [107–109]. Subsequently, DMPG bilayers on a PSS/PAH cushion were utilized to study the interaction of β-amyloid peptide with membranes [107]. In this work, neutron reflectometry demonstrated the presence of a 10 Å amyloid layer adsorbed to the phospholipid bilayer. Similarly, the in situ adsorption of a dense layer of the enzyme, phosphatidylinositol-3-kinase (PI3K), to phosphatidic acid on PAH/PSS, was followed by neutron reflection [109]. In another study, the adsorption of pectin onto charged and uncharged lipid bilayers on PE cushions was investigated [108]. Recently, neutron reflectometry was applied to study the interaction of DNA with PE-supported DMPC and DMPG bilayers in situ [110]. At high phospholipid concentrations ($>0.5\,\mathrm{mg\,ml^{-1}}$), the adsorption of multilamellar stacks of DMPC on PSS/PAH cushions was observed [111]. A very recent study demonstrated the option to deposit further PE layers on top of a DMPC bilayer which in turn was adsorbed to PE multilayer cushions [112]. This approach might lead to PE encapsulated biomembrane mimics with potential application for novel drug delivery systems. A PE cushion terminated with a terpolymer that links stearoyl chains to a styrolsulfonate anchor was created by deposition of positively charged poly-L-lysine and negatively charged alginate. Subsequent transfer of a phospholipid monolayer resulted in the construction of a tethered hybrid bilayer membrane whose structure was in detail assessed by neutron reflectometry [113].

14.7 Conclusions

Neutron reflectometry provides valuable insights into the structure and assembly of complex biological systems at fluid surfaces and solid–liquid interfaces. The examples described in this chapter were selected to demonstrate current capabilities. For a more detailed characterization of biomimetic systems, on the other hand, not only organization perpendicular to the membrane surface is required, but also lateral structures, membrane in-plane organization and dynamic membrane features [1, 114]. Improvements in grazing-incidence neutron diffraction and off-specular neutron reflectivity are requested to tackle these questions. Investigations by these techniques on phospholipid membranes as outlined in the contribution by Salditt et al., in this volume and pioneering studies by Huang and coworkers [115,116] demonstrate the feasibility of these approaches, which currently, however, still lack sufficient neutron flux.

Acknowledgments

M.L. wishes to acknowledge support by the Volkswagen Foundation (Grant no. I/77709), the National Institutes of Health (Grant no. 1 RO1 RR14812) and The Regents of the University of California.

References

1. E. Sackmann, M. Tanaka, Trends Biotechnol. **18** (2000), 58–64
2. M. Lösche, in *Current Topics in Membranes*, Vol. 52, S.A. Simon, T.J. McIntosh (Eds.) (Academic Press, New York, 2001)
3. G. Fragneto-Cusani, J. Phys.: Condens. Matter **13** (2001), 4973–4989

4. C.F. Majkrzak, N.F. Berk, Appl. Phys. A **74** (2002), S67–S69
5. R. Cubitt, G. Fragneto, R.E. Gosh, A.R. Rennie, Langmuir **19** (2003), 7685–7867
6. R. Frahm, J. Weigelt, G. Meyer, G. Materliki, Rev. Sci. Instrum. **66** (1995), 1677–1680
7. J.F. Ankner, C. Rehm, Physica B **336** (2003) 68–74
8. K. Lieutenant, H. Fritzsche, F. Mezei, Appl. Phys. A **74** (2002), S1613–S1615
9. P. Krüger, M. Lösche, in *Lecture Notes in Physics*, Vol. 634, R. Haberlandt, D. Michel, A. Pöppl, R. Stannarius (Eds.) (Springer, Berlin–New York, 2004)
10. J. Als-Nielsen, D. Jacquemain, K. Kjaer, M. Lahav, F. Leveiller, L. Leiserowitz, Phys. Rep. **246** (1995), 251–313
11. M. Lösche, J.P. Rabe, A. Fischer, B.U. Rucha, W. Knoll, H. Möhwald, Thin Solid Films **117** (1984), 269–280
12. M. Lösche, E. Sackmann, H. Möhwald, Ber. Bunsenges. Phys. Chem. **87** (1983) 848–852
13. P. Krüger, M. Lösche, Phys. Rev. E **62** (2000), 7031–7043
14. M. Lösche, H. Möhwald, Rev. Sci. Instrum. **55** (1984), 1968–1972
15. P.S. Pershan, Colloids Surf. A: Physicochem. Eng. Aspects **171** (2000), 149–157
16. C.W. Meuse, S. Krueger, C.F. Majkrzak, J.A. Dura, J. Fu, J.T. Connor, A.L. Plant, Biophys. J. **74** (1998), 1388–1398
17. S. Krueger, C.W. Meuse, C.F. Majkrzak, J.A. Dura, N.F. Berk, M. Tarek, A.L. Plant, Langmuir **17** (2001), 511–521
18. M.C. Wiener, S.H. White, Biophys. J. **59** (1991), 174–185
19. D. Vaknin, K. Kjaer, J. Als-Nielsen, M. Lösche, Biophys. J. **59** (1991), 1325–1332.
20. M. Schalke, P. Krüger, M. Weygand, M. Lösche, Biochim. Biophys. Acta **1464** (2000), 113–126
21. M. Schalke, M. Lösche, Adv. Colloid Interf. Sci. **88** (2000), 243–274
22. R.S. Armen, O.D. Uitto, S.E. Feller, Biophys. J. **75** (1998), 734–744
23. H.L. Brockman, Curr. Opin. Struct. Biol. **9** (1999), 438–443
24. J. Als-Nielsen, H. Möhwald, in *Handbook on Synchrotron Radiation*, Vol. 4, S. Ebashi, M. Koch, E. Rubenstein (Eds.) (Elsevier, North Holland, 1991) pp. 1–53
25. G.S. Smith, J. Majewski, in *Lipid Bilayers, Structure and Interactions*, J. Katsaras, T. Gutberlet (Eds.) (Springer, Berlin, 2001) pp. 127–148
26. G. Büldt, H.U. Gally, A. Seelig, J. Seelig, G. Zacchai, Nature **27** (1978), 182–184
27. G. Büldt, H.U. Gally, J. Seelig, G. Zacchai, J. Mol. Biol. **134** (1979), 673–691
28. T.M. Bayerl, R.K. Thomas, J. Penfold, A.R. Rennie, E. Sackmann, Biophys. J. **57** (1990), 1095–1098.
29. T. Brumm, C. Naumann, E. Sackmann, A.R. Rennie, R.K. Thomas, D. Kanellas, J. Penfold, T.M. Bayerl, Eur. Biophys. J. **23** (1994), 289–295.
30. C. Naumann, C. Dietrich, J.R. Lu, R.K. Thomas, A.R. Rennie, J. Penfold, T.M. Bayerl, Langmuir **10** (1994) 1919–1925.
31. C. Naumann, T. Brumm, A.R. Rennie, J. Penfold, T.M. Bayerl, Langmuir **11** (1995) 3948–3952.
32. S.J. Johnson, T.M. Bayerl, W. Weihan, H. Noack, J. Penfold, R.K. Thomas, D. Kanellas, A.R. Rennie, E. Sackmann, Biophys. J. **60** (1991), 1017–1025
33. B. Demé, D. Hess, M. Tristl, L.T. Lee, E. Sackmann, Eur. Phys. J. E **2** (2000), 125–136

34. C. Naumann, C. Dietrich, A. Behrisch, T.M. Bayerl, M. Schleicher, D. Bucknall, E. Sackmann, Biophys. J. **71** (1996), 811–823
35. J. Strzalka, B.R. Gibney, X. Chen, C.C. Moser, P.L. Dutton, S.K. Satija, B.M. Ocko, J.K. Blasie, Biophys. J. **78** (2000), 325A
36. M. Kent, H. Yim, D. Sasaki, J. Majewski, G.S. Smith, K. Shin, S. Satija, B.M. Ocko, Langmuir **18** (2002), 3754–3757
37. M. Kent, H. Yim, D. Sasaki, S. Satija, J. Majewski, T. Gog, Langmuir **20** (2004), 2819–2829
38. W.K. Fullagar, K.A. Aberdeen, D.G. Bucknall, P.A. Kroon, I.R. Gentle, Biophys. J. **85** (2003), 2624–2632
39. W.K. Fullagar, I.R. Gentle, S.A. Holt, ISIS2003 Science Highlights
40. U.B. Sleytr, M. Sára, D. Pum, B. Schuster, Progr. Surf. Sci. **68** (2001), 231–278
41. B. Wetzer, A. Pfandler, E. Györvary, D. Pum, M. Lösche, U.B. Sleytr, Langmuir **14** (1998), 6899–6706
42. A. Diederich, C. Sponer, D. Pum, U.B. Sleytr, M. Lösche, Colloids Surf. B: Biointerf. **6** (1996), 335–346
43. M. Weygand, B. Wetzer, D. Pum, U.B. Sleytr, N. Cuvillier, K. Kjaer, P.B. Howes, M. Lösche, Biophys. J. **76** (1999), 458–468
44. M. Weygand, M. Schalke, P.B. Howes, K. Kjaer, J. Friedmann, B. Wetzer, D. Pum, U.B. Sleytr, M. Lösche J. Mater. Chem. **10** (2000), 141–148
45. M. Weygand, K. Kjaer, P.B. Howes, B. Wetzer, D. Pum, U.B. Sleytr, M. Lösche, J. Phys. Chem. B **106** (2002), 5793–5799
46. D. Vaknin, J. Als-Nielsen, M. Piepenstock, M. Lösche, Biophys. J. **60** (1991), 1545–1552
47. D. Vaknin, K. Kjaer, H. Ringsdorf, R. Blankenburg, M. Piepenstock, A. Diederich, M. Lösche, Langmuir **9** (1993), 1171–1174
48. M. Lösche, M. Piepenstock, A. Diedrich, T. Grunewald, K. Kjaer, D. Vaknin, Biophys. J. **65** (1993), 2160–2177
49. M. Lösche, C. Erdelen, E. Rump, H. Ringsdorf, K. Kjaer, D. Vaknin, Thin Solid Films **242** (1994), 112-117
50. M. Tristl, I. Haase, S. Marx, L.T. Lee, M. Fischer, E. Sackmann, LLB Sci. Rep. (1999-2000) 100–102; K. Sengupta, L. Limozin, M. Tristl, M. Fischer, E. Sackmann, Phys. Chem. Chem. Phys., submitted
51. C.E. Miller, J. Majewski, R. Faller, S. Satija, T.L. Kuhl, Biophys. J. **86** (2004), 3700–3708
52. T.L. Kuhl, J. Majewski, J.Y. Wong, S. Steinberg, D.E. Leckband, J.N. Israelachvili, G.S. Smith, Biophys. J. **75** (1998), 2352–2362
53. J. Majewski, T.L. Kuhl, M.C. Gerstenberg, J.N. Israelachvili, G.S. Smith, J. Phys. Chem. B **101** (1997), 3122–3129
54. A. Wurlitzer, E. Politsch, S. Hübner, P. Krüger, M. Weygand, K. Kjaer, P. Hommes, O. Nuyken, G. Cevc, M. Lösche, Macromolecules **34** (2001), 1334–1342
55. T. Gutberlet, A. Wurlitzer, U. Dietrich, E. Politsch, G. Cevc, R. Steitz, M. Lösche, Physica B **283** (2000), 37–39
56. E. Politsch, G. Cevc, A. Wurlitzer, M. Lösche, Macromolecules **34** (2001), 1328–1334
57. E. Politsch, J. Appl. Crystallogr. **34** (2001), 239–251
58. C. Hiergeist, R. Lipowsky, J. Phys. II France **6** (1996), 1465–1481
59. A. Baszkin, W. Norde (Eds.), *Physical Chemistry of Biological Interfaces* (Dekker, New York, 1999)

60. G. Fragneto, R.K. Thomas, A.R. Rennie, J. Penfold, Science **267** (1995), 657–660
61. G. Fragneto, T.J. Su, J.R. Lu, R.K. Thomas, A.R. Rennie, Phys. Chem. Chem. Phys. **2** (2000), 5214–5221
62. G. Fragneto, J.R. Lu, A.R. Rennie, ILL Exp. Rep. (1996), 9-16-4
63. R.K. Thomas, A.R. Rennie, J.R. Lu, T.J. Su, ILL Exp. Rep. (1997), 9-10-245
64. T. Nylander, F. Tiberg, T.J. Su, J.R. Lu, R.K. Thomas, Biomacromolecules **2** (2001), 278–287
65. F. Tiberg, T. Nylander, T.J. Su, J.R. Lu, R.K. Thomas, Biomacromolecules **2** (2001), 844–850
66. T. Kull, T. Nylander, F. Tiberg, N.M. Wahlgren, Langmuir **13** (1997), 5141–5147
67. J.R. Lu, T.J. Su, P.N. Thirtle, R.K. Thomas, A.R. Rennie, R. Cubitt, J. Colloid Interf. Sci. **206** (1998), 212–223
68. D. Follows, F. Tiberg, M. Larssen, R. Thomas, G. Fragneto-Cusani, ILL Exp. Rep. (2002), 8-02-275
69. A. Liebmann-Vinson, L.M. Lander, M.D. Foster, W.J. Brittain, E.A. Vogler, C.F. Majkrzak, S. Satija, Langmuir **12** (1996), 2256–2262
70. L.R. Kneller, A.M. Edwards, C.E. Nordgren, N.F. Berk, S. Krueger, C.F. Majkrzak, J.F. Blasie, Biophys. J. **80** (2001), 2248–2260
71. N.F. Berg, C.F. Majkrzak, Phys. Rev. B. **51** (1995), 11296–11309; C.F. Majkrzak, N.F. Berk, Phys. Rev. B **52** (1995), 10827–10830
72. J.A. Chupa, J.P. McCauley, R.M. Strongin, A.B. Smith, J.K. Blasie, L.J. Peticolas, J.C. Bean, Biophys. J. **67** (1994), 336–348
73. A. Menelle, J. Jestin, F. Cousin, Neutron News **14** (2003), 26–30
74. B. Cathala, F. Cousin, A. Menelle, LLB Scientific Highlights 2003, 122–123
75. A.K. Adya, M. Shcherbakov, A. Zarbakhsh, J. Bowers, ISIS 2001 Annu. Rep. (2001), RB no. 11140
76. M. Grunze, R. Dahint, D. Schwendel, F. Schreiber, ILL Exp. Rep. (2002), 8-02-290
77. M. Grunze, R. Dahint, D. Schwendel, F. Schreiber, ILL Exp. Rep. (2000), 7-05-154
78. J.R. Lu, E.F. Murphy, T.J. Su, A.L. Lewis, P.W. Stratford, S.K. Satija, Langmuir **17** (2001), 3382–3389
79. C. Czeslik, G. Jackler, T. Hazlett, E. Gratton, R. Steitz, A. Wittmann, M. Ballauf, Phys. Chem. Chem. Phys. **6** (2004)
80. C. Czeslik, G. Jackler, C. Royer, R. Steitz, HMI Annu. Rep. 2002 (2002) 8
81. A.A. Brian, H.M. McConnell, Proc. Natl Acad. Sci. USA **81** (1984), 6159–6163
82. L. Tamm, H.M. McConnell, Biophys. J. **47** (1985), 105–113
83. S.J. Johnson, T.M. Bayerl, D.C. McDermott, G.W. Adam, A.R. Rennie, R.K. Thomas, E. Sackmann, Biophys. J. **59** (1991), 289–294
84. B.W. Koenig, S. Krueger, W.J. Orts, C.F. Majkrzak, N.F. Berk, J.V. Silverton, K. Gawrisch, Langmuir **12** (1996), 1343–1350
85. T. Gutberlet, R. Steitz, G. Fragneto, B. Kloesgen, J. Phys.: Conden. Matter **16** (2004), S2469–S2476
86. I. Reviakine, A. Brisson, Langmuir **16** (2000), 1806–1815
87. E. Reimhult, F. Höök, B. Kasemo, Langmuir **19** (2003), 1681–1691
88. R. Richter, A. Mukhopadhyay, A. Brisson, Biophys. J. **85** (2003), 3035–3047
89. O. Byron, M. Noellmann, R. Gilbert, M. Sferrazza, ISIS 2001 Annu. Rep. (2001), RB no. 11607

90. F. Tiberg, I. Harwigsson, M. Malmsten, Eur. Biophys. J. **29** (2000), 196–203
91. H. Vacklin, F. Tiberg, G. Fragneto, R.K. Thomas, Biochemstry, **44**, (2005), 2811–2821. ILL Exp. Rep. (2002), 8-02-288
92. O.G. Berg, M.K. Jain, *Interfacial Enzyme Kinetics* (Wiley, New York, 2002)
93. T. Charitat, E. Bellet-Amalric, G. Fragneto, F. Graner, Eur. Phys. J. B **8** (1999), 583–593
94. G. Fragneto, T. Charitat, F. Graner, K. Mecke, L. Perino-Gallice, E. Bellet-Amalric, Europhys. Lett. **53** (2001), 100–106.
95. G. Fragneto, F. Graner, T. Charitat, P. Dubos, E. Bellet-Amalric, Langmuir **16** (2000) 4581–4588
96. C.F. Majkrzak, N.F. Berk, S. Krueger, J.A. Dura, M. Tarek, D. Tobias, V. Silin, C.W. Meuse, J. Woodward, A.L. Plant, Biophys. J. **79** (2000) 3330–3340
97. A.V. Hughes, S.J. Roser, M. Gerstenberg, A. Goldar, B. Stidder, R. Feidenhansl, J. Bradshaw, Langmuir **18** (2002), 8161–8171
98. T. Gutberlet, R. Steitz, J. Howse, I. Estrela-Lopis, B. Kloesgen, Appl. Phys. A **74** (2002), S1262–S1263
99. S. Roser, A. Hughes, B. Strider, G. Fragneto, B. Cubitt, ILL Exp. Rep. (2002), 8-02-277
100. S. Dante, T. Hauß, T. Gutberlet, R. Steitz, BENSC Exp. Rep. (2001), 170; R. Steitz, S. Dante, T. Gutberlet, T. Hauß, B. Klösgen, S. Schemmel, HMI Annu. Rep. (2001), 29
101. S. Dante, T. Hauß, N. Dencher, Biophys. J. **83** (2002), 2610–2616
102. M.L. Wagner, L. Tamm, Biophys. J. **79** (2000), 1400–1414
103. J.Y. Wong, J. Majewski, M. Seitz, C.K. Park, J.N. Israelachvili, G.S. Smith, Biophys. J. **77** (1999) 1445–1457
104. G. Decher, Science **277** (1997), 1232–1237
105. J. Schmitt, T. Grünewald, G. Decher, P.S. Pershan, K. Kjaer, M. Lösche, Macromolecules **26** (1998), 7058–7063
106. M. Lösche, J. Schmitt, G. Decher, W.G. Bouwman, K. Kjaer, Macromolecules **31** (1998), 8893–8906
107. I. Estrela-Lopis, R. Steitz, E. Malzeva, BENSC Exp. Rep. (2000), 170
108. M.H. Ropers, M. Axelos, R. Krastev, BENSC Exp. Rep. (2002), 172
109. M.H. Ropers, G. Brezesinski, R. Krastev, BENSC Exp. Rep. (2002), 171
110. S. Gromelski, R. Krastev, T. Gutberlet, G. Brezesinski, PSI Sci. Rep. 2005, submitted
111. R. Krastev, T. Gutberlet, BENSC Exp. Rep. (2003)
112. C. Delajon, R. Krastev, T. Gutberlet, H. Möhwald, Langmuir, **21**, (2005), 8509–8514
113. U.A. Perez-Salas, K.M. Faucher, C.F. Majkrzak, N.F. Berk, S. Krueger, E.L. Chaikof, Langmuir **19** (2003), 7688–7694
114. E. Sackmann, Science **271** (1996), 43–48
115. L. Yang, T.A. Harroun, W.T. Heller, T.M. Weiss, H.W. Huang, Biophys. J. **75** (1998), 641–645
116. L. Yang, T.M. Weiss, T.A. Harroun, W.T. Heller, H.W. Huang, Biophys. J. **77** (1999), 2648–2656

Part II

Inelastic Techniques

15

Quasielastic Neutron Scattering in Biology, Part I: Methods

R.E. Lechner, S. Longeville

15.1 Introduction

Biological macromolecules and biological systems, in general, are constructed according to well-defined building schemes exhibiting a certain degree of long-range order. But they are also characterized by an appreciable amount of disorder, for several reasons. One is that high structural symmetry is generally absent in native samples, except for the rare cases with integral single-crystalline regions. Another is that long-range order is limited to certain parts of the macromolecules and to part of the degrees of freedom. Furthermore, the ubiquitous presence of water which is generally a prerequisite for the unrestrained performance of biological function plays an important role. The interaction of water molecules with biological surfaces, their diffusion close to and within the hydration layers covering biological macromolecules, provides the latter with the indispensable additional space for the conformational degrees of freedom required for function. The presence of mobile water molecules not only allows or induces additional short-range translational and rotational diffusive motion of parts of biological macromolecules, but also causes damping of low-frequency vibrations in the macromolecules. All these motions which are believed to be essential for biological function, are an important part of the dynamical characteristics of biological matter.

The ensemble of low-energy transfer inelastic (IENS) and quasielastic neutron scattering (QENS) techniques is particularly well-suited for their investigation. In the present chapter, the basic principles of these techniques are outlined from the viewpoints of theory, experiment and analysis, with an emphasis on application to biological problems. The method of QENS[1] focuses

[1]The acronym "QENS" is used for quasielastic neutron scattering in general, and "QINS", for quasielastic neutron scattering, when it is purely incoherent, in order to distinguish this case from the coherent one

on scattering processes involving small amounts of energy exchange,[2] with spectral distributions peaked at zero energy transfer. IENS spectra extend to somewhat higher energies, but are, by principle, also overlapping with the QENS energy region. Both together allow us to study dynamical phenomena in the *time* region of 10^{-13}–10^{-7} s. Atomic and molecular motions are explored in *space*, on length scales comparable with the wavelengths of the neutrons used in the scattering experiments. Typical spatial parameters, such as vibrational displacements, jump distances, diffusion paths, and correlation lengths, are amenable to evaluation in the range from 10^{-9} to 10^{-6} cm. Quasielastic and inelastic neutron scattering experiments on such dynamic processes lead to spectra of energy transfers

$$\hbar\omega = E - E_0 \ , \tag{15.1}$$

in a range from 10 to 10^{-5} meV, where E_0 and E are the neutron energies before and after scattering, respectively. The corresponding momentum transfer $\hbar\boldsymbol{Q}$ in such a process is proportional to the scattering vector

$$\boldsymbol{Q} = \boldsymbol{k} - \boldsymbol{k}_0 \ , \tag{15.2}$$

where $\boldsymbol{k}_0$ and $\boldsymbol{k}$ are the neutron wave vectors before and after scattering, respectively. The wave-vector transfer values Q for elastic scattering, $Q = (4\pi/\lambda)\sin(\varphi/2)$, are typically in the region of 0.1–5 Å^{-1} (λ = neutron wavelength, φ = scattering angle, i.e., the angle between the vectors $\boldsymbol{k}_0$ and $\boldsymbol{k}$), such that $2\pi/Q$ ranges from the order of magnitude of interatomic distances to that of diameters of (e.g., biological) macromolecules. The neutron scattering intensity in such a process is proportional to the so-called *scattering function* or *dynamic structure factor* $S(\boldsymbol{Q},\omega)$, which can be calculated for typical dynamical processes; the calculation and determination of this function is the subject of the following paragraphs.

The purpose of QENS experiments is mainly the study of the details of "quasielastic lines", i.e., of the low-energy spectra which are mostly due to some kind of diffusive or damped vibrational atomic and molecular motions. In the energy domain, this mainly refers to the part of the dynamical function which has a maximum centered around zero energy transfer. In the time domain, it corresponds to the relaxation of the dynamical functions. For instance, in the well-understood case of classical atomic (self-) diffusion, the relaxation function has a single exponential time decay for small Q values, $I_s(Q,t) \sim \exp(-\Gamma t)$ with $\Gamma = 1/\tau = D_s Q^2$, where τ is the decay constant and D_s is the self-diffusion coefficient. This transforms in the energy domain to a Lorentzian function with the width Γ. For larger scattering vectors, Γ depends on the geometric and dynamic details of the diffusion process. Obviously, for a particle *at rest*, $\Gamma = 0$ and $S(\boldsymbol{Q},\omega)$ is a sharp (elastic) line $\delta(\omega)$

[2]In this context, "small" means: (i) the concerned energy transfers are located in the low-energy region of inelastic neutron scattering spectra, and (ii) they cover a region close to zero, but are still resolved by the elastic energy resolution

at $\hbar\omega = 0$. This is analogous to resonance absorption of gamma rays, where this corresponds to the well-known Mössbauer line.

The interpretation of the scattering function $S(\boldsymbol{Q}, \omega)$ in terms of diffusive and/or vibrational processes is relatively simple, if

- such motions can be described by classical physics, i.e., when quantum effects can be completely neglected, and if – as in most practical cases
- the scattering can be treated in first Born approximation. This allows an evaluation and interpretation of $S(\boldsymbol{Q}, \omega)$ by pair correlation functions for the scattering nuclei in space and time, which includes the self-correlation function $S_s(\boldsymbol{Q}, \omega)$ as a special case.

This chapter(Part I) is organized in several sections dealing with the various topics relevant in our context in the following order: (i) Basic theory of neutron scattering: Incoherent and coherent scattering functions and intermediate scattering functions; Van Hove correlation functions; coherent structure factor, elastic incoherent structure factor (EISF); experimental resolution and observation time. (ii) Instruments for QENS spectroscopy in $(\boldsymbol{Q}, \omega)$-space: direct and inverted geometry time-of-flight (TOF), and backscattering (BSC) techniques. (iii) Instruments for QENS spectroscopy in $(\boldsymbol{Q}, t)$-space: NSE and NRSE techniques. For applications of these techniques in biological studies, the reader is referred to Part II in this volume.

15.2 Basic Theory of Neutron Scattering

Information on the dynamic structure of condensed matter is obtained by analyzing the intensity of neutrons, for instance from a monochromatic beam scattered by a sample into a solid angle element $\mathrm{d}\Omega$ and an energy interval $\mathrm{d}(\hbar\omega)$. This is proportional to the double-differential scattering cross-section $\mathrm{d}^2\sigma/\mathrm{d}\Omega\mathrm{d}\omega$ which reads

$$\frac{\mathrm{d}^2\sigma}{\mathrm{d}\Omega\,\mathrm{d}\omega} = \frac{k}{k_0}\frac{\sigma}{4\pi}S(\boldsymbol{Q}, \omega) \,. \tag{15.3}$$

It is factorized in three independent components: (i) The ratio of the wave numbers k and k_0 characterizing the scattering process, (ii) the total scattering cross-section σ for a rigidly bound nucleus (where $\sigma = 4\pi b^2$ and b is the corresponding scattering length of the nucleus), and (iii) the Van Hove scattering function $S(\boldsymbol{Q}, \omega)$ [1]. The latter depends on the scattering vector $\boldsymbol{Q}$ and the energy transfer $\hbar\omega$ as defined by Eqs. 15.1 and 15.2. The structural and dynamical properties of the scattering sample are fully described by $S(\boldsymbol{Q}, \omega)$ which, *for monatomic systems*, does not depend on neutron–nuclear interaction, i.e., on the nuclear cross-sections. The scattering function and its relation with several other functions important for the description of scattering experiments will be discussed below. The details of the derivation can be found in standard text books [2–4].

Before we can consider these relations and the properties of scattering functions, we have to note a complication due to the fact that a nuclear species consists of *isotopes* with different scattering lengths $b_1, b_2, \ldots$ and concentrations $c_1, c_2, \ldots$. Therefore the intensity of scattered neutrons will in general have to be summed over terms with different scattering lengths, randomly distributed over the atomic sites $\boldsymbol{r}_i$. This randomness of the amplitudes destroys part of the interference one would observe due to neutron waves scattered by different nuclei, if they all had identical scattering lengths. A similar effect is caused by the *spin* of the nuclei and that of the neutron, because the scattering length depends on their relative orientations. This leads to scattering lengths b_+ and b_- corresponding to parallel and antiparallel orientation, with fractions $c_+ = (I+1)/(2I+1)$ and $c_- = I/(2I+1)$, respectively, where I is the nuclear spin. If nuclei and/or neutron spins are unpolarized, this gives a random distribution of b_+ and b_-. Randomness *destroys part of the interference* and for ideal disorder the cross-section can be separated into a coherent part with interference terms due to pairs of atoms (including the self-terms) and an incoherent part, where interference between waves scattered by different nuclei has completely cancelled out, such that the double-differential cross-section reads

$$\frac{\mathrm{d}^2\sigma}{\mathrm{d}\Omega\,\mathrm{d}\omega} = \frac{k}{k_0}\left[\frac{\sigma_{\mathrm{coh}}}{4\pi}S_{\mathrm{coh}}(\boldsymbol{Q},\omega) + \frac{\sigma_{\mathrm{inc}}}{4\pi}S_{\mathrm{inc}}(\boldsymbol{Q},\omega)\right] . \tag{15.4}$$

The coherent scattering function, $S_{\mathrm{coh}}(\boldsymbol{Q},\omega)$ in the first term, is due to the atom–atom pair-correlations, whereas the incoherent scattering function, $S_{\mathrm{inc}}(\boldsymbol{Q},\omega)$ in the second term,[3] merely conveys self-correlations of single atoms and, as a consequence, only intensities (and not amplitudes) from scattering by different nuclei have to be added.

One can easily show that the total scattering cross-sections σ_{coh} and σ_{inc} have the following meaning:

$$\sigma_{\mathrm{coh}} = 4\pi\bar{b}^2 \quad \text{with} \quad \bar{b} = \sum c_i b_i \ , \tag{15.5}$$

$$\sigma_{\mathrm{inc}} = 4\pi\left(\overline{b^2} - \bar{b}^2\right) \quad \text{with} \quad \overline{b^2} = \sum c_i b_i^2 \ . \tag{15.6}$$

In subsequent sections we will mainly deal with the *incoherent*, and occasionally with the *coherent* scattering function. The reason is, that the neutron scattering from native biological material is in most cases largely dominated by that of hydrogen atoms which are usually present in large numbers in organic molecules. For the hydrogen nucleus (the proton), the incoherent scattering cross-section is between 10 and 20 times larger than other scattering cross-sections, such that the separation of the incoherent scattering function

---

[3]Note that the following notations for the scattering functions (dynamic structure factors) are customary in the literature: $S(\boldsymbol{Q},\omega)$ or $S_{\mathrm{coh}}(\boldsymbol{Q},\omega)$ for coherent, and $S_{\mathrm{s}}(\boldsymbol{Q},\omega)$ or $S_{\mathrm{inc}}(\boldsymbol{Q},\omega)$ for incoherent scattering

$S_s(\boldsymbol{Q}, \omega)$ is especially easy. In the case of scattering-density fluctuations with sizeable correlation lengths, the coherent scattering contribution becomes significantly higher than the incoherent one in the small angle scattering range where the contrast between *the coherent scattering length densities* of different molecules or molecular subunits contributes as the square of the number of diffusing centers (see Sect. 15.5). The total coherent and incoherent (bound) scattering cross-sections are empirically known and can be found in tables [5]. The fundamental aspects of neutron–nucleus scattering are treated in an excellent review, see [6].

15.2.1 Van Hove Scattering Functions and Correlation Functions

Let us now consider the connection between scattering functions and static structure factors. The momentum-dependent and energy-dependent scattering function $S(\boldsymbol{Q}, \omega)$ was derived by Van Hove starting from the well-known static structure factor $S(\boldsymbol{Q})$, which is the energy-integrated scattering intensity (traditionally called "diffraction pattern"). For the simplest case of N *identical* atoms at positions $\boldsymbol{r}_0, \boldsymbol{r}_1, \ldots, \boldsymbol{r}_N$, this is proportional to the square of the sum of amplitudes, where the phase differences between the waves scattered by pairs of particles (i, j), located at instantaneous positions $(\boldsymbol{r}_i, \boldsymbol{r}_j)$ have been taken into account. The structure factor per atom is then given by

$$S(\boldsymbol{Q}) = N^{-1} \sum_{i=1}^{N} \sum_{j=1}^{N} \mathrm{e}^{-\mathrm{i}\boldsymbol{Q}(\boldsymbol{r}_i - \boldsymbol{r}_j)} = N^{-1} \left| \sum_{i=1}^{N} \mathrm{e}^{\mathrm{i}\boldsymbol{Q}\boldsymbol{r}_i} \right|^2 . \tag{15.7}$$

If, however, with the aid of energy analysis in a scattering experiment, the motion of nuclei (connected e.g., with molecular vibrations and/or diffusion) is observed, the structure factor becomes time-dependent: one gets the so-called intermediate scattering function (or intermediate dynamic structure factor) with time-dependent space coordinates $\boldsymbol{r}(t)$

$$I(\boldsymbol{Q}, t) = N^{-1} \sum_{i=1}^{N} \sum_{j=1}^{N} \left\langle \mathrm{e}^{-\mathrm{i}\boldsymbol{Q}\boldsymbol{r}_i(0)} \mathrm{e}^{\mathrm{i}\boldsymbol{Q}\boldsymbol{r}_j(t)} \right\rangle . \tag{15.8}$$

In the most general (i.e., the quantum-mechanical) case, $\boldsymbol{r}_i$ are operators, and $\langle \cdots \rangle$ is a thermal average of the expectation value for the product enclosed in the brackets. Under classical conditions, i.e., at sufficiently low energies and sufficiently high temperatures, which is usually fulfilled in the context of quasielastic neutron scattering on biological samples in a physiological environment, the quantities $\boldsymbol{r}$ are vectors in space and not operators, and $\langle \cdots \rangle$ is simply the thermal average. If the scattering is incoherent, we need only the self-terms in Eq. 15.8 which leads to

$$I_s(\boldsymbol{Q}, t) = N^{-1} \sum_{i=1}^{N} \left\langle \mathrm{e}^{-\mathrm{i}\boldsymbol{Q}\boldsymbol{r}_i(0)} \mathrm{e}^{\mathrm{i}\boldsymbol{Q}\boldsymbol{r}_i(t)} \right\rangle . \tag{15.9}$$

Here, since all the atoms are assumed to be identical, Eq. 15.9 can be simplified: the index i may be omitted and the sum $\sum$ replaced by N. From these relations Van Hove [1] derived the following expression for the coherent scattering function $S(\boldsymbol{Q},\omega)$:

$$S(\boldsymbol{Q},\omega) = (2\pi)^{-1} \int_{-\infty}^{+\infty} I(\boldsymbol{Q},t)\mathrm{e}^{-\mathrm{i}\omega t}\mathrm{d}t \; . \qquad (15.10)$$

Similarly, for its self-part, namely the so-called "incoherent" contribution (see Eq. 15.11), he obtained

$$S_\mathrm{s}(\boldsymbol{Q},\omega) = (2\pi)^{-1} \int_{-\infty}^{+\infty} I_\mathrm{s}(\boldsymbol{Q},t)\mathrm{e}^{-\mathrm{i}\omega t}\mathrm{d}t \; . \qquad (15.11)$$

Furthermore, he defined the correlation functions,[4] namely

$$G(\boldsymbol{r},t) = (2\pi)^{-3} \int_{-\infty}^{+\infty}\int_{-\infty}^{+\infty} \mathrm{e}^{-\mathrm{i}(\boldsymbol{Qr}-\omega t)} S(\boldsymbol{Q},\omega)\,\mathrm{d}\boldsymbol{Q}\,\mathrm{d}\omega \; , \qquad (15.12)$$

and for the self-part

$$G_\mathrm{s}(\boldsymbol{r},t) = (2\pi)^{-3} \int_{-\infty}^{+\infty}\int_{-\infty}^{+\infty} \mathrm{e}^{-\mathrm{i}(\boldsymbol{Qr}-\omega t)} S_\mathrm{s}(\boldsymbol{Q},\omega)\,\mathrm{d}\boldsymbol{Q}\,\mathrm{d}\omega \; . \qquad (15.13)$$

Finally, by inversion the scattering functions are expressed as time Fourier transforms of these correlation functions, i.e.,

$$S(\boldsymbol{Q},\omega) = (2\pi)^{-1} \int_{-\infty}^{+\infty}\int_{-\infty}^{+\infty} G(\boldsymbol{r},t)\mathrm{e}^{\mathrm{i}(\boldsymbol{Qr}-\omega t)}\mathrm{d}\boldsymbol{r}\,\mathrm{d}t \; , \qquad (15.14)$$

$$S_\mathrm{s}(\boldsymbol{Q},\omega) = (2\pi)^{-1} \int_{-\infty}^{+\infty}\int_{-\infty}^{+\infty} G_\mathrm{s}(\boldsymbol{r},t)\mathrm{e}^{\mathrm{i}(\boldsymbol{Qr}-\omega t)}\mathrm{d}\boldsymbol{r}\,\mathrm{d}t \; . \qquad (15.15)$$

The interpretation of Eqs. 15.8–15.13 is straightforward in the *classical approximation.* The interference of neutron waves scattered by *pairs of different atoms* at positions $\boldsymbol{r}_i$ and $\boldsymbol{r}_j$ at different times 0 and t, respectively, is taken into account by the cross terms with $i \neq j$ in $I(\boldsymbol{Q},t)$. $I_\mathrm{s}(\boldsymbol{Q},t)$, however, does not contain such cross terms, because it is only due to the interference of waves scattered by the *same* nucleus which is in general located at different positions for different times. The classical meaning of the Van Hove correlation

[4] Now called Van Hove correlation functions after this author

functions can therefore be described as follows: $G(\boldsymbol{r}_j - \boldsymbol{r}_i, t)$ is the conditional probability per unit volume to find an atom (nucleus) at a position $\boldsymbol{r}_j$ at time t, if this or another atom has been at a position $\boldsymbol{r}_i$, with a distance vector $\boldsymbol{r} = \boldsymbol{r}_j - \boldsymbol{r}_i$, at a previous time $t = 0$. Analogously, the self-correlation function, $G_\mathrm{s}(\boldsymbol{r}, t)$, is the conditional probability per unit volume to find an atom at $\boldsymbol{r}(t)$ at time t, if the same atom has been at the origin $\boldsymbol{r} = 0$ at $t = 0$.

When using this classical interpretation, one should however not forget its limits. The conditions for the validity of the classical approximation are, that the amounts of energy and momentum exchanged in the scattering process remain sufficiently small to fulfill the following relations with the thermal energy:

$$|\hbar\omega| \ll \frac{1}{2}k_\mathrm{B}T, \tag{15.16}$$

$$\frac{(\hbar Q)^2}{2M} \ll \frac{1}{2}k_\mathrm{B}T\,, \tag{15.17}$$

where $k_\mathrm{B}T$ = Boltzmann's constant and M = atomic mass. Therefore quantum effects are expected for large $\boldsymbol{Q}$ and large ω (or for small $\boldsymbol{r}$ and small t). In the realm of quasielastic neutron scattering concerned with ranges of fairly small Q and ω, the scattering functions may nevertheless be calculated classically, if they are then corrected by the so-called detailed-balance factor, $\exp(-\hbar\omega/2k_\mathrm{B}T)$. Because of the energy dependence of level occupation according to the Boltzmann distribution, the exact (quantum-mechanical) scattering functions for energy gain and energy loss are always related in the following way:

$$S(-\boldsymbol{Q}, -\omega) = \exp\left(+\hbar\omega/k_\mathrm{B}T\right)S(\boldsymbol{Q}, \omega). \tag{15.18}$$

This asymmetry with respect to $\hbar\omega = 0$ distinguishes $S(\boldsymbol{Q}, \omega)$ from the classical function, $S^\mathrm{cl}(\boldsymbol{Q}, \omega)$, which is symmetric in ω. The symmetric function obtained, if both sides of Eq. 15.18 are multiplied by the detailed-balance factor, is a very good approximation of $S^{cl}(\boldsymbol{Q}, \omega)$; from this we obtain the true $S(\boldsymbol{Q}, \omega)$:

$$S(\boldsymbol{Q}, \omega) = \exp\left(-\hbar\omega/2k_\mathrm{B}T\right)S^\mathrm{cl}(\boldsymbol{Q}, \omega) \tag{15.19}$$

and analogously for incoherent scattering:

$$S_s(\boldsymbol{Q}, \omega) = \exp\left(-\hbar\omega/2k_\mathrm{B}T\right)S_s^\mathrm{cl}(\boldsymbol{Q}, \omega). \tag{15.20}$$

Since in this chapter, we will essentially deal only with classical scattering functions, the superscript "cl" for classical functions will be omitted.

The following special cases of Van Hove's correlation functions are of particular interest. At $t = 0$, we have

$$G_\mathrm{s}(\boldsymbol{r}, t = 0) = \delta(\boldsymbol{r}) \quad \text{and} \quad G(\boldsymbol{r}, t = 0) = \delta(\boldsymbol{r}) + g(\boldsymbol{r})\ , \tag{15.21}$$

where $g(\boldsymbol{r})$ is the instantaneous pair correlation function accessible through diffraction experiments. From this one gets

$$S(\boldsymbol{Q}) = \int_{-\infty}^{+\infty} S(\boldsymbol{Q},\omega)\mathrm{d}\omega = 1 + \int_{-\infty}^{+\infty} g(\boldsymbol{r})\mathrm{e}^{\mathrm{i}\boldsymbol{Q}\boldsymbol{r}}\mathrm{d}\boldsymbol{r} \; = I(\boldsymbol{Q}, t=0). \tag{15.22}$$

Furthermore, we note that at $\boldsymbol{r} = 0$, $G_\mathrm{s}(\boldsymbol{r},t)$ is the probability that a certain nucleus which was at $\boldsymbol{r} = 0$ for $t = 0$, is still (or again) at $\boldsymbol{r} = 0$ for a time t.

15.2.2 The Elastic Incoherent Structure Factor

Another important special case concerns the behavior of the correlation functions at very long times. For an atom diffusing in a space which is very large as compared to the interatomic distances, the self-correlation function $G_\mathrm{s}(\boldsymbol{r},t)$ vanishes, if t goes to infinity, whereas, for an atom *bound to a finite volume* (e.g., as part of a rotating molecule fixed in a crystal), $G_\mathrm{s}(\boldsymbol{r},t)$ approaches a finite value $G_\mathrm{s}(\boldsymbol{r},\infty)$ for $\boldsymbol{r}$ varying in the interior of this volume. In fact, very generally, the self-correlation function can be split into its asymptotic value in the long-time limit and a time-dependent term $G'_\mathrm{s}(\boldsymbol{r},t)$, according to

$$G_\mathrm{s}(\boldsymbol{r},t) = G_\mathrm{s}(\boldsymbol{r},\infty) + G'_\mathrm{s}(\boldsymbol{r},t). \tag{15.23}$$

The Fourier transform of this expression reads

$$S_\mathrm{s}(\boldsymbol{Q},\omega) = (2\pi)^{-1} \int_{-\infty}^{+\infty}\int_{-\infty}^{+\infty} \mathrm{e}^{\mathrm{i}(\boldsymbol{Q}\boldsymbol{r}-\omega t)}[G_\mathrm{s}(\boldsymbol{r},\infty) + G'_\mathrm{s}(\boldsymbol{r},t)]\mathrm{d}\boldsymbol{r}\,\mathrm{d}t \;, \tag{15.24}$$

which gives

$$S_\mathrm{s}(\boldsymbol{Q},\omega) = S_\mathrm{s}^\mathrm{el}(\boldsymbol{Q})\delta(\omega) + S_\mathrm{s}^\mathrm{in}(\boldsymbol{Q},\omega). \tag{15.25}$$

It is seen that the incoherent scattering function is decomposed into a purely elastic line, $S_\mathrm{s}^\mathrm{el}(\boldsymbol{Q})\delta(\omega)$, with the integrated intensity $S_\mathrm{s}^\mathrm{el}(\boldsymbol{Q})$, and a nonelastic component, $S_\mathrm{s}^\mathrm{in}(\boldsymbol{Q},\omega)$. The elastic line is the result of diffraction of the neutron on the "infinite time" *distribution in space of a single nucleus spread over a finite volume by its motion*, as already pointed out by Stiller [7]. Therefore, we can derive information about the structure in a very direct way from *incoherent* scattering [8]. This is clearly a rather important result of the Van Hove theory. We now turn to its application which will be further discussed later, in the context of practical examples in the subsequent Sections.

In order to systematically exploit the theoretical fact expressed by Eq. 15.25 in neutron scattering experiments, the concept of the *elastic incoherent structure factor* (EISF) was formulated by Lechner in 1971 (for reviews see [9–11]). The EISF concept provides a method, that permits the extraction of structural information on localized single-particle motions by the determination of the elastic fraction of the measured spectral intensity. The idea is simple: First, by employing sufficiently high energy-resolution, the measured integrals of elastic (I^{el}) and nonelastic (I^{in}) components of the scattering function in Eq. 15.25 – after trivial corrections for the factor k/k_0 (see Eq. 15.3), the sample self-attenuation and the energy-dependent detector efficiency – are determined separately. Then an intensity ratio involving the two integrals can be defined,

$$\mathrm{EISF} = \frac{I^{\mathrm{el}}}{(I^{\mathrm{el}} + I^{\mathrm{in}})} = \frac{A S_{\mathrm{s}}^{\mathrm{el}}(\boldsymbol{Q})}{\left[A \int\limits_{-\infty}^{+\infty} S_{\mathrm{s}}(\boldsymbol{Q}, \omega)\mathrm{d}\omega\right]}, \tag{15.26}$$

where A is a normalization factor proportional to experiment parameters such as the incident neutron flux, the sample size, the detector efficiency, the duration of the measurement, etc. Obviously, the difficulty of an absolute intensity calibration, is avoided in the determination of the EISF by Eq. 15.26: The normalization factor cancels, the integral of the incoherent scattering function is equal to 1 by definition, and we simply have EISF $= S_{\mathrm{s}}^{\mathrm{el}}(\boldsymbol{Q})$. Here, we have used the relations 15.25 and 15.26 as a starting point for obtaining a definition of the EISF. The coefficient $S_{\mathrm{s}}^{\mathrm{el}}(\boldsymbol{Q})$ in these two equations is the EISF in its most general form, since it includes *all the motions* of the scattering atom.

However, the determination of this "global" EISF is generally not the immediate aim of an experiment, for the following reasons. First of all, one is often more interested in specific types of motions, than in all of them. Secondly, an unambigous measurement of a global EISF is not easily achieved in just one single experiment. Every measurement has a well defined energy resolution connected with an effective energy transfer window (see the corresponding discussion in Sect. 15.2.3). Essentially only the dynamics of the specific motions occurring in this energy range are visible, because much slower motions are hidden within the energy resolution function, whereas much faster motions appear only as a flat background. This can be used to experimentally isolate the effect of a specific motion. It is therefore much more interesting to apply the EISF concept to the elastic component of each specific type of motion, e.g., to molecular rotation, rather than to the combined effect of all atomic motions. The expression to be used for the analysis of such specific motions is formally the same as Eq. 15.26. However, we have to replace the nonelastic integral (I^{in}) by the corresponding quasielastic integral (I^{qe}), and keep in mind that the incoherent scattering function, $S_{\mathrm{s}}(\boldsymbol{Q}, \omega)$, must now be replaced by a partial incoherent scattering function, $S_{\mathrm{s}}^{\mathrm{qe}}(\boldsymbol{Q}, \omega)$, consisting merely of an elastic term (measured integral: I^{el}) and a quasielastic term (measured integral: I^{qe}) which correspond to the specific motion under study. The EISF then reads:

$$\mathrm{EISF} = \frac{I^{\mathrm{el}}}{(I^{\mathrm{el}} + I^{\mathrm{qe}})} = \frac{A S_{\mathrm{s}}^{\mathrm{el}}(\boldsymbol{Q})}{\left[A \int\limits_{-\infty}^{+\infty} S_{\mathrm{s}}^{\mathrm{qe}}(\boldsymbol{Q}, \omega)\mathrm{d}\omega\right]}. \tag{15.27}$$

Here, the effect of faster motions has been subtracted as a flat "inelastic" background and only appears as an attenuating Debye–Waller factor bound to be cancelled, because it is included in the normalization factor A. Note that $S_{\rm s}^{\rm qe}(\boldsymbol{Q},\omega)$ by definition has the same normalization as $S_{\rm s}(\boldsymbol{Q},\omega)$. The feasibility conditions are that (i) the energy resolution is adapted to the time scale of the motion of interest, (ii) this motion is sufficiently well separated on the energy-scale from other motions of the same atom, and (iii) the assumption of dynamical independence of the different modes from each-other, e.g. rotations and vibrations, represents an acceptable approximation (see Eqs. 16.1 and 16.2 in Sect. 16.2.1 of Part II in this volume).

It is this possibility of isolating the EISF of *specific modes of motions*, that has proved of most practical importance for the application of the technique. This isolation, of course, means separating the elastic from the quasielastic component of the QENS spectrum. Due to an important sum rule concerning the incoherent scattering function, namely the property that its integral is equal to unity, we have

$$\mathrm{EISF}(\boldsymbol{Q}) + \mathrm{QISF}(\boldsymbol{Q}) = 1, \tag{15.28}$$

where QISF is the quasielastic incoherent structure factor. The latter is the $\boldsymbol{Q}$-dependent spectral weight of the quasielastic component. It obviously contains the same structural information as the EISF, and is sometimes used instead of the latter, for technical reasons (e.g., in case of Bragg contamination of the EISF).

The EISF method represents a strategy for finding the appropriate differential equations and their boundary conditions for the dynamical mechanisms of localized atomic motions in condensed matter. In principle, this method can also be applied to pertinent problems concerning biological systems. This is demonstrated by Fig. 15.1 showing typical EISF curves which are the signatures of specific localized atomic motions and their different geometries in space. The curves have been calculated from the expressions given for various models of motion in Table 15.1. The models are characterized by the symmetry of the motion, its spatial extension represented by the radius R, and the orientation of the momentum transfer vector $\boldsymbol{Q}$ with respect to the atomic displacement vectors. Note that two of these models, "no component of motion along the vector $\boldsymbol{Q}$" (*non* $\parallel$ $\boldsymbol{Q}$) and "continuous diffusion on a circle" (cont. C) [12], are anisotropic, whereas two of them, "continuous diffusion on a spherical surface" (sph. S) [13] and "continuous diffusion in the interior of a spherical volume" (sph. V) [14], are intrinsically isotropic. The three jump model curves shown are presented in a form which is isotropic due to orientational averaging. The full theoretical expressions for these and other models can be found in [4,9].

From the shape of the EISF as a function of the dimensionless parameter QR (Q = momentum transfer, R = radius of rotation or radius of the spherical volume of diffusion, respectively), the mechanism of the concerned motion can be recognized. Early and subsequent experiments have been reviewed

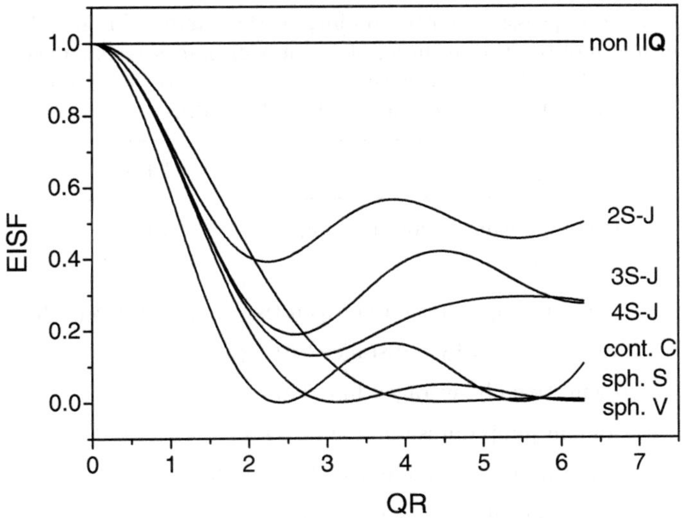

Fig. 15.1. Elastic incoherent structure factor (EISF) for various models of localized diffusive (e.g., rotational) motion of small side-groups of biological macromolecules, or of small solute molecules or even water molecules in an aqueous solution. The corresponding theoretical expressions are given in Table 15.1; the abbreviations have the following meaning: non $\parallel \boldsymbol{Q}$ = no component of motion along the vector $\boldsymbol{Q}$; $2S$–J, random jump diffusion between two sites separated by a distance $2R$; $3S$–J, random jump diffusion between three equidistant sites on a circle of radius R; $4S$–J, random jump diffusion between four equidistant sites on a circle of radius R; cont. C, continuous diffusion on a circle with radius R, with the $\boldsymbol{Q}$ vector in the plane of the circle; sph. S, continuous diffusion on a spherical surface with radius R; sph. V, continuous diffusion in the interior of a spherical volume with radius R. Note that two of these models, non $\parallel \boldsymbol{Q}$ and cont. C, are anisotropic, whereas two, sph. S and sph. V, are intrinsically isotropic. The three jump models are presented in an isotropic form obtained by orientational averaging. The full theoretical expressions for these and other models can be found in [4, 9]

extensively [9, 10, 15]. In [10], the importance of the dynamic independence approximation for the definition of the EISF of specific motions, the relation between Debye–Waller factor, Lamb–Mößbauer factor and EISF, as well as the observation-time dependence of the latter are discussed in detail. In biology-related studies, the method has been applied for instance to isolate the effects of the motion of small side-groups from the total scattering function of a protein containing membrane [16]. More general discussions of Van Hove's theory and its application can be found in text books; see for instance [2, 4, 17].

15.2.3 Experimental Energy Resolution

Let us now turn to the problem of experimental resolution. Eqs. 15.14 and 15.15 represent a Fourier analysis of the neutron scattering functions $S(\boldsymbol{Q}, \omega)$

Table 15.1. EISF expressions corresponding to the curves of Fig. 15.1 for several models of atomic motion; the meaning of the model names is explained. $[J_0(QR)]$ = Bessel function of the first kind, with integer order; $[j_n(QR)]$ = spherical Bessel function of the first kind, with fractional order; R = radius

model name	meaning :	EISF
non $\parallel \boldsymbol{Q}$	no component of motion along the vector $\boldsymbol{Q}$	1.0
$2S$–J	random jump diffusion between two sites separated by a distance $2R$; orientationally averaged function	$\frac{1}{2}[1 + j_0(2QR)]$
$3S$–J	random jump diffusion between three equidistant sites on a circle of radius R; orientationally averaged function	$\frac{1}{3}[1 + 2j_0(QR\sqrt{3})]$
$4S$–J	random jump diffusion between four equidistant sites on a circle of radius R; orientationally averaged function	$\frac{1}{4}[1 + j_0(2QR)$ $+2j_0(QR\sqrt{2})]$
cont. C	continuous diffusion on a circle with radius R, with $\boldsymbol{Q} \parallel$ plane of circle [12]	$[J_0(QR)]^2$
sph. S	continuous diffusion on a spherical surface with radius R [13]	$[j_0(QR)]^2$
sph. V	continuous diffusion in the interior of a spherical volume with radius R [14]	$[3j_1(QR)/QR]^2$

and $S_s(\boldsymbol{Q},\omega)$, respectively, with the Van Hove correlation functions in space and time, $G(\boldsymbol{r},t)$ and $G_s(\boldsymbol{r},t)$, as coefficients. We draw the attention to the fact that the scattering functions S and S_s defined in Eqs. 15.4–15.15, as well as the corresponding correlation functions G, G_s, and the intermediate scattering functions I, I_s, cannot be determined experimentally in their pure forms: For instance, the measured scattering functions are broadened due to convolution with the experimental resolution functions $R(\boldsymbol{Q},\omega)$ in the four-dimensional $(\boldsymbol{Q},\omega)$–space. In the case of incoherent scattering, the $\boldsymbol{Q}$–spread of the resolution can often be neglected, when the studied functions are only slowly varying with $\boldsymbol{Q}$. Then it is sufficient to "deconvolute" (see Sect. 15.5) the measured resolution-broadened "scattering function",

$$[S_s(\boldsymbol{Q},\omega)]_{\text{meas}} = \int_{-\infty}^{+\infty} S_s(\boldsymbol{Q},\omega')R(\omega-\omega')\mathrm{d}\omega' \tag{15.29}$$

merely from the energy resolution function $R(\omega)$. The latter may have a shape close to a Gaussian or a Lorentzian, with an energy width $\Delta(\hbar\omega)$ defined as the half-width at half maximum (HWHM)[5]. Note that this width is connected, by the uncertainty relation, with the experimental *observation time* Δt_{obs} which is the decay time of the observation function $R^*(t)$ in the Fourier time domain [10, 15, 18]:

$$\Delta(\hbar\omega)\Delta t_{\text{obs}} \cong \hbar \tag{15.30}$$

While the resolution function $R(\omega)$ has the effect of broadening the neutron scattering function along the energy transfer coordinate of the experiment, the observation function $R^*(t)$ is a factor, which increasingly attenuates the Van Hove correlation function, as the Fourier time increases: $R^*(t)$ is the Fourier transform of $R(\omega)$ and is therefore in most practical cases a function essentially decaying with increasing time. The net effect is, that the correlation functions are observed in a Fourier time window, with an upper limit controlled by the decay time-constant of the observation function. The low-time limit of this window has a different origin: For instruments working in (Q, ω)-space, it is mainly a consequence of the (always limited) statistical accuracy of the measurement, because quasielastic intensities typically decrease with increasing energy transfer, and therefore counting statistics become the poorer the larger the energy transfers are as compared to the energy-resolution width.

Let us consider an example, in order to illustrate the implications of experimental resolution in the study of dynamic structure in the framework of the Van Hove formalism (Eqs. 15.7–15.15): We assume, for simplicity, that the scattering particle carries out a random motion described by a superposition of several components with n different rates, τ_1^{-1}, τ_2^{-1},..., τ_n^{-1}. The scattering function will then be a sum of Lorentzians centered at zero energy transfer (see the following sections). If this quasielastic spectrum is studied with an instrument resolution $\Delta(\hbar\omega)$, the resulting resolution-broadened spectrum will have a width larger than $\Delta(\hbar\omega)$. This "quasielastic peak" will be dominated by contributions from those motions which have rates $\tau_i^{-1} \cong \Delta(\hbar\omega)$. While much slower motions are hidden within the resolution function, much faster motions will produce only a flat "background" which cannot be easily distinguished from the usual constant background of the experiment. In order to be able to extract information on all relevant motional components, one needs to carry out several measurements with properly chosen resolutions. This procedure may in practice require the application of more than one type of spectrometer. Quasielastic neutron scattering spectra obtained with one single energy resolution only, usually furnish incomplete information. In order to avoid wrong conclusions, it is typically necessary to employ at least three different energy resolutions in the study of a given problem. Figure 15.2 shows

[5]For practical reasons, we prefer here to define the resolution by its HWHM; note, however, that in the literature the resolution width is often represented by its full-width at half maximum (FWHM)

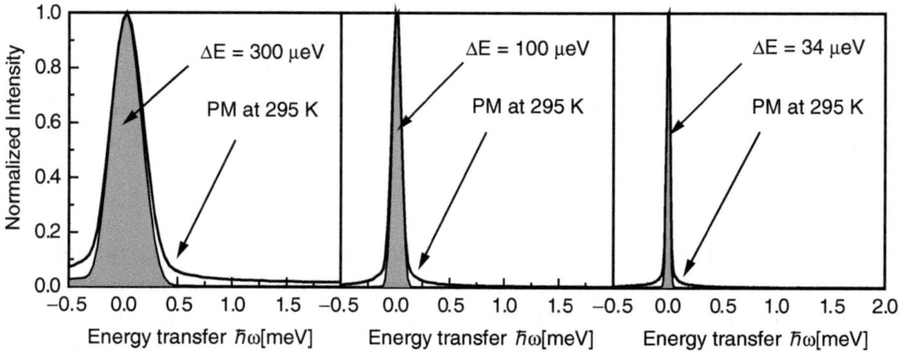

Fig. 15.2. Purple membrane spectra measured with three different energy resolutions (300, 100, and 34 μeV (FWHM)) using incident wavelengths $\lambda = 4.54$, 5.1, and 6.2 Å, respectively. The shaded spectra represent the resolution function obtained from a vanadium standard sample. In spite of identical dynamics the three PM spectra show strong quasielastic components with rather different apparent linewidths. This is of course due to the different observation times (i.e., different energy resolutions) employed, emphasizing dynamical aspects on different time scales of the system under study. Sample: stacks of purple membrane equilibrated at 98% relative humidity (D_2O), at room temperature. Illuminated sample size: $(30 \times 60)\,\mathrm{mm}^2$. Measurement times: 3 h ($\Delta E = 300\,\mu\mathrm{eV}$), 7 h ($\Delta E = 100\,\mu\mathrm{eV}$) and 14.4 h ($\Delta E = 34\,\mu\mathrm{eV}$), respectively. Spectra measured by J. Fitter and R.E. Lechner with the multichopper time-of-flight spectrometer NEAT (see Sect. 15.3.2). Figure from [19]

as an example three spectra from a study of purple membrane [19], which demonstrate the qualitative similarity, but quantitative difference of the spectra in such a series of measurements.

Since $\Delta(\hbar\omega)$ is related in a simple way to the instrumental energy spreads of incident and scattered neutrons, the observation time Δt_{obs} is connected with (although not equal to) the coherence time of the incident neutron wave packet. In spin–echo experiments, where *intermediate* scattering functions are measured (as a function of t), the resolution problem requires a different treatment. Here, instead of the necessity to fold scattering functions with energy resolution functions, the correction for this resolution effect essentially reduces to dividing the measured spectra by the experimental observation function $R^*(t)$.

The principle of experimental observation time, energy and Fourier time windows in quasielastic neutron scattering, and their relevance for the determination of dynamic structure, and especially in problems concerning diffusive atomic and molecular motions in condensed matter, has been discussed extensively in [10, 18]. For further detailed literature related to the Van Hove concept and quasielastic neutron scattering, we refer to the reviews, monographs, and books especially devoted to this topic [4, 9–11, 17, 20–23].

A general review of diffusion studies employing quasielastic neutron scattering techniques is given in [24].

15.3 Instruments for QENS Spectroscopy in $(\boldsymbol{Q}, \omega)$-Space

The role of a neutron scattering spectrometer is to monitor neutron intensity as a function of the wavevector $\boldsymbol{Q}$ (momentum $\hbar\boldsymbol{Q}$) and the energy $\hbar\omega$ exchanged with the sample. If we except the special case of Neutron Spin-Echo spectrometers, the wavelength, velocity or energy of neutrons, or their distribution as a function of these variables, have to be defined before, and analyzed after scattering in the sample. Several techniques are being used: *Crystal Bragg reflection* (XTL) for wavelength definition as a function of the reflection angle, neutron time-of-flight (TOF) selection and measurement using a pulsed incident beam, wavelength-band selection employing a continuous neutron velocity selector. In most of the quasielastic neutron scattering experiments, the scattering function $S(\boldsymbol{Q}, \omega)$ is directly measured by TOF spectrometry with resolutions from about 1 µeV to a few 1000 µeV, or by *backscattering* (BSC) spectroscopy, with resolutions of the order of 0.1–20 µeV. This allows to cover, by QENS, a range of diffusion coefficients between 10^{-12} and $10^{-8}\,\mathrm{m^2s^{-1}}$, or, correspondingly, of characteristic times from 10^{-9} s to 10^{-13} s. Last-not-least, *neutron spin echo* (NSE) spectrometers employ neutron polarization, together with polarization analysis to define and determine the phase of neutron spins precessing in magnetic fields. The NSE method permits the direct determination of the intermediate scattering function, $I(\boldsymbol{Q}, t)$, instead of $S(\boldsymbol{Q}, \omega)$. This technique extends the Fourier time scale up to 10^{-7} s, corresponding to an energy resolution limit in the neV region, where diffusion coefficients of the order of $10^{-13}\,\mathrm{m^2s^{-1}}$ can be measured. This will be discussed in Sect. 15.4. The TOF and BSC methods will be explained in the following.

15.3.1 XTL–TOF Spectrometers

There are different techniques of neutron time-of-flight spectrometry. Basically, XTL–TOF spectrometers[6] [25] use a crystal monochromator to create a continuous monochromatic beam, i.e., the incident neutron beam is monochromatized by Bragg reflection. For a given Bragg angle Θ and a corresponding reciprocal lattice vector $\boldsymbol{G}$, the monochromator selects a certain neutron wave number $k_0 = mv_0/\hbar$, following the Bragg equation:

$$|\boldsymbol{G}| = 2k_0 \sin\Theta \,. \tag{15.31}$$

The monochromatic beam is then periodically chopped by a disk- or Fermi-chopper, before it hits the sample. The energy distribution of the scattered

[6] XTL stands for crystal and TOF for time-of-flight

neutrons is obtained by measuring their time-of-flight from the sample to the detectors which cover a large range of solid angle and consequently of Q-values. A more sophisticated version uses several crystal monochromators (located at slightly different positions and with slightly different orientations on the neutron guide) which reflect several different wavelengths selected. Due to the correlation between the monochromator reflection angle and the wavelength of reflected neutrons (and there velocities), the Fermi-chopper consecutively transmits neutron pulses from these different parts of the incident neutron beam, so that all these neutrons arrive on the detector at the same time. This is the time-focusing principle. The prototype of this spectrometer is IN6 [26–28] at the ILL.

A more recent version of this instrument type is the spectrometer FOCUS at PSI [29] shown by the schematic representation in Fig. 15.3. This spectrometer uses a monochromator covering a large beam area, which is composed of several tens of crystal pieces with horizontal and vertical focusing (variable radius of curvature). The distance between the guide exit and the monochromator can be varied in order to achieve either a high intensity-low resolution mode or one with lower intensity, but higher resolution. This type of time-of-flight instrument is characterized by five main parameters [30]: the lattice spacing d of the crystals, the monochromator Bragg angle θ_{M}, the width W

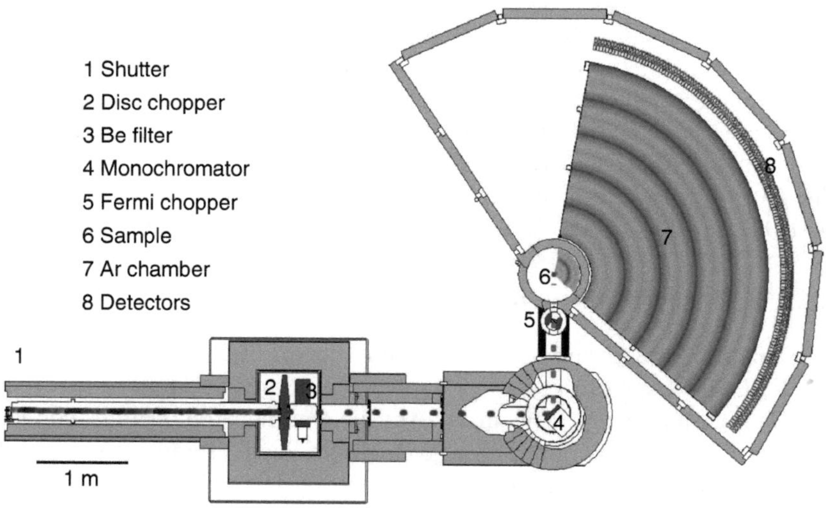

Fig. 15.3. FOCUS spectrometer at Paul-Scherrer Institut (PSI) [29]; FOCUS is a typical XTL–TOF spectrometer, i.e., a time-of-flight instrument with a Bragg monochromator and a TOF analyzer. While the monochromator selects the incident neutron energy E_0, the energy of the scattered neutrons E is determined by measuring the neutron flight time

of the monochromator, the distances from the guide exit to the monochromator and from the monochromator to the sample, d_{GM} and d_{MS}, respectively. The width of the wavelength distribution obtained by this setup is essentially given by:

$$\Delta\lambda = d \sin(\theta_{\mathrm{M}})\, W \cos(\theta_{\mathrm{M}}) \left| \frac{1}{d_{\mathrm{GM}}} - \frac{1}{d_{\mathrm{MS}}} \right| . \qquad (15.32)$$

Further contributions to the energy resolution, i.e., the energy transfer uncertainty at the detector, are the mosaic spread $\Delta\theta_{\mathrm{M}}$ of the monochromator, the incident beam divergence, and the sample–detector time-of-flight spread due to finite thicknesses of sample and detectors. If these contributions are independent of each-other, they may be added quadratically.

When $d_{\mathrm{GM}} \simeq d_{\mathrm{MS}}$, the term given by Eq. 15.32 vanishes, and the highest possible resolution is achieved, but at the expense of beam intensity. On the other hand, the neutron intensity is maximized, when d_{GM} is chosen as small as possible, the resolution is then lower. Applications of the XTL–TOF technique, using IN6, are described in Sects. 16.4.4 and 16.5.2 of Part II in this volume.

15.3.2 TOF–TOF Spectrometers

Alternatively, in the case of a TOF–TOF spectrometer, both $v_0\,(k_0)$ and $v\,(k)$ are selected by time-of-flight using (at least) two choppers in front of the sample, with a mutual phase shift which determines v_0. For thermal neutrons, Fermi-choppers are often used, whereas disk-choppers are preferentially employed in the case of cold neutrons. A multidisk chopper time-of-flight (MTOF) instrument is illustrated schematically in Fig. 15.4 [31]. The two principal choppers, CH1 and CH2, the sample S and the detectors D are separated by the distances L_{12}, $L_{2\mathrm{S}}$, and L_{SD}, respectively, which have values of the order of several meters. CH1 and CH2 create neutron pulses with widths τ_{i} and τ_{ii}, and define the incident neutron wavelength λ_0 and its band width. More precisely, the phase difference between CH1 and CH2 allows the latter to select the "monochromatic" wavelength of the experiment. The scattering processes in the sample then cause neutron wavelength shifts to smaller or larger values of λ. In Fig. 15.5, a neutron flight-path diagram is shown. It explains the method in more detail and demonstrates the filter action of the various disks of the chopper cascade. While the neutron time-of-flight is measured along the horizontal axis, the vertical axis represents the flight-path between the different elements of the chopper cascade. The filter choppers, $\mathrm{CH_P}$ and $\mathrm{CH_R}$ perform pre-monochromatization and pulse-frequency reduction of the beam, respectively, in order to avoid frame-overlap at the monochromator disk CH2 and at the detectors. The spectra shown schematically on the top of the figure, correspond to a study of the rotational motion of OH^- ions [43] with the MTOF spectrometer IN5 at ILL, carried out in 1975 using 4 Å neutrons. The diagram also demonstrates the periodicity of the data acquisition procedure, inherent in a pulsed experiment. Each

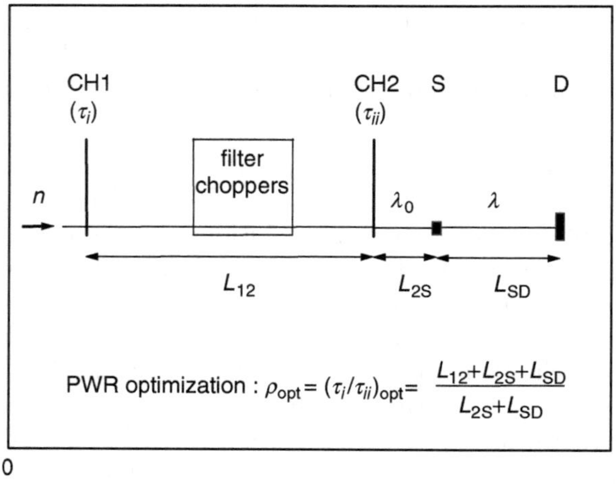

Fig. 15.4. Schematic sketch of a multidisk chopper time-of-flight (MTOF) spectrometer: CH1 and CH2 are the two principal choppers defining the monochromatic neutron pulse and its wavelength bandwidth; S = sample, D = detectors; L_{12}, L_{2S}, and L_{SD} are the distances between these elements of the instrument; τ_i and τ_{ii} are the widths of the pulses created by CH1 and CH2, λ_0, λ the incident and scattered neutron wavelenths (after [31]). *Inset*: the pulse-width ratio (PWR) optimization formula for elastic and quasielastic scattering [32]. Typical instruments of this type are IN5 [33–36] at ILL in Grenoble, MIBEMOL [37] at LLB in Saclay, both France; NEAT [38–41] at HMI in Berlin, Germany; and DCS [42] at NIST in Gaithersburg, USA

TOF period P_{spec} in principle contains one spectrum. But the duration of the measurement must cover a large number of such periods, the spectra of which (about 10^6 for a measurement time of 3 h) are added together, in order to obtain sufficient statistical accuracy. The pulse repetition rate P_{spec}^{-1} of the experiment is limited by the necessity of avoiding frame overlap, which means the superposition of the fastest neutrons (scattered with energy gain) within a time-of-flight period and the slowest neutrons (scattered with energy loss) from the previous period. This requires

$$P_{spec}\,[\mu s] = 252.78\; C\; \lambda_0\,[\text{Å}]\; L_{SD}\,[\text{m}], \tag{15.33}$$

where practical experimental units are indicated. The dimensionless constant C has to be chosen depending on the width of the quasielastic spectrum and the corresponding decay of its intensity on the low-energy side. C is usually in the range of $1.2 \le C \le 1.8$.

For a given incident neutron wavelength, the total intensity at the detectors is essentially governed by the factor $(\tau_i\tau_{ii})$, i.e., by the product of the two chopper opening times [32] (see also [44]). The latter also control the resolution, and thus intensity and resolution are connected through these

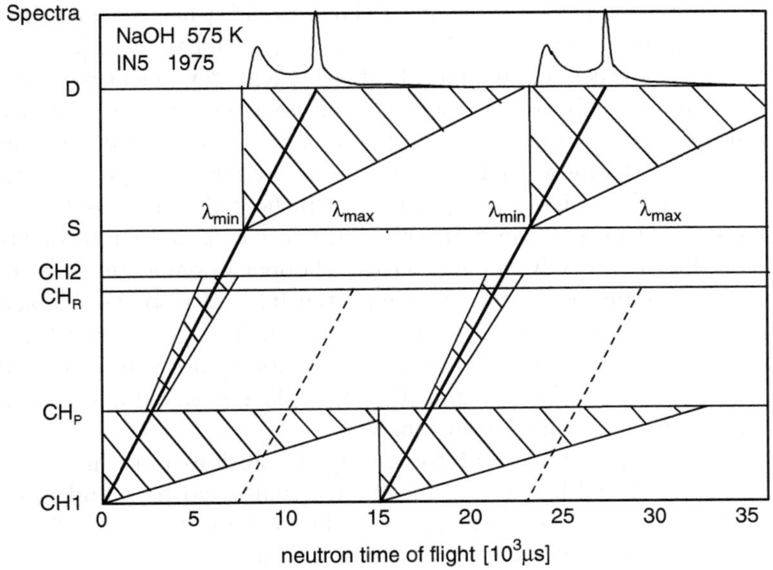

Fig. 15.5. Neutron flight-path diagram: It demonstrates the filter action of the various disks of the chopper cascade. The vertical axis represents the flight-path between the different elements of the chopper cascade. CH1 defines the initial time-distribution of the neutron pulse; CH_P is the pre-monochromator; CH_R is employed for pulse frequency reduction, in order to avoid frame-overlap at the detectors; finally, CH2 selects the "monochromatic" wavelength band for the experiment. The TOF spectra shown schematically on the top of the figure, correspond to a study of the rotational motion of OH^- ions [43] with the MTOF spectrometer IN5 at ILL, carried out in 1975 using 4 Å neutrons. Each spectrum covers one TOF period P_{spec}; see text for more details (from [31])

parameters. The most important and unique feature of this type of instrument is the capability of varying the energy resolution continuously over several orders of magnitude (see Sect. 15.2.3). We therefore give here explicitly an expression for $\Delta(\hbar\omega)$. The energy resolution width (HWHM) at the detector [32], i.e., the uncertainty in the experimentally determined energy transfer $\hbar\omega$, is given by

$$\Delta(\hbar\omega)\,[\mu eV] = 647.2(A^2 + B^2 + C^2)^{1/2}/(L_{12}L_{SD}\lambda^3)/2, \qquad (15.34)$$

where

$$A = 252.78\ \Delta L\ \lambda\ L_{12} \qquad (15.35)$$

$$B = \tau_i(L_{2S} + L_{SD}\lambda^3/\lambda_0^3) \qquad (15.36)$$

$$C = \tau_{ii}(L_{12} + L_{2S} + L_{SD}\lambda^3/\lambda_0^3) \qquad (15.37)$$

ΔL is the uncertainty of the length of the neutron flight path, which is mainly due to beam divergence, sample geometry, and detector thickness. The constant coefficients in Eqs. 15.34 and 15.35 are valid, when the quantities L_{12}, L_{2S}, L_{SD}, ΔL are given in [m], λ_0 and λ in [Å], τ_{i} and τ_{ii} in [μs].

It follows from these expressions, that the energy dependent resolution, for given λ_0, strongly depends on the scattered neutron wavelength λ, whereas the total intensity, as an integral property of the spectrometer, has no such dependence. Furthermore, high resolution is favored by short pulse widths and by large values of the distances L_{12} and L_{SD}. If these distances are fixed, and if sample geometry, λ_0 and energy transfer have been chosen, then total intensity and resolution-width only depend on the chopper opening times τ_{i} and τ_{ii}. Best instrument performance regarding intensity and resolution is achieved not only by selecting suitable values of the individual pulse widths, τ_{i} and τ_{ii}, but also requires the optimization of their ratio (pulse-width ratio (PWR) optimization [32, 38]). The optimization formula for elastic and quasielastic scattering is shown as an inset in Fig. 15.4.

The continuous variation of the energy resolution over three orders of magnitude is achieved by varying the chopper pulse widths τ_{i} and τ_{ii} (which, e.g., in the case of NEAT is possible by a factor between 1 and 40), and by choosing the incident wavelength (yielding another factor, of up to about 30 for the wavelength range from 4 to 12 Å). Applications of this technique using the MTOF spectrometer NEAT at HMI in Berlin, are described in Sects. 16.3.4 and 16.4.3 of Part II, this volume.

15.3.3 XTL–XTL Spectrometers

The energy-transfer regime in the μeV range is covered by *back scattering* (BSC) spectrometry [45–48], which was invented by H. Maier-Leibnitz. BSC-spectrometers are XTL–XTL instruments, i.e., they employ single-crystals as monochromators *and* as analyzers, with Bragg angles close to $\pi/2$ in both cases. For a given incident divergence of the beam, $\Delta\Theta$, the wave number spread produced by reflection from a crystal is given by differentiating Eq. 15.31

$$\frac{\Delta k_{\mathrm{div}}}{k} = \cot\Theta\,\Delta\Theta\ . \tag{15.38}$$

For typical Bragg angles and $\Delta k_{\mathrm{div}}/k \approx 10^{-2}$ rad one achieves an energy resolution in the percent range. However, for Θ approaching $\pi/2$, Δk_{div} from Eq. 15.38 goes to zero and we have to include the curvature of $\sin\Theta$ which leads to a second order contribution,

$$\frac{\Delta k_{\mathrm{div}}}{k} = \frac{(\Delta\Theta)^2}{8} \quad \text{for} \quad \Theta \to \frac{\pi}{2}\ . \tag{15.39}$$

This situation is called "backscattering" which means that the incident and the Bragg reflected beam are practically antiparallel. The square relation Eq. 15.39 replaces the linear relation between Δk_{div} and $\Delta\Theta$: The intensity is proportional to the incident solid angle $(\Delta\Theta)^2$, and we have $(\Delta\Theta)^2 \propto \Delta k_{\mathrm{div}}$

instead of $\Delta\Theta \propto \Delta k_{\text{div}}$, which is valid for Bragg angles other than $\pi/2$. This means that under these conditions resolution and intensity are *decoupled in first order.*

Actually, the wavevector spread is larger than Δk_{div}. Only a finite number of lattice planes contribute to the Bragg line, which causes a finite width of $\boldsymbol{G}$, the so-called Darwin or *extinction width* [49],

$$\frac{\Delta k_{\text{ex}}}{k} = \frac{16\pi N_{\text{c}} F_G}{G^2} , \tag{15.40}$$

where F_G is the structure factor for the Bragg reflection at $\boldsymbol{Q} = \boldsymbol{G}$, N_{c} is the number of lattice cells per unit volume. As an approximation, both contributions can be added such that

$$\frac{\Delta E_0}{E_0} = 2\left[\frac{(\Delta\Theta)^2}{8} + \frac{16 N_{\text{c}} F_G}{G^2}\right] . \tag{15.41}$$

For neutrons from an Ni neutron guide and reflection on an ideal silicon waver, one calculates $\Delta E_0 = (0.24{+}0.08)\,\mu\text{eV}$ as incident neutron energy spread. For the resolution in energy transfer, $\Delta(\hbar\omega)$, of a modern BSC-spectrometer, such as the BSC-spectrometer IN16 (Fig. 15.6) at the ILL high-flux reactor [47,48], one obtains values of 0.09, 0.2, and 0.43 μeV (HWHM), depending on the type of monochromator and analyzer crystals used [48]. So far such values have not been reached by any other kind of crystal spectrometer; they can in principle, however, be achieved also by high-resolution TOF–TOF instruments at future spallation sources [31].

The high energy resolution of IN16 is based on a Bragg angle fixed at 90°; the energy scan is performed by a Doppler drive, moving the monochromator crystal (spherical, perfect Si(111), $450 \times 250\,\text{mm}^2$; label **6** in Fig. 15.6) with a sinusoidally varying speed v_{D}. The resulting energy shift is [50]

$$\frac{\delta E_0}{E_0} = 2\frac{v_{\text{D}}}{v_0} \tag{15.42}$$

which yields an energy window of $\delta E_0 = \pm 14\,\mu\text{eV}$ for maximum speed values of $v_{\text{D}} = \pm 2.5\,\text{ms}^{-1}$ and $\lambda_0 = 6.27\,\text{Å}$. The various components of the instrument are arranged as follows: A double-deflector array (**1** and **5** in Fig. 15.6) selects the useful wavelength band from the cold-neutron guide. The first deflector (**1** in Fig. 15.6), a (broad-band) vertically focusing pyrolythic graphite crystal separates the neutrons to be used from the incident beam and reflects the whole energy-transfer range of about ±14 μeV, covered by the Doppler motion of the monochromator, into a NiTi supermirror guide (**2** in Fig. 15.6). The latter focuses these neutrons vertically and horizontally onto the second deflector (**5** in Fig. 15.6). A Be-filter and a background chopper (**3** and **4**, respectively, in Fig. 15.6) are located in a gap in the middle of this guide. The second deflector (label **5** in Fig. 15.6), made of PG(002) crystals with a wide mosaic, is mounted on a chopper wheel with alternating open and reflecting segments. It sends a neutron pulse towards the monochromator (label **6**

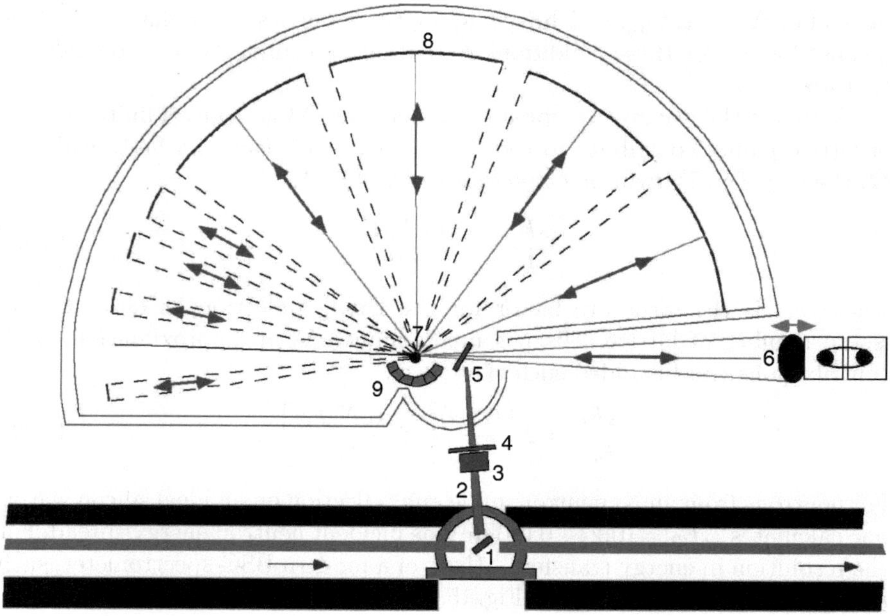

Fig. 15.6. BSC-spectrometer IN16 at the ILL high-flux reactor [47, 48], schematic view: **1** = first graphite deflector crystal; **2** = focusing supermirror guide; **3** = Beryllium-Filter; **4** = background chopper; **5** = stationary graphite-crystal deflector chopper; **6** = Doppler monochromator crystal; **7** = sample; **8** = spherically arranged analyzer crystal array; **9** = multitube detector array. The Bragg angles at the silicon single crystals (i.e., monochromator and analyzer, before and after scattering of the neutrons by the sample) are close to 90°. Other well-known spectrometers of this type are the BSC-spectrometer at the Jülich FRJ-2 reactor, IN10 [46] and IN13 at ILL in Grenoble, France, and HFBS [51] at NIST in Gaithersburg, USA

in Fig. 15.6). The monochromatic backscattered neutron pulse is transmitted through the open segments to the sample. Obviously the system is designed in such a way, that the necessary phase relations between Doppler drive, deflector chopper, and background chopper are observed. To avoid that the scattered neutrons are directly falling onto the detectors (before they have been filtered by the Si analyzers), the incident beam is periodically interrupted by the background chopper, in phase with the Doppler movement. Only when the beam is closed, the consecutively scattered and analyzed neutrons reach the detectors. This ensures that the useful neutrons are reaching the sample, the analyzers, and finally the detectors, while the background that would be caused by neutrons scattered without energy analysis from the sample directly into the detectors, is discriminated. The monochromatic neutrons from the oscillating monochromator crystal, falling onto the sample, being scattered, and finally detected, are individually labelled with the corresponding instantaneous speed of the Doppler drive. The neutrons scattered by the sample are backscattered by spherical shells of Si(111) crystals and thus focussed into a set of ^{3}He detectors. Each detector corresponds to a certain scattering angle or Q-value. When the energy range is too narrow, the reflected neutron energy can be additionally shifted by heating the monochromator, thus increasing the lattice parameter, and/or using monochromators whose lattice parameter is somewhat smaller or bigger than that of the silicon analyzer [52]. In this way, the range of the spectrometer (and also the resolution) can be adapted to the problems.

Sometimes, BSC-spectrometers are used with the Doppler drive *at rest*, i.e., in the so-called *elastic-window scan* mode of operation (see [53], [54] and [23] p. 284). In such a measurement one determines the intensity for the spectrometer set at $\omega = 0$, corresponding to the convolution integral $[S_s(\boldsymbol{Q}, \omega = 0)]_{\text{meas}} = (S_s \otimes R)$, where as an example, S_s is the incoherent neutron scattering

function and R is the energy resolution function (see Eq. 15.29). For scattering functions with Gaussian shape in reciprocal space, when they permit a time-independent atomic mean-square displacement to be defined (e.g., for harmonic vibrations), and when the nonelastic scattering contribution to the elastic channel is negligible, the $\boldsymbol{Q}$-dependence of the scattered intensity at zero energy transfer reads:

$$S_s(\boldsymbol{Q}, \omega = 0) = C \exp\left[- < u^2 > Q^2\right], \tag{15.43}$$

where C is a normalization factor, the exponential is the Debye–Waller factor, and u is the component of the atomic displacement along the vector $\boldsymbol{Q}$. This expression may also be employed for any spatially restricted isotropic motion, as long as $\boldsymbol{Q}$ is small enough (Gaussian approximation; see Sect. 16.3 in Part II of this chapter, in this volume). It has even been used for non-Gaussian probability density distributions, for instance in the context of the so-called "dynamical transition" [55–59] (see Sect. 16.5.2 in Part II ; see also the article by Lehnert and Weik in this volume). This is justified, as long as one keeps in mind, that in such cases the quantity $< u^2 >$ becomes a phenomenological parameter with a less precise meaning than in the harmonic-vibration case. This parameter can be used for the qualitative study of the effects due to the variation of external variables, such as the temperature.

Let us now consider a case, where the scattering function has an appreciable quasielastic component, with a temperature-dependent width of the same order of magnitude as the energy resolution of the instrument. For instance, for a Lorentzian-shaped spectrum $S_s(Q, \omega) = (\Gamma/\pi)/(\Gamma^2 + \omega^2)$ (see for instance Eqs. 15.7 or 15.25 in Part II of this volume), and assuming a Lorentzian shape for the resolution function as well (approximately valid for the classical BSC spectrometer), with a width (HWHM) $\Delta(\hbar\omega)$, one obtains for the measured window-scan intensity at zero energy transfer,

$$I(Q, \omega = 0, T) = \frac{1}{\pi} \frac{1}{(\Gamma(Q, T) + \Delta(\hbar\omega))}. \tag{15.44}$$

We will now, for the purpose of discussion, assume that in our example a single relaxation process (responsible for the Lorentzian line shape) is active over the whole temperature range considered, and that it shows an Arrhenius type behavior. Then, for sufficiently small temperatures, the quasielastic line falls entirely into the energy resolution window. Therefore, it does not cause any measurable quasielastic broadening. Then one gets for the window-scan intensity:

$$I(Q,\omega=0,T)=\frac{1}{(\pi\Delta(\hbar\omega))}. \tag{15.45}$$

With increasing temperature, the line width Γ grows, and finally becomes larger than the window $\Delta(\hbar\omega)$; the measured intensity of the window scan then reveals a "stepwise" decrease. Therefore, a simple temperature scan allows to get a qualitative survey of the diffusion or relaxation processes in the sample as a function of temperature.

The intensity "step" represents a (purely methodical) transition from nonobservability at low T to observability at high T of the relaxation process. From the shape of this step the relaxation time of the single process can easily be determined. Let us consider this problem for the more complex situation of a localized diffusive process, implying an elastic in addition to a quasielastic Lorentzian component. Here the same experimental method can be applied. If the attenuating effect of (harmonic) vibrational motions is described by a classical Debye–Waller factor, the temperature-dependent window-scan intensity, in logarithmic form, is given by [60]

$$\ln(I)=-CTQ^2+\ln(A(Q)/(\pi\Delta(\hbar\omega))+(1-A(Q))/[\pi(2\hbar/\tau_{2S}+\Delta(\hbar\omega))]. \tag{15.46}$$

Here CT is the vibrational mean square displacement (with a temperature coefficient C), $A(Q)$ is the EISF, and τ_{2S} is the relaxation time (for a two-site jump model in our example; see Fig. 15.1 and Table 15.1). It is interesting to note that the observability transition described by Eq. 15.46 can be used not only to yield the relaxation time τ_{2S}, but also for the determination of the EISF, provided that the mechanism does not change in the T-region of the step. For this purpose, the measured low-temperature straight line of ln(I) vs. T is extrapolated to high T and compared with the measured high-temperature line; the latter is obtained, when due to strongly increased line-broadening, the quasielastic contribution to the intensity becomes negligible. A simple division of the intensities yields the EISF according to the equation [60]

$$\ln[A(Q)/(\pi\Delta(\hbar\omega))]=[\ln(I)_{\text{high}\,T}]-\left[\ln(I)_{\text{low}\,T}^{\text{extrapol}}\right]. \tag{15.47}$$

As an example of such a measurement, Fig. 15.7 shows the intensity step due to the effect of OH-group reorientations in CsOH· H_2O [60] appearing in the energy resolution window of the BSC spectrometer IN13.

The elastic-window scan method has been employed in numerous biological experiments, in order to determine the temperature dependence of motional amplitudes ("mean-square displacements"), for instance in the context of the "dynamical transition" (see Sect. 16.5.2 in Part II, this volume).

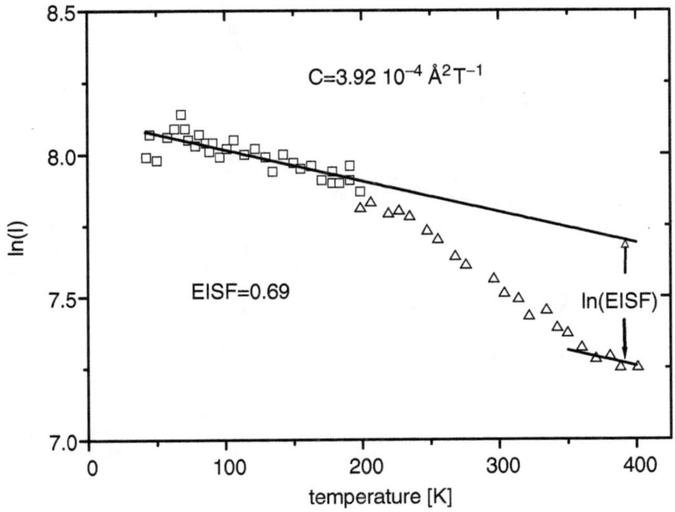

Fig. 15.7. Example of an apparent observability transition (as opposed to a true dynamical transition, where due to a structural phase transition a new type of motion appears at a transition temperature or in a transition region, in case of a higher order transition) exhibited experimentally by the elastic intensity I of $CsOH \cdot H_2O$ measured as a function of temperature: Logarithmic plot of the elastic-window intensity obtained with IN13 (ILL Grenoble) at $Q = 1.89\,\text{Å}^{-1}$. The straight lines show the variation of the Debye–Waller factor in the limits of low T and high T, respectively. The logarithm of the EISF is simply the difference between the values of the two lines at a given temperature [60]

15.3.4 TOF–XTL Spectrometers

Last, but not least, the TOF–XTL technique should be mentioned. This type of hybrid instrument, employs a pulsed polychromatic ("white") incident beam and single-crystals as analyzing filters. It is well adapted to the time-structure of spallation neutron sources. The energies of the incident neutrons are measured with TOF techniques, while the energy of scattered neutrons is fixed by the analyzers. For high energy resolution the crystals are used in BSC or near-BSC geometry ($\theta \simeq \pi/2$), whereas for more moderate resolution $\theta < \pi/2$ is chosen. Since the quality requirements for the crystals depend on these configurations, and for other practical reasons, dedicated instruments have been built for each case. A typical example with analyzer in near-BSC geometry is represented by the inverted geometry instrument IRIS schematically depicted in Fig. 15.8 [61] at RAL in Chilton. Depending on the type of crystals used, several discrete values of energy resolution in the range from about 1 to 55 µeV are achieved. The energy resolution function of such a spectrometer is essentially given by the convolution

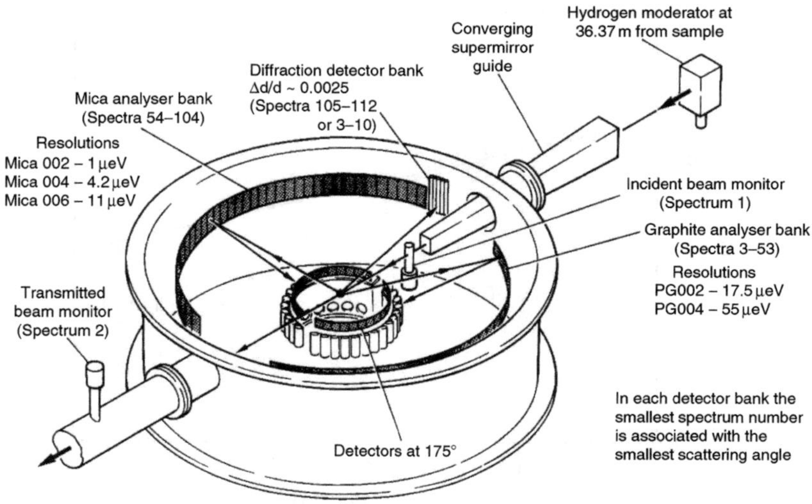

Fig. 15.8. Design of the high-resolution inverted-geometry backscattering spectrometer IRIS [61] on the ISIS pulsed source. The sample is located in the center of the analyzer vacuum vessel, at a distance of 36.5 m from the cold hydrogen moderator. The incident "white" beam reaches the sample through a converging supermirror guide. Graphite (resolutions 17.5, and 55 μeV) and mica (resolutions 1, 4.2, and 11 μeV) analyzer banks are arranged laterally on opposite sides. The detectors are mounted around the sample, in a plane slightly below the latter

of four contributions:

$$R(\omega) \simeq f_t\left(\frac{\Delta t_{\mathrm{mod}}}{t}\right) \otimes f_d\left(\frac{\Delta d}{d}\right) \otimes f_\theta(\cot(\theta)\Delta\theta) \otimes f_s\left(\frac{\Delta t_{\mathrm{SD}}}{t_{\mathrm{SD}}}\right). \quad (15.48)$$

The first function, f_t, arises from the finite pulse width Δt_{mod} produced by the cold moderator of the spallation source or (eventually) by a pulse-shaping chopper; t is the time of flight from the moderator to the detector. f_d is due to the uncertainty of the analyzer crystal's lattice spacing ($\Delta d/d$), and f_θ is the contribution of the Bragg angle uncertainty due to beam divergence and crystal mosaicity, which tends to vanish in perfect BSC geometry. The last function, f_s, is connected with the time-spread Δt_{SD} of the sample–detector flight time t_{SD}, caused by sample and detector thickness. If the four functions are approximated by Gaussian distributions, the elastic resolution $\Delta(\hbar\omega)$ is obtained from the quadratic addition of the individual contributions:

$$\Delta(\hbar\omega) = 2E_0\left[\left(\frac{\Delta t_{\mathrm{mod}}}{t}\right)^2 + \left(\frac{\Delta d}{d}\right)^2 + [\cot(\theta)\Delta\theta]^2 + \left(\frac{\Delta t_{\mathrm{SD}}}{t_{\mathrm{SD}}}\right)^2\right]^{1/2}. \quad (15.49)$$

For IRIS, an example of application is described in Sect. 16.5.1 (see Figs. 16.13 and 16.14 of Part II in this volume).

An example of an inverted-geometry instrument with analyzer scattering configuration significantly deviating from the BSC geometry, is the spectrometer "QENS" at the Intense Pulsed Neutron Source (IPNS) (Argonne, Illinois) [62]. The 22 crystal-analyzer detector-arrays are installed as close as possible to the sample, in order to maximize the analyzed range of solid angle. The crystals are arranged so that they match the "time-focusing-condition" which minimizes the time uncertainty contribution to the resolution. The elastic energy resolution of the spectrometer is around 80 μeV. An interesting feature of these spectrometers is the possibility to measure – simultaneously with the quasielastic spectra – the diffraction pattern of the sample. This is achieved by scattering from the sample directly into additional detectors, over a very wide wavevector range (up to $\simeq$30 $Å^{-1}$ in the case of "QENS"), because of the white beam arriving on the sample.

15.4 Instruments for QENS Spectroscopy in (Q, t)-Space

15.4.1 NSE Spectrometers

Differently from the procedures employed by the QENS techniques discussed in Sect. 15.3, which are based on energy transfer analysis, it is also possible to study scattered neutron intensities by Fourier time analysis. Before discussing neutron spin–echo techniques already well-known for this type of analysis, we briefly mention another method recently proposed [63, 64], which has not yet been widely used. It is based on the measurement of the elastically scattered neutron intensity as a function of observation time. If the latter is properly related to the Fourier time, a direct determination of the intermediate scattering function is achieved, without the detour via Fourier transformation. This method has been proposed as an alternative to energy transfer analysis in the use of TOF–TOF techniques capable of continuous tuning of the energy resolution, a feature so far not available on spectrometers employing crystals as monochromators and/or analyzers.

Let us now turn to the NSE method. It was introduced in 1972 by Mezei [65]. For detailed information on this technique, see [66]. NSE measures the sample dynamics in the time domain, via the determination of the intermediate scattering functions (see Eqs. 15.8 and 15.9). These functions are measured over a range of several orders of magnitude of Fourier time, up to values as large as 10^{-7} s. The drastic reduction of intensity necessarily arising, when in measurements with energy-analyzing QENS techniques the energy resolution is increased by reducing the incident and/or scattered neutron energy band widths, is circumvented here. This property of the NSE method is a consequence of the fact, that the energy transfer occurring in the scattering process, $\hbar\omega = E - E_0$, is causing a phase shift of the neutron spin precession angle for each scattered neutron, which is approximately independent of the neutron wavelength. Therefore the energy resolution depends only weakly on those energy widths of E_0 and E, and a rather large incident wavelength bandwidth can be used with little loss in energy resolution. Obviously, an appreciable intensity gain results from this fact. Of course, the latter is not entirely free of charge: it comes, for example, at the expense of the time-resolution on the Fourier time axis (see Sect. 15.4.3). This can, however, be tolerated, as long as the intermediate scattering functions (relaxation functions) to be measured are slowly varying in time, which is often the case in spin–echo spectroscopy.[7] The phase shift acts for each neutron by its effect on the measured quantity (i.e., the polarization of the scattered beam). Later in the discussion we will return to the reason, why in these measurements the intermediate scattering function is obtained. In the following, we discuss the basic concept of NSE spectroscopy and very briefly introduce the neutron resonance spin–echo (NRSE) method.

Spin 1/2 and Larmor Precession

A basic property of the neutron, relevant for the NSE technique, is its spin 1/2. The description of the behaviour of spin 1/2 particles can be found in standard text books (e.g., [67,68]); detailed instructions for the use of spin–echo spectroscopy are given in [69–72]. When a spin 1/2 is submitted to a magnetic field B_z, its component in the direction collinear to the field is quantized. In the fundamental state, two values of S_z are possible $|+\rangle$ (the spin parallel to the field direction) and $|-\rangle$ (antiparallel). Let us now assume that a neutron spin is oriented in a direction having an angle θ with respect to a homogeneous magnetic field. A precessional motion will occur around the magnetic-field direction. The time evolution of the spin can be determined classically by

$$\frac{\mathrm{d}\mathbf{S}}{\mathrm{d}t} = \gamma \mathbf{S} \times \mathbf{B}. \tag{15.50}$$

This equation describes the precessional motion [73] of $\mathbf{S}$ around $\mathbf{B}$ with a Larmor frequency $\omega_\mathrm{L} = |\gamma|B$. Here, $\gamma = -2\pi\ 2916.4\,\mathrm{G}^{-1}\,\mathrm{s}^{-1}$ is the gyromagnetic ratio of the neutron. The accumulated precession angle, Φ, of the neutron spin depends linearly on the time t it spends in the magnetic field B and on the strength of this field (we assume decoupling of spin state and neutron

[7] This argument is similar to the one justifying an appreciable intensity gain in TOF and BSC spectroscopy, due to the integration over large detector solid angle regions, when scattering functions are slowly varying with Q

position, which is valid for NSE):

$$\Phi = \omega_{\mathrm{L}} t = |\gamma_n| B t = |\gamma| B \frac{l}{v}, \tag{15.51}$$

where l is the length of the flight path in the field, and v is the neutron velocity. Typically, 5 Å neutrons submitted to a field of 10 G will perform around 37 turns per meter, and when the field is increased to 0.5 T (5000 G) this number can be increased up to 18500.

When the direction of the magnetic field changes, two extreme cases can be considered. Let us, for the sake of simplicity, assume that the neutron spin is initially collinear with the field: (i) Slow (or "adiabatic") change of the magnetic field: A change of the field direction, encountered by the neutron on its way along the flight path, is experienced as a rotation of the field in the coordinate frame of the flying neutron. This may be represented by an angular frequency ω_{F} of rotation, which depends not only on the spatial variation of the field in the laboratory coordinate frame, but also on the neutron velocity. The spin follows adiabatically the change of the field when $\omega_{\mathrm{F}} \ll \omega_{\mathrm{L}}$: If it was initially collinear with the magnetic field, it will rotate with the latter, i.e., the spin direction is following the field. (ii) Sudden (or "non-adiabatic") change of the field: this is the limit opposite to the adiabatic change, i.e., $\omega_{\mathrm{F}} \gg \omega_{\mathrm{L}}$; in this case the field variation is so fast that the spin direction can not follow. Such changes are used to initiate the spin precession. This will be explained below.

The Neutron Spin-Echo Principle

The general principle of the neutron spin–echo spectrometer is presented in Fig. 15.9. A neutron of wavelength $\lambda_0^{'}$ moving in direction Z arrives at $t = 0$ in Z_0, with its spin in the $|+\rangle$ state with respect to the Z direction. A $\pi/2$ flipper "suddenly" puts the spin state perpendicular to the direction of the field (the description of these flippers is omitted in the context of this book, but the interested reader is referred to [71]). This action initiates a precessional motion in the counter-clockwise sense (the gyromagnetic ratio γ of the neutron is negative).

Let us now assume a beam of polarized neutrons moving perfectly parallel in Z direction (assuming zero beam divergence), with a wavelength distribution $f(\lambda_0^{'})$, the maximum of the distribution being located at λ_0. We now want to calculate, what will be the precession angle distribution at Z_{A}, if the field path integrals are equal in the two arms. From Z=0 to the sample position Z_{S}, i.e., in the first arm of the spectrometer, the spin of a neutron of wavelength λ_0 will accumulate a total precession angle of

$$\Phi_1^{\lambda_0} = \frac{\gamma m \lambda_0}{h} \int_0^S B_1 \,\mathrm{d}z = 2\pi N_1^{\lambda_0}, \tag{15.52}$$

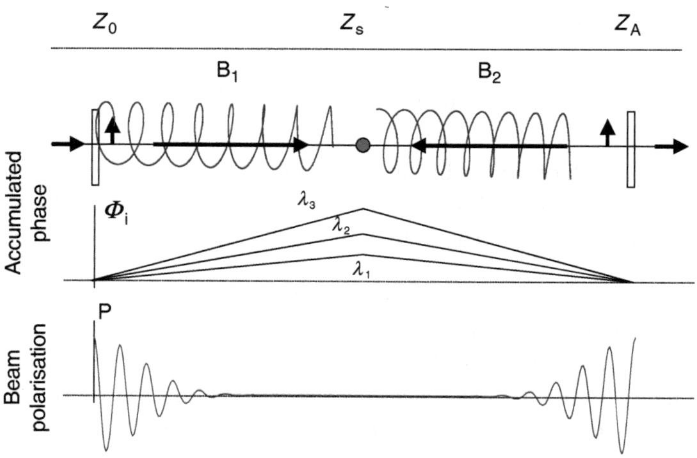

Fig. 15.9. Principle of the neutron spin–echo spectrometer: A neutron of wavelength λ_0' moving in direction Z arrives at $t = 0$ in Z_0, with its spin in the $|+\rangle$ state with respect to the Z direction. Two coils create high magnetic fields of equal strength, but in opposite directions, B_1 and B_2, before and after the sample (*upper part*). A $\pi/2$ flipper "suddenly" puts the spin state of the arriving neutron perpendicular to the direction of the field. This initiates a precessional motion of the neutron spin in the counter-clockwise sense, while this particle is flying along Z. The spin precession stops after a second $\pi/2$ spin flipper placed after the end of the second coil. The accumulated spin phases Φ_i are proportional to the time, wavelength, and distances (*central part*). The beam polarization $P \approx \langle\cos(\Phi_i)\rangle$ is presented in the lower part of the figure. If the sample is a purely elastic scatterer, the neutron wavelength λ_0' is not changed by the scattering process. Because of the two magnetic fields with equal strength but opposite signs, in the case of elastic scattering, the total phase angle accumulated by the neutron spin precession during the flight from Z_0 to Z_A, will be $\Phi(\lambda_0') = 0$. The general case of nonelastic scattering is explained in the text

where

$$N_1^{\lambda_0} = \frac{\gamma m \lambda_0}{2\pi h} \int_0^S B_1 \mathrm{d}z \tag{15.53}$$

is the number of (positive) spin turns between Z_0 and Z_S. For any neutron with a wavelength λ_0' within the incident neutron wavelength distribution, we get:

$$\Phi_1(\lambda_0') = \frac{\gamma m \lambda_0'}{h} \int_0^S B_1 \mathrm{d}z = 2\pi N_1^{\lambda_0} \frac{\lambda_0'}{\lambda_0}. \tag{15.54}$$

The analogous quantities can be calculated for the second arm: Let us assume that the sample is a purely elastic scatterer and thus the neutron wavelength λ_0' is not changed by the scattering process. From Z_S to Z_A the neutron is submitted to a magnetic field in direction opposite to the field before the sample. Then the accumulated precession angle will be

$$\Phi_2^{\lambda_0} = \frac{\gamma m \lambda_0}{h} \int_S^A B_2 \, \mathrm{d}z = -2\pi N_2^{\lambda_0} \tag{15.55}$$

with

$$N_2^{\lambda_0} = \frac{\gamma m \lambda_0}{2\pi h} \int_S^A B_2 \, \mathrm{d}z, \tag{15.56}$$

which is the number of (negative) spin turns between Z_S and Z_A. For any neutron with a wavelength $\lambda_0^{'}$ within the incident neutron wavelength distribution, we get

$$\Phi_2(\lambda_0^{'}) = \frac{\gamma m \lambda_0^{'}}{h} \int_S^A B_2 \, \mathrm{d}z = -2\pi N_2^{\lambda_0} \frac{\lambda_0^{'}}{\lambda_0}. \tag{15.57}$$

Finally, we obtain for the total phase angle accumulated by the neutron spin precession of any neutron in the distribution:

$$\Phi(\lambda_0^{'}) = \Phi_1(\lambda_0^{'}) + \Phi_2(\lambda_0^{'}) = 2\pi \left(N_1^{\lambda_0} \frac{\lambda_0^{'}}{\lambda_0} - N_2^{\lambda_0} \frac{\lambda_0^{'}}{\lambda_0} \right) = \alpha \lambda_0^{'}, \tag{15.58}$$

where α does not depend on $\lambda_0^{'}$

$$\alpha = \frac{2\pi}{\lambda_0} \left(N_1^{\lambda_0} - N_2^{\lambda_0} \right). \tag{15.59}$$

It is easy to realize that at the "echo-point", where the field path integrals in the first and in the second arm of the spectrometer are equal, the total phase angle accumulated by the neutron spin precession will be $\Phi(\lambda_0^{'}) = 0$, and this is true for any neutron velocity; i.e., this result is independent of $\lambda_0^{'}$.

Transmission of Polarizers and Analyzers

Polarizer and analyzer are key elements of the spectrometer. The polarizer is used to prepare a polarized beam (i.e., it selects neutrons with only one of the two quantized neutron spin states). For long-wavelength neutrons ($\lambda > 3\,\text{Å}$), as are generally used in NSE experiments (at least for quasielastic measurements), most of the polarizers employ the principle of magnetic reflection or transmission of supermirrors. For shorter wavelengths, Haeussler-like crystals are used. A potentially very interesting technique for spin analysis is the use of a ^{3}He polarizer and/or analyzer. For the spin–echo technique a ^{3}He polarizing filter unit would have the very interesting property of being simultaneously usable over a wide angular range and independently of the neutron wavelength. At the time of writing the beam polarization achieved with ^{3}He is however, not yet good enough for spin–echo experiments.

The polarizer transmits only neutrons with one of the spin components. The exact fraction of the neutron beam transmitted is not relevant for the principle of the technique, but it should be maximized for the purpose of obtaining good statistical accuracy of the measurements. The number of incident neutrons is usually monitored just before the sample position. For our present purpose, the interesting problems will be: (i) To compute the probability of transmission through the analyzer for a neutron whose spin angle with the static field B_A inside the analyzer is given by θ. (ii) To determine the beam polarization of a neutron population whose spin angles are given by the normalized distribution $F(\theta)$.

These problems require a quantum-mechanical calculation. The spin-state wave function for the neutron spin precessing in the field B_A is

$$|\Psi\rangle = \mathrm{e}^{-\mathrm{i}\tilde{\Phi}} \cos\frac{\theta}{2}\ |+\rangle + \mathrm{e}^{\mathrm{i}\tilde{\Phi}} \sin\frac{\theta}{2}\ |-\rangle, \tag{15.60}$$

where $\tilde{\Phi}$ and θ are the spherical coordinates defining the orientation of the spin with respect to the analyzer field. Each neutron spin will have the probabilities $p_{|+\rangle} = \cos^2\frac{\theta}{2}$ and $p_{|-\rangle} = \sin^2\frac{\theta}{2}$, to be either in the $|+\rangle$ or in the $|-\rangle$ state, respectively.

The solution of problem 1 is that only $|+\rangle$ neutrons will be transmitted, the probability of transmission being given by $p_{|+\rangle} = \cos^2\frac{\theta}{2}$. Thus for a neutron flux N incident on an idealized analyzer (i.e., with no absorption losses), with the normalized spin-precession phase-angle distribution $F(\theta)$, the flux of neutrons transmitted by the analyzer will be

$$N^{|+\rangle} = \int F(\theta) \cos^2\left(\frac{\theta}{2}\right) \mathrm{d}\theta. \tag{15.61}$$

Using a well-known trigonometric relation for the square of the cosine, we can compute the transmission of the analyzer

$$T = \frac{N^{|+\rangle}}{N} = \frac{\int F(\theta)(1+\cos(\theta))\mathrm{d}\theta}{2\int F(\theta)\mathrm{d}\theta} = \frac{1}{2}(1 + \langle\cos(\theta)\rangle). \tag{15.62}$$

It is straightforward to show that $\langle\cos(\theta)\rangle$ corresponds to the polarization of the transmitted beam,

$$P = \frac{p_{|+\rangle} - p_{|-\rangle}}{p_{|+\rangle} + p_{|-\rangle}}, \tag{15.63}$$

which is equivalent to:

$$P = \frac{\int F(\theta)(\cos^2(\frac{\theta}{2}) - \sin^2(\frac{\theta}{2}))\mathrm{d}\theta}{\int F(\theta)\mathrm{d}\theta} = \langle\cos(\theta)\rangle. \tag{15.64}$$

Getting a Spin-Echo, as a Measure of the Polarization

We now return to Eq. 15.58, because we want to compute the polarization of the scattered beam. We will treat the problem for two different cases: (i) The wavelength of neutrons is not changed by the scattering process. (ii) For a quasielastic distribution of energy exchange between the sample and the neutron beam.

First, assuming a *purely elastic scatterer*, $\Phi(\lambda_0')$ does not depend on the sample, but simply on the incident wavelength distribution $f(\lambda_0')$, which is assumed to be Gaussian. The final beam polarization can then be computed as,

$$P = \int_0^\infty f(\lambda_0') \cos(\alpha\lambda_0') \mathrm{d}\lambda_0' \tag{15.65}$$

with

$$f(\lambda_0') = \frac{1}{\sqrt{2\pi\sigma^2}} \mathrm{e}^{-(\lambda_0' - \lambda_0)^2/2\sigma^2}, \tag{15.66}$$

where α was defined by Eq. 15.59, and σ is the standard deviation of the wavelength distribution. The above integral has a simple analytical solution, if the integration is performed from $-\infty$ to $+\infty$, which is possible, because the assumed (in good approximation) Gaussian shape of the distribution to be integrated is practically zero for negative (unphysical) values of the wavelength. The integral yields for the beam polarization

$$P = \cos(\alpha\lambda_0) \mathrm{e}^{-\alpha^2\sigma^2/2}. \tag{15.67}$$

The variation of the beam polarization in the neighborhood of the echo point is shown in Fig. 15.10. It presents a full echo scan around the spin–echo maximum condition, $\int B_1 \, \mathrm{d}l = \int B_2 \, \mathrm{d}l$. In such a scan, the field path integral in the secondary spectrometer (coil 2) is varied by changing the current $\mathbf{i}$ of a small additional solenoid coil. This variation can be computed. Writing for the field path integral of the solenoid,

$$\int B \, \mathrm{d}l = \mu_0 \tilde{N} \mathbf{i}, \tag{15.68}$$

where $\tilde{N}$ is the number of turns of the solenoid wire per unit of length, we get the function (continuous line) plotted together with the measured data points of the full echo scan in Fig. 15.10

$$n(\mathbf{i}) = \frac{N}{2}(1 + P_\mathrm{e} * \cos(\tilde{\alpha}(\mathbf{i} - \mathbf{i}_\mathrm{e})\lambda_0) \mathrm{e}^{-(\tilde{\alpha}(\mathbf{i} - \mathbf{i}_\mathrm{e})\sigma)^2/2} \tag{15.69}$$

with

$$\tilde{\alpha} = \frac{\gamma 10^4 \tilde{N} l m \mu_0}{h}, \tag{15.70}$$

where l is the coil length, m the neutron mass, h the Planck's constant, and μ_0 is the permeability of free space. P_e and $\mathbf{i}_\mathrm{e}$, respectively, are the scattered neutron beam polarization and the current of the NSE tuning coil, at the echo point (where $\int B_1 \, \mathrm{d}l = \int B_2 \, \mathrm{d}l$). N is the number of polarized neutrons per integrated monitor count unit, scattered by the sample towards the analyzer.

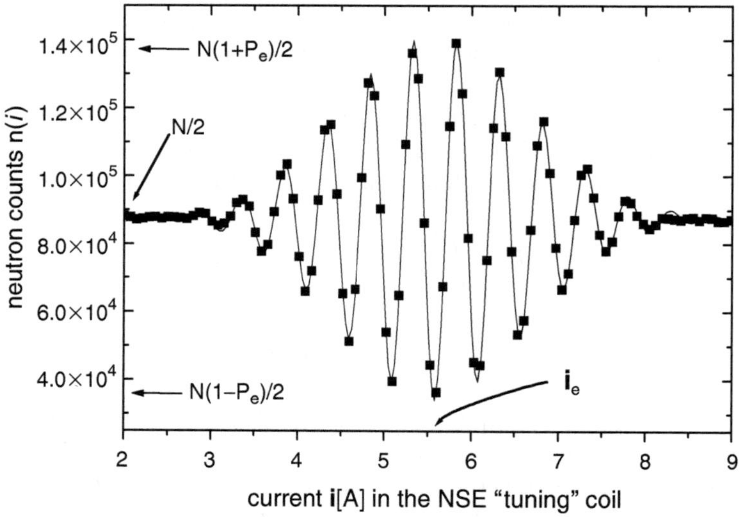

Fig. 15.10. Full echo (asymmetric current scan) measured (*full squares*) with the spectrometer MUSES [74] at the point Z_A, (see Fig. 15.9). The continuous line is the fit of Eq. 15.69 to the measured data points. P_e and i_e, respectively, are the scattered beam polarization and the current of the NSE tuning coil, at the echo point (where $\int B_1\,\mathrm{d}l = \int B_2\,\mathrm{d}l$). N is the number of polarized neutrons per integrated monitor count unit, scattered by the sample towards the analyzer (see text). The maximum and minimum numbers of neutrons that are transmitted by the analyzer, when the current **i** is varied, $N(1+P_\mathrm{e})/2$ and $N(1-P_\mathrm{e})/2$, are also indicated. N/2 is the number of neutron counts, when – outside of the spin–echo region – the beam is completely depolarized

The maximum and minimum numbers of neutrons that are transmitted by the analyzer, when the current **i** is varied, $N(1+P_\mathrm{e})/2$ and $N(1-P_\mathrm{e})/2$, are indicated by horizontal arrows. N/2 is the number of neutron counts, when – outside of the spin–echo region – the beam is completely depolarized.

Measuring Quasielastic Neutron Scattering

We now assume the scattering process to be quasielastic. The energy exchanged between the neutron and the sample during the scattering process is

$$\hbar\omega = \frac{\hbar^2 k^2}{2m} - \frac{\hbar^2 k_0^2}{2m}\,. \tag{15.71}$$

This will induce a change $\delta\lambda$ of the neutron wavelength, so that for incident neutrons with wavelength $\lambda_0^{'}$ the scattered neutron wavelength is $\lambda = \lambda_0^{'} + \delta\lambda$. The polarization of the scattered beam can be expressed as follows:

$$\langle P\rangle = \int_0^{+\infty} I(\lambda_0^{'}) \int_{-\lambda_0^{'}}^{+\infty} p(\lambda_0^{'}, \delta\lambda)\cos(\Phi(\lambda_0^{'}, \delta\lambda))\mathrm{d}(\delta\lambda)\mathrm{d}\lambda_0^{'}. \tag{15.72}$$

Here, $p(\lambda_0^{'}, \delta\lambda)$ is the probability that a neutron scattering process with incident wavelength $\lambda_0^{'}$ and with a change in neutron wavelength $\delta\lambda$ occurs; note that the maximum possible negative value of $\delta\lambda$ is equal to $\lambda_0^{'}$. The function $p(\lambda_0^{'}, \delta\lambda)$, when transformed from the $\delta\lambda$-transfer to the energy-transfer axis, turns into nothing else but the scattering function $S(\boldsymbol{Q}, \omega)$. $\Phi(\lambda_0^{'}, \delta\lambda)$ is the total precession angle. Note also, that writing Eq. 15.72 we have assumed a perfect spectrometer: The incident beam is perfectly polarized, all the neutrons are transmitted through the spectrometer (whatever the wavelength) and with no beam depolarization. We will come back to this assumption, when discussing the different spectral windows and resolutions of quasielastic spectrometers. It is important to realize that the total precession angle $\Phi(\lambda_0^{'}, \delta\lambda)$ represents the energy transfer $\hbar\omega$ which corresponds to the change $\delta\lambda$ in neutron wavelength, caused by the scattering process

$$\Phi(\lambda_0^{'}, \delta\lambda) = 2\pi \left(N_1^{\lambda_0} \frac{\lambda_0^{'}}{\lambda_0} - N_2^{\lambda_0} \frac{\lambda_0^{'} + \delta\lambda}{\lambda_0} \right) = \alpha\lambda_0^{'} - 2\pi N_2^{\lambda_0} \frac{\delta\lambda}{\lambda_0} . \tag{15.73}$$

At the echo point (for QENS experiments, the field path integrals are identical in the first and the second arm) $\alpha = 0$ and hence

$$\Phi(\lambda_0^{'}, \delta\lambda) = \frac{\gamma m}{h} \int_0^S B_1 \mathrm{d}z \delta\lambda . \tag{15.74}$$

For quasielastic scattering, it is assumed that the energy gained or lost by the neutrons remains much smaller than the initial energy of the neutron, i.e., $\delta\lambda/\lambda_0^{'} \ll 1$, hence

$$\delta\lambda \approx -\frac{m(\lambda_0^{'})^3\omega}{2\pi h} \tag{15.75}$$

and thus,

$$\Phi(\lambda_0^{'}, \delta\lambda) = -\frac{\gamma m^2}{2\pi h^2} \int_0^S B_1 \, \mathrm{d}z (\lambda_0^{'})^3 \omega . \tag{15.76}$$

Finally, the scattered beam polarization can be written

$$\langle P \rangle \approx \int_0^{+\infty} I(\lambda_0^{'}) \int_{-\infty}^{+\infty} S(\mathbf{Q}, \omega) \cos(\omega\tau_{\mathrm{NSE}}) \mathrm{d}\omega \, \mathrm{d}\lambda_0^{'} \tag{15.77}$$

with

$$\tau_{\mathrm{NSE}} = \frac{m^2\gamma \int B \mathrm{d}z}{2\pi h^2} (\lambda_0^{'})^3 . \tag{15.78}$$

Note that the second integral in Eq. 15.77 is the real part of the Fourier-transform of the dynamic structure factor $S(\mathbf{Q}, \omega)$. The spin–echo time τ_{NSE} approximates the Fourier time t, and it is important to note that it is proportional to the third power of the wavelength (see also Sect. 15.4.3). Thus, it is now evident that

$$\langle P\rangle \approx I(Q,t) \approx \int_{-\infty}^{+\infty} S(\boldsymbol{Q},\omega)\cos(\omega\tau_{\mathrm{NSE}})\mathrm{d}\omega\,. \qquad (15.79)$$

The $\approx$ symbol stands for the fact that $S(\boldsymbol{Q},\omega)$ is only approximately a symmetric function in ω, and that the integration over the wavelength distribution $I(\lambda_0^{'})$ in Eq. 15.77 has a certain smearing or broadening effect on $I(Q,t)$.

The reason, why NSE spectroscopy measures sample dynamics in the time domain, is that the analyzer transforms a quantity proportional to the time, $\Phi(\lambda_0^{'})$, into a cosine: $\cos(\Phi(\lambda_0^{'}))$; the summation over all the scattered intensity ($\simeq S(\boldsymbol{Q},\omega)$) weighted by this cosine function, Fourier-transforms the dynamic structure factor. One should nevertheless keep in mind, that only the real part of the FT is measured, and thus $\langle P\rangle \simeq I(Q,t)$ only, if $\simeq S(\boldsymbol{Q},\omega)$ is an (almost) even function. This is however, usually the case for NSE experiments, because they probe the very low-frequency part of the dynamic structure factor, where the spectral asymmetry due to the detailed balance factor can be neglected and $S(\boldsymbol{Q},\omega)$ is to a good approximation an even function, as long as the temperature is not too low. For an application of this technique, see Sect. 16.3.2 in Part II of this volume.

15.4.2 Neutron Resonance Spin-Echo Spectrometry

In resonance spin–echo spectrometry [75,76], the two high magnetic-field precession coils are substituted by four radio-frequency coils, two in the first arm and two in the second (see Fig. 15.11). The field geometry in the coils is as follows: a static high field $\mathbf{B}_0$, e.g., oriented vertically, and perpendicular to it a radio-frequency field $\mathbf{B}_1(t)$ rotating in the horizontal plane.

$$\mathbf{B}(t) = \mathbf{B}_0 + \mathbf{B}_1(t)\,. \qquad (15.80)$$

The resonance condition is reached when the frequency of the Larmor precession induced by the static field, $\omega_0 = -\gamma_n B_0$, is equal to the frequency of the rotating field ω_f: then – from the point of view of the neutron spin in the coordinate frame of the rotating field – the magnetic field $\mathbf{B}_0$ vanishes. Under this condition, the motion of the neutron spin in the rotating frame associated with $\mathbf{B}_1(t)$, can be simply reduced to a Larmor precession with a frequency $\omega_1 = -\gamma B_1$. The field $\mathbf{B}_1$ is chosen so that a neutron arriving in a coil with a spin oriented in the scattering plane, will leave it in the same plane, after having performed a precession of π around $\mathbf{B}_1(t)$

$$\pi = \gamma B_1 \frac{d}{v}\,, \qquad (15.81)$$

where d is the coil thickness and v the neutron velocity. One now has to compute the spin orientation along the neutron path through the spectrometer,

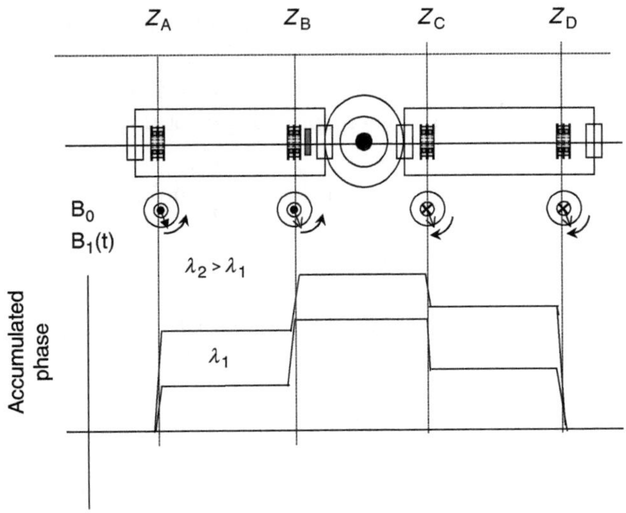

Fig. 15.11. Schematic description of the NRSE principle: Four radio-frequency coils, well-shielded against magnetic contaminations, such as the earth's magnetic field, are placed as two pairs in the two arms of the spectrometer at the positions Z_A, Z_B, Z_C, and Z_D, respectively. A static high field B_0, e.g., oriented vertically, and perpendicular to it, a radio-frequency field $B_1(t)$ rotating in the horizontal plane, are established. In the lower part of the figure the accumulated phase angles of the neutron spin precession are indicated schematically for the cases of elastic scattering for two different wavelengths, respectively

in the scattering plane. Suppose, the spins are polarized in the direction $\boldsymbol{y}$. Let us denote the angle between the spin orientation and $\boldsymbol{y}$ by ϕ, and the four coils by A, B, C, and D; t_A, $t_{A'}$ are the times at which a neutron enters and leaves the coil A, and so on. The angle after the first coil can be written as

$$\phi_{A'} = \omega_f \frac{d}{v} + 2\omega_f t_A - \phi_A \,. \tag{15.82}$$

Only within these coils the neutron spins are submitted to a magnetic field and consequently the remaining neutron path has to be shielded from any magnetic contamination (earth magnetic field, etc.). After computation of the neutron spin phase along the neutron path, and assuming a perfect spectrometer one can obtain the echo condition (i.e., the condition for which all the spins are polarized in the same direction as before entering the spectrometer)

$$\frac{l_{AB} + d}{v} - \frac{l_{CD} + d}{v'} = 0 \tag{15.83}$$

with l_{AB} and l_{CD} being the distances between the coil centers in the first and in the second arm, respectively, which for an elastic or quasielastic process simply reduces to $l_{\mathrm{AB}} = l_{\mathrm{CD}}$. The measured quantity is the polarization of the scattered beam in the $\boldsymbol{y}$ direction (or $\boldsymbol{z}$ after an adiabatic $\pi/2$ turn). If ϕ is the angle $(\boldsymbol{y},S_i)$ between the neutron spin orientation and $\boldsymbol{y}$ after the fourth coil, the contribution of each spin to the total polarization is given by

$$P_z\left(\lambda_0',\delta\lambda\right) \approx \cos\left(2\omega_f \frac{l+d}{h} m\delta\lambda\right). \tag{15.84}$$

Similar to NSE, after summation over all neutron contributions one obtains

$$\langle P\rangle = \int_0^{+\infty} I(\lambda_0') \int_{-\lambda_0'}^{+\infty} p(\lambda_0',\delta\lambda)\cos\left(2\omega_f \frac{l+d}{h} m\delta\lambda\right) \mathrm{d}(\delta\lambda)\,\mathrm{d}\lambda_0'. \tag{15.85}$$

So, within a quasielastic process, we have

$$\langle P\rangle \approx \int_0^{+\infty} I(\lambda_0') \int_{-\infty}^{+\infty} S(\mathbf{Q},\omega)\cos(\omega\tau_{\mathrm{NRSE}})\mathrm{d}\omega\,\mathrm{d}\lambda_0' \tag{15.86}$$

with

$$\tau_{\mathrm{NRSE}} = 2\omega_f \frac{l+d}{2\pi}\frac{m^2}{h^2}(\lambda_0')^3, \tag{15.87}$$

where τ_{NRSE} is the spin–echo time of this method. Again, the spin–echo time is proportional to the third power of the wavelength, and – analogous to Eq. 15.79

$$\langle P\rangle \approx I(Q,t) \approx \int_{-\infty}^{+\infty} S(\mathbf{Q},\omega)\cos(\omega\tau_{\mathrm{NRSE}})\mathrm{d}\omega. \tag{15.88}$$

The consequence of this, and of the use of large incident wavelength bandwidths on the time-resolution will be considered in Sect. 15.4.3

15.4.3 Observation Function, Effect of Wavelength Distribution on Spin-Echo Time

In the spin–echo case, the experimental observation function $R^*(t)$ (compare Sect. 15.2.3) is characterized by the time-dependent decay of the polarization, due to spectrometer imperfections and neutron population distributions (field inhomogeneities, flipper, and polarizer–analyzer efficiencies, spectrometer transmission, wavelength distribution, beam divergence, etc.). In principle, just as in the case of $(\boldsymbol{Q},\omega)$-spectrometers, the decay of the observation function limits the time range, where intermediate scattering functions can be determined. The limit is characterized by the observation time, $\Delta\tau_{\mathrm{obs}}$, i.e., the decay-time of the observation function, which is the inverse of the virtual resolution width (HWHM) the spin–echo spectrometer would have after transformation to the energy axis. Here, there is however a second, independent upper limit, τ_m , of the spin–echo spectrometer's time window, determined by the limit of the ability of the coils to produce high magnetic fields. To compare with other spectrometers and to evaluate the experimental instrument resolution indicating the maximum time up to which a dynamical phenomenon can be measured, one has to compare the value of the observation time, $\Delta\tau_{\text{obs}}$, to τ_m. If $\Delta\tau_{\text{obs}}$ is smaller than τ_m, good statistical accuracy of intermediate scattering functions measured at large times (i.e., close to τ_m) will be difficult to obtain. For the energy-resolution correction, the value of the observation function has to be measured for each spin–echo time and for each wavevector transfer Q, in general,[8] employing a specimen which is a purely elastic scatterer, with a structure factor as close as possible to that of the sample.

We have shown above, that the spin–echo time is given by $\tau = \eta(\lambda_0^{'})^3$ where the coefficients η applying for the two different techniques, are defined by Eqs. 15.78 and 15.87 for NSE and NRSE, respectively. Here, $\lambda_0^{'}$ is the wavelength of the incident neutrons. One of the key points to achieve high resolution in neutron spin–echo spectroscopy is, to first order, the decoupling of intensity and resolution. This allows the use of broad incident wavelength distributions at little cost in energy resolution. The broad wavelength distribution, together with the very strong $\lambda_0^{'}$ dependence of the spin–echo time, are however not without consequence: Since an average over the wavelength distribution has to be taken (see Eqs. 15.77 and 15.86), this amounts to an integration over a corresponding spin–echo time spread. The mean value of the spin–echo time and the time integration window are easily obtained. First, a simple computation of the mean spin–echo time,

$$\langle\tau\rangle = \int \tau(\lambda_0^{'}) f(\lambda_0^{'}) \mathrm{d}\lambda_0^{'} , \tag{15.89}$$

shows that this mean value is slightly different from the time corresponding to the maximum wavelength of the distribution. For a velocity selector with $\Delta\lambda_0/\lambda_0 \simeq 0.15$ (FWHM) one obtains,

$$\langle\tau\rangle \simeq 1.02\ \alpha\lambda_0^3 . \tag{15.90}$$

But, more important is the width of the spin–echo time distribution σ_τ. It can be estimated by the computation of the standard deviation,

$$\sigma_\tau = \sqrt{\langle\tau^2\rangle - \langle\tau\rangle^2} . \tag{15.91}$$

This gives $\sigma_\tau \simeq 0.31\tau_0$, where for a given magnetic field, τ_0 is the spin–echo time corresponding to the wavelength λ_0. Figure 15.12 shows a logarithmical plot of an exponentially decaying intermediate scattering function, $I(\boldsymbol{Q}, t)$, on a linear time-scale. For different spin–echo times the "time integration windows" ($\pm\sigma_\tau$) are indicated by horizontal bars.

---

[8]In the case of NRSE, when S(Q) is a slowly varying function in the Q-range of the experiment, it is often sufficient to do this for only one value of Q

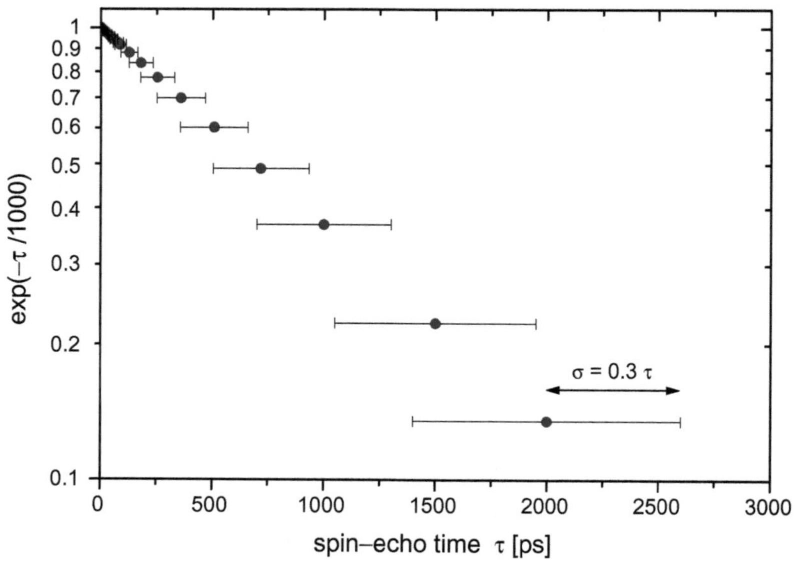

Fig. 15.12. Logarithmical plot of an exponentially decaying intermediate scattering function, $I(\boldsymbol{Q}, t)$, on a linear time-scale, showing typical spin–echo time integration windows of a spin–echo spectrometer with a velocity selector selecting a wavelength band with $\Delta\lambda_0/\lambda_0 \simeq 0.15$ (FWHM). The horizontal bars represent the standard deviation σ of the spin–echo time τ

NSE spectroscopy is often used for the measurement of slowly decaying relaxation functions. In such cases, the consequence of such broad time integration windows is sometimes of little importance and usually neglected.

15.5 Miscellaneous Technical Points: MSC, Calibration, Contrast

Let us finally mention a few technical points concerning the analysis of data originating from QENS experiments. At first, we consider those aiming at the direct determination of Van Hove's scattering functions in (momentum, energy)-space. An important problem in the evaluation of quasielastic neutron scattering spectra is the accurate consideration of the experimental resolution. Although straightforward, this is complex, if the spectrum is a sum of several quasielastic components, eventually superimposed by a purely elastic line, either due to the fact that the diffusive motion is localized, or caused by parasitic incoherent scattering from the host material in which the particles perform a diffusive motion. The measured spectra are convolutions of the scattering function with the resolution function. Since a direct deconvolution generally encounters practical difficulties, in a typical analysis the data are usually not directly corrected for resolution, but an inverse procedure is employed. It consists in folding the theoretical model with the measured resolution function and comparing the result to the measured spectra via nonlinear least-squares fitting calculations. This method, which amounts to an indirect deconvolution, is well-known, and standard software packages are available in the neutron scattering centers.

Another problem is multiple scattering (MSC). In principle, a neutron reaching the detector, may have been scattered in the sample once , twice, or even several times. For a given nominal scattering angle φ or *nominal* Q value ($Q \cong (4\pi/\lambda)\sin(\varphi/2)$) the weighted double-, triple-,... scattering spectra are then superimposed on the single-scattering data. The MSC components should be minimized by using sufficiently thin samples. It is usually assumed that a reasonable sample thickness (a compromise between maximizing single-scattering of neutrons and minimizing the contribution of multiple to total scattering) should correspond to a transmission of $T \simeq 0.9$ (roughly 10% of the incident beam is scattered once and about 1% more than once). This leads for example to about 0.2 mm thickness for $\lambda = 5$ Å in the case of light water and to about 2 mm for deuterated water. Nevertheless, each case should be carefully studied especially with respect to sample absorption and geometry. A numerical correction is often necessary. Analytical methods [77] directly applied to the theoretical models, and model-independent Monte-Carlo techniques (see for instance [78,79]) are employed for this purpose. For the multiple scattering correction of NSE experiments, the same principles as already described apply, except that this has to be applied in the time domain, where instead of the deconvolution a simple division by the observation function has to be performed.

The third problem is *calibration*, which means the absolute determination of the scattered intensity, instead of quoting "arbitrary units". Calibration is performed, for instance, by comparing the scattered intensity of a sample to that of a vanadium scatterer which has the same scattering geometry as the sample itself, and whose incoherent scattering cross-section is well-known (provided that there is no hydrogen contamination). In NSE experiments, usually a relative intensity normalization of the measured intermediate scattering function $I(Q,t)$ is performed. For each spin–echo time, the measured polarization is divided by the polarization measured at $\tau = 0$ (i.e., determined without a magnetic field in the coils). Theoretically, $I(Q,\tau = 0)$ should be equal to $S(Q) = \int_{-\infty}^{\infty} S(Q,\omega)\mathrm{d}\omega$, but due to the fact that the spectrometer transmits only a wavelength band with a limited width, the integral over the scattering function remains incomplete. If a significant amount of the scattered intensity is outside of the spectrometer window, the normalization will therefore not be absolute. For an absolute calibration the required structure factor $S(Q)$ of the sample must be determined independently, for practical reasons on a different instrument which does not have such a limited bandwidth transmission.

For many elements coherent and incoherent scattering coexists, and it is difficult to *separate the contributions* $S(\boldsymbol{Q},\omega)$ and $S_\mathrm{s}(\boldsymbol{Q},\omega)$. Spin-incoherent scattering causes a fraction of $(2/3)[\sigma_\mathrm{inc}/(\sigma_\mathrm{inc}+\sigma_\mathrm{coh})]$ of the neutrons to flip their spin in a polarized neutron beam, whereas coherent and isotope incoherent scattering occurs without spin flip. The high intensity of polarized beams available at the reactor of the ILL in Grenoble allows to apply this property for a direct separation of the coherent and spin-incoherent contributions, even in the study of quasielastic scattering, by orienting spin polarizer and analyzer parallel and antiparallel, respectively ([80]; see also [81]). In NSE experiments, the spin incoherent contribution to the signal can however also induce a significant reduction of the polarization due to its spin-flip component which introduces a negative term (see Eq. 15.63). This adds additional problems of normalization and interpretation, which we will not discuss in the framework of this article.

In biological problems, the interpretation of the results from measurements in (Q, ω)-space is comparatively straightforward, since due to the preponderance of hydrogen atoms in biological matter one can make use of the dominant incoherent scattering of the proton.

The coherent scattering spectrum arises from scattering length density fluctuations in space and time. In the presence of macromolecules in a solution, scattering length fluctuations are characterized by the size of the molecules; their effect dominates the small Q regime of the spectrum. At low concentration (when the interactions between neighboring molecules are negligible) the molecular form factor can be directly determined by experiment. When the concentration increases, macromolecular interaction leads to significant pair-correlations and (via the structure factor) becomes important in the scattered intensity. A relevant factor is the scattering length density contrast between, for example, the solvent and the solute molecules. The coherent scattering lengths of hydrogen and deuterium are very different: $b_c^H = -3.741$ and $b_c^D = 6.674$ Fermi.[9] The deuteration of a solute macromolecule (or of the solvent) allows to strongly increase the coherent part of the spectrum. Note, that this is another important means of separating coherent and incoherent scattering components. In NSE experiments, where spin-flip causes a large background, the possibility of using deuterated samples can be decisive for the feasibility of the measurements.

More information on modern QENS spectrometers can be found in the Instrumentation Brochures and on the Internet home pages of the neutron scattering centers (ILL in Grenoble, France; LLB in Saclay, France; BENSC at HMI in Berlin, Germany; ISIS at RAL in Chilton, UK; NIST in Gaithersburg, USA; FZ Jülich, Germany; FRM-II München, Germany; IPNS at Argonne, USA; PSI Villigen, Switzerland).

15.6 Conclusions

We have given an overview of theory and experimental techniques of quasielastic and low-energy inelastic neutron scattering. Apart from the basics

---

[9] 1 fermi $= 10^{-13}$ cm

of scattering functions and correlation functions, elastic and quasielastic incoherent structure factors have been treated, and the importance of energy resolution and observation time and their variation have been emphasized. A number of selected instruments have been described, which are representative for different ranges of momentum transfer $\boldsymbol{Q}$, energy transfer $\hbar\omega$, and energy resolution $\Delta\hbar\omega$, or Fourier time t and observation time Δt_{obs}. While, in principle, all types of instrument can cover the same or similar Q-ranges depending on the incident neutron wavelength, this is not so for $\hbar\omega$, $\Delta\hbar\omega$, t, and Δt_{obs}. Moderate, high and very high energy resolution are provided by time-of-flight, backscattering and spin–echo techniques, respectively. There is however appreciable overlap between the resolution ranges of different instruments, and this is beneficial, since results from different instruments must often be combined. Furthermore, it is also important to realize, that – because molecular motions occur on time-scales extending over many orders of magnitude – the access to every possible resolution is required, i.e., there is no "bad resolution". As a consequence, the application of *observation-time dependent* quasielastic neutron scattering, in order to investigate dynamic effects due to diffusive atomic and molecular motions has been rapidly expanding and has a very promising future. The high intensities available at modern neutron scattering instruments and especially in the near future at the high-power spallation neutron sources now under construction, will allow the systematic experimental study of such motions on time scales extending over six orders of magnitude at least, and as a function of external parameters.

Regarding the problem, which instrument to choose for a given task, or which spectrometer to start with, it is clear that the relevant temporal and spatial scales are the first criteria to be considered. There are also some general remarks to be made: Once the scale question has been settled, one must decide, whether it is better to employ a $(\boldsymbol{Q},\omega)$- or a $(\boldsymbol{Q},t)$-instrument. Spectrometers working in $(\boldsymbol{Q},\omega)$-space are more efficient for the study of small quasielastic or inelastic effects in the presence of large elastic components. The reason is, that under conditions of sufficient energy resolution, elastic, and nonelastic phenomena are experimentally well separated, whereas in the case of $(\boldsymbol{Q},t)$-techniques the elastic term of the scattering function leads to a time-independent (possibly large) "elastic" background underneath the time-dependent relaxation function to be studied. Without this problem, i.e., when the elastic component is small or absent, $(\boldsymbol{Q},t)$-instruments are very well suited, especially if rather long Fourier times are required. Furthermore, it is known that the latter techniques are generally more suitable for coherent than for spin-incoherent scattering, because the spin-flip scattering component reduces the useful intensity and can appreciably deteriorate the signal-to-noise ratio. The detailed performance (spectral and reciprocal-space regions, i.e., time and energy domains, momentum-transfer ranges, measurement efficiency) of the different QENS techniques, TOF, BSC, NSE, NRSE, in various experimental situations, is best appreciated by comparing different experimental results (for a few examples, see Part II in this volume). The question, how to make best use of the complementarity of the different techniques, in order to obtain a maximum of information, must be addressed systematically on the basis of such a comparison before any experimental study.

References

1. L. Van Hove, Phys. Rev. **95**, 249 (1954)
2. S.W. Lovesey, *Theory of Neutron Scattering from Condensed Matter* (Clarendon Press, Oxford, 1984)
3. G.L. Squires, *Introduction to the Theory of Thermal Neutron Scattering* (University Press, Cambridge, 1978)
4. M. Bée, *Quasielastic Neutron Scattering: Principles and Applications in Solid State Chemistry, Biology and Materials Science* (Adam Hilger, Bristol, 1988)
5. L. Koester, H. Rauch, E. Seymann, At. Data Nucl. Data Tables **49**, 65 (1991)
6. V.F. Sears, Phys. Rep. **82**, 1 (1982)
7. H. Stiller, in *Inelastic Scattering of Neutrons*, Vol. 2 (IAEA Vienna, 1965) p. 179
8. K. Sköld, J. Chem. Phys. **49**, 2443 (1968)
9. A.J. Leadbetter, R.E. Lechner, in *The Plastically Crystalline State*, J.N. Sherwood (Eds.) (J. Wiley and Sons, New York, 1979) pp. 285–320
10. R.E. Lechner, in *Quasielastic Neutron Scattering*, J. Colmenero, A. Alegría, F.J. Bermejo (Eds.), Proceedings of the Quasielastic Neutron Scattering Workshop QENS'93, San Sebastián, Spain 1993 (World Scientific, Singapore, 1994) pp. 62–92
11. M. Bée, Chem. Phys. **292**, 121 (2003)
12. A.J. Dianoux, F. Volino, H. Hervet, Mol. Phys. **30**, 1181–1194 (1975)
13. V.F. Sears, Can. J. Phys. **45**, 237 (1967)
14. F. Volino, A.J. Dianoux, Mol. Phys. **41**, 271 (1980)
15. R.E. Lechner, in *Mass Transport in Solids*, F. Benière, C.R.A. Catlow (Eds.), NATO ASI (1981: Lannion, France) Series B: Physics, Vol. 97 (Plenum Publ. Corp. New York, 1983) pp. 169–226.
16. J. Fitter, R.E. Lechner, G. Büldt, N.A. Dencher, Proc. Natl. Acad. Sci. USA **93**, 7600–7605 (1996)
17. T. Springer, *Quasielastic Neutron Scattering for the Investigation of Diffusive Motions in Solids Liquids*, Springer Tracts in Modern Physics, Vol. 64 (Springer-Verlag, Berlin, 1972)
18. R.E. Lechner, Physica B **301**, 83–93 (2001)
19. B. Rufflé, J. Ollivier, S. Longeville, R.E. Lechner, Nucl. Instr. Methods A **449**, 322 (2000)
20. D.L. Price, K. Sköld, *Neutron Scattering*, Methods of Experimental Physics, Vol. 23, part B, Chapters 8–11 (Academic Press, London, 1987)
21. R.E. Lechner, C. Riekel, in *Neutron Scattering and Muon Spin Rotation*, Springer Tracts in Modern Physics, Vol. 101, (Springer-Verlag, Berlin, 1983) pp. 1–84
22. D. Richter, in *Neutron Scattering and Muon Spin Rotation*, Springer Tracts in Modern Physics, Vol. 101 (Springer-Verlag, Berlin, 1983) pp. 85–222
23. S.W. Lovesey, T. Springer, *Dynamics of Solids and Liquids by Neutron Scattering*, Topics in Current Physics, Vol. 3 (Springer-Verlag, Berlin, 1977)

24. T. Springer, R.E. Lechner, in *Diffusion in Condensed Matter*, P. Heitjans, J. Kärger (Eds.) (Springer Verlag, Berlin, Heidelberg, 2005) pp. 93–164
25. C.G. Windsor, *Pulsed Neutron Scattering* (Taylor & Francis, London, 1981)
26. R. Scherm, C. Carlile, A.J. Dianoux, J. Suck, J. White, ILL Int. Rep. 76S235, (1976)
27. Y. Blanc, ILL Int. Rep. 83BL21G, (1983)
28. see http://www.ill.fr/YellowBook/IN6/
29. J. Mesot, S. Janssen, L. Holitzner, R. Hempelmann, J. Neutron Res. **3**, 293–310 (1996)
30. H. Mutka, Nucl. Instr. Methods Phys. Res. A, **338**, 144–150 (1994)
31. R.E. Lechner, in *Proceedings of ICANS-XV Conference, Tsukuba, 2000*, J. Suzuki, S. Itoh (Eds.) (JAERI, Tokai-mura, Japan, 2001) pp. 357–376
32. R.E. Lechner, in *Neutron Scattering in the 'Nineties* (IAEA, Vienna, 1985) pp. 401–407
33. R. Scherm, Int. Rep., KFA Jülich, (1965) Jül-295-NP
34. F. Douchin, R.E. Lechner, Y. Blanc, Int. Tech. Rep. ITR, ILL Grenoble **26**, 73 (1973)
35. J. Ollivier, H. Casalta, H. Schober, J.C. Cook, P. Malbert, M. Locatelli, C. Gomez, S. Jenkins, I.J. Sutton, M. Thomas, Appl. Phys. A **74**, [Suppl.] S305–S307 (2002)
36. J. Ollivier, M. Plazanet, H. Schober, J.C. Cook, Physica B **350**, 173–177 (2004)
37. S. Hautecler, E. Legrand, L. Vansteelandt, P. d'Hooghe, G. Rooms, A. Seeger, W. Schalt, G. Gobert, in *Neutron Scattering in the 'Nineties* (IAEA, Vienna, 1985) pp. 211
38. R.E. Lechner, in *Proceedings of ICANS-XI*, KEK Report 90–25, M. Misawa, M.Furusaka, H. Ikeda, N. Watanabe (Eds.) (National Laboratory for High-Energy Physics, Tsukuba, March 1991) pp. 717–732
39. R.E. Lechner, Physica B **180 & 181**, 973 (1992)
40. R.E. Lechner, Neutron News **7**, 9 (1996)
41. R.E. Lechner, R. Melzer, J. Fitter, Physica B **226**, 86 (1996)
42. J.R.D. Copley, Nucl. Instr. Methods A **291**, 519–532 (1990)
43. J.G. Smit, H. Dachs, R.E. Lechner, Solid State Commun. **29**, 219–223 (1979)
44. R.E. Lechner, Appl. Phys. A **74**, [Suppl.] S151 (2002)
45. B. Alefeld, M. Birr, A. Heidemann, Naturwissenschaften **56**, 410 (1969)
46. B. Alefeld, T. Springer, A. Heidemann, Nucl. Sci. Eng. **110**, 84 (1992)
47. B. Frick, A. Magerl, Y. Blanc, R. Rebesco, Physica B **234–236**, 1177 (1997)
48. B. Frick, Neutron News **13**, 15 (2002)
49. H. Rauch, in *Neutron Diffraction*, H. Dachs (Eds.) Topics in Modern. Physics, Vol. 6 (Springer-Verlag, Berlin, 1978)
50. C.G. Shull, N.S. Ginrich, J. Appl. Phys. **35**, 678 (1964)
51. P.M. Gehring, D.A. Neumann, Physica B **241–243**, 64–70 (1998)
52. A.J. Dianoux, Physica B **182**, 389 (1992)
53. A. Kollmar, B. Alefeld, in *Proceedings of the Conference on Neutron Scattering, Gatlinburg, Tenn. USA 1976*, R.M. Moon (Eds.), (National Techn. Information Service, US Dept. Of Comm., Springfield, 1976) pp. 330–336.
54. M. Prager, W. Press, B. Alefeld, A. Hüller, J. Chem. Phys. **67**, 5126 (1977)
55. G.P. Singh, F. Parak, S. Hunklinger, K. Dransfeld, Phys. Rev. Lett. **47**, 685–688 (1981)
56. F. Parak, E.W. Knapp, D. Kucheida, J. Mol. Biol. **161**, 177–194 (1982)

57. F. Parak, in *Biomembranes*, Methods in Enzymology, Vol. 127, L. Packer (Eds.) (Academic Press, Inc. London, 1986) pp. 197–206.
58. W. Doster, S. Cusack, W. Petry, Nature **337**, 754–758 (1989)
59. W. Doster, S. Cusack, W. Petry, Phys. Rev. Lett. **65**, 1080–1083 (1990)
60. R.E. Lechner, H.-J. Bleif, H. Dachs, R. Marx, M. Stahn, Solid, State Ionics **46**, 25 (1991)
61. C.J. Carlile, M.A. Adams, Physica B **182**, 431–440 (1992)
62. R.W. Connaster Jr., H. Belch, L. Jirik, D.J. Leach, F.R. Trouw, J.-M. Zanotti, Y. Ren, R.K. Crawford, J.M. Carpenter, D.L. Price, C.-K. Loong, J.P. Hodges, K.W. Herwig, in *Proceedings of ICANS-XVI Conference, Düsseldorf-Neuss, 2003*, G. Mank, H. Conrad (Eds.) (FZ Jülich, Germany, 2003) pp. 279–288
63. W. Doster, M. Diehl, W. Petry, M. Ferrand, Physica B **301**, 65 (2001)
64. W. Doster, M. Diehl, R. Gebhardt, R.E. Lechner, J. Pieper, Chem. Phys. **292**, 487–494 (2003)
65. F. Mezei, Z. Physik **255**, 146–160 (1972)
66. F. Mezei, *Neutron Spin Echo*, Lecture Notes in Physics Vol. 128 (Springer Verlag, Berlin, 1980)
67. C. Cohen-Tannoudji, B. Diu, F. Laloë, *Mécanique Quantique I* (Hermann ed., Paris, 1977)
68. R.P. Feynman, *The Feynman Lectures on Physics*, Quantum Mechanics (Addison-Wesley, 1965)
69. J.B. Hayter, Z. Physik B **31**, 117–125 (1978)
70. J.B. Hayter, J. Penfold, Z. Physik B **35**, 199–205 (1979)
71. O. Schärpf, in *Neutron Spin Echo*, Lecture Notes in Physics Vol. 128, F. Mezei (Eds.) (Springer Verlag, Berlin, 1980) pp. 27–52
72. J.B. Hayter, in *Neutron Spin Echo*, Lecture Notes in Physics Vol. 128, F. Mezei (Eds.) (Springer Verlag, Berlin, 1980) pp. 53–65
73. F. Mezei, in *Coherence and Imaging Processes in Physics*, M. Schlenker, M. Fink, J.P. Goedgebuer, C. Malgrange, J.C. Vienot, R.H. Wade (Eds.), Vol. 112 (Springer-Verlag, Les Houches, 1979) pp. 282–295
74. S. Longeville, W. Doster, M. Diehl, R. Gähler, W. Petry, in *Neutron Spin-Echo Spectroscopy*, Lecture Notes in Physics, Vol. 601, F. Mezei, C. Pappas, T. Gutberlet (eds.) (Springer-Verlag, Berlin, 2003) pp. 325–335
75. R. Gähler, R. Golub, Z. Physik B – Condens. Matter **65** 269–273, (1987)
76. R. Golub, R. Gähler, Phys. Lett. A, **123** 43–48, (1987)
77. V.F. Sears, Adv. Phys. **24**, 1 (1975)
78. R.E. Lechner, G. Badurek, A.J. Dianoux, H. Hervet, F. Volino, J. Chem. Phys. **73**, 934 (1980)
79. M. Russina, F. Mezei, R.E. Lechner, S. Longeville, B. Urban, Phys. Rev. Lett. **84**, 3630 (2000)
80. C. Lamers, O. Schärpf, W. Schweika, J. Batoulis, K. Sommer, D. Richter, Physica B **180 & 181**, 515 (1992)
81. J.C. Cook, D. Richter, O. Schärpf, M.J. Benham, D.K. Ross, R. Hempelmann, I.S. Anderson, S.K. Sinha, J. Phys. Condens. Matter **2**, 79 (1990)

16

Quasielastic Neutron Scattering in Biology, Part II: Applications

R.E. Lechner, S. Longeville

16.1 Introduction

Low-frequency harmonic, damped harmonic, anharmonic, localized diffusive (stochastic) translational, and rotational motions of biological macromolecules or of molecular subunits, are an important part of the dynamical characteristics of biological matter. They are believed to be essential for biological function. The ensemble of low-energy transfer inelastic (IENS) and quasielastic neutron scattering (QENS[1]) techniques, is particularly well-suited for the investigation of dynamic structure. The instrumentation and the basic theory have been treated in Part I in this volume. The present article describes the application of these techniques and gives a brief introduction to the theory of dynamical models employed in the analysis of measured spectra. The QENS method focuses on scattering processes with small energy exchange, where spectral distributions are peaked at zero energy transfer. Such phenomena are mainly due to diffusive (i.e., stochastic, random-walk) atomic and/or molecular motions. IENS spectra, mainly originating from atomic and molecular vibrations, extend to somewhat higher energies, but are often not negligible in the QENS energy region, and therefore have to be dealt with simultaneously in a QENS analysis.

The material to be discussed here, is ordered as follows: Sect. 16.2 deals with the motion decoupling approximation, Sect. 16.3 with the Gaussian approximation, including simple translational diffusion according to Fick's law, vibrational motions, phonon-expansion and Debye–Waller factor (DWF), and the vibrational density of states in a light-harvesting system as an example. In Sect. 16.4, non-Gaussian scattering functions, in particular for atomic jump-diffusion and localized (i.e., confined) diffusive motions, are discussed and

[1]Instead of QENS, the acronym QINS is used for the special case of quasielastic *incoherent* neutron scattering

applications to two different biological systems (lysozyme solvated in glycerol and ligand binding to a protein) are presented. Finally, Sect. 16.5 looks into two-dimensional diffusion on the surface of biological membranes, with purple membrane as an example, and presents results of studies on the so-called dynamical transition of the latter. Many other examples will be mentioned and corresponding references given.

16.2 Dynamical Models

In order to reach an understanding of the atomic motions in the scattering systems and ultimately of the role of such motions in properties and function of biological materials, the theoretical Van Hove scattering functions or the intermediate scattering functions required for a comparison with measured neutron spectra are derived from models for the molecular dynamics. The theoretical formulae are least-squares fitted to the measured spectra, in order to determine the values of the model parameters. In the following, we describe some of the simplest approximations and models and give the pertinent scattering functions for a number of representative cases.

16.2.1 Dynamical-Independence Approximation

Atomic motion in complex systems such as biological macromolecules is generally composed of a number of components occurring on different time-scales. They may be classified according to their frequencies, rates or correlation times, and according to their geometries and spatial extension. Generally, we have to distinguish between periodic (vibrational) and stochastic (diffusive) motions. The latter may be of a long-range type, such as translational diffusion, or have a localized character, like in the case of rotational diffusion. Different motions may be strongly coupled with each other, weakly coupled or uncoupled. In the former cases the equation of motion to be solved, when calculating the Van Hove correlation functions and the scattering functions, must take account of the coupling, while the latter case allows an independent treatment of individual types of motion. This procedure, which is obviously much simpler, permits a modular construction of the Van Hove functions $I_s(\boldsymbol{Q},t)$ and $S_s(\boldsymbol{Q},\omega)$ (see Sect. 15.2.1 in Part I, this volume) obtained by combining together the individual scattering functions. For instance, the calculation is greatly facilitated, if e.g., vibrational, rotational, and translational motions are assumed to be (dynamically) independent (see [1–3]). The intermediate scattering function is then a product, while the scattering function is a convolution of the pertinent modular functions:

$$I_s(\boldsymbol{Q},t) = I_{\mathrm{vib}}(\boldsymbol{Q},t) I_{\mathrm{rot}}(\boldsymbol{Q},t) I_{\mathrm{trans}}(\boldsymbol{Q},t), \tag{16.1}$$

$$S_s(\boldsymbol{Q},\omega) = S_{\mathrm{vib}}(\boldsymbol{Q},\omega) \otimes S_{\mathrm{rot}}(\boldsymbol{Q},\omega) \otimes S_{\mathrm{trans}}(\boldsymbol{Q},\omega). \tag{16.2}$$

Here, $I_{\rm vib}(\boldsymbol{Q},t)$, $I_{\rm rot}(\boldsymbol{Q},t)$, $I_{\rm trans}(\boldsymbol{Q},t)$ are the incoherent intermediate and $S_{\rm vib}(\boldsymbol{Q},\omega)$, $S_{\rm rot}(\boldsymbol{Q},\omega)$, $S_{\rm trans}(\boldsymbol{Q},\omega)$ the incoherent Van Hove scattering functions of the three individual types of motions; the symbol $\otimes$ stands for the convolution in energy transfer $\hbar\omega$. The question, as to when this convolution approximation can be applied, requires some discussion. The vibrational term in this convolution can usually be replaced by a DWF, as long as the study is restricted to the quasielastic region. Strictly speaking, the independence approximation of rotational and translational motions is often invalid, for instance if the two motions are not occurring simultaneously and independently of each other, but as a sequence of alternating steps. The validity of the approximation is, however, recovered, if for instance the rate of rotational steps, $H_{\rm rot}$, is much higher than the rate of translational steps, $H_{\rm trans}$, because then it can be argued that the concerned particle is quasi-continuously participating in the rotational motion. Under the condition that the rates differ by at least an order of magnitude (a situation typical for many real systems), one can study the rotational motion "alone" in a low or medium resolution experiment, by choosing an intermediate energy resolution $\Delta(\hbar\omega)$ such that $\hbar H_{\rm trans} \ll \Delta(\hbar\omega) \leq \hbar H_{\rm rot}$. This simply means that the experimental observation time $\Delta t_{\rm obs}$ [4] (which in practice cannot be made infinite; see the related discussion in Sect. 15.2.3 of Part I in this volume) is larger than the build-up time of the local probability-density distribution (PDD)(for instance due to a reorientational motion), but much smaller than its time of decay due to long-range diffusion. The elastic incoherent structure factor (EISF) ([5] and [6]; see Sect. 15.2.2 of Part I in this volume) of the rotational motion can then be determined from the integral of the only weakly broadened elastic peak, whereas the rotation rate is obtained from the quasielastic line width of the broader spectral component. Vice versa translational diffusion is studied alone in a high resolution experiment with $\Delta(\hbar\omega) \leq \hbar H_{\rm trans} \ll \hbar H_{\rm rot}$, because under this condition the rotational component will contribute only a flat background to the spectrum, whereas the diffusion parameters can be determined from the Q-dependent line width of the central component of $S_{\rm s}(\boldsymbol{Q},\omega)$. This has been the justification for applying the dynamic-independence approximation in many cases, e.g., for a number of superprotonic conductors exhibiting the Grotthuss mechanism [7–9]. The same principles are of course applicable in the case of biological studies.

16.3 The Gaussian Approximation

In early theoretical neutron scattering work, it had already been noticed [10] that in some idealized systems, such as the harmonic solid, the ideal gas and the particle diffusing according to the simple diffusion equation or Langevin's equation, Van Hove's self-correlation function $G_{\rm s}(\boldsymbol{r},t)$ has a simple Gaussian shape in space. The Gaussian has also been employed in the case of less ideal systems, as a model for describing PDDs by elliptical (in 2D) or ellipsoidical (in 3D) functions, with circle and sphere as special cases, respectively. In the isotropic case we have,

$$G_s(\boldsymbol{r},t) = (2\pi w(t))^{-3/2}\exp(-r^2/2w(t)). \qquad (16.3)$$

In reciprocal space, this is obviously reflected by Gaussian shapes in Q of the corresponding incoherent scattering functions, $I_s(\boldsymbol{Q},t)$ and $S_s(\boldsymbol{Q},\omega)$:

$$I_s(\boldsymbol{Q},t) = \exp\left(-\frac{1}{2}Q^2 w(t)\right), \qquad (16.4)$$

$$S_s(\boldsymbol{Q},\omega) = (2\pi)^{-1}\int_{-\infty}^{+\infty} \exp\left(-\frac{1}{2}Q^2 w(t)\right) \mathrm{e}^{-\mathrm{i}\omega t}\mathrm{d}t, \qquad (16.5)$$

where the function $w(t)$ controls the time decay of $G_s(\boldsymbol{r},t)$ and $I_s(\boldsymbol{Q},t)$ and is related to the energy width of $S_s(\boldsymbol{Q},\omega)$. In order to be able to explicitly calculate $S_s(\boldsymbol{Q},\omega)$ by carrying out this Fourier transformation, the (system-dependent) width function $w(t)$ has to be specified. In the following, we give a few examples of Gaussian motion.

16.3.1 Simple Translational Diffusion

In view of the omni-presence of water in living organisms, it is evident that atomic and molecular diffusion play an important role in this context. It is worthwhile to start by considering the most simple diffusion models, even if diffusion processes in the interior of biological materials must be expected to be much more complex.[2] Van Hove's self-correlation function (accessible by incoherent neutron scattering, see Sect. 15.2.1 in Part I, this volume), in the case of simple (continuous) diffusion is derived by solving the macroscopic diffusion equation (e.g., in three dimensions), with the self-diffusion coefficient D_s,

$$\frac{\partial G_s(\boldsymbol{r},t)}{\partial t} = D_s \nabla^2 G_s(\boldsymbol{r},t), \qquad (16.6)$$

with the initial condition $G_s(\boldsymbol{r},0) = \delta(\boldsymbol{r})$. The solution of this equation is well known. It is given by Eq. 16.3, if we write $w(t) = 2D_s\,|t|$ for the width function. The corresponding intermediate scattering function is obtained by making the same substitution in Eq. 16.4. For the incoherent scattering function one obtains from Eq. 16.5 a simple Lorentzian with half-width at half-maximum (HWHM) $f(\boldsymbol{Q})$,

$$S_s(\boldsymbol{Q},\omega) = \frac{f(\boldsymbol{Q})/\pi}{\omega^2 + f(\boldsymbol{Q})^2}, \qquad (16.7)$$

where $f(\boldsymbol{Q}) = D_s Q^2$. We note, however, that this Gaussian behavior of the scattering functions is limited to the regime, where continuous diffusion is a

---

[2]For a more general treatment of diffusion studies employing quasielastic neutron scattering techniques, see [11]

valid picture, namely to sufficiently low Q and long times. This corresponds to large displacements, where in general the diffusing particle is already far away from the origin (where it was at $t = 0$), and has "forgotten" where it came from and what the detailed structure of its neighborhood has been in the early stage of its journey. There is a number of other, more sophisticated, diffusion models proposed by Chudley-Elliott [12], Singwi-Sjölander [13], Oskotskii [14], Egelstaff [15], Rahman-Singwi-Sjölander [16], with Gaussian behavior for low Q and long times. Here, we can merely give reference to some of the original literature, but treat one of them [12] in some detail in Sect. 16.4.1, as a representative of noncontinuum diffusion models taking account of the local structure. Another one (two-dimensional diffusion) is discussed in Sect. 16.5.1.

16.3.2 Three-Dimensional Diffusion of Protein Molecules in Solution (Crowded Media)

For studying the diffusion of molecules as large as proteins in solution with fairly large concentrations, very high energy resolution is required, i.e., very large Fourier times must be reached. For this purpose, neutron spin-echo spectroscopy may be used to study such translational motions of proteins in solution under "crowded" conditions. As an example, we discuss an investigation of myoglobin solutions in vitro as a function of the protein volume fraction $\Phi \simeq 0.05$–0.4 [17]. From a biological point of view, myoglobin is an oxygen storage molecule located in muscles and is believed to assist oxygen transport by diffusion from the high oxygen partial pressure (near the cell membrane) to the low partial pressure regions (the mitochondria). Neutron scattering is the only technique which allows to study both protein interactions and protein motions over intermolecular distances at physiological concentration. In the case of NSE spectroscopy, it is advisable to use coherent neutron scattering (giving access to the Van Hove pair-correlation function, rather than to the self-correlation function). For this reason, and for maximizing the coherent scattering signal at small wave vectors, native hydrogeneous proteins are solved in D_2O. Typical intermediate scattering functions measured on two different spectrometers are shown in Figs. 16.1 and 16.2. The relaxation function follows a single exponential decay and can be refined as

$$I(Q,t) = \exp(-D_c(Q)Q^2\,|t|)\,. \tag{16.8}$$

$D_c(Q)$ is an apparent collective diffusion coefficient, close but not equal to the protein self-diffusion coefficient, because a coherent quasielastic scattering function is measured. The precise interpretation of $D_c(Q)$ in terms of self-diffusion is not straight-forward. The wave vector and volume fraction dependence of $D_c(Q)$ is plotted in Fig. 16.3. At large Q, $D_c(Q)$ tends to a plateau value which corresponds to the short-time self-diffusion coefficient ${}^{s}D_s(\Phi)$. This quantity is very strongly volume fraction dependent, although direct interactions are not relevant over such short distances ($S(Q) \simeq 1$ in

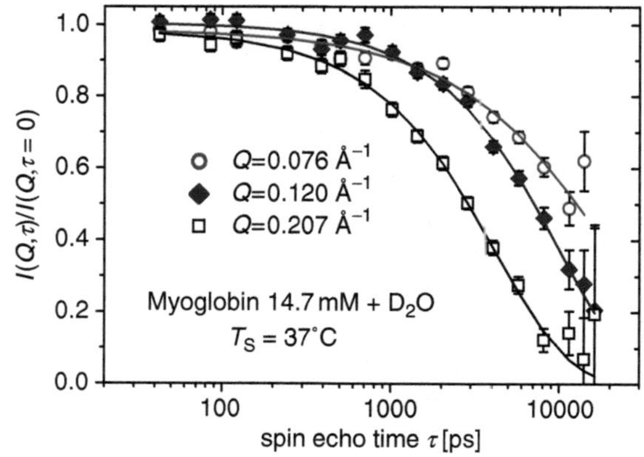

Fig. 16.1. Intermediate scattering function of 14.7 mM myoglobin solution in D_2O, as obtained with the spectrometer MUSES [18] at LLB, Saclay. The instrument was used in its NSE mode of operation (see Sect. 15.4 in Part I, this volume), at three different wavevectors around the structure factor maximum (from [17])

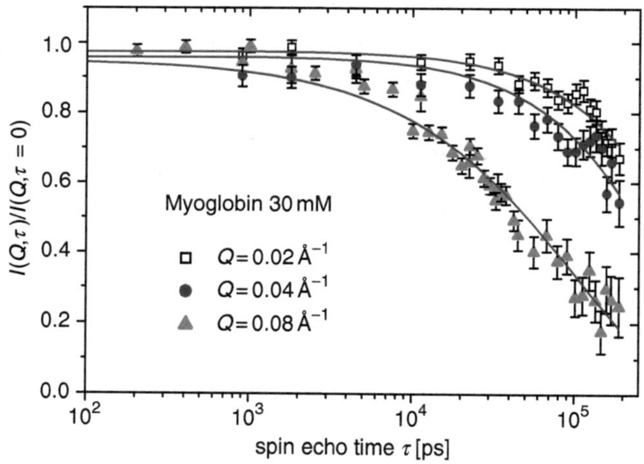

Fig. 16.2. Intermediate scattering function of 30 mM myoglobin solution in D_2O, as obtained with the NSE spectrometer IN15 at ILL, Grenoble (see Sect. 15.4 in Part I, this volume), for three different wavevectors in the very low Q-range (from [17])

this wave vector range). Hydrodynamic interactions are responsible for this strong volume dependence. At smaller wave vectors $D_c(Q)$ increases. It probes pair-correlation relaxations which are accelerated by concentration fluctuation relaxations. In the limit of $Q \Rightarrow 0$,

$$D_c = \frac{1}{f}\frac{\partial \Pi}{\partial C}, \tag{16.9}$$

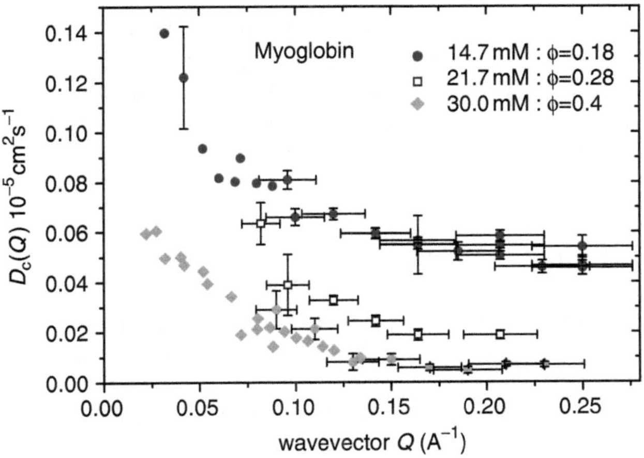

Fig. 16.3. Apparent collective diffusion coefficient $D_c(Q)$ of myoglobin in D_2O solution, for three different concentrations. The wavevector dependence exhibits a plateau at large Q, related to the short-time self-diffusion coefficient ${}^{s}D_s(\Phi)$. Electrostatic interaction is responsible for the increase at small Q (from [17])

where f is a friction coefficient and $\frac{\partial \Pi}{\partial C}$ is the reciprocal osmotic compressibility which acts as a force tending to relax the concentration fluctuations. Generally such studies combined with an analysis of the small-angle neutron scattering (SANS) spectra in terms of direct protein–protein interactions allow to separate hydrodynamic and direct interactions [19]. An interesting property of QENS is that it allows to probe protein dynamics directly inside cells as for example in the case of hemoglobin in red blood cells [20].

16.3.3 Vibrational Motions, Phonon-Expansion and Debye–Waller factor (DWF), Dynamic Susceptibility

In solids, atoms, and molecules perform vibrations around fixed equilibrium positions and orientations. The frequency spectrum of these vibrations is described by the phonon density of states $g(\omega)$. In the simplest case of a long-range ordered crystalline system (cubic crystal in the harmonic approximation, with only one atom per unit cell), the wave-like propagation of vibrations is exactly represented by the scattering function $S_{\text{vib}}(\boldsymbol{Q}, \omega)$, in the form of the phonon-expansion [21, 22]:

$$S_{\text{vib}}(\boldsymbol{Q}, \omega) = \mathrm{e}^{-<u^2>Q^2} \left[\delta(\omega) + \sum_{n=1}^{\infty} \frac{\left[< u^2 > Q^2\right]^n}{n!} S_n(\omega) \right] , \qquad (16.10)$$

where $<u^2>$ is the atomic mean square displacement and $\mathrm{e}^{-<u^2>Q^2}$ is the DWF. The latter obviously stands for the Gaussian behavior in space of the time-averaged atomic probability density distribution in the harmonic solid. The first term of the sum, $\delta(\omega)$, represents the purely elastic scattering, while the terms $S_n(\omega)$ correspond to n-phonon transitions. The normalized one-phonon term $S_1(\omega)$ is the product of the Bose phonon-population factor $n_{\mathrm{B}}(\omega, T)$ with $g(\omega)$,

$$S_1(\omega) = n_{\mathrm{B}}(\omega, T) g(\omega)\,, \tag{16.11}$$

$$n_{\mathrm{B}}(\omega, T) = \frac{1}{\mathrm{e}^{\frac{\hbar\omega}{k_{\mathrm{B}}T}} - 1}\,, \tag{16.12}$$

where T is the temperature, k_{B} and $\hbar$ are the Boltzmann and Planck constants, respectively. Equation 16.12, can also be written in the following form:

$$n_{\mathrm{B}}(\omega, T) = \exp(-\hbar\omega/2k_{\mathrm{B}}T)\,[2\omega \sinh(\hbar\omega/2k_{\mathrm{B}}T)]^{-1} \tag{16.13}$$

Note that now the detailed balance factor, $\exp(-\hbar\omega/2k_{\mathrm{B}}T)$ appears explicitly in this expression (compare Eqs. 16.18 to 16.20, Sect. 15.2.1 in Part I, this volume). Each profile $S_n(\omega)$ $(n \geq 1)$ is obtained by folding $S_1(\omega)$ $(n-1)$ times with itself. Since the detailed-balance factor is invariant against this convolution, it can in practice be extracted from the convolution kernel and put in front of the sum in Eq. 16.10. The normalization of the incoherent scattering function 16.10 follows from the definition:

$$\int_{-\infty}^{\infty} S_1(\omega)\mathrm{d}\omega = 1\,, \tag{16.14}$$

and therefore

$$\int_{-\infty}^{\infty} S_n(\omega)\mathrm{d}\omega = 1\,, \tag{16.15}$$

for all n. In the presence of motions with a diffusive character, in addition to vibrations, the scattering function will also have a quasielastic component that may be combined with the vibrational component 16.10 according to the dynamic independence approximation 16.2, where applicable. If quasielastic scattering is inexistent or unresolved, e.g., in experiments limited to low temperatures ($T \leq 100\,\mathrm{K}$) and/or to large or medium energy resolution widths ($\Delta(\hbar\omega) \geq 90\,\mathrm{meV}$), observed inelastic neutron scattering spectra will be due to vibrational excitations only.

In biological matter, e.g., proteins, the spatial arrangement of atoms and the dynamics are much more complex, than in the above-mentioned type of systems with long-range crystalline order. The unperturbed spatial propagation of vibrational modes is then typically restricted to shorter ranges. An exact theory is usually intractable. But, for practical purposes, we may employ expression 16.10 to extract an effective vibrational density of states from measured neutron scattering spectra. This is a rather useful semi-phenomenological quantity, especially in studies of the evolution of vibrational dynamics as a function of external parameters. The effective density of states

$g(\omega)$ accounts for the individual displacement amplitudes, phonon polarization vectors, and masses of the ensemble of generalized oscillators of the complex protein system. For small values of $<u^2>$ and Q^2 the higher order terms in 16.10 can be neglected, i.e., a one-phonon approximation may be employed, where Eq. 16.10 reduces to

$$S_{\text{vib}}(\boldsymbol{Q},\omega) = \mathrm{e}^{-<u^2>Q^2}\left[\delta(\omega) + < u^2 > Q^2 S_1(\omega)\right] . \tag{16.16}$$

The effective vibrational density of states can be determined using the above expressions in either one of the following two ways: (i) $S_{\text{vib}}(\boldsymbol{Q},\omega)$ according to Eqs. 16.10 and 16.11, is fitted to the spectra measured over an extended Q-range, using suitable model functions for $g(\omega)$, or (ii) $g(\omega)$ is found, independently of model functions, by taking the inelastic part of the measured data to the limit

$$g(\omega) = \lim_{Q\to 0}\left[\frac{2M}{\hbar Q^2}\left[n_{\mathrm{B}}(\omega,T)\right]^{-1} S_{\text{vib}}(\boldsymbol{Q},\omega)\right] , \tag{16.17}$$

where M is the effective molecular mass. The second procedure has sometimes been applied, even when the dynamics of the system under study is not governed exclusively by harmonic or quasiharmonic vibrations. When strongly damped, overdamped or almost purely stochastic motions are also present, the function $g(\omega)$ then obtained by this method, obviously is a "generalized" one, in the sense that it also includes the effects of those additional motions. Their signature is, that the generalized function does not show the ω^2-behavior well-known for harmonic vibrations at low energies, and starts with a finite value at $\omega = 0$ (see Sect. 16.4.4).

Instead of the Van Hove scattering functions the corresponding dynamic susceptibilities are sometimes used. The susceptibility $\chi(\boldsymbol{Q},\omega)$ is defined in the framework of linear response theory [23] as the steady state response of a system to an oscillatory force. In other words, it corresponds to nontransient response and defines the fluctuations in the unperturbed system. The function,

$$\chi(\omega) = \chi'(\omega) + \mathrm{i}\chi''(\omega) , \tag{16.18}$$

is a complex quantity; the real part corresponds to the in-phase and the imaginary part to the out-of-phase response. The fluctuation–dissipation theorem describes, how fluctuations of the system will be dissipated in it at thermal equilibrium. $\chi''(Q,\omega)$ is related to the dynamic structure factor (i.e., the scattering function):

$$S(\boldsymbol{Q},\omega) = n_{\mathrm{B}}(\omega,T)\chi''(\omega). \tag{16.19}$$

Using the imaginary part of the dynamic susceptibility can be particularly useful in some special cases: for instance, in the case of a simple harmonic oscillator and equally for purely harmonic crystals, $\chi''(Q,\omega)$ does not depend on temperature, because here all the temperature effect is contained in the Bose phonon-population factor $n_{\mathrm{B}}(\omega,T)$. $\chi''(Q,\omega)$ is consequently sometimes used to study those effects which do not depend on thermal population.

16.3.4 Vibrational Density of States of the Light-harvesting Complex II of Green Plants

Photosynthesis is a fundamental physiological process in nature representing the transformation of solar radiation into storable chemical energy. One of the important functions of antenna complexes during the primary steps of photosynthesis, is to perform an efficient excitation energy transfer (EET) to photoactive pigments bound in the "reaction center" complexes (for reviews see [24] and [25]). The major antenna complex of green plants is the light-harvesting complex of photosystem II (LHC II) (for a review see [26]). Apart from the energy level structure of the excited electronic states [27–29], the coupling of the purely electronic transitions of pigments to intramolecular and low-frequency vibrational modes of the protein matrix ("electron-phonon" coupling[3]) plays an essential role in light absorption and ultrafast EET [30,31] (for a review see [32]). Because of this coupling and the broadening of electronic states resulting from it, the energy transfer between electronic states of neighboring pigments is facilitated and effectively mediated by the protein matrix's vibrational density of states. A knowledge of the latter is required for the calculation of energy transfer rates. In view of the inherent complexity of the interpretation of phonon structure obtained from optical spectroscopy (hole-burning (HB) and fluorescence line-narrowing (FLN) spectra [33]), inelastic incoherent neutron scattering (IINS) has become a technique of choice as an independent experimental approach to study the vibrational density of states of photosynthetic pigment–protein complexes.

IINS spectra of proteins generally display a peak centered at energy transfers somewhere in the range from 2 to 5 meV, representing an excess of vibrational modes, as compared to the low-frequency region of a Debye-like density of states. The origin of this so-called "Boson peak" known to be characteristic for both, proteins and glassy systems [34–39], is not yet fully understood, but it is believed to be related to the disordered nature of these systems. The effective density of states derived from IINS spectra (see e.g., [34, 37]), which is of a fundamental interest in itself, also permits a comparison to molecular dynamics (MD) simulations of internal protein motions (see e.g., [40]). Here, we discuss results of IINS experiments on the photosynthetic antenna complex LHC II from spinach [41]. Trimeric LHC II was solubilized in a D_2O containing buffer solution, so that the solvent scattering was significantly

[3]In the present context, the term "phonon" is largely employed, when referring to low-frequency protein vibrations (<25 meV). This is customary in studies of pigment–protein complexes in order to distinguish such modes from higher-frequency vibrational modes of the pigment molecules (60–250 meV). However, in biological systems one is generally not dealing with phonons in a rigorous sense. Due to lack of long-range order, vibrational modes, except for long-wavelength sound waves, are expected to travel only over short distances before they are seriously damped

reduced (as compared to a solution in H_2O). The experiments were performed at four temperatures between 5 and 100 K, using the time-of-flight spectrometer NEAT [42–44] at BENSC in Berlin. Incident neutron wavelengths of 2.0 Å ($E_0 \approx 20$ meV) and 5.1 Å ($E_0 \approx 3.2$ meV) were employed, corresponding to (elastic) Q ranges from 1.0 to 5.7 Å^{-1} and 0.3 to 2.3 Å^{-1}, respectively. The elastic energy resolutions were $\Delta\hbar\omega = 0.920$ meV and $\Delta\hbar\omega = 0.093$ meV, respectively. Low temperatures were chosen in order to ensure the validity of the harmonic approximation for the protein vibrations under investigation and a low probability of multiphonon processes (see Eq. 16.10), as well as the absence of fast diffusive motions. For the biologically relevant dynamics to be observed, low T is not prohibitive, since it is well-established that EET in antenna complexes is fully functional for temperatures in the above range (see e.g., reviews [45] and [46]). The measured scattering functions $S_s(Q,\omega)$ of LHC II are shown in Fig. 16.4. The data qualitatively exhibit the temperature dependence expected for a thermal population of vibrational levels increasing with rising temperature. Three prominent features were found to characterize the dynamics of LHC II at temperatures from 5 to 100 K: (a) a linear Q^2-dependence of the logarithm of the elastic intensity (see left frame of Fig. 16.5), (b) a linear Q^2-dependence of the inelastic peak intensity divided by the DWF, and (c) a linear temperature dependence of the average atomic mean square displacement $<u^2>$ (see right frame of Fig. 16.5). The linear increase of $<u^2>$ with temperature is generally predicted for harmonic vibrational motions (see for instance [3]) and has already been reported for PS II membrane fragments [47] and for the bacterial reaction center [48]. Based on these findings, it was concluded that both, the elastic and inelastic contributions of the IINS data presented here, follow the Q-dependence given by

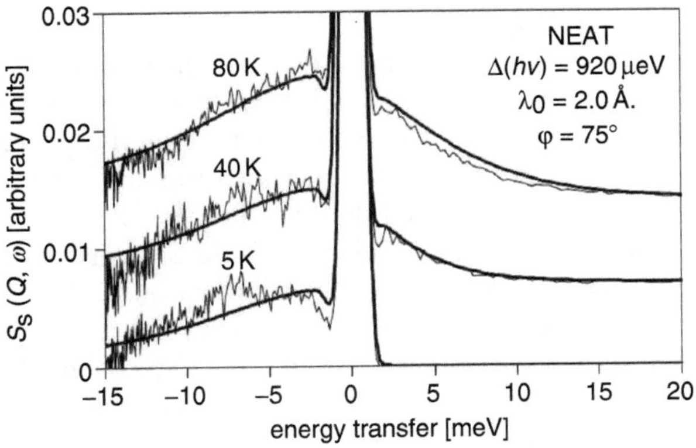

Fig. 16.4. Measured LHC II spectra from IINS experiments using the MTOF spectrometer NEAT (see Sect. 15.3.2 in Part I, this volume). The experiment parameters are indicated in the inset; here, $\Delta(h\nu)$ = elastic energy-resolution, λ_0 = incident neutron wavelength, φ = scattering angle; data taken from [41]

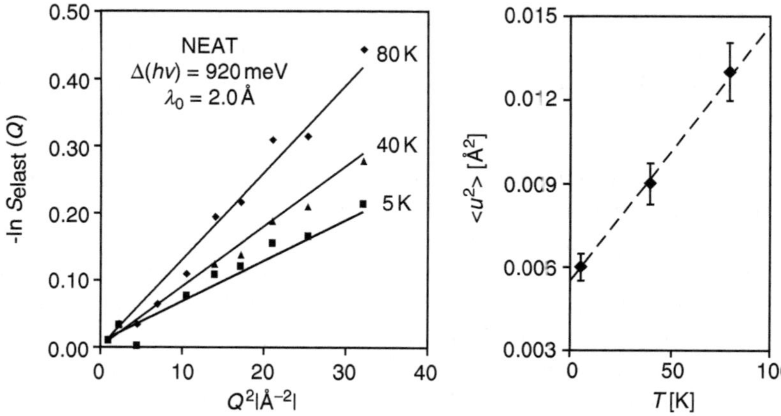

Fig. 16.5. LHC II Debye–Waller factor (*left*) and mean-square displacement $<u^2>$ (*right*) from NEAT experiments; data from [41]

Eq. 16.10. Thus, they display the properties of harmonic vibrational protein dynamics of LHC II in the investigated temperature range. The "Boson peak" feature typical for proteins was here observed around an energy transfer of 2.5 meV. The effective vibrational density of states $g(\omega)$ of LHC II was calculated from $S_s(Q,\omega)$ using Eq. 16.16, i.e., the approximate version of Eq. 16.10, valid at low temperatures. The obtained $g(\omega)$ is similar for all temperatures, within experimental error. Thus, results for $g(\omega)$ obtained separately from the 2.0 Å data taken at 5, 40, and 80 K, were averaged to yield the effective density of vibrational states $g(\omega)$ displayed by diamonds in frame (a) of Fig. 16.6. Independently, the function $g(\omega)$ was also calculated from the data obtained using an incident wavelength of 5.1 Å at 100 K. The 2.0 Å results were fully confirmed (full line in Fig. 16.6). In order to confirm an expected deviation from a Debye-like density of vibrational states, where $g(\omega)$ would be proportional to ω^2, frame (a) of Fig. 16.6 shows a plot of the experimental $g(\omega)$ together with a curve proportional to ω^2, which was fitted to the low-energy part of the experimentally obtained density of states. An inspection of these data reveals, that $g(\omega)$ of LHC II indeed displays an excess of vibrational modes. For comparison, a plot of the $g(\omega)/\omega^2$ function is also shown as a full, smooth line in frame (b) of Fig. 16.6. The ω^2-curve of frame (a), now transformed into a constant, is shown as a dashed horizontal line. This representation makes the excess of vibrational modes extending over a frequency range around 3 ± 2 meV more clearly visible. In conclusion, the density of vibrational states $g(\omega)$ of LHC II exhibits properties typical for proteins with an excess of vibrational modes compared to the low-frequency Debye-behavior due to acoustic vibrations. One may speculate that low-lying intermolecular optical modes are contributing to this component. Similar $g(\omega)$ functions have been determined for other proteins like myoglobin [34] and β-lactoglobulin [37] and

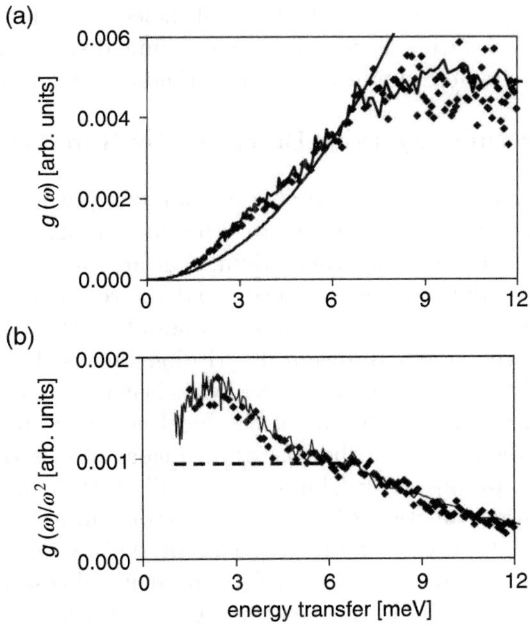

Fig. 16.6. LHC II vibrational density of states (**a**), and the latter divided by ω^2 (**b**), from NEAT experiments; data from [41]

seem to generally reflect the disordered nature of protein molecules. The wide distribution of protein vibrations found for LHC II can be understood in terms of structurally inequivalent protein domains within the LHC II trimer leading to a partial localization of protein phonons. One possible interpretation is, that the low-energy vibrational peak of LHC II at about 2.3 meV corresponds to vibrations fully delocalized over the protein matrix of the LHC II monomer or trimer. In contrast to this, the slight tailing towards higher vibrational frequencies may be constituted by separate contributions from vibrations with smaller amplitudes. These vibrations might e.g., be localized on one of the different α-helices forming the protein matrix of LHC II.

16.4 Non-Gaussian Motion

It must be pointed out that there are many situations where Gaussian behavior of scattering functions is absent. Because of strongly non-Gaussian (space- and time-dependent) atomic probability–density distributions, the Gaussian approximation is then either far from being applicable, or limited to purely phenomenological descriptions.[4] This is probably true in the case of many

[4]These may nevertheless be useful for some practical purposes

of the functionally most relevant motions of molecular subunits in biological macromolecules exhibiting large anisotropic atomic displacements. Here are a few of the most simple examples of models of non-Gaussian motion.

16.4.1 Atomic Jump Motions Described by Rate Equations

The simple diffusion theory is satisfactory for our purpose only, when the locations $(\boldsymbol{r}, t)$ in space and time, where experimental observations are made, are at distances from the space-time origin, that are large as compared to the small elementary diffusive step distance $\boldsymbol{l}$ and the residence time τ_0 between consecutive steps. If, however, such small values of $(\boldsymbol{r}, t)$ are to be covered by the experiment, the simple-diffusion description breaks down, and $S_{\mathrm{s}}(\boldsymbol{Q}, \omega)$ then significantly differs from the above case, because it contains structural information concerning the details of the local geometry directly related to the interatomic distances and the spatial arrangement of atoms.

A model was proposed by Chudley and Elliott [12], to obtain the classical self-correlation function $G_{\mathrm{s}}(\boldsymbol{r}, t)$ for an atom diffusing on an assumed quasicrystalline lattice of a liquid. Because of the assumed regular spatial arrangement of atom sites and its relative simplicity, this so-called CE-model has also become very popular for treating atomic diffusion on interstitial lattices in crystals. We briefly discuss it in the following, in order to provide some insight into how jump models based on rate equations are constructed.

Let us first consider diffusion by atomic jumps on a Bravais lattice. One calculates the probability $P(\boldsymbol{r}_m, t)$ to find the diffusing atom on a site $\boldsymbol{r}_m$ of the lattice at time t, where it spends a time τ on the average. The time τ_j required for the diffusive jump from site to site is neglected. The jumps occur between a given site $\boldsymbol{r}_m$ and its neighbors $\boldsymbol{r}_m + \boldsymbol{d}_\nu$, where $\boldsymbol{d}_\nu$ $(\nu = 1, 2, \ldots, s)$ is a set of jump vectors connecting the given site with the neighbors. The *master equation* for $P(\boldsymbol{r}_m, t)$ is then

$$\frac{\partial P(\boldsymbol{r}_m, t)}{\partial t} = -\frac{1}{\tau} P(\boldsymbol{r}_m, t) + \frac{1}{s\tau} \sum_{\nu=1}^{s} P(\boldsymbol{r}_m + \boldsymbol{d}_\nu, t) \; . \tag{16.20}$$

The first and second term on the right side are the loss and growth rate due to the jumps to and from adjacent sites, respectively. In order to obtain $G_{\mathrm{s}}(\boldsymbol{r}, t)$, the initial condition

$$P(\boldsymbol{r}_m, 0) = \delta(\boldsymbol{r}_m) \tag{16.21}$$

is imposed. Because in a Bravais lattice all sites are equivalent, one directly obtains the self-correlation function as the solution of the master equation: $G_s(\boldsymbol{r}, t) \equiv P(\boldsymbol{r}_m, t)$. Several further assumptions made, when deriving Eq. 16.20 are that consecutive jumps are uncorrelated, and that blocking and mutual interaction effects are neglected. This implies a low site occupancy c for the diffusing particles. The basic CE-model (represented by Eq. 16.20) has been extended to cover many special and more complex situations, for instance to include effects due to atom–atom correlations [49, 50]. A general theory of the master equation is given in Chapter 18 of [51]. Correlated jumps are discussed in Chapters 1, 3, 10, and 18 of [51]. Now we introduce the Fourier transform

$$P(\boldsymbol{r}_m, t) = \int_{-\infty}^{+\infty} \tilde{P}(\boldsymbol{Q}, t) \mathrm{e}^{-\mathrm{i}\boldsymbol{Q}\boldsymbol{r}_m} \mathrm{d}\boldsymbol{Q}, \qquad (16.22)$$

where $\tilde{P}(\boldsymbol{Q}, t) \equiv I_\mathrm{s}(\boldsymbol{Q}, t)$ from Eq. 15.9, Sect. 15.2.1 in Part I, this volume. This leads to a differential equation of first order for $I_\mathrm{s}(\boldsymbol{Q}, t)$ with an exponential decay function as solution,

$$I_\mathrm{s}(\boldsymbol{Q}, t) = \mathrm{e}^{-\Gamma(\boldsymbol{Q})|t|}\ , \qquad (16.23)$$

which fulfils the initial condition Eq. 16.21. From this one gets

$$\Gamma(\boldsymbol{Q}) = \frac{1}{s\tau} \sum_{\nu=1}^{s} (1 - \mathrm{e}^{-\mathrm{i}\boldsymbol{Q}\boldsymbol{d}_\nu})\ . \qquad (16.24)$$

The resulting dynamic structure factor is the Fourier transform of $I_\mathrm{S}(\boldsymbol{Q}, t)$, i.e. the normalized Lorentzian

$$S_\mathrm{s}(\boldsymbol{Q}, \omega) = \frac{\Gamma(\boldsymbol{Q})/\pi}{\omega^2 + \Gamma(\boldsymbol{Q})^2} \qquad (16.25)$$

with energy halfwidth (HWHM) $\Gamma(\boldsymbol{Q})$. For small Q, i.e., for $Q \ll 1/d$, where d = length of jump vector, one gets the limiting case

$$\Gamma(\boldsymbol{Q}) = \frac{d^2}{6\tau} Q^2 = D_\mathrm{s} Q^2\ . \qquad (16.26)$$

i.e., the Gaussian behavior of $S_\mathrm{s}(\boldsymbol{Q}, \omega)$ and $I_\mathrm{s}(\boldsymbol{Q}, t)$ is recovered in this limit (see Eq. 16.4). This relation holds generally for jump diffusion in three dimensions, and it is *independent of the detailed jump geometry.* When diffusion occurs in a crystalline lattice, $\Gamma(\boldsymbol{Q})$ is periodic in reciprocal space. It has a maximum at the Brillouin zone boundary and it is zero, if a reciprocal lattice point $\boldsymbol{G}$ is reached, such that $\Gamma(\boldsymbol{Q} = \boldsymbol{G}) = 0$. This "line narrowing"[5] is related to Bragg diffraction of the neutron wave from the probability density of the proton distributed over the sites of the Bravais lattice (see, for instance, [53–55]).

Finally, we mention here a useful simplification of expression 16.25, which has often been applied in the case of polycrystalline samples requiring orientational averaging of the theoretical scattering function. If, instead of $S_\mathrm{s}(\boldsymbol{Q}, \omega)$, the width function is averaged, one obtains the "isotropic approximation":

$$\Gamma(\boldsymbol{Q}) = \frac{1}{\tau} \left[1 - \frac{\sin(Qr)}{Qr} \right] \qquad (16.27)$$

---

[5]This effect should not be confused with the coherent line narrowing predicted by De Gennes [52]

This relation can be used to determine an estimate of an (average) jump-distance r from the Q-dependence of measured quasielastic line widths [7–9]. An example, where this procedure has been applied to two-dimensional diffusion on the surface of a membrane, is discussed in Sect. 16.5.1 (see Fig. 16.14).

16.4.2 Confined or Localized Diffusive Atomic and Molecular Motions

Diffusive motions of individual atoms or molecules do not necessarily extend over the whole volume of the system under investigation. They may instead be restricted to a certain space of molecular order of magnitude. This occurs for instance, when atoms bound in a molecule – e.g., in a mobile sidegroup of a polypeptide chain – are undergoing rotational diffusion with the latter, or when molecules are confined to diffusion within the space available inside the cages of an inclusion compound. In the context of biological macromolecules, this situation may be encountered for instance by water molecules trapped inside of these molecules, or in cavities or grooves at their surface. Here, the dynamical problem consists in solving a suitable equation of motion applying the appropriate boundary conditions which correspond to the specific spatial limitation of the motion. Then, for instance, when a molecule or molecular subunit performs discrete reorientational steps (rotational jumps) – inspite of the spatial limitation – a rate equation of the type represented by Eq. 16.20 can still be employed, whereas for continuous diffusion Eq. 16.6 may still apply. All molecular motions limited in space have in common, that the atomic probability density within the confinement does not decay to zero at infinite time. As a consequence, the temporal Fourier transform to be calculated for obtaining the Van Hove incoherent scattering function, contains an (elastic) $\delta(\omega)$-term, in addition to nonelastic terms (see Sect. 15.2.2 in Part I, this volume). Generally, the above type of equations for localized or confined diffusive motions yield scattering functions (here for convenience labelled with the subscript for rotation) of the following form [5] and [6]:

$$S_{\mathrm{rot}}(\boldsymbol{Q},\omega) = A_0(\boldsymbol{Q})\delta(\omega) + \sum_{j=1}^{n} A_j(\boldsymbol{Q}) L_j(H_j,\omega). \tag{16.28}$$

Here $A_0(\boldsymbol{Q})$ is the EISF of the localized diffusive motion and the functions $A_j(\boldsymbol{Q})$ are the n corresponding QISFs of the n Lorentzian components, $L_j(H_j,\omega)$, with halfwidths (HWHM) H_j, and with (see Sect. 15.2.2 and Fig. 15.1 in Part I, this volume):

$$\mathrm{QISF} = \sum_{j=1}^{n} \mathrm{QISF}_j(\boldsymbol{Q}) = \sum_{j=1}^{n} A_j(\boldsymbol{Q}) = 1 - \mathrm{EISF}. \tag{16.29}$$

Many models yielding expressions of the type 16.28 have been developed in the past for specific boundary conditions imposed by the geometry of different localized diffusive motions. Several representative examples have been given in Table 15.1 and Fig. 15.1 of Part I, this volume, including the corresponding references to the original literature. When, in addition to the confined diffusive motion, other dynamical phenomena, such as vibrational motions or translational diffusion, are present in the observation time window of the experiment, $S_{\mathrm{rot}}(\boldsymbol{Q},\omega)$ may be combined with the pertinent additional scattering functions by convolution according to Eq. 16.2. Without going into a detailed discussion, we merely note here, that Eqs. 16.28 and 16.29 have been applied to isolate the contribution due to two-site and three-site jump-reorientation of small molecular subunits, such as methyl-groups and short side-chains to the QENS intensity observed in experiments on purple membrane [56]. The classical model of diffusion on the surface of a sphere [57] (corresponding, e.g., to continuous rotational diffusion of molecules in solution) has been formulated for incoherent, as well as for coherent scattering, and applied not only for the former [58–62] (see also [1] and [3]), but also in the latter case [61, 63, 64]. The model for continuous diffusion in the interior of a spherical volume [65] has become popular for the description of localized diffusive (confined molecular) motion in the context of protein dynamics [66–72]. H-bond flip-flop motions based on the assumption of a double-well potential have been treated using a symmetric two-site jump model [73]. A potentially interesting model for dealing with H-bond formation–breaking reaction equilibria in biological macromolecule/hydration–water complexes, which is based on an asymmetric two-site jump model, can be found in [74].

16.4.3 Environment-Dependence of Confined Diffusive Protein Motions: Example Lysozyme

Any functional activity of biological systems involves the quasicontinuous performance of multiple conformational rearrangements of macromolecules such as proteins, enzymes, etc. [75]. The molecular dynamics involved is fully developed only at physiological conditions of the environment. Because of the strong correlation between functionality, dynamics and environment, it is of great interest, to study the dynamics of biological systems as a function of environmental parameters. These may concern physical or biochemical aspects, as temperature, pressure,..., humidity, salinity,..., and – more generally – the kind and concentration of molecular and ionic species surrounding the macromolecular surface. It is well-known that the omnipresence of liquid water in biological systems has a plasticizing effect ensuring the conformational flexibility (e.g., of proteins) required for biological function (see also Sect. 16.5.2). On the other hand, proteins embedded in sugars or glycerol have higher melting temperatures, i.e., they possess an increased thermostability, as compared to the same systems in a state of hydration. The reason is, that under certain conditions well above room temperature, proteins are still stable in anhydrous solvents, when the plasticizing action of water which would give them a conformational flexibility sufficient for denaturing in the hydrated state, is absent [76]. In general, thermal fluctuations of proteins, and in particular stochastic motions occurring on the pico- to nano-second time-scales accessible to neutron scattering, are affected by the character of the environment. Neutron scattering experiments have been carried out, employing the elastic-window scan technique [77] (see Sect. 15.3.3 in Part I, this volume), for studying the molecular mobility of lysozyme solvated in glycerol, with various water contents, as a function of temperature [78, 79]. These studies have revealed the alteration of the protein's internal mobility, when the character of its environment changes from a stabilizer-like to a plasticizer-like nature. Addition of water strongly affects the dynamical behavior of glycerol-solvated lysozyme, which with this technique is monitored via a measured effective mean square displacement of hydrogen atoms in the protein. A quasielastic neutron scattering experiment shows this effect in more detail [39]. As an example, Fig. 16.7 shows IINS spectra of dry lysozyme solvated in glycerol, as obtained at $Q = 1.1\,\text{Å}^{-1}$, at 100 K (open circles) and at 300 K (full circles). The low-T spectrum essentially exhibits a broad Boson-peak type component centered near 4 meV, which is due to quasiharmonic vibrations. There is also a very small quasielastic contribution at energy transfers below about 2 meV. At 300 K, where diffusive motions have become rather important, a large quasielastic component dominates the spectral shape. The continuous line is the elastic energy resolution function measured with a vanadium plate. The inset shows the same spectra, but rescaled by dividing through the Bose

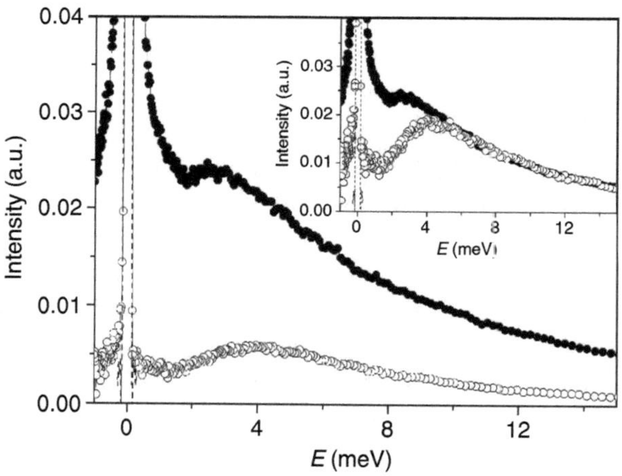

Fig. 16.7. Inelastic neutron scattering spectra of dry lysozyme solvated in glycerol, at 100 K (*open circles*) and 300 K (*full circles*), at $Q = 1.1\,\text{Å}^{-1}$, measured with the MTOF spectrometer NEAT; see Sect. 15.3.2 in Part I, this volume. The continuous line is the elastic energy resolution function measured with a vanadium plate. Inset: for energy transfers above 5 meV, where the quasielastic component is negligible, the inelastic region of the 100 K spectrum overlaps precisely with that of the 300 K spectrum, after both have been rescaled by dividing through the Bose phonon-population factor (after [39])

phonon-population factor, in order to make the vibrational part of the scattered intensity temperature-independent. The result of this procedure is, that the vibrational components observed at the two different temperatures are perfectly overlapping in the energy transfer region above 5 meV, where the (quasielastic) spectral contribution due to diffusive motions is negligible. The spectral weight of the quasielastic component varies strongly not only with the temperature, but also with the degree of hydration. This is shown in Fig. 16.8, where spectra of lysozyme in glycerol at 300 K are presented for three different hydration levels: 0.0h (open circles), 0.2h (full circles), 0.4h (open squares). One observes that, similarly to the temperature effect, increased hydration induces an increased global mobility of lysozyme: both, the inelastic and the quasielastic signals are increased at higher hydration levels. This effect is particularly large in the latter case. In order to investigate the nature of the relaxational dynamics, the quasielastic spectral shape must be studied alone. The vibrational component, as determined at 100 K (see Figs. 16.7 and 16.8), can be fitted by a suitable function, like the phonon-expansion described in Sects. 16.3.3 and 16.3.4 (see Eqs. 16.10–16.16). When this is subtracted from the measured spectra, the sole quasielastic component remains. This is shown in Fig. 16.9 for the room temperature (300 K) spectra at the three hydration levels, 0.0h (open circles), 0.2h (full circles) and 0.4h (open squares). These spectra were quantitatively analyzed by fitting a quasielastic scattering function of the type 16.28, folded with the experimental resolution function. Two Lorentzian components were found to well reproduce the spectra, with the QISFs ($A_j(\boldsymbol{Q})$), the linewidths (H_j) and a trivial global normalization

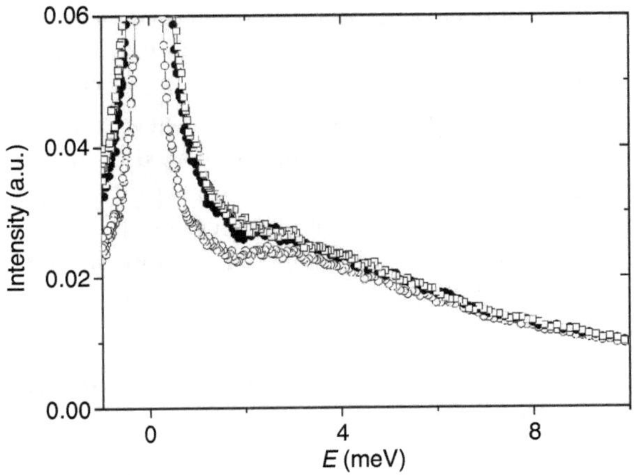

Fig. 16.8. Inelastic neutron scattering spectra of lysozyme solvated in glycerol, at room temperature (300 K), for three different hydration levels, 0.0h (*open circles*), 0.2h (*full circles*) and 0.4h (*open squares*), measured with the MTOF spectrometer NEAT. The momentum transfer value is $Q = 1.1\,\text{Å}^{-1}$ (after [39])

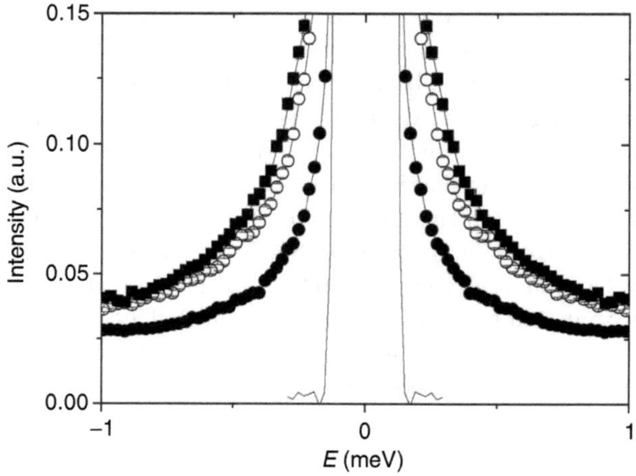

Fig. 16.9. Quasielastic neutron scattering spectra of lysozyme solvated in glycerol for a momentum transfer value of $Q = 1.1\,\text{Å}^{-1}$, at room temperature (300 K), for three different hydration levels, 0.0h (open circles), 0.2h (full circles) and 0.4h (open squares), measured with the MTOF spectrometer NEAT. The vibrational component of the scattered intensity has been subtracted (see text). The narrow continuous line is the elastic energy resolution (after [39])

factor as free parameters. Figure 16.10 shows the fit result for the 0.4h sample. It is seen that the two Lorentzians exhibit rather different widths, namely ($H_1 = 0.135 \pm 0.010$ meV) and ($H_2 = 2.45 \pm 0.70$ meV), which means that two relaxational phenomena occurring on quite different time scales are observed. For different values of the hydration (h), the same pair of widths is found for the two Lorentzians, but the quasielastic signals (i.e., the QISFs) increase with h. This is the signature of a progressive onset of new relaxational degrees of freedom of internal protein motions, probably related to groups located at the protein surface. Apparently, also in the presence of glycerol, water molecules are able to gradually activate the whole lysozyme dynamics by progressively hydrating the protein surface groups. This may be interpreted as a hydration-driven onset of confined diffusive internal protein motions.

16.4.4 Change of Protein Dynamics on Ligand Binding: Example Dihydrofolate Reductase

It is of fundamental importance in biology and medicine to understand how ligands bind to proteins [80–85]. Various types of interactions may be involved in ligand association: van der Waals, electrostatic and hydrophobic interaction, hydrogen bonding, or different intermolecular interactions specific for the protein–ligand complex. The signature of complex-formation is expected to be visible in the spectrum of protein vibrations or – more

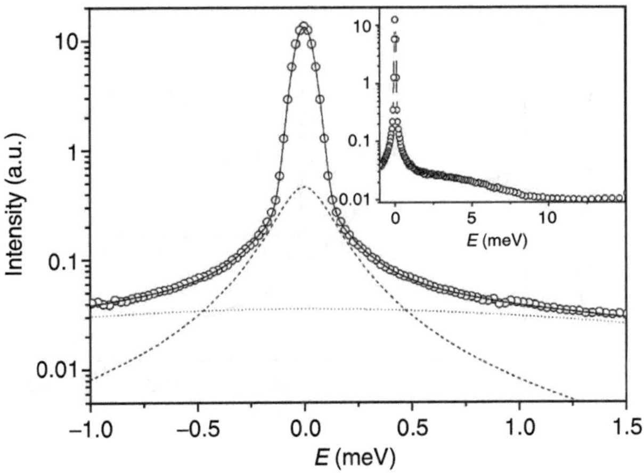

Fig. 16.10. Fit (*full line*) of a scattering function comprising an elastic and two quasielastic (Lorentzian) terms of different widths to the room temperature quasielastic spectrum of lysozyme solvated in glycerol, and hydrated at 0.4h (*open circles*), at $Q = 1.1\,\text{Å}^{-1}$. The dashed and dotted lines are the narrow and the broad Lorentzian components, respectively. The inset shows the quasielastic spectrum over a larger energy range. These spectra (after [39]) were derived from measurements with the MTOF spectrometer NEAT; see Sect. 15.3.2 in Part I, this volume

generally – of protein dynamics, as compared to that of the unbound protein [86–92]. An experimental investigation of the dynamic structure factor modification due to ligand binding was carried out using neutron scattering [93] on the enzyme dihydrofolate reductase (DHFR), an important target for anticancer and antibacterial drugs [94–98]. In the presence of DHFR and of the nicotinamide adenine dinucleotide phosphate (NADPH) cofactor, a catalytic reduction of dihydrofolate to tetrahydrofolate takes place. The ligand used is methotrexate (MTX), a folate antagonist of DHFR employed effectively as a cytotoxic agent in the treatment of cancers [99]. In order to minimize scattering from solvent molecules, the system was exchanged with D_2O. Both the uncomplexed enzyme (DHFR+NADPH) and the complexed enzyme (DHFR+NADPH+MTX) were hydrated to a degree of 30%, i.e., 30 mg of D_2O per 100 mg of dry weight protein. It is important to keep in mind, that this degree of hydration is only valid for the environmental conditions under which it was established, whereas at much lower temperatures, part of the water will be effectively separated from the protein due to a "dehydration by cooling" effect [100], but might still be present within the sample container in the form of crystalline D_2O-ice. But, because of the comparatively low scattering cross-section of D_2O, the latter (with its 23% fraction of the total sample weight) will not contribute more than about 2.5% to the integral of the inelastic scattering spectrum.

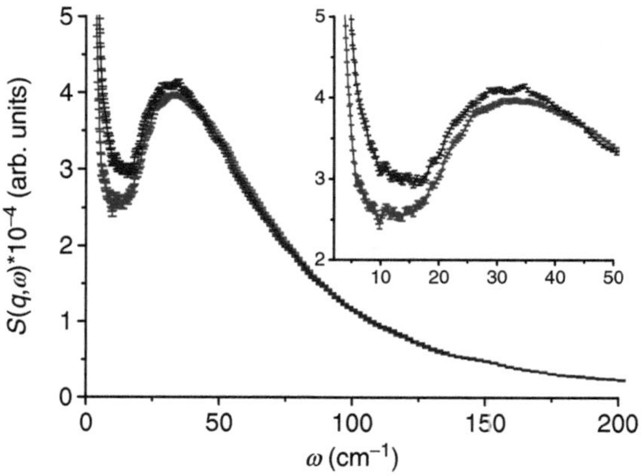

Fig. 16.11. Dynamic structure factor versus frequency ω [cm^{-1}] for uncomplexed DHFR (*lower curve*) and complexed with methotrexate form of DHFR (*upper curve*) at 120 K. Data, taken with the direct-geometry TOF spectrometer IN6 (see Sect. 15.3.1 in Part I, this volume) using an incident neutron wavelength of 5.12 Å, have been summed over all scattering angles of the experiment (elastic Q-range approximately from 0.3 to 2.2 Å^{-1}). Both spectra are normalized to the elastic peak height. Inset: low frequency region of the spectra; after [93]

Measurements were performed with the XTL-TOF spectrometer IN6 at ILL Grenoble (see Sect. 15.3.1 in Part I, this volume), at the fairly low temperature of 120 K, in order to minimize the contribution of diffusive motions. Figure 16.11 shows the measured incoherent scattering functions of the uncomplexed and the complexed form of the enzyme. Both spectra exhibit, in a qualitatively similar manner, two fundamentally different parts: The first one is an inelastic region including the Boson peak (ω above $\simeq 20\,\mathrm{cm}^{-1}$)[6], which is probably exclusively due to harmonic vibrations. The second part is the low-frequency region (ω below $\simeq 20\,\mathrm{cm}^{-1}$), where a careful comparison with the energy resolution function determined by scattering from a vanadium sample (not shown in the figure) suggests that the harmonic vibrational component is overlapping with a quasielastic contribution on top of it. The latter indicates, that at 120 K the spectra still contain a non-negligible effect from strongly anharmonic, or overdamped vibrational, or even from localized diffusive modes of motion, inspite of the low temperature. This is, however, not unusual and has also been found earlier, for instance in the cases of hydrated purple membrane [101, 102], and in alanine dipeptide [103], where anharmonic motions and rotational motions of small side-groups have been observed down to 100 K and below (see also the 100 K spectra of Fig. 16.7 in Sect. 16.4.3). In this low-energy region, a rather significant difference was observed between the spectra of the complexed and the uncomplexed enzyme. The quasielastic component

[6]Note that the energy transfer unit 1 meV corresponds to $8.07\,\mathrm{cm}^{-1}$

of the former is appreciably larger than that of the latter. This may be due to one or several of various conceivable reasons. Let us first consider the possibility, that the additional quasielastic scattering may be due to the mobility of small side-groups located on the ligand, which might be higher than that of side-groups in the main part of the complex. Because of the low weight fraction of the ligand, this is not considered as very probable, although it cannot be completely excluded. Further studies employing a deuterated ligand could however clarify this question. Second, multiphonon scattering can in principle lead to a quasielastic contribution due to phonon–phonon annihilation. This effect should, however, be negligible in the concerned Q-range (below $2.2\,\text{Å}^{-1}$), and especially at the low temperature of the experiment, where it is estimated to be of the order of only 2% of the one-phonon scattering. Furthermore such a component is not likely to be very different for the two kinds of sample. Similar arguments apply for a multiple scattering contribution, which again is negligibly small, because of the large value of the employed sample transmissions (97.7%).

Finally, we come to what is believed to be the main reason for the observed difference in the spectra. The complexation may lead to an increased importance of relaxational modes and possibly of very-low-frequency optical vibrations in this whole system and thus to an increased flexibility of the macromolecular ensemble, as compared to the uncomplexed case. The intensity increase also extends further into the inelastic region up to $\simeq 40\,\text{cm}^{-1}$. It is likely, but cannot be proved at present, that the latter effect might be due to complexation-caused damping of the vibrations in the latter intermediate frequency region. Damping could lead to a shift of modes from higher to lower frequencies. The deviation from purely harmonic behavior of both systems, (DHFR+NADPH) and (DHFR+NADPH+MTX), is demonstrated by calculating, in the limit $Q \to 0$, a "generalized" density of states $g(\omega)$ using Eq. 16.17 in Sect. 16.3.3. This is shown in Fig. 16.12 presenting the $g(\omega)$ curves for the complexed and for the uncomplexed protein. [7] It is seen, that in the frequency region below $\simeq 20\,\text{cm}^{-1}$ the obtained generalized function does not show the well-known ω^2-behavior expected for purely harmonic acoustic vibrations at low energies. Furthermore, it seems to be clear that the curves start with finite values at $\omega = 0$, although measured data are of course not available at zero frequency (see Sect.16.3.3). This is of course the phenomenological signature of the quasielastic component. It is also evident from the figure, that the complexed form of DHFR has an appreciably larger $g(\omega)$ in

[7] Note that the low-frequency part (ω below $\simeq 20\,\text{cm}^{-1}$) of this function must be considered as purely phenomenological: because stochastic fluctuations generally do not obey Bose–Einstein statistics, the division of the scattering function by the Bose phonon-population factor does not have quite the same justification as in the case of harmonic vibrations. But this does not change the main result, namely the observation that complexation of DHFR leads to additional relaxational phenomena causing an increase in the quasielastic scattering component.

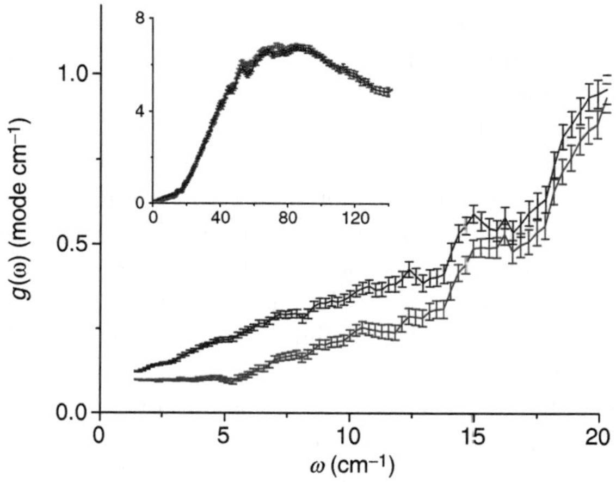

Fig. 16.12. Generalized density of states $g(\omega)$ for uncomplexed DHFR (*lower curve*) and for DHFR complexed with methotrexate (*upper curve*) at 120 K. Data, derived from measurements with the direct-geometry TOF spectrometer IN6 (see Sect. 16.3.1 in Part I, this volume), have been summed over a large range of scattering angles. Inset: enlarged frequency region of the spectra; after [93]

this low-frequency region, suggesting that complexation of DHFR softens the enzyme and makes this macromolecule more flexible, an effect which could be relevant for its biochemical activity. However, we have to note, that we do not know, whether this behavior is also relevant to the same extent at physiological conditions, where the level of hydration established initially (see above) is effective.

Finally, it is interesting to note that, using NMR relaxation experiments, other authors have observed an increased conformational flexibility on binding a hydrophobic ligand to mouse major urinary protein [104]. However, we should note that a flexibility *increase* does not necessarily occur as a rule upon ligand binding, since this depends on the strength of the involved interaction. In other investigations using NMR and crystallography, a flexibility *decrease* was observed on binding small organic ligands to proteins [105–107].

16.5 Low-Dimensional Systems

16.5.1 Two-Dimensional Long-Range Diffusion of Rotating Molecules

Low-dimensional diffusion plays an important role for certain biological objects, such as ion transport channels and membrane surfaces. Obviously, the concept of low-dimensionality should not be understood for instance in the sense of strictly planar motions, since especially on biological membranes there will be appreciable local deviations from this idealized picture. The essential condition is, that the diffusing particles stay within or near a fictive diffusion plane (e.g., representing the membrane surface) at least during the observation time defined by the energy resolution of the experiment (see Sect. 15.3, Eq. 15.30 in Part I, this volume and the related discussion). This also implies that the mean-square displacement parallel to the diffusion layer, the particle has acquired during this time, is much larger than perpendicular to it.

The incoherent neutron scattering function in the low-Q limit for long-range translational diffusion (TD) was already considered above (see Eqs. 16.6 and 16.7. Accordingly, in three dimensions, when the diffusive motion is isotropic, with a diffusion coefficient D_{3D}, we have for this function:

$$S_{TD}^{3D}(\boldsymbol{Q},\omega) = \frac{1}{\pi}\frac{D_{3D}Q^2}{(D_{3D}Q^2)^2+\omega^2} \quad , \tag{16.30}$$

Certain analytic results [108] concerning the anisotropic case, where the diffusion process is restricted to the surface of a plane or to the planes of a layered structure, will be discussed in the following. If single-crystalline samples are available, the dimensionality of the diffusional motion can of course be determined directly by studying the orientation dependence in a QINS experiment. For a single crystal we have [109]:

$$S_{TD}^{2D}(\boldsymbol{Q},\omega) = \frac{1}{\pi}\frac{D_{2D}(Q\sin\theta)^2}{(D_{2D}(Q\sin\theta)^2)^2+\omega^2} \quad , \tag{16.31}$$

where D_{2D} is the coefficient of self-diffusion in two dimensions, $Q\sin\theta$ the component of the scattering vector in the diffusion plane, and θ the angle between **Q** and the normal to this plane. If single-crystals are not available, one has to resort to polycrystalline samples requiring orientational averaging of the above expression. It is known, that the resulting integral over all orientations exhibits a logarithmic singularity at zero energy transfer [109]. This is caused by the fact that diffusion planes which are perpendicular (or close to perpendicular) to the scattering vector **Q** contribute elastic (or almost elastic) scattering to the QINS function. Fortunately, the logarithmic singularity is cancelled by finite resolution. It has been shown that for resolution functions which have the shape of a Lorentzian (with HWHM equal to H) or of a sum of Lorentzians (which can be the case in BSC experiments) the orientationally averaged resolution-broadened QINS function for 2D-diffusion,

$$< S_{TD}^{2D}(\mathbf{Q},\omega) >_{\text{orient.}}^{\text{res}} = \frac{1}{2\pi}\int_0^{2\pi} \mathrm{d}\Phi \, \frac{1}{2}\int_0^{\pi} \sin\theta \, \mathrm{d}\theta \; L(\theta,\omega), \tag{16.32}$$

where

$$L(\theta,\omega) = \frac{1}{\pi}\frac{D_{2D}(Q\sin\theta)^2 + H}{[D_{2D}(Q\sin\theta)^2 + H]^2 + \omega^2} \tag{16.33}$$

can be written in closed form [108]. The explicit expression is lengthy and will not be given here. The determination of such a characteristic line shape is very difficult. It requires an experimental resolution width $H \ll D_{2D}Q^2$, in order to be able to distinguish it from a simple Lorentzian shape. This problem is discussed in detail in [108].

The study of the anisotropy of translational diffusion in strongly anisotropic material, such as biological membranes, is straight-forward, if the membranes are well aligned in the samples. Pertinent investigations with quasielastic neutron scattering techniques have been carried out on purple membrane [110,111] and on the superficial layer of porcine skin (stratum corneum) [112]. As an example we discuss the transport of water molecules on the surface of purple membrane which has a two-dimensional crystalline structure. In biological membranes, proton diffusion connected with conduction of protons provides an important mechanism of energy transduction in living organisms. The purple membrane (PM) of *Halobacterium salinarum*, for instance, contains the protein bacteriorhodopsin (BR) which becomes a one-dimensional stochastically pulsed proton conductor, when activated by light (see Sect. 16.5.2). This is a light-driven proton pump generating an electrochemical gradient across the membrane, which is employed by the bacterium as an energy source. After having been pumped from the cell-interior to the membrane surface, the protons are transported by a mechanism of surface diffusion towards other proteins located within the same membrane: The light-generated electrochemical potential across the cell membrane is utilised by the halobacteria to furnish the driving force for ATP synthesis by energizing the rotation of the turbine-like machinery in ATP synthase. Water molecules near the surface are known to be relevant for this biological function, since they have been shown to assist the proton conductivity [113]. Hydration water, its interaction with the surface of biological macromolecules and macromolecular complexes, and its diffusion generally play an important role in structure, dynamics, and function of biological systems [114, 115]. It is therefore of considerable interest to study the proton diffusion mechanism within and close to the hydration layers of membranes. Similar to the case of bulk water, it is expected that during the diffusion process protons are exchanged between water molecules acting either as acceptors (forming for instance $(H_3O)^+$) or as donors (producing $(OH)^-$ ions). A "solid-like" Grotthuss feature [116–118] is added to the diffusion of protons in the liquid water phase by the presence of fixed protonation sites on the surface of purple membrane. It is worthwhile to note, that these sites are arranged in space in a perfectly regular manner, since the bacteria use the most efficient packing of BR: trimers of BR molecules embedded in a lipid bilayer matrix are aligned in a two-dimensional hexagonal single-crystalline structure [119]. However, the pH-value not being very different from 7, the concentration of charged particles is very low. The protons spend most of the time as part of diffusing neutral water molecules with only rare events of exchange between different "vehicles" [116]. Therefore, for the purpose of analyzing neutron scattering experiments, the whole mechanism of diffusion may to a good approximation be classified as that of molecular diffusion [116].

In the QINS [110] and PFG-NMR [111] measurements concerning the anisotropy of proton diffusion on the membrane, the crystalline order of PM has been exploited. For this purpose about 20,000 layers of purple membranes were stacked approximately parallel to each-other (mosaic spread: about 12° FWHM) at defined relative humidities (r.h.). The membrane stacks had been produced, starting from an aqueous suspension of membrane pieces, by alignment of the membranes through evaporation of water, using aluminium foils as a substrate. At 100% r.h., with water layer thickness of about 10 Å between neighboring membranes, the protons of water molecules were found to participate in a process of two-dimensional long-range translational diffusion parallel to the membrane plane, with a self-diffusion coefficient $D_s = 4.4\times10^{-6}\ \mathrm{cm^2 s^{-1}}$ at room temperature, i.e., about five times smaller than the known bulk value of water. At the same time, they are also participating in a fast localized diffusive motion which can – at least partially – be assigned to the rotation of water molecules. This was observed to be about six times slower than in bulk water. These motions are sufficiently fast to produce quasielastic broadenings clearly seen with an elastic energy resolution of 16 µeV FWHM (see Fig. 16.13). Finally, a translational diffusion jump distance of 4.1 Å was derived from the

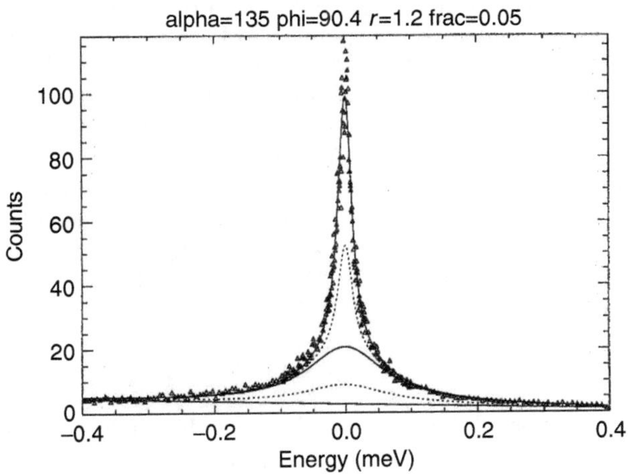

Fig. 16.13. Fit of a model for two-dimensional translational diffusion of water to QINS spectra (data points shown as *triangles*) of hydrated purple membrane stacks oriented with an angle $\alpha = 135°$ with respect to the incident neutron beam. These data were taken with the inverted time-of-flight technique (IRIS, see Sect. 15.3.4 in Part I, this volume) at the scattering angle $\varphi = 90.4°$. The spectra were not corrected for multiple scattering, but MSC was taken into account in the fitted theoretical model: The *solid lines* represent (from *top* to *bottom*) the fit result for the total scattering function including all terms, the sum of all rotation-broadened components of the scattering function (including the backgound B), and B. The dotted lines are the two MSC contributions: a rotation-broadened MSC component sitting on the background line, and an MSC component without rotation broadening on top of the rotation-broadened single-scattering contribution; after [111]

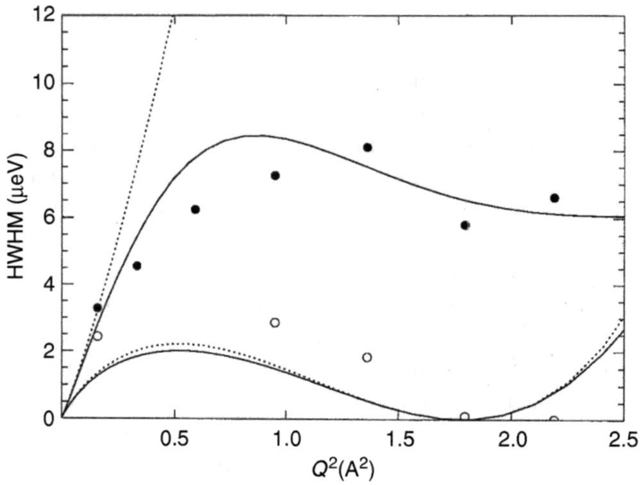

Fig. 16.14. Translational diffusion linewidths (exhibited by spectra as shown in Fig. 16.13) of water on hydrated purple membrane, plotted as a function of Q^2 for two sample orientations: $\alpha = 45°$, open circles; $\alpha = 135°$, full circles. The experimental values are compared with the theoretical width corresponding to an isotropic approximation of the Chudley–Elliott jump-diffusion model in two dimensions. Note that for $\alpha = 135°$, the $\boldsymbol{Q}$ vector is exactly parallel to the membrane plane, when the scattering angle is $\varphi = 90°$, whereas $\boldsymbol{Q}$ is perpendicular to the membrane for the same scattering angle, when $\alpha = 45°$. Therefore the linewidth curve for the latter case approaches zero near $Q^2 = 1.784\,\text{Å}^{-2}$, in agreement with the values observed experimentally. This means that diffusion parallel to the membrane plane has clearly been seen in this experiment, whereas in the direction perpendicular to it the diffusive motion is too slow to be observable with the QINS technique. Dotted lines: behavior of the width according to the D_sQ^2 – law, if this was valid in the whole region of Q^2 shown in the figure (after [111])

Q-dependent behavior of the quasielastic linewidth at large scattering angles (see Fig. 16.14). This distance is three times larger than the corresponding quantity of bulk water.

The relative slowness of the diffusion process may be partly due to the restricted space available within the hydration layers. It might also be caused, together with the large value of the jump distance, by the presence of fixed protonation sites on the membrane surface. These might have nearest-neighbor distances similar to those of neighboring lipid head groups, which are of this order of magnitude. At these protonation sites, hydrogen bonds are expected to be formed, with life times exceeding those of bonds which exist between neighboring H_2O molecules in bulk water.

16.5.2 Dynamical Transition and Temperature-Dependent Hydration: Example Purple Membrane

The study of the relation between the hydration of biological systems and their function has attracted much attention at least over the last 40 years. A large amount of results has been accumulated and summarized in a number of review articles (see for instance [114, 115, 120–124]) and books [125–127]. It is known that the amount of water associated with any specific biological system and the temperature are two of the most important factors controlling the kinetics of the biological function of the involved macromolecules. As an example, we consider studies of the purple membrane (PM) of *Halobacterium salinarum*. This is a lipid bilayer system containing a two-dimensional hexagonal arrangement of trimers of the integral protein bacteriorhodopsin (BR) [128]. BR is a well-known proton pump and the prototype of a membrane protein (see also Sect. 5.1). The absorption of a light quantum by light-adapted BR leads to the conformational transition from the all-*trans* to the 13-*cis* isomer of the retinal chromophore and initiates a sequence of interconversions between a number of spectroscopic intermediates usually designated by the letters K, L, M[8], N, and O, finally leading back to the initial state of BR. Under physiological conditions, as a result of the reactions occurring during this photocycle, a proton is released from the retinal's Schiff base and pumped to the exterior surface of BR, while another proton is taken up on the cytoplasmic side to replace the first proton at its original binding site (for a review, see for instance: [130]). Experimental observations have indicated that the gradual removal of water from PM and thus from the protein modifies the photocycle and has specific consequences on the rates of individual transitions between intermediates. While the processes leading to M-formation are only moderately affected by changes in the amount of water present, the relaxation rate of the M decay strongly depends on the level of hydration. For instance, between 90% and 75% relative humidity (r.h.) the rate decreases by a factor of 10 and slows down by another two orders of magnitude, when the hydration water is removed almost completely by drying at 7% r.h. or less [131]. More recent experimental results [132, 133] in qualitative agreement with [131], suggest that from 100% to 85% r.h. the M-decay rate decreases by a factor of 2, whereas the strong decrease of the M-decay rate essentially starts, when the humidity is decreased below about 85% to 80% r.h. With decreasing humidity less and less light-induced charge translocation occurs across the protein [134]. Upon further lowering of the r.h., the

[8]From X-ray diffraction and FT-IR measurements it is known that the intermediate M splits into the two structurally different states M_1 and M_2, which can not be distinguished spectroscopically [129]

photocycle and the actual vectorial proton transport activity of the proton pump are drastically modified [135–137].

It is interesting to compare the hydration dependence of the photocycle to its variation observed as a function of temperature. While the quantum efficiency of BR is practically temperature independent, the decay of all intermediates relying on thermal activation obviously is not. The photocycle slows down with decreasing temperature and is "frozen in" at some intermediate state, when the time constant of the latter has become practically infinite due to the low value of the temperature. It was found, for instance, that bacteriorhodopsin can be "captured" in the M_2-intermediate state at 260 K [138]. Furthermore, according to [139,140], the photocycle stops at the M-intermediate state below $T \simeq 220\,K$ (at the M_1-intermediate state near 230 K, after [141]), at the L-intermediate below $T \simeq 180\,K$ (near 155 K, after [141]) and at the K-intermediate below $T \simeq 150\,K$ (near 90 K, after [141]). The N and O intermediates were not observed at all by low-temperature spectrophotometry. Below we discuss the question, whether the T- and h-dependences of the photocycle are significantly correlated with the T- and h-dependent effects observed in the static and dynamic structure of multilayered stacks of PM.

In a previous study [100] we have investigated the static structure of multilayered stacks of PM as a function of temperature (T), after having established well-defined levels of hydration (h) by equilibration at fixed relative humidities (r.h.), at room temperature. The lamellar spacing d_l of these systems was measured with neutron diffraction as a function of T and h. The observed large T-dependent variations of d_l indicate that PM is partially dehydrated, when cooled below a "hydration water freezing point." This effect is reversible, but a hysteresis is observed, when PM is rehydrated upon reheating. This phenomenon of dehydration and rehydration, induced by cooling and reheating, respectively, appears to be a general property of biological membranes (see also [142]). It is caused by the presence of hydration forces and by the specific (different) temperature dependences of the chemical potentials of interbilayer water and ice. These forces result in a local freezing point depression and are the main reason, why nucleation of ice crystals first occurs outside of the space between bilayers. The temperature variation of the chemical potentials leads to a difference in vapor pressure, causing water to be successively extracted from the interbilayer space in the presence of these ice crystals, when the temperature is lowered. Vice versa upon heating, the water originating from the melting of the same ice crystals, within the closed system, is "sucked" back into the interbilayer space. This effect is due to the temperature-dependent change in the balance between the hydration forces and the chemical force resulting from ice formation. It is important to note, that this leads to what amounts to an effective spatial separation of the crystallized water from the biological surface.

For the temperature dependence of the lamellar spacing the following qualitative behavior of d_l was observed [100]: During the cooling cycle of samples equilibrated at a given relative humidity (r.h.), starting at room temperature (295 K), d_l stays approximately constant down to a few degrees below the freezing point of bulk water ($T_f = 273.15$ K for pure H_2O and 276.97 K for pure D_2O). Then, at a temperature which we denote by T_{fh}, a discontinuity occurs, which is connected with a large decrease in d_l by a step of the order of 20–30% . The hydration water remaining bound to the purple membrane below about 240 K is nonfreezing. Its amount was found to be $h_{nf} = 0.24(\pm 0.02)$ [g D_2O/g BR] for all PM samples equilibrated at room temperature in the presence of D_2O vapor at ≥ 84 % r.h..

Figure 16.15 shows an example of such a study of a PM specimen [143]. The lamellar lattice constant of the PM multilayer stack is displayed as a function of temperature in a heating cycle; the sample had been equilibrated at room temperature in an atmosphere of 98% r.h. (D_2O vapor). The experimental values (triangles) are shown together with a phenomenological model curve. The vertical arrows labeled with letters, indicate the approximate limiting temperatures, below which – in the course of the photocycle – BR does no longer return to the ground state, but stops at the intermediates K, L, M_1 and M_2, respectively.

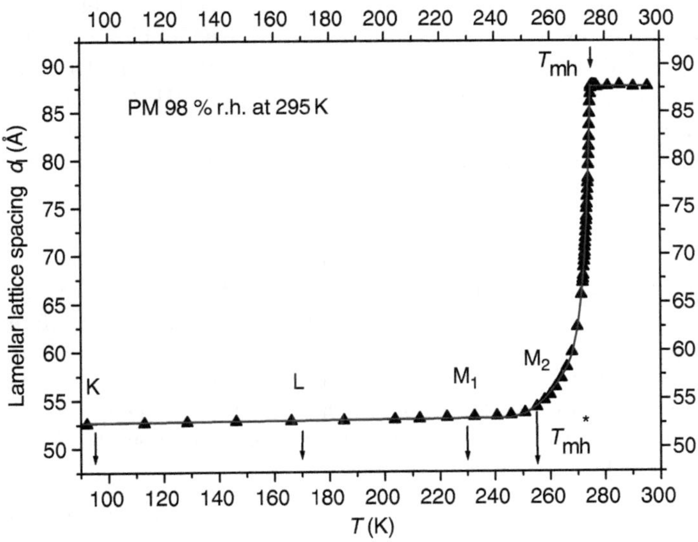

Fig. 16.15. Lamellar lattice constant of a purple membrane multilayer stack as a function of temperature in a heating cycle; the sample (labeled H1) had been equilibrated at room temperature in an atmosphere of 98% r.h. (D_2O vapor). The experimental values (*triangles*) obtained with the membrane-diffractometer V2 at BENSC in Berlin, are shown together with a phenomenological model curve. The vertical arrows indicate the approximate limiting temperatures, below which – in the course of the photocycle – BR does no longer return to the ground state, but stops at the intermediates K, L, M_1, and M_2, respectively [141]. Figure taken from [143]

Let us now discuss possible correlations between the T-dependent dehydration/rehydration behavior of PM and the variation of its dynamic structure with temperature. It has been shown by studies of purple membrane using quasielastic incoherent neutron scattering (QINS), that the ability of bacteriorhodopsin to functionally relax and complete the photocycle initiated by the absorption of a photon, is strongly correlated with the onset of low-frequency, large-amplitude "anharmonic" molecular motions. This manifests itself as the well-known "dynamical transition" [35, 144–147] starting from a low-temperature harmonic toward a high-temperature "anharmonic" regime [36]. More precisely, the "dynamical transition" announces the onset of overdamped vibrational and/or localized (i.e., spatially restricted) diffusive (stochastic) molecular motions at temperatures in the neighborhood of 200 K. It is characterized by the appearance of quasielastic neutron scattering due to these motions[9]. This has also been made visible indirectly by studying the temperature dependence of elastic scattering ("elastic-window scan", see Sect. 15.3.3 in Part I, this volume), which allows to extract an atomic mean-square displacement $<u^2>$. In this case, for fully hydrated PM, the onset of diffusive motions was found near 230 K by an analysis of $<u^2>$. Here, the validity of a Gaussian approximation to the motion of all the hydrogen atoms in PM was assumed, which – for the employed energy resolution (IN13, with FWHM = 10 μeV) – includes the frequencies between $10^{10}\,\mathrm{s}^{-1}$ and $10^{13}\,\mathrm{s}^{-1}$ [36]. In this experiment, a large Q-range (up to $Q = 4.5\,\mathrm{Å}^{-1}$) was used . However at higher energy resolution (FWHM = 1 μeV), and in a lower Q-range (up to $Q = 1.8\,\mathrm{Å}^{-1}$), elastic-window scans with IN16 yielded the onset of a dynamical transition already near 150 K [148, 149][10]. From

[9] Note that it is not possible to give one very precise temperature value for this "transition," for two reasons: (i) there is a multitude of different molecular motions that are gradually activated with rising temperature; the transition is therefore occurring in a continuous way over a certain temperature range; (ii) any newly arising motion can only be detected, when it occurs within the energy window of the experiment; therefore, the temperature dependence of this "transition" phenomenon is also correlated with the resolution-dependent variation of experimental observability; see Sect. 15.3.3, Fig. 15.7, in Part I, this volume

[10] The elastic-window scan method has also been employed in numerous other biological experiments, in order to determine the temperature dependence of motional amplitudes ("mean-square displacements"), concerning for instance the environment-dependence of confined diffusive protein motions (see Sect. 16.4.3), the dynamics of hydrated starch saccharides [150], the effect of myelin basic protein on the dynamics of oriented lipids [151], proton mobilities in crambin and glutathionine S-transferase [152], the dynamic properties of an oriented lipid/DNA complex (proposed as DNA vector in gene therapy) [153], the influence of solvent composition on global dynamics of human butyrylcholinesterase powders [154], the relation between the glass-forming character and the cryoprotective capability of disaccharide–water mixtures [155, 156], and the comparison of the macromolecular dynamics in psychrophile, mesophile, thermophile, and hyperthermophile bacteria [157]

such observations, it has been concluded, that the degree of "softness" of the membrane modulates the function of bacteriorhodopsin by allowing or not allowing large-amplitude molecular motions in the protein [36]. Of course the parameter $< u^2 >$ as defined in [36] represents some kind of an average over three orders of magnitude on the frequency scale. But not all modes of diffusive molecular motions behave exactly in the same way. A more detailed phenomenological analysis of QINS data in terms of quasielastic linewidths and quasielastic incoherent structure factors (QISFs), corresponding to an interpretation by orientational jump-diffusion of molecular subunits, suggests that slower diffusive motions with jump rates of about $10^9\,s^{-1}$ start to be seen already near 190 K, whereas faster motions with rates from about 10^{10} to $10^{12}\,s^{-1}$ appear to be "turned on" in the temperature region from 200 K to about 220 K [101,102]. It should therefore be emphasized, that the "dynamical transition" does not have a unique transition temperature.

Already previously we have pointed out [100], that the dehydration/rehydration behavior of PM is strongly correlated with the temperature-dependent behavior of the dynamic structure factor. Above the "dynamical transition" of PM, reported earlier [36], a second dynamical transition is caused by the temperature-induced rehydration upon heating the PM, which starts near 255 K. Figure 16.16 shows both effects as derived from QINS spectra of the same purple membrane sample [143]. The parameters directly obtained from the measurements are the amount of quasielastic scattering (i.e., the value of the QISF) and its energy width, both as a function of Q. These quantities stand for the macromolecular flexibility. The QISF is related to the spatial extension of the volume required for the confined diffusive motions of H-atoms in the sample, (see also Sect. 15.2.2 in Part I, this volume, and Sects. 16.4.2 and 16.4.3 of this article). It permits a weighted average of the amplitudes of stochastic (localized diffusive) motions of the hydrogen atoms to be calculated, which are bound in the macromolecules. For a qualitative analysis the simplest model was used. It allows to estimate this volume by calculating the QISF for orientationally averaged two-site jumps (see Eqs. 15.27 and 15.28, Table 15.1 and Fig. 15.1 in Part I, this volume), with a jump distance d_{2S} (see for instance [2], p. 193–196) as parameter of the fit. The value of d_{2S} which would yield the measured QISF, was determined as a function of temperature in a heating cycle (Fig. 16.16). The sample had been equilibrated at room temperature in an atmosphere of 98% r.h. (D_2O vapor). The experimental values of d_{2S} (triangles) are shown together with a phenomenological model curve. The vertical arrows indicate the limiting temperatures, near which – in the course of the photocycle – BR does no longer return to the ground state, but stops at the intermediates K, L, M_1, and M_2, respectively [143].

The comparison of Figs. 16.15 and 16.16 suggests the following interesting observations: In the temperature region below $T \simeq 255$ K, the value of d_l is practically constant, i.e., the water layer thickness on the membrane surface essentially does not vary. Above $T \simeq 255$ K, the layer thickness increases drastically, but in a continuous fashion, until the final value is reached at the melting discontinuity point T_{mh} of membrane water. The temperature

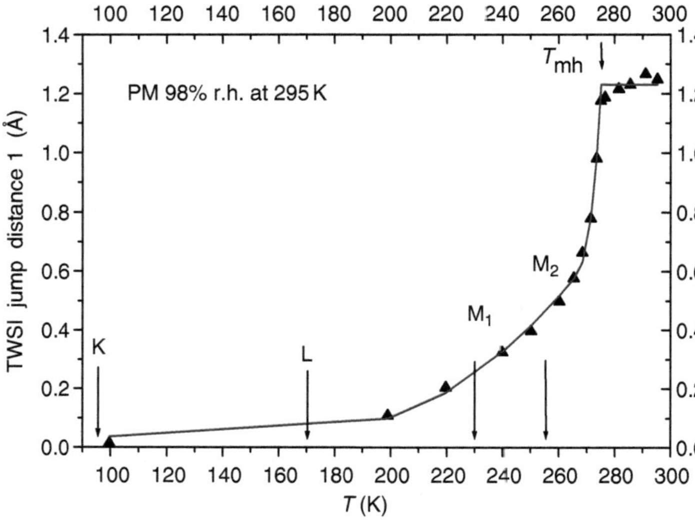

Fig. 16.16. Two-site jump distance d_{2S} of the purple membrane sample H1 as a function of temperature in a heating cycle, as determined from the experimental QISF. This phenomenological parameter, derived from QINS spectra measured with the direct-geometry TOF spectrometer IN6 (see Sect. 15.3.1 in Part I, this volume) at ILL, stands for the molecular flexibility. It measures a weighted average of the amplitudes of stochastic (localized diffusive) motions of the hydrogen atoms bound in the macromolecules (see Eqs. 15.27 and 15.28, Table 15.1 and Fig. 15.1 in Part I, this volume). The sample had been equilibrated at room temperature in an atmosphere of 98% r.h. (D_2O vapor). The experimental values (*triangles*) are shown together with a phenomenological model curve. The vertical arrows indicate the limiting temperatures, near which the photocycle does not reach its end, i.e., BR does not return to the ground state, but stops at the intermediates K, L, M_1, and M_2, respectively [141]. Figure taken from [143]

variation of the jump distance d_{2S} (Fig. 16.16) below $T \simeq 255$ K, is fundamentally different from that of the lamellar lattice constant. The jump distance stays close to zero below 200 K, but starts to grow continuously and significantly beyond this point. Up to $T \simeq 255$ K, this is clearly a pure effect of thermal activation, since the amount of water present in the membrane multilayer is constant in this temperature region. Therefore, this phenomenon must be attributed to the intrinsic dynamical transition of the purple membrane complex including a monomolecular water layer bound to the membrane surface. Above $T \simeq 255$ K, however, the addition of liquid interbilayer water visibly accelerates the increase in flexibility, as demonstrated by a drastic increase of the jump distance. A separation of the pure effect of temperature on the purple membrane, from that of the plasticizing action of water has been achieved here. At the same time, these observations suggest the following interpretation of the results of [138–141]: When $T \simeq 230$ K, because of a lack of intrinsic flexibility of the purple-membrane hydration-shell complex at this and lower temperatures, the photocycle stops at the M_1-intermediate state; at somewhat higher temperatures, near $T \simeq 255\,K$, the photocycle proceeds further, but stops just after the M_2-intermediate state has been reached, because at this and lower temperatures, the required additional flexibility is missing, which is only provided, when water is present at the membrane surface, not exclusively within the first hydration shell, but also beyond it.

16.6 Conclusions

Quasielastic scattering including low-energy-transfer inelastic scattering of neutrons, is an excellent tool to evaluate the elementary steps of diffusive motions and the character of vibrational densities of state of the complex stucture of biological matter, on scales from atomic to mesoscopic dimensions and times. We have reviewed a number of examples of application in this article.

First, we have presented results concerning translational long-range diffusive motions of myoglobin in aqueous solution under "crowded" conditions. This oxygen storage molecule is located in muscles and is believed to assist oxygen transport by diffusion from the cell membrane to the low partial pressure regions in the mitochondria. Neutron scattering is so far the only technique which allows to study both protein interactions and protein motions over intermolecular distances at physiological concentration. Of particular interest is the fact, that QENS allows to probe protein dynamics directly inside cells.

The second topic treated is antenna complexes. One of their important functions during the primary steps of photosynthesis, is to perform an efficient excitation energy transfer. In this, the coupling of the purely electronic transitions of pigments to low-frequency vibrational modes of the protein matrix play an essential role. Apart from optical spectroscopy, IINS has become a technique of choice as an independent experimental approach to study the effective vibrational density of states $g(\omega)$ of photosynthetic pigment–protein complexes. Here, we have discussed results of IINS experiments on the photosynthetic antenna complex LHC II from spinach, leading to the determination of this function.

Because of the strong correlation between functionality, dynamics, and environment, it is of great interest, to study the dynamics of biological systems as a function of environmental parameters. We have discussed quasielastic neutron scattering experiments dealing with the molecular mobility of lysozyme solvated in glycerol, with various water contents, as a function of temperature. These studies have revealed the alteration of the protein's internal mobility, when the character of its environment changes from a stabilizer-like to a plasticizer-like nature. This is the signature of a progressive onset of new relaxational degrees of freedom of internal protein motions, probably related to groups located at the protein surface. Apparently, also in the presence of glycerol, water molecules are able to gradually activate the whole lysozyme dynamics by progressively hydrating the protein surface groups. This may be interpreted as a hydration-driven onset of confined diffusive internal protein motions.

Another interesting subject concerns the modification of protein dynamics due to ligand binding. A pertinent experimental neutron scattering investigation on the enzyme dihydrofolate reductase, an important target for anticancer and antibacterial drugs, was presented. The ligand used is methotrexate (MTX), a folate antagonist of DHFR employed effectively as a cytotoxic agent in the treatment of cancers. The results suggest, that the complexation may lead to an increased importance of relaxational modes (at low frequencies) in this whole system, and to complexation-caused damping of the vibrations (in an intermediate frequency region) and thus to an increased flexibility of the macromolecular ensemble, as compared to the uncomplexed case. Complexation of DHFR apparently softens the enzyme and makes this macromolecule more flexible, an effect which could be relevant for its biochemical activity. From the methodical and application points of view, such results may be considered as promising for the development of medical diagnostics.

Hydration water, its interaction with the surface of biological macromolecules and macromolecular complexes, and its diffusion generally are relevant for structure, dynamics, and function of biological systems. Low-dimensional diffusion plays an important role for certain biological objects, such as ion transport channels and membrane surfaces. Two-dimensional proton diffusion and conduction assisted by water molecules provide an important mechanism of energy transduction in living organisms. The QENS experiments we have discussed here, concern the transport of water molecules on the surface of the purple membrane (PM) of *Halobacterium salinarum*. PM contains the protein bacteriorhodopsin (BR) which becomes a one-dimensional stochastically pulsed proton conductor, when activated by light. This is a light-driven proton pump generating an electrochemical gradient across the membrane, which is employed by the bacterium as an energy source, for instance to furnish the driving force for ATP synthesis. The water molecules were found to serve as vehicles of two-dimensional long-range translational proton diffusion parallel to the membrane plane. Self-diffusion and reorientation of water molecules are about five to six times slower, and the translational jump distance is three times larger than the corresponding quantities of bulk water at room temperature. These facts are related to the presence of protonable sites on the membrane surface, which have distances of this order of magnitude from each-other and slow down the diffusion process.

The amount of water associated with any specific biological system, and the temperature are two of the most important factors controlling the kinetics of the biological function of the macromolecules involved. We have discussed the question, how these environmental conditions are correlated with the macromolecular dynamics, and what this correlation can tell us about the role of dynamics for function. It is a known fact, that the photocycle slows down with decreasing temperature and is "frozen in" at specific intermediate states, when the growing time constant of the latter has become practically infinite. This phenomenon has been related with the dynamical transition of PM, announcing the onset (upon heating) of localized diffusive molecular motions at temperatures in the neighborhood of 200 K, and with the temperature-induced rehydration of PM upon heating, which starts near 255 K. A strong correlation between the temperature-dependent behavior of the dynamic structure factor observed with neutron scattering and the course of the photocycle was discovered. Both the temperature and the T-dependent level of hydration are important parameters of this correlation. A separation of the pure effect of temperature from that of the plasticizing action of water on the purple membrane with its integral protein, the proton pump BR, and its ability to function, has been achieved. This clearly demonstrates that the photocycle, in order to be complete (and thus the proton pumping function of bacteriorhodopsin to be operational), requires molecular dynamics in the picosecond to nanosecond range and is turned off when the corresponding molecular motions are frozen or dried out.

Some of the dynamical features which are specific for the complex molecular systems encountered in biology, are connected with the fact that hydrogen bonding plays an important role. In particular, it is the process of formation and breaking of H-bonds, which represents an important ingredient of related diffusive mechanisms. The observability by neutron scattering of the chemical reaction equilibrium underlying the H-bond formation and breaking process has already been established in other fields [74], but not yet extensively exploited for biological problems.

In general, the synthesis between the results of quasielastic and inelastic neutron scattering, Mößbauer spectroscopy, nuclear resonance scattering of synchrotron radiation, nuclear magnetic resonance studies, frequency- dependent conductivity spectroscopy using electromagnetic waves from the microwave to the (far) infrared region, and ultrasonic measurements, computer simulation and the development of new analytical approaches, is fruitful and necessary for the further development of the field. Future efforts in this sense will contribute to a more comprehensive and sophisticated understanding of the low-frequency and diffusive dynamics in biological matter.

Acknowledgments

We would like to thank our colleagues, J. Pieper, A. Paciaroni, J.C. Smith, M. Tehei, W. Doster, J. Fitter, and N. A. Dencher for a number of clarifying discussions concerning the experimental results presented.

References

1. A.J. Leadbetter, R.E. Lechner, in *The Plastically Crystalline State*, J.N. Sherwood (Eds.) (J. Wiley and Sons, New York, 1979) pp. 285–320
2. R.E. Lechner, in *Mass Transport in Solids*, F. Benière, C.R.A. Catlow (Eds.), NATO ASI, 1981: Lannion, France, Series B: Physics, Vol. **97** (Plenum Publ. Corp., New York, 1983) pp. 169–226
3. M. Bée, *Quasielastic Neutron Scattering: Principles and Applications in Solid State Chemistry, Biology and Materials Science* (Adam Hilger, Bristol, 1988)
4. R.E. Lechner, Physica B **301** 83–93, (2001)
5. R.E. Lechner, in *Quasielastic Neutron Scattering*, J. Colmenero, A. Alegría, F.J. Bermejo (Eds.), Proceedings of the Quasielastic Neutron Scattering Workshop QENS'93, San Sebastián, Spain, 1993 (World Scientific, Singapore, 1994) pp. 62–92
6. M. Bée, Chemical Physics **292** 121, (2003)
7. R.E. Lechner, Th. Dippel, R. Marx, I. Lamprecht, Solid State Ionics **61**, 47 (1993)
8. M. Pionke, T. Mono, W. Schweika, T. Springer, H. Schober, Solid State Ionics **97**, 497 (1997)
9. A.V. Belushkin, C.J. Carlile, L.A. Shuvalov, J. Phys.: Condens. Matter **4**, 389 (1992)
10. G.H. Vineyard, Phys. Rev. **110**, 999 (1958)
11. T. Springer, R. E. Lechner, *Diffusion Studies of Solids by Quasielastic Neutron Scattering*, in *Diffusion in Condensed Matter*, P. Heitjans, J. Kärger (Eds.) (Springer Verlag, Berlin, Heidelberg, 2005) pp. 93–164
12. C.T. Chudley, R. J. Elliott, Proc. Phys. Soc. **77**, 353 (1961)
13. K.S. Singwi, Phys. Rev. **119**, 863 (1960)
14. V.S. Oskotskii, Soviet Phys. Solid State **5**, 789 (1963)
15. P.A. Egelstaff, Advanc. Phys. **11**, 203 (1962)
16. A. Rahman, K.S. Singwi, A. Sjölander, Phys. Rev. **126**, 986–997 (1962)
17. S. Longeville, W. Doster, G. Kali, Chem. Phys. **292**, 413–424 (2003)
18. S. Longeville, W. Doster, M. Diehl, R. Gähler, W. Petry, in *Neutron Spin-Echo Spectroscopy*, Lecture Notes in Physics, Vol. **601**, F. Mezei, C. Pappas, T. Gutberlet (Eds.) (Springer Verlag, Berlin, 2003) pp. 325–335
19. S. Longeville, W. Doster, to be published
20. W. Doster, S. Longeville; to be published
21. A. Sjölander, Arkiv för Fysik **14**, 315–371 (1958)
22. R.E. Lechner, C. Riekel, in *Neutron Scattering and Muon Spin Rotation*, Springer Tracts in Modern Physics, Vol. **101**, (Springer Verlag, Berlin, 1983) pp. 1–84
23. R. Kubo, Rep. Progr. Phys. **29**, 255 (1966)
24. R. Van Grondelle, J.P. Dekker, T. Gillbro, V. Sundstrom, Biochim. Biophys. Acta **1187**, 1 (1994)
25. G. Renger, in *Concepts in Photobiology and Photomorphogenesis*, G.S. Singhal, G. Renger, K. Sopory, K.-D. Irrgang, Govindjee (Eds.) (Narosa Publishing House, New Delhi, India, 1999) p. 52
26. H. Paulsen, Photochem. Photobiol. **62**, 367 (1995)
27. J. Voigt, T. Renger, R. Schödel, T. Schrötter, J. Pieper, H. Redlin, Phys. Stat. Sol. (b) **194**, 333 (1996)

28. T. Renger, J. Voigt, V. May, O.J. Kühn, Phys. Chem. **100**, 15654 (1996)
29. T. Renger, V. May, Phys. Rev. Lett. **84**, 5228 (2000)
30. S.V. Kolaczkowski, J.M. Hayes, G.J. Small, J. Phys. Chem. **98**, 13418 (1994)
31. J. Pieper, J. Voigt, G.J. Small, J. Phys. Chem. B **103**, 2319–2322 (1999)
32. O. Kühn, T. Renger, V. May, J. Voigt, T. Pullerits, V. Sundström, Trends in Photochem. Photobiol. **4**, 213 (1997)
33. J. Pieper, J. Voigt, G. Renger, G.J. Small, Chem. Phys. Lett. **310**, 296–302 (1999)
34. S. Cusack, W. Doster, Biophys. J. **58**, 243 (1990)
35. W. Doster, S. Cusack, W. Petry, Phys. Rev. Lett. **65**, 1080–1083 (1990)
36. M. Ferrand, A.J. Dianoux, W. Petry, G. Zaccai, Proc. Natl. Acad. Sci. USA **90**, 9668–9672 (1993)
37. B. Frick, D. Richter, Science **267**, 1939 (1995)
38. A. Orecchini, A. Paciaroni, A.R. Bizzarri, S. Cannistraro, J. Phys. Chem. B **105**, (48) 12150 (2001)
39. A. Paciaroni, A. Orecchini, S. Cinelli, G. Onori, R.E. Lechner, J. Pieper, Chemical Physics **292**, 397 (2003)
40. A. Paciaroni, A.R. Bizzarri, S. Cannistraro, J. Mol. Liq. **84**, 3 (2000)
41. J. Pieper, K.-D. Irrgang, G. Renger, R.E. Lechner, J. Phys. Chem. B **108**, 10556 (2004)
42. R.E. Lechner, Physica B **180 & 181**, 973 (1992)
43. R.E. Lechner, Neutron News **7**,(4) 9 (1996)
44. R.E. Lechner, R. Melzer, J. Fitter, Physica B **226**, 86 (1996)
45. N.R.S. Reddy, P.A. Lyle, G. Small, J. Photosyn. Res. **31**, 167 (1992)
46. R. Jankowiak, G. Small, J. Chem. Res. Toxicol. **4**, 256 (1991)
47. A. Garbers, F. Reifarth, J. Kurreck, G. Renger, F. Parak, Biochemistry **37**, 11399 (1998)
48. F. Parak, Rep. Prog. Phys. **66**, 103 (2003)
49. K. Funke, Prog. Solid St. Chem. **22**, 111 (1993)
50. K. Funke, Z. Phys. Chem. **188**, 243 (1995)
51. P. Heitjans, J. Kärger, *Diffusion in Condensed Matter* (Springer Verlag, Berlin, Heidelberg, 2005)
52. P.G. De Gennes, Physica **25**, 825 (1959)
53. R.E. Lechner, Solid State Ionics **61**, 3 (1993)
54. M.H. Dickens, W. Hayes, P. Schnabel, M.T. Hutchings, R.E. Lechner, B. Renker, J. Phys. C **16**, L1 (1983)
55. P. Schnabel, W. Hayes, M.T. Hutchings, R.E. Lechner, B. Renker, Radiation Effects **75**, 73 (1983)
56. J. Fitter, R.E. Lechner, G. Büldt, N.A. Dencher, Proc. Natl Acad. Sci. USA **93**, 7600–7605 (1996)
57. V.F. Sears, Can. J. Phys. **45**, 237 (1967)
58. J. Teixeira, M.-C. Bellissent-Funel, S.H. Chen, A.J. Dianoux, Phys. Rev. A **31**, 1913–1917 (1985)
59. A. Deriu, F. Cavatorta, D. Cabrini, C. J. Carlile, H.D. Middendorf, Europhys. Lett. **24**, 351–357 (1993)
60. D. Di Cola, A. Deriu, M. Sampoli, A. Torcini, J. Chem. Phys. **104**, 4223–4232 (1996)
61. S. Longeville, R.E. Lechner, Physica B **276–278**, 183–184 (2000)
62. V. Calandrini, A. Deriu, G. Onori, R.E. Lechner, J. Pieper, J. Chem. Phys. **120**, 4759–4767 (2004)

63. D.A. Neumann, J.R.D. Copley, R.L. Cappelletti, W.A. Kamitakahara, R.M. Lindstrom, K.M. Kreegan, D.M. Cox, W.J. Romanow, N. Coustel, J.P. McCauley, Jr, N.C. Maliszewskyj, J.E. Fischer and A.B. Smith, III: Phys. Rev. Lett. **67**, 3808 (1991)
64. D. Wilmer, K. Funke, M. Witschas, R.D. Banhatti, M. Jansen, G. Korus, J. Fitter, R.E. Lechner, Physica B **266**, 60 (1999)
65. F. Volino, A.J. Dianoux, Mol. Phys. **41**, 271 (1980)
66. M.-C. Bellissent-Funel, J. Teixeira, K.F. Bradley, S.H. Chen, J. Phys. I France **1**, 995–1001 (1992)
67. M.-C. Bellissent-Funel, K.F. Bradley, S.H. Chen, J. Lal, J. Teixeira, Physica A **201**, 277–285 (1993)
68. V. Receveur, P. Calmettes, J.C. Smith, M. Desmadril, G. Coddens, D. Durand, Proteins: Struct., Funct., Genet. **28**, 380–387 (1997)
69. J. Pérez, J.-M. Zanotti, D. Durand, Biophys. J. **77**, 454–469 (1999)
70. M.-C. Bellissent-Funel, S.H. Chen, J.-M. Zanotti, Phys. Rev. E **51**, 4558–4569 (1999)
71. J.-M. Zanotti, M.-C. Bellissent-Funel, J. Parello, Biophys. J. **76**, 2390–2411 (1999)
72. J. Fitter, J. Phys. IV France **10**, 265–270 (2000)
73. Th. Steiner, W. Saenger, R.E. Lechner, Mol. Phys. **72**, 1211–1232 (1991)
74. R.E. Lechner, Solid State Ionics **145**, 167 (2001)
75. J.A. McCammon, S.C. Harvey, *Dynamics of Proteins and Nucleic Acids* (Cambridge University Press, New York, 1987).
76. A.M. Klibanov, Nature **409**, 241 (2001)
77. A. Kollmar, B. Alefeld, in *Proceedings of the Conference on Neutron Scattering*, Gatlinburg, Tenn. USA 1976, R.M. Moon (Eds.) (National Techn. Information Service, U.S. Dept. of Comm., Springfield, 1976) pp. 330–336.
78. A. Paciaroni, S. Cinelli, G. Onori, Biophys. J. **83**, 1157 (2002)
79. S. Cinelli, A. De Francesco, G. Onori, A. Paciaroni, Phys. Chem. Chem. Phys. **6**, 3591–3595 (2004)
80. I.M. Klotz, Q. Rev. Biophys. **18**, 227 (1985)
81. S.J. Benkovic, C.A. Fierke, A.M. Naylor, Science **239**, 1105 (1988)
82. L.Y. Lian et al., Methods Enzymol. **239**, 657 (1994)
83. M.K. Gilson et al., Biophys. J. **72**, 1047 (1997)
84. M.L. Lamb, W.L. Jorgensen, Curr. Opin. Chem. Biol. **1**, 449 (1997)
85. W. Wang et al., Annu. Rev. Biophys. Biomol. Struct. **30**, 211 (2001)
86. I.Z. Steinberg, H.A. Scheraga, J. Biolumin. Chemilumin. **238**, 172 (1963)
87. M.I. Page, W.P. Jencks, Proc. Natl Acad. Sci. USA **68**, 1678 (1971)
88. J.M. Sturtevant, Proc. Natl Acad. Sci. USA **74**, 2236 (1977)
89. H.P. Erickson, J. Mol. Biol. **206**, 465 (1989)
90. A.V. Finkelstein, J. Janin, Protein Eng. **3**, 1 (1989)
91. B. Tidor, M. Karplus, J. Mol. Biol. **238**, 405 (1994)
92. S. Fischer, J.C. Smith, C.S. Verma, J. Phys. Chem. B **105**, 8050 (2001)
93. E. Balog, T. Becker, M. Oettl, R.E. Lechner, R. Daniel, J. Finney, J.C. Smith, Phys. Rev. Lett. **93**, 028103–1 (2004)
94. S.R. Stone, J. F. Morrison, Biochemistry **23**, 2753 (1984)
95. E.E. Howell et al., Science **231**, 1123 (1986)
96. D.M. Epstein, S.J. Benkovic, P.E. Wright, Biochemistry **34**, 11037 (1995)
97. M.R. Sawaya, J. Kraut, Biochemistry **36**, 586 (1997)

98. T. Kamiyama, K. Gekko, Biochim. Biophys. Acta **1478**, 257 (2000)
99. F.M. Huennekens, Adv. Enzyme Regul. **34**, 397 (1994)
100. R.E. Lechner, J. Fitter, N.A. Dencher, T. Hauß, J. Mol. Biol. **277**, 593–603 (1998)
101. J. Fitter, R.E. Lechner, G. Büldt, N.A. Dencher, Physica B **226**, 61–65 (1996)
102. J. Fitter, R.E. Lechner, N.A. Dencher, Biophys. J. **73**, 2126–2137 (1997)
103. G.R. Kneller, W. Doster, M. Settles, S. Cusack, J.C. Smith, J. Chem. Phys. **97**, 8864–8879 (1992)
104. L. Zidek, M.V. Novotny, M.J. Stone, Nat. Struct. Biol. **6**, 1118 (1999)
105. J.W. Cheng, C.A. Lepre, J.M. Moore, Biochemistry **33**, 4093 (1994)
106. C. Rischel et al., Biochemistry **33**, 13997 (1994)
107. D. Fushman,, O. Ohlenschlager, H. Ruterjans, J. Biomol. Struct. Dyn. **11**, 1377 (1994)
108. R.E. Lechner, Solid State Ionics **77**, 280 (1995)
109. A.J. Dianoux, F. Volino, H. Hervet, Mol. Phys. **30**, 1181–1194 (1975)
110. R.E. Lechner, N.A. Dencher, J. Fitter, G. Büldt, A.V. Belushkin, Biophys. Chem. **49**, 91 (1994)
111. R.E. Lechner, J. Fitter, Th. Dippel, N.A. Dencher, Solid State Ionics **70/71**, 296 (1994)
112. J. Pieper, G. Charalambopoulou, Th. Steriotis, S. Vasenkov, A. Desmedt, R.E. Lechner, Chem. Phys. **292**, 465–476 (2003)
113. G. Careri, M. Geraci, A. Giansanti, J.A. Rupley, Proc. Natl. Acad. Sci. USA **82**, 5342 (1985)
114. N.A. Dencher, J. Fitter, R.E. Lechner, *Hydration of Biological Membranes: Role of Water in Structure, Dynamics, and Function of the Proton Pump Bacteriorhodopsin* in: *Hydration Processes in Biology*, M.-C. Bellissent-Funel (Eds.), Proceedings ASI Les Houches, France 1998 (IOS Press, Amsterdam, 1999) pp. 195–217
115. J. Fitter, R.E. Lechner, N.A. Dencher, J. Phys. Chem. B **103**, 8036–8050 (1999)
116. R.E. Lechner, Ferroelectrics **167**, 83 (1995)
117. E. Hückel, Z. Elektrochem. **34**, 546 (1928)
118. K.D. Kreuer, in *Proton Conductors, Chemistry of Solid State Materials 2*, Ph. Colomban (Eds.) (Cambridge University Press, 1992)
119. R. Henderson, P.N.T. Unwin, Nature **257**, 28 (1975)
120. I.D. Kuntz, Jr., W. Kauzmann, *Hydration of proteins and polypeptides*, in: *Advances in Protein Chemistry*, Vol. **28**, C.B. Anfinsen, J.T. Edsall F.M. Richards (Eds.) (Academic Press, New York, London, 1974) pp. 239–345
121. J.T. Edsall, H.A. McKenzie, Adv. Biophys. **16**, 53–183 (1983)
122. W. Saenger, Annu. Rev. Biophys. Biophys. Chem. **16**, 93–114 (1987)
123. J.A. Rupley, G. Careri, in *Protein Hydration and Function*, Advances in Protein Chemistry, Vol. **41**, C.B. Anfinsen, J.T. Edsall, F.M. Richards, D.S. Eisenberg (Eds.), (Academic Press, New York, London, 1991) pp. 37–172
124. W. Doster, M. Settles, Biochim. Biophys. Acta **1749**, 173–186 (2005)
125. F. Franks, *Water, A Comprehensive Treatise*, Vols. **4** and **6** (Plenum Press, New York and London, 1982)
126. G.A. Jeffrey, W. Saenger, *Hydrogen Bonding in Biological Structures* (Springer-Verlag, Berlin, 1994)

127. M.-C. Bellissent-Funel, *Hydration Processes in Biology*, Proceedings ASI Les Houches, France 1998 (IOS Press, Amsterdam, 1999)
128. D. Oesterhelt, W. Stoeckenius, Nature New Biol. **233**, 149–152 (1971)
129. H.J. Sass, I.W. Schachowa, G. Rapp, M.H. Koch, D. Oesterhelt, N.A. Dencher, G. Büldt, EMBO J. **16**, 1484–1491 (1997)
130. D. Oesterhelt, J. Tittor, Trends Biochem. Sci. (TIBS) **14**, 57–61 (1989)
131. R. Korenstein and B. Hess, Nature **270**, 184–186 (1977)
132. T. Hauß, G. Papadopoulos, S.A.W. Verclas, G. Büldt, N.A. Dencher, Physica B **234–236**, 217–219 (1997)
133. J. Fitter, S.A.W. Verclas, R.E. Lechner, H. Seelert, N.A. Dencher, FEBS Lett. **433**, 321–325 (1998)
134. G. Váró, L. Keszthelyi, Biophys. J. **43**, 47–51 (1983)
135. G. Váró, J.K. Lanyi, Biophys. J. **59**, 313–322 (1991)
136. G.U. Thiedemann, J. Heberle, N.A. Dencher, in *Structures and Functions of Retinal Proteins*, J.L. Rigaud (Eds.), Colloque INSERM, Vol. **221** (John Libbey Eurotext Ltd., 1992) pp. 217–220
137. G.U. Thiedemann, *Zeitaufgelöste optische Spektroskopie an Bacteriorhodopsin: Photozyklusintermediate, Einfluß von Wasser*. PhD-thesis, FU-Berlin FB Physik (1994)
138. P. Ormos, Proc. Natl. Acad. Sci. USA **88**, 473–477 (1991)
139. T. Iwasa, F. Tokunaga, T. Yoshizawa, Biophys. Struct. Mech. **6**, 253–270 (1980)
140. T. Iwasa, F. Tokunaga, T. Yoshizawa, Photochem. Photobiol. **33**, 539–545 (1981)
141. N.A. Dencher, H.J. Sass, G. Büldt, Biochim. Biophys. Acta **1460**, 192–203 (2000)
142. J. Fitter, O.P. Ernst, T. Hauß, R.E. Lechner, K.P. Hofmann, N.A. Dencher, Eur. Biophys. J. **27**, 638–645 (1998)
143. R.E. Lechner, J. Fitter, N.A. Dencher, T. Hauß, to be published
144. G.P. Singh, F. Parak, S. Hunklinger, K. Dransfeld, Phys. Rev. Lett. **47**, 685–688 (1981)
145. F. Parak, E.W. Knapp, D. Kucheida, J. Mol. Biol. **161**, 177–194 (1982)
146. F. Parak, in *Biomembranes, Methods in Enzymology*, Vol. **127**, L. Packer (Eds.) (Academic Press, Inc., London, 1986) pp. 197–206
147. W. Doster, S. Cusack, W. Petry, Nature **337**, 754–758 (1989)
148. U. Lehnert, V. Réat, M. Weik, G. Zaccai, C. Pfister, Biophys. J. **75**, 1945–1952 (1998)
149. U. Lehnert, V. Réat, B. Keßler, D. Oesterhelt, G. Zaccai, Appl. Phys. A **74**[Suppl.] S1287-S1289, (2002)
150. M. Di Bari, A. Deriu, G. Albanese, F. Cavatorta, Chem. Phys. **292**, 333–339 (2003)
151. F. Natali, A. Relini, A. Gliozzi, R. Rolandi, P. Cavatorta, A. Deriu, A. Fasano, P. Riccio, Chem. Phys. **292**, 455–464 (2003)
152. U.N. Wanderlingh, C. Corsaro, R.L. Hayward, M. Bée, H.D. Middendorf, Chem. Phys. **292**, 445–450 (2003)
153. F. Natali, C. Castellano, D. Pozzi, A. Congiu Castellano, Biophys. J. **104**, (2004) Nov. 12
154. F. Gabel, M. Weik, B.P. Doctor, A. Saxena, D. Fournier, L. Brochier, F. Renault, P. Masson, I. Silman, G. Zaccai, Biophys. J. **86**, 3152–3165 (2004)

155. S. Magazù, G. Maisano, F. Migliardo, C. Mondelli, J. Phys. Chem. B **108**, 13580–13585 (2004)
156. S. Magazù, G. Maisano, F. Migliardo, C. Mondelli, Phys. Chem. Chem. Phys. **6**, 1962–1965 (2004)
157. M. Tehei, B. Franzetti, D. Madern, M. Ginzburg, B.Z. Ginzburg, M.-T. Giudici-Orticoni, M. Bruschi, G. Zaccai, EMBO Reports **5**, 66–70 (2004)

17

Conformational Dynamics Measured with Proteins in Solution

J. Fitter

17.1 Introduction

Biological macromolecules, such as enzymes or transport proteins, share a structural complexity which is also reflected in a complex dynamical behavior. Therefore, a complete understanding of these biomolecules is possible only if we know the three-dimensional structures of the ground state and the intermediate states, the kinetics of conformational changes, and thermal equilibrium fluctuations occurring in the structures. In particular, fast stochastic fluctuations on a picosecond timescale (a typical task for neutron spectroscopy) are relevant to overcome energy barriers which are given by the energy landscape in the biomolecules [1–3]. These fluctuations can determine the kinetics of transitions between conformational (intermediate) states (see for example [4–7]). In addition, these fluctuations contribute significantly to the conformational entropy of a biopolymer and are therefore important for protein stability, protein folding, and protein–ligand interactions [8–11]. The variation of environmental parameters, such as temperature, hydration level, pressure, salinity, or pH, offers a possibility to elucidate mechanisms how fluctuations control important properties of biomolecules, like catalytic activities or protein stability.

The ability to study proteins under most physiological conditions, which in most cases is provided by a solution, is a challenge for any experimental technique. Due to problems with strong solvent scattering (70–90% of the scattered intensity in a protein solution is related to solvent scattering), neutron spectroscopy was applied mainly to hydrated powder samples in the past. In some cases this type of sample with hydration levels in the order of 0.2–0.5 g solvent per gram biomaterial exhibits almost physiological characteristics (e.g., hydrated stacks of purple membranes, hydrated powder of myoglobin [5, 12, 13]). In a living cell, the volume fraction of biopolymers (mainly proteins) is in the order of up to 30%, but in general the concentration of one type of protein (there are about a few thousand different species in one single cell) is in the submicromolar regime (a notable exception from this can be found in the case of red blood cells with very high concentration of oxygen carriers such as hemoglobin). There is evidence that proteins already show many functional properties when hydrated with only 1–2 layers of water [14], which corresponds to a hydrated powder owing a hydration level of ~0.3 g solvent per gram protein. However, in most cases globular proteins need much more surrounding solvent to ensure sufficient diffusion rates of substrates in order to display full functionality.

Other properties of interest, such as folding/unfolding transitions of proteins or specific protein–ligand interactions, are generally strongly perturbed by pronounced unspecific protein–protein interactions such as protein aggregation at very high protein concentrations (like in hydrated powders). Therefore, a protein solution (still with concentrations $\geq 20\,\mathrm{mg\,ml^{-1}}$) is a step in the right direction, at least if a broad spectrum of applications with rather different proteins is in the focus of interest. Recently, an increasing number of neutron spectroscopy studies investigating protein solutions are emerging in the literature. Approaches on how to gain information on structural dynamics of proteins in solution and how to investigate biologically relevant question with this type of samples are the purpose of this chapter.

17.1.1 Dynamics in Proteins: Types of Motions and Their Biological Relevance

It is quite obvious to any considerate observer that a fundamental feature of living matter is "motion." Rest of atoms or absolute rigidity of biological structures is indicative for death or at least for the absence of life. For a large scope of timescales and length scales living and motion are indispensable related to each other. In the case of biological macromolecules which consist of a few thousand atoms, the involved timescales of motions range from 10^{-14} s to seconds. Possible biological relevance of these motions would strongly depend on the particular functional role of the considered protein. Dynamical features of proteins having structural roles (constituents of fibrous tissue or muscle) are expected to play a different role as compared to enzymes or transport proteins. Our knowledge about these motions and their specific roles for function and for other aspects, like the structural stability, is far from complete. However, a promising step toward understanding these roles is to determine the nature of these structural fluctuations themselves. This includes the study of the amplitudes of motions, the probability of motions (how often they occur) and the timescales of motions. Various techniques, such as NMR spectroscopy [6], fluorescence spectroscopy [15,16], molecular dynamics (MD) simulations [17], Mössbauer spectroscopy [18], vibrational spectroscopy [19,20], time-resolved diffraction techniques [21], and neutron spectroscopy [12,13,22–25] were already and are still employed to study these tasks. From a vast body of experimental and theoretical studies, a picture of protein dynamics emerged which is schematically summarized in Fig. 17.1.

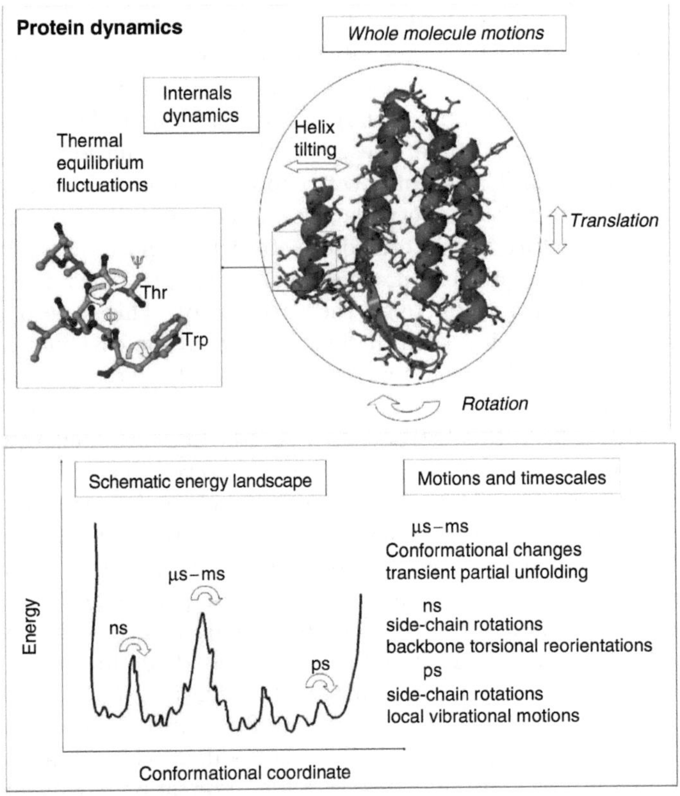

Fig. 17.1. *Top*: Various examples of dynamical features related to proteins are summarized in this picture. With respect to neutron spectroscopy, mainly thermal equilibrium fluctuations of side-chains are observed. *Bottom*: Schematic display of the energy landscape of a protein. The local minima correspond to the so-called conformational substates which are characterized by nearly the same energy level and slightly different conformations (degenerated states). The native folded state of a protein is determined by a dynamic conversion (in the picosecond to nanosecond time regime) between these conformational substates. Less frequent motions, such as conformational changes or folding/unfolding transitions, correspond to overcoming higher energy barriers

In principle, a distinction between structural *equilibrium fluctuations* on the one hand, and *conformational changes* on the other hand is a helpful approach to elucidate the hierarchy and cooperativeness of different internal motions. Structural equilibrium fluctuations in proteins are omnipresent motions of individual atoms or groups of atoms (e.g., polypeptide side chains) which are activated by the surrounding solvent molecules. In general, all parts of the protein structure are affected by these motions; more pronounced in flexible regions and less pronounced in rigid regions of the protein. The interplay between flexibility and rigidity also determines the characteristic of these motions; either fast harmonic vibrational motions with rather small amplitudes or diffusive, liquid-like motions (characterized by an energy barrier crossing process; see Fig. 17.1) with larger amplitudes up to few Ångstroms. In the subnanosecond time regime, these equilibrium fluctuations are the main topic of investigations if neutron spectroscopy is employed. Conformational changes are often transitions between two or more well-defined different 3D structures (ground state and intermediate states). In most cases, these structural differences between the conformational states are restricted to specific parts of the protein structure (e.g., the active site). This type of motion is specific for a given protein, often directly related to the catalytic function of the protein, and generally occurs in a time regime of microsecond to milliseconds. Conformational changes do not occur spontaneously,[1] but need a special interaction with a ligand (substrate) or a specific activation energy.

Various techniques, such as NMR [6], time-resolved diffraction studies [21] (see also contribution by May et al., this volume), and time-resolved vibrational spectroscopy [26], were successfully employed to elucidate details of conformational changes and their role during catalysis. As the conformational changes are generally directly related to functional properties of proteins, the study of these motions is in the focus of most biophysical studies. Nevertheless, there is evidence from various studies that picosecond equilibrium fluctuations are essential for much slower conformational changes [4,13,27]. These studies suggest that the omnipresent fast equilibrium fluctuations act as a "lubricant" and let the conformational changes occur as frequent as necessary for proper kinetics and turn over rates in catalysis. Some interesting case studies in this direction are discussed in more detail by Lechner et al., Part II, Doster, and Lehnert et al., in this volume.

Another important feature of picosecond equilibrium fluctuations is their relation to the structural stability of proteins. The measure of equilibrium fluctuations can give valuable information about how the structural stability of proteins is achieved under extreme environmental conditions (for example in thermophiles, halophiles, barophiles [28–31]). Comparing the equilibrium fluctuations of the folded, partially unfolded, and totally unfolded states provides a direct measure of conformational entropy changes during folding/unfolding transitions. The impact of neutron spectroscopy studies on tasks, such as thermal adaptation and folding/unfolding transition, will be discussed in more detail later in this chapter.

Besides internal structural dynamics, proteins perform whole molecule motions in a solution. In the cell the rotational and translational diffusion of proteins, characterized by correlation time ranging from nanoseconds to

[1]In principle, the conformational changes might occur spontaneously, but they are very infrequent and therefore spontaneous conformational changes in general do not have an impact on catalysis

milliseconds, play a role for transport, signaling, and specific protein–protein interactions (see for example [32]).

17.2 Samples in Neutron Spectroscopy: Sample Preparation, Sample Characterization, and Sample Environment

As discussed by Lechner et al., Part I in this volume, the measured neutron scattering of biological samples is dominated by incoherent scattering from hydrogen nuclei for the most scattering angles (i.e., when Bragg reflections are not excited in the observed Q-space). The high density of hydrogen nuclei in all biological macromolecules results in a total scattering function $S(Q,\omega)$ which is dominated by the incoherent scattering function $S_{\mathrm{i}}(Q,\omega)$. The Fourier-transform in space and time of the latter provides a direct interpretation of dynamical properties by means of the so-called Van Hove self-correlation function $G_{\mathrm{s}}(r,t)$. Because the bound hydrogens are distributed more or less homogeneously in the protein structure, they serve perfectly as local probes to monitor internal structural fluctuations of the protein. Most neutron spectroscopy applications to biological samples deal with this type of measurements. Here, the measured spectra are utilized to estimate the trajectories of all hydrogens in a specific time and space window given by the instrumental settings (see the contribution by Lechner et al., Part I this volume). Unfortunately, trajectories of individual hydrogens cannot be resolved in a spectrum where in general a few thousand hydrogens, as bound in average to a protein, contribute to the total scattering function. However, a sum over the scattering from all these hydrogens is nevertheless a feasible measure of important dynamic properties of a biomolecule (see Doster, Lehnert et al., Sokolov et al., this volume and Sect. 17.4, this contribution).

In order to obtain appropriate samples for neutron spectroscopy, one has to consider all contribution (predominantly from hydrogens) of ingredients (buffer, solvent, etc.) with respect to their impact on the total scattering. Therefore, in general D_2O is used as solvent instead of H_2O, because the total cross-section of deuterium is smaller by a factor of 11 as compared to hydrogen (the incoherent cross-section of deuterium is even smaller by a factor of $\sim$40 as compared to hydrogen). Due to their specific chemical environment various hydrogens in a protein structure are not bound covalently to the protein structure but can form hydrogen bonds either with other structural parts of the protein or with solvent molecules. Many groups in the protein, like back-bone NH-groups and various side-groups, such as tryptophan, arginine and asparagine, glutamine (the latter under specific pH condition), can act as proton donors in transient hydrogen bonds. As a consequence, extensive hydrogen/deuterium exchange has to take into account when deuterated solvents are used. In order to prepare proper samples with an uniform ensemble of well-characterized proteins, a stable and well-defined number of hydrogens bound to the protein exhibiting a scattering background as low as possible, the following procedures and precautions are helpful:

- The large amount of protein (∼100 mg, which is generally required for neutron spectroscopy studies) should be transferred in a deuterated buffer. An extensive washing and purifying of the protein sample is recommended because all exchanged hydrogens in the protein solution need to be removed. Further, quite often lyophilized protein powders (as received from the producer) contain salts or other ingredients which can give rise to unwanted scattering. For example, a threefold to fourfold use of a desalting gel filtration column gives reasonable results.
- If certain ingredients in the buffer are used in very high concentrations and if they contain hydrogens, it is necessary to use these ingredients in a deuterated form. The rather high concentrations (1–5 $\mathrm{mol\,l^{-1}}$) of urea or guanidine hydrochloride (unfolding ingredients) are examples for which these considerations have to be taken into account.
- In the case of studies with unfolded proteins, one has to consider the fact that hydrogens in the interior of a folded protein might exchange only after a long time (sometimes after month) if ever. In the unfolded state the H/D exchange will be accelerated. In order to achieve proper results, one has to make sure that the H/D exchange of the folded protein is already (more or less) completed before performing the washing and purifying procedures (see above). Increased temperatures or other reversible "prefolding" conditions can promote a more complete H/D exchange.
- A proper characterization of the proteins in our sample before (and in some cases after) the neutron scattering measurements is a prerequisite for data analysis and any reasonable interpretation of results. Depending on the accessibility to various spectroscopic techniques and the specific topic of investigations the following measurements might be helpful or are necessary: (i) Vibrational IR-absorption of amide protons in the protein is indicative for the completeness of H/D exchange (i.e., the comparison of amide II band (1550 $\mathrm{cm^{-1}}$) and of amide II′ band (1450 $\mathrm{cm^{-1}}$, see for example [30]). (ii) Tryptophan fluorescence emission spectroscopy and circular dichroism (CD) spectroscopy gives information about structural integrity of the 3D structure and of secondary structure elements (see for example [33, 34]). (iii) Dynamic light scattering (DLS) [35], PFG-NMR [36], Fluorescence correlation spectroscopy (FCS) [37] and SANS/SAXS [38] are well-suited techniques to measure diffusion coefficients of proteins diffusing in solutions. These measurements are helpful in order to find proper condition for samples without unspecific aggregation.
- As mentioned earlier, in most cases the solvent is D_2O. On the other hand, most published studies on enzymatic activity or thermodynamic properties (e.g., melting temperature T_m or C_p and H_m of the folding/unfolding transition) of the protein of interest, have been performed in H_2O. Therefore,

one has to check whether D_2O solvent changes some of these properties significantly.

The average beam size of a neutron spectrometer is in the order of 10–20 cm^2. In a reasonable neutron scattering experiment, one would like to illuminate as much sample as possible in order to obtain a good signal-to-noise ratio. The limiting factor (besides availability of sample material) is multiple scattering, caused by too high densities of scattering nuclei. As a rule of thumb the sample transmission T for neutrons should be in the order of 0.9. The sample transmission

$$T = \mathrm{e}^{-\rho_\mathrm{s}/m_\mathrm{s}\cdot N_\mathrm{a}\cdot(\sigma_\mathrm{s}+\sigma_\mathrm{a})d} \tag{17.1}$$

is a function of the density of scattered nuclei (ρ_s in g cm^{-3}), the molecular mass (m_s in g) of the scatterer, the scattering and absorption cross-sections (σ_s and σ_a both in 10^{-24} cm^2, i.e., barns), and the effective travelling length of neutrons through the sample (d in cm). A rather thin sample, covering an area in the order of the beam size, is appropriate with respect to our purpose. Slab-shaped samples meet theses requirements. In the case of hydrated powders samples of biological macromolecules exhibit a thickness of a few tenths of a millimeter. For protein dissolved in D_2O buffer the thickness can be around 1 mm. The sample containers (often made from aluminum because of its low neutron scattering and absorption cross-sections) must be sealed perfectly in order to prevent the sample from dehydration during the experiment. A typical measuring time is in the order of hours and the sample containers are often placed in vacuum. Therefore, sealing is a crucial point in sample preparation, and it is recommended to check proper sealing by weighing the sample before and after each measurement.

17.3 From Spectra to Results: Data Acquisition, Data Analysis, and Data Interpretation

The obtained experimental data in a quasielastic neutron scattering experiments using a time-of-flight or a backscattering spectrometer is a set of energy-transfer spectra measured at different scattering angles. For a proper data analysis and interpretation not only a sufficient counting statistics of the sample measurements is essential but also an additional measurements of a vanadium sample (detector calibration and determination of the elastic energy resolution) and of the empty sample container (scattering background measurements) are required. If protein solutions are under investigation, an additional measurement of the pure buffer solutions can be used to subtract solvent scattering in order to isolate scattering solely from the protein. In this case, measurements with excellent counting statistic are required because the spectra of interest are difference spectra (see below). Before more details about the data analysis of studies dealing with protein solutions are discussed, some general properties of scattering spectra from proteins are introduced first. As already shown in previous chapters, a typical energy transfer spectrum is characterized by a component of elastic scattering, several components of quasielastic scattering, and a continuum-like set of inelastic components occurring as a broad background scattering. The dominant quasielastic scattering from proteins at physiological conditions (room temperature and sufficient hydration) is caused by a diffusive type of motion and can be parameterized by Lorentzian-shaped curves (see Fig. 17.2a for hydrated powder samples).

This type of motion is specified by a characteristic correlation time (how often that type of motion occurs) and by an "amplitude" of the motion (a jump distance for example, depending on a specific model used for data interpretation). Characteristic correlation times of motions are obtained from the widths of the Lorentzian-shaped curves. The amplitudes are given by the Q-dependence of the elastic incoherent scattering amplitude (EISF or A_0) (see Fig. 17.2b).

As mentioned earlier, we deal with a few thousand hydrogen nuclei bound to the protein, every single one in a slightly different environment, contributing to the measured total scattering function. Therefore, we would not expect a perfectly uniform type of motion for all contributing hydrogens. However, within the limit of the given energy and Q resolution ($Q_{\mathrm{max}} \sim 2\,\text{Å}^{-1}$), obtained spectra are characterized by the following properties:

- All spectra can be fitted reasonably well with only one or two quasielastic Lorentzian-shaped components (besides an elastic component and a flat background component).
- The obtained quasielastic components give line-widths which do not change significantly over the measured Q-range.
- The above described features are at least observable for different energy resolutions between 2 and 500 μeV (FWHM). The line width of the major quasielastic component strongly depends on the elastic energy resolution and is, in average, two times larger as compared to the elastic energy resolution.

These observations, which are visible from the so-called phenomenological fits without using a specific model (see contribution by Lechner et al., Part II), give rise to the following picture: The protein dynamics is characterized by broad continuum of different correlation times (at least in the time regime from nanoseconds to picoseconds) of localized diffusive, jump-like motions (e.g., polypeptide side-group reorientation). In this time regime, the "average amplitudes" of motion are in the order of 0.4–1.2 Å for native proteins. Because of the heterogeneous dynamic character of hydrogen movements in a complex structure like a protein, the obtained parameters (correlation times, amplitudes of motion) must be understood as averaged values. Nevertheless, one would like to use these parameters to compare protein properties at different environmental conditions, to compare these properties for

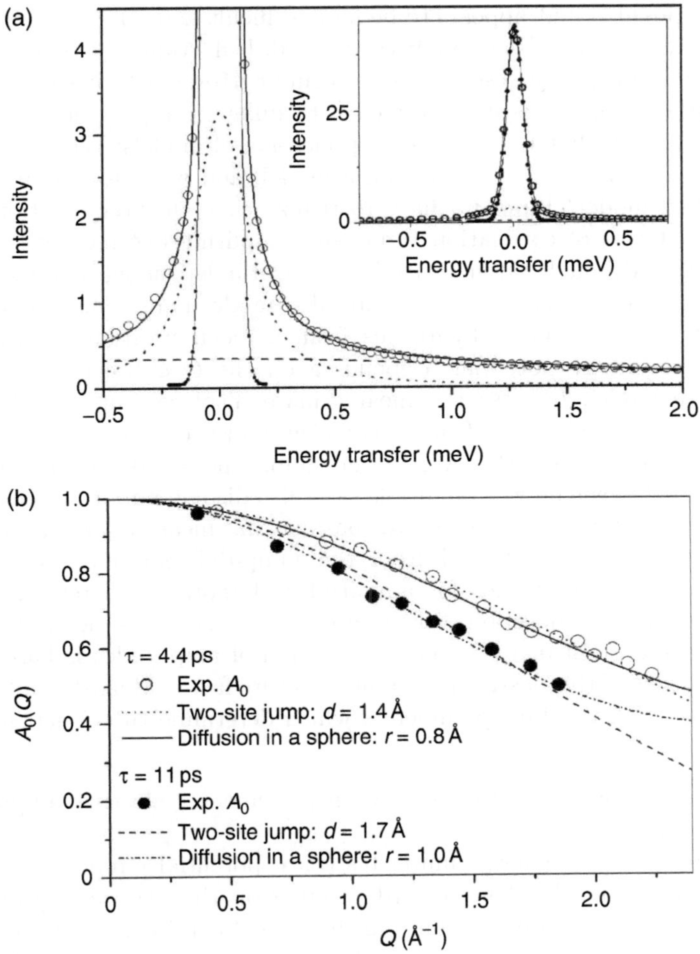

Fig. 17.2. (**a**) A typical energy-transfer spectrum of a hydrated powder sample (protein) and a separation of spectral components, as demonstrated by a fit, are shown. Two quasielastic components (*dotted line* with a line-width of 120 μeV (HWHM), *dashed line* with a line width of 2 meV), and an elastic component (*small full circles*) add up to the total scattering function (*open symbols*). Inset shows the spectrum in full extension while the main figure shows a frame which corresponds to 10% of the elastic peak intensity. (**b**) The symbols give elastic incoherent structure factors (A_0) as a function of Q as results from phenomenological fits of data measured with two different energy resolutions (*open symbols*: 100 μeV (FWHM), *solid symbols*: 34 μeV). Reasonable results from fitting procedures were obtained by using different jump models. It is evident from this figure that a distinction between both models is only possible with reliable data at higher Q-values

different proteins, and to compare results obtained with different techniques (e.g., B-factors from crystallography, NMR, MD simulations). In principle this is always possible, but appears to be more difficult and not straightforward if different approaches of data analysis are used. Unfortunately, this is generally the case when different researchers perform neutron scattering studies.

In order to judge results obtained with different approaches for the data analysis, we will shortly discuss the application of models: What is a reasonable model, what can we learn from a model, and what is the limitation of the applied model? First, we have to realize that all interpretations of dynamics in terms of correlation times and amplitudes of motions are based on models. Although often termed as model-independent parameters, the mean square displacements $<u^2>$ are also model parameters, describing a transient displacement of hydrogens from a mean position determined by a harmonic energy potential. Depending on the Q–ω-range this can be a reasonable model, because the mean square displacement is a descriptive parameter for amplitudes of motions which is also used for results obtained with other techniques. However, other simple models describing successive jumps of hydrogens between multiple sites also display important characteristics of protein dynamics. One has to consider that mean square displacements and jump distances obtained from the jump-diffusion models in general give rather different values for amplitudes of motion. Therefore, it is not easy to answer the question of how large the effective average amplitude of motion in a protein really is. The calculation of amplitudes is based on the Q-dependence of the elastic incoherent structure factor. Realistic models have to consider at least three parameters which determine the Q-dependence of the EISF:

- Jump distance or mean square displacement (effective amplitude of motion).
- The number of adjacent jump-sites (local potential minima) which characterize the localized motion (depending on the observed Q-range this number can be finite or infinite in the case of continuous diffusion).
- A fraction of hydrogens bound to the protein scatter purely elastic because their motions are outside the Q–ω-window of our measurement.

In a complex structure of a protein with a few thousand hydrogens as part of the protein structure, we cannot determine these factors with a sufficient reliability because these factors are strongly coupled in the fits. In general, only data at higher Q-values can help to deconvolute these parameters in a fit and give a more detailed and precise picture of the characteristic motions in the protein. In practice such data for protein samples is not easy to measure (problems with counting statistics, more difficult separation of quasielastic components because higher orders of the Bessel functions have to take into account, for more details see Bee [17]). Therefore in the most cases the measured data is not sufficient to overcome this problem of ambiguity and absolute values of effective amplitudes of motions cannot be determined (see Fig. 17.2b). This is the reason why rather different values of amplitudes of motions were published for proteins by different authors using different procedures of data analysis [13,30,40–42]. However, comparative studies (applying one and the same type of data analysis) give nevertheless reasonable results if dynamical properties of a protein are investigated as a function of hydration, temperature, solvent composition for example.

As earlier mentioned, one of the difficulties with neutron scattering studies on protein solutions is a strong scattering contribution from the solvent. Even at rather high protein concentrations (30–50 mg ml^{-1}) approximately 80% of the total scattering is caused by solvent molecules (D_2O). Due to a "peak-like" coherent scattering of D_2O which is caused by O–O distance correlations, the contribution from the solvent scattering is even higher at Q-values ranging from 1.5 to 2.5 Å^{-1} (see Fig. 17.3). As a consequence, difference spectra obtained from measurements performed at this Q-range show a reduced counting statistics and are therefore less trustable. More reliable difference spectra are obtained for Q-values below 1.5 Å^{-1} (see Fig. 17.4).

Difference spectra are calculated by the use of spectra measured with pure D_2O buffer (S_{sol}) and of spectra measured with protein solutions (S_{PS}). In this simple subtraction procedure, an "excluded volume fraction" – factor f_{ex} is applied which considers the volume of the total amount of dry protein (typical specific volume of a medium sized protein is ~0.75 ml g^{-1}) in the volume of the protein solution

$$S_{\text{diff}}(Q,\omega) = S_{\text{PS}}(Q,\omega) - (1 - f_{\text{ex}})S_{\text{sol}}(Q,\omega). \tag{17.2}$$

It can be assumed that the dominant contribution of scattering in difference spectra is due to internal dynamics of nonexchanged hydrogens bound to the protein. However, the data interpretation in terms of internal protein dynamics has to consider to following sources of errors:

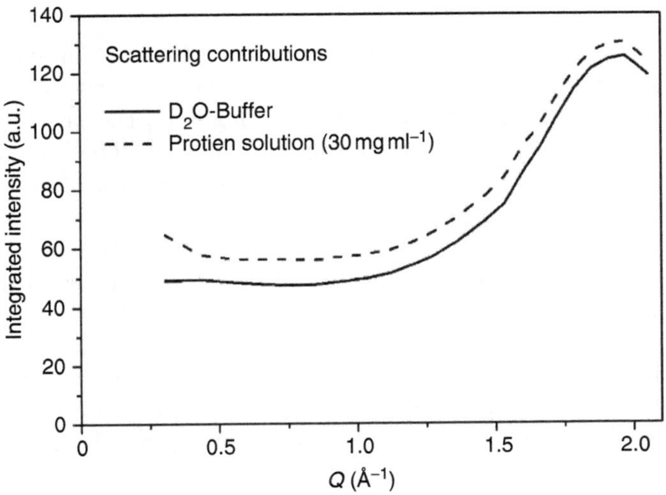

Fig. 17.3. Integrated intensities (energy transfer range from −1 to 3 meV) as measured at 30°C for D_2O-buffer and for a protein solution with 30 mg protein per millimeter solvent (from [57])

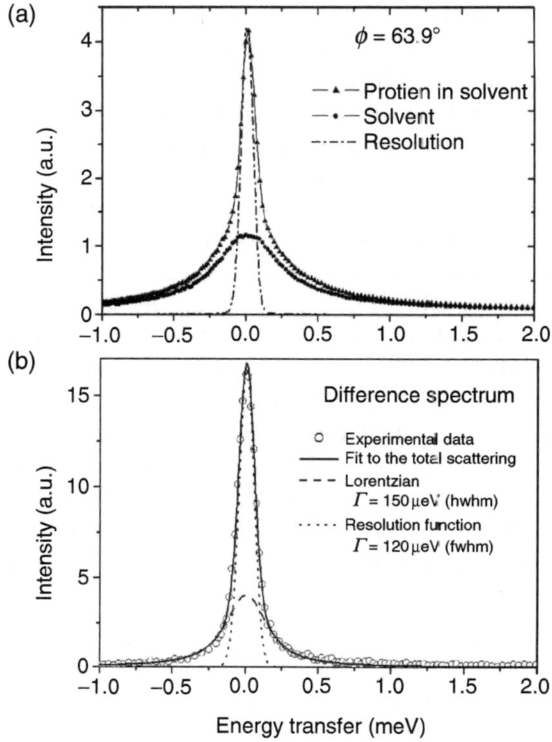

Fig. 17.4. (**a**) Comparison of spectra measured with the protein solution and with solvent buffer only at a Q-value of $\sim 1.3\,\text{Å}^{-1}$ (data from [30]). (**b**) Difference spectrum as obtained from subtractions procedure applied to spectra shown in (**a**)

– We have to consider not only scattering from the protein and from bulk D_2O-solvent but also a contribution from hydration water which is less mobile as compared to the bulk water. Therefore, the procedure of subtracting pure D_2O-solvent scattering to eliminate the D_2O signal from scattering of our protein solution, decreases the quasielastic scattering which is supposed to represent protein dynamics. Assuming that a protein is more or less fully hydrated at $h = 0.3$–0.5 g solvent per gram protein [14] and that mainly this water shows a pronounced difference in the dynamical behavior with respect to bulk water [43–47] we have to consider in our case a contribution of hydration water to the total amount of D_2O which is about 1.4%. Thus, the scattering from the hydration water is in the order of 5% compared to the scattering of the protein. It is possible to reduce this contribution from hydration water in difference spectra below 1% if one subtracts not pure D_2O-buffer, but a protein solution with deuterated protein. Other approaches which use H_2O solvent and focus their data analysis on a Q–ω-space where fast solvent dynamics do not superimposed (see for example [31]) with internal protein dynamics will have problems with this separation of solvent and protein dynamics when dynamics of the slower hydration water contributes significantly to the scattering.

– Proteins in solution perform rotational and translational diffusion as a whole unit which may contribute to nonelastic scattering in the Q–ω regime of our measurements. For small and medium-sized globular proteins the apparent diffusion coefficient (including translational and rotational diffusion of the whole protein) is in the order of 10^{-6} to $10^{-7}\,\text{cm}^2\,\text{s}^{-1}$. The translational diffusion is proportional to $1/R$ (R: hydrodynamic radius of the protein) and the rotational diffusion scales with $1/R^3$, while for both types of diffusion the diffusion is proportional to the viscosity of the protein solution. In the case of medium-sized proteins (with R in the order of 30 Å) additional line-broadening due to whole molecules motions is in the order of 4–6 μeV (at a Q-value of about $1\,\text{Å}^{-1}$). With respect to a moderate energy resolution of about 100 μeV (FWHM) this leads to an underestimation of the EISF in the order of 10%. This estimation is based on diffusion coefficients (see above) measured at protein concentrations in the order of $2\,\text{mg}\,\text{ml}^{-1}$ (and lower). In much higher concentrations as used for neutron scattering studies, the translational diffusion drastically slows down [36], which will decrease the contribution of whole molecule motions in our spectra. For an adequate data treatment one has to consider the whole molecule motions by including an appropriate scattering component in the fitting procedure (see for example [41, 48]). In order to include this component in the data analysis, one has really to measure the apparent diffusion coefficients for the given protein at those conditions (i.e., at the given concentrations, and for example in the folded and unfolded state) as used in the neutron scattering studies. For very high protein concentrations PFG-NMR or FCS are well-suited techniques to determine diffusion coefficients.

The majority of published studies dealing with neutron scattering applied to protein solutions presents a data analysis employing simple jump-diffusion models (diffusion inside a sphere, three-site jump diffusion; for details see the contribution by Lechner et al., Part I this volume). A comparison of theses studies exhibits rather different results. If we only focus on those studies with native proteins which apply the same model fits (diffusion inside a sphere, see Sect. 17.4.1), the obtained radii vary from 1.2 to 5.5 Å (see for example [30, 40–42]). In most cases, the EISF reaches a plateau at higher Q-values which is in principle indicative for a certain fraction of hydrogens which do not diffuse in the observed Q–ω-space (consider however the less reliable data quality of difference spectra at Q-values above $1.5\,\text{Å}^{-1}$, see Fig. 17.3). The fraction of nondiffusive hydrogens in the protein differs significantly in the considered studies. Although different researchers performed studies with different proteins, it is unlikely that these rather large differences mainly occur from a very different protein dynamics. It is more likely that mainly the way of sample preparation and the applied solvent subtraction procedures give rise to very different resulting parameter. However, for a comparative analysis of different species (e.g., folded/unfolded protein or homologous proteins) treated with the same procedures, these studies can nevertheless permit reasonable results and interpretations. Some applications and recent results from studies will be presented in Sect. 17.4.

17.4 Applications and Examples

It is well known that environmental conditions like temperature, hydration, salinity, or pressure have a profound influence on the stability of proteins as well as on the kinetics of enzymatic reactions. A promising approach to investigate the underlying mechanisms of protein stability by concurrent functionality, is to measure dynamical properties of proteins under extreme condition (high temperatures, high salt concentrations, high static pressure; see for example [31, 40, 42, 49–51]). One goal of studies in this direction is to figure out to which extent and how the dynamic properties contribute to the free energy which stabilizes the native folded state. For these purposes the protein structures of interest have not only to be studied in the folded state but also in the unfolded state.

17.4.1 Comparison of Folded and Unfolded States

In this section, results from measurements on an amylolytic enzyme, α-amylase from *B. licheniformis* (BLA), are presented. α-Amylases are subjects in numerous studies on protein stability and thermal adaptation [30, 52–54], because various species of this protein show a nearly identical 3D overall structure but rather different thermostabilities (homologues proteins). Samples of thermostable BLA dissolved in D_2O-buffer were measured with the time-of-flight spectrometer IN6 (ILL, Grenoble). The obtained data (see Fig. 17.5) were analyzed using a simple jump-diffusion model (diffusion inside a sphere [55]).

The obtained difference spectra (see Eq. 17.2) were fitted using one quasi-elastic (A_1) and one elastic (A_0) component

$$S_{\text{diff}}(Q,\omega) = \mathrm{e}^{-<u^2>Q^2}[A_0(Q)\delta(\omega) + A_1 L_1(H_1,\omega)]. \tag{17.3}$$

Here A_0 is given by

$$A_0(Q) = \left[3\frac{\sin(Qr) - (Qr)\cos(Qr)}{(Qr)^3}\right]^2 , \tag{17.4}$$

and is parameterized by the radius of the sphere r (for more details see [11, 57]).

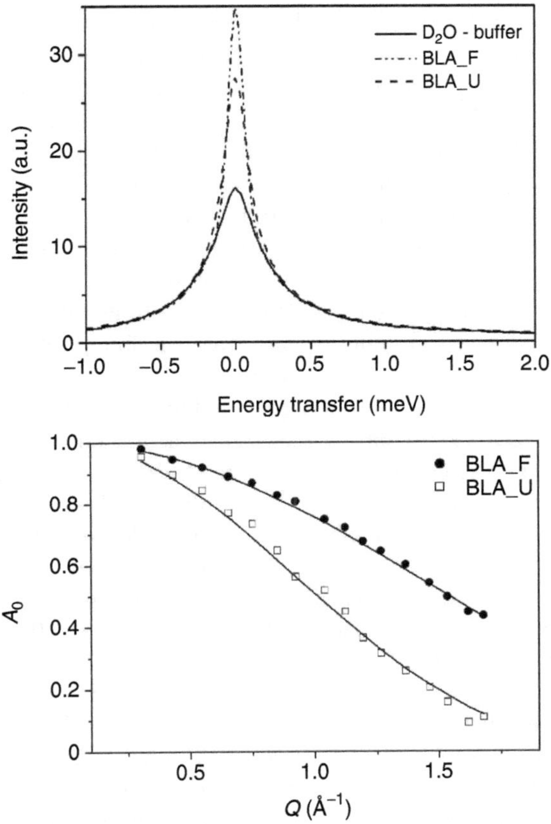

Fig. 17.5. *Top*: A comparison of measured spectra from BLA in D_2O-buffer solution (BLA_F: folded state at pH 7.8; BLA_U: unfolded state at pH 13) and from D_2O-buffer solely (data from [57]). Each displayed spectrum represents a sum of seven individual spectra measured at scattering angles between 60° and 65° which corresponds to an average Q-value of 1.27 Å^{-1}. The spectra represent measured raw data and, for a better comparability of quasielastic contributions, they are plotted as lines. *Bottom*: For both samples (folded and unfolded) the elastic incoherent structure factor A_0 is given as a function of Q. The structure factors were obtained from phenomenological fits of difference spectra (line width H_1 of the quasielastic component is 150 μeV).The experimental errors of the structure factors are about 5%. The solid lines represent the Q-dependence of A_0 as given by the model. The following parameter values of r (radius of the sphere) were obtained: $r = 1.2 \pm 0.06$ Å (folded protein) and $r = 1.8 \pm 0.11$ Å (unfolded protein)

The major effect of unfolding a protein is already apparent in the raw data and is characterized by a strong increase of quasielastic scattering as shown in spectra of Fig. 17.5 (upper part). This figure shows grouped raw data from samples (protein solution with protein in the folded and unfolded state) without solvent subtraction. We observe a large and mainly quasielastic contribution of D_2O scattering in the spectra of protein solutions. In comparison to spectra measured with the unfolded state, spectra obtained from the folded state clearly show more pronounced elastic scattering accompanied by a reduced quasielastic scattering. This feature was analyzed in more detail by applying fits to the difference spectra. Considering three spectral contributions, an elastic component, a single lorentzian ($H_1 = 150 \pm 10$ µeV, the same for both states and constant over the Q-range used here), and a constant background, provide a sufficient fit quality for the data within an energy transfer range from -1 to 2 meV. Such a result is consistent with the predictions of the applied model, where in the low Q-range the line width is constant with increasing Q [55].

The resulting elastic incoherent structure factors (A_0) are shown as a function of Q in Fig. 17.5 (lower part). This figure clearly shows smaller A_0-values and a steeper decrease of these values with Q for the unfolded state as compared to the folded state. In terms of the applied "diffusion inside a sphere" – model we obtain a radius value of $r = 1.2$ Å for the folded protein and a clearly larger value of $r = 1.8$ Å for the unfolded protein. From this analysis the following picture of internal structural fluctuations emerges: Most of the nonexchangeable hydrogens are localized in the side groups (~80%), and therefore the observed dynamics represents mainly side group diffusive reorientations with an average correlation time of about 4.4 ps confined in the volume given by the radii. In accordance with other experimental results (e.g., [40,42]), the increase of the radius in our model due to protein unfolding reflects a situation, where the proteins have less defined and more heterogeneous structures with a larger degree of freedom for structural fluctuations as compared to the folded state. A demonstrative description of this property is given by the concept of energy landscapes and the *protein folding funnel* [56].

Conformational fluctuations in proteins as measured with neutron spectroscopy can contribute significantly to the conformational entropy. In particular, measurements as a function of temperature, of denaturant concentrations, or comparative studies between homologues proteins can reveal important information about thermodynamic parameter. In contrast to the role of dynamics with respect to functional properties, where spatial resolution of dynamical features in principle is desirable (for example at the active site), the use of dynamical features for thermodynamic interpretations do not depend on spatial resolution. Therefore, neutron spectroscopy, which in general is not able to give information about the localization of observed structural fluctuations in a complex structure of a protein (for some exceptions see the contribution by Lehnert et al., this volume) is an ideal technique for deriving thermodynamic parameter, such as conformational entropy. An approach how to interpret neutron scattering data in this framework is described in the Sect. 17.4.2.

17.4.2 Conformational Entropy Calculation from Neutron Scattering Data

The stability of the folded state of a protein, which is the native and functional state under physiological conditions, is operated by a subtle balance of enthalpic and entropic contributions ($\Delta G_{\mathrm{unf}} = \Delta H - T\Delta S$). Both contributions consist of opposing fractions which either stabilize or destabilize the folded state (see for example [58]). The conformational entropy of the polypeptide chain is larger for unfolded state compared to the more compact folded state characterized by much more restricted conformational space (see Sect. 17.4.1). Therefore, this contribution stabilizes the unfolded state ($\Delta S_{\mathrm{conf}} = S_{\mathrm{U}} - S_{\mathrm{F}} > 0$). The interaction of solvent water with nonpolar side chains of the protein stabilizes the folded state, because solvation of these side groups induces ordering of water which is unfavorable. Since nonpolar groups are exposed to the solvent mainly in the unfolded state and not in the folded state, this contribution stabilizes the folded state ($\Delta S_{\mathrm{hyd}} < 0$). As an approach to elucidate more details about the role of the conformational entropy during thermal unfolding, neutron spectroscopy can be utilized to measure structural fluctuations as function of temperature and to calculate conformational entropy changes from the data.

Proper entropy calculations are difficult, because these calculations require knowledge of the complete conformational space and are strongly influenced by correlations between the motions. However, on the basis of the applied analytical model an estimation of a contribution to the conformational entropy related to the observed picosecond dynamics is given here. Due to the fact that hydrogens located mainly in the protein side-groups (more than 80%) contribute to the observed scattering, we assume side-groups or parts of side-groups as single individual "moving units", which explore a certain part of the conformational space during their confined diffusive motions. For these units we can calculate the conformational entropy by

$$S_{\mathrm{conf}} = R \ln(Z). \tag{17.5}$$

Here R is the gas constant (8.3144 J K^{-1} mol^{-1}) and Z is the partition function with $Z = V/\lambda_{\mathrm{D}}^3$ determined by the de Broglie wavelength $\lambda_{\mathrm{D}} = h/\sqrt{2\pi m k_{\mathrm{B}} T}$ and by V, the accessible volume of the conformational space occupied by the corresponding state (see Fig. 17.6).

For amino-acid residues moving at the given temperatures (30–70°C), the thermal deBroglie wavelength is 0.1 Å. Therefore, $\lambda_{\mathrm{D}}^3 \ll V$ and Boltzmann statistics is valid in the classical approximation (see [59]). The conformational entropy change during unfolding with a radius of confinement for motions in the folded state r_{f} and in the unfolded state r_{u} is given by

$$\Delta S_{\mathrm{conf}} = S_{\mathrm{u}} - S_{\mathrm{f}} = 3R \ln\left(\frac{r_{\mathrm{u}}}{r_{\mathrm{f}}}\right). \tag{17.6}$$

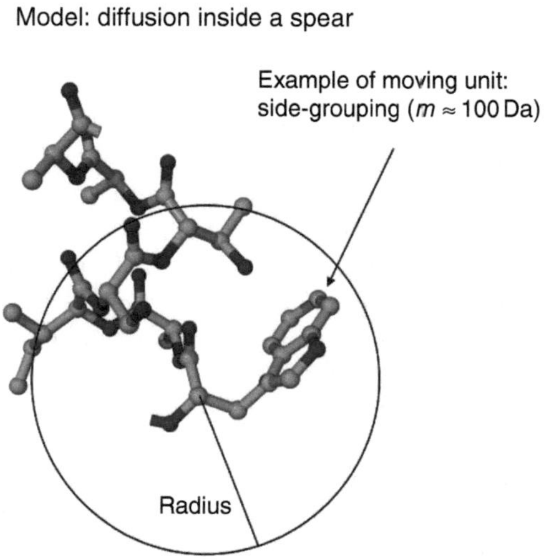

Fig. 17.6. Schematic representation of a polypeptide side-group diffusing in the volume of a sphere. With respect to the observed time regime the "moving units" are not only side groups (as shown here) but also part of side-groups (e.g., methyl groups) or parts of the polypeptide backbone. The given radius r simply determines the conformational volume $V = 4\pi r^3/3$, which is explored by the "moving unit"

Using radius values as determined in Sect. 17.4.1 (at 30°C) we obtain for ΔS_{conf} a value of $10.85\,\mathrm{J\,K^{-1}\,mol^{-1}}$ per residue (or $2.6\,\mathrm{cal\,K^{-1}\,mol^{-1}}$ per residue). Compared to estimated and measured ΔS_{conf} values given in the literature (see for example [8,9,60]) our value is in average smaller by 30–60%. Due to fact that our analysis covers mainly the picosecond time regime and contributions to the conformational entropy are also effected by fluctuations in other time regimes, this deviation is pretty reasonable.

17.5 Conclusions and Outlook

It was demonstrated in this chapter that dynamical features of internal structural fluctuations in proteins can be extracted from measurements with proteins dissolved in buffer solutions. A number of useful procedures and precautions which have to be considered for a reasonable data analysis and interpretation were discussed. As already demonstrated in the presented and numerous published examples [30, 31, 40, 42, 50, 51], the challenge of neutron scattering studies applied to protein solutions is to increase the scope of biological relevant topics, where internal fluctuations of proteins play an important role. With the accessibility of next generation high-flux neutron sources (SNS, ESS) biological neutron spectroscopy will benefit significantly. Further applications on proteins with trapped folding intermediates, functional intermediate states, or on protein–protein and on protein–ligand interactions are of general interest in structural molecular biology and biophysics and are certainly in the scope of neutron spectroscopy.

Acknowledgments

The author would like to thank R.E. Lechner, N.A. Dencher, and G. Büldt for a long standing and fruitful collaboration. The presented work was partly supported by grants from the German Government (03DE4DAR-1 and 03DE5DA1-8).

References

1. H. Frauenfelder, S.G. Silgar, P.G. Wolyens, Science **254**, 1598 (1991)
2. M. Karplus, J. A. MacCammon, Sci. Am. **254**, 42 (1986)
3. A.E. Garcia, Phys. Rev. Lett. **68**, 2696 (1992)
4. B.F. Rasmussen, A.M. Stock, D. Ringe, G.A. Petsko, Nature **357**, 423 (1992)
5. J. Fitter, S.A.W. Verclas, R.E. Lechner, H. Seelert, N.A. Dencher, FEBS Lett. **433**, 321 (1998)
6. L.E. Kay, Nat. Struct. Biol. **5**, 513 (1998)
7. G. Zaccai, Science **288**, 1604 (2000)
8. M. Karplus, T. Ichiye, B.M. Pettit, Biophys. J. **52**, 1083 (1987)
9. A.J. Doig, J.E. Sternberg, Protein Sci. **4**, 2247 (1995)
10. D. Yang, L.E. Kay, J. Mol. Biol. **263**, 369 (1996)
11. J. Fitter, Biophys. J. **84**, 3924 (2003)
12. W. Doster, S. Cusack, W. Petry, Nature **337**, 754 (1989)
13. M. Ferrand, A.J. Dianoux, W. Petry, G. Zaccai, Proc. Natl. Acad. Sci. USA **90**, 9668 (1993)
14. J.A. Rupley, G. Careri, Protein hydration and function, in: *Advances in Protein chemistry*, vol. **41**, ed. by C.B. Anfinsen, J.T. Edstall, F.M. Richards, D.S. Eisenberg (Academic Press, New York, 1991) pp. 37–172
15. J.R. Lakowicz, Photochem. Photobiol. **72**, 421 (2000)
16. N.L. Thompson, A.M. Lieto, N.W. Allen, Curr. Opp. Struct. Biol. **12**, 634 (2002)
17. J. A. MacCammon, S.C. Harvey, in: *Dynamics of Proteins and Nucleic Acids* (Cambridge University Press, New York, 1987)
18. F. Parak, E.W. Knapp, D. Kucheida, J. Mol. Biol. **161**, 195 (1995)
19. A. Barth, C. Zscherp, Q. Rev. Biophys. **35**, 369 (2002)
20. H. Deng, R. Callender, Methods Enzymol. **308**, 176 (1999)
21. B.L. Stoddard, Curr. Opp. Struct. Biol. **8**, 612 (1998)
22. J.C. Smith, Q. Rev. Biophys. **24**, 227 (1991)
23. J. Fitter, R.E. Lechner, G. Büldt, N.A. Dencher, Proc. Natl Acad. Sci. USA **93**, 7600 (1996)
24. J. Fitter, R.E. Lechner, N.A. Dencher, Biophys. J. **73**, 2126 (1997)
25. F. Gabel, D. Bicout, U. Lehnert, M. Tehei, M. Weik, G. Zaccai, Q. Rev. Biophys. **35**, 327 (2002)
26. J. Heberle, Biochim. Biophys. Acta **1458**, 135 (2000)

27. J. Heberle, J. Fitter, H.J. Sass, G. Büldt, Biophys. Chem. **85** 229, (2000)
28. T. Lazaridis, I. Lee, M. Karplus, Protein Sci. **6**, 2589 (1997)
29. P. Zavodszky, J. Kardos, F. Svingor, G. Petsko, Proc. Natl Acad. Sci. USA **95**, 7406 (1998)
30. J. Fitter, J. Heberle, Biophys. J. **79**, 1629 (2000)
31. M. Tehei, D. Madern, C. Pfister, G. Zaccai, Proc. Natl Acad. Sci. USA **98**, 14356 (2001)
32. E. Haustein, P. Schwille, Methods **29**, 153 (2003)
33. F.X. Schmidt, in: *Protein Structure: A Practical Approach*, ed. by T.E. Creighton (IRL Press, Oxford, 1989), pp. 299–321
34. C.N. Pace, J.M. Scholz, in: *Protein Structure: A Practical Approach* , ed. by T.E. Creighton (IRL Press, Oxford, 1989), pp. 261–289
35. K. Gast, G. Damaschun, R. Misselwitz, D. Zirwer, Eur. Biophys. J. **21**, 357 (1992)
36. I. Nesmelova, V.D. Skira, V.D. Fedotov, Biopolymers **63**, 132 (2002)
37. C. Zander, J. Enderlein, R.A. Keller, *Single Molecule Detection in Solution* (Wiley-VCH, Berlin, 2002)
38. D. Russo, J. Perez, J.M. Zanotti, M. Desmadril, D. Durand, Biophys. J. **83**, 2792 (2002)
39. M. Bee, in: *Quasi-Elastic Neutron Scattering* (Hilger, Bristol (UK), 1988)
40. V. Receveur, P. Calmettes, J. Smith, M. Desmadril, G. Coddens, D. Durand, Proteins. **28**, 380 (1997)
41. J. Perez, J.M. Zanotti, D. Durand, Biophys. J. **77**, 454 (1999)
42. Z. Bu, D.A. Neumann, S.H. Lee, C.M. Brown, D.M. Engelmann, C.C. Han, J. Mol. Biol. **301**, 525 (2000)
43. M.C. Bellissent-Funel, J.M. Zanotti, S.H. Chen, Farraday Discuss. **103**, 281 (1996)
44. R.E. Lechner, N.A. Dencher, J. Fitter, T. Dippel, Solid Sate Ionics **70/71**, 296 (1994)
45. J. Fitter, R.E. Lechner, N.A. Dencher, Phys. Chem. B **103**, 8036 (1999)
46. M. Tarek, D.J. Tobias, J. Am. Chem. Soc. **121**, 9740 (1999)
47. C. Bon, A.J. Dianoux, M. Ferrand, M.S. Lehmann, Biophys. J. **83**, 1578 (2002)
48. J.A. Hayward, J.L. Finney, R.M. Daniel, J.C. Smith, Biophys. J. **85**, 679 (2003)
49. J. Fitter, R. Herrmann, T. Hauss, R.E. Lechner, N.A. Dencher, Physica B **301**, 1 (2001)
50. Z. Bu, J. Cook, D. Callaway, J. Mol. Biol. **312**, 865 (2001)
51. W. Doster, R. Gebhardt, Chem. Phys. **292**, 383 (2003)
52. J.E. Nielsen, T.V. Borchert, Biochim. Biophys. Acta **1543**, 253 (2000)
53. J. Fitter, R. Herrmann, N.A. Dencher, A. Blume, T. Hauss, Biochemistry **40**, 10723 (2001)
54. D. Georlette, V. Blaise, T. Collins, S. D'Amico et al., FEMS Microbiol. Rev. **28**, 25 (2004)
55. F. Volino, A.J. Dianoux, Mol. Phys. **41**, 271 (1980)
56. J.N. Onuchic, Z. Luthey-Schulten, P.G. Wolynes, Annu. Rev. Phys. Chem. **48**, 545 (1997)
57. J. Fitter, Chem. Phys. **292**, 405 (2003)
58. A. Fersht, in: *Structure and Mechanism in Proteins Science* (W.H. Freeman and Company, New York, 1999)
59. R. Becker, in: *Theorie der Wärme* (Springer Verlag, Berlin, 1985)
60. T.P. Creamer, Proteins **40**, 443 (2000)

18

Relating Protein Dynamics to Function and Structure: The Purple Membrane

U. Lehnert, M. Weik

18.1 Introduction

Biological macromolecules have a highly complex structure and this is reflected in their dynamical behavior. The dynamical spectrum covers a time domain of several orders of magnitude from 10^{-14} to 1 s and longer [1]. High frequency motions such as vibrations of bonded atoms are localized over a few atoms. Their energy varies from 35 to about 350 meV and the amplitudes are in a range from 0.01 to 0.15 Å. The low frequency modes such as vibrations of globular regions, side-chain rotations at the surface, or torsional motions may involve more sizeable masses and can be considered as rigid body motions. They can be thermally excited at 300 K in an energy range from 0 to 25 meV. Neutron scattering probes such kind of rigid body motions. The investigated energy range (μeV–meV) corresponding to the nano-second to pico-second time-range, reflects motions of the atomic groups to which the H-atom is associated [2,3].

Fluctuations of local or global character are believed to play an important role in protein function [4]. A description of protein motions has been suggested by Frauenfelder and coworkers [5,6], in which the potential energy is described by a multidimensional energy landscape as a function of conformational coordinates. A protein can assume a large number of nearly isoenergetic conformations with only small structural differences. Each of the various protein conformations, represented by a valley, is divided into several conformational substates. Each conformational substate is organized in a hierarchy of a number of tiers. At low temperature, the protein atoms vibrate harmonically within their substate. As temperature increases, sufficient energy is provided to cross from one well to another, which gives rise to diffusive, nonvibrational dynamics. These motions can be investigated by quasielastic neutron scattering. Molecular dynamics simulations of myoglobin have shown that these diffusive motions as seen in neutron scattering experiments result from rigid-body motions of the protein side-chains attached to the protein backbone [7]. A model has been developed based on a hydrated powder of C-phycocyanin, which decomposes protein dynamics into three components: global motions of the residues, side-chain rigid-body motions and internal deformation [8,9]. It has been demonstrated that the global motions of the residues correspond to vibrations in local energy minima and jumps between minima, whereas the rigid-body side-chain motions mainly describe the nonharmonic component.

18.1.1 Elastic Incoherent Neutron Scattering

Harmonic and nonharmonic protein motions can be explored by elastic incoherent neutron scattering (for a review see [3]). We recall that in the Gaussian approximation (for further information see contributions by Lechner et al., Part I and Doster, this volume) the scattering function can be described as

$$S_{\text{in}}(\mathbf{Q},0) = \exp\left\{\frac{<u^2>}{6}Q^2\right\}. \tag{18.1}$$

Two factors determine, which motions can be observed in an elastic scattering experiment:

- The accessible Q-range. The wave-vector Q is proportional to the inverse of a distance. The higher the studied Q-value, the smaller the amplitudes of the observed motions.
- The energy resolution of the instrument. The better the energy resolution, the lower the frequencies which can be investigated, i.e., slower motions become accessible on the spectrometer.

If only the elastic component is studied, all motions in the time interval corresponding to the energy resolution contribute to the scattering, but are not resolved into different correlation times or frequencies. Only motions faster than the corresponding time limit are observed.

Mean square amplitudes, $<u^2>$, are accessible as a function of temperature from the intensity of elastically scattered neutrons. These $<u^2>$ correspond to the full extend of the motion which has been swept out by the atom during a short time t. At low temperature, where only harmonic vibrations are present, the $<u^2>$ show a linear increase with temperature. A break in the increase of the mean square amplitude vs. temperature signals the onset of nonharmonic motions. This feature is termed a dynamical transition and has been observed in neutron scattering for hydrated protein powders and membranes. A first transition arises typically around 200 K and has been interpreted with several physical models, which will be detailed further in Sect. 18.2.2.

In the following sections we will elucidate how dynamics investigated by incoherent neutron scattering can be related to functional and structural characteristics of the biological system under study. We will focus on dynamics extracted from the elastic scattering intensity. In particular, we will highlight the membrane protein bacteriorhodopsin and exemplify the relationship between protein structure, dynamics, and hydration on one hand and protein function and activity on the other.

18.2 Methods of Investigation

18.2.1 Elastic Incoherent Neutron Scattering on Powder Samples

Protein dynamics was first studied in powder samples. Powder samples consist of dried proteins or proteins in various hydration states. In general, hydration (h) varies from 0 to 0.5 g of solvent per g of protein ($\mathrm{g\,g^{-1}}$). At a hydration value of $h = 0.4\,\mathrm{g\,g^{-1}}$ a smaller protein is completely covered by one hydration shell and is in an active state [10]. The dynamics of powder samples can be investigated over a large temperature range. Owing to the absence of crystalline ice formation, there is no need to employ a cryoprotectant for studies at low temperatures as there is for liquid samples. Myoglobin is a widely studied system and a dynamical transition of Fe motions has first been observed by Mössbauer spectroscopy [11]. Later, neutron scattering revealed a dynamical transition for global motions at 180 K (Fig. 18.1) [12]. Further, the effects of trehalose on myoglobin dynamics have been studied [13]. Trehalose is a disaccharide, which can protect biomolecules under extreme conditions (high temperature and low hydration) against dehydration and denaturation. When myoglobin becomes coated in trehalose it behaves like a harmonic solid up to room temperature. The dynamical transition formerly observed, is absent (Fig. 18.1). Moreover, it has been reported that characteristic motions as observed by Mössbauer, optical absorption spectroscopy [14] and obtained from MD simulations [15] are hindered. Therefore, it has been concluded that trehalose traps the protein in a conformational substate, inhibiting larger nonharmonic motions, which might lead to protein unfolding.

18.2.2 Models for Describing Thermal Protein Dynamics

Several models have been developed to describe protein motions. We focus here on models aiming to describe motions measured in energy resolved elastic incoherent neutron scattering.

A double-well model has been applied to explain the dynamical behavior of myoglobin (inset Fig. 18.1) [12]. At temperatures below the dynamical transition (part A of triangles in Fig. 18.1), the atoms are trapped in a harmonic well. The dynamical transition was observed at 180 K, which was interpreted as the onset of jumps of hydrogen atoms between states of different energy (part B of triangles in Fig. 18.1). These nonharmonic motions were modeled by a two site-jump diffusion, where the sites are at a distance d with a free energy difference ΔG. This model allows to extract parameters like jump distance d, enthalpy ΔH and entropy $\Delta S/R$ difference from the mean square amplitude vs. temperature data.

The concept of force constants has been introduced by Zaccai and coworkers [16, 17] in order to quantify the molecular resilience of a protein structure. From the slope of the $<u^2>$ values with respect to the temperature,

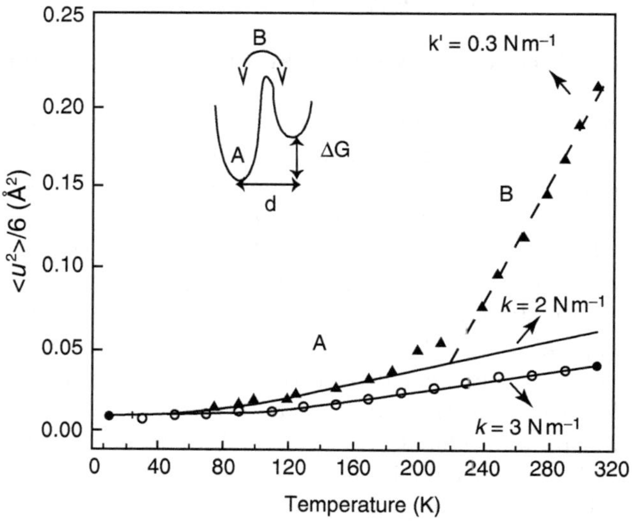

Fig. 18.1. Mean square amplitudes $<u^2>$ of a hydrated myoglobin powder in D_2O (*triangles*, redrawn from [12]) and of myoglobin in trehalose (*circles*, redrawn from [13]). The inset shows the double-well model of Doster and coworkers [12]. Hydrated myoglobin molecules are trapped in the lower well at temperatures below the dynamical transition (**a**). Above this transition (**b**), they can sample both wells. Force constants $<k>$ and effective force constants $<k'>$ are given for myoglobin in trehalose and hydrated myoglobin below and above the dynamical transition (Reprinted figure with permission from [16]. Copyright 2000 AAAS)

a force constant $<k>$ can be calculated in the harmonic region (part A of Fig. 18.1) and an effective force constant $<k'>$ in the nonharmonic region (part B of Fig. 18.1). Myoglobin is most resilient (least "soft") in trehalose ($<k>= 3\,\mathrm{N\,m^{-1}}$), and least resilient as a hydrated powder above the dynamical transition ($<k'>= 0.3\,\mathrm{N\,m^{-1}}$). The underlying model describes the atomic potential by two concentric wells of different depths and a hard outer sphere. It is valid in all dynamical regimes of the mean square amplitudes as a function of temperature, and assumes that two classes of conformational fluctuations essentially control atomic motion in a protein. The first class consists of movements about the equilibrium position, which are limited by the local environment of the atoms. The second one is due to interaction-mediated fluctuations of the molecule that allow larger excursions of the particle. These motions are restrained to a larger cage formed by the neighboring molecules. Concretely, for a protein atom this can be regarded as two conformational cages, i.e., a small and a large one, fitting together. The force constants extracted from the harmonic regime as well as the effective force constants can be regarded as a measure of the degree of flexibility of the molecule. They can be used for comparison of different systems having similar transition temperatures. In this model, the dynamical transition temperature observed in the $<u^2>$ vs. temperature plot corresponds to the system gaining sufficient energy to explore the upper well. The dynamical transition has been defined as the temperature where 10% of the atomic population is located in the outer potential of higher energy and displays larger amplitude motions [17].

18.2.3 H/D Labeling Techniques

As described in various chapters in this book, neutrons discriminate between different isotopes of an element leading to a different scattering cross-section for hydrogen than for deuterium. This particular specificity is not only employed in diffraction experiments to localize hydrogen atoms or to minimize the solvent contribution to a scattering spectrum. Owing to the fact that the incoherent scattering cross-section of a hydrogen atom exceeds by far the one of deuterium and of other atoms in biological samples, a specific H/D labeling allows the study of local dynamics in the protein part where the hydrogens have been preserved while the deuterated part is masked. For neutron dynamics studies, only a few examples on specifically labeled samples exist.

For instance, a study focussing on hydration water used deuterated phycocyanin [18]. Another example deals with specifically labeled purple membranes [19]. In the latter case, only two types of amino acids have been labeled to investigate local core motions of the protein. This example will be presented in further detail in Sect. 18.3.3. An often discussed question is whether or not deuterium labeling alters the "real" motions of the system under study. One could argue that as neutron scattering experiments are sensitive to motions of the amino acid side-chains, deuteration could alter the dynamical spectrum. However, a typical contribution to the scattering spectrum is related to CH_2 or CH_3 groups and deuteration would only cause a minor change in mass. Moreover, a comparison of $<u^2>$ of fully deuterated PM with a natural abundance PM sample showed similar results [19]. However, to date not enough data are available for definitely excluding that deuterium labeling does not affect the dynamics probed by neutron scattering and care should be taken when conceptualizing a project.

A facility for partial or full deuteration of biological molecules, such as proteins, nucleic acids and lipids has been set up by the Institut Laue-Langevin (ILL) and the EMBL Outstation in Grenoble. It aims to support the development and improvement of labeling techniques for biological systems and provides a user service for the production of deuterated samples for neutron scattering and NMR experiments (http://www.ill.fr/deuteration).

18.3 Relating Thermal Motions in Purple Membranes to Structural and Functional Characteristics of Bacteriorhodopsin

In the following we will present dynamical studies of different powder samples. In particular, we have chosen to highlight the relation between dynamics, structure and function of a biological system on the example of bacteriorhodopsin (BR), and the purple membrane (PM).

18.3.1 Thermal Motions in Bacteriorhodopsin and the Purple Membrane

Purple membranes from halophilic Archaea display a 2D-crystalline arrangement in their native state and are composed of lipids and of only one type of protein, the light-driven proton pump bacteriorhodopsin [20]. Absorption of a photon by the chromophore retinal in BR triggers a cyclic series of ground and intermediate states (photocycle), namely bR, J, K, L, M, N, O, and bR [21,22]. As a result, one proton per photocycle is transferred from the cytoplasmic (CP) to the extracellular (EC) part of the membrane. The photocycle as well as the proton pump activity and their dependence on hydration and temperature have been thoroughly studied. Therefore, the relationship between structure – BR activity and protein dynamics can be explored on this system. BR and PM dynamics can be studied at several levels, each associated with interesting biological properties:

- Dynamical properties of the entire membrane as a function of temperature and hydration.
- Local motions by combining the powerful method of $H/^2H$ labeling and neutron scattering.
- The influence of the lipid environment on protein motions.

A protein is only active if it adopts its particular three-dimensional structure and if all internal motions attached to its function can be accomplished. Based on this, Zaccai [23] has suggested a hypothesis that the close packing of PM in a cooled or dry state would inhibit the motions necessary to complete the BR photocycle and the proton pumping across the membrane. A first set of incoherent neutron experiments, performed on PM in two extreme hydration states (dry: 0% relative humidity (r.h.), hydrated: 93% r.h), substantiated this hypothesis [24]. The investigation of mean square amplitudes, $<u^2>$, as a function of temperature by elastic incoherent neutron scattering has revealed a transition around 230–250 K between harmonic and nonharmonic motions only for the hydrated PM sample (Fig. 18.2a, [24,25]). Below this temperature the hydrated sample, following a harmonic behavior, is not active, whereas above the transition temperature nonharmonic diffusive motions appear and the BR photocycle can be completed. In contrast, the dry PM sample exhibits

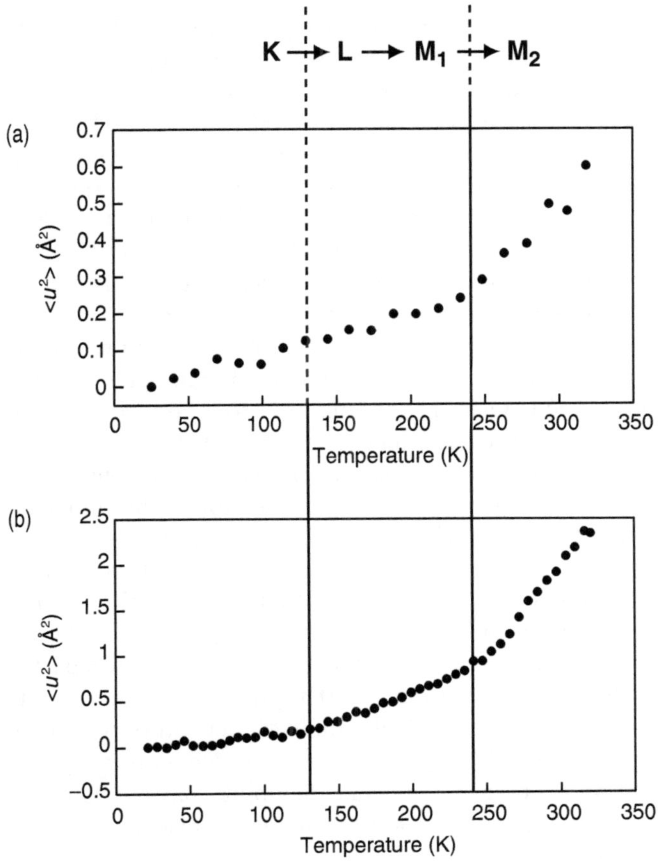

Fig. 18.2. Dynamical transitions and trapping of BR intermediate states. Mean square amplitudes of small-amplitude motions (**a**) and high-amplitude motions (**b**) in purple membranes at 93% r.h. were measured on IN13 and IN16 at the Institut Laue Langevin, Grenoble, France, respectively. Small- and high-amplitude motions both show a dynamical transition at around 240 K. High-amplitude motions show an additional dynamical transition at about 130 K. IN13 data were recorded in the wave-vector range $Q = [1.99 ; 3.71]$ Å^{-1} and IN16 data in the range $Q = [0.43 ; 1.25]$ Å^{-1}. The K, L, M1 and M2 intermediate states of BR are usually trapped by illumination at 110, 170, 230, and above 240 K, respectively [22,26,27]. The dynamical transition at 130 K may, therefore, be correlated with the K to L transition and the one at 240 K with the M1 to M2 transition. The figure is based on [26]

harmonic motions up to room temperature and BR is not functional. The fact that nonharmonic motions are not necessarily liberated at high temperatures indicate that temperature alone is not sufficient to obtain high amplitudes of motion.

These results have supported significantly the hypothesis that a global soft environment obtained through temperature and hydration, which allows larger diffusive thermal motions, is required to ensure the functioning of BR. Complementary studies of the elastic incoherent scattering for smaller scattering vectors have demonstrated a dynamical transition at about 130–150 K, which is present for both the dry and the hydrated membrane (Fig. 18.2b [19, 28]). This has led to a division of PM dynamics into two different populations of motions: "high" amplitude motions exhibiting smaller frequency (maximum $<u^2>$ of 2 Å^2 at 300 K for the hydrated PM sample), and "small" amplitude motions at a higher frequency (maximum $<u^2>$ of 0.5 Å^2 at 300 K for the hydrated PM sample). The two dynamical transitions can be compared with the ability of BR to proceed in its photocycle (Fig. 18.2). The dynamical transition at 130 K (Fig. 18.2b) might be correlated with the ability of BR to convert from the K into the L intermediate. Large-scale structural changes that accompany the transition from an early to a late M state (i.e., from M1 to M2) correlate with the dynamical transition at 240 K (Fig. 18.2a and b).

A different approach using quasielastic incoherent neutron scattering on oriented PM samples has yielded information about the diffusive motions of different spatial extent and with different correlation times in the nano-second to picosecond time scale [29]. This technique has allowed characterization of the main components of diffusive motions such as three-site jump diffusion of methyl group rotations and two-site jump diffusion [30]. The onset of quasi-elastic scattering has been found around 200 K with a particular strong increase of this component around 260 K [31]. Moreover, the different categories of diffusive motion have been found to depend on the hydration state of PM [29], supporting the hypothesis that a decrease in hydration results in a solidification in internal molecular flexibility of the protein structure. It has been concluded that the observed diffusive motions are essential for BR photocycle activity.

18.3.2 Hydration Dependence of Thermal Motions

The hydration dependence of the thermal motions has been investigated further in a broad temperature range [25, 30]. "Fast" and "slower" motions as defined by the correlation time from quasielastic measurements have shown to be influenced differently by hydration. In general, the slower local diffusive motions become much more pronounced at a high hydration state, which led to the suggestion that they correspond to motions of groups close to the surface of the membrane [30]. The behavior of the $<u^2>$ values, as a function of temperature and hydration degree, has shown that the two populations of high- and small-amplitude motions respond differently to hydration of the membrane [25]. These results have quantified the dynamical heterogeneity of the membrane and its dependence on hydration and temperature. The high amplitude motions in the nonharmonic region above the transition at 250 K are not very sensitive to changes in hydration until a high degree is reached where a sharp increase of the $<u^2>$ values is seen. In contrast, the small amplitude motions rise continuously with increasing hydration above a threshold of 60% r.h. and above the transition temperature. The hydration dependence of photocycle and pumping characteristics of BR have been correlated with the thermal motions, i.e., below 60% r.h. the BR photocycle can not be completed [32, 33] and proton pumping ceases [27]. It has been shown that not only a global fluidity is necessary for the functioning of BR, but that the particular response to hydration of the different populations of motions can be correlated to the functional behavior of BR.

The examination of the lamellar spacing of PM stacks as a function of temperature by neutron diffraction has shown that cooling of a PM sample leads to a decrease in the lamellar spacing [34]. The authors suggested that this corresponds to dehydration of PM that takes place at a temperature between 245 and 265 K, depending on the hydrational level of PM. Upon heating, this effect is reversible and rehydration starts at around 255 K. The initial spacing is reached again at 265–275 K. This process has been put in the context of the dynamical behavior and the transition temperatures of the membranes. In particular, the rehydration has been correlated to the transition temperature around 250 K and the solvent melting effect observed by Réat and coworkers [19]. If a hydrated PM stack is flash-cooled, a fraction of the amorphous inter-membrane water turns liquid-like at 200 K during subsequent heating and leaves the intermembrane space without triggering a dynamical transition [35].

18.3.3 Local Core Motions

Local dynamics of the protein core elucidated in more detail the dynamical heterogeneity of BR [19]. For this purpose, a completely deuterated membrane with only the retinal and the amino acids Met and Trp being hydrogenated, has been prepared. These selected groups are located in the retinal binding pocket and the extracellular half of the protein (Fig. 18.3).

The retinal binding pocket is functionally particularly important in the uni-directional proton transfer during the photocycle. The mean square amplitudes $<u^2>$ of such a specifically labeled and hydrated PM sample show significantly decreased flexibility compared to the global membrane, with the motions revealed as being more restrained and shielded from solvent melting effects. Moreover, the investigation of core dynamics in different motion regimes (small- and high-amplitude motions) as a function of hydration revealed a particular behavior for the core motions different to the one discovered in the membrane globally [36]. This has quantified the dynamical heterogeneity of BR and the results have been interpreted in terms of a more rigid protein core, which meets the functional requirements of the protein.

The retinal binding pocket has to act as a valve between the extracellular and cytoplasmic parts of BR, and, therefore, a certain environmental rigidity is required to control stereo-specific selection of retinal conformations to accomplish the valve function, whilst a certain global fluidity is required which allows large conformational changes during the photocycle.

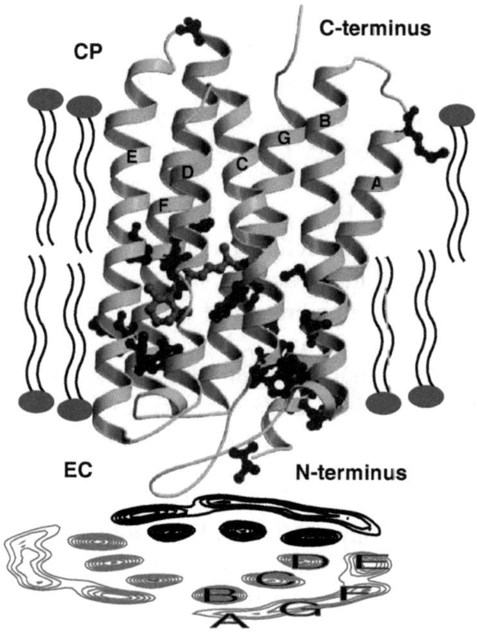

Fig. 18.3. Schematic view of selectively labeled BR in the PM and the projection structure of a BR trimer in the PM lattice. The entire PM has been deuterated, except for the Met and Trp residues and the chromophore retinal (shown in ball-and-stick mode) that have been kept hydrogenated. Structural data are from PDB entry 1qhj [38] and the figure has been prepared with MOLSCRIPT [39] and RASTER3D [40]

18.3.4 Lipid Environment

A point of discussion is always to which extent lipid motions contribute to the observed PM motions. Using partly delipidated PM samples, which are still crystalline and containing BR trimers, Fitter and coworkers [37] started addressing the role of the lipid environment on PM dynamics. A reduced flexibility of the delipidated membrane has been observed and the difference in flexibility was more pronounced for a wet than for a dry membrane. This has led the authors to the conclusion that it is the hydration of the lipids which plays a major role for the dynamical behavior of the global membrane as suggested by Zaccai [23]. In addition, photocycle kinetics is partly slowed down upon delipidation and a correlation between the BR photocycle and the dynamical behavior of the protein–lipid complex in different conditions has been made [37].

A comparison between motions explored by quasielastic neutron scattering on PM and disk membranes, which contain rhodopsin and lipids in a weight ratio of 45% rhodopsin and 50% of lipids, indicated higher internal flexibility of the disk membranes [41]. From this piece of work a suggestion of a linear dependence between the amplitudes of motions (in this case extracted from a two-site jump diffusion model or diffusion inside a sphere model) and the relative amount of lipids has been emerged.

What might be the impact of packing in the membrane on protein and lipid flexibility? Packing can be perceived as a lateral pressure on the protein, which maintains the molecule in a certain conformational state. It is important to note that BR is not inserted in a symmetric way in PM, but that the extracellular part of the protein is more deeply buried in the lipids. Therefore, the lateral pressure might not be identical on both sides of the protein. This agrees with the more rigid dynamics of the specifically labeled PM sample presented above. From neutron diffraction experiments it has been concluded that BR needs a soft environment in order to be functional, which allows large conformational changes of the protein. In dry membranes or at low temperature, conditions in which BR is not functional, the close-packed environment would inhibit the required amplitudes for accomplishing the photocycle and the pump activity [23]. It has been pointed out that thermal dynamics serves as a kind of lubricant to enable large-scale displacements necessary to ensure correct protein activity [19, 25, 42]. Thus, it can be suggested that protein packing in the membrane participates in the creation of the necessary soft environment including the required well-defined rigidity to maintain the dynamical heterogeneity, which influences functionally important conformational changes during the photocycle. The different kinetics of the rise and decay of the photointermediates in monomeric BR compared to BR in PM have been attributed to the packing of BR in PM [43]. However, one has to be careful, because it is difficult to estimate which of the changed parameters such as monomerisation or disruption of lipid protein interaction produce the most prominent perturbation in the photocycle and of proton pumping. For lactose permease from *E. coli*, it has been suggested that the critical lateral lipid packing pressure is an essential prerequisite for correct membrane insertion and function of the protein [44]. The permease seems to be an extremely flexible, metastable molecule that requires external stabilization by the surrounding membrane and, therefore, it is difficult to crystallize such a membrane protein without its lipids. Recently, the first high-resolution structure became available [45].

18.3.5 Relation between PM Dynamics and Crystallographic B-factors

Electron and X-ray diffraction data of BR provide temperature-factors (B-factors) as one result of the structure refinement process, which reflect both static and dynamic disorder. They have indicated that the protein is most rigid in its center near the retinal binding pocket and that it becomes more flexible at the surfaces [46]. Furthermore, it has been found that the temperature-factors are asymmetrically distributed over the protein, i.e., larger in the cytoplasmic (CP) than in the extracellular (EC) region [38], supporting the neutron scattering results from Réat et al. [19] discussed in Sect. 18.3.3. In molecular dynamics simulations, the fluctuations of the helix extremities in the CP half of BR appeared twice as large than those in the EC half [47].

The question of the physical significance of these temperature factors, and how they are related to the thermal motions observed in neutron scattering, can be addressed. The isotropic temperature-factor or B-factor is described by $B = 8\pi^2 <x^2>$ with a mean square deviation $<x^2>$ from a mean atomic position. The overall $<x^2>$ of an atom originates from several sources including the internal static disorder (different configurations in different unit cells), internal dynamic disorder (vibrations within molecules), lattice defects and lattice vibrations. In the case of the BR ground-state structure determined at 100 K, the average $<x^2>$ value is 0.3 Å^2 (see for example the B-factors from the PDB entry codes 1c3w [48] and 1cwq [49]). They are about four times larger compared to $<u^2>$ values from neutron scattering at 100 K ($<u^2>$ values are around 0.07 Å^2). The motions, which enter in the B-factor calculation, are assigned to single atoms and reflect the vibrations of backbone and side-chain atoms. Vibrations of bonded atoms are of the order of 0.1 Å in a high frequency range (10^{13}–10^{14} s^{-1}) [1], whereas the $<u^2>$ reflect the motions of the amino acid side-chains (obtained from the scattering of the H-atoms in these groups) in a frequency range of 10^9–10^{12} s^{-1} [2, 7]. In contrast to the mean square amplitudes $<u^2>$, the B-factors are not directly measured and only arise from the structure refinement. They mainly reflect a general tendency of the protein flexibility. They do not supply quantitative information in contrast to neutron scattering, which provided evidence for the dynamical heterogeneity. A comparison between temperature factors and root mean square deviation (RMSD) from an averaged BR structure of six models has revealed that the RMSD are mostly lower in the helices than the corresponding B-factors whereas the RMSD loop values are mostly higher. This has led to the suggestion that the B-factors are highly dependent on characteristics associated with crystal nature and quality and that they are not useful as absolute values [50].

18.3.6 Comparison of Force Constants with Forces Measured by AFM

The force constants underlying thermal motions (20–200 pN/Å) are of the same order of magnitude as forces explored by atomic force microscopy (AFM) on PM samples [51, 52]. The surface structure of BR has been explored by AFM and an applied force of 200 pN was necessary to bend away the E–F loop by 2 Å [53]. Effective force constants of thermal fluctuations represent an average value of the entire protein or the labeled part and are certainly higher than what can be expected for the flexible loop regions.

BR helices have been pulled off the membrane in pairs by applying forces around 350 pN [52]. The ruptured bonds correspond to interhelix and intrahelix hydrogen bonding and interaction with lipids and are, therefore, considerably higher than the bending forces.

To pull off lipids, a force of only 25 pN is sufficient, as the interactions are smaller due to the fluid-like lipid behavior. This suggests that the force constants derived from thermal motions of lipids should be much smaller than those of the protein. Indeed, quasielastic incoherent neutron scattering on lipid bilayers have characterized fast in-plane and out-of-plane motions of lipid chains and molecules with amplitudes rising up to 3 Å [54].

18.4 Protein Dynamics and Function in Some Other Proteins

As outlined in the preceding sections, biological activity of BR (i.e., proton pumping) is temperature-dependent. In particular, the ability of BR to complete its photocycle is abolished below the dynamical transition [24]. A correlation between the onset of biological function and the dynamical transition has been established in other proteins as well. In myoglobin, e.g., escape of the CO molecule from the heme pocket after flash photolysis is only possible at temperatures above the dynamical transition determined by Mössbauer spectroscopy [55]. Evidence for a dynamical transition at 220 K in Ribonuclease A has been obtained from the temperature-dependence of atomic B-factors [56]. Substrate-binding to Ribonuclease A is possible above, yet not below 220 K as revealed by temperature-controlled protein crystallography [57], thus suggesting a link between the dynamical transition and enzyme activity. Protochlorophyllide oxidoreductase is a light-activated enzyme and intermediates in the reaction pathway can be monitored by temperature-dependent absorption and fluorescence spectroscopy after initiating catalysis at cryo-temperatures (e.g., 100 K, [58]). The final step in the reaction pathway can only proceed above 200 K [59], a temperature close to the dynamical transition of many proteins. Despite these examples, a possible link between the dynamical transition and onset of biological activity remains a controversially discussed issue. Daniel and collaborators [60] conducted parallel measurements of dynamics and activity of glutamate dehydrogenase in a cryo-solution. Incoherent elastic neutron scattering experiments in the temperature range from 80 to 300 K revealed a dynamical transition of motions faster than 100 ps at 220 K. Enzymatic activity of glutamate dehydrogenase in the same cryo-solvent did not deviate from Arrhenius behavior in the measured temperature range from 280 down to 190 K. The authors concluded that the observed dynamical transition is decoupled from the rate-limiting step in the reaction pathway. However, the dynamical transition observed at 140 K for slower motions (i.e., faster than 5 ns) might, or might not be coupled to enzymatic activity, which cannot be measured below 190 K because of cryo-solvent freezing [61]. Not only fast motions probed by Mössbauer and neutron spectroscopy, but also much slower protein motions (µs–ms) measured by NMR have been identified as being involved in enzyme catalysis [62,63]. Even though protein dynamics is generally acknowledged to be crucial for activity, it remains unclear for proteins less well-studied than BR, in how far motions on different timescales are coupled to each other and which ones are essential for biological activity.

18.5 Conclusions

A general picture of the relation between thermal motions, structure and function emerges from the study of PM samples. The thermal motions reflect the protein flexibility that is necessary to access different conformational substates, but do not represent the conformational change in itself for which additional light energy is required to cross the barrier. The energy that is necessary for large conformational changes to occur during the photocycle has been provided by the light absorption and the thermal relaxations lead to the different photointermediates. Thermal motions help to overcome barriers between the different intermediates, but they do not directly reflect the conformation of the protein. Thermal fluctuations have been described as "...searching out the path or paths along which the transition takes place." [4]. They are needed to prepare a "soft" environment in which the large conformational changes can take place and where the photointermediates are maintained in a stable configuration.

Acknowledgments

We thank G. Zaccai and F. Gabel for many fruitful discussions and for critical reading of the manuscript. U. Lehnert thanks the German Academic Exchange Service (DAAD) for a Postdoctoral fellowship. Support by the European Union under contract numbers HPRI-CT-2001-50035 and RII3-CT-2003-505925 is gratefully acknowledged.

References

1. J.A. McCammon, S.C. Harvey, *Dynamics of Proteins and Nucleic Acids* (Cambridge University Press, Cambridge, 1987)
2. J.C. Smith, Quat. Rev. Biophys. **24**, 227–291 (1991)
3. F. Gabel, D. Bicout, U. Lehnert, M. Tehei, M. Weik, G. Zaccai, Q. Rev. Biophys. **35**, 327–367 (2002)
4. M. Karplus, J.A. McCammon, Ann. Rev. Biochem. **53**, 263–300 (1983)
5. H. Frauenfelder, P.J. Steinbach, R.D. Young, Chem. Scripta **29A**, 145–150 (1989)
6. H. Frauenfelder, S.G. Sligar, P.G. Wolynes, Science **254**, 1598–1603 (1991)
7. G.R. Kneller, J.C. Smith, J. Mol. Biol. **242**, 181–185 (1994)
8. G.R. Kneller, Chem. Phys. **261**, 1–24 (2000)
9. K. Hinsen, A.-J. Petrescu, S. Dellerue, M.-C. Bellissent-Funel, G.R. Kneller, Chem. Phys. **261**, 25–37 (2000)

10. J.A. Rupley, G. Careri, Adv. Protein Chem. **41**, 37–172 (1991)
11. F. Parak, E.W. Knapp, D. Kucheida, J. Mol. Biol. **161**, 177–194 (1982)
12. W. Doster, S. Cusack, W. Petry, Nature **337**, 754–756 (1989)
13. L. Cordone, M. Ferrand, E. Vitrano, G. Zaccai, Biophys. J. **76**, 1043–1047 (1999)
14. L. Cordone, P. Galajda, E. Vitrano, A. Gassmann, A. Ostermann, F. Parak, Eur. Biophys. J. **27**, 173–176 (1998)
15. G. Cottone, L. Cordone, G. Ciccotti, Biophys J. **80**, 931–938 (2001)
16. G. Zaccai, Science **288**, 1604–1607 (2000)
17. D.J. Bicout, G. Zaccai, Biophys. J. **80**, 1115–1123 (2001)
18. M.C. Bellissent-Funel, J.M. Zanotti, S.H. Chen, Faraday Discuss. **103**, 281–294 (1996)
19. V. Réat, H. Patzelt, M. Ferrand, C. Pfister, D. Oesterhelt, G. Zaccai, Proc. Natl. Acad. Sci. USA **95**, 4970–4975 (1998)
20. A.E. Blaurock, J. Mol. Biol. **93**, 139–158 (1975)
21. U. Haupts, J. Tittor, D. Oesterhelt, Annu. Rev. Biophys. Biomol. Struct. **28**, 367–399 (1999)
22. J.K. Lanyi, B. Schobert, Biochemistry **43**, 3–8 (2004)
23. G. Zaccai, J. Mol. Biol. **194**, 569–572 (1987)
24. M. Ferrand, A.J. Dianoux, W. Petry, G. Zaccai, Proc. Natl. Acad. Sci. USA **90**, 9668–9672 (1993)
25. U. Lehnert, V. Réat, M. Weik, G. Zaccai, C. Pfister, Biophys. J. **75**, 1945–1952 (1998)
26. R. Neutze, E. Pebay-Peyroula, K. Edman, A. Royant, J. Navarro, E.M. Landau, Biochim. Biophys. Acta **1565**, 144–167 (2002)
27. N.A. Dencher, H.J. Sass, G. Büldt, Biochim. Biophys. Acta **1460**, 192–203 (2000)
28. V. Réat, G. Zaccai, M. Ferrand, C. Pfister, Functional dynamics in purple membrane, in *Biological Macromolecular Dynamics* (Adenine, Guilderland, Newyork, 1997) pp. 117–122
29. J. Fitter, R.E. Lechner, G. Büldt, N.A. Dencher, Proc. Natl. Acad. Sci. USA **93**, 7600–7605 (1996)
30. J. Fitter, R.E. Lechner, N.A. Dencher, Biophys. J. **73**, 2126–2137 (1997)
31. J. Fitter, R.E. Lechner, N.A. Dencher, Physica B **226**, 61–65 (1996)
32. H.J. Sass, I.W. Schachowa, G. Rapp, M.H.J. Koch, D. Oesterhelt, N.A. Dencher, G. Büldt, EMBO J. **16**, 1484–1491 (1997)
33. M. Weik, G. Zaccai, N.A. Dencher, D. Oesterhelt, T. Hauss, J. Mol. Biol. **275**, 625–634 (1998)
34. R.E. Lechner, J. Fitter, N.A. Dencher, T. Hauss, J. Mol. Biol. **277**, 593–603 (1998)
35. M. Weik, U. Lehnert, G. Zaccai, Biophys. J. **29**, 29 (2005)
36. U. Lehnert, V. Réat, B. Kessler, D. Oesterhelt, G. Zaccai, Appl. Phys. A **74**, S1287-S1289 (2002)
37. J. Fitter, S.A. Verclas, R.E. Lechner, H. Seelert, N.A. Dencher, FEBS Lett. **433**, 321–325 (1998)
38. H. Belrhali, P. Nollert, A. Royant, C. Menzel, J.P. Rosenbusch, E.M. Landau, E. Pebay-Peyroula, Struct. Fold Des. **7**, 909–917 (1999)
39. P. Kraulis, J. Appl. Cryst. **24**, 946–950 (1991)
40. E.A. Merritt, D.J. Bacon, Methods Enzymol. **277**, 505–524 (1997)
41. J. Fitter, O.P. Ernst, T. Hauss, R.E. Lechner, K.P. Hofmann, N.A. Dencher, Eur. Biophys. J. **27**, 638–645 (1998)

42. C.L. Brooks, M. Karplus, B.M. Pettitt, Proteins: a theoretical perspective of dynamics, structure, and thermodynamics, in *Advances in Chemical Physics*, I. Rice, S.A. Prigogine (Eds.) (John Wiley & Sons, New York, 1988) pp. 94–95.
43. G. Váró, J.K. Lanyi, Biochemistry **30**, 5008–5015 (1991)
44. J. Le Coutre, L.R. Narasimhan, C.K. Patel, H.R. Kaback, Proc. Natl. Acad. Sci. USA **94**, 10167–10171 (1997)
45. J. Abramson, I. Smirnova, V. Kasho, G. Verner, H.R. Kaback, S. Iwata, Science **301**, 610–615 (2003)
46. N. Grigorieff, T.A. Ceska, K.H. Downing, J.M. Baldwin, R. Henderson, J. Mol. Biol. **259**, 393–421 (1996)
47. O. Edholm, O. Berger, F. Jähnig, J. Mol. Biol. **250**, 94–111 (1995)
48. H. Lücke, B. Schobert, H.T. Richter, J.P. Cartailler, J.K. Lanyi, J. Mol. Biol. **291**, 899–911 (1999)
49. H.J. Sass, G. Büldt, R. Gessenich, D. Hehn, D. Neff, R. Schlesinger, J. Berendzen, P. Ormos, Nature **406**, 649–653 (2000)
50. J.B. Heymann, D.J. Müller, E.M. Landau, J.P. Rosenbusch, E. Pebay-Peyroula, G. Büldt, A. Engel, J. Struct. Biol. **128**, 243–249 (1999)
51. D.J. Müller, G. Büldt, A. Engel, J. Mol. Biol. **249**, 239–243 (1995)
52. F. Oesterhelt, D. Oesterhelt, M. Pfeiffer, A. Engel, H.E. Gaub, D.J. Müller, Science **288**, 143–146 (2000)
53. D.J. Müller, H.-J. Sass, S.A. Müller, G. Büldt, J. Mol. Biol. **285**, 1903–1909 (1999)
54. S. König, W. Pfeiffer, T. Bayerl, D. Richter, E. Sackmann, J. Phys. II France 2 **8**, 1589–1615 (1992)
55. H. Lichtenegger, W. Doster, T. Kleinert, A. Birk, B. Sepiol, G. Vogl, Biophys. J. **76**, 414–422 (1999)
56. R.F. Tilton, J.C. Dewan, G.A. Petsko, Biochemistry **31**, 2469–2481 (1992)
57. B.F. Rasmussen, A.M. Stock, D. Ringe, G.A. Petsko, Nature **357**, 423–424 (1992)
58. D.J. Heyes, A.V. Ruban, H.M. Wilks, C.N. Hunter, Proc. Natl. Acad. Sci. USA **99**, 11145–11150 (2002)
59. D.J. Heyes, A.V. Ruban, C.N. Hunter, Biochemistry **42**, 523–528 (2003)
60. R.M. Daniel, J.C. Smith, M. Ferrand, S. Hery, R. Dunn, J.L. Finney, Biophys. J. **75**, 2504–2507 (1998)
61. R.M. Daniel, J.L. Finney, V. Réat, R. Dunn, M. Ferrand, J.C. Smith, R.V. Dunn, J. Finney, Biophys. J. **77**, 2184–2190 (1999)
62. F.A. Mulder, A. Mittermaier, B. Hon, F.W. Dahlquist, L.E. Kay, Nat. Struct. Biol. **8**, 932–935 (2001)
63. E.Z. Eisenmesser, D.A. Bosco, M. Akke, D. Kern, Science **295**, 1520–1523 (2002)

19

Biomolecular Spectroscopy Using Pulsed-Source Instruments

H.D. Middendorf

19.1 Introduction

Neutron scattering techniques have already been applied to a wide range of problems in biomolecular dynamics, from anhydrous building blocks to a variety of partially or fully hydrated biomolecules and their assemblies [1–7]. However, much of what has been published so far deals with data sets that are very limited in relation to the structural and dynamical complexities of the systems studied, and the impact of this work in the life sciences has been small.

There is a broad concensus that increases in data collection rates by 1–2 orders of magnitude are required if the full potential of neutron scattering as a more widely used tool is to be realized [8–10]. Only high-intensity pulsed sources can deliver increases of this magnitude. A factor of 10–20 will be needed to bring sample sizes down into the 20–50 mg region. The 1/2 g quantities required for present-day neutron spectroscopy are much too large. Another factor of 10 could be "spent" either on higher-quality data sets from experiments designed to reveal functionally significant difference spectra, or on more extensive coverage of the large parameter domain of biomolecular systems.

19.2 Why Pulsed Sources?

It has been recognized since the 1980s [11,12] that there is little room for flux increases from fission reactors beyond the level achieved by the world's most powerful steady-state neutron source at the Institut Laue-Langevin (ILL) in Grenoble. In Europe, the only new research reactor built during the past 20 years (FRM-II at Munich) requires highly enriched uranium for optimal performance, a politically contentious design option that is unlikely to be adopted ever again.

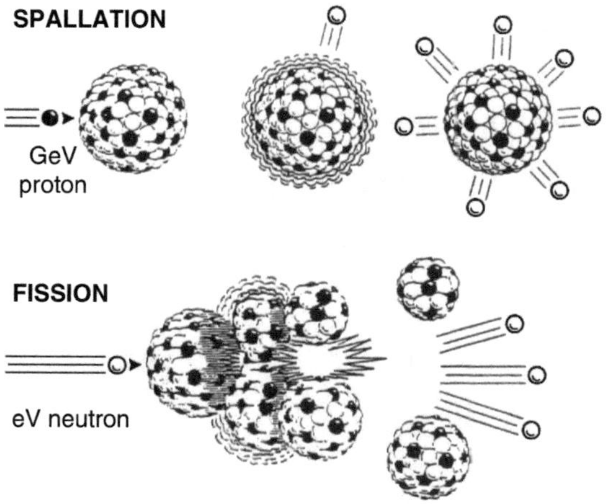

Fig. 19.1. A GeV proton incident on a heavy-metal nucleus leads to an excited state that emits 15–25 spallation neutrons, but the nucleus does not undergo fission. In the fission case, a uranium nucleus produces on average only 2.5 neutrons, one of which is needed to sustain the chain reaction

An alternative way of producing intense neutron beams is by bombarding a block of heavy metal with GeV proton pulses [11]. Each incident proton "spalls" between 15 and 25 neutrons off a target nucleus (Fig. 19.1). In addition to cold and thermal neutron beams from appropriate moderators, spallation sources provide epithermal neutrons, which open up possibilities for electron volt spectroscopy and medical applications. Following pioneering work at research centers with low-intensity accelerators, the technology of such sources was advanced significantly by the commissioning in 1986 of the first high-intensity spallation source, the ISIS facility at the Rutherford Appleton Laboratory near Oxford, England. New facilities are under construction currently at Tokai (JSNS [13]), Japan, and at Oak Ridge (SNS [14]) USA, and a second target station is being built at ISIS (TS2 [15]).

Pulsed-source instruments can achieve large gain factors even with inferior time-averaged fluxes. By cutting out a series of spikes from a continuous white beam and monochromatising them, conventional choppers always "throw away"' more than 99.9% of incident neutrons. With a pulsed beam, on the other hand, neutrons already arrive "concentrated" as spikes separated by 20–100 ms according to the duty cycle of the accelerator. Since the source is effectively "switched off" between successive pulses, scattered neutrons can be energy-analyzed with excellent signal/noise ratios. A drawback of fixed-frequency pulsed sources is that the gain factors of instruments covering different energy regions cannot be optimized simultaneously. By constructing a second target station running at a lower frequency, ISIS is the only facility aiming for the best of both worlds.

Table 19.1. Spallation neutron sources

Country	Name	Institution	Location	Current (μA)	Frequency (Hz)
UK	ISIS[a]	Rutherford Appleton Lab	Chilton	200	50/10
CH	SINQ[b]	Paul Scherrer Institut	Villigen	1,300	qu.cont.
USA	IPNS	Argonne National Lab.	Argonne	15	30
USA	LANSCE	Los Alamos National Lab	Los Alamos	100	20
USA	SNS[a]	Oak Ridge National Lab	Oak Ridge	1,300	60
Japan	KENS	KEK	Tsukuba	6	20
Japan	JSNS[a]	J-PARC (KEK-JAERI)	Tokai	330	25

[a]under construction: SNS, JSNS, ISIS-TS2 (10-Hz target station of ISIS)
[b]quasi-continuous (50 MHz)

At present there are five neutron research centers operating spallation sources, and four of these are pulsed sources (Table 19.1). Instruments at SINQ (Paul Scherrer Institute, Switzerland) are not able to exploit the inherent time structure of a pulsed beam since the source is quasi-continuous. In the following the discussion will be limited to instruments and experiments at the four spallation sources producing pulsed beams.

19.3 Pulsed Source vs. Reactor Instruments

Neutron scattering experiments using pulsed-source instruments are not in principle different from those on reactor-source instruments. With the exception of the spin-echo technique [16, 17], the objective in both is to measure double-differential cross-sections $\mathrm{d}^2\sigma/\mathrm{d}\Omega\,\mathrm{d}E$ from which dynamic structure factors $S_{\mathrm{inc}}(Q,\omega)$ and $S_{\mathrm{coh}}(Q,\omega)$ for incoherent and coherent scattering can be extracted [11] ($\hbar Q$ = momentum transfer, $\hbar\omega$ = energy transfer). However, there are significant differences in instrument design and operation that may affect experimental strategy, and there are also instrument types covering Q,ω-regions that are not or only partly accessible by spectrometers on conventional sources. Most applied-science users familiar with instruments on the latter will appreciate the larger dynamic range and (for backscattering) better Q-resolution of pulsed-source spectrometers.

A comparison of the Q,ω-envelopes of spectrometers at ILL and ISIS (Fig. 19.2) shows an area of overlap flanked by regions of lower and higher energy transfers, which at present are accessible only by reactor-source (low $\hbar\omega$) or pulsed-source instruments (high $\hbar\omega$). While this reflects the fact that the energy distribution of raw (i.e., undermoderated) spallation neutrons is centered at higher energies than that of fission neutrons, the complementarity

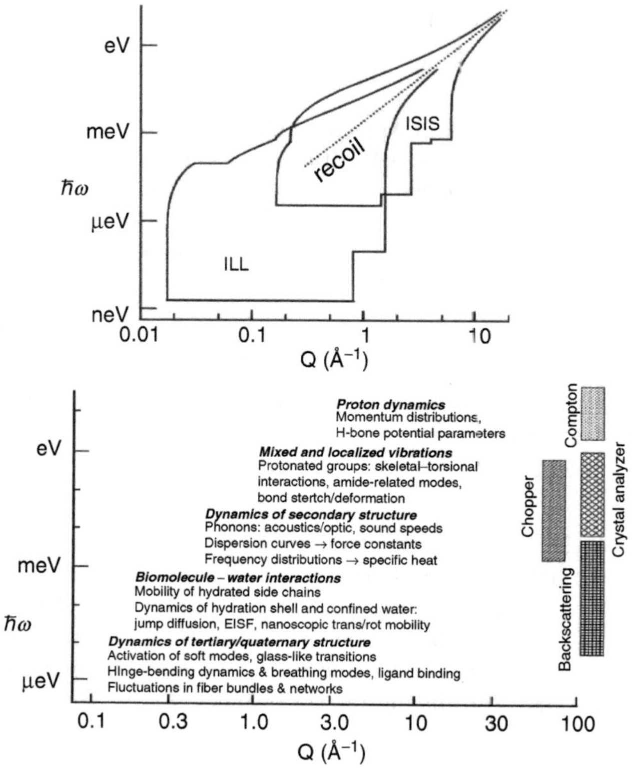

Fig. 19.2. *Top*: Envelopes of the Q, ω-domains of commonly used spectrometers at ILL (IN5/6/10/13/16) and at ISIS (IRIS, OSIRIS, MARI, HET) for typical operating parameters. TOSCA and VESUVIO at ISIS operate on or very close to the recoil asymptote given by $\hbar\omega$ (μeV) $= 2.072\, Q^2$ Å^{-2} (see Sect. 19.5.2). *Bottom*: Synopsis of effects observed and information derived from experiments in different regions of the Q, ω-plane. Vertical bars give energy ranges of the four pulsed-source instrument types discussed in the text

apparent from Fig. 19.2 is deceptive. Until the late 1980s it was thought generally that there would not be enough cold neutrons in beams from spallation sources to allow μeV spectroscopy with useful intensities. However, the success of instruments on the liquid H_2 moderator at ISIS [18] has led to a sea change in thinking about cold-neutron capabilities, and the new sources under construction will all be fully instrumented to cover the $\hbar\omega < 10$ μeV region. Apart from advances in moderator optimization for low-frequency sources, two important developments are relevant here: (i) At ILL, Grenoble, much work has gone into building a multi-angle spin-echo spectrometer on a chopped beam mimicking pulsed-source operating conditions, and this instrument is essentially ready for implantation at one of the new sources [17]. (ii) Designs for direct-geometry multichopper spectrometers on 10–25 Hz pulsed sources have evolved significantly since the first such instrument was proposed [19,20], and versions optimized for ISIS-TS2, JSNS, and SNS are under construction. All offer resolutions down to a few μeV, wide dynamic range, and large gain factors relative to comparable reactor instruments.

Four types of instrument are being used for biomolecular spectroscopy at the research centers listed in Table 19.1. The interaction processes and effects observable are given in Fig. 19.2 as a function of their approximate location in Q, ω-space. The following sections provide outlines of the main characteristics of these spectrometers along with selected results from problem areas that illustrate what can be achieved with existing instrumentation.

19.4 Backscattering Spectrometers

Backscattering spectrometers on pulsed sources, by virtue of their wide dynamic range and high Q-resolution, are versatile instruments for studying the motions of protons in and around biomolecules and their building blocks [11,21,22]. Sets of temperature and/or hydration dependent spectra with energy transfers $\hbar\omega$ from a few μev to several meV reflect the rich dynamics of such systems, from nearly free or restricted translational diffusion (central peak broadenings at low Q) to segmental fluctuations and various hindered rotational motions at intermediate and high momentum transfers (Fig. 19.3).

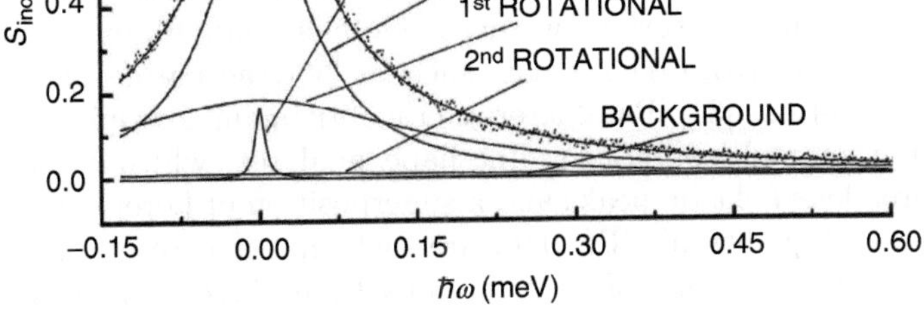

Fig. 19.3. Example of the decomposition of a quasielastic lineshape from IRIS data for a 1 % agarose–H_2O gel ($Q = 1.01\,\text{Å}^{-1}$) into elastic, translational, and two rotational components. For any measured spectrum in the quasielastic region, the momentum transfer $\hbar Q$ is approximately independent of $\hbar\omega$ and equal to the elastic momentum transfer $\hbar Q_0 = \hbar(4\pi/\lambda_0)\sin\theta$ (λ_0 = incident wavelength, 2θ = full scattering angle)

IRIS and OSIRIS are currently the two most advanced pulsed-source backscattering instruments [23]. In the "inverse" scattering geometry employed here (see Fig. 19.8 in Lechner et al. Part 1, this volume), the energy scan is achieved by the spread in velocity as neutron pulses from the moderator travel through a long guide toward the sample. Two disk choppers cut out a wavelength band that can be shifted over a wide interval by appropriate phasing. Neutrons scattered by the sample are energy-analyzed in near-backscattering (175°) by two semicircular crystal arrays: a 51-element pyrolytic graphite (PG) array, and a similar mica array on the opposite semicircle. Backscattered neutrons are detected by 2×51 scintillator counters on a 0.55-m circle just below the analyzer-sample plane. By resetting chopper phasings and acquisition windows, it is possible to configure instruments of this kind for resolutions between $\approx$1 and 100 μeV with energy windows ranging from around 100 μeV to several meV.

Although spectrometer configurations offering resolutions <10 μeV are already available at ISIS, they are rarely used because of the demands on beam time. This will change dramatically with IRIS-type spectrometers on the new spallation sources because of both higher flux levels and the fact that these instruments are well suited to multianalyzer operation. The backscattering spectrometer DYANA [24] at JSNS, for example, features a stack of three analyzer banks, which can be moved into and out of the scattering plane. Since the emphasis in future work will increasingly be on complex and challenging biomolecular systems requiring measurement of difference structure factors $\Delta S(Q,\omega)$ with high accuracy, the potential of instruments such as DYANA for leading-edge experiments and for opening up new applications of neutron spectroscopy is very substantial indeed.

19.4.1 Hydration Dynamics

Hydration experiments using IRIS and OSIRIS have been concerned mostly with the dynamics of sorbed water in globular and fibrous proteins [25–28], and in fully hydrated systems such as polysaccharide gels [29, 30] or "crowded" solutions [31]. Parameters characterizing nanoscopic mobilities are extracted from multicomponent lineshape analyses, which typically reveal slightly broadened elastic peaks and a superposition of Lorentzian-like wings. The range and quality of IRIS data are such that it is often possible to analyze lineshapes in terms of contributions from three dynamically distinct proton populations: (i) tightly bound backbone protons and other protons that belong to very slowly fluctuating structural elements, both contributing an effectively elastic fraction $\sim\delta(\omega)$; (ii) protons belonging to closely associated waters and mobile side chains, with highly restricted rotational and librational degrees of freedom; (iii) protons of water molecules in the secondary hydration shells, with reduced translational and rotational mobilities that are qualitatively similar to those of supercooled H_2O. The Q-dependence of the relevant width parameters are key quantities here, together with geometrical information contained in the elastic incoherent structure factor (EISF [2]).

Spectra obtained from aqueous polysaccharide gels at IRIS have provided insight into fundamental aspects of water–biopolymer interactions [29,30]. The polymer matrix dynamics and solvent coupling in agarose and hyaluronate gels differ substantially as a result of very different Coulomb interactions along their hydrated chains. For gels in the 1–20% concentration range, the spectra have been analyzed in terms of two mobile proton populations plus an irrotationally bound fraction. Agarose forms well-defined gels down to low concentrations, allowing a "perturbative" approach: spectra can be measured for a concentration series covering the transition from pure H_2O to a dilute gel and then in steps to more concentrated gels. The data reveal changes in the solvent dynamics introduced by a random, increasingly dense 3D network of hydrophilic fibers carrying OH groups that mesh with and perturb transiently hydrogen-bonded water clusters. The Q-resolution made it possible to observe an oscillatory component in $\Gamma_t(Q)$, the translational water proton mobility. This component appears to be the dynamic signature of density fluctuations with a correlation length of about 6 Å. MD simulations of pure H_2O have predicted density fluctuations involving clusters of 20–30 waters [32]. The spatial wavelength extracted from $\Gamma_t(Q)$ via Fourier transformation is close to the most probable cluster size determined by simulations. This suggests that superimposed on the dominant nearest-neighbour jump diffusion there is a nanoscopic process associated with the breaking and reforming of water aggregates with average diameters of 10–12 Å.

19.4.2 Low-Temperature Dynamics and Glass-Like Transitions

Observing the evolution of the energy landscape of biomolecules from very low to near-ambient temperatures has been a focus of much neutron work since the pioneering studies of Rorschach et al. at Oak Ridge [33]. Due to their size and complex architecture, biomolecular systems possess a large number of conformational substrates, which may be described by a potential energy hypersurface (or "energy landscape") resembling that of small-molecule glass formers. Oligopeptides and proteins hydrated with H_2O or D_2O at specific hydrations h up to ≈ 0.7 (h = g water/g anhydrous protein) have been used in IRIS and OSIRIS experiments on the "unfreezing" of softer degrees of freedom and activation of anharmonic modes around 200 K [25–28, 34, 35]. The aim here is to obtain data either in the form of window-integrated intensities, $S_{qe}(Q,T)$, or as fully resolved dynamic structure factors that quantify the spectral evolution of low-frequency fluctuations and localized diffusive modes. Since for H_2O-hydrated and slightly D_2O-hydrated biomolecules the scattering is incoherent to a good approximation, $S_{qe}(Q,T)$ is essentially proportional to a proton-weighted Debye–Waller (DW) factor in regions where the dynamic behavior is adequately described by a system of harmonic or quasiharmonic oscillators. Up to ≈180 K a mean-square displacement $<u_p^2>$ averaged over all protons can be derived from the slope of $\ln S_{qe}(Q,T)$ vs. Q^2 or, more accurately, from fitting expressions involving Debye-type density of states functions. Around and above 200 K, significant nonlinear increases in $<u_p^2>$ are observed for all hydrated biomolecules (Fig. 19.4), and described analytically as products of effective DW factors and barrier-jump terms [36], Plazcek-type expansions $\sim\exp(-u_p^2Q^2 + bQ^4 +)$, or integrals over a distribution of mean-square displacements [37].

Experiments at using instrument IN13 have concentrated on deriving $<u_p^2>$ from $S_{qe}(Q,T)$ scans for 20–25 temperature points between ≈30 and 300 K [7]. By contrast, the objective of some IRIS and OSIRIS experiments has been to examine spectral changes across the 200 K region and to look for any dependence on thermal history [27, 35]. In experiments on hydrated hemoglobin samples, for example, one was quenched rapidly in liquid N_2 and spectra were recorded starting at 100 K then going up in steps to 260 K, the other was "warm" initially (260 K) and spectra were taken down to 100 K.

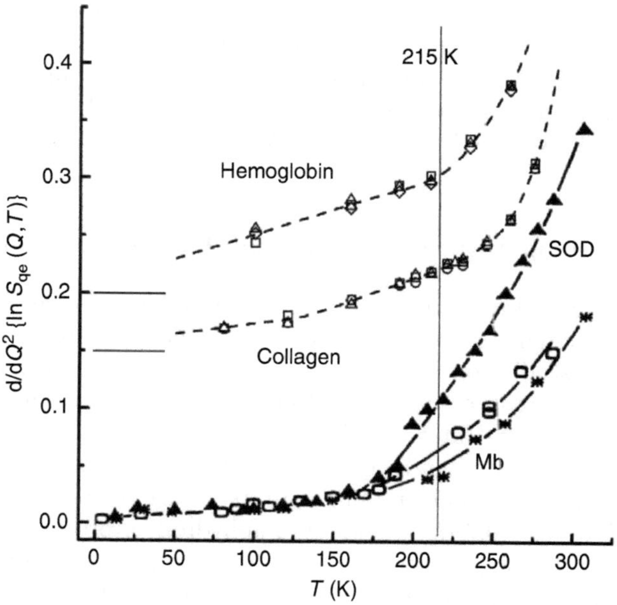

Fig. 19.4. Transition temperatures and slopes $|\ \mathrm{d}\ln S_{qe}(Q;\omega)/\mathrm{d}Q^2\ |$ for superoxide dismutase (SOD), myoglobin (Mb), hemoglobin (Hb), and collagen (for clarity, Hb and collagen curves are y-shifted by 0.2 and 0.15, respectively). Data from IRIS (SOD [25], hemoglobin [27]), IN6 at ILL (myoglobin [36]), and FOCUS at PSI (collagen [28]). In the harmonic and quasiharmonic (for collagen, between ≈110 and 220 K) temperature regions where the ordinate may be identified with $<u_p^2>/3$, the resulting proton displacements range from 0.15 to 0.4 Å

Three conclusions have emerged so far: (i) $<u_\mathrm{p}^2>$ values increase substantially from around 200 K, as expected, but in addition the low-Q behavior of $S_\mathrm{qe}(Q,T)$ exhibits some "up" vs. "down" differences that may be ascribed to differential freezing vs. unfreezing of low-frequency modes over 15–20 Å (Fig. 19.5); (ii) the build-up of small shoulders in the wings at the onset of anharmonic and mode-coupling processes allow Q-dependent broadenings to be determined (Fig. 19.5); (iii) both the window-integrated and the spectrally resolved structure factors show small but significant differences at and above 235 K that depend on thermal history. Experiments of this kind shed light on the molecular basis of glass-like transitions, especially by providing data constraining the parameters of studies modeling proteins as hierarchically organized multilevel systems.

19.4.3 Enzyme Dynamics and Folding–Unfolding Processes

$S(Q,\omega;T)$ functions derived from neutron spectra provide a basis for interpreting the "unperturbed" dynamics of biomolecules. For a fuller understanding of functionally relevant properties, it is necessary to probe interactions with other (usually smaller) molecules, or to probe the response to externally applied electric fields or irradiation, e.g., by laser pulses. The redistribution of low-frequency modes as a function of ligand binding, changes in enzymatic activity, or folding/unfolding processes is of particular interest here [38]. This is typically done by measuring spectra for two states A and A' of a solution sample in consecutive runs, plus additional runs for buffer and ligand alone. Data of this kind may be analyzed (i) to quantify changes in the wings of quasielastic peaks and their merging into damped low-frequency modes, and (ii) to derive proton-weighted frequency distributions $\Delta Z_p(\omega)$ from spectral changes in the 1–50 meV region (8–400 cm^{-1}) with the aim of relating them to macroscopic quantities characterizing the $A \rightarrow A'$ transition.

Very few studies along these lines have been undertaken so far, reflecting the difficulty of separating spectral contributions due to the dynamics of interest from those due to the D_2O buffer. It is essential to minimize "excess" solvent scattering that is unlikely to affect the macromolecular dynamics. Although the total specific cross-section of a hydrogenous D-exchanged protein is larger than that for D_2O by a factor of about 6, volumetrically the protein signal gets "diluted" quickly as the concentration is lowered. Since the information in nearly all such experiments is carried by the predominant incoherent scattering, the problem is that toward lower concentrations the useful signal becomes smaller, whereas at higher concentrations the scattering from "excess" water and from the shell of water closely associated with the protein globule become comparable. In the latter situation, reliable difference structure factors $\Delta S(Q,\omega)$ characterizing the macromolecular dynamics proper cannot be extracted just by subtracting spectra measured for the buffer alone.

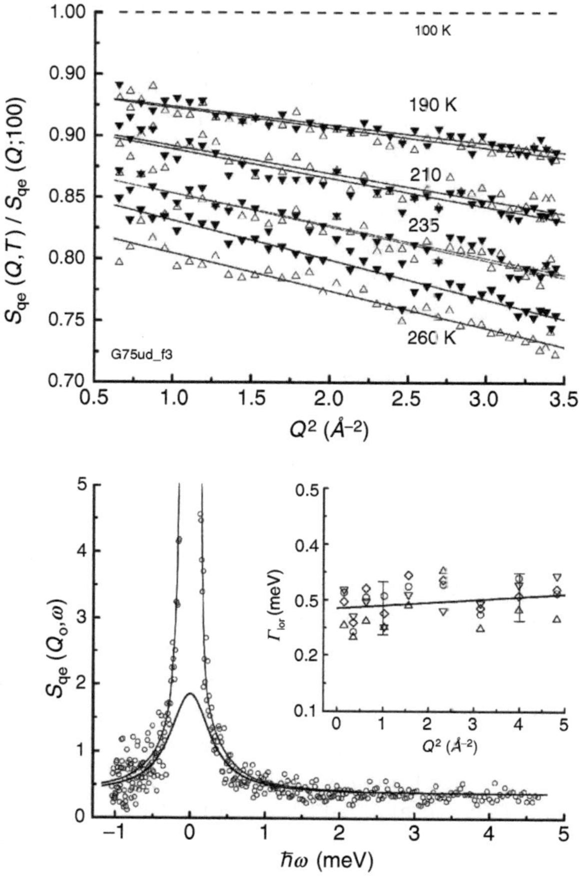

Fig. 19.5. *Top*: $S_{qe}(Q,T)$ for 29% H_2O-hydrated hemoglobin powders for "down" (*solid triangles*) and "up" (*open triangles*) temperature series relative to $S_{qe}(Q,T = 100) = 1$, with slopes from linear regression fits. No significant differences are observed up to and including 235 K, whereas at 260 K intensities and slopes differ by 5–10% (from [27]). *Bottom*: Example of a FOCUS spectrum (*circles*) measured for 50% H_2O-hydrated collagen fibers at T = 260 K, showing onset of wing broadening (single spectrum from [51], $Q_0 = 1.77\,\text{Å}^{-1}$, not smoothed). Full curves represent results from three-parameter EISF analysis [2, 27] convoluted with the near-Gaussian resolution function, on top of a flat background. Insert shows deconvoluted Lorentzian wing broadenings $\Gamma_{lor}(Q^2)$ for the 260 K data set grouped into nine spectra. The approximate Q-independence suggests that this onset is due to effectively localized excitations. To bracket errors, fit intervals and weights were varied. Shown are four sets of Γ_{lor} values obtained in this way

In early experiments at ILL [1,39], attempts were made to observe changes due to "hinge-bending" modes, i.e., low-frequency motions of domains around the active site of an enzyme. Difference spectra for lysozyme revealed that the lysozyme–$(NAG)_3$ complex (NAG = N-acetylglucosamine) is "tighter" than the free enzyme, and that the differences are larger at Q values corresponding to distances of 10–15 Å. The only experiment on enzyme–ligand interactions using pulsed-source spectrometers was performed at KEK [40]. In spectra from Ca-saturated calmodulin with and without bound mastoparan (a tetradecapeptide), Izumi et al. observed difference bands centered on 4 and 50 meV (30 and 400 cm^{-1}).

Significant redistributions of low-frequency modes accompanied by increases in diffusive degrees of freedom occur when a protein begins to unfold. In experiments at KEK, Kataoka et al. [41,42] compared neutron spectra measured for staphylococcal nuclease (SNase) at 100 and 300 K with those from a deletion mutant. This mutant lacks 13 residues at the C-terminal (SNase-13) and is known to assume a compact but denatured form. Relative to native SNase, the SNase-13 sample gave enhanced quasielastic but less inelastic scattering. To circumvent problems associated with solvent subtraction, slightly hydrated protein powders ($h = 0.39$) were used in these experiments. These few early results on ternary systems and functionally relevant interactions can only give a glimpse of what will be possible in the future once fast and versatile pulsed-source spectrometers capable of producing high-quality spectra from small samples become available.

19.5 Inelastic Scattering at 1 meV < $\hbar\omega$ < 1 eV ($8 < \hbar\omega < 8{,}000$ cm^{-1})*

At higher energy transfers, overlapping with backscattering instruments in the 1–10 meV region (8–80 cm^{-1}), two types of pulsed-source spectrometer cover about three decades in $\hbar\omega$: (i) direct-geometry chopper spectrometers provide large sets of Q-dependent spectra within the accessible Q,ω-domain, whilst (ii) time-focusing crystal-analyzer or filter-difference instruments produce only one or two spectra constrained to fixed trajectories in Q,ω-space. For (i), typical areas of application involve mapping out $S(Q,\omega)$ functions characterizing collective processes, which are either intrinsically Q-dependent (such as phonons or Brillouin modes), or for which a Q-dependence could give clues to their nature (such as bands in the "Boson peak" region). Instruments of type (ii), on the other hand, are high-throughput spectrometers designed to resolve effectively nondispersive excitations over a wide $\hbar\omega$ range at Q values reflecting short-range interactions.

*Optical spectroscopy units are commonly used in this area of neutron scattering: 1 meV = 8.06554 cm^{-1}, 1 cm^{-1} = 0.12398 meV.

Table 19.2. Chopper (direct geometry) and crystal or filter analyzer (indirect geometry) spectrometers at pulsed sources. Highest energies accessible are between 0.5 and 1 eV

Facility	Current μA	Chopper instruments	Crystal or filter analyzer instruments
ISIS	190	MARI, HET, MAPS	TOSCA (TFXA) (crystal)
LANSCE	100	HRMECS	FDS (Be-filter)
IPNS	15	PHAROS	CHEX (crystal)
KEK	6	INC	CAT (crystal)

19.5.1 Chopper Spectrometers

Direct-geometry chopper spectrometers on pulsed sources (Table 19.2) are straightforward adaptations of reactor instruments. Since the time structure of pulses is already "prepared" by the source, a single Fermi chopper (a fast-rotating assembly of curved slits) plus one or two slower auxiliary disk choppers (to suppress background) are sufficient to select a wide range of near-monochromatic slices from the incident pulses. For geometrical reasons and the fact that the sample is not exposed to some fraction of the full spectrum of incident neutron pulses, direct-geometry spectrometers are easier to equip with low-angle arrays. This is an attractive feature for macromolecular work where low-Q coverage down to 0.1 or 0.05 $Å^{-1}$ is desirable. MARI at ISIS, the fastest and best equipped instrument, uses a Fermi chopper to give incident energies in the range 9–1000 meV with resolutions of 1–2% [43]. It is similar in design to its sister machine HET, but with detector banks (920 detectors in total) providing full angular coverage from $2\theta = 3°$ to $135°$. The range and quality of MARI data allow excellent 2D or 3D visualization of dynamic structure factors $S(Q,\omega)$, a data presentation option that is highly developed for this instrument.

19.5.2 Crystal-Analyzer and Filter-Difference Spectrometers

Of the four instruments of this kind listed in Table 19.2, the time-focusing crystal-analyzer spectrometer TOSCA (Fig. 19.6) is by far the most refined and powerful instrument [44]. TOSCA is an upgraded version of an instrument called TFXA (to avoid confusion, no explicit distinction will be made between TFXA and TOSCA in the following). Time focusing is achieved by positioning the elements of two sets of analyzer crystals (one for backscattering, one for forward scattering) at slightly different Bragg angles in such a way that the differences in reflected neutron energies are compensated for by small differences in the total neutron flight paths. The effective scattering angles are 135° (backscattering) and 45° (forward scattering) so that the conservation equations for momentum and energy yield $Q^2 = (2m_n/\hbar^2)E_f[2+\widetilde{\omega}\pm(1+\widetilde{\omega})^{1/2}\sqrt{2}]$ where $\widetilde{\omega} = \hbar\omega/E_\mathrm{f}$ and $E_\mathrm{f} = 3.95\,\mathrm{meV}$ ($31.86\,\mathrm{cm}^{-1}$) is the fixed analyzer

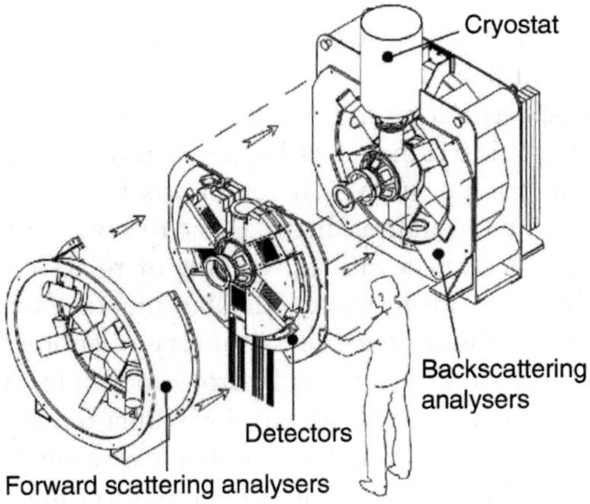

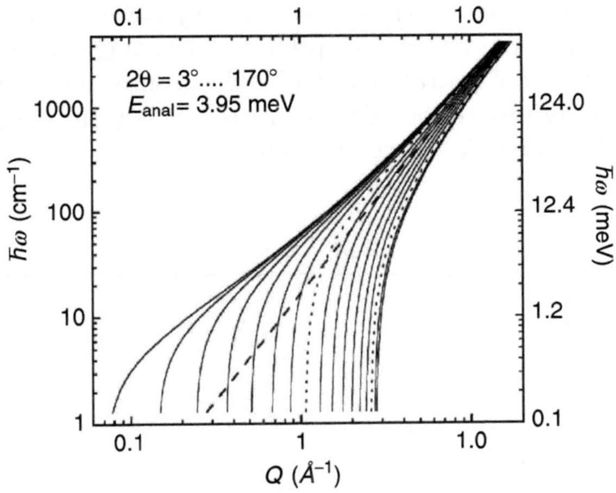

Fig. 19.6. *Top*: Exploded view of the focusing crystal-analyzer spectrometer TOSCA at ISIS. *Bottom*: Kinematic Q, ω-dependence of spectra recorded by an array of $2\theta = \text{const}$ detectors ($2\theta = 3°, 6°, 10°, 15°, ..., 170°$). Each curve corresponds to a single nonlinear 'slice' through the Q, ω-domains shown in Fig. 19.2. Except for the lowest angles, $Q \approx \text{const}$ up to $\hbar\omega \approx 1\,\text{meV}$ (or 8 cm^{-1}) whence it gradually veers toward higher Q and approaches the recoil asymptote (dashed line). The dotted curves, for $2\theta = 45°$ (forward scattering) and 135° (backscattering), represent the $\omega(Q)$ trajectories of TOSCA spectra

energy. TOSCA thus produces two spectra per run, each with a nonlinear $Q(\omega)$ dependence such that Q increases from $1.4\,\text{Å}^{-1}$ (forward scattering) or $3.1\,\text{Å}^{-1}$ (backward) at $\hbar\omega = 4\,\text{meV}$ ($32.3\,\text{cm}^{-1}$) to around $15\,\text{Å}^{-1}$ at the highest bond-stretch frequencies (see Fig. 19.6). At larger $\hbar\omega$ the Q-dependence approaches the recoil asymptote $Q^2\,\text{Å}^{-2} = 0.483\,\hbar\omega\,\text{meV} \simeq 0.0598\,\hbar\omega\,\text{cm}^{-1}$. Instead of crystal analyzers it is possible to use a difference method employing low-energy band-pass filters (Be or BeO) between sample and detector.

The spectrometer FDS at Los Alamos works in this way, achieving an energy resolution $\Delta E/E_o$ of 3–5%.

Biomolecular spectroscopy in the energy transfer regions covered by crystal or filter analyzer instruments is largely limited to the low temperature and low hydration regime, unless the emphasis is on water or ice dynamics "perturbed" by a biomolecular additive. Basic reasons for this are: (i) the scattering kinematics, which necessitates working at relatively high Q (Fig. 19.6) and consequently small mean-square displacements $<u^2>$ to give tolerable DW factors, and (ii) the absence of symmetry selection rules. The latter is an asset in studies of small, nearly anhydrous molecules where neutron spectra can provide unique data on optically forbidden modes. In experiments on more complex systems, however, the number of variously broadened excitations together with their overtones and combination lines becomes large and they merge into bands carrying weak, ill-differentiated features. It is possible, in principle, to introduce contrast selectively by covalent H/D substitution, but at present this is not a realistic option for the sample sizes required. H/D exchange of the water of hydration and of labile groups (mostly OH, some NH) remains as the only way to introduce contrast. This is illustrated in Fig. 19.7 by spectra in the amide A+B region of collagen.

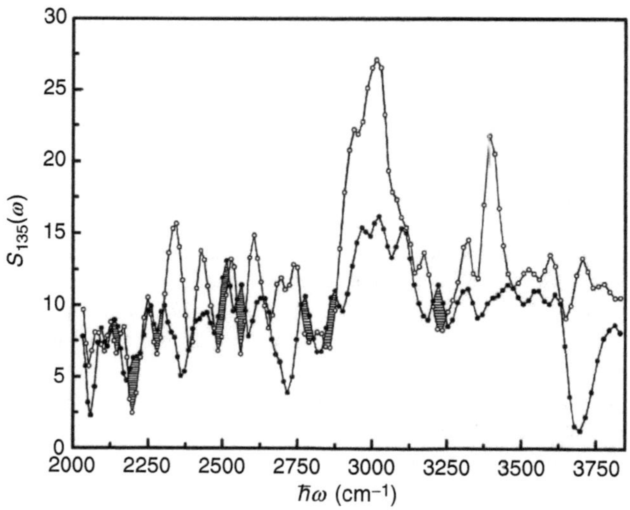

Fig. 19.7. High-frequency TOSCA spectra from collagen fibers: Hydrogenous (*open circles*) and exchange-deuterated (*full circles*) samples of aligned fibers (rat tail collagen) at 25 K. Prominent peaks at 3000 (CH, NH) and 3400 cm^{-1} (OH) are due to stretch modes. Hatched are spectral features "inversely" affected (i.e., where deuteration leads to enhanced rather than decreased intensities) or shifted due to differences in effective oscillator masses

19.5.3 Building Blocks and Model Compounds

Neutron studies of the dynamics of biological macromolecules need to be underpinned by similar work on their building blocks and on functionally important small molecules they interact with. Much current work focuses on saccharide–water interactions, in particular on understanding the molecular mechanisms that determine the remarkable effectiveness of trehalose as a bioprotectant in nature and as a cryo-preservation agent in various applications [45, 46]. The problems addressed in this work relate closely to the role played by the glass transition phenomenon and its molecular characterization in H-bonded systems that are classed as more or less "fragile" in the terminology of physicochemical studies. Molecular dynamics (MD) simulations of saccharides in dilute aqueous solutions have shown that water molecules can be highly localized at polar and hydrophilic positions, and this effect is expected to become more pronounced at higher concentrations. It is here that neutron data on proton mobilities and the concentration-dependent transition from jump to continuous diffusion in saccharide–water systems provide valuable data that are difficult to obtain otherwise. Experiments using MARI and TOSCA have emphasized the role of intramolecular stretching and bending modes together with intermolecular librational modes [46].

Amino acids and peptides are another group of important building blocks for which neutron experiments are beginning to be performed more frequently. Proton translocation across potential barriers has been studied for a number of model systems, principally carboxylic acids and the model peptides *N*-methylacetamide (NMA) [47, 48], acetanilide (ACN) [49], and alanine dipeptide (AdP) [50]. NMA is the simplest of these as it consists of a single peptide group flanked by two CH_3 groups. In ACN one of these is replaced by a phenyl ring; AdP is composed of two peptide groups and three CH_3 groups. Synthetic polypeptides have also been examined: polyglycine (in parallel with NMA) and some triple-helical polypeptides (see Sect. 19.5.5).

Of the model peptides, ACN is by far the best characterized system as a result of anomalous optical properties that triggered a great deal of work on nonlinear mode coupling. TOSCA spectra for its hydrogenous form and two isotopomers have been analyzed in considerable detail by atomic trajectory simulations using the CHARMM package, incorporating anharmonic potential functions [49]. The low and intermediate frequency regions of ACN spectra are well developed (Fig. 19.8) and similar to that of some amino acids because of the aromatic ring in addition to the methyl group. The aim of neutron studies has been to simulate the full frequency and amplitude dependence for comparison with $S(Q,\omega)$ data on both isotopomers (ACN-d_8 and ACN-d_5), in order to characterize better the interaction of low-frequency lattice modes with amide proton vibrations. The phonon dispersion diagram has been calculated from CHARMM simulations, but further experiments providing Q-dependent data over a larger temperature range are required for conclusive interpretations.

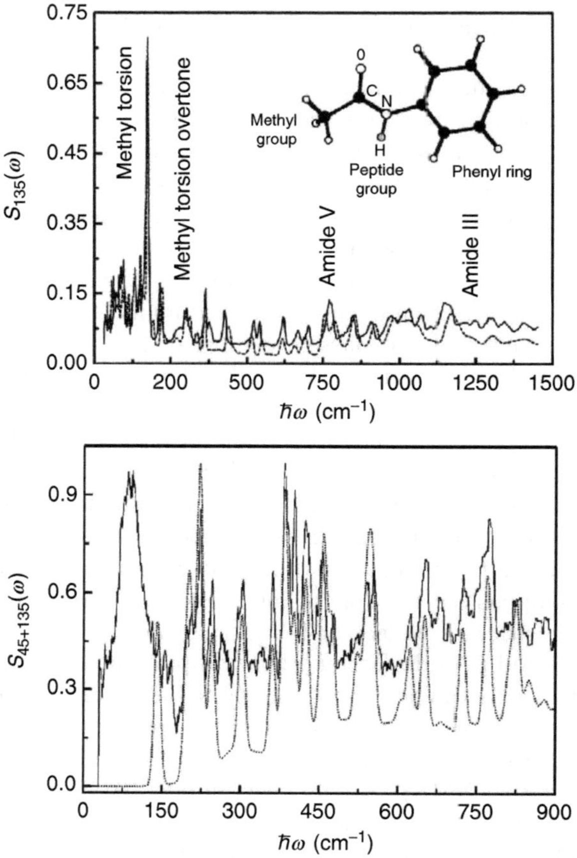

Fig. 19.8. *Top*: TOSCA spectrum of ACN-h$_9$ at 25 K (*solid line*), compared with CHARMM simulation (*hatched*). Details of $\approx$35 mode assignments up to 1500 cm^{-1} and force field parameters are given in [49]. *Bottom*: TOSCA spectrum of anhydrous glutathione at 30 K (*solid line*, *y*-shifted for clarity) compared with a simulation (*dotted*) using GAUSSIAN 98 [51]. Note that a full trajectory simulation package (such as CHARMM) is needed to simulate spectra in the acoustic phonon region $\lesssim$150 cm^{-1}

Two peptides of biomedical interest are currently being studied using TOSCA in parallel with backscattering spectrometers [51]. Carnosine (CAS, β-alanyl-L-histidine) is an antioxidative and bioregulatory dipeptide, glutathione (GSH, L-(-glutamyl-L-cysteinyl-glycine)) is a ubiquitous tripeptide serving essential cellular functions. Both the CAS and the GSH spectra show numerous well-resolved peaks up to the Amide I and II region. The 100–900 cm^{-1} region of a TOSCA spectrum from GSH is shown in Fig. 19.8 together with a result from simulations in progress. The spectra obtained provide a good basis for comparison with and refinement of CHARMM simulations of GSH in the context of drug-binding studies involving glutathione *S*-transferase [52].

19.5.4 Interpretational Aspects

Spectra from biomolecular systems pose challenging interpretational problems. Neutron techniques go beyond optical spectroscopy in that they furnish not only frequency distributions with emphasis on proton motions but also amplitude information. This puts greater constraints on the quantitative interpretation of neutron spectra, offsetting to some extent the advantage (relative to optical techniques) of dealing with a much simpler elementary scattering process. Data from TOSCA and MARI are often analyzed by means of CLIMAX [53], a normal-mode spectral simulation program based on the Zemach–Glauber expression for scattering from a harmonic oscillator [11]. Apart from the three well-established molecular mechanics packages (CHARMM, GROMOS, AMBER), more sophisticated ab initio simulations incorporating density functional theory (DFT) have been used increasingly in recent years. In the context of neutron studies, DFT simulations were first employed by Ulicny et al. [54] to interpret low-temperature vibrational features of two antiviral agents (anthrone and bianthrone) belonging to the hypericin group of molecules.

During the 1990s, in CLIMAX simulations of spectra from NMA and polyglycine, Fillaux et al. [47, 48] found that satisfactory agreement between experiment and simulation could be reached only by discarding long-established force fields and assuming instead that the proton and backbone dynamics were largely decoupled, leading to reassignment of the optically well-established N–H stretch frequency. Recent work by Kearley et al. [55] using DFT methods for the simulation of NMA spectra has gone some way towards resolving much debate and controversy about this subject. These authors concluded that previous CLIMAX simulations are seriously in error, and that the description of vibrations in molecular crystals such as NMA cannot be based on single-molecule dynamics but has to take into account intermolecular hydrogen-bonding.

19.5.5 Proteins and Biomaterials

In experiments on the dynamics of complex molecules, a strong case can often be made for using TOSCA and MARI (or HET) in parallel. Their energy ranges overlap to a large degree, and TOSCA data can already give a rough idea of the scattering response of a sample by providing "slices" through $S(Q,\omega)$ distributions. This approach was adopted in several experiments on biomolecules and model compounds.

As an example of MARI data, Fig. 19.9 shows sets of low-frequency spectra for D_2O-hydrated samples of C-phycocyanin below and above the 200 K region [56]. As the only in vivo deuterated protein available in gram quantities (from algae grown in perdeuterated media), the light-transducing chromoprotein C-phycocyanin has been a prominent "guinea pig" protein for neutron studies ever since the first such experiments at Harwell and Grenoble

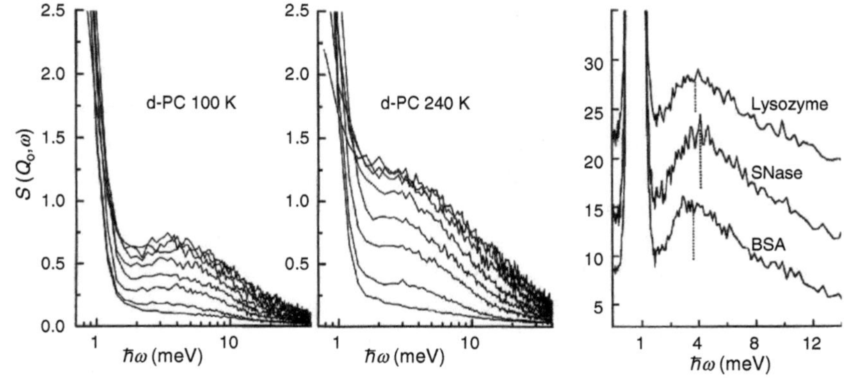

Fig. 19.9. *Right*: MARI spectra of D_2O-hydrated d-phycocyanin binned in eight groups with $<Q_0>$ from 0.284 (lowest) to 2.87 Å^{-1} (highest traces) [54]. Distinct Q-dependent bands centered on 3–4 meV (24–32 cm^{-1}) are observed at 100 K. Above the dynamic transition region they merge with rapidly increasing quasi-elastic broadening. *Left*: Spectral shape of "Boson peak" region with approximate band centroids (*dotted lines*) for lysozyme (3.80 meV), staphylococcal nuclease (4.03 meV), and bovine serum albumin (3.57 meV), measured using LAM-40 at KEK. Spectra averaged over $0.3 < Q < 2$ Å^{-1}. Adapted from [42]

in the 1970s [1, 21]. The bands shown in Fig. 19.9 and the way they change across the 200 K region relate to much-discussed questions about the significance of the ubiquitous "Boson peaks" observed in spectra from nearly all biopolymers. Three features of well-resolved data from MARI experiments are of particular interest: (i) substantially higher quasielastic wing intensities ($0.5 \lesssim \hbar\omega \lesssim 2$ meV) at 240 K relative to 100 K, probably due to enhanced merging of low-frequency α-relaxation processes with a broad distribution of β-type relaxations at somewhat higher $\hbar\omega$; (ii) disproportionate intensity increases at low-to-medium Q values relative to the high-Q spectra; (iii) red-shift of the Boson peak centroids from 3.3 (100 K) to 2.8 meV (240 K). A more quantitative interpretation of these first results requires data at intermediate temperatures and full MD simulations.

In studies of natural biocomposites like bone or teeth, or of part-synthetic composites developed for medical applications, the dominance of modes due to protonated groups may be exploited to contribute decisive information to the interpretation of optical and NMR spectra. Bone is a dense composite of calcium phosphate crystallites and type I collagen fibers in roughly equal proportions, with about 15% of the volume occupied by living cells in channels and cavities. The structural and spectroscopic properties of variously substituted hydroxyapatite nanocrystals are of central interest here. In IR and Raman spectra, the vibrational modes of OH groups and hydrogen phospate are often masked in the presence of substituent anions, whereas in neutron spectra they show up prominently. Both the OH librational band at 660 cm^{-1} and

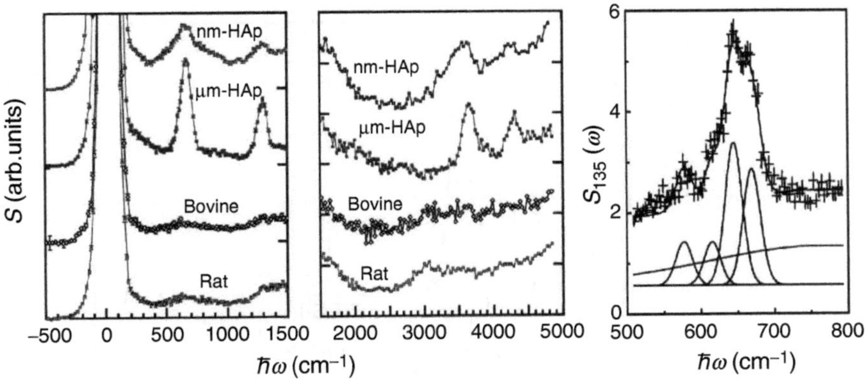

Fig. 19.10. *Left*: Neutron spectra of micrometer-sized and nanometer-sized hydroxyapatite (HAp) crystalline powders, compared with bovine and rat bone crystallites (y-shifted for clarity) (from [58]). *Right*: Decomposition of the OH librational band of HAp into components assigned to the monoclinic and the hexagonal form (from [57])

the high-energy bond-stretch region (3000–3600 cm^{-1}), together with overtone and combination bands, provide valuable data relating in particular to questions about the degree of hydroxylation and the extent to which adsorbed water may have "contaminated" certain preparations. These problems are being addressed in current work at ISIS using TOSCA and HET [55], and at IPNS using the chopper spectrometer HRMECS [57] (Fig. 19.10). The best-resolved data come from TOSCA experiments [58]. Shown in Fig. 19.10 is an example of a decomposition of the OH libration band into four components is shown.

19.5.6 Biopolymers

Collagen [59], amylose [60], and DNA [61] are the only natural biopolymers for which wide-range vibrational neutron spectra have been studied. The low-frequency region typically consists of two broad bands each carrying a number of distinct lines, the first due primarily to dispersive phonon modes and the second to weakly dispersive skeletal deformation modes. In the absence of MD simulations, the semiquantitative interpretation of collagen data from TOSCA and MARI requires comparison with spectra and normal-mode calculations for structurally related triple-helical polypeptides: polyglycine II (PGII), polyproline II (PPII), and (Pro-Pro-Gly)$_{10}$ (PPG10). Theoretical results for PGII give a maximum due to longitudinal modes at 13 cm^{-1} in the isolated molecule, but this is shifted upward to 40 cm^{-1} when H-bonding to neighboring chains is taken into account. The intensity maximum appears to be due to an inflection in the longitudinal mode branch, induced by softening of modes with wavevectors matching the pitch of the crystallographic helix. The 45 cm^{-1} peak in collagen and PPG10 spectra can be identified with

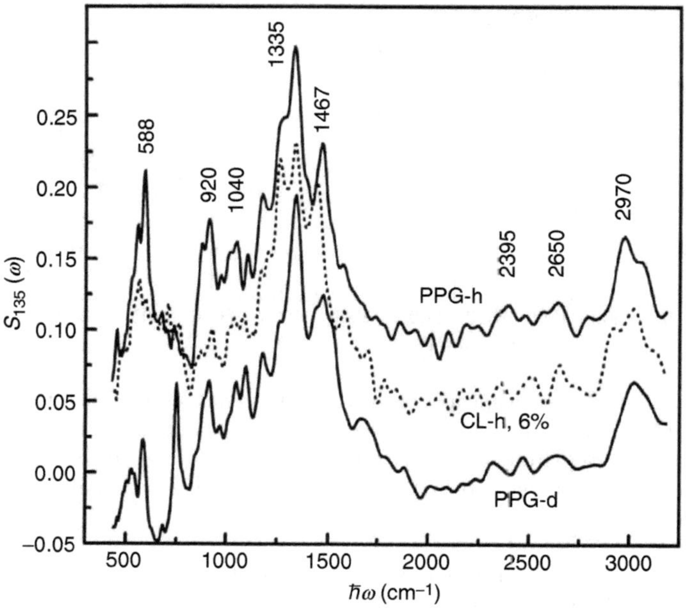

Fig. 19.11. Comparison of TOSCA spectra at intermediate and high frequencies (all at 30 K): PPG-h and PPG-d, hydrogenous and deuterated $(\text{Pro-Pro-Gly})_{10}$; CL-h, collagen fibers oriented at right angles to the scattering plane, "dry" (6% H_2O). Spectra shifted by $\Delta S = 0.05$ relative to CL-h (from [59])

this maximum for longitudinal acoustic modes in PGII-like chains that are H-bonded and supercoiled to form the collagen triple helix (Fig. 19.11). The longitudinal acoustic phonon frequency in PGII matches that of its associated interhelical ice shell, suggesting efficient coupling between modes in the latter and the polypeptide backbone.

At higher energy transfers, the modes observed involve predominantly stretching or bending of C–H, N–H and O–H bonds, together with torsional deformations. The Amide V band is one focus of attention here. Triple-helical supercoiling leads to a downward shift of the Amide V mode of the Gly-Pro linkage, together with skeletal deformation and C=O in-plane bending modes of the Pro-Pro linkages. Examples are the various amide modes between 700 and 1550 cm^{-1}, the in-plane and out-of-plane bending modes of aromatic ring hydrogens between 600 and 1600 cm^{-1}, and the methylene and methyl deformations between 950 and 1450 cm^{-1}. These major features of collagen spectra, and the assignment of several other bands and peaks on the basis of optical data and normal-mode calculations for triple-helical polypeptides, have been discussed in considerable detail [59].

19.5.7 Nucleotides and Nucleosides

Low-temperature neutron spectra for dry polycrystalline powder samples of the purines (adenine, guanine), pyrimidines (cytosine, thymine, uracil), and the nucleoside uridine were measured at ISIS using TOSCA [62], whereas similar data for the nucleosides adenosine, guanosine, and cytidine were collected at ILL using the beryllium-filter spectrometer IN1-BeF [62, 63]. Following earlier, less refined simulations, most of the nucleotide and nucleoside spectra were reanalyzed during the past 4 years at the DFT level, by Gaigeot et al. [64] for isolated molecules and by Plazanet et al. [65] for hydrogen-bonded unit cells with periodic boundary conditions. In 15 K spectra of the bases, for example, a large number of peaks in the interval 25–200 meV (200–1600 cm^{-1}) could be assigned to modes involving CH, NH, NH_2, and ring deformations (Fig. 19.12). As in the peptide work discussed in Section 19.5.2, neutron spectra at $T < 100$ K reflect closely the normal modes, and state-of-the-art simulations can provide a wealth of detail on both frequencies and intensities when analyzed in conjunction with crystallographic data. It is an open question, however, to what extent the information gained from anhydrous systems at very low temperatures will be transferable to, and hopefully improve, the empirical potentials of the commonly used biodynamics simulation packages. TOSCA data for peptides [51] and DNA [61] show how features that are well-resolved below 100 K quickly coalesce into broad undifferentiated bands as the temperature is raised, due to the combined effect of higher DW factors, enhanced overtone and combination lines, and the onset of anharmonic interactions.

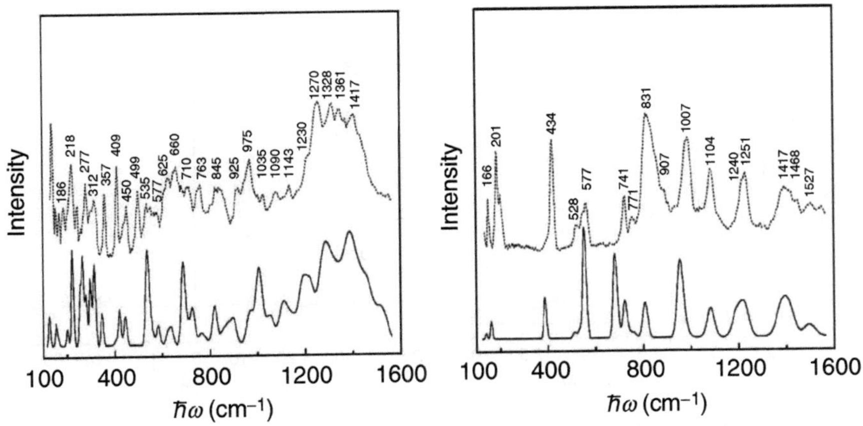

Fig. 19.12. Neutron spectra for anhydrous uridine (*left*) and uracil (*right*) at 15 K compared with DFT simulations for isolated molecules. Adapted from [64]

19.6 Neutron Compton Scattering (NCS)

For energy transfers beyond the highest excitations observable by the spectrometers discussed in Section 19.5.2, but below the region of ionizing interactions, neutron–nuclei collisions are dominated by a neutron analogue of the Compton effect [66]. Neutrons with eV energies "see" target nuclei as almost free and interact with them over distances shorter than 0.1 Å. A key concept here is the impulse approximation (IA) according to which the dynamic structure factor $S(Q,\omega)$ (effectively incoherent) consists of a superposition of Gaussians centered on the recoil energy $E_{\mathrm{rec},i} = (\hbar Q)^2/m_i$ of each atom species i. The relevant partial structure factor $S_i(Q,\omega)$ can be expressed as an integral over the product of $n_i(\mathbf{p})$ and $\delta(\omega + p^2/2m_i - (\mathbf{p}\cdot\mathbf{Q})^2/2m_i)$, where $\mathbf{p}$ is the nuclear momentum and $n_i(\mathbf{p})$ the momentum distribution of nuclei with mass m_i. The total intensity of each recoil contribution is proportional to the fraction of nuclei of mass m_i in the sample weighted by the scattering cross-section, whilst its width and shape are related analytically to $n_i(\mathbf{p})$ along $\mathbf{Q}$. The zero-order interpretation of NCS spectra, valid for $Q \longrightarrow \infty$, is based on the IA. In this limit, $S_i(Q,\omega)$ is proportional to the neutron Compton profile, $J_i(y)$, i.e., the projection of the $n_i(\mathbf{p})$ along the direction of $\mathbf{Q}$. For large but finite Q, deviations from the IA are of great interest since they provide a means to measure two important quantities: Δ^2V, the Laplacian of the interatomic potential V, and $<F^2>$, the mean-square force on the target atom.

At ISIS, the eV spectrometer VESUVIO [67] (formerly called eVS) has been used in several mostly exploratory experiments. Compton profiles have been measured for H and D in ACN crystals (Fig. 19.13) [68], in paracrystals

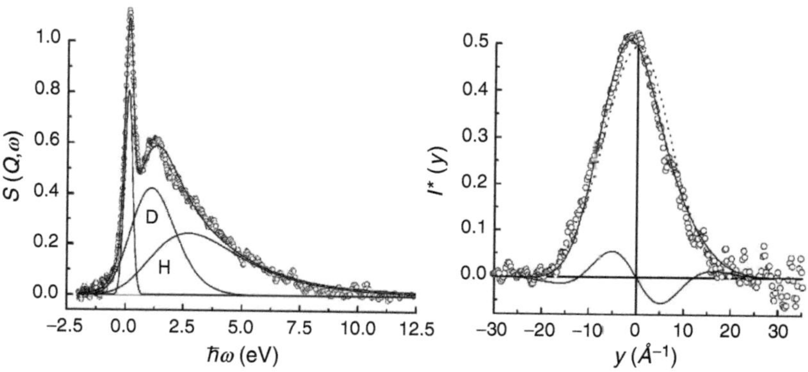

Fig. 19.13. *Left*: NCS spectrum from ACN at $2\theta = 33°$, showing decomposition into near-elastic (C,O,N), deuterium (D), and proton (H) recoil lines (from [68]). *Right*: Proton Compton profile analyzed as superposition of pure Gaussian (impulse approximation, dotted line) plus asymmetric component representing the first term of Sears expansion (from [68])

of DNA [69], and in aligned arrays of collagen fibers [70]. In a recent experiment on phospholipid bilayer stacks, the aim was to detect anisotropies in the mean kinetic energy of protonated groups due to the incorporation of a functionally important protein [71]. Although improvements in resolution and data reduction methods are needed before the full potential of NCS can be realized, it seems clear that we have a valuable new source of information here. This technique will be useful mainly for the study of highly ordered, minimally hydrated biomolecules and their building blocks. Exploiting orientation, H/D occupancies and temperature as principal variables, it should be feasible in favorable cases to unscramble well-resolved sets of neutron Compton profiles with the aim of obtaining direct information on potential well shapes and force fields.

19.7 Conclusions and Outlook

The instrumental developments and range of applications reviewed in this chapter demonstrate amply that neutron techniques using pulsed sources have an increasingly important role to play in work on biomolecular dynamics. Although at present only about one in six to seven biodynamics projects use pulsed-beam spectrometers, it is clear that this is bound to change substantially in the years following the commissioning of the new facilities under construction.

It is equally clear, however, that significant increases in data collection rates, while crucial for bringing sample sizes down to reasonable levels, are but one of three or four factors determining the overall impact of biodynamics work using neutrons. Paramount among these will be efforts to ease access and coordinate scheduling for users from the life sciences who are willing to embark on longer-term projects addressing problems for which neutron experiments can provide key data. Projects of this kind are likely to require substantial resources for producing selectively or fully covalently deuterated samples, which even for 10–20 mg quantities can be very considerable. While the value of neutron experiments in providing complementary or unique data is recognized by many biology users, the low level of access, haphazard system of beam time allocation and piecemeal data collection combine to make neutron techniques unattractive compared with NMR and optical techniques, which are widely available and seen as more user-friendly. Experiments addressing questions of more "direct" relevance to research in the life sciences have decreased markedly during the 1990s, whereas physics-dominated projects have increased steadily. This shift is largely due to experimental and simulation work focusing on efforts to understand glass-like transitions in globular proteins and fibrous biopolymers at $T < 270\,\mathrm{K}$.

The scientific programmes for the third-generation sources to be commissioned between 2006 and 2008 all contain strong commitments to prioritize applications of neutron scattering to soft-matter physics, materials science, nanotechnology, and the biosciences. If these aims will be realized, the future looks bright indeed!

Acknowledgments

I wish to thank colleagues at ISIS, ILL, PSI, KEK, and ANL for information and helpful comments on draft versions of this chapter. C. Corsaro, A. Deriu, M. Kataoka, C.K. Loong, S.F. Parker, K. Shibata, M.G. Taylor, and M.T.F. Telling kindly provided figures.

References

1. H.D. Middendorf, Annu. Rev. Biophys. Bioeng. **13** (1984), 425
2. M. Bée, *Quasielastic Neutron Scattering* (Adam Hilger, Bristol, 1988)
3. P. Martel, Progr. Biophys. Mol. Biol. **57** (1992), 129
4. H.D. Middendorf, A. Miller, in *Neutrons in Biology*, B.P. Schoenborn, R.B. Knott (Eds.), (Plenum, New York, 1996), pp. 239–265
5. S. Cusack, H. Büttner, M. Ferrand, P. Langan, P. Timmins (Eds.), *Biological Macromolecular Dynamics* (Adenine Press, NewYork, 1997)
6. H.D. Middendorf, in *Neutron Scattering from Novel Materials*, A. Furrer (Ed.) (World Scientific, Singapore, 2000), pp. 141–158
7. F. Gabel, D. Bicout, U. Lehnert, M. Tehei, M. Weik, G. Zaccai, Qu. Rev. Biophys. **35** (2002), 327
8. H.D. Middendorf, R.L. Hayward, J. Neutron Res. **10** (2002), 123
9. N. Niimura, K. Shibata, H.D. Middendorf, J. Neutron Res. **10** (2002), 163
10. H.D. Middendorf, J. Neutron Res. **13** (2005), 79
11. C.G. Windsor, *Pulsed Neutron Scattering* (Taylor & Francis, London, 1981)
12. J. Newport, B.D. Rainford, R. Cywinski (Eds.), *Neutron Scattering at a Pulsed Source* (Adam Hilger, Bristol, 1988)
13. S. Ikeda, Appl. Phys. A **74** (2002), S15
14. T.E. Mason, R.K. Crawford, G.J. Bunick, A.E. Ekkebus, D. Belanger, Appl. Phys. A **74** (2002), S11
15. A. Taylor, Physica B **276–278** (2000), 36
16. F. Mezei, Neutron News **5** (1994), 2
17. B. Farago, in *Time-of-Flight Neutron Spin Echo, in Neutron Spin Echo Spectroscopy*, F. Mezei, C. Pappas, T. Gutberlet (Eds.) (Springer, Berlin, 2003)
18. C.J. Carlile, J. Penfold, Neutron News **6** (1995), 5
19. R.E. Lechner, Appl. Phys. A **74** (2002), S151
20. R. Bewley, R. Eccleston, Appl. Phys. A **74** (2002), S218
21. H.D. Middendorf, J. Phys. V-C3 France **5** (1995), 387; Physica B **226** (1996), 113
22. M.-C. Bellissent-Funel (Ed.): *Hydration Processes in Biology* (IOS Press, Amsterdam, 1999)
23. M.T.F. Telling, S.I. Campbell, D.D. Abbley, D.A. Cragg, J.J.P. Balchin, C.J. Carlile, Appl. Phys. A **74** (2002), S61

24. K. Shibata, I. Tamura, K. Soyama, M. Arai, H.D. Middendorf, N. Niimura, High Efficiency Indirect Geometry Crystal Analyzer TOF Spectrometer: DYANA, in *Proc. ICANS-XVI* (ICANS-XVI, Düsseldorf-Neuss, 2003)
25. C. Andreani, A. Filabozzi, F. Menzinger, A. Desideri, A. Deriu, D. DiCola, Biophys. J. **68** (1995), 2519
26. U.N. Wanderlingh, M. Cutroni, L. De Francesco, in *Biological Macromolecular Dynamics*, S. Cusack , H. Büttner, M. Ferrand, P. Langan, P. Timmins (Eds.) (Adenine Press, New York, 1997), pp. 171–174
27. C. Tengroth, L. Börjesson, W.W. Kagunya, H.D. Middendorf, Physica B **266** (1999), 27
28. L. Foucat, J.-P. Renou, C. Tengroth, S. Janssen, H.D. Middendorf, Appl. Phys. A **74** (2002), S1290
29. A. Deriu, F. Cavatorta, D. Cabrini, C.J. Carlile, H.D. Middendorf, Europhys. Lett. **24** (1993), 351
30. H.D. Middendorf, D. DiCola, F. Cavatorta, A. Deriu, C.J. Carlile, Biophys. Chem. **47** (1994), 145
31. A. Faraone, C. Branca, S. Magaz, G. Maisano, H.D. Middendorf, P. Migliardo, V. Villari, Physica B **276–278** (2000), 524
32. I. Ohmine, H. Tanaka, P.G. Wolynes, J. Chem. Phys. **89** (1998), 5852
33. H.E. Rorschach, D.W. Bearden, C.F. Hazlewood, D.B. Heidorn, R.M. Nicklow, Scanning Microscopy **1** (1987) 2043.
34. A. Deriu, Neutron News **11** (2000), 26
35. H.D. Middendorf, U.N. Wanderlingh, M.T.F. Telling, OSIRIS Exp. RB14292 (2004)
36. W. Doster, S. Cusack, W. Petry, Phys. Rev. Lett. **65** (1990), 1080
37. H. Nakagawa, H. Kamikubo, I. Tsukushi, T. Kanaya, M. Kataoka, J. Phys. Soc. Jap. **73** (2004), 491
38. D. Kern, E.R. Zuiderweg, Curr. Opin. Struct. Biol. **13** (2003), 748
39. H.D. Middendorf, Physica B **182** (1992), 415
40. Y. Izumi, K. Sakai, H. Oshino, M. Kataoka, Physica B **213–214** (1995), 772
41. M. Kataoka, M. Ferrand, A.V. Goupil-Lamy, H. Kamikubo, J. Yunoki, T. Oka, J.C. Smith, Physica B **266** (1999), 20
42. M. Kataoka, H. Kamikubo, J. Yunoki, F. Tokunaga, T. Kanaya, Y. Izumi, K. Shibata, J. Phys. Chem. Solids **60** (1999), 1285
43. M. Arai, Adv. Colloid Interface Sci. **71–72** (1997), 209
44. S.F. Parker, J. Neutron Res. **10** (2002), 173
45. S. Magazù, V. Villari, P. Migliardo, G. Maisano, M.T.F. Telling, H.D. Middendorf, Physica B **301** (2001), 130
46. C. Branca, S. Magazù, G. Maisano, S.M. Bennington, B. Fåk, J. Phys. Chem. B **107** (2003), 1444
47. F. Fillaux, J.P. Fontaine, M.-H. Baron, G. Kearley, J. Tomkinson, Chem. Phys. **176** (1993), 249
48. G.J. Kearley, F. Fillaux, M.H. Baron, S.M. Bennington, J. Tomkinson, Science **264** (1994), 1285
49. R.L. Hayward, H.D. Middendorf, U. Wanderlingh, J.C. Smith, J. Chem. Phys. **105** (1995), 5525
50. J. Baudry, R.L. Hayward, H.D. Middendorf, J.C. Smith, in *Biological Macromolecular Dynamics*, S. Cusack , H. Büttner, M. Ferrand, P. Langan, P. Timmins (Eds.) (Adenine Press, New York, 1997), pp. 46–49

51. M.T.F. Telling, C. Corsaro, U.N. Wanderlingh, H.D. Middendorf, Eur. Biophys. J., (2005), submitted.
52. U.N. Wanderlingh, C. Corsaro, R.L. Hayward, M. Bée, H.D. Middendorf, Chem. Phys. **292** (2002), 445
53. G.J. Kearley, Spectrochimica Acta **48A** (1992), 349
54. J. Ulicny, M. Ghomi, H. Jobic, P. Miskovsky, A. Aamouche, J. Molec. Struct. **410** (1997), 497
55. G.J. Kearley, M.R. Johnson, M. Plazanet, E. Suard, J. Chem. Phys. **115** (2001), 2614
56. H.D. Middendorf, W. Montfrooij, S.M. Bennington, MARI Exp. RB5527 (unpubl. data).
57. M.G. Taylor, S.F. Parker, P.C.H. Mitchell, J. Molec. Struct. **651–653** (2003), 123
58. C.-K. Loong, C. Rey, L.T. Kuhn, C. Combes, Y. Wu, S.-H. Chen, M.J. Glimcher, Bone **26** (2000), 599
59. H.D. Middendorf, R.L. Hayward, S.F. Parker, J. Bradshaw, A. Miller, Biophys. J. **69** (1995), 660
60. F. Cavatorta, A. Deriu, N. Angelini, G. Albanese, Appl. Phys. A **74** (2002), S504
61. H.D. Middendorf, S.F. Parker, TFXA Exp. RB9726 (unpubl. data)
62. Z. Dhaouadi, et al., J. Phys. Chem. **97** (1993), 1074
63. M. Ghomi, A. Aamouche, H. Jobic, C. Coulombeau, O. Bouloussa, in *Biological Macromolecular Dynamics*, S. Cusack , H. Büttner, M. Ferrand, P. Langan, P. Timmins (Eds.) (Adenine, New York, 1997), pp. 73–78
64. M.-P. Gaigeot, N. Leulliot, M. Ghomi, H. Jobic, C. Coulombeau, O. Bouloussa, Chem. Phys. **261** (2000), 217
65. M. Plazanet, N. Fukushima, M.R. Johnson, Chem. Phys. **280** (2002), 53
66. J. Mayers, Phys. Rev. B **41** (1991), 41; Phys. Rev. Lett. **71** (1993), 1553
67. R. Senesi, et al, Physica B **276** (2000), 200
68. U.N. Wanderlingh, A.L. Fielding, H.D. Middendorf, Physica B **241–243** (1998), 1169
69. H.D. Middendorf, R.L. Hayward, U.N.Wanderlingh, Nuovo Cimento **20 D** (1998), 2215
70. U.N. Wanderlingh, F. Albergamo, R.L. Hayward, H.D. Middendorf, Appl. Physics A **74** (2002), S1283
71. A. Deriu et al, VESUVIO Exp. RB14697 (unpubl. data)

20

Brownian Oscillator Analysis of Molecular Motions in Biomolecules

W. Doster

20.1 Introduction

Dynamic analysis of biomolecules often works by the principle of difference spectroscopy: What is the qualitative difference in structural flexibility of a protein with and without ligand? This method, illustrated elsewhere in this book, is quite useful considering the complexity of biomolecules. Sometimes, however, differences between different samples are easier to obtain than reproducable identical results. This chapter is addressed to students of biophysics, who would like to proceed further. We present a modern statistical analysis of neutron scattering data applied to biomolecules. We start from the simple model of the harmonic oscillator, introduce the visco-elastic oscillator and conclude with a model-independent moment expansion of the density correlation function. To illustrate the method, a number of recent results on protein dynamics are presented. The power of neutron scattering is that it provides, both spectral and spatial, information from which one can reconstruct in principle the microscopic trajectory of labeled particles on a picosecond time scale. Such results can be used to test molecular dynamic simulations of biomolecules, and simulations can be used to interpret the neutron scattering spectra. Since protein–water interactions belong to the most interesting questions that can be approached with neutron scattering, we start with a brief outline on this topic.

20.2 Dynamics of Protein–Solvent Interactions

The nature of protein–solvent interactions is central to most basic questions in molecular biophysics ranging from protein folding, protein–ligand association to desease-related formation of protein aggregates. Biological structures owe their existence to a delicate balance of weak hydrophilic and hydrophobic forces, which are mediated by the solvent [1]. Moreover, proteins are dynamical structures, which undergo continuous thermal motion induced by the solvent. Dynamic neutron scattering experiments allow probing the protein–water fluctuations on the relevant time scales. In the absence of solvent or in a rigid environment functionally relevant motions are arrested. Water thus acts as a lubricant to protein dynamics. Figure 20.1 shows a neutron scattering spectrum of myoglobin, exposed to three types of environment: vacuum (dehydrated), fully hydrated with D_2O ($0.4\,\mathrm{g\,g^{-1}}$) and vitrified in a perdeuterated glucose glass [2].

The wings of the protein spectra appear to be broadened with respect to the resolution function, which is the signature of structural fluctuations on a picosecond time scale. It is obvious that the spectral broadening is more pronounced with the hydrated sample. The excess broadening derives from water-plasticized translational motions of side chains. That the spectra of the dehydrated and the glucose-vitrified protein display a finite and similar width, indicates rotational transitions of side chains, which persist irrespective of the protein environment. From a dynamical point of view, liquids and proteins are radically different, liquids exhibit short-range order and long-range translational diffusion. Molecular displacements in liquids are continuous and isotropic. Proteins in contrast are long-range ordered, but molecular diffusion is short-ranged. Internal displacements are discontinuous, rotational, and anisotropic. The protein–water interaction introduces liquid aspects to otherwise solid-like molecules. Molecular displacements in dense liquids are dominated by short-range repulsive interactions. For a molecule to move also requires that the nearest neighbors have to move. This is a collective phenomenon resembling more a continuous search for escape out of a cage rather than a discontinuous jump across an energetic barrier. By-passing the barrier by collecting sufficient free volume instead of barrier crossing appears to be the dominant diffusion mechanism in the liquid state [3]. The protein–water

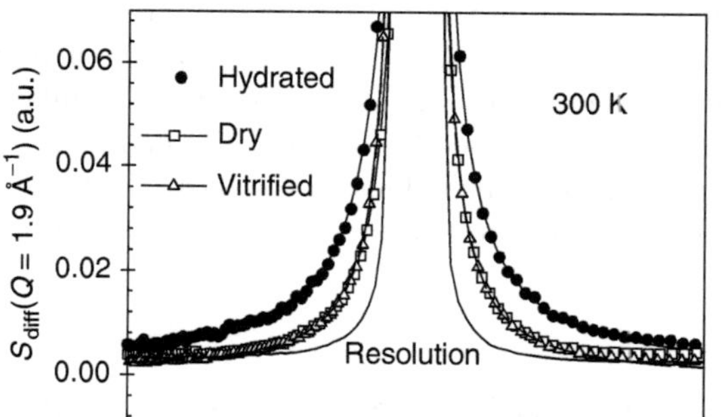

Fig. 20.1. Neutron scattering spectra of diffusive motions at $Q = 1.9\,Å^{-1}$ of myoglobin embedded in three environments as indicated ($1\,\mathrm{meV} = 8\,\mathrm{cm^{-1}}$). *Full line*: instrumental resolution function (IN6, ILL). A vibrational background was subtracted

interaction causes the protein torsional barriers to fluctuate. The analysis of these motions by dynamic neutron scattering is a classical topic, which has been discussed by many authors, just to cite a small sample [4–6]. The relation between liquids and proteins was discussed in [7–11]. A recent review was published in [12]. The neutron scattering method has the particular advantage to probe the very low frequency range of a few terahertz, where vibrational and relaxational motions overlap. The corresponding spectral features cannot be assigned to vibrations of a particular group. Instead it is dominated by collective motions of many particles. To illustrate the application of neutron scattering to protein–water dynamics we follow a simple physical concept: Protein structural fluctuations are spatially constrained by covalent and van der Waals forces. As a general dynamic model of protein variables. we consider a set of harmonic oscillators which are driven by the random forces of a heat bath, which is essentially the solvent. The generalized protein coordinates X_α then obey a Langevin equation driven by the random forces R_α

$$M_\alpha \ddot{X}_\alpha + f_\alpha \dot{X}_\alpha + \omega_\alpha^2 X_\alpha = R_\alpha(t). \tag{20.1}$$

We thus pick up the basic idea of a normal mode analysis of proteins, complemented by an appropriate frictional force [13,14]. Neutron scattering provides the tools to study the frictional force. In a dynamic neutron scattering experiment, the protein–water hydrogen atoms serve as a monitor to record the trajectory of the X_α in space and time. The theoretical aspects of the application of the Brownian oscillator model to neutron scattering has been discussed by Kneller [15].

20.3 Properties of the Intermediate Scattering Function

An insightful article on the neutron scattering process was written by Mezei [16]. Neutrons are scattered by the nuclei of the atoms, which are point-like entities. The scattered beam pattern is thus determined by the superposition (interference) of spherical waves emitted by the individual atoms. The respective scattering amplitudes depend on the individual nuclear cross-sections, which for C, N, O, H amount to σ_c = 5.5, 11.5, 4.2, and 1.76 b ($1\,\mathrm{b} = 10^{-24}\,\mathrm{cm}^2$), respectively [17]. The coherent cross-sections, which specify phase-preserving processes, contribute generally less than 10% to the total scattering intensity with protein samples. Three types of disorder generate an incoherent background (a) chemical disorder, neutron waves scattered by different types of atoms (N, C, H) or isotopes do not interfere; (b) positional disorder, protein powder samples or protein solutions are rotationally disordered, thus waves scattered by identical atoms in different proteins exhibit a random phase relationship; and (c) spin disorder, the neutron cross-section of hydrogen fluctuates depending on whether the respective neutron–proton spins are parallel or antiparallel. The spin-disorder in combination with the negative scattering length of the proton, lead to a large incoherent cross-section, $\sigma_{\text{inc}} = 80.2\,\text{b}$ [17]. Therefore, roughly 90% of the combined scattering cross-section of D_2O-hydrated protein samples is incoherent. The scattering function $S_{\text{inc}}(Q, \omega)$ thus should be interpreted as the sum of intensities and not as a square of the sum of amplitudes. Finally only those waves can interfere, which originate from one and the same hydrogen atom. This is called "self-interference" and refers to the average behavior of single particles: The motion of the hydrogen atom modulates the phase of the scattered wave (Doppler shift). The self- scattering trace in time thus reflects the trajectory of individual particles. The coherent fraction corresponds to relative displacements of two distinct atoms. Single particle motions are generally easier to interpret than the relative motion of two distinct particles. One can thus derive most of the relevant dynamical information from incoherent scattering, which has the further advantage to be much more intense. Dynamic analysis of coherent scattering is important with perdeuterated proteins or solutions with D_2O. The neutron scattering experiment determines the statistical average of the phase factors at different times.

This is the self-intermediate scattering function $I_{\text{s},i}(\boldsymbol{Q}, t)$ defined for each atom (i) by

$$I_{\text{s},i}(\boldsymbol{Q}, t) = \langle \exp(\mathrm{i}\boldsymbol{Q}\boldsymbol{r}_i(0)) \cdot \exp(-\mathrm{i}\boldsymbol{Q}\boldsymbol{r}_i(t)) \rangle \,. \tag{20.2}$$

We omit the cross-sections, since we consider a system of hydrogen atoms, dominating the scattering intensity. The scattering vector $\boldsymbol{Q}$ is an adjustable parameter, which allows to modify the spatial scale probed by the scattering process. $Q = 4\pi/\lambda_n \sin(\theta/2)$ defines its length, λ_n denotes the wavelength of the incident neutrons, and θ is the scattering angle. $I(\boldsymbol{Q}, t)$ can be expressed as the Fourier transform of the van Hove self-correlation function in space, $G_{\text{s},i}(\boldsymbol{r}, t)$

$$G_{\text{s},i}(\boldsymbol{r}, t) = \int \frac{\mathrm{d}^3 Q}{(2\pi)^3} \exp(-\mathrm{i}\boldsymbol{Q}\boldsymbol{r}) \cdot I_{\text{s},i}(\boldsymbol{Q}, t) \,. \tag{20.3}$$

This even function in space (and time) [17] describes the single particle dynamics of a system averaged over the possible starting points in space. It denotes the probability density, that atom (i) which is initially at $\boldsymbol{r}_0$ moves to a position $\boldsymbol{r}$ within a time interval t. For a classical system it can be written as

$$G_{\text{s},i}(\boldsymbol{r}, t) = \int \mathrm{d}^3 r_0 \, p(\boldsymbol{r}_0 + \boldsymbol{r}, \boldsymbol{r}_0, t) \cdot p_0(\boldsymbol{r}_0) \,, \tag{20.4}$$

with the equilibrium distribution

$$p_0(\boldsymbol{r}) = p(\boldsymbol{r}, \boldsymbol{r}_0, t = \infty) \,. \tag{20.5}$$

Thus the Q-dependence of $I_{\text{s},i}(\boldsymbol{Q},\ t_{\text{res}})$ contains the complete information about the single particle dynamics at any fixed instant of time $t = t_{\text{res}}$ (often defined by the energy resolution of the instrument). This argument is stressed because the intermediate scattering function is usually introduced as a correlation function versus time or spectrum versus frequency.

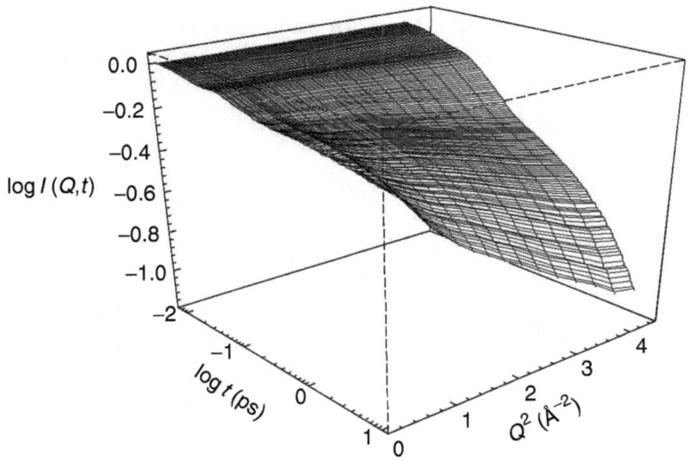

Fig. 20.2. Intermediate scattering function, $I(Q,t)$ of hydrated myoglobin at 300 K, derived by Fourier-deconvolution of the dynamical structure factor $S(Q,\omega)$ (IN6, ILL)

Figure 20.2 shows the experimental Q–t profile of the intermediate scattering function of D_2O-hydrated myoglobin averaged over all atoms (i). The oscillations at short times reflect damped low frequency proteins modes (Boson peak). Two processes with correlation times at 0.3 and 5 ps are visible at high Q correspond to vibrational dephasing and fast water-coupled motions. The variation with Q contains the geometry of the displacements. The decay at $Q = 0$ is due to second-scattering processes, generating a Q-independent quasielastic background (see below).

The interpretation of experimental data becomes most transparent in the case of small Q and/or short times, as can be seen from the expansion of $I_{s,i}(\boldsymbol{Q},t)$ in powers of Q^2

$$I_{s,i}(\boldsymbol{Q},t) = \int \mathrm{d}^3r \, \exp(-\mathrm{i}\boldsymbol{Qr}) \cdot G_s(\boldsymbol{r},t) \tag{20.6}$$

$$= 1 - \frac{1}{2} \cdot Q^2 \cdot \langle(\hat{Q}\boldsymbol{r})^2\rangle(t) + \frac{1}{24} \cdot Q^4 \cdot \langle(\hat{Q}\boldsymbol{r})^4\rangle(t) - O(Q^6) \tag{20.7}$$

with $\boldsymbol{Q} = Q \cdot \hat{Q}$. This is the moment expansion of the displacement distribution function $G(r,t)$. In the limit

$$Q^2 \cdot \langle(\hat{Q}\boldsymbol{r})^2\rangle(t) \ll 1, \tag{20.8}$$

only the first term of the expansion contributes and two essential consequences arise:

– The correlation function and spectrum are completely determined by the mean squared displacement.

– The intermediate scattering function and dynamic structure factor factorize into a term $\propto Q^2$ and a purely time- or frequency-dependent function, respectively,[1]

$$I(\boldsymbol{Q},t) = 1 - \frac{1}{6} \cdot Q^2 \cdot \langle r^2(t) \rangle, \tag{20.10}$$

$$S(\boldsymbol{Q},\omega) = \delta(\omega) + \frac{1}{6} Q^2 \cdot \mathrm{FT}\left\{-\langle r^2(t)\rangle\right\}(\omega). \tag{20.11}$$

In the following we consider only the self-scattering functions averaged over all atoms. $S(\boldsymbol{Q},\omega)$ denotes the so-called dynamical structure factor, which is the quantity determined by most spectrometers (time-of-flight and backscattering). It is obtained from constant angle cuts at fixed frequency and involves interpolation (see below). The intermediate scattering function is then derived by numerically transforming $S(Q,\omega)$ to the time domain

$$I(Q,t) = \int S(Q,\omega) \exp(\mathrm{i}\omega t) \mathrm{d}(\hbar\omega). \tag{20.12}$$

If the inequality of Eq. 20.8 holds up to a certain time $t_{\max}$, then the Fourier transform is only valid in the set of discrete points

$$\omega_n = n \cdot \frac{\pi}{t_{\max}} \quad \text{for } n \geq 1\,. \tag{20.13}$$

Note also that the measured linewidth of a localized process with time constant τ_{loc} less than $t_{\max}$, i.e.,

$$\langle r^2(t_{\max})\rangle_{\mathrm{loc}} \approx \langle r^2(\infty)\rangle_{\mathrm{loc}} \tag{20.14}$$

is independent of Q and is given by its actual value $\Gamma_{\mathrm{loc}} = 1/\tau_{\mathrm{loc}}$. In addition the squared amplitude of such a process is directly given by the integral over the corresponding quasielastic spectrum. A localized motion or glassy state leads to a long-time plateau of the intermediate scattering function, $\mathrm{EISF}(Q) = I(Q, t \to \infty)$, the so-called elastic incoherent structure factor. Then a purely elastic component (δ-function) arises in $S(Q,\omega)$ at $\omega = 0$. The $\mathrm{EISF}(Q)$ is the Fourier transform of the displacement distribution $G(r, t \to \infty)$. A finite elastic fraction also obtains, if the correlations do not vanish within the time defined by the energy resolution of the spectrometer. With Eqs. 20.4 and 20.5 one obtains the useful relation

$$\mathrm{EISF}(\boldsymbol{Q}) = \left| \int \mathrm{d}\boldsymbol{r} \cdot \exp(\mathrm{i}\boldsymbol{Q}\boldsymbol{r}) \cdot p(\boldsymbol{r} - \boldsymbol{r}_0) \right|^2 . \tag{20.15}$$

The elastic fraction, $\mathrm{EISF}(Q)$, thus represents the orientationally averaged (single particle) displacement distribution at infinite time.

[1]Let us – for notational simplicity – assume the usual case of an isotropic sample (not necessarily isotropic dynamics!, see below) which leads to the orientational average of the scalar products in the displacement moments

$$\frac{1}{4\pi} \int \mathrm{d}\Omega \, \langle (\hat{Q}\boldsymbol{r})^{2n} \rangle(t) =: \frac{1}{2n+1} \cdot \langle r^{2n}(t) \rangle. \tag{20.9}$$

20.4 Relevant Time and Spatial Scales

What are the relevant spatial and temporal scales of molecular motions? The shortest time scale is given by the ballistic flight: a particle with mass m is moving with the thermal velocity $v_{\rm th} = \sqrt{k_{\rm B}T/m}$ across a distance δ, which requires a time $\tau_{\rm mic}$:

$$\tau_{\rm mic}^2 = (\delta/v_{\rm th})^2 = m/(k_{\rm B}T \cdot Q^2). \tag{20.16}$$

Here the length of the wavevector Q is an experimental parameter, which defines a length scale $Q = 1/\delta$ across which the trajectory of the particle is observed. Ballistic flight is a basic feature of motion in gases, but applies also to molecular motions in liquids and proteins over short times. The corresponding intermediate scattering function is Gaussian [18]

$$I(Q,t) = \exp[-(t/\tau_{\rm mic})^2] = \exp[-(k_{\rm B}T/m) \cdot Q^2 t^2]. \tag{20.17}$$

However, liquids and solids unlike gases exhibit a characteristic length scale, the interparticle distance δ. Thus also a characteristic time scale exists

$$\tau_{\rm mic}^2 = \delta^2 m/(k_{\rm B}T), \tag{20.18}$$

which is independent of the probing length scale $1/Q$ as in Eq. 20.15. $\tau_{\rm mic}$ is typically in the range of 0.5 ps for proteins depending on the mass of the molecular fragments. For times larger than $\tau_{\rm mic}$ collective interparticle correlation becomes important, the correlation function deviates from a single exponential decay and the correlation time depends on Q. This regime of complex many-particle interactions is crucial to water-assisted protein flexibility. At long times and large distances in isotropic liquids one enters the hydrodynamic (Gaussian) limit: The correlation time depends on the chosen scale: $\tau^{-1} = D \cdot Q^2$, where D is the diffusion constant of the particle. The corresponding intermediate scattering function is single exponential in time, the quadratic Q-dependence reflects the Gaussian distribution of displacements

$$I(Q,t) = \exp[-Q^2 \cdot D \cdot t]. \tag{20.19}$$

Protein internal coordinates are constrained by covalent and weak forces stabilizing the native state. Long-range diffusion of side-chains is thus prohibited. In Sect. 20.5 we investigate a model of Brownian motion constrained by a harmonic potential.

20.5 The Brownian Oscillator as a Model of Protein-Residue Motion

The harmonic oscillator, driven by random forces of the solvent, has been frequently used as model of protein-residue motion. It is the basic concept underlying the normal mode analysis of proteins [13, 14]. Mössbauer resonance spectra recording the motion of the heme group were analyzed using a "rugged" Brownian oscillator (BO) model [19]. Surprisingly the BO is not very popular among neutron experimentalists. Instead it is often assumed that protein residues perform free diffusion inside a rigid sphere [20]. The oscillator model does not involve specific assumptions (which are hard to test) and provides a more general perspective. It allows us to pick up the discussion on time and length scales of chap. 19. An example of a Brownian mode analysis was discussed for phycocyanin by Hinsen et al. [21]. Neutron scattering probes the hydrogen atoms attached to the side chains and the main chain. We assume that the amino acid positional fluctuations follow approximately Eq. 20.1 and start with a one-dimensional model: $V(x) = Kx^2$. The harmonic potential implies for the deviations a Gaussian probability distribution. The mean square displacements evolve at high damping monotonically toward an equilibrium value, the average thermal amplitude δ^2 [21, 22]

$$\langle \Delta x^2 \rangle = \delta^2 \cdot (1 - \exp[-2\Gamma_0 \cdot t]). \tag{20.20}$$

The thermal amplitude amounts to $\delta^2 = k_{\mathrm{B}}T/K$. The relaxation rate is given by $\Gamma_0 = K/f$ and $D = k_{\mathrm{B}}T/f$ may be interpreted as a diffusion coefficient. Inserting Eq. 20.20 into Eq. 20.10, yields the following intermediate scattering function of the overdamped BO

$$I_{\mathrm{B}}(Q,t) = \exp[-Q^2\delta^2 \cdot (1 - \mathrm{e}^{-\Gamma_0 t})]. \tag{20.21}$$

The correlation function, $I_{\mathrm{B}}(Q,t)$, exhibits several interesting features shown in Fig. 20.3. It is nonexponential in time, its effective relaxation time depends on Q and it decays toward a finite plateau at long times the so-called elastic incoherent structure factor, EISF(Q).

$$\mathrm{EISF}(Q) = \exp[-Q^2 \cdot \delta^2] \tag{20.22}$$

Figure 20.3 shows an experimental intermediate scattering function derived from D_2O-hydrated myoglobin [7, 8, 23] covering a wide Q-range. This experiment thus reflects essentially protein structural fluctuations. The harmonic oscillator model fits the data below $Q = 2\,\mathring{A}^{-1}$. But the observed long-time plateau is much higher than predicted by the model at large Q. The spatial constraints of the protein-displacements are thus more severe than those of an isotropic harmonic potential.

The dashed line in Fig. 20.3 was obtained by assuming a two-dimensional harmonic oscillator (see below). The EISF(Q) decreases with increasing spatial resolution (Q) at fixed relaxation rate Γ_0. Since $I_{\mathrm{E}}(Q,t)$ is not a single exponential one obtains a Q-dependent average relaxation rate according to

$$\frac{1}{\Gamma} = \int_0^\infty \mathrm{d}t(I_{\mathrm{B}}(Q,t) - I_{\mathrm{B}}(Q,\infty))/I_{\mathrm{B}}(Q,\infty). \tag{20.23}$$

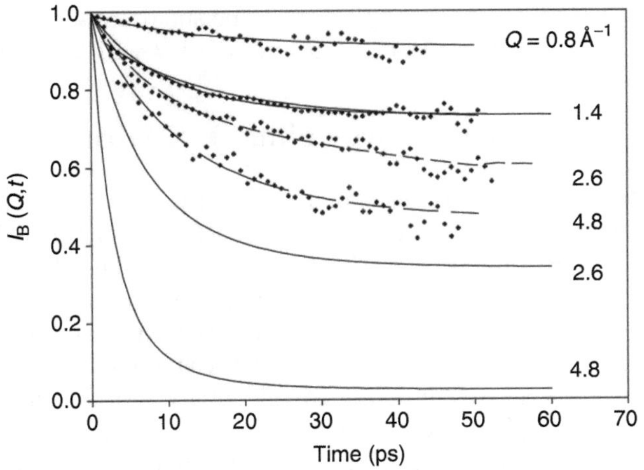

Fig. 20.3. Intermediate scattering function of the BO versus Q (*full line*), Eq. 20.21, and experimental data of D_2O-hydrated myoglobin ($0.35\,g\,g^{-1}$) based on backscattering data (IN13, ILL). Fits (*dashed line*): $\delta = 0.15\,Å^2$, $\Gamma_0^{-1} = \tau_0 = 10\,ps$

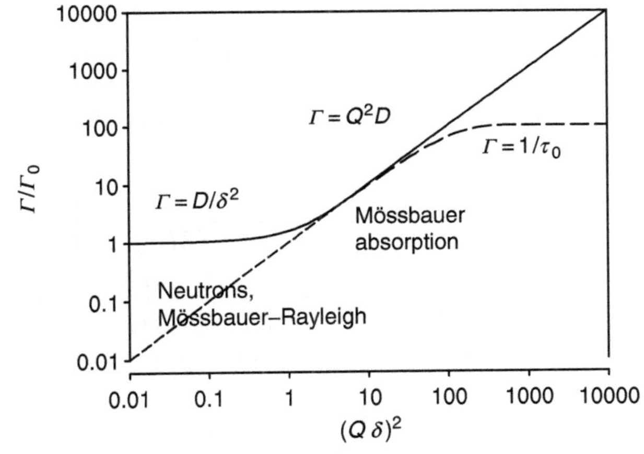

Fig. 20.4. Average relaxation time of the damped harmonic oscillator versus $Q\delta$ Eq. 20.24

Figure 20.4 shows the resulting average rate versus Q. At low Q, the relaxation rate is independent of the spatial scale, the oscillator has explored its entire phase space within a time $\tau = \delta^2/D$. At intermediate Q the displacements are diffusive, $\Gamma \propto Q^2$, while at high Q a limiting rate is achieved reflecting the finite time required for discrete molecular steps. Also indicated in the figure is the experimental range achievable by neutron scattering, Mössbauer absorption spectroscopy and Rayleigh scattering assuming $\delta^2 = 0.2\ Å^2$. So with Mössbauer spectroscopy at $Q = 7\,\text{Å}^{-1}$ one should observe a much larger rate than for the same oscillator motion at $Q = 2\,\text{Å}^{-1}$ with neutron scattering.

20.6 The Visco-Elastic Brownian Oscillator

In our simple model we consider conformational fluctuations coupled to fast motions of the solvent, which acts as a heat bath. The BO model of Eq. 20.1 assumes a clear cut separation of time scales between heat bath coordinates and Brownian motion. This allows treating the random force as a δ-correlated Gaussian process (white noise) and the friction coefficient to be time independent. Protein structural fluctuations observed with neutron scattering occur, however, on time scales comparable to solvent dynamics. Thus one has to take into account explicitly the spectrum of random forces, which leads to a time-dependent friction according to the fluctuation-dissipation theorem. The Mori–Zwanzig theory provides an algorithm for the equation of motion, which yields for the density correlator $\Phi_Q(t)$ a generalized oscillator equation containing a time dependent friction kernel $m(t)$, reflecting the slow force correlations [3, 24, 25]. Ω and γ_0 denote a generalized frequency and a regular damping coefficient:

$$\ddot{\Phi}_Q(t) + \Omega^2 \Phi_Q(t) + \gamma_0 \dot{\Phi}_Q(t) + \Omega^2 \int_0^t \mathrm{d}t' m(t-t') \dot{\Phi}_Q(t') = 0\,. \tag{20.24}$$

A general solution to Eq. 20.24 can be given in the frequency domain for the spectrum $S(Q, \omega)$ [3, 18]

$$S(Q,\omega) = -\mathrm{Im}\left\{ \frac{\omega + \Omega^2 M(Q,\omega)}{\omega^2 - \Omega^2 + \omega \Omega^2 M(Q,\omega)} \right\}. \tag{20.25}$$

"Im" denotes "imaginary part of" and $\mathrm{i} = \sqrt{-1}$. $M(Q, \omega)$ represents a generalized friction kernel, which can be decomposed into a Newtonian friction γ_0 (collisions) and a slow relaxing part $m_{\mathrm{ps}}(Q, \omega)$ due to protein solvent coupling [25]

$$M(Q,\omega) = \mathrm{i}\gamma_0 + m_{\mathrm{ps}}(Q,\omega). \tag{20.26}$$

To illustrate the basic features of visco-elasticity, we introduce Maxwell's relaxation model. The spectrum of force correlations is Lorentzian in shape, corresponding to an exponential decay of the force correlations [25, 26]

$$m_{\mathrm{ps}}(\tau,\omega) = -F(Q)/(\omega + \mathrm{i}/\tau). \tag{20.27}$$

F is the amplitude and τ denotes the characteristic time of the relaxing friction, which is proportional to the viscosity $\eta = G_\infty \cdot \tau$. G_∞ is the high frequency shear modulus of the liquid. The real part of $\mathrm{m_{ps}}(\omega\tau)$ contributes to the elastic modulus, while the imaginary part yields a frequency-dependent friction coefficient, $f(\tau, \omega) \propto m''_{\mathrm{ps}}(\tau, \omega)$:

$$\omega m''_{\mathrm{ps}}(\tau,\omega) = F(Q) \cdot \frac{\omega\tau}{1 + \omega^2\tau^2}. \tag{20.28}$$

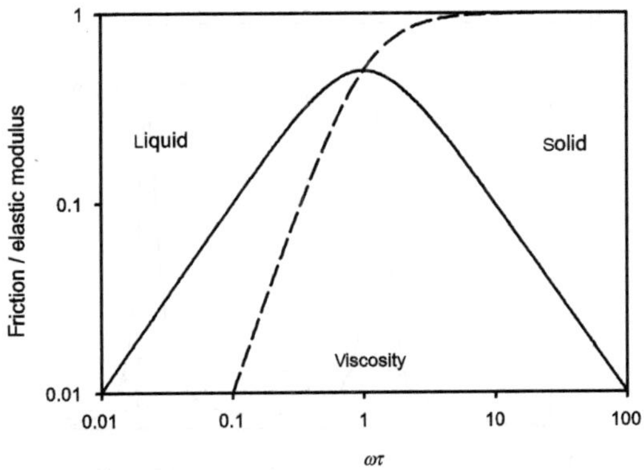

Fig. 20.5. Real (*dashed*) and imaginary part of $\omega m_{\mathrm{ps}}(\omega)$ according to Eq. 20.28

The dissipation rate thus assumes a maximum at $\omega\tau = 1$. The friction coefficient vanishes in both limits, $f(\tau \to \infty) \propto \tau^{-1}$ and $f(\tau \to 0) \propto \tau$ at fixed ω as shown in Fig. 20.5. The visco-elastic system thus turns into an elastic solid if the correlation time and thus the viscosity of the heat bath diverges. The same arguments apply to a variation of the frequency at fixed τ. For the simple BO increasing friction will always enhance the viscous properties. As shown in Fig. 20.6a, increasing the damping constant γ leads to a downshift and broadening of the resonance maxima. In the overdamped regime only a narrow central line will remain. In the visco-elastic case, the friction coefficient declines with increasing frequency. Thus even at high viscosity or large τ, an oscillation will persist since $\tau \gg \omega_0^{-1}$.

Experimental neutron scattering spectra behave very much like those in Fig. 20.6b. The vibrational feature near 3 meV has been termed the "boson peak," and involves low frequency oscillations of the protein structure. It is most prominent for large τ (low temperature and high viscosity) and becomes "overdamped" at high temperatures, when the viscoelastic relaxation times are comparable to ω_0^{-1}. A visco-elastic analysis was performed with neutron scattering data of myoglobin [8, 26]. The low-temperature spectrum (150 K) was adjusted to Eq. 20.25, assuming two oscillators as shown in Fig. 20.7. Then by decreasing only the visco-elastic relaxation time τ, the spectrum at 300 K could be simulated. Note that the maximum of Ω_2 is nearly independent of τ in contrast to Ω_1. Note also the difference in position of the boson peak at low temperature between dry and hydrated myoglobin [27].

In this particular experiment we were interested whether the energy of photons absorbed by the heme group of myoglobin was channeled into low frequency vibrations near 30 cm^{-1}. Following a laser pulse at 532 nm, we found a simultaneous emission in the far infrared below 50 cm^{-1}, suggesting exactly this [26]. The self-friction kernel $\mathrm{m}_{\mathrm{ss}}(Q,\omega)$ of simple liquids has been calculated on a molecular basis by mode-coupling theory (MCT) [3]. MCT predicts a self-induced structural arrest of the liquid, when the density, depending on

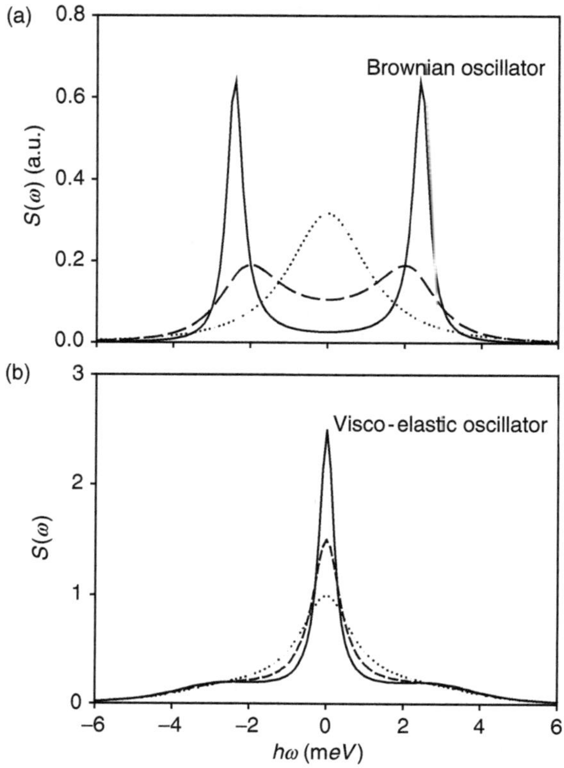

Fig. 20.6. (**a**) Spectrum of a BO: underdamped regime (*full line*), intermediate (*dashed*) and the overdamped case (*dotted*), parameters in meV: $\omega_0 = 2.4, \gamma = 0.5$, 3.0, 6.0. (**b**) spectrum of visco-elastic oscillator (Eq. 20.24), the relaxation time τ increases from *dotted*, *dashed* to *full line*, $F = 1.7$, $\gamma = 0.5$, $1/\tau = 0.5$, 1.0, 2.0 meV. The broadening at $\omega = 0$ is sometimes called the Mountain line

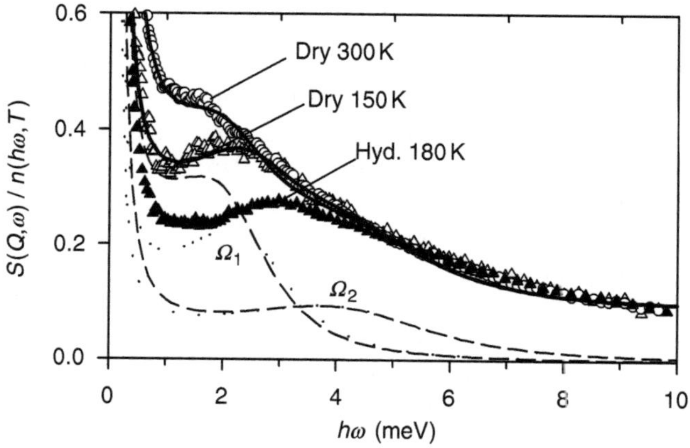

Fig. 20.7. Spectrum of dry and hydrated myoglobin and visco-elastic analysis assuming two oscillators, $\omega_0 = \Omega_1, \Omega_2$ [26]

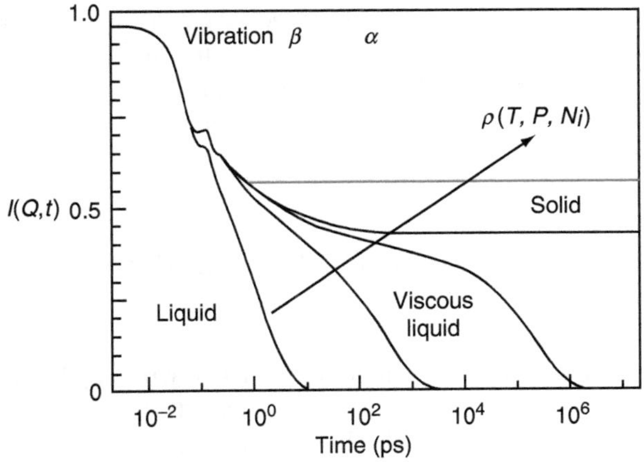

Fig. 20.8. Schematic intermediate scattering function of the generalized MCT-oscillator showing a two-step decay due to fast local (β-process) and slow collective motions (α-process). The oscillatory feature at short times generates the boson-peak in the frequency domain

temperature, pressure, and number concentration, exceeds a critical value. The signature of the glass-transition are nonvanishing density correlations and thus a finite value of the long time intermediate scattering function as shown in Fig. 20.8. MCT approximates the so-called cage effect, each particle is surrounded by a cage of nearest neighbors. Escape out of the cage (the α-process) constitutes the first step leading to long-range diffusion, which is the essence of the liquid state.

Fast local motions (β-processes) can also occur in the glass, but their amplitude (not their rate!) decreases when the density increases. In water, β-processes include hydrogen bond fluctuations and reorientational motions while the α-process involves translation. The glass transition is accompanied by a diverging α-relaxation time, the cage becomes a trap. This reasoning also applies to protein hydration water, which forms a glass at low temperatures [28]. As mentioned in Sect. 20.1, protein residues in a native structure are highly constrained and cannot perform long-range translational displacements. A native protein is not in a liquid state and thus cannot exhibit a liquid to glass transition. However, the protein residues are frictionally coupled to hydration water, which performs a self-induced glass transition at low temperatures [28]. Consequently, water-coupled protein motions will also be arrested because the plasticizer is arrested. Vitrification can also be achieved by cosolvents which seems to be equivalent to dehydration, as Fig. 20.1 shows. Sticking to the generalized oscillator concept, one may introduce the dynamic protein–solvent interactions at the level of the friction kernels

$$m_{\mathrm{ps}}(\omega) \approx m_{\mathrm{p}}(\omega) + m_{\mathrm{ss}}(\omega). \tag{20.29}$$

The internal friction due to water-decoupled structural motions is represented by m_p. With this kernel, water-coupled structural fluctuations will freeze in parallel with the hydration water and behave liquid-like at high temperature. Proceeding in this direction, we have to study the properties of hydration water, which is a true liquid.

20.7 Moment Analysis of Hydration Water Displacements

In the following we analyze experiments performed with H_2O-dehydrated myoglobin in the range of $0.35\,\mathrm{g\,g^{-1}}$ degree of hydration. At this level, most of the water is in close contact with the protein surface. The hydration water does not freeze forming ice at low temperatures, instead it vitrifies forming an amorphous structure [28]. Figure 20.9 shows the intermediate scattering function $I(Q,t)$ of water in the hydration shell of H_2O-hydrated myoglobin at various Q-values. It is obtained by Fourier-transforming the spectral function, $S(Q,\omega)$. The amplitude of the fast component at 0.3 ps increases with Q indicating that this component is highly localized. It reflects damped translational oscillations of water molecules. The rate constant of the second process increases with Q (spatial resolution) [29].

This is a characteristic feature of translational diffusion. One expects an exponential correlation function with a characteristic rate $1/\tau = Q^2 \cdot D$ with diffusion constant D. Instead we observe a stretched-exponential decay where

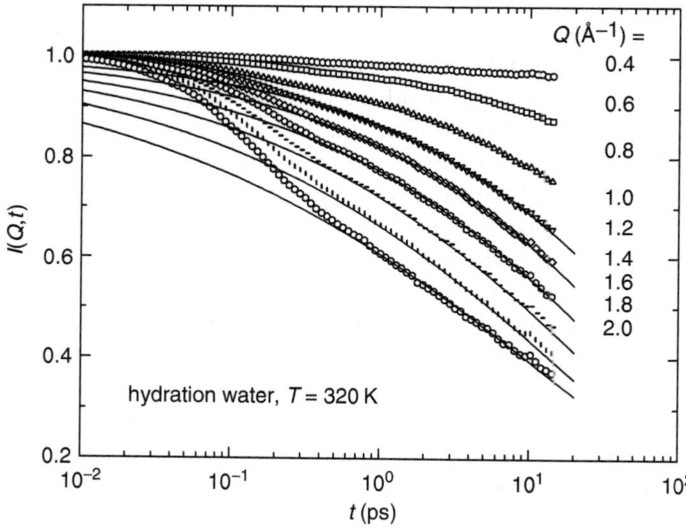

Fig. 20.9. Intermediate scattering function of myoglobin hydration water at $0.35\,\mathrm{g}\,H_2O$ per g protein and fits to a stretched exponential function (IN6, ILL) [29]

the diffusion coefficient is Q-dependent [29]

$$I(Q,t) = \exp[-Q^2 \cdot D(Q) \cdot t]^\beta = \exp[-Q^2 \cdot \langle r^2(Q,t)\rangle/6]. \qquad (20.30)$$

The stretched behavior, ($\beta < 1$) results from partially localized water molecules at the protein surface. The second equality in Eq. 20.30 refers to a more general property of $I(Q,t)$, which may be expanded in terms of the second and higher moments of the displacement distribution function according to Eq. 20.6

$$I(Q,t) = 1 - A_0(t) - \frac{1}{6}Q^2 \left\langle \boldsymbol{r}^2(t)\right\rangle + \frac{1}{24\cdot 5}\cdot Q^4 \left\langle \boldsymbol{r}^4(t)\right\rangle - \dots . \qquad (20.31)$$

Adjusting the data to a fourth-order polynom yields the second and the fourth moment of the displacement distribution function $G(r,t)$. In practice one has to account for multiple scattering corrections $A_0(t)$ [29,30]. Figure 20.10 shows the mean square displacements of protein interfacial water in comparison with bulk water [29]. The data on bulk water by Brockhouse and collaborators provided the first information on fast motions in a liquid using neutron spectroscopy [31]. After the initial rise due to vibrational dephasing, the displacements of bulk water reach the limiting diffusion region ($\langle \Delta x^2\rangle = 2D\cdot t$), within 10 ps. The slope of the dashed line corresponds to the long time diffusion coefficient of water, $D = 2.45 \cdot 10^{-5}\,\mathrm{cm^2\,s^{-1}}$.

Note the sublinear regime near 1 ps, which reflects the motion of water molecules inside the cage formed by its nearest neighbors. For interfacial water the sublinear range is drastically extended, it takes about 100 ps to reach the

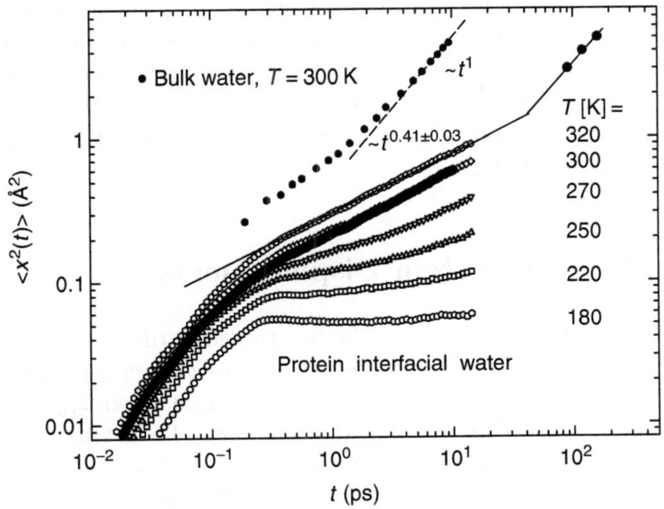

Fig. 20.10. Time-resolved displacements of bulk water and myoglobin interfacial water at 0.4 g D_2O per g protein versus temperature (IN6, ILL; *full circles*: IN15, ILL)

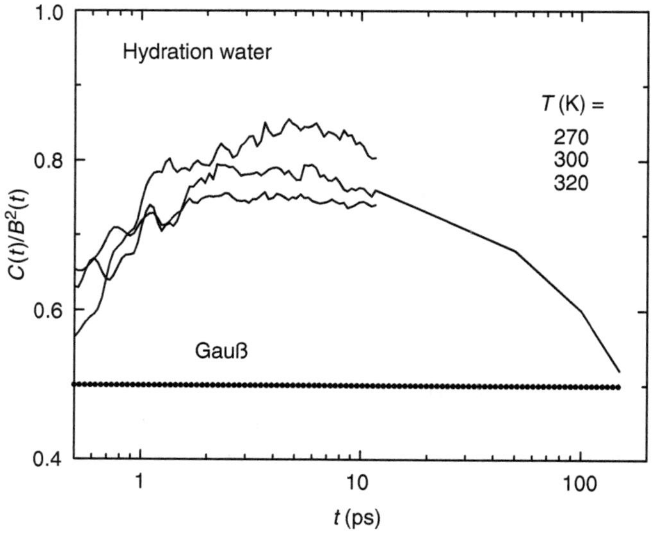

Fig. 20.11. Gauss deviation (fourth moment) of the hydration water displacement distribution versus time of myoglobin (0.35 g g^{-1})

regime of regular diffusion. The protein thus enhances the cage effect leading to partially localized water states. With decrease in temperature the cage becomes a trap. The extended plateau at 180 K is the signature of a solid state. Structural arrest is achieved continuously as the temperature decreases, which excludes discontinuous transitions such as ice formation.

Figure 20.11 shows the time-dependent fourth moment of the water displacement distribution which is larger than the Gaussian value of 0.5. This indicates that water displacements near the protein surface occur preferentially along a preferred direction. However, at times above 100 ps, the Gaussian value is reestablished, consistent with the observed linear time dependence of the second moment, Fig. 20.10.

20.8 Analysis of Protein Displacements

We now turn to protein motions. As mentioned earlier the relevant spatial information is contained in the Q-dependence of the intermediate scattering function $I(Q,t)$. Figure 20.12 shows this function at fixed $t_{\mathrm{res}} = 50$ ps versus temperature for myoglobin in three environments of Fig. 20.1.

A linear (Gaussian) behavior of $\ln[I(Q)]$ versus Q^2, is observed at low temperatures reflecting vibrational displacements. But above 200 K nonGaussian deviations, first described in [7] become significant. Nearly identical scattering functions are found for the dry and vitrified sample, pointing to similar intramolecular motions. The addition of water has a significant effect on the

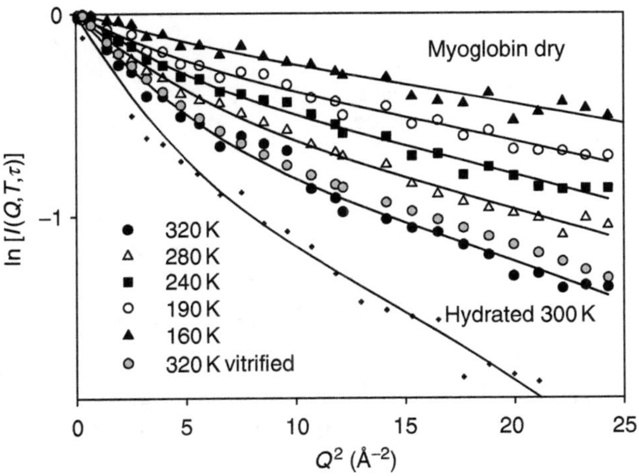

Fig. 20.12. Long time value of $I(Q, t \approx 50\,\text{ps})$ versus Q and temperature for dry myoglobin. Selected data of vitrified and hydrated myoglobin are also shown, *line*: fit to Eq. 20.18, instrument: IN13, ILL

density correlations, indicating water-assisted protein displacements. The simplest model, which fits the data, is a displacement distribution composed of two Gaussians [32, 33]

$$I(Q, T, t_{\text{fix}}) = A_1 \cdot \exp(-Q^2 \left\langle x_1^2 \right\rangle /2) + A_2 \cdot \exp(-Q^2 \left\langle x_2^2 \right\rangle /2). \qquad (20.32)$$

Figure 20.13 shows the second moments of the two Gaussian components determined on absolute scale [7, 10].

The first component has a plateau at low temperatures given by the protein zero point vibrations (0.014(±0.002) Å^2. This component shows harmonic behavior across the whole temperature range in the case of the dehydrated and the vitrified system. However, in the hydrated system an additional increase occurs above 240 K due to water-assisted motions. In contrast the second component emerges above 150 K and leads to a nonlinear enhancement in the total displacements independent of the protein environment. Figure 20.14 shows that the corresponding displacement distribution of hydrated myoglobin at fixed time versus temperature, $4\pi\, r^2 G(r, T, t = 50\,\text{ps})$. Component 1 broadens with increasing temperature due to vibrational motions. But above 240 K the maximum is shifting and the width increases. This effect points to small scale continuous motions. It is seen only with hydrated samples. In contrast, component 2 is observed in all samples independent of the protein environment. It has its maximum near 1.5 Å, indicating large scale excursions.

What is the molecular nature of the two types of motions? Several authors have emphasized the relevance of dynamical heterogeneity in the context of neutron scattering experiments [33, 34]. With neutron scattering we probe, because of their large cross-section, the trajectories of nonexchangeable

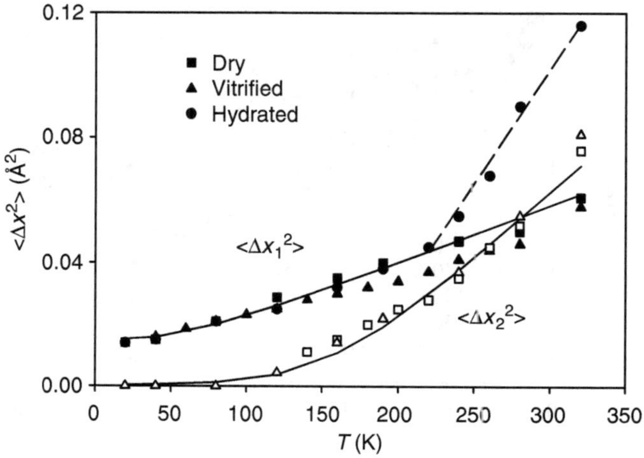

Fig. 20.13. Second moment of the displacement distribution (myoglobin, dry, d-glucose-vitrified and hydrated ($0.4\,\mathrm{g\,g^{-1}}$) at a fixed time of 50 ps, derived from data in Fig. 20.12, *closed symbols*: harmonic component, *open symbols*: total displacements, lines: from fits in Fig. 20.12, (instrument: IN13, ILL)

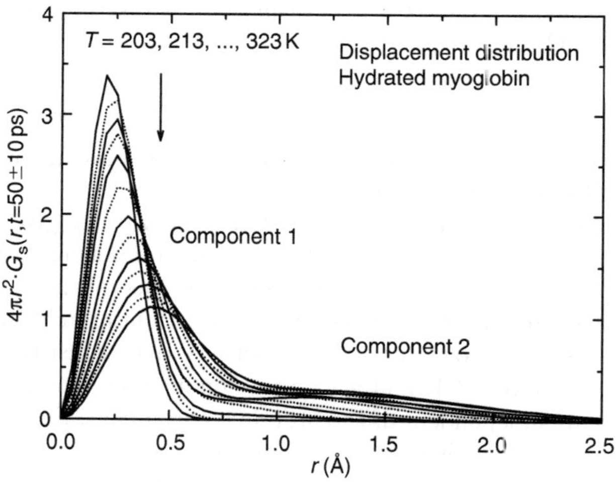

Fig. 20.14. Displacement distribution (myoglobin, D_2O-hydrated ($0.4\,\mathrm{g\,g^{-1}}$) at a fixed time of 50 ps versus temperature derived from data in Fig. 20.12, (instrument: IN13, ILL)

hydrogens, which are attached to the protein side-chains and, to a minor fraction, to the main chain. As discussed above structural changes of the protein chain involve mostly rotational jumps, while displacements in liquids are more continuous and on a small scale. Rotational jumps of methyl groups and of heavy atom dihedral transitions are the most natural modes of motion of a polypeptide chain. Moreover, the partial cross-section due to methyl groups is about 25%, which is the most significant individual contribution. Since we know its partial cross-section, the structure factor and the barrier to rotation from energy-resolved experiments one can calculate the scattering function without adjustable parameters. The results are shown as the solid lines in Fig. 20.12 and 20.13. The average rotational barrier of methyl groups in myoglobin amounts about 10 kJ mol^{-1}. The increase in the apparent displacements due to component 2, $\langle \Delta x^2 \rangle_2$, in Fig. 20.13 most likely results from an increasing rotational rate at fixed instrumental resolution (50 ps). A low temperatures the apparent displacement of component 2 is zero because the transitions are too slow to be resolved by the instrument. The slight discrepancy between the data and the theoretical curve at low temperatures may indicate a distribution of rates. The close agreement with experimental data produced by this model suggests strongly that rotational transitions, essentially of methyl groups, are the origin of the nonGaussian displacement distribution. Dynamical heterogeneity seems to be of minor importance. Rotation of methyl groups occurs in the hydrated, the dry as well as in the vitrified state. Water induces additional small scale protein displacements, which appear as anharmonic enhancements of the vibrational component 1.

20.9 Data Analysis

High resolution instruments collect data at constant angle and not at constant Q in contrast to three-axis spectrometers. As a result one has to transform the experimental data from a "constant angle" to a "constant Q" format: The momentum exchange, $\Delta \boldsymbol{p} = \hbar \boldsymbol{Q}$, at fixed angle, varies with the energy transfer $\hbar\omega$, according to [17]

$$Q^2 = \frac{2m_\mathrm{n}}{\hbar^2}(2E_0 + \hbar\omega - 2\cos(\theta) \cdot \sqrt{E_0(E_0 + \hbar\omega)}). \tag{20.33}$$

To determine the physically relevant quantity $S(Q,\omega)$ based on $S(\theta,\omega)$ data involves extrapolation and interpolation as indicated in Fig. 20.15. Only the area of the kinetic plane enclosed by the lines contains experimental data. Figure 20.16 illustrates for a particular case, how the "constant angle" data differ from the final interpolated "constant-Q" spectrum.

The low-Q regime, which is quite important to data analysis, is hampered by two difficulties; the necessity to extrapolate finite Q data and the relevance of multiple scattering. Multiple scattering tends to generate a Q-independent inelastic background [35]. Figure 20.17 shows that even for a thin sample with 91% transmission about 15% of the incident neutrons are multiply scattered.

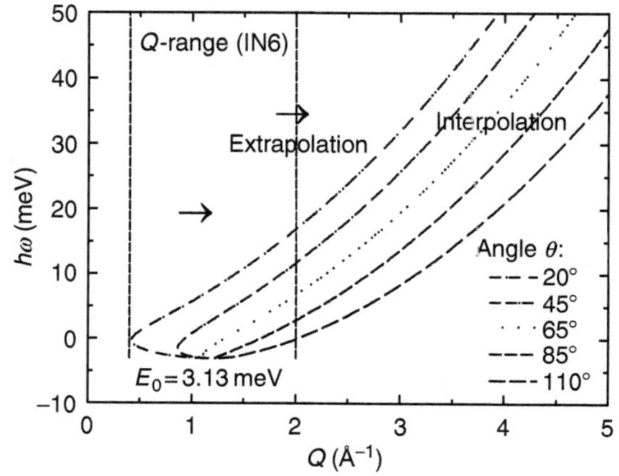

Fig. 20.15. Kinetic plane for an incident neutron wavelength of 5.1 Å or $E_0 =$ 3.13 meV. Only the area enclosed by the lines is accessible to the experiment. Constant-Q data can be calculated from constant angle experiments by interpolation. To reach the low Q range, extrapolation is required [30]

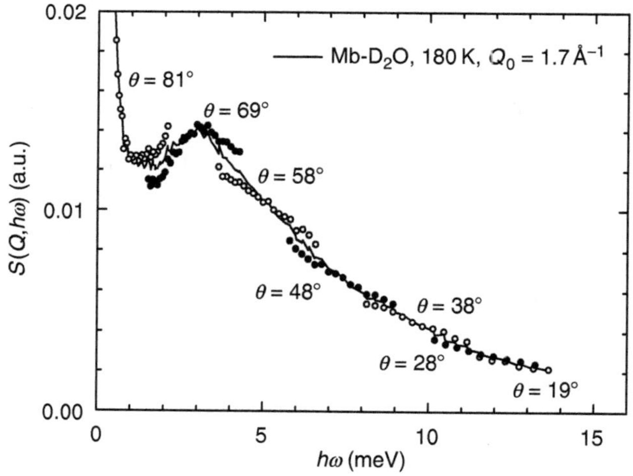

Fig. 20.16. Experimental "constant angle" (θ) neutron scattering spectra of myoglobin (IN6, ILL) and interpolated "constant Q" spectrum. The "elastic" $Q = Q_0$ at $E = E_0$ was 1.7 Å^{-1} [30]

A very detailed analysis always requires consideration of multiple scattering. An initial estimate is the Q-independent background extrapolated to $Q = 0$. Then the corrected spectrum can be used to calculate the multiple scattering corrections iteratively. This is particularly relevant to high-frequency data [30].

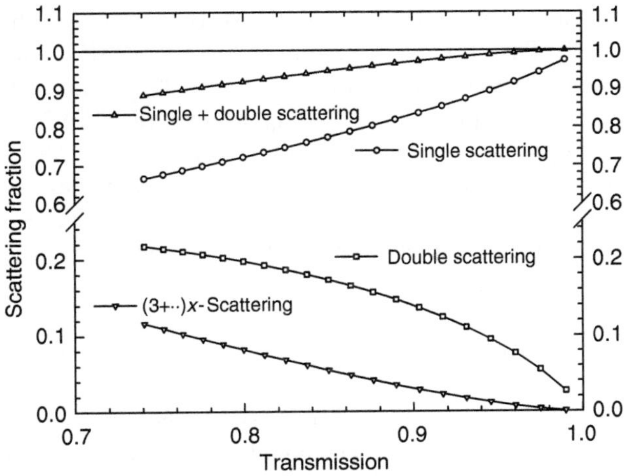

Fig. 20.17. Multiple scattering fraction versus sample transmission calculated for an infinite, thin slab sample holder inclined at 135° relative to the beam [17,30]

20.10 Conclusions

Dynamic neutron scattering provides a unique tool to probe the statistical properties of picosecond motions in biomolecules. To understand the effect of water on structure and stability of proteins requires to study the interactions on the time scale where hydrogen bonds are broken and formed. INS thus complements other local methods like NMR and fluorescence emission. We suggest to assume a general perspective without resigning too early to particular models: A moment analysis of scattering data leads to model-independent insight into molecular mechanisms. It requires, however, high quality data. Two classes of protein displacements could be discriminated: torsional transitions and water-assisted motions. The latter are composed of fast H-bond fluctuations and slower small scale displacements. The dynamical transition is driven by the H-bond dynamics of hydration water [10,28,36], which has, however, minor effects on the rotational transitions. The transition temperature thus varies with H-bond strength and viscosity [37]. Within a glass-transition scenario our results support the notion of water-plasticized β-processes in proteins proposed by Green and Angell [11]. A quite interesting perspective is to study the real time protein–water coupling processes using generalized harmonic oscillator models. Structural rearrangements correlated with motions of water molecules occur on a time scale of picoseconds. In what sense is then the fast motion of water molecules relevant to the observed slow substrate conversion in enzymes? For myoglobin, it could be shown, that the solvent modulates the barrier which controls ligand entry and escape according to Kramers law of activated escape [37,38]. The solvent generates the "seascape" of fluctuating barriers. This does not imply, that protein activity occurring on a time scale of seconds vanishes, because of reduced picosecond structural fluctuations as suggested in [39]. The essential dynamical quantity is the solvent viscosity (Eq. 20.1), which was not taken into account in [39]. The displacements of molecules on a microscopic scale including water molecules in the active site are discontinuous and always fast. The millisecond time scales come about by high energetic or entropic barriers which prevent particular rearrangements for long-time intervals. Enzymes are thus devices, which select by construction a small fraction of events out of a large number of fast structural fluctuations.

Acknowledgments

The author is grateful for technical support by the instrument responsibles of the Institut Laue Langevin and the collaboration with many colleagues, in particular with Marcus Settles, who rediscovered the moment method for biology. Financial support by the Bundesministerium für Bildung und Forschung (grant 03DOE2M) is gratefully acknowledged.

References

1. K.A. Dill, Biochemisby **29** (1990), 7132–7155
2. W. Doster, M. Diehl, H. Leyser, W. Petry, H. Schober, in *Spectroscopy of Biological Molecules* J. Greve, G.J. Puppels, C. Otto (Eds.) (Kluwer, Dordrecht, 1999) pp. 655–659
3. W. Göetze, L. Sjögren, Rep. Prog. Phys. **55** (1992), 241–376
4. J. Fitter, R. Lechner, G. Büldt, N.A. Dencher, Proc. Natl Acad. Sci. USA **93** (1996), 7600–7605
5. U. Lehnert, V. Reat, M. Weik, G. Zaccai, C. Pfister, Biophys. J. **75** (1998), 1945–1952
6. M.C. Bellissent-Funel, J.M. Zanotti, S.H. Chen, Faraday Discuss. **103** (1996), 281–294
7. W. Doster, S. Cusack, W. Petry, Nature (London) **337** (1989), 754
8. W. Doster, S. Cusack, W. Petry, Phys. Rev. Lett. **65** (1990), 1080
9. M. Tarek, D. Tobias, Mod. Phys. Lett. B **5** (1991), 1407
10. W. Doster, M. Settles, in *Hydration Processes in Biology, Protein Dynamics, The Role of Hydrogen Bonds*, Nato Science Series A Life Science, Vol. **305** (IOS Press, Amsterdam, 1998), pp. 177–194
11. J.L. Green, J. Fan, A. Angell, J. Phys. Chem. **98** (1994), 13780–13790
12. W. Doster, M. Settles, Biochim. Biophys. Acta **1749** (2005), 173–186
13. A. Kitao, F. Hirata, N. Go, Chem. Phys. Lett. **158** (1991), 447–472
14. S. Hayward, A. Kitao, F. Hirata, N. Go, J. Mol. Biol. **234** (1993), 1207–1217
15. G. Kneller, Chem. Phys **261** (2000), 1–24
16. F. Mezei, in *Liquids, Freezing and The Glass Transition*, J.P. Hansen, D. Levesque, J. Zinn-Justin (Eds.) (Elsevier, Amsterdam, 1991, Les Houches, 1989), pp. 632

17. M. Bee, *Quasielastic Neutron Scattering* (Adam Hilger, Bristol, Philadelphia, 1988), p. 16
18. W. Marshall, S.W. Lovesey, *Theory of Thermal Neutron Scattering* (Clarendon, Oxford, 1971)
19. F. Parak, E.W. Knapp, D. Kucheida, J. Mol. Biol. **161** (1982), 177–194
20. D.J. Bicout, Phys. Rev. E. **62** (2000), 261–271
21. K. Hinsen, A.J. Petrescu, S. Dellerue, M.C. Bellissent-Funel, G.R. Kneller, Chem. Phys. **261** (2000), 25–37
22. G.E. Uhlenbeck, L.S. Orstein, Phys. Rev. **36** (1930), 823–829
23. W. Doster, M. Settles, Biochim. Biophys. Acta 1749 (2005), 173–186
24. J.B. Boon, S. Yip, *Molecular Hydrodynamics*, (McGraw-Hill, New York, 1980)
25. W. Göetze, L. Sjögren, Trans. Theory Stat. Phys. **24** (1995), 801–853
26. H. Leyser, W. Doster, M. Diehl, Phys. Rev. Lett. **82** (1999), 2897
27. M. Diehl, W. Doster, W. Petry, H. Schober, Biophys. J. **73** (1997), 2726–2732
28. W. Doster, T. Bachleitner, M. Hiebl, E. Lüscher, A. Dunau, Biophys. J. **50** (1986), 213–219
29. M. Settles, W. Doster, Faraday Discuss. **103** (1996), 269–279
30. M. Settles, W. Doster, in *Biological Macromolecular Dynamics*, H. Büttner et al. (Eds.) (Adenine, New York, 1996), pp. 307–331
31. M. Sakamoto, B.N. Brockhouse, R.G. Johnson, N.K. Pope, J. Phys. Soc. Jpn **17** (1962), 370–376
32. M. Settles, Thesis (Technical University Munich, Munich, 1997)
33. H. Nakagawa, H. Kamikubo, I. Tsukushi, T. Kanaya, M. Kataoka, J. Phys. Soc. Jpn **73** (2004), 491–495
34. J.A. Hayward. J. Smith, Biophys. J. **82** (2002), 1216
35. S. Cusack, W. Doster, Biophys. J. **58** 1990, 243–251
36. M. Tarek, D. Tobias, Phys. Rev. Lett. **88** (2002), 138101
37. H. Lichtenegger, W. Doster, T. Kleinert, B. Sepiol, G. Vogl, Biophys. J. **76** (1999), 414–422
38. T. Kleinert, W. Doster, H. Leyser, W. Petry, V. Schwarz, M. Settles, Biochemistry **37** (1998), 717–733
39. R.M. Daniel, J. Smith, M. Ferrand, S. Hery, R. Dunn, J. Finney, Biophys. J. **75** (1998), 2504

21

Internal Dynamics of Proteins and DNA: Analogy to Glass-Forming Systems

A.P. Sokolov, R.B. Gregory

21.1 Introduction

One of the main goals of protein research is to understand the relationship between protein structure, dynamics and function. Methods of structural analysis (X-ray diffraction and NMR spectroscopy) are well developed and, as a consequence, the structures of thousands of proteins are now known. Describing the dynamic properties of proteins, i.e., exploring the potential energy landscape, and relating dynamics to function is a far more difficult task. The importance of protein and DNA motions for an understanding of their function, and the advantages of neutron spectroscopy for analysis of protein dynamics have already been emphasized in previous chapters. In this chapter we focus our attention on the internal dynamics of proteins and DNA. Their dynamic behavior exhibits many similarities to the dynamics of glass-forming systems or, speaking more generally, to dynamics of complex systems. Free translational diffusion and rotation of a molecule as a whole is excluded from our discussion. These kinds of motions do contribute to neutron scattering spectra of protein solutions, but are usually easily distinguished from internal dynamics [1,2].

Neutron scattering spectra of hydrated lysozyme and DNA are shown in Fig. 21.1. These spectra are typical for biological macromolecules at ambient conditions and are dominated by quasielastic scattering. The dynamics of proteins and DNA span an enormous time (frequency) range from picoseconds (THz) to seconds, and involve local atomic fluctuations, collective motions of side-chains and loops, segmental motions and a variety of rigid body motions such as hinge-bending [5]. Neutron spectroscopy easily covers the range from Terahertz down to $\nu \sim 100\,\mathrm{MHz}$, i.e., from picoseconds to nanoseconds. Using the neutron spin-echo (NSE) technique, the time scale can be extended to hundreds of nanoseconds. Because of some technical difficulties, NSE has been used for analysis of protein dynamics in only a few cases [6]. Above $\nu \sim 3\,\mathrm{THz}$ ($\sim 100\,\mathrm{cm}^{-1}$) high-frequency atomic vibrations of small groups of atoms are present. This frequency range can be easily analyzed using modern IR or Raman spectroscopy. At frequencies below $\sim 3\,\mathrm{THz}$ collective excitations of many atoms (amino-acid residues and base-pairs) are present which are amenable to study by neutron spectroscopy.

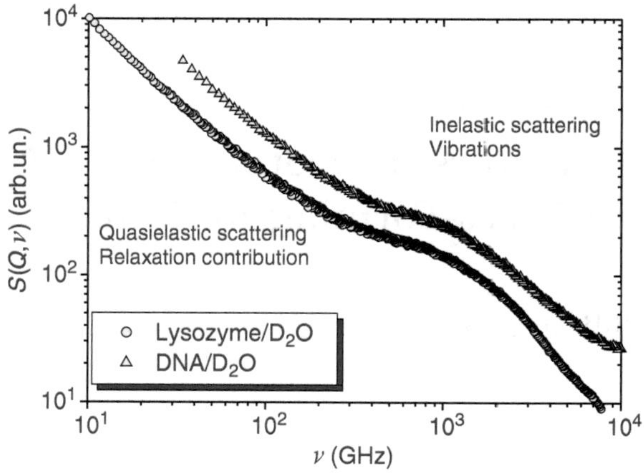

Fig. 21.1. Neutron scattering spectra of hydrated lysozyme (~0.4 g of D_2O per gram of lysozyme) at $T = 290\,K$ and DNA (~0.46 g of D_2O per gram of DNA) at $T = 320\,K$ presented as the dynamic structure factor $S(Q, \nu)$. Data are from [3,4]. The spectra include time-of-flight (TOF) data only and are summed over all detectors (all Q).

Neutron scattering spectra at $\nu < 3\,THz$ exhibit at least three main components in the quasi-elastic and inelastic regions of the spectrum [7–9]: (i) low-frequency vibrations, the so-called boson peak, at energies $h\nu \sim$ 2–6 meV (~15–50 cm^{-1} or 0.5–1.5 THz) have been observed in spectra of all proteins and DNA analyzed so far. Results of computer simulation for hydrated Ribonuclease-A [10] show that vibrational modes at the boson peak involve the whole protein, side groups and backbone, polar and nonpolar groups, and also water molecules. Thus these vibrations are collective motions of many amino-acid residues (base-pairs in the case of DNA). (ii) Fast conformational fluctuations in the picosecond time range are present in hydrated and dry proteins and exist even at very low temperatures. (iii) A slow relaxation process is observed in hydrated proteins and DNA which exhibits a strong temperature dependence. The same three components are observed in the dynamics of glass-forming liquids. Moreover, temperature dependencies of these components are similar in biological macromolecules and in glass-forming liquids. An overview of neutron scattering data related to the two relaxation components (ii and iii) of the dynamics of proteins and DNA is presented below. The nature of the dynamic transition in biological macromolecules is also discussed.

21.2 Analysis of Relaxation Spectra: Susceptibility Presentation vs. Dynamic Structure Factor

Conformational changes in proteins occur at all temperatures [11]. They show up in neutron scattering spectra as a quasielastic contribution to the dynamic structure factor, $S(Q,\omega)$ (Fig. 21.1). The width of the quasielastic spectrum provides an estimate of the characteristic relaxation time. The quasielastic neutron scattering spectra are often approximated by a sum of a few Lorentzians from which the wave-vector (Q) and temperature dependence of the Lorentzian line width Γ are analyzed [1,8]. It is known, however, that relaxation in complex systems, including biological macromolecules, usually cannot be represented by a single exponential decay, but instead involves a number of processes, each of which is strongly stretched [3,7,9,12].

An alternative way to analyze quasielastic neutron scattering spectra is through their representation as the imaginary part of the dynamic susceptibility, $\chi''(Q,\nu)$, instead of the dynamic structure factor, $S(Q,\nu)$. The dynamic susceptibility is related to $S(Q,\nu)$ through the Bose occupation number $n_{\mathrm{B}}(\nu) = [\exp(h\nu/kT) - 1]^{-1}$

$$\chi''(Q,\nu) \propto S(Q,\nu)/n_{\mathrm{B}}(\nu). \tag{21.1}$$

The susceptibility presentation has several advantages: (i) trivial temperature variations are taken into account; (ii) scattering data can be directly compared to dielectric $\varepsilon''(\nu)$ or mechanical $G''(\nu)$ loss spectra; (iii) a relaxation process in the susceptibility spectrum appears as a maximum at $\nu_{\max} \sim (2\pi\tau)^{-1}$, where τ is a characteristic relaxation time; relaxation processes with well-separated τ appear as separated peaks; (iv) the spectral shape of the susceptibility maximum provides information on the stretching of the relaxation process, usually high- and low-frequency tails can be approximated by a power law $\chi''(Q,\nu) \propto \nu^a$; where values of $a = 1$ or -1 correspond to a single exponential relaxation, while $-1 < a < 1$ corresponds to a stretched relaxation process or a broad distribution of relaxation times. In this respect the dynamic susceptibility presentation simplifies analysis of complex relaxation spectra.

Figure 21.2 shows neutron scattering spectra of DNA and lysozyme using the susceptibility presentation. The spectra clearly show two well-separated relaxation processes in hydrated biomolecules, while only one, the fast process, remains in the spectra of dry DNA and lysozyme. Both processes are strongly stretched and an attempt to fit them by Lorentzian functions can give incorrect estimates of the characteristic relaxation times.

21.3 Slow Relaxation Process

In this section we focus on analysis of the slow relaxation process, its temperature and hydration dependence and its relationship to the dynamic transition. Susceptibility spectra of hydrated samples reveal a slow relaxation process with a maximum in the Gigahertz frequency range (Fig. 21.2). Time-of-flight spectrometers do not have sufficient energy resolution to explore slow relaxation processes and so a combination of time-of-flight and backscattering spectrometers is required to provide reasonable spectra of the slow process in a

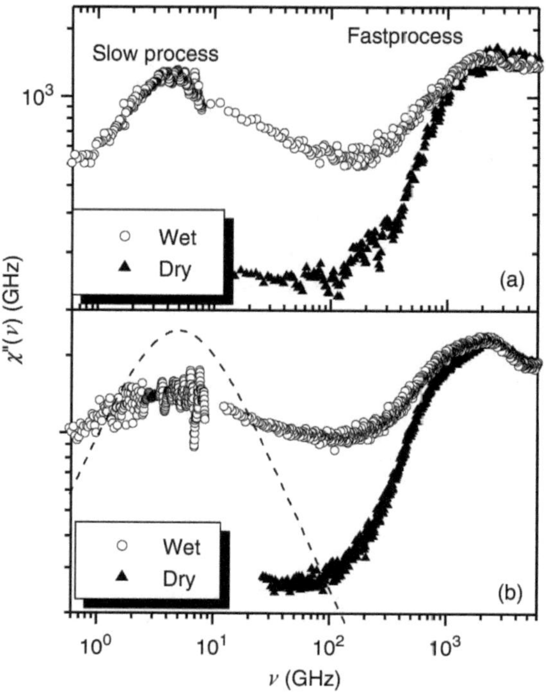

Fig. 21.2. The same spectra shown in Fig. 21.1 of hydrated DNA at $T = 320$ K (**a**) and lysozyme at $T = 290$ K (**b**) presented as the imaginary part of the susceptibility. The spectra are extended at lower frequencies using data from a backscattering spectrometer. The dashed line shows the spectrum expected for a single exponential relaxation (Lorentzian). Significant stretching is observed in the spectra of the slow process of both biomolecules. Spectra of dry samples at the same temperatures are shown for comparison. No slow process is observed in the spectra of dry samples. Data are from [3, 4]

broad enough frequency range. The slow relaxation process in hydrated DNA at T 320 K has a maximum at $\nu_{\max} \sim 4$ GHz (Fig. 21.2a) corresponding to a relaxation time $\tau_S \sim (2\pi\nu_{\max})^{-1} \sim 40$ ps. The maximum for the slow process in fully hydrated lysozyme at T 290 K appears at $\nu_{\max} \sim 3$ GHz, which corresponds to $\tau_S \sim 50$ ps (Fig. 21.2b). For comparison, Fig. 21.2 includes the spectrum that would be obtained for an intermediate scattering function, $I(Q, t)$, consisting of a single exponential relaxation, i.e., $I(Q, t) = \exp(-t/\tau_S)$ (a Lorentzian peak in the susceptibility spectrum). The slow relaxation peak observed for both protein and DNA is significantly broader than the Lorentzian peak, indicating that the slow relaxation process is strongly stretched in these systems (Fig. 21.2) [3, 9, 12] and better described by a Kohlrauch–Williams–Watts relaxation function: i.e., $I(Q, t) = \exp[-(t/\tau_S)^{\beta}]$. The slope of the susceptibility spectrum ($\chi''(\nu) \propto \nu^{-b}$) provides an estimate of the stretching exponent $b \sim 0.3 - 0.35$ that appears to be rather temperature independent

(Fig. 21.3). The stretching exponents β (stretching in the time domain) and b (stretching in the frequency domain) are not identical, but they are directly related [13]. The value of the exponent b is smaller than in most of glass-forming systems, i.e., the stretching of the slow relaxation process in proteins and DNA is stronger than in molecular liquids or polymers, suggesting a more complex relaxation process or a broader distribution of relaxation times. The contribution of the slow process to the susceptibility spectrum changes strongly with temperature (Fig. 21.3). The changes of the spectra are usually ascribed to temperature variations of the characteristic relaxation time $\tau_S(T)$. Analysis of the spectra (Fig. 21.3a) reveals a strong temperature dependence of $\tau_S(T)$ in DNA at $T > 250$ K (Fig. 21.4) such that the slow process moves out of the accessible frequency range at $T < 210$–230 K and makes estimates of τ_S unreliable. A strong (usually nonArrhenius) temperature dependence of the relaxation time and stretching of the relaxation spectrum are characteristic of structural relaxation in glass-forming liquids. Thus the slow process in

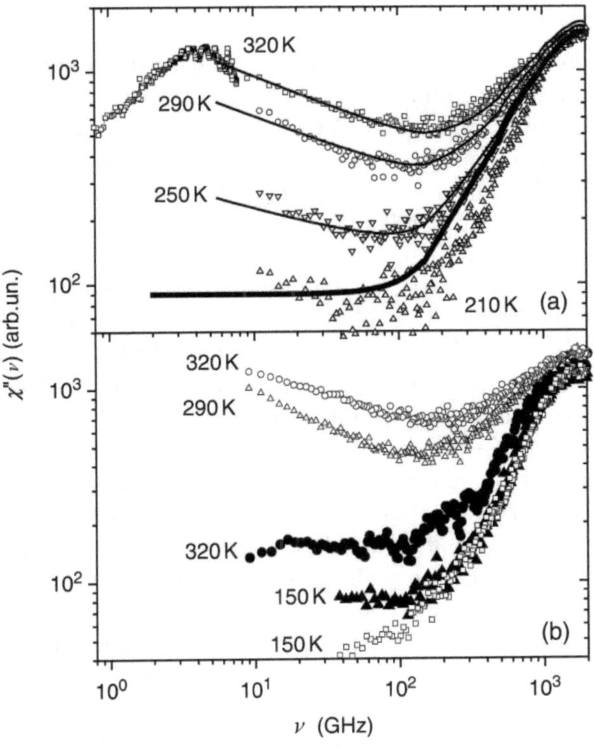

Fig. 21.3. Temperature variations of susceptibility spectra for DNA at various hydration levels: (**a**) 0.46 g of D_2O per gram of DNA, (**b**) 0.66 g of D_2O per gram of DNA (*open symbols*) and 0.03 g of D_2O per gram of DNA (*closed symbols*) (for details, see [3]). Thin *solid lines* show the fit to Eq. 21.4. The thick *solid line* shows the spectrum of the fast process $\chi''_{\text{fast}}(\nu)$

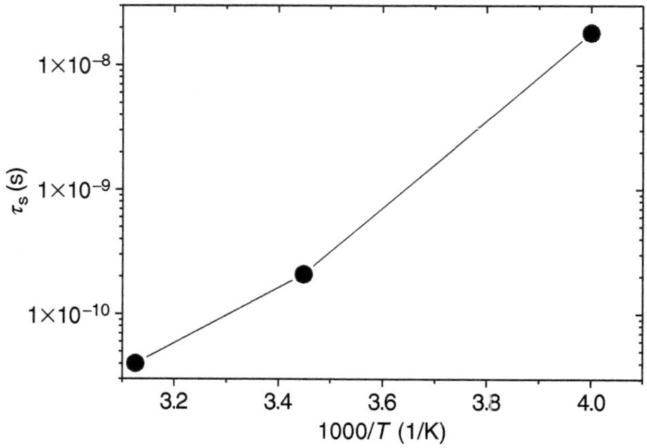

Fig. 21.4. Temperature dependence of the characteristic relaxation time τ_S of the slow relaxation process in hydrated DNA (Fig. 21.3a) obtained from a fit to Eq. 21.4.

biological macromolecules appears to be similar to the main structural relaxation process in glass-forming liquids [14–16].

There is, however, one significant difference between the structural relaxation (α-relaxation) in glass-forming liquids and the slow process in biological macromolecules. It appears in the Q-dependence of the characteristic relaxation time. The structural relaxation time, τ_α in glass-forming systems has a strong dependence on Q, i.e., $\tau_\alpha \propto Q^{-c}$, with $c \sim 2$–4 [17, 18] indicative of diffusive motion. For particles undergoing free translational diffusion, the van Hove correlation function obeys Fickian diffusion and the corresponding intermediate scattering function is given by $I(Q,t) = \exp(-t/\tau_\alpha)$ with $\tau_\alpha \propto 1/DQ^2$ where D is the diffusion coefficient. By contrast, analysis of τ_S in proteins reveals no significant Q-dependence [1, 4], consistent with the slow relaxation process in proteins involving localized diffusive motions with atomic displacements smaller than ~ 3 Å. If diffusion is confined to a sphere of radius r, the relaxation time is independent of Q for $Q < \pi/r$ and varies as $1/DQ^2$ for Q values $> \pi/r$ [19]. Using a two-site jump model, Doster and Settles [16] estimated the jump distance to be ~1.5 Å. Such confined diffusive motion of groups of atoms in the interior of proteins is to be expected given the compact folding of the polypeptide backbone in the native state.

The microscopic picture behind the slow process remains unclear. However, results of recent computer simulations [20] provide additional insight. The authors found that a few collective modes distributed over the whole protein molecule are activated above T 180–220 K. These modes represent small displacements of secondary structures (α-helices) of the protein. Because the slow process is also activated at this temperature range, we tend to ascribe it to these kinds of modes. That will explain weak Q-dependence of the relaxation time.

There is much evidence for water acting as a plasticizer of proteins including ESR studies of spin labels [21–23], Mössbauer spectroscopy [22, 24], phosphorescence [25,26], solid-state ^{13}C cross polarization/magic angle sample spinning (CP/MASS) NMR [27], hydrogen isotope exchange [28,29], dielectric relaxation [30, 31], Rayleigh scattering of Mössbauer radiation (RSMR) [32], light-scattering [12] and incoherent neutron scattering [1, 3, 4, 33]. The main conclusion to be draw from these studies is that at 300 K the internal dynamics of proteins recovers at hydration levels of 0.1–0.2 g water g protein^{-1}, much lower than the amounts of water required for full hydration (0.4–0.5 g water g protein^{-1}). In fully hydrated systems, internal dynamics recover at temperatures of about 180–220 K.

The detailed mechanism by which water plasticizes proteins and the coupling of the solvent to the protein interior are not fully understood. It has been suggested that water plasticizes proteins in the same way it plasticizes polyamides such as Nylon 6,6 and Nylon 4,6 by providing alternate, mobile hydrogen bond donors and acceptors for peptide groups [34]. This may involve the migration of buried water molecules in the protein interior via "Mobile Defects" in the hydrogen bonding network [35] or could be due to water at the protein surface. Several recent simulations [36–38] suggest that the dynamics of water at the protein surface activates protein internal dynamics. Recent neutron scattering studies of the effect of hydration on proteins and DNA clearly show that the slow relaxation process occurs in the hydrated systems but not in the dry systems (Figs. 21.2, and 21.3b) [3, 4, 12], suggesting that water enables the slow relaxation process.

There is also evidence that the slow relaxation process occurs in proteins in glycerol (Fig. 21.5). The contribution of the slow process is weaker in the case of lysozyme in glycerol and nearly disappears from the accessible frequency range at T 250 K. There has been much discussion of the effects of cosolvents such as glycerol and sugars on proteins. The relationship between the solvent viscosity and kinetics of biochemical reactions has been discussed in many papers [39–46]. Increases in solvent viscosity can increase internal friction associated with molecular motions and slow down protein dynamics. But cosolvents also have effects on the thermodynamics of proteins. It is well-established that polyols such as glycerol and sugars increase protein chemical potentials. As the protein–water–polyol system adjusts to relieve the unfavorable increase in free energy, the polyols are preferentially excluded from the vicinity of the protein (the protein is preferentially hydrated) [47] and the protein favors more compact conformations in order to minimize its contact with the polyol. The later effect may have a large influence on protein internal dynamics as has been suggested for the effect of glycerol on protein hydrogen isotope exchange [48].

We should emphasize that the appearance of the slow relaxation process in the presence of glycerol, shown in Fig. 21.5, occurs in the absence of water. These neutron scattering results in glycerol [4] are also supported by parallel light scattering measurements [12] and clearly indicate that glycerol itself can

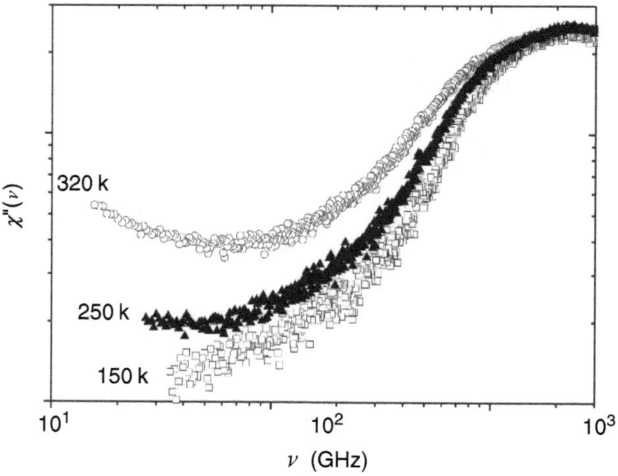

Fig. 21.5. Susceptibility spectra of lysozyme/glycerol sample (1 g of deuterated glycerol per gram of lysozyme) at different temperatures (data from [4]). The slow process is strongly suppressed already at T 250 K

plasticize the protein. Unlike water, there is no evidence that glycerol penetrates the protein interior so its plasticizing effect must be achieved through interactions with protein surface groups. Like water, glycerol can function as a mobile hydrogen bond donor and acceptor, and it is possible that the slow relaxation process is enabled by the translational diffusion of glycerol at the protein surface through transient formation and breaking of hydrogen bonds.

21.4 The Nature of the Dynamical Transition in Proteins and DNA

Analysis of mean-squared displacements of atoms $<x^2>$ in hydrated proteins and DNA as a function of temperature reveals a transition at temperatures around 200–230 K [11, 49–52] changing from relatively weak nearly linear dependence on T at lower temperature to much stronger rise at higher temperature (Fig. 21.6). This behavior of $<x^2>$ has been observed in neutron scattering, X-ray and Mössbauer spectroscopic measurements [11, 41, 49–51] and is usually interpreted as a dynamical transition from nearly harmonic motion at $T < T_{\rm D}$ to strongly anharmonic motion at $T > T_{\rm D}$. Several studies have shown that proteins are only active at temperatures above $T_{\rm D}$ [41, 50] and it is widely accepted that proteins require internal flexibility in order to function, although one enzyme (glutamate dehydrogenase) is known which remains active below the dynamical transition temperature [53]. There is therefore a great deal of interest in understanding the mechanism of the dynamical transition and its relationship to protein activity. The dynamical transition

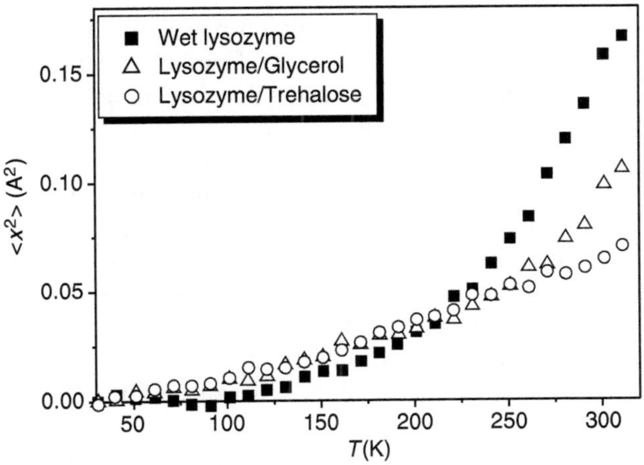

Fig. 21.6. Mean-squared displacements of atoms $<x^2>$ in lysozyme dissolved in various solvents (data from [4]). The dynamical transitions for hydrated lysozyme appears at T_D 220 K while it appears in lysozyme embedded in glycerol at T_D 270 K. No dynamic transition appears in lysozyme embedded in trehalose up to T 320 K

does not appear in dry proteins and DNA even at temperatures as high as $T \sim 320$ K [11,51,52]. Moreover, the transition can be strongly influenced by solvent (Fig. 21.6), shifting to higher $T \sim 270$–280 K for proteins embedded in glycerol [4,51] and being suppressed at least up to $T \sim 320$ K in proteins formulated in solid trehalose [4,54]. All these results suggest that solvents play a crucial role in the dynamical transition and activity of proteins.

The mean square displacement, $<x^2>$, is an integrated quantity that usually includes contributions from vibrations, conformational changes, diffusion and other molecular motions. Analysis of time or frequency resolved spectra might therefore help to understand which processes are activated as the temperature is increased above T_D. Neutron spectroscopy is an extremely useful tool for these measurements because it provides information over a wide range of frequencies. These kinds of measurements have been performed on only a few proteins and on DNA [3,7,9,12] and indicate that the dynamical transition is associated with activation of the slow relaxation process that enters the Gigahertz frequency window at temperatures $T > T_D$ 200–230 K (Fig. 21.3).

The microscopic mechanism of the dynamical transition remains a subject of much discussion. Mode coupling theory (MCT) developed in the mideighties [55] to describe the dynamics of simple liquids, supercooled liquids and glass-forming systems may provide a useful framework for understanding the dynamic behavior of proteins and DNA. MCT predicts a dynamic crossover in glass-forming liquids marking a transition from liquid-like to solid-like behavior on molecular time and length scales at temperatures much above conventional glass transition temperatures. Many experimental results support the

MCT predictions [56], and the theory gives a reasonably good qualitative description of relaxation spectra in many glass-forming systems at temperatures above the crossover temperature T_C, although other experimental observations are not in accord with idealized MCT [57].

MCT predicts that relaxation in glass-forming systems at high temperatures ($T > T_C$) occurs by a two-step process involving an initial fast decay corresponding to relaxation of a particle (a molecular unit) confined ("rattling") in a cage formed by its neighbors; which decays to a particular level but does not lead to a complete relaxation; followed by a slow second step that corresponds to escape of the particle from the cage. The theory predicts that both relaxation processes are described by stretched decays. Details of MCT predictions and its experimental tests can be found in various reviews (see, for example [55, 56]).

On a qualitative level, MCT predicts that at $T > T_C$ the fast process is essentially temperature independent and the main variation is a strong slowing down of the slow relaxation process [55]. Moreover, the characteristic relaxation time of the slow relaxation process should exhibit a critical temperature dependence

$$\tau_{\text{slow}}(T) \propto (T - T_C)^{-\gamma} \qquad (21.2)$$

where the value of the critical exponent γ is directly related to the stretching exponent b [55]. This provides a reasonably good qualitative description of data from various ionic and van der Waals molecular systems, polymers, as well as covalent and hydrogen bonded liquids [56, 58, 59], including water and glycerol [60, 61].

Doster et al. [7] applied MCT to describe neutron scattering spectra for myoglobin. They showed that the theory can describe the spectra and their temperature variations, and estimated a crossover temperature T_C 195 K. Sokolov et al. performed an analysis of neutron scattering spectra of DNA at various levels of hydration [3, 9] and demonstrated (Fig. 21.3) that MCT describes well the qualitative features of the dynamics of DNA down to temperatures of about 230 K. The authors fit the susceptibility spectra by a sum of a slow and a fast relaxation process

$$\chi''(\nu, T) = \chi''_{\text{slow}}(\nu, T) + \chi''_{\text{fast}}(\nu) = A[\nu\tau_S(T)]^b + \chi''_{\text{fast}}(\nu). \qquad (21.3)$$

Here the slow process was approximated by a power law with a stretching exponent $b \sim 0.3$. All parameters, except $\tau_S(T)$ were assumed to be temperature independent and $\tau_S(T)$ was the only free fitting parameter [3]. Its temperature dependence is presented in Fig. 21.4. This high-temperature MCT description works well down to T 230 K. Below this temperature, Eq. 21.4 cannot provide a good description of the spectra because the fast process becomes temperature dependent [3, 9]. This corresponds to the dynamic crossover and gives an estimate of T_C 230 K [3, 9]. As noted earlier, MCT also predicts temperature variations of $\tau_S(T)$ (Eq. 21.3), so the temperature dependence of the only free fitting parameter can be verified. The exponent γ in Eq. 21.3 is related to the

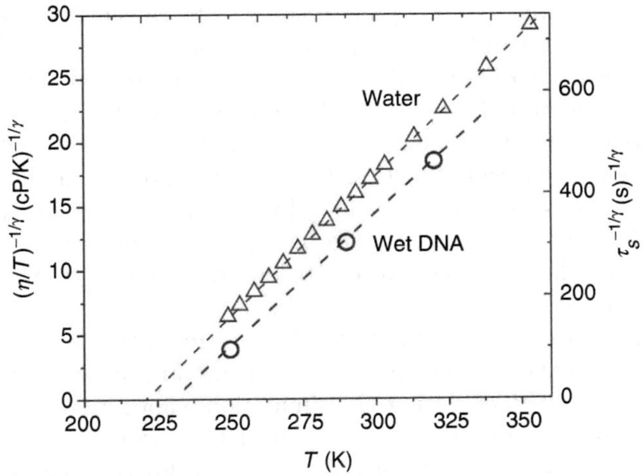

Fig. 21.7. Critical temperature behavior for τ_S in hydrated DNA. The critical behavior for the viscosity time scale of water (data from [60]) is shown for comparison

exponent b, and it is $\gamma \sim 3.9$ when $b \sim 0.3$ [3]. Analysis of the results given in [3] reveals (Fig. 21.7) the predicted critical behavior of $\tau_S(T)$. It gives an estimate of T_C 230 K [3] consistent with the earlier analysis of the spectra. Thus, neutron spectra of proteins and DNA and their temperature variations can be consistently described in the framework of MCT [3, 7, 9, 16]. In that respect the dynamical transition corresponds to a crossover from liquid-like to solid-like dynamics on a molecular scale. The solid-like dynamics corresponds to a very limited motion of amino-acid residues (or base-pairs in the case of DNA).

As noted above, the temperature, T_D, at which the dynamical transition occurs in proteins depends on the solvent and is approximately the same, T_D 200–230 K, for different hydrated proteins and for DNA [9, 11, 50–52]. This observation together with the existence of the dynamic crossover in bulk water at T 225 K [60] (Fig. 21.7) led Sokolov and coworkers [3, 9] to speculate that the dynamic crossover of water is responsible for the dynamical transition observed in hydrated biological macromolecules. Analysis of data for the protein lysozyme in glycerol provides strong support for this idea (Fig. 21.6) [12, 51]. The dynamical transition temperature $T_D = 270$ K estimated for lysozyme/glycerol samples from the temperature dependence of $< x^2 >$ [4, 51] (Fig. 21.6) and from an analysis of light scattering spectra [12] is very similar to the crossover temperature of bulk glycerol $T_C \sim 270 - 290$ K [62].

These results clearly demonstrate that the dynamic crossover in a solvent controls the dynamical transition in proteins and DNA. Moreover, the analysis suggests that the dynamical transition might be considered as an arrest of the slow relaxation process. This arrest leads to suppression of the biochemical activities of proteins. What is the microscopic picture behind the correlation between T_{D} of a protein and T_{C} of a solvent? Results of several recent computer simulations provide some insight. In one case, the diffusion of water molecules was constrained by placing the oxygen atoms in a very deep energy minimum. The water molecules could rotate, and the lifetime of hydrogen bonds remains similar to the unrestricted case, however, analysis of the intermediate scattering function $I(Q,t)$ of the protein in an environment with restricted water diffusion was found to resemble $I(Q,t)$ for the dry protein [36]. The slow process appears to be suppressed and restriction of water diffusion places the hydrated protein below its dynamic transition even at room temperature [36].

A recent molecular dynamics simulation employing a dual heat bath method, in which the temperature of the solvent and protein could be maintained at different values during the simulation, shows that the protein dynamical transition is driven by a dynamical transition in the translational motion of water [37]. The latter enables increased fluctuations of the protein especially in the motions of side chains at the protein surface. According to MCT, the dynamic crossover is essentially a dynamic arrest on a molecular time and length scale. In this case, decreasing the temperature below T_{C} of a solvent is equivalent to restricting translational diffusion of the solvent molecules. This restriction brings the protein below its dynamical transition. This picture provides a microscopic explanation for the observed correspondence of the solvent dynamic crossover temperature T_{C} and the protein dynamical transition temperature T_{D}. It agrees with the idea proposed some time ago [43] that proteins are "slaves" of the solvent. The dynamics of water bonded to the protein surface are known to differ from the dynamics of bulk water, which might affect the T_{C} of solvent molecules at the protein interface, but this effect might be of secondary importance.

21.5 Fast Picosecond Relaxation

In this section we focus our attention on the fast relaxation process. It is present in the spectra of all proteins even at very low temperatures. It overlaps with the boson peak vibrations and has a stretched low-frequency tail (Fig. 21.2). Changes in temperature lead to a change in amplitude rather than a change in relaxation time (Fig. 21.3b). All these properties are known for the fast relaxation process in glass-forming systems [63]. Nevertheless, the microscopic picture of the fast relaxation in proteins and DNA remains unclear. The small mean square atomic displacements $<x^2>$ associated with the fast relaxation process (see for example, the data for $<x^2>$ shown in Fig. 21.6) suggest it is a local process, and we interpret it as involving fast conformational fluctuations of side-chains, and perhaps peptide groups. These fast conformational fluctuations might be the necessary precursors for global conformational transitions in proteins that happen at much longer times.

It has been emphasized in many papers that at ambient temperature the conformational relaxation contribution in hydrated biomolecules is stronger than in dry ones. It is observed as a higher quasielastic scattering, and is primarily associated with the slow relaxation process that appears in the Gigahertz frequency window (Fig. 21.6). The situation, however, changes at low temperatures, when the slow relaxation process is out of the frequency window of neutron spectroscopy: The fast process in the spectra of dry samples appears to have a higher intensity than in hydrated samples as seen for example between 40 and 200 GHz in the spectra for DNA at 150 K [3] (Figs. 21.3b and 21.8). This also is clearly seen for lysozyme [4], myoglobin [32] and α-amylase [64], but has not been emphasized. A similar behavior is observed in molecular dynamics simulations of ribonuclease A in which translational diffusion of water is suppressed [36]: the amplitude of the fast decay is larger in the dry protein than in the hydrated one. This observation suggests that fast conformational fluctuations in dry protein are stronger than in the hydrated protein. Apparently, water of hydration suppresses these fast fluctuations.

The microscopic nature of the fast relaxation in proteins remains unclear. It is not well understood even in the case of glass-forming systems. In the latter it is usually ascribed either to the rattling of an atom in a cage formed by its neighbors [55, 56], or to small local jumps between substates separated by low-energy barriers inside a megabasin in the systems energy landscape [65]. In the former case, it would correspond to librational ("rattling") motions of amino-acid residues in a cage formed by their neighbors and the water of hydration. This idea can explain the observed decrease of fast fluctuations in hydrated biomolecules: Water molecules, being rather small, fill free volume, and reduce the cage size available for rattling motions of amino-acid residues

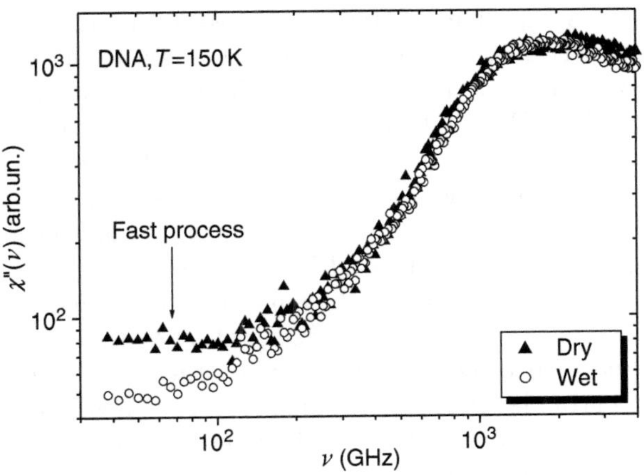

Fig. 21.8. Susceptibility spectra of dry (~0.03 g D_2O per gram of DNA) and hydrated (~0.46 g of D_2O per gram of DNA) DNA at low temperature (data from [3]). Fast process appears in the dry sample appears to be stronger than in the wet sample

or base-pairs. Fast fluctuations of peptide groups may also contribute to the fast relaxation process. ^{2}H NMR studies of spin–lattice relaxation times of ^{2}H-labeled lysozyme provide evidence of small amplitude (9°–12°) torsional motions with correlation times of about 15 ps [66].

The energy landscape approach is similar to the idea of hierarchical states proposed for proteins by Frauenfelder and coworkers [45]. Using this approach, one can estimate the distribution of barrier heights $g(V)$ between substates. This approach has been used for analysis of susceptibility spectra of some glass formers [63, 67]. In a simple approximation one can estimate $g(V)$ from the measured $\chi''(\nu)$ [63, 67]

$$g(V) \propto \chi''(\nu)/T; \text{ where } V = -T\ \ln(2\pi\nu\tau_0). \tag{21.4}$$

The nature of the fast relaxation can be analyzed using coherent neutron scattering which can provide microscopic information on the molecular motions involved, and the cooperativity of motions of amino-acid residues.

The role of the fast relaxation in the function of biological molecules is not understood. Fast fluctuations are thought to be a necessary precursor for any slower processes, including those important for protein function. Suppression of the fast relaxation process in proteins by putting them in various sugar formulations is known to greatly increase their deactivation time [68, 69].

21.6 Conclusions and Future Prospects

The dynamics of proteins and nucleic acids have been the subject of intense study with a variety of experimental methods and computational modeling. Neutron spectroscopy occupies a special position in such studies (along with techniques such as Rayleigh Scattering of Mössbauer radiation) because the experimental observable, the double differential scattering cross-section related to the dynamic structure factor, combines time- and space-correlations in molecular motion. The combination of neutron spectroscopy with computer simulations, in our view, is emerging as one of the best approaches to study the dynamics of complex systems such as proteins. This combination helps to verify parameters of computer modeling and to visualize the motions that appear in scattering functions probed by neutrons.

One very promising direction is the analysis of the dynamics of selectively deuterated proteins [70, 71]. Analysis of selectively deuterated synthetic polymers is a well-developed method that provides rather precise information on the motion of particular parts of the molecule. It is now possible to synthesize proteins with cell-free protein synthesis systems that will allow specific deuterated amino acids to be incorporated into the protein or for all amino-acid residues in the protein to be deuterated. The main problem is in the quantities of sample required for modern neutron scattering spectroscopy – usually ∼100 mg or more. Production of this amount of deuterated protein is usually beyond the resources of most laboratories. However, the European Community has recently established a laboratory for large-scale production of deuterated molecules [72]. These problems with sample size should also be alleviated when new spallation neutron sources become available. It is expected that they will provide signals ~100 (in the most optimistic estimate even ~1,000) times stronger than current instruments. With these sources samples required for scattering measurements might be reduced to 1–10 mg, i.e., to the amount available to many biochemical laboratories.

Precise selective deuteration of biological macromolecules will open tremendous possibilities for analysis of motion of active sites, relative motions of secondary structures, for studies of dynamic heterogeneity in proteins (see, for example, ideas formulated in [73, 74]) and many other questions. Completely deuterated proteins will open the possibility for analysis of coherent scattering that will allow the analysis of pair-correlation functions.

We also expect that significant advantages in understanding protein and DNA dynamics might be achieved by using ideas developed in the field of dynamics of complex systems. Significant progress in this field has been achieved during the last two decades. It was based on developments of two approaches – the energy landscape and mode coupling theory. In particular, it emphasizes that the dynamics of liquid systems can be separated into three regimes [65]: (i) simple liquid behavior at very high temperatures; (ii) onset of some collective (or coupled) dynamics, which happens below some onset temperature T^*. This region is usually called "energy landscape influenced" and extends down to the dynamic crossover temperature T_C; (iii) crossover from liquid-like to solid-like dynamics on the molecular scale which occurs at T_C; the liquid becomes nonergodic on a short time scale. The region at $T < T_C$ is called "energy landscape dominated". The first regime has no interest for biological macromolecules because of its high temperature. In our view, biomolecules in their native environment are in the second regime. In this regime, relaxation over large barriers does not dominate the dynamics. The hierarchy of conformational substates available to the native protein gives rise to a rough energy landscape with a broad distribution of barrier heights. The current view is that the system can relax or undergo conformational changes by negotiating its way around the high-energy barriers in the multidimensional energy landscape. It does not have to jump over these barriers. Below T_D, however, the system becomes trapped in deep energy minima of the landscape. As a result, harmonic vibrations dominate $S(Q, \nu)$ and over-barrier relaxation becomes the dominant mechanism of conformational motion in the third regime. Application of these ideas to analysis of dynamics of proteins and DNA might be very helpful in understanding microscopic mechanisms of molecular motions. Neutron spectroscopy continues to play a key role in the analysis of dynamics of complex systems and with the developments in new neutron sources and the production of deuterated proteins should receive increased attention from the biophysical and biochemical community in future.

Acknowledgements

AS is grateful to NSF (DMR-0315388) for financial support, and to ILL and NIST Neutron Center for assistance with neutron measurements.

References

1. J. Perez, J.M. Zanotti, D. Durand, Biophys. J. **77** (1999), 454
2. M. Tarek, D.A. Neumann, D.J. Tobias, Chem. Phys. **292** (2003), 435
3. A.P. Sokolov, H. Grimm, A. Kisliuk, A.J. Dianoux, J. Biol. Phys. **27** (2001), 313
4. J.H. Roh, G. Caliskan, A.P. Sokolov, R. Gregory, et al. (unpublished)
5. C.L. Brooks III, M. Karplus, B.M. Pettitt, in *Proteins: A Theoretical Perspective of Dynamics, Structure, and Thermodynamics* (Wiley Interscience, 1988) p. 19
6. S. Longeville, W. Doster, G. Kali, Chem. Phys. **292** (2003), 413
7. W. Doster, S. Cusack, W. Petry, Phys. Rev. Lett. **65** (1990), 1080
8. J. Fitter, R.E. Lechner, N.A. Dencher, Biophys. J. **73** (1997), 2126
9. A.P. Sokolov, H. Grimm, R. Kahn, J. Chem. Phys. **110** (1999), 7053
10. M. Tarek, D.J. Tobias, J. Chem. Phys. **115** (2001), 1607
11. F.G. Parak, Rep. Prog. Phys. **66** (2003), 103
12. G. Caliskan, A. Kisliuk, A. Sokolov, J. Non-Crystallogr. Sol. **307–310** (2002), 868
13. C.P. Lindsey, G.D. Patterson, J. Chem. Phys. **73** (1980), 3348
14. J.L. Green, J. Fan, C.A. Angell, J. Am. Chem. Soc. **98** (1994), 13780
15. C.A. Angell, Science **267** (1995), 1924
16. W. Doster, M. Settles, in *Workshop on Hydration Processes in Biology: Theoretical and Experimental Approaches*, M.C. Bellissent-Funel (Ed.) (Les Houches, IOS Press, 1999) p. 177
17. D. Richter, et al., J. Phys. Condens. Matter **11** (1999), A297
18. J. Colmenero, et al., J. Phys. Condens. Matter **11** (1999), A363
19. F. Volino, A.J. Dianoux, Mol. Phys. **41** (1980), 271
20. A.L. Tournier, J.C. Smith, Phys. Rev. Lett. **91** (2003), 208106
21. J.A. Rupley , P. Yang, G. Tollin, in *Water in Polymers*, S.P. Rowland (Ed.) (Am. Chem. Soc., Washington, DC, 1980) pp. 111–132
22. O.V. Belonogova, E.N. Frolov, S.A. Krasnopol'skaya, B.P. Atanasov, V.K. Gins, E.N. Mukhin, A.A. Levina, A.P. Andreeva, G.I. Likhtenshtein, V.I. Goldanskii, Dokl. Akad. Nauk USSR **241** (1978), 219
23. G.I. Likhtenshtein, V.R. Bogatyrenko, A.V. Kulikov, Appl. Magn. Reson. **4** (1993), 513
24. F. Parak, Methods Enzymol. **127** (1986), 196
25. G.B. Strambini, E. Gabellieri, Photochem. Photobiol. **39** (1984), 725
26. N.K. Shah, R.D. Ludescher, Photochem. Photobiol. **58** (1993), 169
27. R.B. Gregory, M. Gangoda, R.K. Gilpin, W. Su, Biopolymers **33** (1993), 513
28. P.L. Poole, J.L. Finney, Int. J. Biol. Macromol. **5** (1983), 308
29. J.E. Schinkel, N.W. Downer, J.A. Rupley, Biochemistry **24** (1985), 352
30. S. Bone, R.J. Pethig, Mol. Biol. **181** (1985), 323

31. R. Pethig, in *Protein–Solvent Interactions*, R.B. Gregory (Ed.) (Marcel Dekker, New York, 1995) pp. 265–288
32. V.I. Goldanskii, Y.F. Krupyanskii: Rev. Biophys. **22** (1989), 39; V.I. Goldanskii, Y.F. Krupyanskii, in *Protein-Solvent Interactions*, R.B. Gregory (Ed.) (Marcel Dekker, New York, 1995) pp. 289–326
33. M. Diehl, W. Doster, W. Petry, A. Schulte, Biophys. J. **73** (1997), 2726
34. R.B. Gregory, in *Protein–Solvent Interactions*, R.B. Gregory (Ed.) (Marcel Dekker, New York, 1995) pp. 191-264
35. R.W. Lumry, A. Rosenberg, Colloques. Int. CNRS **246** (1975), 53
36. M. Tarek, D.J. Tobias, Phys. Rev. Lett. **88** (2002), 138101
37. A.L. Tournier, J. Xu, J.C. Smith, Biophys. J. **85** (2003), 1871
38. D. Vitkup, D. Ringe, G.A. Petsko, M. Karplus, Nat. Struct. Biol. **7** (2000), 34–38
39. A. Ansari, C.M. Jones, E.R. Henry, J. Hofrichter, W.A. Eaton, Science **256** (1992), 1796
40. T. Kleinert, et al., Biochemistry **37** (1998), 717
41. H. Lichtenegger, et al., Biophys. J. **76** (1999), 414
42. B. Gavish, M.M. Werber, Biochem. **18** (1979), 1269
43. D. Beece, L. Eisenstein, H. Frauenfelder, D. Good, M.C. Marden, L. Reinish, A.H. Reynolds, L.B. Sorensen, K.T. Yue, Biochem. **19** (1980), 5147
44. M. Settles, W. Doster, F. Kremer, F. Post, W. Schirmacher, Philos. Mag. **65** (1992), 861
45. R.H. Austin, K.W. Beeson, L. Eisenstein, H. Frauenfelder, I.C. Gunsalus, Biochem. **14** (1975), 5355
46. B. Gavish, S. Yedgar, in *Protein–Solvent Interactions*, R.B. Gregory (Ed.) (Marcel Dekker, New York, 1995) pp. 343–373
47. S.N. Timasheff, in *Protein–Solvent Interactions*, R.B. Gregory (Ed.) (Marcel Dekker, New York, 1995) pp. 445–482
48. R. Gregory, Biopolymers **27** (1988), 1699
49. W. Doster, S. Cusak, W. Petry, Nature **337** (1989), 754
50. B.F. Rasmussen, et al., Nature **357** (1992), 423
51. A.M. Tsai, D.A. Neumann, L.N. Bell, Biophys. J. **79** (2000), 2728
52. M. Ferrand, A.J. Dianoux, W. Petry, G. Zaccai, Proc. Natl Acad. Sci. USA **90** (1993), 9668
53. R.M. Daniel, J.C. Smith, M. Ferrand, S. Hery, R. Dunn, J. Finney, Biophys. J. **75** (1998), 2504
54. L. Cordone, M. Ferrand, E. Vitrano, G. Zaccai, Biophys. J. **76** (1999), 1043
55. W. Gotze, L. Sjogren, Rep. Prog. Phys. **55** (1992), 241
56. see, for example, Transp. Theory Stat. Phys. **24** (1995), Special issue devoted to Relaxation Kinetics in Supercooled Liquids – Mode Coupling Theory and Its Experimental Tests
57. W. Stefen, A. Patkowski, H. Glaser, G. Meier, E.W. Fischer, Phys. Rev. E **49** (1994), 2992
58. A.P. Sokolov, W. Steffen, E. Rössler, Phys. Rev. E **52** (1995), 5105
59. A.P. Sokolov, Science **273** (1996), 1675; Endeavour **21** (1997), 109
60. A.P. Sokolov, J. Hurst, D. Quitmann, Phys. Rev. B **51** (1995), 12865
61. E. Rössler, A.P. Sokolov, A. Kisliuk, D. Quitmann, Phys. Rev. B **49** (1994), 14967
62. A.P. Sokolov, J. Non-Crystallogr. Solids **235–237** (1998), 190

63. N.V. Surovtsev, J. Wiedersich, V.N. Novikov, E. Rössler, A.P. Sokolov, Phys. Rev. B **58** (1998), 14888
64. J. Fitter, Biophys. J. **76** (1999), 1034
65. P.G. Debenedetti, F.H. Stillinger, Nature **410** (2001), 259
66. J.W. Mack, M.G. Usha, J. Long, R.G. Griffin, R.J. Wittebort, Biopolymers **53** (2000), 9
67. G. Caliskan, A. Kisliuk, V.N. Novikov, A.P. Sokolov, J. Chem. Phys. **114** (2001), 10189
68. G. Caliskan, A. Kisliuk, A. Tsai, C. Soles, A.P. Sokolov, J. Chem. Phys. **118** (2003), 4230
69. M.T. Cicerone, A. Tellington, L. Trost, A. Sokolov, Bioprocess Int. **1** (2003), 2
70. V. Reat, H. Patzelt, C. Pfister, M. Ferrand, D. Oesterhelt, G. Zaccai, PNAS USA **95** (1998), 4970
71. A. Orecchini, A. Paciaroni, A.R. Bizzarri, S. Canistraro, J. Phys. Chem. B **106** (2002), 7348
72. Information about the ILL-EMBL Deuteration Laboratory can be found on the web: www.ill.fr/deuteration
73. Y. Zhou, D. Vitkup, M. Karplus, J. Mol. Biol. **285** (1999), 1371
74. S. Dellerue, A.J. Petrescu, J.C. Smith, M.C. Bellissent-Funel, Biophys. J. **81** (2001), 1666

22

Structure and Dynamics of Model Membrane Systems Probed by Elastic and Inelastic Neutron Scattering

T. Salditt, M.C. Rheinstädter

22.1 Introduction

Phospholipid membranes are intensively studied as simple model systems to understand fundamental structural and physical aspects of their much more complex biological counterparts [1]. The lateral structure of membranes, including both height and compositional fluctuations, remains an important experimental challenge of present-day biophysics, concerning in particular the biologically relevant fluid L_α state, where the material softness compromises the use of scanning probe microscopy. Neutron scattering can contribute to the elucidation of the molecular structure, as is well documented in the literature (see, e.g., [2]).

Dynamical properties are often less well understood in biomolecular systems, but are important for many fundamental biomaterial properties, e.g., elasticity properties and interaction forces. Furthermore, lipid membrane dynamics on small molecular length scales determines or strongly affects functional aspects, like diffusion and parallel or perpendicular transport through a bilayer. The specific advantages of neutron scattering to study fluctuations of phospholipid membranes on lateral length scales between several micrometer down to a few Ångströms can give unique insights.

The present chapter concentrates on mainly two neutron scattering techniques, which give information on very different types of dynamics: (i) nonspecular neutron reflectivity (NSNR) as a tool to probe thermal fluctuations of lipid bilayers on mesoscopic length scales, and (ii) inelastic neutron scattering (INS) for studies of the short-range collective motions in the acyl chains. The methods can also be applied to more complex model systems, including lipid – peptide and lipid–protein mixtures, as well as in some cases to real biological membranes like purple membranes.

The chapter is so organized that at first sample preparation and sample environment for multilamellar lipid phases on solid support is presented. Next, specular neutron reflectivity (SNR) and NSNR from lipid membranes is described. Mainly work on pure lipid phases is presented and reviewed, a few examples relate to structure and interaction of the antimicrobial peptide Magainin 2 in phosphocholine bilayers, as an example of how the methods presented can be extended and applied to probe lipid – peptide interaction or more generally the interaction of bilayers with membrane-active molecules (such as stereols, peptides, and proteins). Afterward recent advances are highlighted in the application of classical INS at triple-axis spectrometers (TAS) to study the short-range dynamics of lipid bilayers, e.g., the collective motion of the acyl chains. The chapter closes with a summary and conclusions.

22.2 Sample Preparation and Sample Environment

Highly oriented multilamellar bilayers for neutron reflectivity can be easily prepared on cleaned silicon wafers ((1 1 1)-orientation or (1 0 0)-orientation) by spreading from organic solution [3]. The wafers are cleaned by subsequent washing in methanol and ultrapure water (specific resistivity $\geq 18\,\mathrm{M\Omega\,cm}$), and made hydrophilic by washing in a 5-molar solution of KOH in ethanol for about a minute, or alternatively by plasma etching. Lipid and/or peptide components are codissolved in the desired ratio (molar ratio P/L) in trifluoroethanol (TFE) or (1:1) TFE–chloroform mixtures at concentrations between 10 and $40\,\mathrm{mg\,ml^{-1}}$, depending on the total mass to be deposited.

After a slow evaporation process avoiding film rupture, remaining traces of solvent in the sample are removed by exposing the samples to high vacuum overnight. Then the films are rehydrated in a hydration chamber and tempered above the main phase transition. The orientational alignment of such multilamellar stack with respect to the substrate (mosaicity) is typically better than 0.01°. A very low mosaicity is a prerequisite to apply interface-sensitive scattering techniques. The lateral domains sizes are in the range of $100\,\mu\mathrm{m}$, exhibiting a broad distribution in the total number N of the bilayers.

For measurements on samples immersed in water thick (1 cm), polished silicon blocks are used, where the neutron beam can be coupled into the substrate without refraction from the side. The beam then impinges at grazing incidence onto the sample with the incoming and reflected beam path in silicon rather than in water, avoiding incoherent scattering in H_2O or D_2O/H_2O mixtures (Fig. 22.1b). The low mosaicity of the samples prepared was preserved after immersing the samples in water.

For INS experiments highly oriented membrane stacks were prepared essentially in the same way as presented earlier [4]. However, ten or more such wafers separated by small air gaps are combined and aligned with respect to each other to create a "sandwich sample" consisting of several thousands of highly oriented lipid bilayers (total mosaicity of about 0.6°), with a total mass of several hundred milligrams of deuterated phospholipid, to maximize the scattering volume. Figure 22.1c shows a photograph of a sample for the inelastic neutron experiments.

During the neutron experiments, the solid-supported multilamellar films are kept in temperature and hydration controlled chambers. For measurements

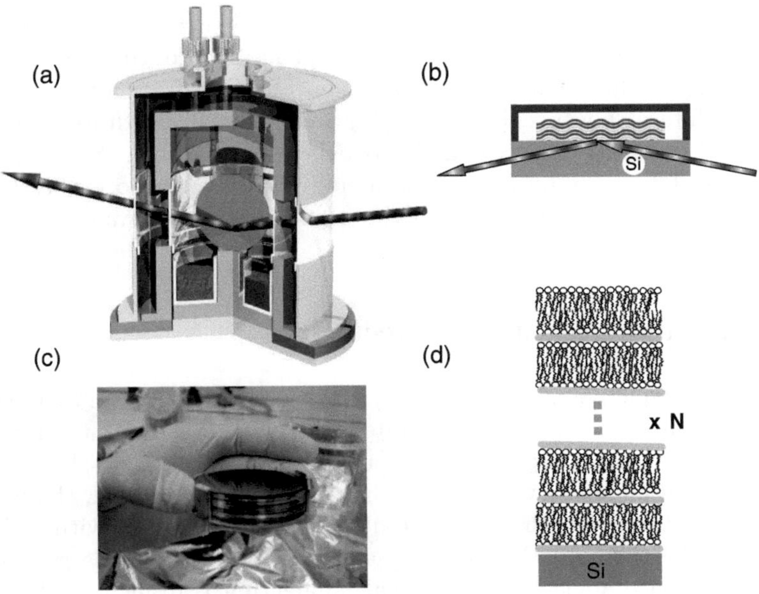

Fig. 22.1. (**a**) Schematic of the chamber used for the neutron experiments to control temperature and humidity of the bilayers. (**b**) Samples immersed in water were applied on 1 cm thick Si wafers. The neutron beam was then coupled into the wafer from the side. (**c**) Photograph of the "sandwich sample" prepared for the inelastic neutron experiments. (**d**) Hydrated multilamellar sample with layers of water (H_2O/D_2O) between the bilayers, respectively

carried out at partial hydration with the bilayers facing D_2O vapor, a chamber with two concentric high-purity aluminum cylinders was used, see Fig. 22.1a. The inner cylinder was heated or cooled by a flow of oil, connected to a temperature-controlled reservoir. The space between the two cylinders was evacuated to minimize heat conduction. The temperature was measured close to the sample holder by a Pt100 sensor, indicating a thermal stability of better than 0.02 K over several hours. At the bottom of the inner cylinder, a water reservoir was filled with salt-free Millipore water, such that the sample was effectively facing a vapor phase of nominally 100% relative humidity.

Despite the nominally full hydration condition, bilayer samples of DMPC were typically swollen only up to a repeat distance of only $d \simeq 50$–55 Å in the fluid L_α-phase, i.e., were only partially hydrated. This limited swelling of solid-supported lipid films is well known as the so-called vapor-pressure paradox, and has recently been explained by Katsaras and coworkers on the basis of small temperature gradients in the chamber [5,6]. Accordingly, it was demonstrated that chambers of suitable design do not show this effect, and that bilayers can be swollen to equilibrium (full hydration) from the vapor phase.

Alternatively, a chamber operating at full hydration has been used, where the bilayers are immersed in water, and the beam impinges through a thick

Si block as shown in Fig. 22.1b. This setup is the neutron analog of an X-ray chamber used in a recent temperature-dependent study of the specular reflectivity on aligned lipid bilayers [7]. Conditions of full hydration are important, since the diffuse (nonspecular) scattering from partially hydrated films may be affected by static defects and the associated strain fields [8], rather than by thermal diffuse scattering. Furthermore, it is desirable to probe the elasticity and fluctuation properties in the physiologically relevant state of full hydration.

22.3 Specular Neutron Reflectivity

SNR offers unique possibilities of studying the structure of thin biomolecular films and membranes on solid substrates or at the air–water interface and is widely used for this purpose complementing X-ray techniques by means of contrast variation. As is well known, the vertical structure, i.e., the laterally averaged scattering length density profile $\rho(z)$, can be derived with molecular resolution. To this end the SNR signal is measured over a large range of grazing incidence angles α_i, or vertical momentum transfer q_z. The lateral interface structure (variations of the scattering length density in the xy plane) is contained in the nonspecular scattering (NSNR) measured at angles of exit $\alpha_\mathrm{f} \neq \alpha_\mathrm{i}$ [9, 10], as discussed in Sect. 22.4

For the study of model membrane systems, an appropriate control of the sample temperature and hydration is essential. Figure. 22.2 illustrates the effect of hydration on the structure of DMPC bilayers, with two curves at partial and full hydration. Both curves correspond to the L_α phase. At full hydration the high-density (deuterated) water region becomes almost as thick as the low-density (hydrogenated) hydrophobic chain region, leading to a cancellation of

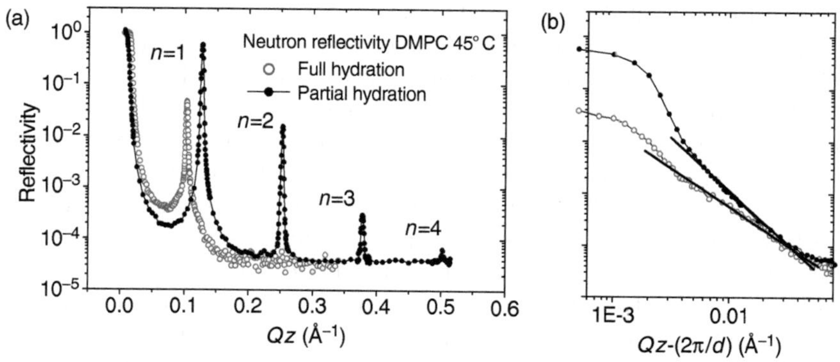

Fig. 22.2. The effect of full (*solid symbols*, measurement with sample immersed in D_2O) versus partial hydration (*open symbols*, measurement in vapor (humidity chamber)) on (**a**) the complete reflectivity curves and (**b**) of the Bragg peak tail in double logarithmic scale. Both curves correspond to DMPC in the L_α phase [4, 11]

the second-order Bragg peak. At the same time the increase in thermal fluctuations at full hydration leads to a significant damping of higher-order peak, so that only one (strong) Bragg peak is observed. At partial hydration more than four Bragg peaks can be measured, depending on the exact relative humidity (osmotic pressure).

The correct analysis of X-ray and neutron reflectivity relies on very low mosaicity (narrow orientational distribution of domains). A necessary condition is a clear distinction between specular and nonspecular scattering components. The data analysis and modeling of the measured reflectivity should be based on an appropriate scattering theory, such as the fully dynamical Parratt algorithm (taking into account multiple reflections) or on the semikinematical reflectivity pioneered by Als-Nielsen [12]. The observation of a region of total external reflection and hence of the critical angle α_c allows for the determination of the scattering length density profile on an absolute scale. Moreover, since the full q_z-range can be used for data analysis by fitting the reflectivity curve to a parametrized model of the density profile [13], a reasonable resolution in $\rho(z)$ can also be reached for fully hydrated systems. Furthermore, the phase problem is reduced, since the change of sign in the bilayer form factor (real valued due to centro-symmetry) is often accompanied by an observable cusp in the (continuously measured) reflectivity curve. Alternatively, phasing can be performed by the so-called swelling method. The advantage of full q_z-fitting has also been demonstrated in bulk (SAXS) studies, see for example [14].

In most published studies of oriented bilayers, however, only the integrated Bragg peaks of the multilamellar samples are used for data analysis, and the one-dimensional density profile $\rho(z)$ is computed by Fourier synthesis using a discrete set of Fourier coefficients f_n as described in [15, 16]. In this approach, the exact relation between Bragg peak intensity and the Fourier coefficients f_n is an open problem for which there may not even exist a general solution. Taking into account effects of absorption, polarization, specular and nonspecular Fresnel reflectivity components, illumination correction, etc., a widely used correction factor is $I_n = |f_n|^2/q_z$, where q_z^{-1} is termed a Lorentz factor for oriented bilayers. In the absence of a rigorous theoretic derivation, such a correction is at best empirical. Furthermore, the two scattering contributions of specular and nonspecular scattering are often measured in the same scan, adding up the background and making quantitative analysis questionable, since both contributions are governed by a different q_z dependence. However, the peak-to-peak distance of a reconstructed bilayer profile is luckily relatively stable against variations in data analysis like different choices of Lorentz factors.

Figure 22.3a shows DMPC reflectivity curves measured at a rather low resolution on a TAS, along with the results for the density profiles obtained by a Fourier synthesis approach (Fig. 22.3b). The data have been recorded as function of T below and above the main phase transition, simultaneously with inelastic data discussed in Sect. 22.5. The temperature-dependent structural

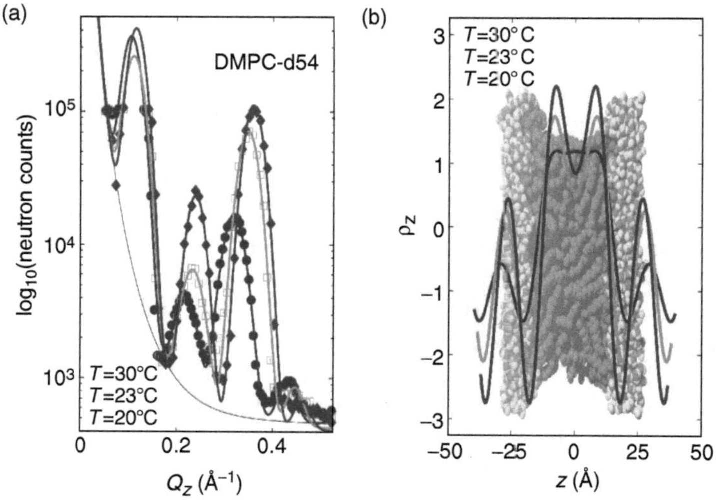

Fig. 22.3. (**a**) Reflectivity curves and (**b**) scattering length density profiles as obtained from Fourier synthesis of integrated peak intensities (d54-DMPC, hydrated from D_2O, IN12/ILL, from *top* T =30°C to *bottom* T = 20°C)

parameters of the bilayer profile are shown in Fig. 22.4. The two different phases show distinct differences in ρ_z. Whereas the scattering length density in the center of the bilayer in the gel phase (T = 20°C) is almost box shaped, it softens in the more loosely packed and dynamic fluid phase (T = 30°C).

The density profile $\rho(z)$ has been calculated using the relation $\rho_0(z) = \frac{2}{d_z}\sum f_n \cos\left(\frac{2n\pi}{d_z}z\right)$. Insertion of $|f_n|^2 = I_n q_z = I_n 2\pi n/d_z$ gives

$$\rho_0(z) = \frac{2}{d_z}\sum_{n=1}^{M}\sqrt{\frac{2n\pi}{d_z}}\sqrt{I_n}\,v_n \cos\left(\frac{2n\pi}{d_z}\,z\right), \quad z \in \left[-\frac{d_z}{2}, \frac{d_z}{2}\right] \tag{22.1}$$

with I_n the integrated intensity of the nth Bragg reflection, v_n the corresponding phase of f_n and d_z the periodicity of the layers in z direction. The sum goes over all orders of reflection.

In an attempt to go beyond the conventional Fourier synthesis approach, we have developed a reflectivity model in the framework of the semikinematical scattering theory, in which both the structure factor of the stack and the bilayer form factor can be suitably chosen [13], e.g., according to the given experimental resolution. This is possible since the lipid bilayer density profile $\rho_1(z)$ and the associated form factor $F(q_z)$ is parametrized by a variable number N of Fourier coefficients taken to describe the model density profile $\rho(z)$, where N is adapted to the resolution of the measurement. In contrast to conventional box models the total number of parameters can thus be kept small, while still fitting to reasonable density profiles. Moreover, structural constraints can be easily implemented by a transformation to N independent structural parameters.

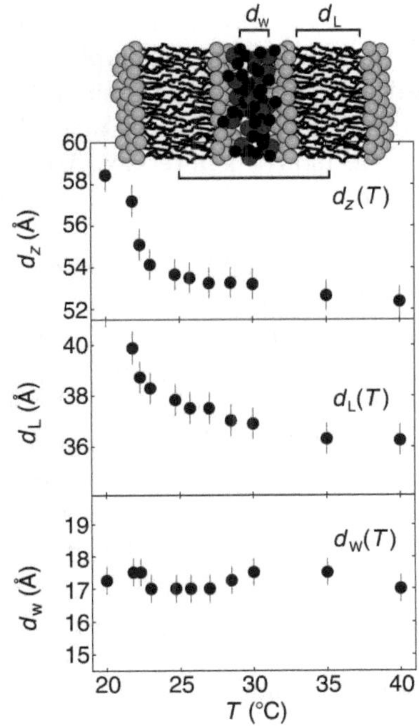

Fig. 22.4. Temperature dependence of the periodicity d_z, water layer d_{w}, and hydrophobic layer d_{L} (deuterated acyl chains) thickness of DMPC in the vicinity of the main phase transition. The values are evaluated from the density profiles as shown in Fig. 22.3

Starting point for this treatment is the so-called master equation of reflectivity from a structured interface in the semikinematic approximation [12]. There, the reflectivity from an interface with its normal along z is characterized by the (laterally averaged) scattering length density profile $\rho(z)$ (electron density profile for X-rays) between a medium 1 (air or water) with scattering length density ρ_1 and a medium 2 (solid substrate) with density ρ_2, thus

$$R(q_z) = R_{\mathrm{F}}(q_z)\,|\Phi(q_z)|^2 = R_{\mathrm{F}}(q_z)\left|\frac{1}{\Delta\rho_{12}}\int \frac{\partial\rho_{\mathrm{e}}(z)}{\partial z}\mathrm{e}^{-\mathrm{i}q_z z}\mathrm{d}z\right|^2 , \qquad (22.2)$$

where R_{F} is the Fresnel reflectivity of the ideal (sharp) interface between the two media. $\Delta\rho_{12}$ is the scattering length density contrast. ρ is obtained by the combination of the solid surface and a step train of lipid bilayers, convolved with a function describing the positional fluctuations and multiplied by a coverage function. The critical momentum transfer or the critical angle in R_{F} is directly related to the density contrast by $q_z = 4\pi/\lambda\,\sin(\alpha_{\mathrm{c}}) \simeq 4\sqrt{\pi\Delta\rho_{12}}$. Absorption can be accounted for by an imaginary component of the wave vector.

The solid–membrane interface is the only relevant interface for the R_{F} term due to the following reason: (i) in many cases the membrane sample impinges through the water phases and there is almost no contrast between membrane and water (for X-rays), so no refraction takes place at the water–membrane interface, (ii) due to a decreasing coverage with N the water–membrane interface is much broader and less well defined, again leading to vanishing reflectivity and refraction effects at this interface. We assume a monotonously decreasing coverage function $c(N)$ with $c(1) = 1$ and $c(N) = 0$. More details of this approach are discussed in [17].

The thermal fluctuations for a solid-supported stack of lipid bilayers have been calculated in [18] and can be easily included in the structure factor, as well as a decreasing coverage function, or defect densities [19]. In practice, the range of the reflectivity determines the number N_o of orders, which should be included in the model. The parametrization of n Fourier coefficients can be easily changed by a linear transformation into a parametrization of n (independent) structural parameters of the bilayer, such as bilayer thickness (headgroup peak-to-peak), density maximum in the headgroup maximum, density in the bilayer center plane, density of the water layer, etc.

22.4 Nonspecular Neutron Reflectivity

Thermal fluctuations of lipid membranes on molecular to mesoscopic scales reflect fundamental physical properties of the lipid bilayer, related to thermodynamic stability, elasticity, interaction potentials, and phase transitions [1]. Reciprocally, fluctuations may strongly influence these phenomena, as well as different self-assembly properties of lipid membranes. If this interplay is quantitatively understood in simple model systems, more complex biomimetic membrane systems with membrane-active peptides and membrane proteins can be addressed. Thermal fluctuations in lipid bilayer phases have been probed by X-ray and neutron scattering with lineshape analysis, carried out on aqueous bulk suspensions [20, 21]. Multilamellar phases of stacked lipid bilayers exhibiting smectic liquid–crystalline symmetry have received particular attention [20–23]. They allow for a tremendous increase in scattering volume over single bilayer or more dispersed phases. Interaction potentials between neighboring membranes can also be inferred from such studies [20, 21, 24].

However, the full information on the characteristic displacement correlation functions is lost due to powder averaging over the isotropic distribution function of domains in solution. Correspondingly, model assumptions have to be introduced in the data analysis, such as the Caillé or modified Caillé models derived from linear smectic elasticity theory. In this model two elastic constants B and K govern the compressional and bending modes of the smectic phase, respectively. For large bilayer bending rigidity $\kappa \gg kT$ typical for phospholipid systems, only the combination $\sqrt{KB}$ of the two Caillé moduli B and $K = \kappa/d$ can be inferred from the measured lineshape exponent $\eta = \pi kT/2d^2\sqrt{KB}$. The second fundamental parameter, the smectic penetration length $\Lambda = \sqrt{K/B}$ is usually not accessible, unless the bending rigidity becomes small $\kappa \simeq kT$ [20, 21].

In aligned (oriented) lamellar phases on solid substrates the study of fluctuations does not suffer from the above restrictions. In such systems, specular and diffuse (nonspecular) X-ray (NSXR) and neutron reflectivity (NSNR) yield model independent information on the height–height correlation functions [8,25]. In both cases the nonspecular (diffuse) intensity distribution can be described by a structure factor $S(q_z, q_{\|})$, measured as a function of the parallel and vertical component of the momentum transfer [26], $q_{\|} = \sqrt{q_x^2 + q_y^2}$ and q_z, respectively.

Using NSNR in addition to NSXR or alone offers several specific advantages. The transparency of the solid substrate for neutrons gives access to a continuous range of parallel momentum transfer $q_{\|} = \sqrt{q_x^2 + q_y^2}$ at fixed vertical momentum transfer (e.g., for constant $q_z = 2\pi/d_z$), see Fig. 22.5, opening up the possibility of studying fluctuations, on length scales between a few Ångstroms up to several micrometer, essentially in one scan [4]. In time-of-flight (TOF) mode the simultaneous collection of neutrons over a

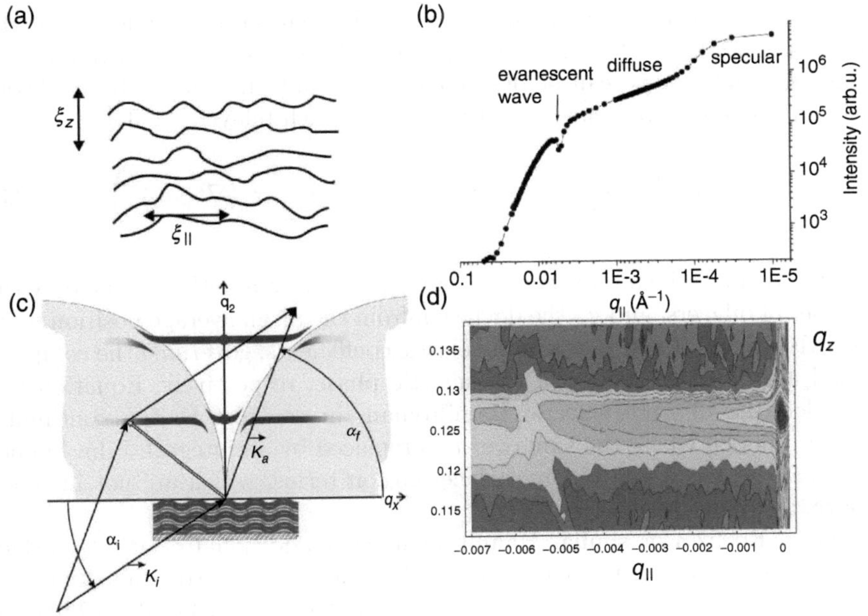

Fig. 22.5. (**a**) Schematic of fluctuating membranes causing diffuse (nonspecular) scattering. The precise nature of the height–height correlation functions, and in particular the characteristic correlation length in parallel, $\xi_{\|}$, and perpendicular direction, ξ_Z, determine the scattering distribution, as shown in (**b**) and (**d**) (DMPC, L_α phase, hydrated from D_2O [4]). (**b**) Intensity decrease with $q_{\|}$ after integration along q_z of the two-dimensional intensity mapping shown in (**d**). Refraction effects and intensity modulation due to evanescent waves are observed when the incident or exit beam is close to 0 (**c**)

range of wavelengths $\lambda \in 2, \cdots, 20$ Å becomes possible, significantly reducing accumulation time. Thus time-dependent white-beam measurements can be carried out without the risk of radiation damage, as is often the case in X-ray scattering. For specular reflectometry, the advantages of TOF, in particular the small accumulation times, have already been well established, e.g., see [27] and references therein. The contrast variation, which is widely used in specular neutron reflectometry [28, 29], allows to label specific molecular groups.

Up to now, only a few studies have presented quantitative analysis of nonspecular (diffuse) scattering in TOF mode, see e.g., [10, 30–32], and some technical aspects related to resolution and data treatment have to be discussed in more detail or are presently still a matter of debate.

22.4.1 Models of Bilayer Undulations

Multilamellar membrane fluctuations can be described by the linearized free energy density of a 3D smectic liquid crystal [20, 21, 33–36]. On large length scales the finite size effects and the presence of film boundaries (e.g., at the substrate) become apparent and limit the fluctuation amplitudes σ_n of the bilayers [18, 37]. On short length scales the bilayer undulations (bending and compressional modes) can be described by the bulk smectic elasticity theory using a discrete displacement field $u_n(r, z)$ for each bilayer [34–38],

$$H = \int_A \mathrm{d}^2 r \sum_{n=1}^{N-1} \left(\frac{1}{2} \frac{B}{d} \left(u_{n+1} - u_n\right)^2 + \frac{1}{2} \kappa \left(\nabla_{xy}^2 u_n\right)^2 \right) , \qquad (22.3)$$

where κ denotes the bilayer bending rigidity, A the area in the xy-plane, N the number of bilayers, and u_n the deviation from the mean average position nd of the nth bilayer. B and $K = \kappa/d$ are elastic coefficients, governing the compressional and bending modes of the smectic phase, respectively. Equation 22.3 is called the discrete smectic Hamiltonian, in contrast to the continuum (Caillé) model, where the sum over n is replaced by an integral. Film boundaries can be accounted for by surface tension terms, which are not included above.

From Eq. 22.3 or similar Hamiltonians that additionally include surface terms, the characteristic height–height (displacement) correlation functions $g_{ij}(r) = \langle [u_i(\mathbf{r}') - u_j(\mathbf{r}' + \mathbf{r})]^2 \rangle$ can be calculated [18, 34, 35], describing the self and the cross-correlations of the bilayers labeled by i and j [9, 39].

The height–height self-correlation functions $g(r) = g(r, j = i)$ are a special case of particular interest. The correlation functions are characterized by several length scales: (i) the maximum lateral wavelength of fluctuations $\xi_{\max}$ (finite only in thin films, infinite in bulk systems), (ii) the rms fluctuation amplitude of the nth bilayer σ_n, measured on the lateral length scale $\xi_{\max}$, and (iii) the vertical length scale ξ_z over which the fluctuations of wavelength

$\xi_{\max}$ are correlated, defining the conformality (Fig. 22.5a). The conformality or cross-correlations of the bilayer undulations leads to the presence of diffuse Bragg sheets in reciprocal space, as can be directly verified from Eq. (22.4).

On length scales $r > \xi_{\max}$ the bilayers are essential flat, since the associated smectic damping length $\xi_z \simeq \xi_{\max}^2/\Lambda$, with $\Lambda = \sqrt{K/B}$ excludes a corresponding relaxation of the profile within the film thickness D. The presence of a sharp specular Bragg peak on top of the diffuse Bragg sheet reflects the flatness of the bilayers on the macroscopic length scale. On length scales $r \ll \xi_{\max}$ the fluctuations are not affected by the the film boundaries and should be described by bulk smectic theory. While $\xi_{\max}$ and ξ_z depend on Λ, σ scales with d^2 and the dimensionless constant $\eta = \pi kT/2d^2\sqrt{KB}$. The diffuse scattering measured in the plane of reflection can be written as a unique transformation of the $g_{ij}(r) = 2\sigma_i\sigma_j - 2c_{ij}(r)$, by [9,39]

$$S(q_x, q_z) = \frac{L_x L_y}{q_z^2} \sum_{i,j}^{N} \Delta\rho^2 \, \mathrm{e}^{-q_z^2\sigma_i\sigma_j} \, \mathrm{e}^{-\mathrm{i}q_z(h_i-h_j)} \, \epsilon_{ij}(\mathbf{q}) \tag{22.4}$$

with

$$\epsilon_{ij}(\mathbf{q}) = \int \mathrm{d}r r \left(\mathrm{e}^{q_z^2 c_{ij}(r)} - 1\right) \cos(q_x r) \; , \tag{22.5}$$

where $\Delta\rho$ is the effective scattering length contrast between the bilayer and D_2O, $L_x L_y$ the illuminated area. The diffuse scattering is integrated over the direction perpendicular to the plane of incidence (hence over q_y), as in the present experiments. For a point-like detector slit setting, the term $\cos(q_x r)$ has to be replaced by $J_o(q_{\|}r)$ with $q_{\|}^2 = q_x^2 + q_y^2$. By insertion of modeled or derived functions for $g_{ij}(r)$ in Eq. 22.4, the scattering distribution can be calculated.

In the following, we mainly consider two characteristic quantities: (i) the cross-correlation length $\xi_z(q_{\|})$ defining the length scale over which a thermal mode is correlated as a function of the corresponding wave vector $q_{\|}$, and (ii) the decay of the q_z-integrated diffuse scattering with $q_{\|}$, i.e., the structure factor of the fluctuations. The cross-correlation function and the parameter ξ_z can be deduced from the peak lineshape and width (HWHM) along q_z. A Lorentzian lineshape indicates an exponential decrease of the cross-correlations along z with a characteristic length scale $\xi_z = 1/\mathrm{HWHM}$. Since the HWHM increases with $q_{\|}$, ξ_z depends on the wave vector of the height fluctuations, or conversely, the lateral length scale of the fluctuation. The linear smectic elasticity predicts $\xi_z = 1/(q_{\|}^2\Lambda)$.

22.4.2 Monochromatic NSNR Experiments

Using monochromatic neutrons the diffuse scattering from thermal fluctuations in phospholipid model systems has been investigated in a series of experiments as described in [4,40]. A typical mapping of the reciprocal space in the region of the diffuse Bragg sheet is shown in Fig. 22.5d for fully hydrated DMPC in the fluid L_α phase at $T = 45°$. The modulations at values of q_x, which correspond to the transition from reflection to transmission geometry, is illustrated in Fig. 22.5c and discussed in detail in [4, 40]. With the beam near the sample horizon, the scattering is modified by refraction effects and evanescent waves. The mapping can be evaluated with respect to the self-correlation and the cross-correlation functions. Most important is the analysis of the decay of the q_z-integrated Bragg sheet intensity $\tilde{S}(q_{\|}) = \int_{\mathrm{BZ}} \mathrm{d}q_z S(q_z, q_{\|})$, BZ denotes one Brillouin zone $2\pi/d \pm \pi/d$. It can be shown for stationary fluctuation amplitudes $\sigma_n =$ const. that the contributions of the cross-correlation terms in Eq. 22.4 vanish, and that one is left with a curve, which corresponds to the transform of an average height–height self-correlation function or the effective single-bilayer structure factor $\tilde{S}(q_{\|})$ [41,42]. In many cases, a power law behavior is observed at high $q_{\|}$, indicating a corresponding algebraic regime of the real space correlation function at small r.

In Fig. 22.5b the integrated Bragg sheet intensity for DMPC is plotted in double logarithmic scale, illustrating the large range in intensity and parallel momentum transfer, which can be achieved in these measurements. The curve is related to the power spectral density (PSD) of the average fluctuations. In the limit of small $q_z\sigma$ the curves are strictly proportional to the SPD of the average bilayer. At finite $q_z\sigma$, the intensity decay still contains the information on the PSD, but in addition to the diffuse component there is a specular peak observed at small q_x with a Gaussian lineshape. The specular beam is clearly separated from the diffuse Bragg sheet. The systematic deviations of the observed lineshape (in particular the asymptotic power law slope) at high $q_{\|}$ from the predictions of the Caillé model have independently been found in diffuse X-ray studies of oriented bilayers, measured at full hydration, and points to the limits of the smectic model. Physical reasons for this effect are unclear. It is speculated that protrusion modes governing the fluctuations at small distances in the plane of the bilayer could explain this observation.

22.4.3 White-Beam NSNR Experiments

Neutrons in a broad range of wavelengths λ are recorded simultaneously in TOF-NSNR, and registered as a function of their respective TOF, as well as scattering angle on a two-dimensional multiwire detector. Intensity distributions can be obtained without moving any motors as a function of q_z and q_x. The TOF-NSNR data analysis includes the corrections of the raw data for detector sensitivity, wavelength distribution of the primary beam, resolution, the nontrivial transformation of the detector counts to the intensity matrix I(θ,λ), and/or to the reciprocal space mapping to I(q_x,q_z) [25, 43].

Figure 22.6 shows a representative 2D data set I(2θ, λ) of a multilamellar stack of DMPC in the fluid L_α phase at $T = 40°$C. The different columns of the multiwire detector matrix correspond to different scattering angles 2θ, with

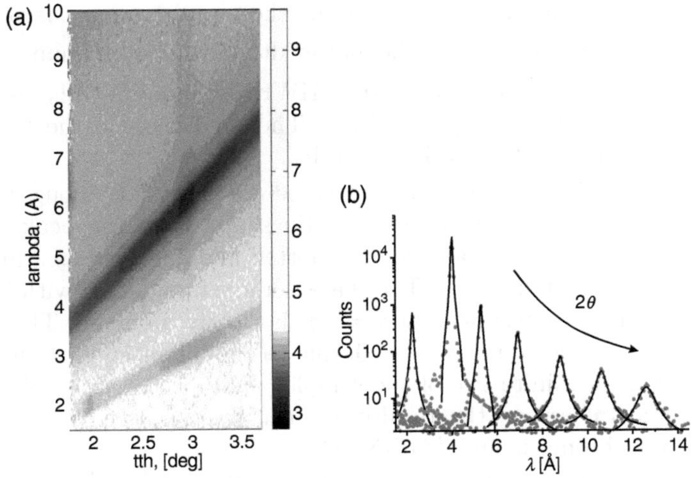

Fig. 22.6. Typical intensity distribution (logarithmically scaled gray shades) observed on the detector at D17 in TOF mode for a DMPC sample in the fluid L_α phase at T = 40°C, as a function of scattering angle 2θ (calculated from the horizontal pixel number), and neutron wavelength λ calculated from the time of flight (*left*). The angle of incidence was kept constant at $\alpha_\mathrm{i} = 2.94°$. The inclined streaks correspond to the first (*top*) and second (*bottom*) diffuse Bragg sheets. The vertical line at $2\theta = 2\alpha_\mathrm{i}$ corresponds to the specular reflectivity. Only a relatively weak specular intensity enhancement is observed over the strong diffuse signal at the Bragg peak position. The vertical cuts of the detector matrix show the intensity distribution of the first Bragg sheet as a function of λ for constant 2θ (*right*). The cuts can be fitted to a Lorentzian with HWHM values increasing with the distance away from the specular condition. The resulting HWHM values increase according to the prediction of the smectic elasticity model

$2\theta = \alpha_\mathrm{i} + \alpha_\mathrm{f}$ where α_i is the angle of incidence and α_f the exit angle. Enhanced diffuse scattering stemming from correlated thermal fluctuations is observed at the position of the diffuse Bragg sheets, for all wavelength λ and detector angles 2θ, satisfying the position of the first two diffuse Bragg sheets at $q_z = 2\pi/d$ and $q_z = 4\pi/d$ (higher-order Bragg sheets are not observed in the angular and wavelength range covered). The diffuse Bragg sheets appear as a straight line at oblique angles. The column of the matrix corresponding to $2\theta = 2\alpha_\mathrm{i}$ defines the specular axis. Its intersection with the diffuse Bragg sheets defines the specular Bragg peaks, which are enhanced over the diffuse Bragg sheets due to the specular component. From these positions, a lamellar periodicity of $d = 59$ Å is obtained. Quantitative information on the height–height correlation functions can now be obtained by evaluating the intensity matrix along the different principal axis, e.g., along the horizontal $2\theta/2$ and vertical λ axis.

As known from monochromatic X-ray and neutron scattering [44, 45], the vertical cross-correlation length $\xi_z(q_\|)$ of thermal fluctuations can be inferred from the analysis of the Bragg sheet width in q_z as a function of $q_{\|} = \sqrt{q_x^2 + q_y^2}$, i.e., from the half-width at half-maximum (HWHM) $\text{HWHM}_{qz}(q_{\|})$. The conversion between HWHM_λ and HWHM_{qz} is straightforward according to $\mathrm{d}q_z/q_z = -\mathrm{d}\lambda/\lambda$. The prediction of the the smectic Hamiltonian in Eq. 22.3 is $\text{HWHM}_{qz} = \Lambda q_{\|}^2$.

Cuts along λ for different constant values 2θ (Fig. 22.6) along with least-square fits to the predicted Lorentzian lineshape yield the peak width for each angle 2θ. The results can then be plotted and analyzed as a function of $\theta = 2\theta/2$ or correspondingly $q_{\|}$. The characteristic broadening with increasing θ away from the specular peak at $\theta = 2\alpha_{\mathrm{i}}$ is clearly observed. The curves of $\text{HWHM}(q_x)$ never go to 0 for $q_x \to 0$, due to the intrinsic resolution limited width (instrument, finite size of the sample). Within the limits of resolution and experimental errors, the results of such analysis were shown to agree between monochromatic and TOF-NSNR.

Recently, the propagation of layer perturbation induced by lithographic surface gratings has been mapped by TOF-NSNR to compare thermal to static perturbations, and to have a direct control of the corresponding lateral length scales. In this case characteristic satellites occur in the diffuse Bragg sheets at the Fourier components of the surface grating [46].

22.4.4 Change of Fluctuations by Added Antimicrobial Peptides

The diffuse scattering in multilamellar systems changes upon the insertion of membrane-active molecules, such as the antibiotic peptide Magainin 2. In the system DMPC/Magainin 2 we have observed significant changes with changing the peptide-to-lipid ratio P/L, indicating corresponding changes in the fluctuation and elasticity parameters, or perhaps also the defect structure of the lamellar phase. Reciprocal space mappings $I(q_x, q_z)$ of the first Bragg sheet were measured in monochromatic mode [11] at partial hydration in the fluid L_α phase for samples of different molar ratios $P/L = 0, 0.02, 0.01, 0.033, 0.05$. Figure. 22.7 shows the reciprocal space mappings for $P/L = 0.02, 0.01$, and 0.033, from top to bottom (logarithmically scaled). The intensity of the Bragg sheet decreases with increasing P/L and the width (HWHM_{qz}) of the Bragg sheet increases with P/L. The decay of the intensities is also evident in the reflectivity curves. A corresponding disordering of the lamellar structure, possibly due to both thermal fluctuations and static defects, is observed at high P/L.

The decay of the specular Bragg peaks and the diffuse (nonspecular) Bragg sheets is accompanied by an increase in d of the lamellar stack with increasing P/L. This is probably due to electrostatic repulsion of the lamellae stemming from the increasing surface charge density, since each peptide carries about 4–5 net charges at neutral pH. It is interesting to quantify the decrease of lamellar ordering. The obvious approach would be to evaluate the parameters

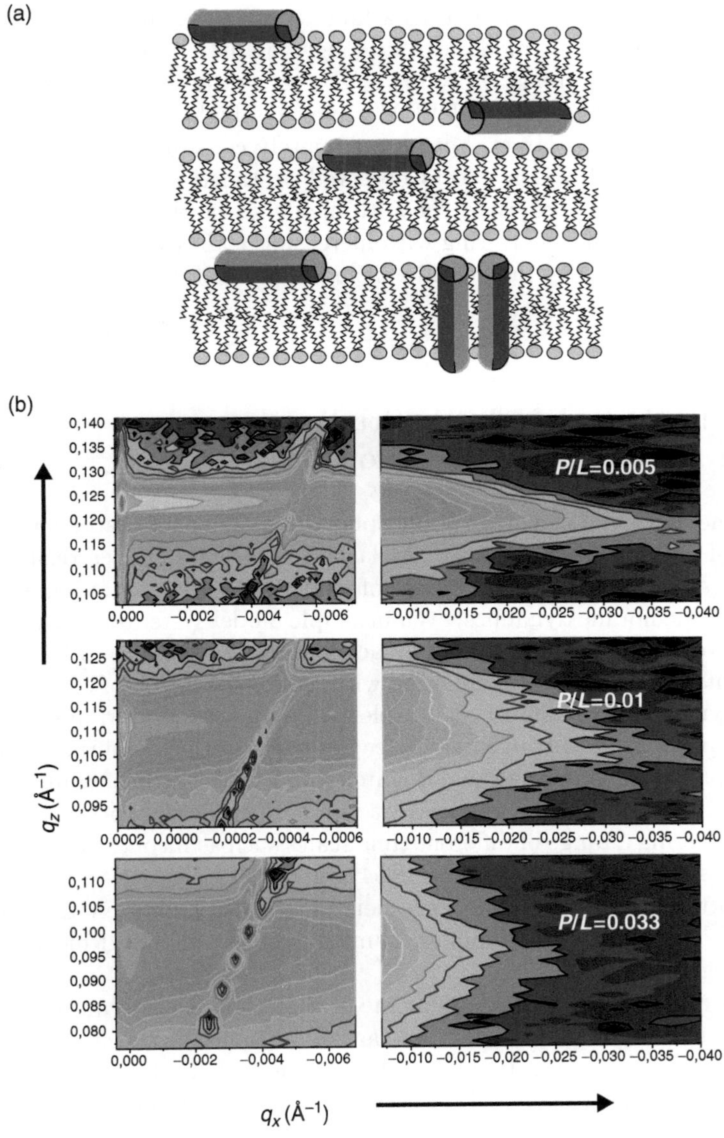

Fig. 22.7. (**a**) Schematic illustration of different states, which the peptides can adopt when bound at the bilayer. The surface (S) state versus the inserted (I) state according to the notation of Huang and coworkers, who have studied the concentration dependent transition S to I in several different amphipathic peptide systems [47–49]. (**b**) Diffuse scattering intensity in DMPC/Magainin 2 covering the relevant P/L range of the S to I transition. Reciprocal space mappings of the first Bragg sheet are shown for P/L ratios of 0.005, 0.01, and 0.033, from *top* to *bottom*. A distinct broadening of the Bragg sheet in q_z is observed, reflecting significant changes in the fluctuations spectrum with increasing peptide concentration

of the smectic model, B and Λ, as a function of P/L. However, the analysis shows that the smectic model can no longer be used to describe the data of the peptide–lipid systems. Only for $P/L = 0$, we can observe the characteristic parabolic increase in the HWHM values. For higher P/L, the width of the Bragg sheet becomes larger, but approximately constant as a function of q_x, apart from the refraction effects observed at the transition zone where α_i changes sign. An appropriate theoretic model is lacking to account for the changes in the diffuse scattering with increasing P/L, which reflect the lamellar disorder induced by the peptide, including both static defects and thermal fluctuations.

22.5 Elastic and Inelastic Studies of the Acyl Chain Correlation Peak

While the molecular structure of phospholipid model membranes has been the object of many investigations in the last three decades and is relatively well studied (see, e.g., [1]), the knowledge of membrane dynamics and in particular collective membrane dynamics, even in simple model systems as DMPC, is still scarce. Nevertheless it is now widely acknowledged that several key functions of a membrane cannot be understood without consideration of collective membrane dynamics [50]. The short wavelength dynamics is attributed to play a key role in the transport of small molecules through the membrane [51]. Molecular vibrations, conformational dynamics and "one particle" diffusion in the plane of the bilayer can be studied by a number of different spectroscopic techniques covering a range of different time scales such as incoherent inelastic and quasieleastic neutron scattering [52–54] or nuclear magnetic resonance [55]. The short-range collective motions mentioned earlier can be elucidated only by a few experimental techniques, namely coherent INS and inelastic X-ray scattering.

Figure 22.8 shows examples of some of the motions that can be probed by coherent neutron scattering, as there are bilayer undulation modes with typical length scales of several hundred Ångströms and short wavelength density fluctuations on nearest neighbor distances of the hydrocarbon acyl chains in the plane of the bilayer, which we discuss in the following. Recently Chen et al. made a seminal inelastic measurements in phosphocholine model membranes using IXS techniques [56]. They could determine the dispersion relation in the gel and the fluid phase of DLPC bilayers, finding a minimum at Q_0, the maximum of the static structure factor $S(Q)$.

22.5.1 Inelastic Neutron Scattering

We applied INS for the study of the collective dynamics of the hydrocarbon acyl chains in lipid bilayers [57]. The main differences with respect to inelastic

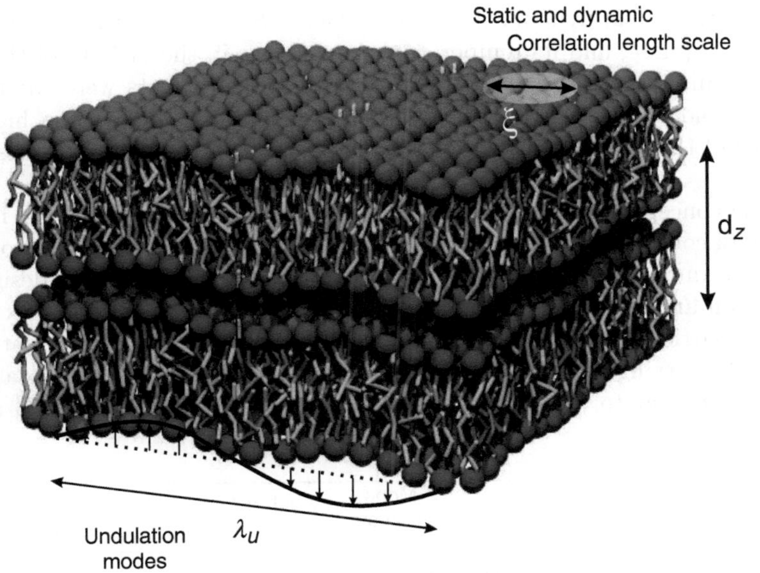

Fig. 22.8. Schematic of a double bilayer with some elementary excitations and the corresponding length scales. Apart from undulation modes with typical wavelengths of several hundred Ångströms, the short wavelength correlations and dynamics in the plane of the membrane can be probed by coherent neutron scattering. The corresponding length scale ξ is thereby in the order of 20 Å. d_z is the bilayer thickness

X-ray scattering are related to the energy–momentum relation of the neutron versus the photon probe, strongly affecting energy resolution, and accessible (Q,ω) range. At high Q the energy of the incident neutrons is in the range of the excitations (some microelectron volts) resulting in a high energy resolution (up to ~300 µeV), in comparison to 1.5 meV of the inelastic X-ray experiment. A better energy resolution in combination with a smaller ratio between central peak and Brillouin amplitudes leads to very pronounced satellites, which are easier to evaluate. This is of particular advantage for the identification of peaks in the central part of the dispersion relation, as well as for the experimental verification of a predicted nondispersive mode at high energies, as shown later. Due to the dispersion relation of the neutron itself ($\sim Q^2$), the range at low Q and high ω values is difficult to access by INS. The determination of the exact speed of sound is therefore a domain of inelastic X-ray scattering.

The collective dynamics of the lipid acyl chains in the model system DMPC (-d54, deuterated 1,2-dimyristoyl-*sn*-glycero-3-phosphatidylcholine), were studied by INS. By selective deuteration of the chains, the respective motions are strongly enhanced over other contributions to the inelastic scattering cross-section. The dynamical structure factor $S(Q_r, \omega)$ in the gel (P_β) and fluid (L_α) phase, and its temperature dependence in the vicinity of the main phase transition has been investigated. The measurements were carried out on the cold TAS IN12 and the thermal spectrometer IN3 at the high flux reactor of the ILL in Grenoble, France, the principles of which are described next.

The concept of TAS has undoubtedly been very successful in the investigation of collective excitations in condensed matter physics, i.e., phonons and magnons in crystals. Advantages of TAS are their relatively simple design and operation and the efficient use of the incoming neutron flux to the examination of particular points in (Q, ω) space. Figure 22.9 shows a schematic of a TAS. By varying the three axes of the instrument, the axes of rotation of the monochromator, the sample and the analyzer, the wavevectors $\boldsymbol{k}_{\mathrm{i}}$ and $\boldsymbol{k}_{\mathrm{f}}$

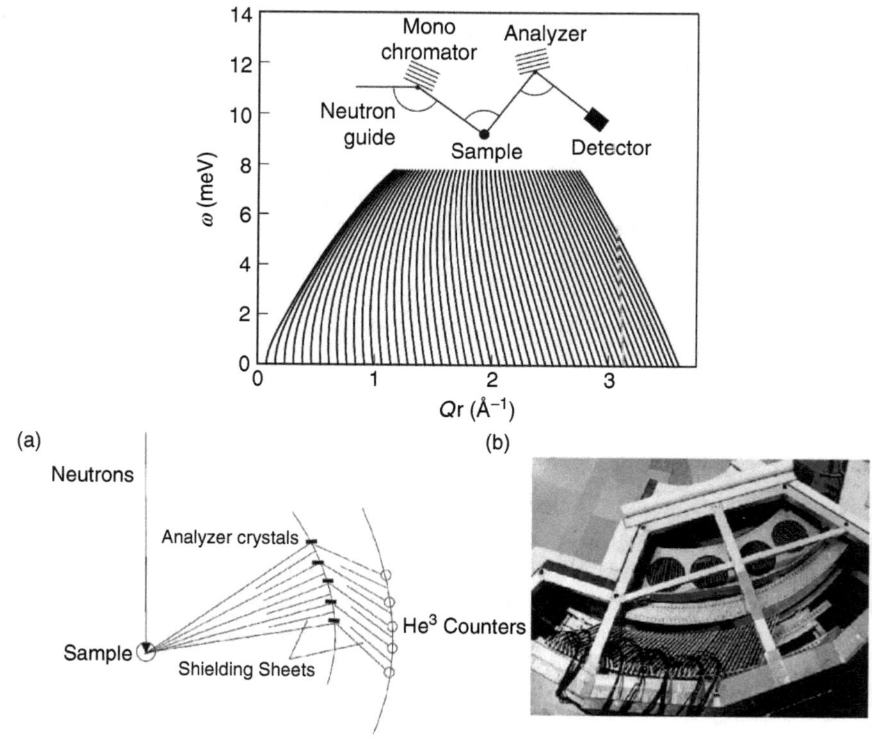

Fig. 22.9. Accessible (Q,ω) range of the cold TAS IN12 (*top*). The inset shows a schematic of a triple-axis instrument with monochromator, sample and analyzer. Principle of IN3's multianalyzer detector (*bottom* (**a**)). 32 Cu-analyzer blades with a dedicated ^{3}He counter (from [59]) each covering 1° in scattering angle 2θ (**b**). All analyzers are arranged on a circle (with a radius of 1031 mm) and aligned to the same final energy of $E_{\mathrm{f}} = 31$ meV

and the energies E_i and E_f of the incident and the scattered beam, respectively, can be determined. **Q**, the momentum transfer to the sample, and the energy transfer, ω, are then defined by the laws of momentum and energy conservation:

$$\mathbf{Q} = \boldsymbol{k}_\mathrm{f} - \boldsymbol{k}_\mathrm{i} \quad \text{and} \quad \omega = E_\mathrm{i} - E_\mathrm{f}. \tag{22.6}$$

The accessible ($\mathbf{Q}$,ω) range of IN12 for a fixed energy of the scattered beam E_f of 10 meV is shown in Fig. 22.9 and covers very well the range of the excitations expected in phospholipid model membranes. It is limited by the range of incident neutron energies offered by the neutron guide as well as by mechanical restrictions of the spectrometer. The instrumental energy resolution in this configuration is $\Delta\omega = 500\,\mu\mathrm{eV}$. By choosing smaller incident energies and energy transfers the energy resolution can be enhanced. A detailed description of the experimental set-up and method can be found elsewhere [57, 58].

With a conventional TAS, single points in (Q,ω) space are scanned one by one. A more efficient use of the triple-axis technique is achieved by *multiplexing*, i.e., the use of several independent analyzer blades and position-sensitive detectors to investigate multiple (Q,ω) points at the same time. This option is implemented on the spectrometer IN3 by the usage of a multianalyzer detector. A schematic of the set-up is shown in Fig. 22.9a; Fig. 22.9b shows a photograph of the detector unit with analyzers and ^{3}He counter tubes (for a detailed description of the set-up see Demmel et al., [59]).

The use of the multidetector is especially useful in systems with restricted dimensions. Considering the membranes as stacked two-dimensional layers, only two directions in space can be differentiated, i.e., the plane of the bilayer, Q_r, or the normal to it, Q_z. Because of the missing periodicity in the third direction in space the scattered intensity is distributed rod-like in reciprocal space. The scattering is independent of one of the reciprocal axis and this allows to measure different **Q** vectors of the sensitive reciprocal axis at the same time in the different channels of the multi analyzer. The dispersion relation can therewith be measured for several Q-points at the same time. Considering elastic scattering, large areas of reciprocal space can be measured (or *mapped*) simultaneously.

22.5.2 Elastic Neutron Scattering

The use of a TAS offers the possibility of measuring the static structure factor in the plane of the membranes, $S(Q_r)$, the in-plane dynamics, $S(\mathrm{Q}_r,\omega)$, and the reflectivity on the same instrument in the same run without changing the setup (diffraction is measured at energy transfer $\omega = 0$). This is an invaluable advantage as the thermodynamic state of the lipid bilayer not only depends on temperature and relative humidity, but also on cooling and heating rates, preparation and thermal history. The combination of elastic and inelastic measurements (see later) leads to a complete picture of structure and (collective) dynamics of model membranes on a molecular length scale.

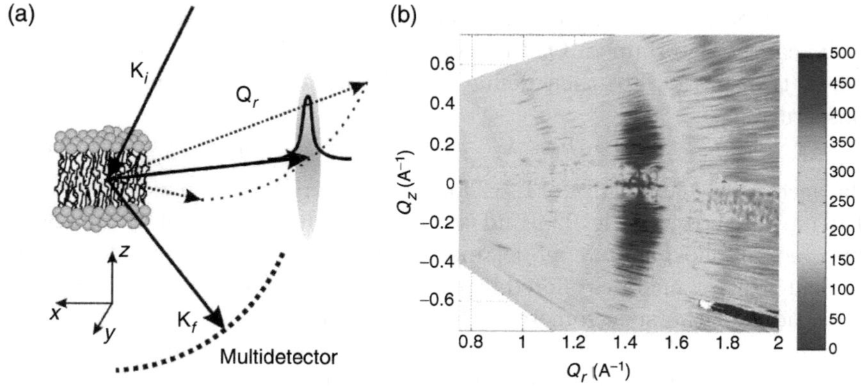

Fig. 22.10. (**a**) A typical multidetector Q-scan in reciprocal space. The incident (k_{i}) and the outgoing beam (k_{f}) and the resulting momentum transfers Q for the different analyzer blades are given in the figure. (**b**) By rotating (rocking) the sample, the reciprocal (Q_r, Q_z) space around the chain correlation peak can be mapped. The measurement was done in the gel phase of the phospholipid bilayer at $T = 20°$

Figure 22.10 shows a (Q_r,Q_z) mapping of the interacyl chain correlation peak in the gel phase of the deuterated DMPC bilayer at $T = 20°$C. Because of the quasi two dimensionality of the system, the peak is sharp in Q_r, but the intensity is smeared out in the perpendicular Q_z-direction. The data show excellent agreement with results of molecular dynamics (MD) simulations on the structure of lipid bilayers [60].

On IN12 temperature-dependent Q-scans through the interacyl chain peak of deuterated DMPC in the temperature range from $T = 20°$C to $T = 40°$C were performed to probe the static correlations in the plane of the bilayer. Figure 22.11a shows Q-scans of the interchain correlation peak at temperatures of $T = 20$, $T = 23$, and $T = 30°$C, respectively. The deuterated compound undergoes the phase transition from the more rigid gel phase into the liquid-like fluid phase at about T_c=21°C, some degrees lower than in the protonated compound. When going from the gel to the fluid phase, the peak position changes to smaller Q_r-values (larger average nearest neighbor distances) and the peaks broaden, indicating a decreasing correlation length ξ_r in the plane of the membranes.

Figure 22.11b gives Q_0, the amplitude I and the correlation length ξ_r as obtained from the peak position, amplitude and width of the peak for all measured temperatures from $T = 20$–$40°$C. The average nearest neighbor distance, as calculated from $2\pi/Q_0$, enlarges with temperature. Although the phase transition is of first order [22, 23, 61], the values point to a critical behavior (anomalous swelling) of the bilayer. As the analyzer cuts out only the elastically scattered neutrons and the quasielastic contribution to the background is reduced, the signal-to-noise ratio is drastically improved.

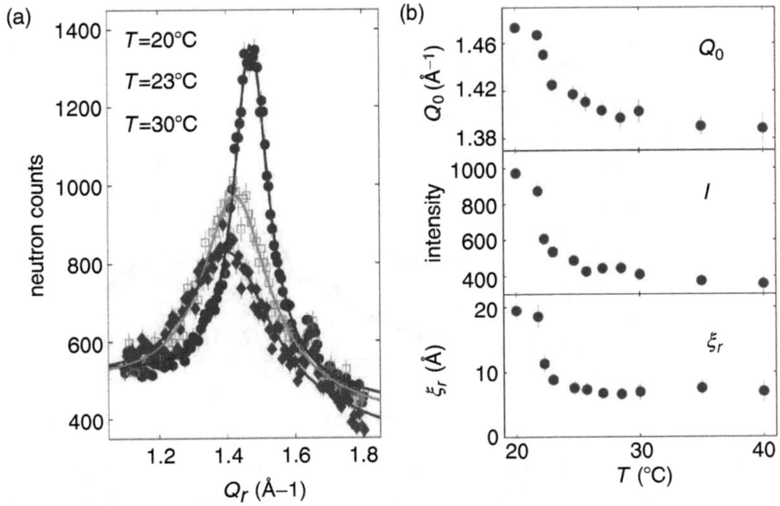

Fig. 22.11. (**a**) Q-scans of the acyl chain peak for temperatures from *top* to *bottom* $T = 20$, $T = 23$, and $T = 30$°C through the main phase transition of the DMPC-d54 bilayer. (**b**) Q_0, intensity and correlation length ξ_r as extracted from peak position, amplitude and width for all measured temperatures between $T = 20$ and 40°

22.5.3 Collective Dynamics

A typical energy scan of deuterated DMPC collected at $T = 20$°C, in the gel phase of the bilayer, measured on IN12 at $Q = 1.0\,\text{Å}^{-1}$ is shown in Fig. 22.12. The inset shows the excitations of the bilayer in the gel and the fluid phase ($T = 30$°C, nine degrees above the phase transition temperature), exhibiting well-pronounced peaks the position and width of which can be easily determined. The inelastic scans can be evaluated by the generalized three-effective eigenmode theory (GTEE) [56, 62, 63], using the following function for least-square fitting:

$$\frac{S(Q,\omega)}{S(Q)} = \frac{1}{\pi}\left(A_0\frac{\Gamma_h}{\omega^2+\Gamma_h^2} + A_s\left[\frac{\gamma_s + b(\omega+\omega_s)}{(\omega+\omega_s)^2+\gamma_s^2} + \frac{\gamma_s - b(\omega-\omega_s)}{(\omega-\omega_s)^2+\gamma_s^2}\right]\right).$$

The model consists of a heat mode, centered at $\omega = 0$ meV (Lorentzian with a width Γ_h), two sound modes, represented by Lorentzians at $\omega = \pm\omega_s$ and a damping γ_s [62, 63]. From the width of the central mode, and the width and position of the Brillouin lines, the thermal diffusivity, the sound frequency, and the sound damping can be determined, respectively, within the framework of a hydrodynamic theory. To fit the neutron data an additional Lorentzian component is added describing the broad quasielastic contribution presumably

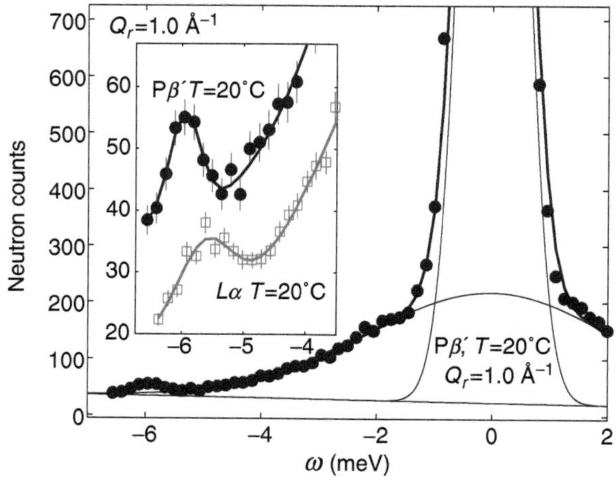

Fig. 22.12. Energy scan at a $Q_r = 1.0\,\text{Å}^{-1}$ in the gel phase of the bilayer. The inset shows the excitations of the bilayer in the gel (*top*) and the fluid phase (*bottom*)

associated with intramolecular degrees of freedom and incoherent scattering, not seen by inelastic X-ray scattering. The solid line in Fig. 22.12 is a fit by the three-effective eigenmode model with an additional Lorentzian component.

Figure 22.13a shows the dispersion relation in the gel and the fluid phase as measured by several constant Q-scans at Q-values ranging from $Q = 0.7$ to $3.0\,\text{Å}^{-1}$. The fluid dispersion has been measured far in the fluid phase of the DMPC bilayer. At small Q_r, longitudinal sound waves in the plane of the bilayer are probed and give rise to a linear increase of $\omega \propto Q_r$, saturating at some maximum value ("maxon"), before a pronounced minimum Ω_0 ("roton") is observed at $Q_0 \simeq 1.4\,\text{Å}^{-1}$, the first maximum in the static structure factor $S(Q_r)$ (the interchain correlation peak). Qualitatively, this can be understood if Q_0 is interpreted as the quasi-Brillouin zone of a two-dimensional liquid. Collective modes with a wavelength of the average nearest neighbor distance $2\pi/Q_0$ are energetically favorable, leading to the found minimum. At Q_r values well above the minimum, the dispersion relation is dominated by single particle behavior. The inelastic neutron data are less noisy and cover a wider range as compared to inelastic X-ray scattering [56]. A quantitative theory that predicts the absolute energy values of "maxon" and "roton" on the basis of molecular parameters is absent so far. However, the dispersion relation can be extracted from MD simulations by temporal and spatial Fourier transformation of the molecular real space coordinates [64] which shows excellent agreement.

Figure 22.13b shows corresponding energy scans taken at $Q = 1.5\,\text{Å}^{-1}$ in the dispersion minimum for temperatures T between 20 and 40°C. In the gel phase an excitation at $\omega_s = -1\,\text{meV}$ is found. At higher temperatures, the gel excitations decreases and a new excitation at energy values of $\omega_s = -2\,\text{meV}$ grows associated with the fluid phase. Clearly the excitations are not spurious

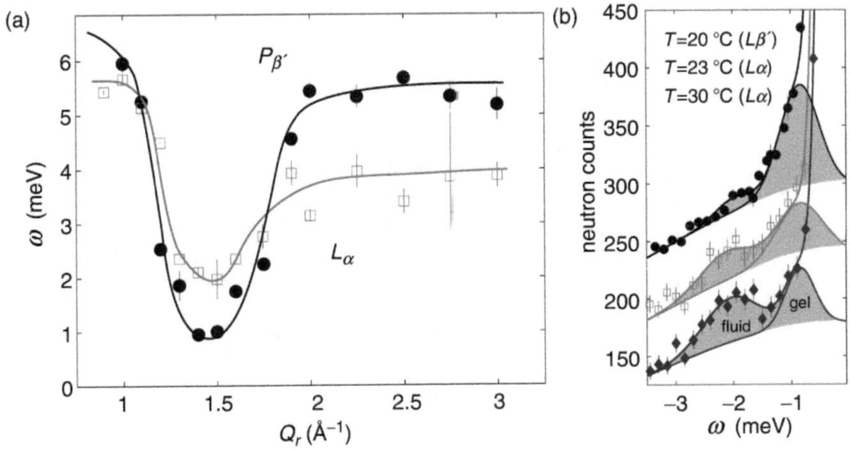

Fig. 22.13. (**a**) Dispersion relations in the gel and the fluid phase of the DMPC bilayer as measured by several constant Q-scans at Q values ranging from Q=0.7 to 3.0 Å^{-1}. (**b**) Energy scans in the dispersion minimum at $Q = 1.5$ Å^{-1} for temperatures $T = 20°$C, $T = 23°$C, and $T = 30°$C (from *top* to *bottom*)

effects as they are symmetric around the central peak. The assignment of the excitations to the particular phases is justified by their temperature dependence as in each phase there is a dominant excitation. Both modes are clearly dispersive and change its energy position when moving out of the minimum (Fig. 22.13a). Small traces of the "fluid excitation" are already present in the gel phase. At $T = 30°$C, far in the fluid phase, the "gel excitation" is still present, indicating coexistence of fluid-phase and gel-phase domains. This coexistence is only observed in the range of the dispersion minimum, which coincides with the maximum of the static structure factor. One has to note that at the same time, the elastic scans do not show coexistence of two phases. While the transition is of first order, a pseudocritical swelling, i.e., a continuous change of the interlamellar distance in the range of T_c, is observed for DMPC and other lipids [22, 61, 65]. The changes in d_z (see Fig. 22.3) are accompanied by corresponding changes in the mean distance between the acyl chains (Fig. 22.11b).

A crucial point is the size of the domains. The coexistence of macroscopic domains with sizes larger than the coherence length ξ_n of the neutrons in the sample would lead to a peak splitting of the acyl chain peak. Therefore, the domain sizes must be smaller than a few hundred angstroms estimated for ξ_n. Fluid domains in the gel phase and vice versa with sizes smaller than 0.01 μm^2 have been reported in a recent AFM study [66], and have been related to lateral strain resulting from density differences in both phases. While the "maxon" and the high-Q range are energetically higher in the gel than in the fluid phase (due to stiffer coupling between the lipid

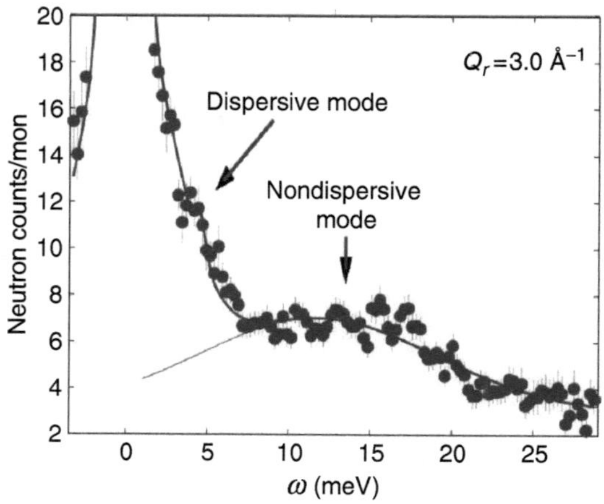

Fig. 22.14. Scan over a higher energy transfer range at Q_r=3.0 Å^{-1} (IN3 data), showing both the dispersive excitation and the nondispersive (optical) excitation predicted by MD calculation

chains in all-*trans* configuration), Ω_0, the energy value in the dispersion minimum, is actually smaller in the gel phase, roughly analogous to soft modes in crystals.

Figure 22.14 shows an energy scan in the gel phase at Q_r=3.0 Å^{-1}, up to an energy transfer of 30 meV. Aside from the dispersive excitation due to in-plane density waves, a second nondispersive (optical) mode is observed at about $\omega = 14$ meV with a width (FWHM) of about 13 meV, corresponding quite well to the predictions by Tarek et al., [64], and can be attributed to the motions of the methyl ends of the acyl chains. This mode was not observed by Chen et al., possibly due to the low signal-to-noise ratio in the inelastic X-ray measurements at high energy transfers.

22.6 Conclusions

Only a few methods are currently available to determine the fundamental smectic length scale Λ or the bilayer bending rigidity κ. While light scattering or optical microscopy techniques determine κ from thermal fluctuations on much larger length scales, which may lead to different values, X-ray powder diffraction and line shape analysis is sensitive to Λ only in the limit of very soft and strongly undulating systems, untypical for phospholipids. By nonspecular neutron and X-ray scattering from aligned phases the bilayer structure and fluctuation is accessible over a wide range both for relatively stiff and soft systems, covering length scales from the molecular scale up to a few 100 nm.

Nonspecular neutron scattering is a very useful tool to elucidate phospholipid membrane interaction on the basis of changing fluctuation and elasticity properties as described in this chapter. To this end, an appropriate model is needed. Turner and Sens have explored the statistical physics of particle inclusions in smectic liquid crystals [67,68], providing a quantitative description of the deformation fields around static defects, as caused by inclusions. Their model, however, is based on the smectic model, and the inclusions lead to effective smectic parameters, while the present data show that the scattering distribution can no longer be described by Caillé theory.

Furthermore, the following problems are associated with the verification of the parabolic law $\mathrm{HWHM}_{qz}(q_x)$ predicted for smectic systems: (i) the width saturates for $q_x \geq 0.01\,\text{Å}^{-1}$, indicating contributions from collective molecular motions that are distinct from bending, (ii) the initial increase in the curve shows some spread and may also be explained by functions other than parabolas. The first point is accompanied by corresponding deviations from the smectic model in the curve $S(q_x)$ also observed at high q_x. We speculate that collective protrusion of peristaltic modes of the bilayers is at the origin of this observation. The present results have been verified by different samples and show consistency between two completely different modes of the experiments, monochromatic and TOF. However, suitable techniques of measuring the resolution, i.e., the intrinsic width of the cross-sections (cuts) along λ for constant angles 2θ have to be developed, in order to verify the resolution model according to [43].

Measurements of the collective short wavelength dynamics in lipid bilayers are a new but promising field, because the collective dynamics is likely to play a crucial role for different biological functions. The use of a triple-axis spectrometer allows to measure structure and dynamics, i.e., reflectivity, acyl chain correlation peak and in-plane dynamics, of model membranes on a molecular length scale in the same run without changing set-up. The dispersion relation measured in the gel and the fluid phase of the DMPC model system and the interpretation of the temperature dependent experiments point to a new interpretation of the gel–fluid phase transition and the collective excitations in lipid bilayers. Further inelastic investigations will address the influence of different head and tail groups to the collective dynamics.

On the interaction of peptides with model membranes the results presented also show that pronounced changes of the fluctuation spectrum occur already at moderate peptide concentration, e.g., at $P/L = 0.005$, where the changes in the specular reflectivity are still quite small. Furthermore the influence of cholesterol and membrane-active proteins will be studied, hopefully giving new insight into the functionality of these systems. The investigation of more and more complex systems might once lead to a better understanding of real biological membranes.

Acknowledgments

We thank R. Cubitt for helpful comments and for his great achievements in the design and construction of the novel D17 reflectometer, C. Münster and

C. Ollinger for enjoyable collaborations on some of the original work discussed here, R. Siebrecht and V. Leiner for excellent support at ADAM, F. Demmel for help on the IN3 multidetector, G. Fragneto for experimental support and discussions and G. Brotons for helpful discussions regarding the data analysis. Financial aid by the Deutsche Forschungsgemeinschaft (DFG) through grants SA-772/3 and SA-772/4 and by the German Research Ministry under contract numbers 05300CJB6 and 05KS1TSA7 is gratefully acknowledged.

References

1. R. Lipowsky, E. Sackmann (eds.), *Structure and Dynamics of Membranes*, Vol. 1, *Handbook of Biological Physics* (Elsevier, North-Holland, Amsterdam, 1995)
2. G. Büldt, H.U. Gally, J. Seelig, G. Zaccai, J. Mol. Biol. **135** (1979), 673
3. M. Seul, M.J. Sammon, Thin Solid Films **185** (1990), 287
4. C.M. Münster, T. Salditt, M. Vogel, R. Siebrecht, J. Peisl, Europhys. Lett. **46** (1999), 486
5. J. Katsaras, Biophys. J. **75** (1998), 2157
6. J.F. Nagle, J. Katsaras, Phys. Rev. E **59** (1999), 7018
7. M. Vogel, C. Münster, W. Fenzl, T. Salditt, Phys. Rev. Lett. **84** (2000), 390
8. T. Salditt, C. Münster, J. Lu, M. Vogel, W. Fenzl, A. Souvorov, Phys. Rev. E **60** (1999), 7285
9. S.K. Sinha, E.B. Sirota, S. Garoff, H.B. Stanley, Phys. Rev. B **38** (1988), 2297
10. R. Pynn, Phys. Rev. B **45** (1992), 602
11. C. Münster, PhD thesis, Sektion Physik, Universität München, 2000
12. J. Als-Nielsen, D. McMorrow, *Elements of Modern X-Ray Physics* (John Wiley & Sons, New York, 2001)
13. T. Salditt, C. Li, A. Spaar, U. Mennicke, Eur. Phys. J. E **7** (2002), 105
14. G. Pabst, M. Rappolt, H. Amenitsch, P. Laggner, Phys. Rev. E **62** (2000), 4000
15. A.E. Blaurock, Biochem. Biophys. Acta **650** (1982), 167
16. J. Katsaras, Biochem. Cell. Biol. **73** (1995), 209
17. T. Salditt, Curr. Opin. Coll. Interface Sci. **19** (2000), 232
18. D. Constantin, U. Mennicke, C. Li, T. Salditt, Eur. Phys. J. E **12** (2003), 283
19. U. Mennicke, PhD thesis, Institut für Röntgenphysik, Georg-August-Universität Göttingen, 2003
20. C.R. Safinya, E.B. Sirota, D. Roux, G.S. Smith, Phys. Rev. Lett. **57** (1986), 2718
21. C.R. Safinya, E.B. Sirota, D. Roux, G.S. Smith, Phys. Rev. Lett., **62** (1989), 1134
22. J.F. Nagle, H.I. Petrache, N. Gouliaev, S. Tristram-Nagle, Y. Liu, R.M. Suter, K. Gawrisch, Phys. Rev. E, **58** (1998), 7769
23. H.I. Petrache et al., Phys. Rev. E **57** (1998), 7014
24. R.P. Rand, V.A. Parsegian, Biochim. Biophys. Acta **988** (1989), 351
25. T. Salditt, C. Münster, U. Mennicke, C. Ollinger, G. Fragneto, Langmuir **19** (2003), 7703
26. G. Smith, E.B. Sirota, C.R. Safinya, N.A. Clark, Phys. Rev. Lett. **60** (1988), 813
27. R.K. Thomas, J. Penfold, Curr. Opin. Coll. Interface Sci. **1** (1995), 23
28. T.M. Bayerl, R.K. Thomas, A. Rennie, J. Penfold, E. Sackmann, Biophys. J. **60** (1991), 1

29. T.L. Kuhl, J. Majewski, J.Y. Wong, S. Steinberg, D.E. Leckband, J.N. Israelachvili, G.S. Smith, Biophys. J. **75** (1998), 2352
30. G.P. Felcher, R.J. Goyette, S. Anastasiadis, T.P. Russell, M. Foster, F. Bates, Phys. Rev. B **50** (1994), 9565
31. S. Langridge, J. Schmalian, C.H. Marrows, D.T. Dekadjevi, B.J. Hickey, Phys. Rev. Lett. **85** (2000), 4964
32. Z.X. Li, J.R. Lu, R.K. Thomas, A. Weller, J. Penfold, J.R.P. Webster, D.S. Sivia, A.R. Rennie, Langmuir **17** (2001), 5858
33. W. Helfrich, Z. Naturforsch. **28c** (1973), 693
34. L. Ning, C.R. Safinya, R. Bruinsma, J. Phys. II France **5** (1995), 1155
35. N. Lei, PhD thesis, Rutgers, 1993
36. A. Caillé, C. R. Acad. Sci. **274** (1971), 891
37. R. Holyst, D.J. Tweet, L.B. Sorensen, Phys. Rev. Lett. **65** (1990), 2153
38. J. Als-Nielsen, J.D. Litster, R.J. Birgenau, M. Kaplan, C.R. Safinya, Phys. Rev. B **22** (1980), 312
39. S.K. Sinha, J. Phys. III (France) **4** (1994), 1543
40. A. Schreyer, R. Siebrecht, U. Englisch, U. Pietsch, H. Zabel, Physica B **248** (1998), 349
41. T. Salditt, T.H. Metzger, J. Peisl, Phys. Rev. Lett. **73** (1994), 2228
42. T. Salditt, D. Lott, T.H. Metzger, J. Peisl, G. Vignaud, P. Høghøj, O. Schärpf, P. Hinze, R. Lauer, Phys. Rev. B **54** (1996), 5860
43. R. Cubitt, G. Fragneto, Appl. Phys. A **74** (2003), 329
44. M. Vogel, PhD thesis, Universität Potsdam, 2000
45. T. Salditt, M. Vogel, W. Fenzl, Phys. Rev. Lett. **90** (2003), 178101
46. C. Ollinger, D. Constantin, J. Seeger, T. Salditt, Europhys Lett. **71** (2005), 311
47. S. Ludtke, K. He, H. Huang, Biochemistry, **34** (1995), 16764
48. S.J. Ludtke, K. He, W.T. Heller, T.A. Harroun, L. Yang, H.W. Huang, Biochemistry **35** (1996), 13723
49. L. Yang, T.M. Weiss, R.I. Lehrer, H.W. Huang, Biophys. J. **79** (2000), 2002
50. T. Bayerl, Curr. Opin. Coll. Interface Sci. **5** (2000), 232
51. S. Paula, A.G. Volkov, A.N. Van Hoek, T.H. Haines, D.W. Deamer, Biophys. J. **70** (1996), 339
52. W. Pfeiffer, Th. Henkel, E. Sackmann, W. Knorr, Europhys. Lett. **8** (1989), 201
53. S. König, T.M. Bayerl, G. Coddens, D. Richter, E. Sackmann, Biophys. J. **68** (1995), 1871
54. E. Endress, H. Heller, H. Casalta, M.F. Brown, T.M. Bayerl, Biochemistry **41** (2002), 13078
55. A.A. Nevzorov, M.F. Brown, J. Chem. Phys. **107** (1997), 10288
56. S.H. Chen, C.Y. Liao, H.W. Huang, T.M. Weiss, M.C. Bellisent-Funel, F. Sette, Phys. Rev. Lett. **86** (2001), 740
57. M.C. Rheinstädter, C. Ollinger, G. Fragneto, F. Demmel, T. Salditt, Phys. Rev. Lett. **93** (2004), 108107
58. M.C. Rheinstädter, C. Ollinger, G. Fragneto, T. Salditt, Physica B **350** (2004), 136
59. F. Demmel, A. Fleischmann, W. Gläser, Nucl. Instrum. Methods Phys. Res., Sect. A **416** (1998), 115
60. A. Spaar, T. Salditt, Biophys. J. **85** (2003), 1576
61. F.Y. Chen, W.C. Hung, H.W. Huang, Phys. Rev. Lett. **79** (1997), 4026
62. C.Y. Liao, S.H. Chen, F. Sette, Phys. Rev. Lett. **86** (2001), 740
63. T.M. Weiss, P.-J. Chen, H. Sinn, E.E. Alp, S.H. Chen, H.W. Huang, Biophys. J. **84** (2003), 3767

64. M. Tarek, D.J. Tobias, S.-H. Chen, M.L Klein, Phys. Rev. Lett. **87** (2001), 238101
65. P.C. Mason, J.F. Nagle, R.M. Epand, J. Katsaras, Phys. Rev. E **63** (2001), 030902(R)
66. A.F. Xie, R. Yamada, A.A. Gewirth, S. Granick, Phys. Rev. Lett. **89** (2002), 246103
67. P. Sens, M.S. Turner, J. Phys II. France **7** (1997), 1855
68. P. Sens, M.S. Turner, Phys. Rev. E **55** (1997), R1275

23

Subnanosecond Dynamics of Proteins in Solution: MD Simulations and Inelastic Neutron Scattering

M. Tarek, D.J. Tobias

23.1 Introduction

Complete understanding of protein function requires knowledge of their structure, energetics, and dynamics at the atomic level. To probe dynamics, inelastic neutron scattering (INS) has emerged as a powerful tool allowing the space and time-resolved study of a wide spectrum of motions in biomolecules taking place in the subpicosecond to multinanosecond timescales [1–8]. The ranges of energy and momentum transfers accessible on presently available neutron spectrometers correspond closely to the lengths and sizes of Molecular Dynamics (MD) simulations that are feasible nowadays for biological molecules (cf. Fig. 23.1). Therefore, simulations are a potentially valuable tool for interpreting neutron data on biomacromolecules, and similarly neutron scattering experiments can be used to verify results from MD calculations in a previously inaccessible time regime, and this has a direct impact on the rapidly growing area of computational biology.

Due to experimental limitations, neutron scattering was initially used primarily to study protein dynamics in powder environments in order to balance the trade-off between maximal scattered neutron flux and minimal data collection time. Despite their widespread usage in the food and pharmaceutical industries [9–11] or in research applications, very little is known about the organization and interactions of the protein molecules in these powders. Hydrated samples are usually prepared by adding a prescribed amount of water to a dry lyophilized powder. Lyophilization consists of freeze–drying an aqueous solution of the protein under vacuum. Although reversible by full hydration, lyophilization is known to induce changes in protein structure and make proteins more rigid [12]. Thus there is some concern that partially hydrated powder samples may contain a certain amount of nonnative structure and therefore, the scattering data taken on powders may not be truly representative of native functioning proteins.

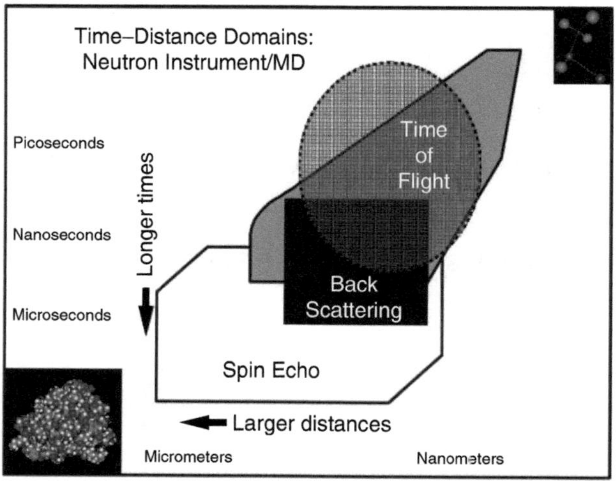

Fig. 23.1. Time and length scales accessible by typical neutron spectrometers and by molecular dynamics simulations (*shaded area*)

We used molecular dynamics simulations in comparison to INS experiments on model protein powders, and showed, for the first time, that it is possible to reproduce quantitatively the neutron data through careful consideration of the environment; hence assessing the ability of current generation force fields to capture the essence of protein internal dynamics as well as that of their hydration water on the $\simeq$100 ps timescale. Our previous results show clearly that in order to obtain quantitative agreement with INS experiments performed on powder samples, simple representations of the protein monomers surrounded by a prescribed amount of water molecules in vacuum are not appropriate [13–16]. Such a representation leads to an overestimation of the protein motion as well as a much faster dynamics of the hydration water in the vicinity of the protein surface, and therefore does not guarantee that the dynamics is well described by the MD simulation.

Recently, neutron scattering experiments have begun to be applied to probe the dynamics of proteins in solution. Pérez et al. [17] showed that it is possible through careful data processing to get information on the internal motions of globular proteins in solution samples. Not long after, other authors used neutrons to probe the dynamics of unfolded proteins [18,19] and of protein complexes in solution [20]. Pérez et al. [17] proposed an original treatment of quasielastic neutron scattering data (QENS) from proteins in solution. The authors take into account the overall motion of the protein i.e., its overall translation and rotation, and assume that it is decoupled from the internal motion. The analysis of the data appears to indicate that at the length scale ($0.5 \leq Q \leq 2.0\,\text{Å}^{-1}$) and timescale ($t \leq 100$ ps) covered by the experiment, the scattering contains a substantial contribution from the motion of the protons due to the global motion of the protein. While such treatment of the INS data has been considered in some studies [19], the validity of the assumptions made remains unclear. The contribution from the overall motion has simply been ignored in several other studies. This stems mainly from the fact that it is counterintuitive that the overall translational and rotational motion of globular proteins in solution, known from light scattering and NMR measurements to be appreciable only on the nanosecond timescale, have an effect on motion probed with neutrons on the tens of picoseconds timescale.

Here we extend our MD simulation studies to investigate the dynamics of proteins in solutions, at low to moderate concentrations. We have two objectives. First, we use the simulation results to provide support to the assumptions made in experimental data reduction. Second, we use the atomistic modeling to characterize the changes in the internal dynamics of a protein upon going from low hydration to high hydration. To these ends we present results obtained from a series of simulations of four globular proteins in solution: Ribonuclease A (RNase A), myoglobin (Myo), lysozyme (Lys), α-lactalbumin (α-Lact) (Fig. 23.2). In the case of RNase A, we compare results from the solution simulation with previous simulations in powder environments both at very low and intermediate hydrations.

The purpose of this chapter is not to present a comprehensive review on probing dynamics of proteins with neutron scattering (up-to-date references are reported in several chapters of this book) and MD simulations. Rather, we focus on a few examples to demonstrate the utility of combining neutron scattering experiments and MD simulations in order to elucidate the dynamical

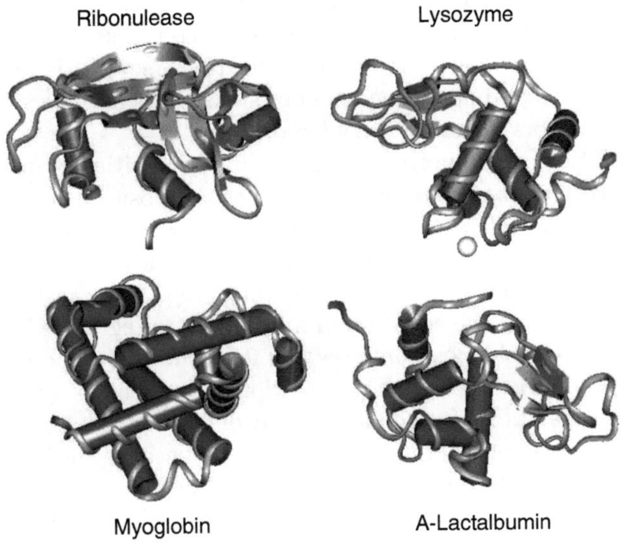

Fig. 23.2. Ribbon representation of the four proteins studied (Protein Data Bank access codes in parentheses): Bovine pancreatic Ribonuclease (7RSA) [21], sperm whale Myoglobin (1MBC) [22], human α-Lactalbumin (1HML) [23], and hen egg white Lysozyme (194L) [24]

behavior of proteins in solution at the atomic level. The chapter is organized as follows. First, we describe the computational approaches used in the simulations, including the system setup, simulation protocols, and the methods used for analysis of the results. Next, we discuss the MD simulations results in comparison to neutron scattering measurements performed on proteins in powder and solution environments, and conclude with a detailed description of the internal dynamics of proteins in the different environments.

23.2 MD Simulations

Molecular dynamics refers to a family of computational methods aimed at simulating macroscopic behavior through the numerical integration of the classical equations of motion of a microscopic many-body system. Macroscopic properties are expressed as functions of particle coordinates and/or momenta, which are computed along a phase space trajectory generated by classical dynamics. The underlying assumption is that over long times the system will reach an equilibrium state, in which time averages can be equated to statistical ensemble averages. When performed under conditions corresponding to laboratory scenarios, molecular dynamics simulations can provide a detailed view of the structure and dynamics at the atomic and mesoscopic levels that is not presently accessible by experimental measurements. They can also be used to perform "computer experiments" that could not be carried out in the laboratory, either because they do not represent natural behavior, or because the necessary controls cannot be achieved. Lastly, MD simulations can be implemented as simple test systems for condensed matter theory. In order to realize this wide spectrum of applications, several issues concerning system modeling, dynamical generation of statistical ensembles, and efficient numerical integrators, must be considered.

MD simulations use information (positions, velocities or momenta, and forces) at a given instant in time, t, to predict the positions and momenta at a later time, $t + \Delta t$, where Δt is the time step, usually taken to be constant throughout the simulation. Numerical solutions to the equations ofmotion are thus obtained by iteration of this elementary step. The most popular algorithms for propagating the equation of motion, or "integrators," are based on Taylor expansions of the positions (see [25] for a survey). A popular approach to increasing the time step in MD simulations of molecular systems is to use holonomic constraints to freeze the motion of the highest frequency bonds, or all of the bonds altogether. This is acceptable because the bond stretching motion is effectively decoupled from the other degrees of freedom that are typically of greater interest. There are a variety of algorithms (e.g., SHAKE, RATTLE) for solving the appropriate constraint equations and integrating the equations of motion (see [25]).

The utility of the constraint method rests on the fact that the fastest degrees of freedom place an upper bound on the time step, and the time step can therefore be increased if the fastest degrees of freedom are eliminated.

A more general approach is based on the premise that the forces associated with different degrees of freedom evolve on different timescales, and the slower forces do not need to be computed as often as the faster forces. It turns out, for molecular systems, that the fastest forces, those associated with intramolecular interactions, require the least computational effort to evaluate ($\mathcal{O}(N)$), while the slowest forces, those associated with nonbonded interactions, require the most computational effort ($\mathcal{O}(N^2)$). Thus, in the interest of computational efficiency, it is clearly advantageous to calculate the rapidly evolving forces at each elementary time step, and calculate only the slowly evolving forces at a larger time interval that is several times the elementary time step. This is the essence of multiple time step MD, which is used in most MD simulations of biomolecules these days (see [26] for a review and technical details).

To date, most MD simulations have been driven by forces that were derived from assumed, empirical potentials (for a detailed overview of commonly used empirical potentials, see [27]). Although the number and nature of the potential energy terms varies from one application to the next, the potential for an arbitrarily complicated molecule such as a biopolymer is generally written as the sum of "bonded" and "nonbonded" terms. The bonded terms include energy penalties, usually represented by harmonic potentials, for deforming chemical bonds and the angles between bonds from their equilibrium values as well as periodic potentials to describe the energy change as a function of torsion (dihedral) angle about rotatable bonds. The bonded terms are usually taken to be diagonal, i.e., there are no terms describing coupling between deformations of bonds and angles, bonds, and torsions, etc. These off-diagonal terms are important for accurate reproduction of vibrational properties, as are anharmonic terms (e.g., cubic, quadratic). The nonbonded terms, as the name suggests, describe the interactions between atoms in different molecules, or interactions within a molecule that are not completely accounted for by the bonded terms, i.e., they are separated by more than two bonds. The nonbonded terms are typically assumed to be pairwise-additive functions of interatomic separation, r, and include van der Waals interactions and Coulomb interactions in polar or charged molecules in which the charge distribution is represented by partial charges (usually placed on the atoms). The van der Waals interactions, commonly modeled by the Lennard–Jones potential, include a term that is strongly repulsive (exponential or inverse 12th power of r), and a weakly attractive dispersion term (e.g., inverse sixth power of r, representing the induced dipole-induced dipole interaction).

An important aspect of performing simulation studies is to calculate experimentally measurable properties for comparison with measurements made at particular thermodynamic state points. Most experiments are carried out under the conditions of constant temperature and either constant pressure or volume, e.g., in the isobaric–isothermal or canonical statistical mechanical ensembles. However, integration of Newton's equations of motion generates trajectories at constant volume and energy, i.e., in the microcanonical ensemble. Several techniques for generating trajectories in ensembles other than the microcanonical ensemble have been developed (see [25] for an overview). The most powerful techniques are based on the concept of an "extended system" (ES), in which the atomic positions and momenta are supplemented by additional dynamical variables representing the coupling of the system to an external reservoir (see [26] for a review and technical details).

23.2.1 Systems Set-up and Simulations

The simulations were all initiated from crystal structures. For each system, the protein monomers were first solvated in a well-equilibrated water box. After energy minimization, a constant volume and temperature (300 K) equilibration run was followed by constant pressure (1 atm.) and constant temperature (300 K) runs. The powder simulations have been previously described in length [13–16]. We therefore only briefly outline here the set up procedure and run parameters. The powder representations of RNase A each contain eight protein molecules replicated by periodic boundary conditions (so that they are actually polycrystalline) constructed from four unit cells (a 2a × 2b × c lattice) of the monoclinic crystal with the water molecules removed. First, a constant volume MD run at 500 K was used to produce non-native, disordered configurations on the surfaces of the protein molecules. This was followed by a constant pressure run at 300 K during which the system contracted, enabling the protein molecules to interact with their neighbors and periodic images, followed this. The system was then hydrated up to hydration levels of $h = 0.05\ g$ and $h = 0.42\ g$ D_2O per gram protein to correspond to low- and high-hydration experiments by adding 280 and 2,188 water molecules, respectively.

A summary of the simulated protein in solution systems is given in Table 23.1. The CHARMM22 force field [28] was used. Three-dimensional periodic boundary conditions were applied and the Ewald sum was used to calculate the electrostatic energies, forces, and virial in all of the simulations. The Lennard–Jones interactions and the real-space part of the Ewald sum were smoothly truncated at 10 Å, and long-range corrections to account for the neglected interactions were included in the energies and pressures [25]. The reciprocal space part of the Ewald sum was calculated using the smooth particle mesh method [29]. The extended system Nosé–Hoover chain method [30] was used to the control the temperature in all of the simulations, with separate thermostat chains for the water and protein molecules. The constant

Table 23.1. Characteristics of the simulated systems

protein	PDB	no. of protein residues/atoms	no. of α domains β strands	box/ dimensions (Å^3)	no. of water molecules
Ribonuclease A	7RSA	124/1,144	3/9	57 × 52 × 40	3,453
Lysozyme	194L	129/1,141	4/3	56 × 48 × 78	6,461
Myoglobin	1HBC	153/1,404	9/0	57 × 52 × 40	3,290
α-Lactalbumin	1HML	142/1,102	4/3	58 × 50 × 80	4,655

pressure simulations were carried out in a fully flexible simulation box by using the extended system algorithm of Martyna et al. [31]. A multiple time step algorithm [32] was used to integrate the equations of motion with a 4 fs time step. The lengths of bonds involving H/D atoms were held fixed by using the SHAKE/RATTLE algorithm [33, 34].

23.2.2 Generating Neutron Spectra

MD simulations produce phase-space trajectories that consist of the positions and momenta of all the atoms in the system as a function of time. Here, we focus on quantities related to incoherent neutron scattering measurements that probe motions of hydrogen atoms on picosecond timescales. Neutron spectroscopy experiments essentially measure the total dynamic structure factor, $S^{\mathrm{meas}}(\boldsymbol{Q},\omega)$, in which $\hbar\boldsymbol{Q}$ and $\hbar\omega$ are the momentum and energy transfers, respectively. Because the incoherent scattering length of hydrogen is an order of magnitude larger than the scattering lengths of all the other atoms in proteins and water (D_2O) molecules, for the systems of interest here, the scattering is primarily incoherent, and hence $S^{\mathrm{meas}}_{\mathrm{tot}}(\boldsymbol{Q},\omega) = S^{\mathrm{meas}}_{\mathrm{inc}}(\boldsymbol{Q},\omega)$. The incoherent dynamical structure factor may be written as the Fourier transform of a time correlation function, $I_{\mathrm{inc}}(\boldsymbol{Q},\mathrm{t})$ the intermediate scattering function

$$S_{\mathrm{inc}}(\boldsymbol{Q},\omega) = \frac{1}{2\pi}\int_{-\infty}^{\infty} I_{\mathrm{inc}}(\boldsymbol{Q},t)\mathrm{e}^{-\mathrm{i}\omega t}\,, \tag{23.1}$$

$$I_{\mathrm{inc}}(\boldsymbol{Q},t) = \frac{1}{N}\sum_j \left\langle \mathrm{e}^{-\mathrm{i}\boldsymbol{Q}\cdot\mathbf{r}_j(0)}\mathrm{e}^{\mathrm{i}\boldsymbol{Q}\cdot\mathbf{r}_j(t)} \right\rangle . \tag{23.2}$$

Here $\boldsymbol{r}_j$ is the position operator of atom j, or, if the correlation function is calculated classically, as in an MD simulation, $\boldsymbol{r}_j$ is a position vector, and the angular brackets denote an average over time origins and scatterers. $I_{\mathrm{inc}}(\boldsymbol{Q},\omega)$ can be directly computed from an MD trajectory and Fourier transformed (FT) to afford $S_{\mathrm{inc}}(\boldsymbol{Q},\omega)$. Note that we have left out the prefactor corresponding to the square of the scattering length. This is convenient in the case of a single dominant scatterer because it gives $I(Q,0) = 1$ and normalized to unity. The $I_{\mathrm{inc}}(\boldsymbol{Q},t)$ (and their corresponding spectra) reported in this paper are "powder averages" computed at eight randomly chosen scattering vectors with $|\boldsymbol{Q}| = Q$.

Intermediate scattering functions in proteins do not decay completely on the timescale accessible to most experiments. A direct Fourier transform is therefore not appropriate. A window function is generally used in such cases, and the results may still be accurately compared to experiment if one considers the proper mathematical form for the window function. The best is that corresponding to the resolution function of the instrument.

In practice, the spread in energies of the neutrons incident on the sample results in a finite energy resolution, and the measured spectrum $S^{\prime\,\mathrm{meas}}_{\mathrm{inc}}(Q,\omega)$

is a convolution of the true spectrum, $S_{\rm inc}(Q,\omega)$, and the instrumental resolution function, $R(\omega)$

$$S_{\rm inc}^{\rm meas}(Q,\omega) = R(\omega) \otimes S_{\rm inc}(Q,\omega)\,, \qquad (23.3)$$

where $\otimes$ denotes a convolution product. The width of the resolution function determines the timescale of the dynamics probed by the instrument in a non-trivial way, with narrower widths (higher resolution) corresponding to longer observation times.

Instrumental resolution functions are generally represented by a Gaussian or other peaked function, centered at $\omega = 0$, with width $\delta\omega$ (or $\delta E = \hbar\,\delta\omega$). Noting that a convolution in energy space is equivalent to a product in the time domain, we compute resolution broadened spectra, $S_{\rm inc}^{\rm meas}(Q,\omega)$, by Fourier transforming the product, $I_{\rm inc}(Q,t)R(t)$, where $R(t)$ is the Fourier transform of $R(\omega)$. Typical spectra generated from the Fourier transform of intermediate scattering functions calculated at different Q values from an MD simulation trajectory are depicted in Fig. 23.3. It is clearly evident that as a consequence of the shape of the resolution function, here considered as a Gaussian of half width at half maximum $\sigma \simeq 180$ ps, corresponding to that of an instrumental

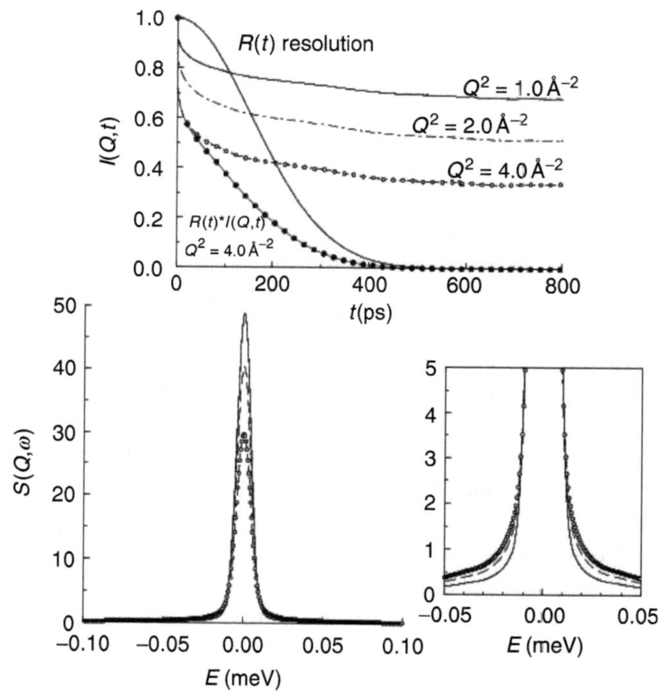

Fig. 23.3. Generating $S_{\rm inc}^{\rm meas}(Q,\omega)$ from MD simulations. *Top*: intermediate scattering functions at several Q values, the resolution function in time and an its product with $I_{\rm inc}(Q=2,t)$ *bottom*: corresponding dynamical structure factors

resolution of $\simeq$100 µeV, the dynamical structure factor contains contributions from motions that are longer the σ. Thus, it is inappropriate to use σ as a fixed time interval for computing time averages if the objective is a quantitative comparison with neutron scattering data.

Interpretation of $S_{\text{inc}}^{\text{meas}}(Q,\omega)$ in terms of atomic dynamics requires models for the diffusive motions. QENS spectra for proteins and other disordered, condensed phase systems are often interpreted in terms of diffusive motions that give rise to an elastic line with a Q-dependent amplitude, and a series of Lorentzian quasi-elastic lines with Q-dependent amplitudes and widths, i.e.,

$$S_{\text{inc}}^{\text{diff}}(Q,\omega) = A_0(Q)\delta(\omega) + \sum_{i=1}^{n} A_i(Q) L_i(\Gamma_i(Q),\omega)\,, \tag{23.4}$$

where $L_i(\Gamma_i(Q),\omega)$ is a Lorentzian centered at $\omega = 0$ with half-width-at-half-maximum $\Gamma_i(Q)$

$$L_i(\Gamma_i(Q),\omega) = \frac{1}{\pi}\frac{\Gamma_i(Q)}{\Gamma_i(Q)^2 + \omega^2}\,. \tag{23.5}$$

The amplitudes of the elastic scattering, $A_0(Q)$, the elastic incoherent structure factor (EISF), provides information on the geometry of the motion, while the line widths are related to the time scales (broader lines correspond to shorter times). The Q and ω dependence of these spectral parameters are commonly fitted to dynamical models for which analytical expressions for $S_{\text{inc}}^{\text{diff}}(Q,\omega)$ have been derived (e.g., jump diffusion, diffusion-in-a-sphere, etc.), affording diffusion constants, jump lengths, residence times, etc. characterizing the motion described by the models [35]. Such models can be scrutinized using the exquisite detail contained in MD simulation trajectories.

23.3 Overall Protein Structure and Motion in Solution

To assess the ability of the simulations to maintain the correct internal structure of the protein molecules, we have computed the root-mean squared deviations (r.m.s.ds) of the C^{α} positions in the simulations from the corresponding crystal structure. The C^{α} atoms define the backbone of the protein molecule. In each case the r.m.s.ds had converged (to values between 1.0 and 1.5 Å) before the averaging period, in the sense that they exhibited small fluctuations in time about their averages, which were not drifting. The results indicate that the overall protein structure is reasonably well maintained during the simulations. When making such comparisons it is important to keep in mind that some deviation from the crystal structure is expected because the intermolecular contacts present in the crystal are absent in solution. We have also computed the time evolution of the radius of gyration R_{g} of the molecules during the simulations, where R_{g} is computed using all the protein atoms according to the standard formula: $R_{\text{g}} = \sum_i^N m_i(r_i - r_{\text{com}})^2/\sum m_i$, r_{com} being the protein center of mass position, r_i is the position of the atom i and m_i its mass. For all proteins under study, after the equilibration period, the radius of gyration show plateau values at 14.5, 14.2, 15.3, and 14.5 Å, respectively, for RNase A, Lys, Myo, and αLact.

In order to estimate the contribution from the overall motion of the protein to the total scattering measured on the 100 ps time scale, we have calculated, for a wide range of Q values, $I_{\rm inc}^{\rm tot}(Q,t)$, the intermediate scattering functions computed directly from the trajectories, and $I_{\rm inc}^{\rm int}(Q,t)$, the intermediate scattering functions computed after removing the translational and rotational motion of the protein in the solvent (by rigidly rotating and translating the whole molecule so that the backbone is optimally superposed on that of a reference structure in a least-squares sense), i.e., singling out the internal motion. The Fourier transforms of $I_{\rm inc}^{\rm tot}(Q,t)$ and $I_{\rm inc}^{\rm int}(Q,t)$ correspond, respectively, to the spectra measured by INS experiments, and to that resulting from the relative motion of the protons with respect to the protein center of mass (heretofore referred to as the internal motion). The results reported in Fig. 23.4 show clearly that, on the hundred ps timescale, $I_{\rm inc}^{\rm tot}(Q,t)$ decays much more rapidly than $I_{\rm inc}^{\rm int}(Q,t)$, and the corresponding structure factors are broader.

Assuming that the MD simulations reproduce qualitatively the overall motion of the protein in the solution, the present results are the first direct evidence that the scattering from a protein in solution, on the length, and timescale studied here, contains a nonnegligible contribution due to the overall

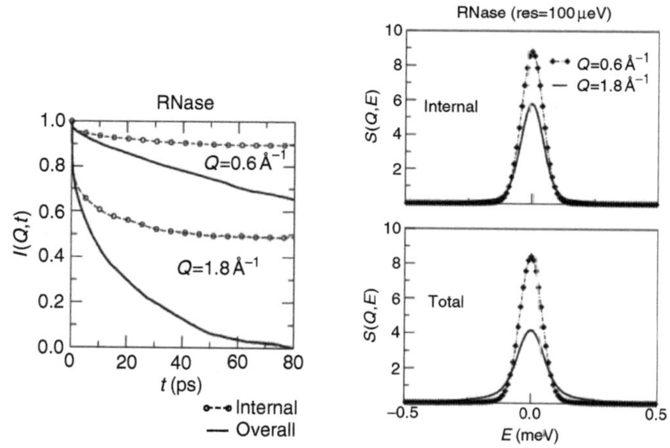

Fig. 23.4. Overall (total) and internal intermediate scattering functions (*left*) and corresponding dynamical structure factors (*right*) computed from MD simulations of RNase A in solution at 300 K, at $Q = 0.6$ and $1.8\,\text{Å}^{-1}$

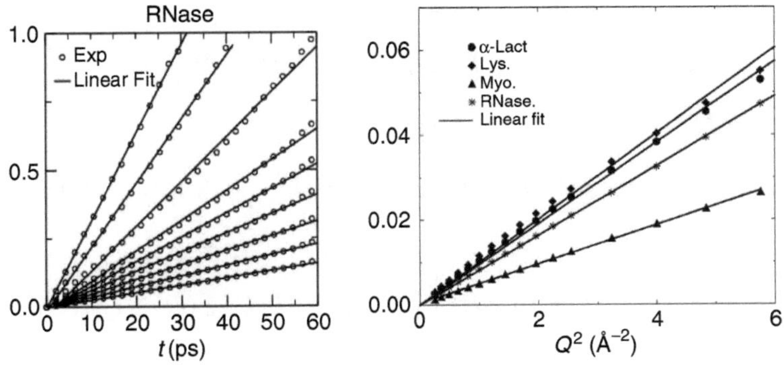

Fig. 23.5. Analysis of the overall motion of the proteins in solution. *Left*: plots of $\ln(1/I_{\text{inc}}^{\text{glob}}) = \ln(I_{\text{inc}}^{\text{int}}(Q,t)/I_{\text{inc}}^{\text{tot}}(Q,t)$ at several values of the wave vector transfer Q, up to $Q = 2.5\,\text{Å}^{-1}$, for RNase A in solution. *Right*: corresponding slopes, ν (cf. text), as a function of $Q^2(\text{Å}^{-2})$ for all the studied proteins

diffusion of the protein. The contribution from this motion may be analyzed by considering the ratio $I_{\text{inc}}^{\text{glob}} = I_{\text{inc}}^{\text{tot}}(Q,t)/I_{\text{inc}}^{\text{int}}(Q,t)$, reported in Fig. 23.5 (left) for RNase A at several Q values in a range accessible by time of flight spectrometers. The results indicate that in the 100 ps time scale, the overall motion may be described by a simple exponential decay i.e., $I_{\text{inc}}^{\text{glob}} = \exp(-\nu t)$. Figure 23.5 (left) shows that in the Q range studied, ν displays a linear dependence on Q^2. One may therefore write

$$I_{\text{inc}}^{\text{tot}}(Q,t) = I_{\text{inc}}^{\text{int}}(Q,t)\exp\left(-D_{\text{eff}}Q^2 t\right) = I_{\text{inc}}^{\text{int}}(Q,t) I_{\text{inc}}^{\text{glob}}(Q,t)\,, \tag{23.6}$$

where D_{eff} is the slope of the linear fits to the curves reported in Fig. 23.5 (right). By Fourier transform one obtains

$$S_{\text{inc}}^{\text{meas}}(Q,\omega) = S_{\text{inc}}^{\text{int}}(Q,\omega) \otimes S_{\text{inc}}^{\text{glob}}(Q,\omega)\,, \tag{23.7}$$

where $S_{\text{inc}}^{\text{glob}}(Q,\omega)$ is the Fourier transform of $I_{\text{inc}}^{\text{glob}}(Q,t)$.

The right-hand side of each of the previous two equations may be considered as the contribution from the global motion of the proteins in solution (i.e., overall rotation and translation of the protein). Direct evidence from our simulations shows that the internal and the global motions are, within the length and timescales of the analysis, decoupled.

Turning now back to fitting the data from an INS experiment, our data support the use of a model in which the measured dynamical structure factor is fitted considering the expression in Eq. 23.7, where the component $S_{\text{inc}}^{\text{glob}}(Q,\omega)$, is a Lorentzian $L(\Gamma_{\text{glob}}(Q))$ of width $\Gamma_{\text{glob}}(Q) = \nu = D_{\text{eff}}Q^2$, i.e.,

$$S_{\text{inc}}^{\text{glob}}(Q,\omega) = 1/\pi \frac{D_{\text{eff}}Q^2}{\omega^2 + (D_{\text{eff}}Q^2)^2}\,. \tag{23.8}$$

Assuming now that the internal motion may be decomposed as

$$S_{\mathrm{inc}}^{\mathrm{int}}(Q,\omega) = A_0(Q)\delta(\omega) + (1 - A_0(Q)L(\Gamma_{\mathrm{int}}(Q)) , \tag{23.9}$$

one may write

$$S_{\mathrm{inc}}^{\mathrm{meas}}(Q,\omega) = L(\Gamma_{\mathrm{glob}}(Q)) \otimes [A_0(Q)\delta(\omega) + (1 - A_0(Q)L(\Gamma_{\mathrm{int}}(Q)))] \tag{23.10}$$

and

$$S_{\mathrm{inc}}^{\mathrm{meas}}(Q,\omega) = A_0(Q)L(\Gamma_{\mathrm{glob}}(Q)) + [1 - A_0(Q)]L(\Gamma_{\mathrm{glob}}(Q) + \Gamma_{\mathrm{int}}(Q)) \tag{23.11}$$

where $L(\Gamma_{\mathrm{glob}}(Q))$and $L(\Gamma_{\mathrm{int}}(Q))$ are Lorentzians representing the overall and internal motion, respectively.

At this stage, our analysis supports the model used by Pérez et al. [17], in which the dynamical structure factor is fitted with two Lorentzians, one with a narrow width $\Gamma_{\mathrm{glob}}(Q)$ corresponding to the overall motion of the protein, and one with a broader one, $\Gamma_{\mathrm{int}}(Q)$, corresponding to the diffusive internal motion of the protons. The constraint on the intensities of the two components given by the above equation affords a direct estimate of the EISF of the internal motion.

Estimates of D_{eff} extracted from the MD results are reported in Table 23.2. These are in satisfactory agreement with the estimates by Pérez et al. for myoglobin and lysozyme solutions, i.e., $8.2 \pm 0.2 \times 10^{-7}\,\mathrm{cm}^2\,\mathrm{s}^{-1}$ and $9.1 \pm 0.2 \times 10^{-7}\,\mathrm{cm}^2\,\mathrm{s}^{-1}$ respectively, in light of the fact that values extracted from the fit of the INS spectra may contain a large uncertainty resulting from the subtraction of the buffer scattering from the raw data.

While such an analysis of the scattering from a solution sample is appropriate for the timescales corresponding to time of flight spectrometers ($t \leq 100\,\mathrm{ps}$), the accuracy of the models fails at much longer time scales, i.e., for data collected with high resolution backscattering and spin echo spectrometers, and at high Q values. Indeed, at longer time scales and/or high Q, the overall motion of the protein dominates in dilute samples. The corresponding correlation functions decay faster than the times corresponding to the experimental resolution. Extracting information about the internal dynamics of the protein protons in such cases would likely be inaccurate.

Table 23.2. Effective diffusion coefficient from MD simulations

protein	molecular weight	$D_{\mathrm{eff}}(10^7\,\mathrm{cm}^2\,\mathrm{s}^{-1})$
Ribonuclease A	13,674	12.43
Lysozyme	14,296	15.33
Myoglobin	17,184	7.2
α-Lactalbumin	13,674	14.52

23.3.1 Internal Protein Dynamics

To compare results from the simulations to available experimental data, the internal dynamics may be analyzed using the same Gaussian/Lorentzian models generally adopted to fit the structure factor. In the following cases, $I_{\rm inc}^{\rm int}(Q,t)$, the intermediate scattering functions computed from the MD trajectories, are Fourier transformed as described above by considering a 100 μeV resolution (similar to that of IN6 at the Institut Laue Langevin) to generate $S_{\rm inc}^{\rm int}(Q,\omega)$, which is fitted according to Eq. 23.9. The EISF corresponding to the intensity of the Gaussian (localized motion) component and the width of the Lorentzian component corresponding to the diffusive motion are reported in Fig. 23.6. The results shown here for RNase A are again in satisfactory agreement with the Pérez et al. data on myoglobin.

In order to highlight the effect of inappropriate data analysis, and to investigate the effect of the resolution shape on the results, we show in Fig. 23.7 an example where the same fitting model is used to extract the EISF and the width of the diffusive component. The analysis shows again that the scattering contains a significant contribution from the overall motion of the protein.

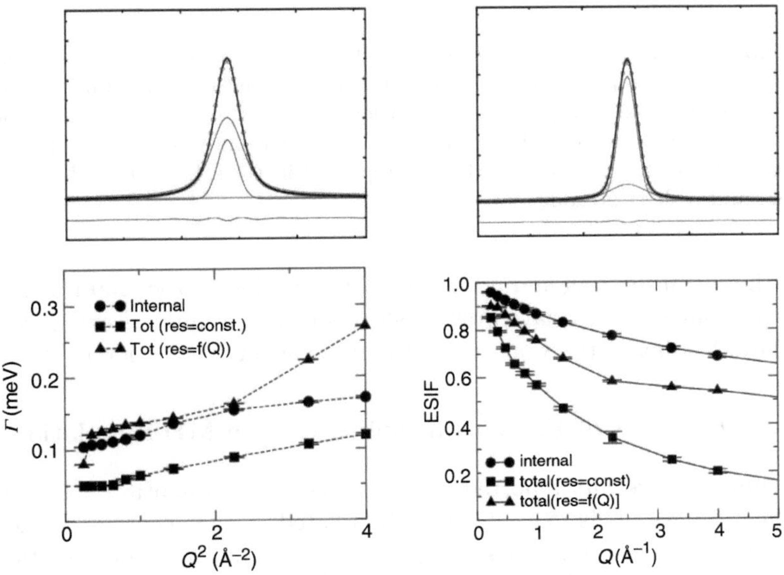

Fig. 23.6. Data analysis of $S(Q,\omega)$ for RNase A in solution. Top: fit with Eq. 23.9 of both $S_{\rm inc}^{\rm tot}(Q,\omega)$ (*left*) and $S_{\rm inc}^{\rm int}(Q,\omega)$ (*right*) calculated from the MD trajectory at a 100 μeV resolution. Bottom: corresponding parameters, i.e., half width at half maximum of the Lorentzian component and the EISF. The triangle symbols represent the parameters extracted from the fit of $S_{\rm inc}^{\rm tot}(Q,\omega)$ considering a typical Fermichopper instrument resolution varying between 120 and 260 μeV for $0.5\,\text{Å}^{-1} \leq Q \leq 2.4\,\text{Å}^{-1}$

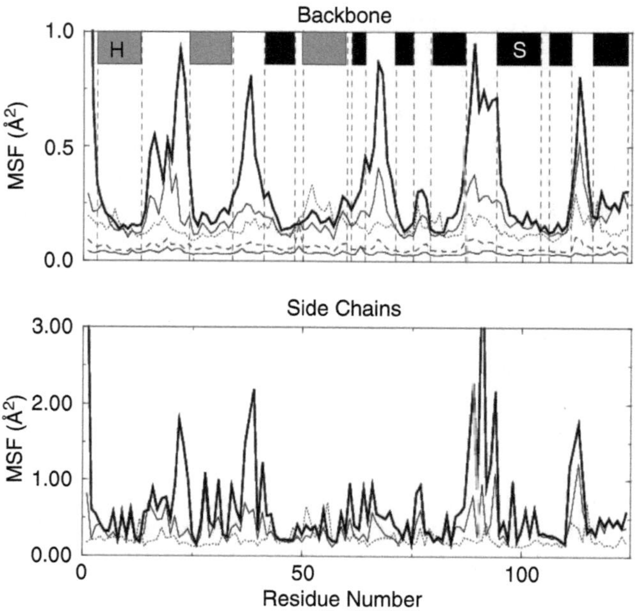

Fig. 23.7. Mean-squared fluctuations (MSFs) of the backbone (*top*) and the side chains atoms (*bottom*) of Rnase A as a function of the residue number along the chain. From bottom to top the results for low temperature (150 K) and low powder hydration simulations to high hydration powders at room temperature. The results for the solution simulations (300 K) are reported in thick lines. H and S stand for α-helix and β-sheet strands

More importantly, this and additional artifacts due to the characteristics of the spectrometers may have drastic effects on the parameters extracted from the data, leading in the worst cases to erroneous interpretation of the data.

23.3.2 Dynamics of Proteins in Solution from MD Simulations

The simulation results reported above did not agree quantitatively with experimental data. This may of course be related to the accuracy of the force field used, or to the simulation setup (sampling of multiple conformations for the protein). One should also keep in mind, however, the experimental limitations. Errors due to data treatment may contribute equally to the discrepancy. For example, it is important to recall the measured spectra result from the scattering from the protein and from the solvent. It turns out that subtraction of the latter is rather complicated and often user-dependent. Bearing in mind this shortcoming, and based on our previous results obtained for low- and high-hydration powders, where contributions from the solvent and from the overall protein diffusion are not an issue, one can claim that the simulations are rather satisfactory.

The next step is now to provide a "real space" description of the motions of the protein atoms, and examine how appropriate the models adopted by experimentalists are to describe such motions. In a typical experiment, it is not possible to label specific protons to monitor independently their motion. It follows that one probes the motion of all nonexchangeable protons (the protein is immersed in a D_2O bath) simultaneously. Moreover, the structure factor is an intensive quantity representing scattering intensity per atom. Therefore, most if not all models used to describe the protein dynamics assume that mobile protons (those that give rise to a quasielastic signal in the time domain corresponding to the experimental resolution) have similar amplitudes and time scales of motion. The analysis of motion from MD simulations shows clearly that such models are inappropriate. Indeed, for proteins in solution, at room temperature, one finds that the protein motion is characterized by a very large heterogeneity. In particular, as shown in Fig. 23.7, the amplitudes of motion (mean-squared fluctuations) of different residues along the protein sequence can be as much as five- to tenfold different. This is consistent with the sequence dependence of B-factors determined by X-ray crystallography [36] and previous analysis of MD simulations [37].

One interesting feature emerging from the simulation is the connection between the secondary structure and the amplitude of motions. Indeed, the results show that, as expected, the atoms belonging to those residues in highly structured regions of the protein (α-helices and β-sheets) are much less mobile that those attached to unstructured parts (e.g., loops) of the backbone, whether located at the surface of the protein or not. It is this kind of observation that should somehow be feed back into the experimental data analysis.

Inelastic neutron data are often interpreted in terms of a model of nonexchangeable hydrogen atoms diffusing in a sphere. The complexity of the models used to fit the data is limited by the small number of parameters that are extractable from the spectral lineshapes. Thus, it is important to keep in mind that, while the model of diffusion-in-a-sphere fits QENS data reasonably well, it is clearly an approximation to ascribe a single sphere radius to hundreds or thousands of hydrogen atoms in a protein molecule.

MD trajectories may be used to quantify the dispersion in the amplitudes of nonexchangeable hydrogen motion on the timescale probed by current neutron spectrometers. In Fig. 23.8 we show the distributions of the mean-squared fluctuations, $\langle \Delta r_i^2 \rangle = \langle (r_i - \langle r_i^2 \rangle)^2 \rangle$ of the nonexchangeable hydrogen atoms in the MD simulations of the native α-lactalbumin in solution, computed as averages over blocks of 100 ps. The mean-squared fluctuations were calculated after removing the overall translational motion of the protein and hence represent the amplitudes of the internal motion. It is immediately evident that there is a broad distribution of H atom amplitudes on the 100 ps time scale. The distribution is sharply peaked at values near 0.5 Å, with pronounced asymmetry on the higher amplitude side.

To make contact with the diffusion-in-a-sphere model we identify the average root-mean-squared fluctuation, $\langle \langle \Delta r_i^2 \rangle \rangle^{1/2}$, where the outer angular brackets denote an average over H atoms, with the average effective radius of

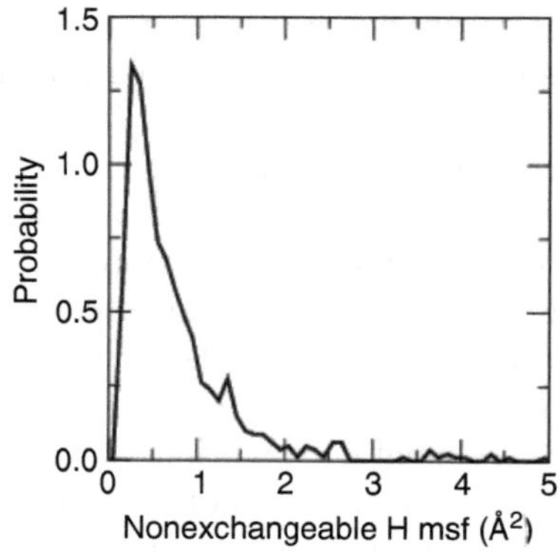

Fig. 23.8. Distribution of mean-squared fluctuations of nonexchangeable H atoms from an MD simulation of α-lactalbumin in solution

the sphere in which the H atoms diffuse, obtaining a sphere radius of 0.89 Å. Our estimates of the radii are about half of those obtained for phosphoglycerate kinase [38] and almost five times smaller than those obtained for α-lactalbumin [19]. Part of the discrepancy can be attributed to the fact that the global translational and rotational motion of the protein were not taken into account in the analysis of the QENS data, while our values include only internal motions. Indeed, when we do not remove the global motion, we obtain a value of 1.80 from the simulation which agrees very well with the corresponding value for phosphoglycerate kinase.

Application of the diffusion-in-a-sphere model with a single radius for all H atoms is clearly an oversimplification. Indeed, heterogeneity in the diffusion-in-a-sphere model may be accounted for by using a Gaussian distribution of sphere radii centered at zero [17,39]. In principle, the shape of the distributions in Fig. 23.8 could provide the basis for the development of a more realistic model that includes a distribution of amplitudes, which could be used to fit QENS data. However, it is not clear whether or not such a model could be formulated in terms of the small number of fitting parameters that are available from QENS data.

23.4 Conclusions

In summary, while classical MD simulations using current generation force fields have allowed us to reproduce quite well the dynamics of proteins in a variety of environments, as probed by neutron scattering data, the full potential of such calculations has not yet been fully exploited. We have illustrated some examples where simulations can be used to provide support for the models used by experimentalists, and others where it is clear that further work is needed to extract the maximum information from QENS spectra. At any rate, it is crucial that simulations and experiments on such complex systems go hand in hand, so that "raw data" may be compared side-by-side, and all the pitfalls, both in the simulation protocols and the experimental data analysis, may be identified and overcome.

Acknowledgment

This work has benefited from support from a collaborative grant between the University of Pennsylvania and the National Institute of Standards and Technology. We are grateful to Dr. Michael Klein for support and to Dr. Taner Yildirim for very helpful discussions and for providing software for scattering data analysis. This work was partially supported by a grant from the National Science Foundation (CHE-0417158).

References

1. W. Doster, S. Cusack, W. Petry, Nature **337**, 754–756 (1989)
2. W. Doster, W. Petry, Phys. Rev. Lett. **65**, 1080–1083 (1990)
3. Smith, J.C. Quart. Rev. Biophys. **24**, 227–291 (1991)
4. C. Andreani, F. Menzinger A. Desidiri A. Deriu D. Di Cola, Biophys. J. **68**, 2519–2523 (1995)
5. J. Fitter, R.E. Lechner, N.A. Dencher, Biophys. J. **73**, 2126–2137 (1997)
6. G.R. Kneller, J.C. Smith, J. Mol. Biol. **242**, 181–185 (1994)
7. J.M. Zanotti, S.H. Chen, Phys. Rev. E **59**, 3084–3093 (1999)
8. A. Paciaroni, C. Arcangeli, A.R. Bizzarri et al., Eur. Biophys. J. **28**, 447–456 (1999)
9. S. Timasheff, in *Stability of protein pharmaceuticals, Part B: In vivo pathways of degradation and strategies for protein stabilisation.* (Plenum, New York, 1992)
10. M.J. Hageman, in *Stability of protein pharmaceuticals, Part B: In vivo pathways of degradation and strategies for protein stabilisation.* (Plenum, New York, 1992)
11. B. Hancock, G. Zografi, J. Pharm. Sci. **86**, 1–12 (1997)
12. R.L. Remmele Jr., J.F. Carpenter, Pharm. Res. **14**, 1548–1555 (1997)
13. M. Tarek, D.J. Tobias, J. Am. Chem. Soc. **121**, 9740–9741 (1999)
14. M. Tarek, D.J. Tobias, J. Am. Chem. Soc. **102**, 10450–10451 (2000)
15. M. Tarek, D.J. Tobias, Biophys. J. **79**, 3244–3257 (2000)
16. M. Tarek, D.J. Tobias, J. Chem. Phys. **115**, 1607–1612 (2001)
17. J. Pérez, J.M. Zanotti, D. Durand, Biophys. J. **77**, 454–469 (1999)
18. Z. Bu, S-H. Lee, C.M. Brown, D.M. Engelman, C.C. Han, J. Mol. Biol. **301**, 525–536 (2000)
19. Z. Bu, J. Cook, D.J.E. Callaway, J. Mol. Biol. **312**, 865–873 (2001)
20. A. Gall, B. Robert, M.C. Bellissent-Funel, J. Phys. Chem. B. **106**, 6303–6309 (2002)
21. A. Wlodawer, L. Sjolin, G. Gilliland, Biochemistry **27**, 2705–2717, (1988)
22. J. Kuriyan, M. Karplus, G.A. Petsco, J. Mol. Biol. **192**, 133 (1986)
23. J. Ren, K.R. Acharya, J. Biol. Chem. **266**, 19292 (1993)

24. M.C. Vaney, M. Riess-Kautt A. Ducruix, Acta. Cryst. D. Biol. Crystallogr. **52**, 501 (1996)
25. M.P. Allen, D.J. Tildesley, *Computer simulation of liquids* (Clarendon Press, Oxford, 1989)
26. M.E. Tuckerman, G.J. Martyna, J. Phys. Chem. B **104**, 159–178 (2000)
27. A.R. Leach, *Molecular Modelling, Principles and Applications* (Addison-Wesley Longman Limited; Singapore, 1996)
28. A.D. MacKerell Jr., M. Bellott, R.L. Dunbrack Jr. et al., J. Phys. Chem. B **102**, 3586–3616 (1998)
29. U. Essmann, M.L. Berkowitz, T. Darden, L.G. Pedersen, J. Chem. Phys. **103**, 8577–8593 (1995)
30. G.J. Martyna, M.L. Klein, J. Chem. Phys. **97**, 2635–2643 (1992)
31. G.J. Martyna, M.L. Klein, J. Chem. Phys. **101**, 4177–4189 (1994)
32. G.J. Martyna, D.J. Tobias, M.L. Klein, Mol. Phys. **87**, 1117–1157 (1996)
33. J.-P. Ryckaert, H.J.C. Berendsen, J. Comp. Phys. **23**, 327–341 (1977)
34. H.C. Andersen, J. Comp. Phys. **52**, 24–34 (1983)
35. M. Beé, *Quasielastic neutron scattering: Principles and applications in solid state chemistry, biology, and materials science* (Adam Hilger, Bristol, 1988)
36. E.D. Chrysina, K.R. Acharya, J. Biol. Chem. **47**, 37021–37029 (2000)
37. E. Paci, C.M. Dobson, M. Karplus, J. Mol. Biol. **306**, 329–347 (2001)
38. V. Receveur, J.C. Smith, M. Desmadril et al., Proteins: Struct. Funct. Genet. **28**, 380–387 (1997)
39. D. Russo, J.M. Zanotti, M. Desmadril et al., Biophys. J. **83**, 2792–2800 (2002)

Index

MIX
Papier aus verantwortungsvollen Quellen
Paper from responsible sources
FSC® C105338

If you have any concerns about our products,
you can contact us on
ProductSafety@springernature.com

In case Publisher is established outside the EU,
the EU authorized representative is:
Springer Nature Customer Service Center GmbH
Europaplatz 3, 69115 Heidelberg, Germany

Printed by Libri Plureos GmbH
in Hamburg, Germany